Guard Your health with Wise Food safety Routines

ISBN:979-8-9959180-0-4

Guard Your Health with Wise Food Safety Routines

Dr. Udo F. Ufomadu

ISBN:979-8-9959180-0-4

PREFACE

Knowledge translates to capability, and understanding empowers positive actions. Information allows one to make better decisions. In the context of food safety, lack of basic safe food knowledge or refusal to follow food safety advice can expose individuals and communities to severe illness, suffering, and even death.

In many parts of the world, some dishonest individuals adulterate food and drinks for profit. Harmful substances may be added to food products to increase weight, improve appearance, extend shelf life, or reduce production costs. Examples include the addition of dangerous chemicals, contaminated water, artificial colors, or low-quality ingredients, all of which can seriously harm consumers. When people fail to understand food safety principles or ignore warnings from food safety officials, they become more vulnerable to foodborne diseases, poisoning, allergic reactions, and long-term health complications.

Hosea 4:6 reminds society that knowledge is essential for protection. Consumers must therefore educate themselves about safe food-handling practices, proper labeling, hygiene, risks of adulteration, and reputable food sources. Food manufacturers, vendors, and handlers also have a moral responsibility to protect public health rather than pursue profit at the expense of human lives.

Smart food safety practices such as reading labels carefully, checking expiration dates, avoiding suspicious products, maintaining cleanliness, cooking foods thoroughly, and reporting unsafe food practices can help prevent illness and save lives. Governments and food safety agencies must continue to educate the public while enforcing strict regulations against food adulteration and unsafe food-handling practices.

Ultimately, food safety knowledge is a form of protection. When people value and apply sound food safety advice, they guard not only their own health but also the well-being of their families and communities. As a result, this book focuses on contamination, adulteration, counterfeiting, substitution, and mislabeling in the food we eat. Contamination is the presence of harmful substances or microorganisms in food that make it unsafe for human consumption. It can happen at any stage of the food supply chain, including production, processing, transport, or cooking.

Adulteration is the process of making a product impure or inferior by adding a cheaper substitute chemical or other substance. This is very common in food and beverages (e.g., diluting milk or bulking up spices with cheaper adulterants). Counterfeiting frequently involves the use of nonfood additives in human food. Substitution is the direct replacement of a premium ingredient or material with a cheaper chemical or other alternative. Simulation is the use of chemicals to artificially mimic the flavor, color, or scent of a natural product (such as artificial vanillin instead of real vanilla extract), and Mislabeling is selling a product under a false name or description, which often involves using cheaper substitute chemicals without listing them.

Knowledge of biological, chemical, and physical hazards, as illustrated in this book, helps ensure the proper prevention of foodborne illness. Types of Biological hazards include bacteria, viruses, and parasites. Types of chemical hazards are pesticides and cleaning agents. Physical hazards are dirt, glass, and metal fragments. Causes of hazards are poor hygiene, cross-contamination, and unsafe storage. As a result, efficient and effective Food Safety Management is of utmost significance to all of us and the food industry for these six reasons:

Public Health: Ensuring food safety is critical for protecting public health. Contaminated or improperly handled food can lead to foodborne illnesses, which can range from mild discomfort to severe illness and even death. By preventing foodborne diseases through smart food safety measures, you safeguard the well-being of your consumers and everybody.

Consumer Confidence: Maintaining high standards of food safety helps to build and maintain consumer trust. When individuals have confidence in the safety of the food supply, they are more likely to purchase and consume a wide variety of foods, supporting the success of the food industry.

Legal and Financial Implications: Failure to comply with food safety regulations can have serious legal and financial consequences for food businesses. Violations can result in fines, legal action, and reputational damage. Outbreaks of foodborne illness can also lead to costly recalls, litigation, and a loss of consumer trust.

Economic Impact: Food safety is crucial for the economic stability of the food industry and its communities. When foodborne outbreaks occur, they can have far-reaching economic consequences, affecting not only the companies directly involved but also the broader food supply chain and related industries.

Regulatory Compliance: Compliance with food safety regulations is mandatory for food businesses. Adhering to these regulations helps to ensure that food is produced, processed, and distributed in a manner that minimizes the risk of contamination and illness.

Reputation and Brand Protection: Maintaining high standards of food safety protects the reputation and brand integrity of food businesses. A strong track record in food safety practices can differentiate a company in the marketplace and enhance its long-term success.

Yes, Food safety is a critical part of public health and well-being. According to the World Health Organization (WHO), an estimated 600 million people in the world, which translates to practically

1 in 10 people in the world, get sick after eating contaminated food, and 420,00 die yearly. Unsafe foods that contain harmful bacteria, viruses, parasites, chemicals, and physical hazards cause more than 200 diseases, ranging from diarrhea to cancer. 40 percent of food-borne disease burdens are carried by children 5 years and under. Foodborne diseases undermine socioeconomic growth by straining health care systems and harming national economies, tourism, and trade. From personal experience, I firmly affirm that Food safety is important to the food producers, the consumers, and other stakeholders. Safe food protects companies and stakeholders from expensive penalties and legal actions. Yes! Meaningful collaboration among governments, producers, and consumers is always needed to help ensure our food is safe. Overall, food safety is essential for protecting public health, maintaining consumer confidence, complying with regulations, and sustaining the economic viability of the food industry. It is a shared responsibility that requires diligence and commitment from all stakeholders involved in the production, processing, distribution, and preparation of food. Bear in mind that there is nothing more fulfilling than a tasty meal that is also safe; know thee also that when a banquet is secured from illness, it makes every bite bloom with delight; and be aware that

a palatable meal resides where flavor and safety are in control.

Dr. Udo Fosby Ufomadu

www.ufomaduconsulting.com

TABLE OF CONTENTS

CHAPTER ONE

All foods Require Food Safety Precautions

All foods, no matter how tasty or familiar they seem, require proper safety precautions because risk can exist at every stage from production to consumption. Here are the key reasons:

1. Microbial contamination can occur in any food
Even foods that look clean can carry harmful microorganisms like bacteria, viruses, or parasites. Raw produce, meats, dairy, and even cooked foods can become contaminated through handling, water, or surfaces.

2. Cross-contamination is always a risk
Foods can become unsafe when they come into contact with contaminated utensils, surfaces, or other foods (especially raw meats). Without precautions, even safe food can quickly become hazardous.

3. Improper storage leads to spoilage
All foods have a limited shelf life. Incorrect temperature control can promote bacterial growth or degrade food quality, increasing the risk of foodborne illness.

4. Food handling practices affect safety
Poor hygiene (e.g., not washing hands) or unsafe preparation methods can introduce contaminants at any stage—whether at home, in restaurants, or during processing.

5. Chemical contamination is possible
Foods can be exposed to pesticides, cleaning agents, or packaging chemicals. Proper washing, storage, and handling help reduce these risks.

6. Allergens can be present or transferred
Even trace amounts of allergenic foods (like nuts, dairy, or fish) can be dangerous for sensitive individuals. Safety precautions prevent accidental exposure.

7. Food can be misidentified or improperly prepared
Some foods (such as certain mushrooms or undercooked meats) can be harmful if not properly identified or cooked to safe internal temperatures.

8. Time and temperature abuse increase danger
Leaving food out too long or not cooking/reheating it properly allows harmful bacteria to multiply rapidly.

9. Global supply chains increase exposure risks
Food often travels long distances and passes through multiple handlers, increasing the chances of contamination along the way.

10. Vulnerable populations are at higher risk
Children, older adults, pregnant individuals, and people with weakened immune systems are more susceptible to foodborne illnesses, making it essential to take precautions with all foods.

CHAPTER TWO

Interpreting Grocery Stores' Codes or Stickers on Food

Understanding codes like PLU and UPC on food isn't just trivia; it gives customers practical control over food safety and quality and helps them make informed choices. These are the small-number stickers usually attached to meat or produce packages. Knowing them helps you distinguish conventional, organic, and genetically modified produce. Make safer choices if you're trying to avoid certain farming practices or pesticide exposure. Verify that what you're paying for matches what you're getting to reduce the risk of mislabeling. The barcodes are used for scanning packaged foods. For customers, they help identify the exact product and manufacturer, which is crucial during food recalls. Allow you to check product details (ingredients, origin) using apps or databases. Improve traceability, which is important in the event of contamination issues. Date/handling codes like sell-by, use-by, and lot numbers. These are critical for safety: Indicate freshness and shelf life, helping you avoid spoiled or unsafe food. Provide batch/lot information, which is essential during recalls or outbreaks. Guide proper storage, especially for refrigerated or perishable items, to reduce the risk of foodborne illnesses. Why it matters overall: When customers understand these codes, they can avoid expired or mishandled food, respond quickly to recalls, make healthier, safer purchasing decisions, and reduce food waste by buying and using items appropriately. In short, these codes turn you from a passive buyer into an informed and safety-conscious consumer.

Interpreting the Codes:

1) PLU codes (Price Look-Up codes)- What they look like: 4–5 digits printed on a sticker attached to the fruit (sometimes directly on the skin as a number, especially for small fruits).

- What they mean: - 4 digits: Conventional (standard) produce. The first digit is 3 or 4 (e.g., 4011 for bananas). 5-digit codes starting with 9 are Organic. The numeric structure is the same as the conventional one, but with a 9 added in front (e.g., 94011 for organic bananas).

- 5 digits starting with 8: Genetically modified (GM) or bioengineered. Note: This is not widely used in practice for consumer labeling, and many retailers have stopped using 8-digit codes; some stores no longer indicate GM status on PLUs due to regulatory and labeling changes. If you see a PLU starting with 8, treat it as an alert that the fruit is genetically modified, but verify with the store if you need explicit confirmation.

2) UPC codes (Universal Product Code) on packaging

- What they look like: 12-digit barcode on packaging, not typically on loose fruit.

- What they mean: They identify the product for checkout, inventory, and pricing in the store's system. They are retailer-specific and not designed for consumer interpretation beyond the product ID.

3) Country of Origin Labeling (COOL)

- What they look like: Often a small sticker or printed label stating the country of origin (e.g., "Product of Mexico," "Produced in USA").

- What do they mean? They indicate where the fruit was grown or produced. Some items may list multiple origins if they were imported or partially produced abroad.

4) Certification and quality marks (less common on fresh fruit, more on packaged)

- Examples: Organic certification logos (USDA Organic, EU organic), fair trade on certain fruit products or packaged items, non-GMO labels.

- What they mean: Organic indicates compliance with organic farming standards. Non-GMO indicates the product was produced without genetically engineered organisms.

Genetically Engineered Food Products (GE foods)

GE foods, also called genetically modified (GM) foods, are foods produced from organisms whose DNA has been altered using modern biotechnology. This process allows scientists to introduce specific traits that do not occur naturally through traditional breeding. Genetic engineering involves directly modifying an organism's genetic material. Scientists may add genes from another organism (even from a different species), modify existing genes, or turn certain genes "on" or "off". Common Examples of GE Foods are Corn (e.g., insect-resistant varieties), Soybeans (herbicide-tolerant), Cottonseed oil, Canola oil, Sugar beets, Papaya (virus-resistant), and many processed foods contain ingredients derived from these crops. The purpose of Genetic Engineering in Food is to increase crop yield, enhance nutritional value (e.g., vitamin-enriched rice), improve resistance to pests, diseases, and drought, reduce the need for chemical pesticides, extend shelf life, and improve food security. Consequently, Higher productivity helps feed growing populations and reduces pesticide use. Some crops are engineered to resist insects. Enhanced nutrition: For example, biofortified crops have economic benefits, including lower production costs for farmers.

Health concerns: Potential allergens or long-term effects, though current evidence shows that approved GE foods are safe. Environmental impacts include gene transfer to wild plants and biodiversity loss. Ethical issues relate to tampering with nature. Economic issues relate to dependence on biotech companies for seeds. In terms of Regulation and Safety, in the United States, GE foods are regulated by the Food and Drug Administration (FDA), the United States Department of Agriculture, and the Environmental Protection Agency. These agencies ensure that GE foods are tested for safety before they enter the market. Are GE Foods Safe? Scientific consensus from organizations like the World Health Organization and the National Academy of Sciences is that currently approved GE foods are as safe as conventional foods. Under U.S. labeling rules, GE foods may be labeled "bioengineered," allowing consumers to make informed choices. Genetically engineered foods are a powerful tool in modern agriculture, offering solutions to food security and nutrition challenges. At the same time, they raise important environmental, ethical, and economic questions that require careful regulation. It is important to read the label for organic, non-GMO, GMO, or other certifications, and scan the UPC for price/identity at checkout if you're curious about the exact product. Whether GMO or not, it is essential to check whether the prices of the products you are buying are reduced for the wrong reasons.

-Sell-By Date (Store Focus): Tells the store how long to display the product for sale. It is not a safety date; food is usually still good for several days or weeks after this date, including milk, which lasts about a week longer.

-Use-By Date (Consumer Focus): This is the last date recommended for the use of the product while at peak quality. It is not necessarily safe, except when used in infant formulas.

Best if Used By/Before (Quality Focus): Indicates when a product will be at its best in terms of flavor or quality.

Another way to gauge what you are getting for your health is to pay attention to the order in which the product's ingredients are listed. Ingredients are listed in descending order by weight, so the first three ingredients are the most prevalent. Watch for Hidden Sugar: Food labels often “hide” sugar under many different names. Recognizing them helps you spot how much added sugar you’re actually consuming. Basic sugars: Glucose, Fructose, Sucrose (table sugar).Lactose (milk sugar), Maltose, Syrups often highly processed and quickly absorbed, High-fructose corn syrup (HFCS), Corn syrup/corn syrup solids, Rice syrup / brown rice syrup, Maple syrup, Agave nectar, “Natural” or less obvious sweeteners, Honey, Molasses, Cane sugar/cane juice / evaporated cane juice, Coconut sugar, Date sugar, Dextrose, Galactose, Invert sugar, Malt syrup, Barley malt, Fruit juice concentrate. Be careful if a label lists multiple types of sugar; it's often a way manufacturers spread them out so sugar doesn't appear as the first ingredient, even though the total sugar content may be high.

Salt compounds don't always appear as salt compounds in the ingredient statement; they could be listed as Sodium chloride, Sodium bicarbonate (baking soda), Sodium benzoate (preservative Monosodium glutamate (MSG), and Sodium nitrate or nitrite. These contribute to total sodium intake but may not be obvious to consumers.

Fats and Oils: Fats and oils can be disguised in different forms, such as vegetable oils (soybean, palm, and canola); hydrogenated or partially hydrogenated oils; trans fats; shortening; lard; butterfat; and Palm kernel. Listing different oils and fats separately can make the total appear lower.

Starches and carbohydrate fillers are often used as thickeners or fillers, including modified food starch, Corn starch, Tapioca starch, Potato starch, and maltodextrin. These can contribute to carbohydrate intake and affect the glycemic response.

Milk Derivatives or Milk ingredients appear under many names, such as Casein/caseinates, Whey, Lactose, Milk solids, and Butter oil. Proper information on milk ingredients is important for people with allergies or lactose intolerance.

Gluten-containing ingredients are not always labeled as gluten but could be listed as Wheat flour, Barley malt/malt extract, Rye, or hydrolyzed wheat protein, which are critical for individuals with Celiac disease.

Flavor Enhancers / Additives have multiple identities (monosodium glutamate), Hydrolyzed vegetable protein (HVP), Autolyzed yeast extract, Natural flavors (can include many compounds)

Preservatives can be listed under many chemical names, such as BHA (butylated hydroxyanisole), BHT (butylated hydroxytoluene), TBHQ, and sulfites (e.g., sodium sulfite).

Why This Matters (Food Safety & Consumer Awareness)? This matters because Manufacturers may use different names to spread ingredients across the list (since ingredients are listed by weight). This can make products seem healthier than they are, and it's not necessarily illegal, but it can be misleading. Also, read the label to check for allergens if you or your family members are allergic to certain ingredients.

Allergens

Allergens are a food safety concern that deserves attention in this book. For a food buyer, identifying allergens requires scanning ingredient lists for the "Big 9": milk, eggs, fish, crustaceans, shellfish, tree nuts, peanuts, wheat, and soybeans. Always check for "Contains" statements, scrutinize ingredient lists for hidden names, and read labels every time, as formulas change. If you are in doubt, contact manufacturers or store management directly. Be familiar with alternative names for allergens, e.g., albumin for eggs, casein for milk.

All countries should have rules and regulations on allergens because many people are still unaware of their dangers, and in some parts of the world, they can be mistaken for something else. If an allergen enters the body after food consumption, the reaction can be rapid and potentially more severe than with typical ingestion alone because it bypasses some digestive barriers and enters the circulation more directly.

The response is driven by an immune mechanism called an allergic reaction and, in severe cases, can progress to Anaphylaxis. The Typical reactions, ranging from mild to severe, are Skin reactions, often the first signs, Hives (urticaria), Itching, Flushing (redness), swelling - especially lips, eyelids, face. 2. Gastrointestinal symptoms are Nausea, Vomiting, Abdominal cramps, and diarrhea. 3. Respiratory symptoms (serious warning signs) are wheezing, Shortness of breath, Throat tightness, Hoarseness, and difficulty swallowing. These indicate airway involvement and require urgent attention. 4. Cardiovascular symptoms involve a drop in blood pressure, which could be dangerous, Dizziness or fainting, and a rapid or weak pulse. 5. Full systemic reaction: Anaphylaxis. This is the most severe outcome and can occur within minutes, and the Key features are a combination of skin + breathing + circulation symptoms, swelling of the throat, airway blockage, Sudden collapse or loss of consciousness. This is a medical emergency. Act immediately when you have trouble breathing, swelling of the tongue or throat, dizziness, or fainting.

Allergen labeling rules are broadly similar worldwide; most countries require disclosure, but they differ in which allergens are covered, how they're displayed, and whether precautionary warnings are mandatory. Here's a clear comparison of major allergens. In the United States, allergens are governed by the Food Allergen Labeling and Consumer Protection Act (FALCPA) and updates under the FASTER Act. There are 9 major allergens: milk, eggs, fish, shellfish, tree nuts, peanuts, wheat, soy, and sesame, and they must be declared in plain English, e.g., "Contains: Milk". For the European Union, allergens are regulated under the EU Food Information Regulation (FIR). Relative to the European

Union, 14 allergens must be declared, including cereals containing gluten, crustaceans, eggs, fish, peanuts, soyabeans, milk, tree nuts, celery, mustard, sesame seeds, sulfur dioxide/sulfites, lupin, and mollusks. Allergens must be visually emphasized in bold, color, or font. It applies to packaged and non-packaged foods, e.g., restaurants and bakeries. If the information is not on the restaurant's menu, a sign must be clearly visible directing you to where it is held, such as a special binder or other easily accessible location.

The United Kingdom has the Same 14 allergens as the EU. Mandatory labeling, even for prepacked foods prepared on-site; strong emphasis on clear highlighting and accessibility; and, obviously, among the strictest systems, especially for takeaway and prepared foods. Canada is regulated by the Canadian Food Inspection Agency. Their 11 priority allergens include allergens not in the U.S., such as mustard; require plain-language labeling similar to that in the U.S.; and cover both ingredients and certain derivatives. Australia / NZ New Zealand: Very strict; similar to the EU. Mandatory labeling of key allergens, including sesame and gluten. Strong enforcement and clarity standards. Japan requires labeling of specific mandatory allergens, e.g., egg, milk, wheat, and peanuts. Additional allergens are recommended but not mandatory.

Africa: At least 5 countries with enacted allergen labeling regulations are explicitly documented in regional summaries, e.g., South Africa, Kenya, Uganda, Tanzania, and Rwanda via the EAC framework; additional individual laws are noted across the continent. The literature emphasizes that African allergen labeling laws are emerging.

India: India has considered and moved toward mandatory disclosure; it is frequently listed among Asian jurisdictions with allergen requirements or active proposals and regulations.

Asia (including East, Southeast, and parts of South and Central Asia): At least 10+ countries with specific allergen labeling regulations are included, comprising China, Japan, South Korea, Malaysia, Singapore, Hong Kong, Kuwait, and others within the region that have enacted laws aligned with Codex or local lists. Notably, South Korea (21 allergens).

The Middle East, often grouped under Gulf/GSO and wider West Asia, have At least 6 Gulf Cooperation Council countries applied GSO allergen labeling standards (Saudi Arabia, UAE, Kuwait, Bahrain, Oman, Qatar), with Yemen also referenced under GSO coverage; additional Middle Eastern jurisdictions like Israel and Jordan have broader food labeling laws, and Kuwait is repeatedly cited with explicit allergen disclosure.

Know Hidden Names for Your Allergen

Regardless of location, hidden allergens are a real risk because ingredients aren't always listed by their simple names, such as "milk" or "peanuts." Here's a practical, food-safety-focused guide to common hidden names you should watch for on labels. Hidden names for milk and milk proteins are Casein / Caseinates (e.g., sodium caseinate), Whey (whey protein, whey solids), Lactalbumin / Lactoglobulin, Lactose (milk sugar), Butterfat / Butter oil, Ghee, Cream / Curds, Milk solids / Dry milk powder, Cheese powder, Caramel flavoring (sometimes contains milk derivatives), artificial butter flavor, protein powders, Some processed meats (use casein as a binder). Pay attention, if you see "casein" or "whey," it is = milk. Yes, Casein and whey are two primary proteins found in cow's milk.

Hidden names for peanuts can be especially tricky due to the use of oils and international terms. Common hidden terms are Groundnuts (common outside the U.S.), Arachis oil (the scientific name for peanut oil), Peanut flour/peanut protein, and Beer nuts (often peanuts). Highly refined peanut oil may not always trigger allergies, but it's still risky unless confirmed safe.

Hidden names for tree nuts like Almond (marzipan, nougat), Walnut, Pecan, Cashew, Pistachio, Hazelnut (filbert), Nut butters/nut pastes, Nut oils (e.g., walnut oil), Pesto (often contains pine nuts), Cereals, granola, snack bars. Liqueurs and flavoring.

Hidden names for eggs include Albumin (egg white protein), Globulin, Lecithin (can be egg-based, though often soy-based), Lysozyme, Ovalbumin. Baked goods, Pasta (especially fresh pasta), Mayonnaise, dressings, Foams (like meringue).

Hidden names for wheat/gluten are Semolina, Durum, Farina, Kamut, Bulgur, Triticale, Soy sauce (often contains wheat), Modified food starch (sometimes wheat-based in certain countries).

Hidden names for soy include soy protein isolate/concentrate, Textured vegetable protein

(TVP), Edamame, Miso, Tempeh, Natto, Lecithin (often soy-based), Vegetable oil blends, and Processed foods.

Hidden names for fish & shellfish. Fish: Anchovy (in sauces like Worcestershire), Fish sauce, Surimi (imitation crab), Shellfish: Crab, lobster, shrimp, Prawn, Krill. Watch sauces, soups, and international cuisine that you are not familiar with,

I recommend that you learn the names of your allergens (e.g., casein, whey for milk) and always read the full ingredient list, not just the “Contains” statement. Be food safety smart, enjoy, and always be aware that a banquet so safe from illness makes every bite bloom with delight; a feast consumed safely makes every bite sparkle with memorable flavor; and a meal grounded in safety produces unfailing satisfaction.

CHAPTER THREE

FOOD SAFETY OFFICIALS

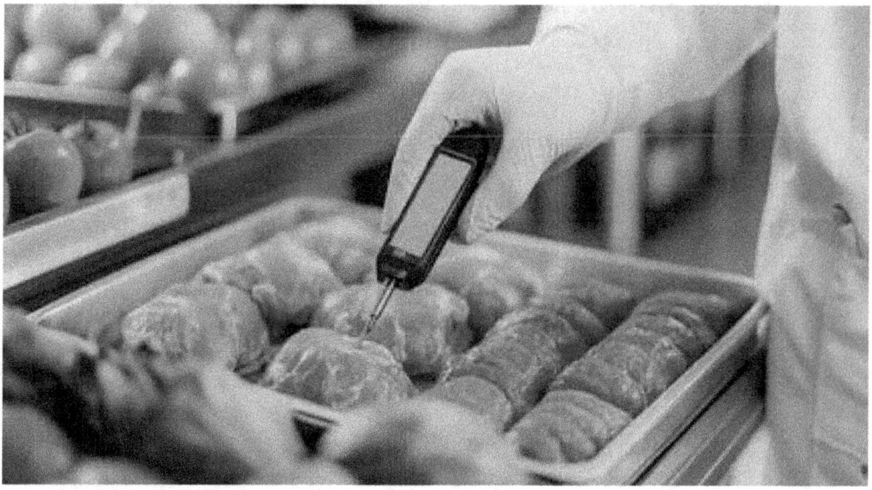

Working in food safety, whether with a state like Alabama or at the federal level with the United States Department of Agriculture, is one of those careers where your paycheck and your purpose line up in a very real way. Both state and federal programs are highly equipped to tackle food safety issues facing the nation. You will become part of a system that directly protects millions of people worldwide every single day. The work was deeply fulfilling, knowing my vigilance protected families, supported honest businesses, and strengthened the public's faith in our food system.

To the younger generation, here's the truth: this job feels good.

There's quiet satisfaction in knowing that what you do prevents illness before it happens. You may not always see the people you help, but every safe meal on someone's table is part of your impact. When you get paid, it's not just a paycheck; it's compensation for safeguarding public health, supporting proper growth and businesses, and maintaining trust in the food supply.

Why the job is rewarding

- You protect lives by preventing foodborne illnesses.
- You contribute to national food security and public confidence.
- The work is stable, respected, and needed everywhere.

- You gain specialized knowledge that makes you an expert in a critical field.
- There's room to grow in inspection, policy, research, training, and leadership.

Other aspects that make it a good career

- **Job stability:** Food safety is essential; there's always demand.
- **Government benefits:** Health insurance, retirement plans, and structured career progression.
- **Variety of work:** One day you're inspecting facilities, another day you're doing hands-on, another day you are reviewing industry records, and another day you are having a meeting with the Plant Managers.
- **Pride in service:** You're part of something bigger than yourself.

How to prepare for the job
Start early and be intentional:

- **Education:** Study fields like food science, biology courses, public health, nutrition, agriculture, and veterinary medicine.
- **Understand regulations:** Learn the basics of ROP (Rules of Practice), SPS(Sanitation Performance Standards), SSOP (Sanitation Standard Operating Procedures), HACCP

(Hazard Analysis and Critical Control Points), and Food Safety Regulatory Essentials.

- **Develop attention to detail:** This job requires sharp observation and integrity.
- **Get experience:** Work in food production or Industry, Internships in food production, labs, or public health departments help a lot.
- **Build communication/relationship skills:** You'll need to explain rules clearly and work with different people. You must be able to respect others. The more you respect others, the more you are respected.

A word of advice from experience

Take the responsibility seriously; it's about protecting families, children, and entire communities. It is about ensuring that the food on people's tables is wholesome, safe, and free of adulteration. When you approach work with integrity, the job gives back to you in pride, stability, and purpose. If you're looking for a career where you can earn a living and make a real difference every single day, I assure you this is one worth pursuing.

GOVERNMENT FOOD SAFETY OFFICIALS ARE YOUR FRIENDS, NOT ENEMIES

A government food safety official is not an enemy of the food industry; rather, the official is a public health partner whose responsibility is to help food businesses succeed while protecting consumers from foodborne illness, contamination, adulteration, allergens, and unsafe practices. The relationship between regulators and the food industry should be based on cooperation, education, transparency, and mutual respect.

Food safety officials understand that restaurants, processors, manufacturers, farmers, distributors, and retailers are essential to the economy and to public well-being. Their goal is not to punish businesses unnecessarily but to help operators comply with food safety laws, improve food-handling practices, prevent outbreaks, and maintain consumer confidence. When consumers trust that food is safe, the industry benefits from stronger sales, fewer recalls, fewer lawsuits, and an improved reputation.

A professional food safety official demonstrates friendship toward the industry in several important ways:

- By educating food operators on safe food handling, sanitation, allergen control, temperature control, labeling, and contamination prevention.

- By explaining regulations clearly instead of simply issuing violations.
- By helping businesses correct problems before they become serious public health hazards.
- By encouraging preventive systems such as Hazard Analysis and Critical Control Points (HACCP).
- By recognizing and praising businesses that maintain high food safety standards.
- By treating owners and employees fairly, respectfully, and professionally during inspections.
- By understanding the financial and operational challenges faced by food businesses while still protecting public health.
- By serving as a technical advisor and resource during emergencies, outbreaks, recalls, and compliance questions.

For example, when a food establishment has improper cold-holding temperatures, the official does not simply seek punishment. Instead, the official explains why temperature control matters, how bacteria such as Salmonella, Listeria, and Campylobacter can grow in unsafe conditions, and how corrective actions can protect both customers and the business itself.

Similarly, during food recalls or contamination events, food safety officials work closely with industry partners to quickly identify affected products, remove unsafe food from commerce, and restore consumer confidence. This cooperative relationship protects lives and helps businesses recover more effectively.

I strongly believe that a good food safety official also recognizes that most food operators want to do the right thing. Many violations result from a lack of training, misunderstanding, equipment failure, or resource limitations rather than intentional misconduct. Therefore, the official focuses on prevention, training, and continuous improvement rather than on fear and intimidation.

Ultimately, both the food industry and government food safety agencies share the same mission:

- Safe food
- Healthy consumers
- Strong businesses
- Public confidence in the food supply

When food safety officials and the food industry work together as partners rather than adversaries, communities become healthier, outbreaks decrease, businesses prosper, and the entire food system becomes stronger and more trusted.

EXPLAIN FOOD SAFETY TO CHILDREN

Food safety inspectors are those superheroes who check the food we eat. Picture the food we buy from stores and restaurants as our favorite snacks or meals. You want it to be safe to eat, just like you would expect other superheroes to defend you from danger. Food safety inspectors are like those superheroes. They go to places where food is prepared, such as restaurants and food-processing facilities, to ensure everything is clean and safe. They make sure the food doesn't contain harmful germs or anything that could make you sick.

These superhero inspectors use elite tools and their super senses to look at the food, the kitchen, the processing room, the cooking or smokehouse areas, the packaging room, and the entire premises. They want to make sure that the food is good to eat and won't hurt you. If they notice any problems, they inform the people who make the food so they can correct them before anyone eats it.

Yes, food safety inspectors are necessary because they keep us safe from getting sick when we eat our favorite foods. Just as superheroes guard us from bad people, food safety inspectors protect us from harmful things in the food we eat.

CHAPTER FOUR

GET THE POISON OFF YOUR FOOD

You can't realistically eliminate the risk of food poisoning *100% all the time*; that's the first thing to be clear about. But an ordinary person can reduce the risk to nearly zero by consistently following a few core food safety habits. These are the same principles used by professionals, just applied at home.

1. Keep everything clean

- Wash hands with soap before and after handling food
- Clean cutting boards, knives, and countertops
- Rinse fruits and vegetables under running water
- Avoid preparing food when you're sick

2. Separate raw and cooked foods

- Keep raw meat, poultry, and seafood away from ready-to-eat foods
- Use separate cutting boards if possible
- Store raw meat on the lowest shelf in the fridge to prevent dripping

3. Cook food thoroughly

- Undercooking is one of the biggest causes of food poisoning
- Make sure foods—especially meat, poultry, eggs, and seafood—are fully cooked
- When possible, use a food thermometer (not just appearance)

4. Keep food at safe temperatures

- Refrigerate perishable foods promptly (within 2 hours, or 1 hour if it's hot)
- Keep your fridge cold (around 40°F / 4°C or below)
- Don't leave cooked food sitting out too long

5. Store food properly

- Follow "first in, first out" (use older food before newer items)

- Pay attention to expiration or "use-by" dates
- Store leftovers in sealed containers and eat within a few days

6. Reheat food safely

- Heat leftovers until they are steaming hot all the way through
- Avoid reheating food multiple times

7. Be cautious with high-risk foods

- Raw or undercooked eggs, unpasteurized milk, and raw seafood carry a higher risk
- Be extra careful with these, especially for children, elderly people, or pregnant women

The bottom line:
Food safety isn't about one big action—it's about consistent small habits. If you clean, separate, cook, and chill food properly every day, you dramatically reduce the chance of food poisoning and protect both yourself and others. The rationale behind the core food safety practices—cleaning, separating, cooking/cooling, and proper storage—is grounded in one simple goal: to prevent the growth and spread of harmful microorganisms that cause foodborne illness.

Cleaning (food, hands, and surfaces)
Cleaning removes dirt, bacteria, and other contaminants that can cause illness. Harmful microorganisms such as Salmonella, Escherichia coli, and Listeria monocytogenes can easily transfer from hands, utensils, or surfaces onto food. Rationale

Rationale: Reduce contamination at the source and stop pathogens from entering the food in the first place.

2. Separating raw and cooked foods
Raw foods—especially meat, poultry, and seafood—often carry bacteria that are destroyed during cooking. If they come into contact with cooked or ready-to-eat foods, they can reintroduce harmful microbes.

- **Rationale:** Prevent **cross-contamination**, which is one of the leading causes of food poisoning.

3. Cooking and cooling properly
Cooking kills most harmful microorganisms, but if food is not cooled correctly afterward, surviving bacteria can multiply rapidly. Bacteria grow

fastest in the "danger zone" (about 40°F–140°F / 4°C–60°C).

- **Rationale:**
    - Cooking → **destroys pathogens**
    - Cooling → **prevents rapid bacterial growth.**

4. Proper storage of food

Even after safe cooking, improper storage can allow bacteria to grow again. Refrigeration slows bacterial growth, while proper sealing prevents contamination and spoilage.

- Yes, it is pertinent that you ensure that your food is free of pathogens. A pathogen is a bacterium or other microorganism that can cause disease. Regardless of how tasty, nutritious, or organic your food is, it is not a healing food if it is loaded with pathogens. There are five major types of pathogens: bacteria, fungi, parasites, protozoa, and viruses. The most common Foodborne Pathogens are 1. Norovirus: Leading cause of acute gastroenteritis. 2. Campylobacter: A common cause of bacterial diarrhea. 3. Salmonella: Frequently linked to contaminated poultry and produce. 4. Clostridium perfringens: Often found in inadequately heated food. 5. Shiga toxin-producing Escherichia coli (STEC): Known for causing severe stomach cramps and diarrhea. Food pathogens are living microorganisms that fall under the biological hazard category, but this book discusses all three hazards.

Identify Food Hazards

The best way for food industries, cafeterias, restaurants, and other food outlets to utilize to prevent these illness-causing agents is to first identify food hazards. The categories of hazards to be discussed here are biological, chemical, and physical. Foodborne illness is most commonly caused by biological hazards, but serious harm can also come from chemical and physical hazards. A complete food safety system must address all three categories simultaneously.

CHAPTER FIVE

BIOLOGICAL HAZARDS

Biological hazards can cause illnesses. Potential biological hazards include bacteria, viruses, toxins, protozoa, and parasites. The most critical of all these biological hazards are bacteria, which cause either infections or intoxications. Bacteria are responsible for the majority of all food-borne illnesses. So, whenever you see the word "pathogen" or "pathogenic," it simply implies that the bacteria cause illness. According to the United States Department of Agriculture, “Most food-borne illness can result in diarrhea, nausea, vomiting, cramping, and in more serious cases, organ damage, stillborn births, and death.”

The pathogens commonly found in cattle, sheep, swine, chickens, and turkeys are Salmonella, Campylobacter, Listeria monocytogenes, and Escherichia coli O157:H7. There are other forms of E. coli that are not pathogenic, but for emphasis, E. coli 0157:H7 causes human illness. The main source of all these pathogens is healthy animals. Pathogens like

Listeria are widespread in the environment, water, soil, and feed.

Some bacteria produce toxins in Ready-to-Eat or Not Ready-to-Eat foods during their growth. Toxins we should be concerned about are produced by Clostridium botulinum, Clostridium perfringens, Bacillus cereus, and Staphylococcus aureus, and these result from bacterial growth and toxin production in foods that were not properly handled. When toxin like C. botulinum toxin is consumed, they can cause headaches, neurological damage, disorientation, paralysis, or even death. These toxins can affect nerve endings and disrupt nerve impulses. Other examples of these toxins include enterotoxins, neurotoxins, and hemotoxins, which are produced by organisms. Salmonella, E. coli 0157:H7, and Campylobacter may be controlled effectively by cooking your meat to an internal temperature of 158 degrees Fahrenheit, and poultry to 160 degrees Fahrenheit. It is possible for the food to get contaminated after it is cooked. Proper cooling and storage of food will avert bacterial growth and, more significantly, the production of their toxins. Nevertheless, cooking will not destroy some of these toxins once they are formed in food.

Our Cafeterias, food-processing industries, and other food businesses are supposed to ensure that the food we eat is free of pathogens. However, one strain of Listeria species is of great concern: **Listeria Monocytogenes (LM)**. For the reason that it could be very pathogenic to pregnant women, newborns, the elderly, and those with weakened immune systems, food businesses and kitchens must be mindful of the pathogen. Listeria monocytogenes is in the intestines of healthy animals, and in the setting in which food manufacturing animals are raised and processed, for instance, soil, water, and vegetation, it can be successively introduced into the food processing atmosphere. Lm can contaminate the surfaces of food processing equipment, floors, walls, drains, overhead structures, and more.

It is important to know that Listeria monocytogenes can easily spread through direct contact with a contaminated surface. Amazingly, it can survive and grow in cool or damp areas. As a result, it is important to remove food debris from the floor and any area where food particles are attached, so as not to provide nutrients and support for bacterial multiplication.

It is of immense importance to follow strict sanitary measures to combat Lm, as it is more heat-resistant than most foodborne pathogens and can survive freezing or drying. Listeria is a very stubborn pathogen that can grow in refrigerated, packaged, ready-to-eat products, resists salt, nitrite, and acidity, and can even grow in vacuum-packed food products. The fact that organisms require moisture to grow is a good reason to keep your storage and production areas and equipment as dry as possible because it will minimize the chances of Listeria monocytogen's reproduction. You must ensure that product-contact utensils, equipment, and floors in the processing areas are kept free of standing water whenever possible.

Some foodborne illnesses and deaths have been linked to recalled products adulterated with Lm. The conclusion is that adulteration occurred through cross-contamination from environmental sources after the food was cooked. Cross-contamination occurs when cooked food is exposed to contaminated surfaces, employees, home food preparers, or environmental reservoirs. The best way to prevent cross-contamination is for cafeterias and other food industries to ensure that effective sanitation and employee hygiene are maintained in areas where food is processed and stored after cooking. You can also mess up your food if you prepare it in an area undergoing reasonable construction. Certain types of construction and building conditions are unsafe for food preparation because they harbor contaminants, are difficult to clean, or allow pests and moisture to accumulate. The core principle is simple: if a surface or structure can't be cleaned, sanitized, and maintained, it's unsafe. Here are the main types of unsafe construction: 1. Porous and absorbent materials. Examples: unfinished wood, raw plywood, carpet, cardboard. These materials absorb moisture, grease, and bacteria. They cannot be properly cleaned or sanitized. Risk: growth of pathogens like Salmonella and mold. 2. Cracked, broken, or damaged surfaces, Examples: chipped tiles, cracked countertops, peeling paint. Cracks trap food particles and moisture, becoming breeding grounds for bacteria and pests. Risk: hidden contamination and difficult cleaning. 3. Rusted or corroded metals. Rust flakes can contaminate food. Corroded surfaces are no longer smooth or cleanable. Risk: physical and microbial contamination. 4. Poorly sealed joints and gaps like open seams, unsealed wall-floor junctions, which allow entry of pests (rodents, insects), trap dirt and moisture. Risk: contamination from pests and bacteria. 5. Inadequate drainage and standing water. Water accumulation promotes bacterial growth and attracts pests. Increases risk of slips and unsanitary conditions. Risk: growth of organisms like Listeria monocytogenes. 6. Low ceilings or overhead contamination risks like exposed pipes, dripping, or flaking ceilings, which can drip water, dust, or debris onto food. Risk: direct contamination of food and surfaces. Poor ventilation systems cause the buildup of heat, smoke, grease, and moisture, leading to mold growth and condensation. Risk: contamination and unsafe working conditions. The use of toxic or non–food-grade materials, such as lead-based paint, treated wood, and industrial chemicals, can cause harmful substances to leach into food, posing a risk of chemical contamination.

Whether you are a small or large food-processing business, a school or church cafeteria, or a personal kitchen, work such as breaking walls, removing floor coatings, or other activities that are bound to generate dust must be done with caution. Sometimes, dust from construction can be hard to detect or control, so emphasis should be placed on ensuring that food is properly handled, stored, and kept free of dust during construction. The establishments I covered usually built demarcations to trap dust in areas under construction. Once construction is complete, detailed cleanup and sanitization will be carried out in the construction area. After the construction, USDA and establishment's samples usually come back negative for Listeria monocytogenes because effective precautionary measures were taken. While it will cost you a little bit to constructively block the dust with a temporary demarcation, it will cost you hugely if you don't block the dust and a USDA sample comes back positive, or your food eaten by consumers gets some people sick because it had the pathogen, a deadly bacterium.

The sanitizers shown to be effective against Lm are quaternary ammonium compounds, chlorine solutions, iodophors, and peracetic acid-containing products. You must apply your sanitizer after the equipment and environment have been thoroughly cleaned. Heat applied as hot water could also be used as a sanitizer. Rotating Sanitizer is generally an effective method, as it offers effective protection against bacteria. Rotating a sanitizer could mean using a particular type for a period and then switching to another. An additional method of rotating sanitizers could entail using different sanitizers in different areas of the establishment. For example, you could use one on the equipment and another on the drains. Also, alternating between alkaline- and acid-based detergents could systematically shift the pH, thereby preventing bacterial adaptation to a particular environment. Consequently, rotating sanitizers can help prevent biofilm formation. A biofilm is a thin layer of microorganisms that principally attach themselves together to stick to the surface of equipment and other structures. Since the biofilm is stubborn, the bacteria are hard to remove and are sometimes protected from the chemicals used to clean and sanitize surfaces.

Be mindful of reservoirs in the environment.

Many processes in the food processing industry can expose ready-to-eat products to environments that can contaminate them. For example, moist conditions boost microbial growth that can contaminate food. Splash from cleaning materials can contaminate products, and drips from overhead cooling units and other structures can also be sources of contamination. Wet insulation, maintenance tools, pallets, and forklifts can contaminate products. Even switch boxes, motor housing, cracked water hoses, cracked walls, and ice makers can contaminate products if not properly managed.

Structural Design

The location of the ready-to-eat products is critical. You must ensure that your structural design does not encourage constant traffic between raw and ready-to-eat areas. Minimize overhead structures that can harbor dirt or moisture. Install overhead structures that discourage condensation. You must ensure that the raw processing area is clearly separated from the ready-to-eat processing area. Furthermore, make sure that air is not flowing from raw areas to RTE areas. Airflow can be affected by the location of ventilation and refrigeration equipment. If not guarded, airflow from a raw product area into the ready-to-consume area could carry Listeria monocytogenes, contaminating products and product-contact surfaces.

Salmonella: Another biological hazard and bacterium commonly found in raw or improperly cooked poultry, eggs, and contaminated meat products. Preventing Salmonella contamination in food is critical due to the potential health risks associated with this bacterial infection. Salmonella is a common cause of foodborne illness worldwide and can lead to a condition called salmonellosis. Here are some reasons why it is crucial to prevent Salmonella in your food:

(a) Human health impact: Salmonellosis can cause a range of symptoms, including diarrhea, abdominal cramps, fever, vomiting, and nausea. While most people recover without medical treatment, severe cases can lead to dehydration, hospitalization, and even death, particularly in vulnerable populations such as infants, young children, elderly individuals, and those with weakened immune systems.

(b) Widespread occurrence: Salmonella bacteria can contaminate a wide variety of foods, including poultry, eggs, meat, dairy products, fruits, vegetables, and even processed foods. It can be present in raw or undercooked animal products, and cross-contamination can occur if proper food handling and preparation practices are not followed.

(c) Public health burden: Salmonella outbreaks can have significant public health implications, leading to many illnesses, hospitalizations, and even deaths. These outbreaks can strain healthcare systems and result in economic losses from medical costs, reduced productivity, investigation, and management costs.

(d) Food industry consequences: Foodborne illness outbreaks caused by Salmonella can have severe consequences for the food industry. Companies may face legal actions, product recalls, reputational damage, and financial losses. Ensuring food safety and preventing Salmonella contamination is crucial for maintaining consumer trust and protecting the integrity of food businesses.

Preventing transmission: Salmonella can be easily transmitted from contaminated food to humans. By implementing preventive measures throughout the food production chain, including proper hygiene practices, through cooking, effective sanitation, and regular testing, the risk of Salmonella contamination can be significantly reduced, preventing the spread of the bacteria and subsequent illnesses. Overall, preventing Salmonella in food is essential to safeguard public health, minimize the risk of foodborne illness, and maintain the safety and integrity of the food supply chain. It requires a collective effort involving government agencies, food producers, processors, retailers, community leaders, and consumers to ensure proper food safety practices are followed at every stage.

Escherichia Coli (E. Coli): Certain strains of E. coli, such as E. coli O157:H7, can cause severe foodborne illness. They are commonly associated with undercooked ground beef and contaminated water. Preventing E. coli O157:H7 contamination of human food is critical because it can cause severe illness and even death. E. coli O157:H7 is a strain of Escherichia coli bacteria that produces a toxin called Shiga toxin, which can lead to a condition called hemolytic uremic syndrome (HUS). Here are a few key reasons why it is crucial to prevent E. coli O157:H7 contamination:

(a) Human health risks: E. coli O157:H7 is a foodborne pathogen that can cause gastrointestinal symptoms such as severe abdominal cramps, diarrhea (often bloody), and vomiting. In more severe cases, it can lead to HUS, a potentially life-threatening condition characterized by kidney failure, anemia, and low platelet count. Children, the elderly, and individuals with weakened immune systems are particularly vulnerable to these complications.

(b) Outbreaks and public health: E. coli O157:H7 has been associated with numerous foodborne illness outbreaks worldwide, often linked to contaminated ground beef, raw milk, fresh produce (such as spinach or lettuce), and other food products. These outbreaks can result in significant public health impacts, including hospitalizations, economic losses, and even fatalities. Preventing contamination at the source helps to minimize the risk of outbreaks and protect public health.

(c) Food safety regulations and consumer confidence: Preventing E. coli O157:H7 contamination is a fundamental aspect of food safety regulations implemented by government agencies and industry standards. By adhering to these regulations and implementing preventive measures, food producers and processors can maintain consumer confidence in the safety and quality of their products. Public awareness of foodborne illness risks has increased in recent years, making it essential for food businesses to prioritize safety and prevent contamination.

(d) Economic implications: Foodborne illness outbreaks associated with E. coli O157:H7 can have significant economic consequences for the food industry. These outbreaks often lead to product recalls, market withdrawals, and temporary or permanent closure of food production facilities, resulting in financial losses for companies. Additionally, the negative publicity and loss of consumer trust can have long-term effects on brand reputation and customer loyalty.

Prevention of E. coli O157:H7 contamination: Food safety practices such as good hygiene, proper sanitation, thorough cooking, and effective supply chain management are crucial. Regular testing and monitoring throughout the food production process, from farm to fork, can help identify and mitigate potential contamination risks, ensuring the safety of the food supply and protecting public health.

Campylobacter: a bacterial pathogen commonly associated with poultry, but it can also contaminate other meats. Improper handling and undercooking of meat products can lead to Campylobacter infection. Preventing Campylobacter from contaminating human food is crucial for several reasons:

(a) Illness and Health Risks: Campylobacter is one of the leading causes of bacterial foodborne illness worldwide. Consuming contaminated food, especially raw or undercooked poultry, unpasteurized milk, or contaminated water, can result in Campylobacteriosis. This illness can cause symptoms such as diarrhea, abdominal pain, fever, and vomiting. Although most cases are self-limiting and resolve within a week, severe cases can lead to hospitalization and potentially life-threatening complications, especially in vulnerable populations such as young children, the elderly, and individuals with weakened immune systems.

(b) Public Health Burden: Campylobacteriosis is a significant public health concern due to the high number of cases reported annually. According to the World Health Organization (WHO), Campylobacter is estimated to cause over 96 million cases of gastroenteritis worldwide each year. By preventing Campylobacter contamination in food, the incidence of illness can be reduced, resulting in lower healthcare costs and improved overall public health.

(c) Economic Impact: Campylobacter contamination in food can have substantial economic consequences. Outbreaks of Campylobacteriosis can lead to increased healthcare expenses, lost productivity, and economic losses for the food industry. Additionally, food recalls and loss of consumer confidence can significantly impact the sales and reputation of food producers and retailers.

(d) Antibiotic Resistance: Campylobacter strains have shown increasing resistance to antibiotics commonly used to treat bacterial infections. The inappropriate or excessive use of antibiotics in both human and animal health contributes to the development and spread of antibiotic-resistant Campylobacter strains. Contaminated food can be a source of antibiotic-resistant Campylobacter infections. By preventing Campylobacter contamination, the selective pressure for antibiotic resistance can be reduced, helping to preserve the effectiveness of antibiotics for treating bacterial infections.

Preventing Campylobacter: To prevent Campylobacter contamination in food, it is important to implement good hygiene practices during food production, processing, and handling in personal kitchens, and during preparation in school, church, and business restaurants. This includes properly cooking meat and poultry, avoiding cross-contamination between raw and cooked foods, practicing good hygiene, and ensuring the safety and quality of water sources used in food production.

Clostridium perfringens: a bacterium that can cause food poisoning when food is improperly handled or stored. Preventing Clostridium perfringens contamination in human food is critical due to the associated health risks. Clostridium perfringens is a bacterium commonly found in the environment, including soil and the intestines of humans and animals. It can also be present in raw or undercooked meat and poultry, as well as in foods prepared in large quantities and kept warm for extended periods, such as at buffets or catering events. When Clostridium perfringens contaminates food, it can multiply rapidly, particularly in low-oxygen environments like sealed containers or vacuum-sealed packages. The bacteria produce a toxin that causes illness when ingested. The main concern with Clostridium perfringens is its ability to cause foodborne illness, known as Clostridium perfringens food poisoning or gastroenteritis.

The symptoms of Clostridium perfringens food poisoning typically appear within 6 to 24 hours after consuming contaminated food. They include abdominal pain, cramps, diarrhea, and occasionally nausea and vomiting. Although the illness is usually self-limiting and resolves within 24 to 48 hours, it can cause significant discomfort and dehydration, particularly in vulnerable populations such as the elderly, young children, and individuals with weakened immune systems.

Preventing Clostridium: Preventing Clostridium perfringens contamination involves implementing proper food safety practices. This includes:

(a) Maintaining proper food handling and storage: Food should be stored at appropriate temperatures to prevent bacterial growth. Hot foods should be kept above 140 F (60 C), and cold foods should be refrigerated below 40 F (4 C).

(b) Thorough cooking: Adequate cooking temperatures and times should be followed, especially for meat and poultry, to ensure the destruction of any potential bacteria.

(c) Rapid cooling: Cooked foods should be cooled quickly to prevent bacterial growth. They should be refrigerated or stored in shallow containers to promote faster cooling.

(d) Avoiding prolonged temperature abuse: Foods should not be held at room temperature for extended periods, as this allows bacteria to multiply rapidly.

(e) Practicing good personal hygiene: Food handlers should maintain proper hand hygiene, including washing hands thoroughly with soap and water before handling food.

By following these practices, the risk of Clostridium perfringens contamination in food can be minimized, thus reducing the occurrence of foodborne illness and protecting public health.

Staphylococcus aureus: a bacterium that can produce toxins causing food poisoning. It can be found in processed meats and other food products when handled by individuals with poor hygiene. Preventing Staphylococcus aureus from contaminating human food is important for several reasons:

(a) Foodborne illness: Staphylococcus aureus can cause food poisoning when the bacteria or their toxins are ingested. The symptoms of Staphylococcus aureus food poisoning include nausea, vomiting, abdominal cramps, and diarrhea. While most cases are self-limiting and resolved within a few days, severe cases can lead to dehydration and require medical attention.

(b) Toxin production: Staphylococcus aureus produces heat-stable toxins that are not destroyed by cooking. One of the best-known heat-resistant toxins produced by Staphylococcus is called a Staphylococcal enterotoxin. Yes! These toxins are not easily destroyed by heat. These toxins can be present in contaminated food and cause illness even if the bacteria themselves are killed. This means that even if the food is cooked properly, it can still cause food poisoning if contaminated with Staphylococcus aureus.

(c) Cross-contamination: Staphylococcus aureus can be present on the skin, in the nose, and on other surfaces of humans. If proper hygiene practices are not followed during food preparation, the bacteria can be transferred from these sources onto the food, leading to contamination. This can occur when food handlers do not wash their hands properly, when contaminated utensils or equipment are used, or when raw and cooked foods come into contact.

(d) Rapid growth: Staphylococcus aureus can multiply rapidly in certain conditions, particularly when food is kept at temperatures between 40 f and 140 f degrees. This temperature range is often referred to as the “danger zone” for bacterial growth. If contaminated food is left at these temperatures for an extended period, the bacteria can multiply to high levels, increasing the risk of foodborne illness.

To prevent Staphylococcus aureus contamination in food, it is important to follow good food hygiene practices, such as regular handwashing, proper cooking, and storage temperatures, avoiding cross-contamination, and maintaining clean food preparation surfaces and utensils; these measures help reduce the risk of foodborne illness and ensure the safety of the food we consume.

Shigella: It is important to prevent Shigella, a type of bacteria that can cause gastrointestinal illness, from contaminating human food for several reasons:

(a) Health Risks: Shigella is a major cause of diarrheal disease worldwide, particularly in developing countries with poor sanitation practices. Infection with Shigella can lead to symptoms such as diarrhea (often bloody), abdominal pain, fever, and vomiting. In severe cases, it can cause complications like dehydration, malnutrition, and even death, especially in vulnerable populations such as young children, the elderly, and individuals with weakened immune systems.

(b) Transmission: Shigella is highly contagious and can spread easily from person to person through the fecal-oral route. Contaminated food and water are common sources of transmission, particularly when proper hygiene and sanitation practices are lacking. Preventing Shigella from entering the food supply helps reduce the risk of outbreaks and the subsequent transmission of the bacteria to individuals who consume contaminated food.

(c) Foodborne Outbreaks: Shigella outbreaks associated with contaminated food can have significant public health implications. These outbreaks can affect many people within a short period, leading to increased healthcare costs, productivity losses, and potential long-term health consequences for those affected. By preventing Shigella contamination in food, the likelihood of such outbreaks can be minimized.

(d) Food Safety Regulations: Government authorities and regulatory agencies have established food safety regulations and guidelines to protect consumers from foodborne illnesses. Compliance with these regulations is crucial for food businesses to ensure the safety of their products. Failure to prevent Shigella contamination can result in legal consequences, damage to the business's reputation, and loss of consumer trust.

(e) Preventive Measures: Preventing Shigella contamination in food involves implementing good hygiene and sanitation practices throughout the food production and handling processes. This includes proper handwashing, maintaining clean food preparation surfaces and utensils, avoiding cross-contamination between raw and cooked foods, ensuring adequate cooking temperatures, and practicing safe food storage and handling. By adhering to these preventive measures, the risk of Shigella contamination can be significantly reduced, safeguarding public health. Overall, preventing Shigella contamination of human food is crucial to protecting public health, minimizing the spread of infection, preventing foodborne outbreaks, complying with food safety regulations, and ensuring consumer safety and well-being.

Bacillus Cereus: Preventing its contamination of human food is important due to its potential health risks. Bacillus cereus is a common cause of foodborne illness and can produce toxins that cause two main types of food poisoning: emetic and diarrhea.

(a) Emetic Food Poisoning: Bacillus cereus can produce a heat-stable toxin that causes emetic food poisoning. This type of poisoning typically results from consuming contaminated rice or other starchy foods. Symptoms usually appear within 1 to 5 hours and include nausea, vomiting, and occasionally abdominal cramps. Although emetic food poisoning is generally self-limiting and resolves within 24 hours, it can cause significant discomfort and illness.

(b) Diarrheal Food Poisoning: Bacillus cereus can also produce toxins that cause diarrheal food poisoning. This type of poisoning occurs when the bacteria multiply in food and release toxins that affect the gastrointestinal system. Symptoms typically include watery diarrhea, abdominal cramps, and occasionally nausea and vomiting. Diarrheal food poisoning caused by Bacillus cereus usually lasts for about 24 to 48 hours.

While most healthy individuals recover fully from Bacillus cereus food poisoning, certain vulnerable populations, such as children, the elderly, and individuals with weakened immune systems, are at a higher risk of developing severe complications. In rare cases, severe infections can occur, leading to systemic illnesses such as septicemia and pneumonia.

Clostridium Botulinum: Preventing botulism from contaminating human food is crucial because it is a serious, potentially life-threatening illness caused by the toxin produced by this bacterium. This bacterium can thrive in low-oxygen environments, such as improperly canned or preserved foods, and, when conditions are right, produce a toxin that, if ingested, can cause botulism. Here are a few reasons why it's important to prevent botulism from getting into human food:

(a) Health Risks: Botulism can cause severe paralysis, difficulty breathing, and can be fatal if not treated promptly. Even small amounts of botulinum toxin can cause serious illness, so preventing its presence in food is a matter of public health and safety.

(b) Food Safety: Ensuring that food is free from botulism helps maintain food safety standards and prevents the risk of contaminated products reaching consumers. This is particularly important for commercially produced and distributed foods, as well as for homemade or locally produced goods.

(c) Economic Impact: Contamination of food products with botulism can lead to costly recalls, loss of consumer confidence, and potential legal liabilities. Preventing botulism helps protect the food industry from financial losses and reputational damage.

To prevent botulism in food, it's important to follow food safety and preservation practices, including using appropriate canning methods, maintaining proper hygiene during food preparation, and exercising caution with home-canned or fermented foods. Additionally, commercial food producers should adhere to strict quality control measures and government regulations to ensure product safety.

Another biological hazard of interest is **the virus**. Viruses can cause gastroenteritis and diseases like hepatitis and polio in people.

a. **Norovirus** is a common food-borne sickness which causes a medical condition depicted by diarrhea, vomiting, and abdominal pain. In most cases, the presence of viruses in food or even water is linked to sick food workers. Healthy animals are not necessarily considered carriers of viruses. This hazard does not grow or reproduce in meat and poultry products. There are 3 main ways these microorganisms can enter human bodies. They can enter through the respiratory system, be transmitted via the infected person's body fluids, or be transmitted via a contaminated object. An infected worker or a cook can easily transfer it to another person. This fact is a good reason to ensure that people who work in food-processing areas, restaurants, or personal kitchens avoid being around food if their illnesses cause them to sneeze or cough.

b. **Hepatitis A virus**: Hepatitis A virus can cause liver inflammation and is transmitted through the fecal-oral route. Infected food handlers who do not practice proper hygiene can contaminate food and beverages, leading to outbreaks. Raw or undercooked shellfish, fruits, and vegetables, as well as contaminated water, are common sources of the hepatitis A virus.

c. **Rotavirus**: Rotavirus primarily affects young children and causes severe diarrhea and vomiting. Although it is commonly transmitted through person-to-person contact, contaminated food and water can also contribute to its spread.

d. **Hepatitis E virus**: Hepatitis E virus is prevalent in developing countries and is transmitted through contaminated water and undercooked or raw pork and other meats. It can cause acute hepatitis, like the hepatitis A virus.

e. **Enteroviruses**: Enteroviruses, including coxsackievirus and echovirus, can cause gastrointestinal illnesses. These viruses can contaminate food and water and spread through the fecal-oral route.

It's essential to note that proper food handling, preparation, and hygiene practices, such as thorough cooking, proper handwashing, and avoiding cross-transmission of these viruses through food.

Another biological hazard of concern is **parasites.** Parasites are usually found in raw meat. a.

Trichinellosis or trichinosis can be gotten from eating raw or not well-cooked meat of animals that are infected with the larvae of the worm species known as Trichinella. The symptoms and signs of Trichinellosis include diarrhea, vomiting, nausea, fever, fatigue, abdominal discomfort, headache, and muscle pain. A good reason to cook your meat properly. Ensure that your grocer or family members separate your raw meat from ready-to-eat foods and from foods that could be eaten raw, such as fruits and vegetables. If you have only one sink, it may be important to process your fruits and vegetables first: wash, rinse, and sanitize them before handling raw meat. Also, ensure that your water is free of contamination.

b. **Toxoplasma gondii**: This parasite can also be found in undercooked raw meat. Toxoplasma gondii is a parasitic protozoan that can cause the disease toxoplasmosis in humans. It is important to prevent Toxoplasma gondii from contaminating human food because the parasite can pose a risk. Here are a few reasons why it's important to prevent Toxoplasma gondii from getting into human food.

* Health Risks: Toxoplasmosis can cause mild flu-like symptoms in healthy individuals, but it can be more severe in people with weakened immune systems. Pregnant women who become infected can pass the parasite to their unborn child, leading to potential birth defects or other serious health problems.

* Food Safety: Contamination of food with Toxoplasma gondii can compromise food safety. This can occur through the consumption of raw or undercooked meat containing the parasite's cysts, or through contact with contaminated soil, water, or surfaces.

Preventing Toxoplasma gondii from contaminating human food involves several steps, including:

- Cooking Meat Thoroughly: As stated earlier, Toxoplasma gondii can be present in raw or undercooked meat, especially lamb, pork, and venison. To prevent contamination, it is important to cook meat to a safe internal temperature (usually 145 F for whole cuts of meat and 160 F for ground meats) to kill any parasites present.

- Washing Fruits and Vegetables: Toxoplasma gondii can be present in soil, so it's important to thoroughly wash fruits and vegetables, especially if they are consumed raw.

- Good Hygiene Practices: Practicing good hygiene, such as washing hands and kitchen surfaces thoroughly after handling raw meat, can help prevent the spread of Toxoplasma gondii and other foodborne pathogens.

- Avoiding Cross-Contamination: It's important to prevent cross-contamination between raw and cooked foods, as well as between foods that will be eaten raw and those that will be cooked.

- Avoiding Unpasteurized Products: Consuming unpasteurized dairy products or juices can also pose a risk of Toxoplasma gondii contamination, so it's important to choose pasteurized products to reduce the risk.

c. **Taenia solium (pork tapeworm) and Taenia saginata (beef tapeworm):** These tapeworms can be transmitted to humans through the consumption of undercooked or raw pork or beef, respectively. Once ingested, the tapeworms can cause a condition known as cysticercosis, which can affect the muscles, eyes, and brain. As a result, it is important to prevent the transmission of parasites such as Taenia solium and Taenia saginata into the human food supply, as these parasites can cause serious health issues when ingested. Here's why it's important to prevent them and steps to help avert their transmission:

* Health Risks: Consumption of undercooked or raw pork or beef contaminated with the larvae of these tapeworms can lead to the development of tapeworm infection in humans. These infections can cause symptoms such as abdominal pain, nausea, weakness, and weight loss. In severe cases, the larvae can migrate to other parts of the body, including the brain, leading to a condition known as neurocysticercosis, which can result in seizures, neurological damage, and even death.

* Economic Impact: In addition to the health risks, the presence of these parasites in the food supply can have economic implications due to the costs associated with treating infected individuals and

the potential for reduced consumer confidence in pork and beef products.

Steps to Prevent Transmission:

1. Regulation and Inspection: Governments and regulatory bodies can play a crucial role in preventing the transmission of these parasites by implementing and enforcing regulations related to food safety and hygiene. This can include regular inspections of slaughterhouses and meat-processing facilities to ensure that proper hygiene and sanitation practices are being followed.

2. Meat Inspection: Proper inspection of meat by qualified and certified inspectors at all stages of the supply chain is essential to identify and remove contaminated products from reaching consumers. This includes visual inspection of meat products for cysts and larvae, as well as ensuring that meat is properly cooked before consumption.

3. Education and Awareness: Public education campaigns can help raise awareness about the risks associated with consuming undercooked or raw pork and beef products. Providing information on the

importance of thorough cooking and proper meat-handling practices can help consumers make informed choices to protect themselves from tapeworm infections.

4. Collaboration between Public Health Protectors: Collaboration between food safety officials and communities is essential to prevent tapeworm infections in both animals and humans. This can include strategies such as deworming programs for livestock and surveillance programs to monitor the prevalence of these parasites in animal populations

d. **Anisakis simplex**: This parasite is found in undercooked or raw fish and seafood. Infection with Anisakis can lead to a condition called anisakiasis, which causes severe abdominal pain, nausea, and vomiting. Anisakis simplex is a parasitic worm found in marine fish and squid. When humans consume raw or undercooked seafood contaminated with Anisakis larvae, it can lead to a condition known as anisakis, which can cause severe abdominal pain, nausea, vomiting, and, in some cases, allergic reactions. Therefore, it is important to prevent Anisakis simplex from getting into human food to avoid potential health risks.

Preventing Anisakis simplex from contaminating human food involves several measures:

-Freezing: One of the most effective methods to kill Anisakis larvae in fish is freezing. The United States Food and Drug Administration (FDA) recommends freezing fish at -4 F (20 C) or below for 7 days (total time), or at -31 F (-35 C) or below until solid, and then storing at -31 F (-35 C) or below for 15 hours.

-Cooking: Proper cooking of seafood to the recommended internal temperature can also kill Anisakis larvae. It's important to ensure that fish and squid reach an internal temperature sufficient to destroy any parasites that may be present.

-Inspection and Quality Control: Strict inspection and quality control measures in the fishing, processing, and distribution of seafood can help to prevent contaminated products from reaching consumers.

-Education and Awareness: Educating consumers, food handlers, and foodservice professionals about the risks of consuming raw or undercooked seafood and the importance of proper handling and cooking can help prevent anisakiasis.

By implementing these preventive measures, the risk of Anisakis simplex contamination in human food can be significantly reduced, safeguarding public health and ensuring the safety of seafood consumption.

e. **Giardia lamblia**: This parasite is found in contaminated water and food, particularly in areas with poor sanitation. Infection with Giardia can cause gastrointestinal symptoms such as diarrhea, abdominal cramps, and nausea. Giardia lamblia is a microscopic parasite that causes the diarrheal illness known as giardiasis. It is important to prevent Giardia lamblia from contaminating human food to avoid the spread of this infection and to protect public health. Here are a few reasons why preventing Giardia lamblia from contaminating food is important:

(1) Health risks: Giardiasis can cause symptoms such as diarrhea, abdominal cramps, bloating, nausea, and dehydration. In some cases, the symptoms can be severe and long-lasting, especially in young children, the elderly, and individuals with weakened immune systems.

(2) Public health impact: Giardiasis is a significant public health concern, especially in settings where there is potential for outbreaks of illness. Preventing contamination of food with Giardia lamblia helps to reduce the risk of widespread illness and associated healthcare costs.

To prevent Giardia lamblia from contaminating human food, several measures can be taken:

(a) Practice good personal hygiene: Individuals handling food should wash their hands thoroughly with soap and water before and after handling food, especially after using the bathroom, changing diapers, or having contact with animals.

(b) Ensure food safety: Proper food handling, storage, and preparation practices are essential to prevent contamination. This includes thoroughly washing fruits and vegetables before consumption, cooking food to safe internal temperatures, and avoiding cross-contamination between raw and cooked foods.

(c) Purify water: Since Giardia lamblia can also be transmitted through contaminated water, it's important to drink only safe, treated water. Boiling, filtering, or chemically treating water are effective methods to make it safe for consumption.

(d) Maintain clean food preparation surfaces: Keep kitchen surfaces and utensils clean to prevent the spread of Giardia lamblia and other pathogens.

(e) Educate food handlers and the public: Providing education and training to food handlers and the public about the risks of Giardia lamblia contamination and the importance of proper food hygiene practices can help prevent the spread of infection.

By following these preventive measures, the risk of Giardia lamblia contamination of human food can be significantly reduced, thereby contributing to the overall safety of the food supply and public health.

CHAPTER SIX

CHEMICAL HAZARD

Another food safety hazard of interest is chemical hazards. Exposure to chemicals in food can have serious health effects. When food additives are used in amounts exceeding government-regulated limits, they can pose food safety risks. Overconsumption of nitrites is risky because when nitrites react with certain compounds in the stomach, they can form nitrosamines, some of which are proven to be carcinogenic(cancer-causing). Many chemicals can find their way into our food and cause illness. Some allergens enter our food as chemicals. Chemicals are likely to negatively affect Not Ready to Eat/Ready to Eat foods, which commonly originate from five sources. I will list these sources so you can guard against them in your restaurants, cafeteria, kitchen, or other food business.

1. **Food Chemicals:** preservatives, acids, food additives, sulfating agents, & food processing aids like Nitrites/Nitrates are needed as food safety measures because they prevent C. botulinum spores from germinating and forming the botulism toxin, but excesses of them in food beyond the government-recommended amount could be dangerous. Yes, Nitrites and nitrates have antimicrobial properties that help inhibit bacterial growth. By inhibiting bacterial growth, nitrites and nitrates help ensure the safety of processed meats and prevent foodborne illnesses. Also, Nitrites react with natural proteins in meat, which contribute to the characteristic flavor and color of processed meats. These compounds give products like bacon and hot dogs their distinctive taste and appealing pink color.
2. **Agricultural Chemicals** are pesticides, herbicides, animal medicine, & fertilizers. Preventing pesticides from entering food products requires implementing strict regulations and practices throughout the entire food production process. Here are some measures commonly taken to

minimize pesticide contamination: Regulatory standards: Governments establish maximum residue limits (MRLs) for pesticides in food products. These MRLs dictate the permissible levels of pesticides in meat and other food items. I encourage businesses to regularly monitor and test to ensure compliance with these standards. Pesticide selection and application: Farmers should follow approved pesticide lists and adhere to recommended application techniques. Integrated Pest Management (IPM) strategies that emphasize minimal pesticide use are encouraged. Farmers should carefully follow the pesticide manufacturer's instructions regarding application rates, timing, and pre-harvest intervals. Pre-harvest intervals: the minimum waiting period between pesticide application and harvesting. This allows sufficient time for pesticide residues to break down to safe levels before the crop is harvested. Implementing sound safety practices ensures the safe and responsible use of pesticides. This includes proper storage, handling, and disposal of pesticides, as well as maintaining accurate records of pesticide use.

Quality control and testing: Meat food products must undergo regular quality control and testing procedures to identify and quantify pesticide residues. This includes both random sampling and targeted testing to ensure compliance with regulatory standards.

Animal feed regulations: Livestock raised for meat are often fed feed that may contain pesticide residues. Implementing regulations and monitoring systems for animal feed production helps ensure that feed ingredients are free from excessive pesticide contamination.

Traceability and labeling: Establishing traceability systems allows for the identification and tracking of meat products from farm to fork. Proper labeling ensures that consumers are aware of any pesticide-related information or precautions.

Education and training: Ongoing programs for farmers, food handlers, and agricultural workers are essential to raise awareness of pesticide risks and best practices for safe, responsible use.

It's important to note that while these measures aim to minimize pesticide contamination, it may not be possible to eliminate all traces entirely. However, strict adherence to regulations and best practices significantly reduces the likelihood of harmful pesticide levels in food products.

Animal drugs are also chemical hazards that could be prevented from getting into food products. This fact is all the more reason that you are encouraged to buy meat food products that are inspected by the USDA or other government-approved food regulatory agencies. Preventing animal drugs from entering the human food supply is a critical aspect of food safety and public health. Regulatory agencies and food producers implement several measures to minimize the presence of animal drugs in meat and food products. Here are certain key strategies:

Veterinary Drug Approval: Before a drug is approved for use in food-producing animals, it undergoes rigorous evaluation by regulatory agencies such as the U.S. Food and Drug Administration (FDA) or the United States Department of Agriculture (USDA). These agencies assess the drug's safety and effectiveness and set limits on residues in food.

Veterinary Prescription and Supervision: Animal drugs used in food-producing animals are typically prescription-only, meaning they can be administered only by or under the supervision of a licensed veterinarian. Veterinarians play a critical role in ensuring proper drug use, appropriate dosages, and adherence to withdrawal periods to minimize residues in meat.

Withdrawal Periods: Withdrawal periods are the required time between the last administration of a drug to an animal before it can be slaughtered for human consumption. During this period, the animal’s body eliminates the drug residues. Withdrawal periods are strictly enforced to ensure that meat products do not contain excessive drug residues.

Monitoring and Testing: Regulatory agencies conduct regular monitoring and testing of animal tissues and food products to detect drug residues. Samples are taken from animals at slaughterhouses and analyzed in laboratories to ensure compliance with safety standards. Additionally, food producers may also implement their own testing programs to ensure the safety and quality of their products. This includes proper animal husbandry, hygiene, and nutrition, as well as disease prevention strategies such as vaccinations and biosecurity measures. By maintaining the animals' overall health, the need for therapeutic drug treatments can be reduced.

Compliance and Enforcement: Regulatory agencies enforce strict regulations and standards to ensure compliance with requirements for drug use and withdrawal periods. Non-compliance can result in penalties and legal consequences for producers who fail to adhere to the regulations.

It's imperative to note that although these measures significantly reduce the risk of animal drug residues in meat food products, it may be impossible to eliminate. However, implementing these strategies helps ensure that residue levels in meat remain within safe limits established by regulatory agencies to protect public health. Another chemical hazard of interest is fertilizer. Preventing fertilizer contamination in food products is crucial for food safety and human health. Here are some common strategies and practices that you can employ to minimize the risk of fertilizer contamination:

Good Agricultural Practices (GAPs): Implementing and following them is essential for preventing fertilizer contamination. This includes proper handling, storage, and application of fertilizers in accordance with recommended guidelines. Farmers should receive training on proper fertilizer use, including accurate application rates, timing, and methods.

Soil Testing and Nutrient Management: Conducting regular soil testing helps determine crop nutrient requirements and enables farmers to apply fertilizers more precisely. By understanding soil nutrient levels, farmers can avoid over-applying fertilizers, reducing the risk of excess nutrients leaching into water sources or being taken up by plants.

Proper Application Techniques: Using appropriate application techniques can minimize fertilizer runoff and leaching. These techniques include precision application methods, such as banding or injecting fertilizers directly into the soil, rather than surface-dissemination. Additionally, timing the fertilizer application to coincide with the crop's nutrient uptake can enhance its absorption and minimize losses.

Buffer Zones and Setbacks: Establishing buffer zones and setbacks around water bodies, wells, and sensitive areas can help protect these areas from potential fertilizer runoff. These zones act as a barrier, reducing the likelihood of fertilizers reaching water sources and minimizing the risk of contamination.

Compliance with Regulations: Governments and regulatory bodies established regulations and guidelines regarding fertilizer use. Farmers and agricultural producers must comply with these regulations, including restrictions on the types, quantities, and timing of fertilizer applications. Regular inspections and enforcement by regulatory agencies help ensure compliance.

Education and Training: Providing education and training to farmers and agricultural workers is crucial in promoting proper fertilizer management practices. This can include workshops, seminars, and educational materials that emphasize the importance of responsible fertilizer use and the potential risks associated with improper application.

Monitoring and Testing: Regular monitoring and testing of soil, water, and crops can help identify potential fertilizer contamination issues. These tests can detect excessive nutrient levels or contaminants, enabling prompt corrective action. Organic and Sustainable Farming Practices: Organic farming practices, which limit or eliminate synthetic fertilizers, can reduce the risk of fertilizer contamination. Sustainable farming methods, such as crop rotation, cover cropping, and the use of compost or manure as natural fertilizers, can help maintain soil fertility while minimizing the need for synthetic fertilizers.

By implementing these strategies and practices, farmers and agricultural producers can minimize the risk of fertilizer contamination in food products, safeguarding both consumer health and the environment.

3. **Industrial Chemicals**: cleaners, paint, sanitizers, oils, & lubricants can contaminate food products if not managed properly. Preventing industrial chemicals from contaminating food in food-processing areas is crucial to ensuring food safety. Here are several measures that can be taken to prevent such contamination:

   Implement proper storage: Store industrial chemicals in dedicated areas, separate from food storage and processing. Ensure that they are locked in secure cabinets or rooms to prevent accidental access or spillage. Use appropriate containers and store chemicals in clearly labeled, sealed containers designed for chemical storage. Avoid using containers that resemble food containers to prevent accidental mixing or confusion. Establish clear segregation by maintaining a physical separation between chemical storage areas and food processing areas. This can be achieved through physical barriers, such as walls, partitions, or separate rooms, to minimize the risk of cross-contamination. Implement strict handling procedures: Train employees on proper handling, storage, and use of industrial chemicals. Emphasize the importance of following safety protocols, including wearing appropriate personal protective equipment (PPE), such as gloves and goggles, and avoiding contact between chemicals and food. Use designated equipment: Provide separate tools and equipment for handling chemicals and food processing. This includes using separate utensils, containers, and cleaning equipment to avoid cross-contamination. Maintain good hygiene practices: Implement rigorous cleaning and sanitation procedures throughout food-processing areas. Regularly clean and sanitize surfaces, equipment, and utensils to remove any residual chemicals. Conduct routine inspections: Regularly inspect storage areas, equipment, and containers for signs of damage, leaks, or spills. Address any issues promptly to prevent chemical contamination. Implement effective pest control measures: Pests can carry chemicals or contribute to the spread of chemicals. Implement a robust pest control program to prevent pests from entering food processing areas and potentially contaminating food. Monitor and test: Regularly assess both the food and the environment for chemical contaminants. Implement a comprehensive quality control program that includes routine sampling and testing to ensure compliance with safety standards. Stay updated on regulations: Keep abreast of USDA, FDA, state & international regulations related to food safety and chemical handling. Ensure compliance with all applicable regulations and standards. Remember, it is essential to consult with food safety experts, regulatory agencies, and industry-specific guidelines to develop and implement appropriate measures tailored to your food processing environment.

CHAPTER SEVEN
PHYSICAL HAZARD

Another food safety hazard of concern is the Physical Hazard. You can define a physical hazard as any physical object that is not usually found in food, and which can lead to sickness or injury to an individual eating the food. Physical hazards can enter food through contaminated raw products, poorly designed or maintained facilities and equipment, flawed processing methods, and inadequate employee training and practices.

There are many types of physical hazards that can enter human food. Fragments from nail, screw, needles, knives, equipment parts, jewelry, hair pins, broken bottle or jar pieces, light bulbs, thermometers, windows, eyeglasses, pieces of wood from pallets or other, pieces of board, plastic pieces, packaging materials, food wrappers, bone fragments, pens, etc.

Common public health concerns related to physical hazards in food include broken or damaged teeth and tears to the mouth, esophagus, stomach, and intestines. These hazards can obstruct the airways or the intestines. In some cases, it may lead to death due to suffocation and infections. In most cases, children are more susceptible to the effects of physical hazards because their bodies are smaller and physical objects have a greater impact. Ready-to-eat food can be risky when a physical object is embedded in it, and a consumer may not be able to identify and remove the hazard before eating.

Preventing physical hazards

Preventing physical hazards from entering foods in the processing area or kitchen is critical to ensure food safety. Here are some measures you can take to minimize the risk of physical contamination:

You must implement good personal hygiene practices: All staff members should maintain high standards of personal hygiene, including wearing clean uniforms, head coverings, and appropriate footwear. Regular handwashing and the use of gloves are essential to prevent the transfer of physical contaminants.

Control access to the processing area: Limit access to the processing area or kitchen to authorized personnel only. Implement a controlled-entry system to prevent unauthorized individuals from entering the area and to prevent the introduction of physical hazards.

Maintain a clean, organized workspace: Keep the processing area free of clutter. Regularly clean and sanitize all surfaces, equipment, and utensils to remove any potential physical contaminants.

Inspect and maintain equipment: Regularly inspect and maintain all equipment used in food processing to ensure it is in good working condition. Repair or replace any damaged or worn-out equipment that could potentially introduce physical hazards.

Use appropriate packaging materials: Ensure the packaging used to store and transport food products is high-quality and free of physical contaminants. Inspect packaging materials for any signs of damage or contamination before use.

Implement proper storage practices: Store raw materials, ingredients, and finished products in designated areas, away from potential physical hazards, such as chemicals, cleaning supplies, or equipment that could cause damage.

Train and educate staff: Provide comprehensive training on food safety practices, including identifying and preventing physical hazards. Regularly reinforce the importance of following proper procedures and reporting any potential issues or incidents related to physical contamination.

Implement quality control measures: Establish quality control procedures to regularly inspect and monitor the processing area and kitchen for potential physical hazards. These may include visual inspections, metal detection, sieving, or other appropriate methods to detect and remove physical contaminants.

Establish a hazard reporting system: Encourage all staff members to report any incidents, near misses, or concerns related to physical hazards. Establish a system to promptly report and investigate such incidents, identify the root cause, and take necessary corrective actions.

Stay informed and up to date: Keep up with the latest food safety regulations, guidelines, and best practices to ensure your food-processing area or kitchen meets required standards.

By implementing these measures, you can significantly reduce the risk of physical hazards entering foods in the processing area or kitchen, thereby enhancing food safety and protecting consumer health.

CHAPTER EIGHT PROCESSED FOOD & SAFETY

There are many types of processed foods. During processing, a wide range of ingredients are added to enhance taste, flavor, color, and the overall character of the finished products. These ingredients must be prepared in a sanitary environment, must be unadulterated, and must be harmless to consumers. The safety of these ingredients is paramount, and information about their allergenic potential is of immense importance to consumers and inspectors/regulators. Some of these ingredients are utilized as antimicrobial agents, binders, extenders, curing agents, curing accelerators, spices, and flavorings.

Let's start with antibacterial agents added to processed meat. Antibacterial agents are added to processed food products to inhibit the growth of microorganisms, such as bacteria, fungi, and yeast, thereby increasing the food's shelf life and safety. Here are some common antimicrobial agents used in processed food products:

1. **Sodium nitrite**: Used in processed meats, such as bacon, ham, deli meats, and hot dogs, to prevent the growth of bacteria and to preserve the meat's color. While sodium nitrite serves a valuable purpose in preventing the growth of harmful bacteria and maintaining the color of processed meats, its use has raised concerns regarding potential health risks. Some of the primary food safety concerns associated with sodium nitrite include:

   (a) Formation of Nitrosamines: When sodium nitrite is exposed to high heat during cooking or processing, it can react with compounds naturally present in meat to form nitrosamines. Nitrosamines are known carcinogens and have been linked to an increased risk of cancer, particularly in the digestive system.

(b)Causing Methemoglobinemia: In high doses, sodium nitrite can lead to methemoglobinemia, a condition in which the oxygen-carrying capacity of the blood is reduced. This can be particularly dangerous in infants and young children.

To regulate the use of sodium nitrite and mitigate these safety concerns, food safety authorities and regulatory bodies implement several measures, including:

(a) Maximum Residue Limits: Sodium Nitrite is regulated by the USDA, and you can only add a certain percentage in order to ensure safety. Establishing maximum residue limits for sodium nitrite in meat products ensures that its use remains within safe levels.

(b) Good Manufacturing Practices: Enforcing good manufacturing practices to ensure that sodium nitrite is used in compliance with regulations and in a manner that minimizes the formation of nitrosamines.

(c) Labeling Requirements: Requiring clear and accurate labeling of products containing sodium nitrite to inform consumers of its presence and provide guidelines for safe consumption.

(d) Research and Monitoring: Continuously monitoring scientific research to assess the safety of sodium nitrite and its potential health effects and adjusting regulations accordingly.

(e) Alternative Preservation Methods: Encouraging the development and use of alternative preservation methods, such as natural extracts and other preservatives, to reduce reliance on sodium nitrite in processed meats.

By implementing these measures, regulatory authorities aim to ensure that sodium nitrite is used to minimize potential health risks while still allowing its beneficial effects in food preservation.

(2) **Sodium benzoate**: It is a commonly used preservative in the food and beverage industry, and, like any food additive, it comes with its own set of safety concerns. Some of the main safety concerns associated with sodium benzoate include:

(a) Formation of Benzene: When sodium benzoate is exposed to heat and light in the presence of vitamin C (ascorbic acid), it can form benzene, a known cancer-causing agent. This can occur in certain soft drinks, particularly those with added vitamin C.

(b) Allergic Reactions: Some individuals may be sensitive or allergic to sodium benzoate, and consumption of products containing this preservative can lead to adverse reactions.

To prevent and regulate these concerns, several measures can be taken:

(a) Proper Formulation: Manufacturers can ensure that the formulation of products containing sodium benzoate is carefully controlled to minimize the potential for benzene formation. This can involve adjusting pH levels and avoiding the combination of sodium benzoate with certain other ingredients, such as ascorbic acid.

(b) Storage and Handling: Proper storage and handling of products containing sodium benzoate are crucial. Exposure to heat and light should be minimized to prevent benzene formation.

(c) Labeling and Consumer Education: Clear and accurate labeling of products is essential to inform consumers about the presence of sodium benzoate. Additionally, educating consumers about the potential risks associated with the preservative can help individuals make informed choices.

(d) Regulatory Oversight: Government regulatory agencies, such as the United States Department of

Agriculture (USDA), the Food and Drug Administration (FDA) in the United States, state agencies, and similar organizations in other countries, play a crucial role in establishing and enforcing safety standards for food additives. You are not allowed to add any amount that you want to add. These agencies set maximum allowable levels of sodium benzoate in various food and beverage products to ensure safety.

(e) Monitoring and Testing: Regular monitoring and testing of food products are important to ensure that manufacturers comply with regulations and that products on the market are safe for consumption.

By implementing these measures, it is possible to mitigate the food safety concerns associated with sodium benzoate and ensure that its use in food and beverages is safe and regulated.

(3)Potassium sorbate: Potassium sorbate is a common food preservative used to inhibit the growth of mold, yeast, and fungi in a wide variety of food and beverage products. While it is generally considered safe for consumption when used within recommended limits, there are some food safety concerns associated with potassium sorbate.

(a) Allergic Reactions: Some individuals may be sensitive or allergic to potassium sorbate, leading to adverse reactions such as hives, itching, or respiratory issues. It's important for food manufacturers to clearly label products containing potassium sorbate to alert consumers with allergies.

(b) Potential Mutagenicity: There is some evidence to suggest that potassium sorbate may have mutagenic properties under certain conditions. However, the concentration at which this occurs is typically much higher than that found in food products. Nonetheless, it's important for manufacturers to adhere to recommended usage levels to minimize potential risk.

To prevent these concerns, food manufacturers must adhere to strict regulations and guidelines established by food safety authorities, such as the United States Department of Agriculture (USDA), the U.S. Food and Drug Administration (FDA), and the European Food Safety Authority (EFSA). These regulations dictate the maximum allowable concentrations of potassium sorbate in various food products to ensure its safe consumption.

Additionally, manufacturers should clearly label products containing potassium sorbate to inform consumers and help individuals with allergies avoid them. It's also crucial for producers to conduct thorough safety assessments and quality control measures to ensure that potassium sorbate is used responsibly and within safe limits.

Consumers can minimize their risks by being aware of the ingredients in the foods and beverages they consume, and by following any allergen warnings on product labels. If there are concerns about potassium sorbate or any other food additive, individuals should consult with healthcare professionals for personalized advice.

(4)Nisin: Nisin is a natural antimicrobial peptide used as a food preservative to inhibit the growth of certain bacteria, particularly those that cause food spoilage. However, there are several food safety concerns associated with nisin, including potential allergic reactions in sensitive individuals and the development of antimicrobial resistance in bacteria.

To prevent and regulate these concerns, several measures can be taken:

(a) Allergen Labeling: Nisin should be clearly labeled on food products to alert consumers who may be sensitive or allergic to it.

(b) Maximum Residue Limits: Regulatory authorities may establish maximum residue limits (MRLs) for nisin in certain food products to ensure that its use remains within safe levels.

(c) Good Manufacturing Practices (GMP): Adherence to GMP, including proper dosing and application of nisin in food production, can help minimize potential risks.

(d) Risk Assessment: Concluding thorough risk assessments to evaluate the safety of nisin in specific food applications can provide valuable insights for regulatory decision-making.

(e) Monitoring and Surveillance: Implementing monitoring and surveillance programs to track the use of nisin in food products and to detect any potential adverse effects of emerging issues.

(f) Regulation and Oversight: Government agencies and regulatory bodies should oversee the use of nisin in food products to ensure compliance with safety standards and regulations.

By implementing these measures, it is possible to mitigate the food safety concerns associated with nisin and ensure its safe and responsible use as a food preservative.

(5)Natamycin: Natamycin is a natural antifungal agent commonly used as a food preservative, particularly in dairy products such as cheese and yogurt. While it is generally considered safe when used within recommended limits, there are some food safety concerns associated with its use.

One concern is the potential development of resistance in fungi, which could compromise the effectiveness of natamycin as an antifungal agent over time. Additionally, some people may be sensitive to or allergic to natamycin, though such cases are rare.

To prevent and regulate these concerns, several measures can be taken:

(a) Usage Limits: Regulating authorities, such as the USDA & the FDA in the United States and the EFSA in Europe, have established specific limits on the amount of natamycin that can be used in food products. Adhering to these limits helps mitigate the risk of resistance development and ensures consumer safety.

(b) Monitoring and Surveillance: Regular monitoring and surveillance of natamycin use and its effects on food products can help detect any emerging issues, such as resistance development, and allow for timely intervention

(c) Allergen Labeling: Food products containing natamycin should be clearly labeled to indicate its presence, allowing consumers with allergies or sensitivities to make informed choices.

(d) Research and Development: Continued research into the use of natamycin as a food preservative, including its potential impact on resistance development and allergenicity, can inform regulatory decisions and risk management strategies.

Overall, ensuring that natamycin is used within recommended limits, combined with effective monitoring and clear labeling, can help address food safety concerns associated with its use as a preservative.

(6)Propionic acid: Propionic acid is a naturally occurring carboxylic acid that is commonly used as a preservative in food products. While it is generally recognized as safe (GRAS) by the U.S. Food and Drug Administration (FDA), there are still some food safety concerns associated with its use.

Some of the potential concerns regarding propionic acid include:

(a) Allergic reactions: Some individuals may be sensitive or allergic to propionic acid, leading to adverse reactions if consumed in large quantities.

(b) Potential impact on beneficial bacteria: Propionic acid, like other preservatives, can potentially affect the balance of beneficial bacteria in the gut if consumed in excessive amounts.

To prevent and regulate these concerns, the following measures can be taken:

(a) Adherence to regulations: Food regulatory agencies, such as the USDA & FDA in the United States and the European Food Safety Authority (EFSA) in the European Union, have established guidelines that are maximum allowable limits for the use of propionic acid in food products. Adhering to these regulations helps to ensure that its use remains safe.

(b)Accurate labeling: Food products containing propionic acid should be accurately labeled to inform consumers of its presence. This allows individuals who are sensitive or allergic to the compound to make informed choices about the foods they consume.

(C) Good manufacturing practices: Food manufacturers should follow good manufacturing practices to ensure that propionic acid and other additives are used safely and appropriately, minimizing the potential for adverse effects.

(d)Monitoring and research: Continued monitoring and research into the safety of propionic acid and its potential health effects can help to identify any emerging concerns and inform regulatory decisions.

By following these measures, the food industry can help to prevent and regulate potential food safety concerns associated with the use of propionic acid.

7. **Sorbic acid**: Sorbic acid is a commonly used preservative in the food industry, and there are several food safety concerns associated with its usage. Here

are some of the key concerns and methods for prevention and regulation:

(a) Allergic Reactions: Some individuals may be sensitive or allergic to sorbic acid, which can lead to adverse reactions. To prevent this, food manufacturers should clearly label products containing sorbic acid to alert consumers with allergies.

(b) Microbial Contamination: While sorbic acid is effective in inhibiting the growth of mold, yeast, and some bacteria, it is not effective against all types of microorganisms. As a preventive measure, proper sanitation and good manufacturing practices should be followed to minimize the risk of microbial contamination in food products containing sorbic acid.

(c) Regulatory Limits: To ensure food safety, regulatory agencies such as the United States Department of

Agriculture & the Food and Drug Administration (FDA) in the United States and the European Food Safety Authority (EFSA) in the European Union have established maximum limits for the use of sorbic acid in various food products. Adherence to these limits is crucial to prevent overuse and potential health risks.

(d) Formation of Harmful Byproducts: Under certain conditions, such as exposure to heat and light, sorbic acid may form harmful byproducts. To prevent this, food manufacturers must carefully consider the storage and processing conditions for products containing sorbic acid and take steps to minimize byproduct formation.

(e) Monitoring and Testing: Regular monitoring and testing of food products containing sorbic acid can help ensure that they meet safety standards. This may involve periodic analysis of sorbic acid levels and the presence of any harmful byproducts. By addressing these concerns and adhering to established regulations and best practices, the food industry can mitigate the potential safety risks associated with the use of sorbic acid as a food preservative.

(8)Ethylenediaminetetraacetic acid (EDTA): Added to some processed foods as a preservative and to prevent the growth of certain bacteria. Ethylenediaminetetraacetic acid (EDTA) is a widely used chelating agent in the food industry, primarily for its ability to bind metal ions, thereby preserving the color, flavor, and texture of food products. However, there are some food safety concerns associated with EDTA:
(a)Chelation of essential minerals: EDTA's ability to chelate metal ions can also result in the removal of essential minerals such as calcium, magnesium, and iron from the body, potentially leading to nutritional deficiencies.

(b)Formation of potentially harmful compounds: In certain food processing and storage conditions, EDTA can react with metal ions to form potentially harmful compounds, such as metal-EDTA complexes, which may pose health risks.

To prevent and regulate the potential risks associated with EDTA, the following measures can be taken:

(a) Maximum Residue Limits (MRLs): Regulatory authorities may establish MRLs for EDTA in food products to ensure that the levels of EDTA do not exceed the safe limits.

(b) Good Manufacturing Practices (GMP): Food manufacturers should adhere to GMP to ensure that EDTA is used in safe amounts and under appropriate conditions.

(c) Risk assessment and monitoring: Regular risk assessments and monitoring of EDTA levels in food products can help identify and address potential safety concerns.

(d) Labeling and consumer education: Clear labeling of food products containing EDTA can help consumers make informed choices, and educational initiatives can raise awareness about the potential risks associated with EDTA consumption.

It's important to note that regulations and recommendations may vary by country, and it's essential to consult local regulatory authorities and scientific literature for the most current information on the use and regulation of EDTA in food products.

(9)Sodium nitrate. Sodium nitrate is different from sodium nitrite. Sodium nitrate is also commonly used as a preservative, color fixative, and flavoring in processed meats such as bacon, ham, and hot dogs. However, there are some food safety concerns associated with its use.

One primary concern is the formation of nitrosamines, which are potentially carcinogenic compounds, when sodium nitrate is exposed to high heat, such as during cooking or processing. Nitrosamines have been linked to an increased risk of certain cancers, so minimizing their formation is a key regulatory concern.

To regulate and prevent the formation of nitrosamines from sodium nitrate, several measures can be taken:

(a) Limiting the use of sodium nitrate: Government regulators/inspectors ensure that the scientifically determined limits on the amount of sodium nitrate that can be used in processed meats to reduce the potential for nitrosamine formation are adhered to.
(b) Using alternative ingredients: Food manufacturers can explore the use of alternative preservatives or natural curing agents that do not pose the same risk of nitrosamine formation.

(c) Adding antioxidants: Antioxidants such as vitamin C (ascorbic acid) can be added to processed meats containing sodium nitrate to inhibit the formation of nitrosamines.

(d) Controlling processing conditions: Manufacturers can implement processing methods that minimize the potential for nitrosamine formation, such as using lower cooking temperatures and shorter cooking times.

(e) Labeling and consumer education: Providing clear and accurate labeling about the presence of sodium nitrate in processed meats can help consumers make informed choices about their food consumption. By implementing these measures, regulators and food manufacturers can work to mitigate the potential health risks associated with sodium nitrate in processed meats and enhance food safety for consumers.

(10)Essential Oils: Essential oils are highly concentrated substances extracted from plants and often used for their aromatic properties and potential health benefits. When it comes to processed foods, essential oils are sometimes added for flavoring or as natural preservatives, and if ingested in large quantities or used improperly, they can pose health risks. Common essential oils used in processed foods include Lemon oil, orange oil, Peppermint oil, Spearmint oil, Cinnamon oil, Clove oil, Oregano oil, Rosemary oil, Thyme oil, and Bergamot oil. Despite their benefits, essential oils should be used with caution in food products due to their potency. Food safety areas of concern are:

(a)Ingestion: Essential oils are not intended for ingestion in large quantities. When consumed in excess, they can cause adverse reactions and toxicity. Ingesting certain essential oils can lead to nausea, vomiting, and even organ damage.

(b)Purity and Contamination: The purity of essential oils is crucial for food safety. If not properly processed, contamination with pesticides, heavy metals, or other toxic substances can occur during production and packaging, posing health risks to consumers.

©Allergens and Sensitivities: Some individuals may have allergies or sensitivities to certain essential oils. Improper labeling or inadequate information about potential allergens in essential oils can lead to allergic reactions in sensitive individuals.

To regulate and prevent these food safety concerns, several measures can be taken:

(a)Regulation and Labeling: Government agencies implement regulations that ensure that essential oils intended for food use meet specific purity and safety standards. Clear and accurate labeling should also be mandated to provide consumers with information about proper usage and potential allergens.

(b)Quality Control: Manufacturers should adhere to strict quality control measures during the production, processing, and packaging of essential oils. Regular testing for purity and contamination can help ensure the safety of these products.

©Education and Awareness: Public education campaigns can help raise awareness of proper essential oil use and the potential risks of ingestion. Consumers should be informed about safe usage guidelines.

(d)Industry Standards: Industry organizations and associations must continue to have standards for the production and labeling of food-grade essential oils, promoting best practices and ensuring consumer safety. By implementing these measures, regulators and industry stakeholders can work together to mitigate food safety concerns associated with essential oils and promote their safe and responsible use in food and beverages.

(11)**Acetic Acid**: Acetic acid, which is the main component of vinegar, is generally considered safe for consumption when used in food products in appropriate quantities. However, there are some food safety concerns associated with acetic acid that should be considered:

(a)Concentration: Acetic acid is a strong acid, and in its concentrated form, it can be corrosive and cause skin and eye irritation. Food-grade acetic acid used in food products is typically diluted to safe levels, but improper handling of concentrated acetic acid can pose risks.

(b)Allergies and Sensitivities: Some individuals may be sensitive or allergic to acetic acid, and consuming foods containing acetic acid can potentially cause foodborne illnesses if used in food products.

©Inhalation: While not a concern in food products, inhalation of concentrated acetic acid fumes can be harmful and irritating to the respiratory system.

(d)Interaction with Other Substances: Acetic acid can react with certain substances to produce potentially harmful compounds. For example, when mixed with certain types of cleaning agents or chemicals, acetic acid can produce toxic gases.

Overall, when used appropriately and in accordance with food safety regulations, acetic acid is considered safe for use in food products. However, it's important for food manufacturers to adhere to proper handling, storage, and labeling practices to ensure the safety of products containing acetic acid.

12. Lactic Acid: Lactic acid is generally recognized as safe (GRAS) by the United States Department of Agriculture (USDA) and the U.S. Food and Drug Administration (FDA) when used in accordance with good manufacturing practice. It is commonly used as a food additive for its acidity-regulating, flavor-enhancing, and preservation properties. However, there are some food safety concerns associated with lactic acid, particularly when it is used in food processing and preservation. Here are some of the key concerns:

(a)Microbiological Contamination: Lactic acid itself has antimicrobial properties and can inhibit the growth of certain bacteria, including pathogens. However, in certain food-processing environments, there may be concerns about microbiological contamination if lactic acid is not used properly or is present at high concentrations.

(b)Residue Levels: Residues of lactic acid and its salts may be a concern in certain foods, especially if they are used in high concentrations or if there are inadequate controls in place during food processing. Regulatory authorities set limits on the amount of lactic acid that can be present in food to ensure safety.

(c)Allergen Concerns: While lactic acid itself is not a common allergen, there may be concerns about the presence of allergenic substances in lactic acid produced through fermentation processes. People with specific allergies or intolerances may need to be cautious about consuming foods containing lactic acid.

(d)Chemical Interactions: Lactic acid can react with other compounds in food, potentially leading to the formation of undesirable byproducts. Food manufacturers must consider the potential for such interactions and ensure that lactic acid is used safely alongside other ingredients.

(e)Occupational Safety: While not a direct food safety concern, it's important to note that lactic acid, like many acids, can pose risks to workers involved in its production and handling. Proper safety protocols must be placed to protect workers from potential hazards.

It's important to note that when used in accordance with regulations and best practices, lactic acid is generally considered safe for consumption. However, as with any food additive, it's crucial for food manufacturers to comply with applicable regulations and guidelines to ensure the safety of the food supply.

13. Citric Acid: Citric acid is a common food additive that is used as a preservative, flavoring agent, and acidulant in various food and beverage products. While it is generally recognized as safe (GRAS) when used in accordance with good manufacturing practices, there are some food safety concerns and considerations associated with its use:

(a)Allergies: Some individuals may be allergic to citric acid, although true allergies to citric acid are rare. However, people with known citric acid allergies should avoid products containing this ingredient.

(b)Dental Health: Citric acid is acidic and can contribute to dental erosion when consumed in high amounts or in highly concentrated forms. This is particularly relevant for sour candles or acidic beverages.

(c)Potential Contaminants: Citric acid itself is not a significant source of concern for contaminants, but it can be produced from mold fermentation of sugars, which could introduce mycotoxins if not properly controlled during production.

To minimize these concerns, food manufacturers and consumers can take the following steps:

(a)Quality Control: Manufacturers should ensure that the citric acid used in their products meets quality standards and is free from contaminants. This includes selecting reputable suppliers and performing regular testing for purity and potential contaminants.

(b)Labeling and Allergen Information: Food products containing citric acid should clearly be labeled by the producers, and any potential allergen information should be clearly communicated to consumers, as with any other allergen.

(c)Formulation and Usage: Manufacturers should use citric acid in accordance with regulatory guidelines and industry best practices to ensure that it is used at safe levels and does not contribute to product safety concerns.

(d)Consumer Education: Consumers should be aware of the potential dental health implications of consuming products containing citric acid and should practice good oral hygiene to minimize any negative effects.

By addressing these concerns through proper handling, labeling, and consumer education, the risks associated with citric acid in food products can be effectively minimized.

14. Benzoic Acid: Benzoic acid is commonly used as a food preservative to inhibit the growth of bacteria, yeast, and molds in acidic products such as fruit juices, soft drinks, pickles, and salad dressings. However, there are some food safety concerns associated with benzoic acid, particularly when it is used in combination with ascorbic acid (vitamin C) or in low-pH products. The main food safety concerns regarding benzoic acid include:

(a)Formation of benzene: When benzoic acid is combined with ascorbic acid (vitamin C) in the presence of heat and light, it can lead to the formation of benzene, a known human carcinogen. Benzene formation is of particular concern in beverages, as it has been associated with certain soft drinks.

(b)Allergic reactions: Some individuals may be sensitive or allergic to benzoic acid, which can lead to adverse reactions such as hives, asthma, or rhinitis.

To minimize the food safety concerns associated with benzoic acid, the following measures can be taken:

(a)Use of alternative preservatives: Consider using alternative preservatives or preservative combinations that do not pose the risk of benzene formation, especially in acidic beverages.

(b)Proper formulation: Formulate products containing benzoic acid and ascorbic acid in a way that minimizes the potential for benzene formation. This may involve controlling factors such as temperature, light exposure, and pH levels during production and storage.

©Regulatory compliance: Ensure compliance with regulatory limits and guidelines for the use of benzoic acid in food products. Different countries have specific regulations regarding the maximum permitted levels of benzoic acid in food and beverages.

(d)Allergen labeling: Clearly label products containing benzoic acid to alert consumers who may have allergies or sensitivity to the substance.

(e)Good manufacturing practices: Adhere to good manufacturing practices (GMP) to ensure the safe handling, storage, and use of benzoic acid in food production facilities.

(15)Calcium Propionate: Calcium propionate is a food additive that is used as a preservative to inhibit the growth of mold and prevent spoilage in baked goods and other food products. While it is generally recognized as safe by regulatory authorities when used in appropriate quantities, there are some food safety concerns associated with its use. Here are some of the concerns and ways to prevent and regulate them:

(a)Allergic Reactions: Some individuals may be sensitive or allergic to calcium propionate, which could lead to adverse reactions. To prevent this, food manufacturers should clearly label products containing calcium propionate and comply with labeling regulations to warn consumers of its presence.

(b)Health Effects: There is some evidence to suggest that excessive consumption of calcium propionate may be associated with negative health effects in certain individuals, such as migraines or other adverse reactions. To regulate this, authorities should continue to set and enforce maximum allowable levels of calcium propionate in foods and conduct ongoing research to assess its safety.

(c)Microbial Resistance: There is a concern that the use of calcium propionate in food products may contribute to the development of microbial resistance, potentially reducing the effectiveness of preservatives and leading to food safety issues. To prevent this, food producers should use calcium propionate judiciously and in accordance with recommended guidelines to minimize the risk of microbial resistance.

(d)Regulatory Oversight: To ensure the safe use of calcium propionate, regulatory authorities should closely monitor its use in food products, establish and enforce safety standards, and regularly review scientific evidence to update regulations and guidelines as necessary.

While calcium propionate is generally considered safe when used appropriately, it is important for food producers and regulatory authorities to remain vigilant in monitoring its use, setting, and enforcing safety standards, and educating consumers about its presence in food products to ensure food safety and protect public health. It's also important to note that while these antimicrobial agents are generally recognized as safe by regulatory authorities when used within specified limits, consumers should still be mindful of their intake and consider consuming processed foods in moderation.

CHAPTER NINE

SAFETY of SPICES and SEASONINGS

There are several reasons why some people may be less attentive to the food safety of spices and seasonings. First is lack of Awareness: Many people may not be fully aware of the potential risks associated with some spices and seasonings. Second may be minimal oversight: In some areas, spices and seasonings are well regulated and inspected by the responsible government agencies. In some areas, spices and seasonings are less regulated and inspected than other food products. This can create a false sense of security among consumers, leading them to be less vigilant about food safety with these products. Third, there is a lack of awareness of the importance of proper handling and storage of seasonings and spices. Improper handling and storage of seasonings can contribute to contamination. Since these products often retain their aromas and are stored for longer periods, people may pay less attention to proper storage practices, which can lead to safety issues. Fourth is Cultural Practices. In many cultures, traditional methods of processing and using these spices may not align with modern food safety standards. Even when illnesses arise from such unsafe standards, there is no way to identify the source of the illness as a lack of food safety practices. Fifth is Risk Perception. Because relatively few spices are added, some individuals may perceive the risk of contamination as low, leading to complacency about food safety practices for these products.

It is important to note that while these factors may contribute to a lack of attention to food safety regarding species and food seasonings, it's essential for consumers to understand that these products can indeed pose food safety risks and cause illness if not handled, stored, and used properly. This includes being mindful of expiration dates, avoiding cross-contamination, and following proper food safety guidelines when using spices and seasonings so you can enjoy your food and stay healthy. Certainly! There are numerous spices and seasonings commonly used in processed food industries, restaurants, and personal kitchens. Here are some of the most widely used ones:

1. **SALT**: Used to enhance flavor in almost all types of dishes. Food safety concerns related to salt primarily center on its potential to contribute to health problems when consumed in excess. However, from a food safety perspective, salt itself is typically not a significant concern for contamination or spoilage. Instead, the focus is on its impact on overall diet and health. Here are some food safety concerns related to salt and how they could be regulated and prevented:

   (a) Excessive Sodium Intake: High levels of dietary sodium, primarily in the form of salt, are associated with health issues such as high blood pressure, heart disease, and stroke. To address this concern, regulatory and preventive measures include:

-Labeling regulations: Many countries have implemented regulations requiring food manufacturers to label the sodium content of their products. Clear and easily understandable labeling can help consumers make informed choices about the salt content of the foods they consume.

-Public education: Governments and health organizations often provide educational campaigns to raise awareness about the health risks of excessive salt intake and encourage individuals to make healthier food choices.

-Reformulation of processed foods: Some governments and health organizations work with the food industry to encourage the reformulation of processed foods to contain less salt. This can be done through voluntary agreements or, in some cases, through regulatory mandates.

(b) Contamination in Salt Products: While salt itself is resistant to microbial contamination due to its high osmotic pressure, there can be concerns about contamination in salt products due to inadequate processing or storage. To regulate and prevent contamination in salt products:

-Good manufacturing practices: Regulatory agencies often establish and enforce good manufacturing practices (GMPs) for salt production to ensure that it is processed and handled in a sanitary manner.

-Quality control measures: Regular testing for contaminants, such as heavy metals and microbial pathogens, can help ensure the safety of salt products. Regulatory agencies may set maximum allowable limits for specific contaminants in salt. The primary food safety concerns related to salt revolve around its impact on public health when consumed in excessive amounts and potential contamination in salt products.

Through a combination of labeling regulations, public education, industry collaboration, and quality control measures, these concerns can be regulated and prevented.

2. **Black pepper**: Adds heat and depth to dishes. Black pepper, like any other food product, can pose certain food safety concerns. Here are some of the potential issues and how they could be regulated and prevented:

   (a) Microbial Contamination: Black pepper can be contaminated with bacteria, mold, or other microorganisms during growing, harvesting, processing, or packaging. This can cause foodborne illness if consumed.

   Regulation and Prevention: Good agricultural and manufacturing practices, along with proper hygiene and sanitation during processing and packaging, can help prevent microbial contamination. Regular testing for microbial pathogens, such as Salmonella or E. coli, can also help ensure the safety of black pepper. Regulatory bodies such as the United States Department of Agriculture & Food and Drug Administration (FDA) in the United States or the European Food Safety Authority (EFSA) in Europe establish guidelines and regulations for microbial limits in spices.

(b) Adulteration: Black pepper can be adulterated with lower-cost substances, such as other spices or filters, to increase profits; this can compromise the quality and safety of the product. Regulation and Prevention: Regulatory agencies and food safety organizations often establish standards for the purity and quality of spices, including black pepper. Regular testing and inspections can help detect and prevent adulteration. Traceability systems and quality control measures throughout the supply chain can also help ensure the authenticity of black pepper.

(c) Pesticide Residues: Pesticides used during the cultivation of black pepper can leave residues on the final product, which may pose health risks if consumed in high amounts.

Regulation and Prevention: Regulatory agencies set maximum residue limits (MRLs) for pesticides in food products, including black pepper. Compliance with these limits is enforced through testing and inspections. Good agricultural practices and integrated pest management can help minimize pesticide use and reduce residues in black pepper.

(d) Foreign Objects: Contamination of black pepper with foreign objects such as metal, plastic, or glass during processing and packaging can pose safety hazards.

Regulation and Prevention: Implementing good manufacturing practices and quality control measures, such as metal detectors and sieves, can help prevent foreign-object contamination. Regulatory standards often specify acceptable limits for foreign objects in food products.

Overall, ensuring the safety of black pepper and other spices requires a combination of regulatory oversight, industry standards, good agricultural and manufacturing practices, quality control measures, and testing protocols. Consumers can also contribute to food safety by purchasing spices from reputable sources and storing them properly to maintain quality and safety.

3. **Sucrose**: Sucrose, commonly known as table sugar, is a disaccharide composed of glucose and fructose. While it is a common and generally safe food ingredient, there are some food safety and health issues associated with its consumption:

   (a) Dental Health: Consuming excessive amounts of sucrose can contribute to tooth decay and cavities, especially if oral hygiene is not maintained.

   (b) Obesity and Metabolic Health: Overconsumption of sucrose, especially in the form of sugary beverages and processed foods, can contribute to obesity and related metabolic health issues such as type 2 diabetes and cardiovascular disease.

(c) Blood Sugar Levels: Sucrose can cause rapid spikes in blood sugar levels, which may be of concern for individuals with diabetes or those at risk of developing the condition.

To prevent and regulate the potential health issues associated with sucrose, the following measures can be taken:

(a) Dietary Guidelines: Encourage balanced diets that limit the consumption of added sugars, including sucrose. National dietary guidelines can provide recommendations for limiting sugar intake.

(b) Nutrition Labeling: Require clear and informative nutrition labeling on food products to help consumers make informed choices about the sugar content of the foods they consume.

(c) Public Awareness Campaigns: Educate the public about the health risks associated with excessive sucrose consumption and promote healthier alternatives.

(d) Sugar Taxes or Regulations: Some jurisdictions have implemented sugar taxes or regulations targeting sugary beverages and foods high in added sugars to discourage their consumption.

(e) Product Reformulation: Encourage food and beverage manufacturers to reformulate products to reduce their sucrose content and offer healthier alternatives.

(f) Oral Health Education: Promote good oral hygiene practices and educate the public about the relationship between sucrose consumption and dental health.

(g) Research and Monitoring: Support ongoing research into the health effects of sucrose consumption and monitor population-level data to assess trends and potential health impacts.

By implementing these measures, it's possible to mitigate the food safety and health issues associated with sucrose and promote healthier dietary habits within the population.

4. **Molasses**: Molasses is a byproduct of the sugar-refining process and is commonly used in cooking and baking. While molasses itself does not pose significant food safety risks, there are a few considerations to keep in mind:

(a) Contamination: Molasses can become contaminated with pathogens if they are not handled or stored properly. Bacterial contamination can occur if molasses comes into contact with unclean surfaces or is stored in a warm, moist environment.

(b) Mold: Molasses can also be susceptible to mold growth if it is stored in a damp or humid environment. Mold can produce mycotoxins, which can pose health risks if consumed in large quantities.

To prevent and regulate food safety and health issues associated with molasses, consider the following measures:

(a) Storage and handling: Store molasses in a cool, dry place and keep them tightly sealed to prevent contamination and mold growth. Ensure that storage containers are clean and free from any residues that could contaminate the molasses.

(b) Quality control: Regularly inspect molasses for signs of contamination or mold growth. If any irregularities are observed, the affected product should be discarded, and the storage area should be thoroughly cleaned and sanitized.

(c) Good manufacturing practices: If molasses is being used in a commercial food production setting, adhere to good manufacturing practices (GMP) to ensure the safe handling and processing of molasses. This includes maintaining a clean production environment, training employees in proper food-handling techniques, and implementing quality-control procedures.

(d) Regulatory compliance: Ensure that all relevant food safety regulations and guidelines are followed when handling, processing, and storing molasses. This may include compliance with USDA & health department regulations, as well as any specific regulations related to the production and sale of molasses.

By following these measures, the risks associated with molasses can be effectively managed, ensuring their safe use in food production.

5. **Honey**: Honey is generally considered safe to consume, but there are some food safety and health issues associated with it, as well as regulations in place to ensure its safety.

Here are some of the key issues and measures to prevent and regulate them:

(a) Microbial contamination: Honey can sometimes contain spores of Clostridium botulinum, a bacterium that can produce botulinum toxin under certain conditions. To prevent this, honey should not be given to infants under 1 year old, as their digestive systems are not fully developed enough to inhibit bacterial growth. Additionally, honey should be stored in a cool, dry place to prevent bacterial growth.

(b) Adulteration: There have been cases of honey being adulterated with added sugars or other substances. To prevent this, regulatory measures such as testing and certification of honey's authenticity can be implemented. Consumers can also look for reputable brands and sources of honey.

(c) Pesticide residues: Honeybees can come into contact with pesticides used in agriculture, and traces of these pesticides can potentially be found in honey. To regulate this, governments can establish maximum residue limits for pesticides in honey and enforce monitoring programs to ensure compliance.

(d) Allergen contamination: People with allergies to pollen may experience allergic reactions to honey. To prevent this, honey should be properly labeled to indicate its source and potential allergies, and individuals with known allergies should avoid consuming honey from those sources.

Regulatory measures to ensure the safety of honey can include:

-Establishing and enforcing standards for honey production, processing, and labeling.

-Monitoring and testing honey for contaminants, including pesticides and microbial pathogens.

-Implementing traceability measures to track the source of honey and ensure its authenticity.

-Providing guidelines for the proper storage and handling of honey to prevent contamination.

By implementing and enforcing these measures, the food safety and health issues associated with honey can be effectively managed, ensuring the safety of this popular natural sweetener.

6. **Dextrose**: Dextrose, also known as glucose, is a simple sugar that is often used as a sweetener in food products. It is generally recognized as safe (GRAS) by the United States Department of Agriculture & the U.S. Food and Drug Administration (FDA), but, like any food ingredient, it has some associated food safety and health issues.

    (a) Dental Health: Dextrose, like other sugars, can contribute to dental cavities and tooth decay if consumed in excessive amounts or if proper oral hygiene is not maintained.

(b) Obesity and Diabetes: Consumption of large amounts of dextrose, especially in the form of high-calorie, low-nutrient foods, can contribute to obesity and increase the risk of type 2 diabetes.

(c) Allergies and Sensitivities: Some individuals may have sensitivity or allergies to dextrose, although this is relatively rare.

To prevent and regulate the potential food safety and health issues associated with dextrose, the following measures can be taken:

(a) Nutritional Labeling: Clearly label products containing dextrose to inform consumers about its presence and the potential health implications of consuming products with added sugars.

(b) Moderation: Encourage moderation in the consumption of foods containing dextrose and other added sugars. Providing clear serving size information on packaging can help consumers make informed choices.

(c) Education: Provide education and information to consumers about the potential health effects of excessive dextrose consumption and the importance of a balanced diet.

(d) Regulation: Government agencies can regulate the use of dextrose in food products and set limits on its use in certain categories of products to promote public health.

(e) Reformulation: Encourage food manufacturers to reformulate their products to reduce the size of added sugars, including dextrose, and to develop healthier alternatives.

(f) Dental Hygiene: Promote good dental hygiene practices to mitigate the risk of dental issues associated with dextrose consumption.

By implementing these measures, it is possible to help mitigate potential food safety and health issues associated with dextrose and other added sugars while still allowing their safe use in food products.

7. **Cumin**: Cumin is a widely used spice in cuisines around the world, and like any food product, it can be associated with food safety and health issues. Here are some potential issues associated with cumin and how they can be prevented and regulated:

   (a) Microbial contamination: Cumin, like other spices, can be contaminated with bacteria, molds, and yeast. This can occur during the growing, harvesting, processing, or storage of the spice. To prevent microbial contamination, good agricultural and manufacturing practices should be followed, including proper sanitation, drying, and storage conditions. Regulatory agencies can enforce these practices through inspections and quality control standards.

(b) Allergen contamination: Cross-contamination with allergens such as peanuts, tree nuts, or gluten can occur during the processing and packaging of cumin. To prevent allergen contamination, manufacturers should implement proper cleaning procedures and allergen control measures in their facilities. Regulatory agencies can require allergen labeling and enforce strict allergen control practices.

(c) Adulteration: Cumin is sometimes subject to adulteration with cheaper substitutes or fillers. This can compromise the quality and safety of the spice. Adulteration can be prevented through quality control measures, including purity and authenticity testing. Regulatory bodies can set standards for cumin purity and enforce penalties for adulteration.

(d) Pesticide residues: Pesticide residues on cumin can pose health risks if they exceed safe levels. To prevent this, farmers should follow approved pesticide application practices and adhere to pre-harvest intervals. Regulatory agencies can monitor and enforce pesticide residue limits through testing and certification programs.

Regulation of cumin safety can be achieved through several means:

(a) Setting standards: Regulatory bodies can establish specific standards for cumin quality, purity, and safety. These standards may include maximum allowable limits for microbial contamination.

(b) Inspection and certification: Regulatory agencies can conduct inspections of cumin processing facilities to ensure compliance with food safety regulations. They can also implement certification programs to verify that cumin meets safety and quality standards.

(c) Labeling requirements: Clear and accurate labeling of cumin products can help consumers make informed choices. Regulatory agencies can mandate labeling requirements for allergens, country of origin, and other relevant information.

(d) Import/export controls: Governments can continue to enforce regulations on the import and export of cumin to ensure that it meets safety and quality standards. This can include inspection of imported cumin shipments and adherence to international food safety standards.

By implementing and enforcing these measures, the safety and quality of cumin can be better assured, helping to protect consumers from potential health risks associated with the spice.

8. **Protein Hydrolysates**: Protein hydrolysates are products of protein hydrolysis, a process in which proteins are broken down into smaller peptides or amino acids by the action of enzymes, acids, or other methods. This process yields protein hydrolysates, which are used in a variety of food and nutritional products for their unique functional and nutritional properties.

Food Safety and Health Issues of Protein Hydrolysates:

(a) Allergenicity: Protein hydrolysates can still retain allergenic properties, which may pose a risk to individuals with allergies to the source protein. It's important to properly label products containing protein hydrolysates to alert consumers with allergies.

(b) Contaminant Control: During the production process, there is a risk of microbial contamination, which can lead to foodborne illnesses if not properly controlled.

(c) Quality and Nutritional Integrity: The quality and nutritional integrity of hydrolysates can be compromised if not manufactured and stored under appropriate conditions. This can impact on the effectiveness and safety of the products in which they are used.

Prevention and Regulation:

(a) Good Manufacturing Practices (GMP): Adherence to GMP in the production of protein hydrolysates is crucial to ensure the safety and quality of the products. This includes proper sanitation, personnel training, and facility maintenance.

(b) Allergen Labeling: Clear and accurate labeling of products containing protein hydrolysates is essential to inform consumers of potential allergens and prevent allergic reactions.

(c) HACCP (Hazard Analysis and Critical Control Points): Implementing a HACCP plan can help identify and control potential hazards in the production process of protein hydrolysates, thereby ensuring food safety.

(d) Regulatory Oversight: Government agencies, such as the USDA or FDA in the United States and similar bodies in other countries, should continue to regulate the production and labeling of food products, including those containing protein hydrolysates, to ensure consumer safety and product quality.

By following these preventive measures and regulatory guidelines, the food industry can help mitigate the food safety and health issues associated with protein hydrolysates, ensuring that these products are safe and beneficial for consumers.

9. **Monosodium Glutamate**: Monosodium glutamate (MSG) is the sodium salt of glutamic acid, one of the most abundant naturally occurring non-essential amino acids. It is commonly used as a flavor enhancer in many processed and restaurant foods. MSG is known for enhancing the savory (umami) taste of food and is often used in savory snacks, canned soups, and fast food.

There has been significant debate and concern about the safety of MSG and its potential health effects. Some people report symptoms such as headache, flushing, sweating, facial pressure or tightness, numbness, tingling, palpitations, chest pain, and nausea after consuming foods containing MSG. However, scientific evidence linking MSG to these symptoms is limited and inconsistent. Many studies have failed to show a consistent relationship between MSG ingestion and these symptoms, leading many experts to conclude that MSG is safe for the general population when consumed in typical amounts.

To prevent the negative effects of excessive MSG and regulate its use, food safety authorities in various countries have established guidelines and regulations. These regulations often include labeling requirements specifying the maximum allowable amount of MSG in food products and ensuring that its use is clearly disclosed to consumers. Additionally, some individuals may choose to avoid food containing them, even in moderation, if they believe they are sensitive to them.

As with any food additive, it's important for consumers to be aware of what they are eating and to pay attention to food labels. If you are concerned about consuming MSG, you can look for products labeled as "MSG-free" or "no added MSG." Additionally, cooking meals from fresh ingredients at home can help you better control what goes into your food. If you have specific health concerns related to MSG, it's best to consult with a healthcare professional for personalized advice.

10. **Saltpeter**: Saltpeter, also known as potassium nitrate (KNO3), is a chemical compound that has been historically used for various purposes, including as a food

preservative and a component of gunpowder. In the context of food, saltpeter has been used to cure and preserve meats, particularly in the preparation of bacon, ham, and sausages.

Food Safety and Health Concerns:

(a) Nitrate and Nitrite Content: Saltpeter contains nitrate, which can be converted into nitrite by bacteria in the human body. Nitrites can react with compounds in the stomach to form nitrosamines, some of which are known to be carcinogenic. Excessive consumption of nitrites has been associated with an increased risk of certain types of cancer.

(b) Methemoglobinemia: Nitrites can also cause a condition called methemoglobinemia, especially in infants, where the ability of the blood to carry oxygen is reduced, leading to a bluish discoloration of the skin and potentially serious health effects.

Prevention and Regulation:

(a) Usage Limits: In many countries, the use of saltpeter in food is regulated, and there are strict limits on the amount that can be used in food products. These limits are designed to minimize health risks associated with nitrate and nitrite consumption.

(b) Alternative Preservatives: Many food manufacturers have shifted away from using saltpeter in favor of other preservatives, such as sodium nitrite. Nitrite is often used in combination with salt to cure meat and is regulated to ensure that it is used within safe limits.

(c) Consumer Awareness: Consumers can reduce their exposure to nitrite and nitrite by being aware of the foods they consume. Eating a balanced diet that includes plenty of fruits and vegetables can help mitigate the potential health risks associated with nitrite and nitrite consumption.

(d) Labeling and Monitoring: Food regulatory agencies often require products containing saltpeter or other nitrate and nitrite sources to be clearly labeled, and there are monitoring programs in place to ensure that manufacturers comply with regulations.

It's important to note that while there are concerns associated with the use of saltpeter and other nitrate/nitrite compounds in food, when used within regulatory limits, they can be safe and effective for preserving food products.

11. **Milk Products**: Milk and milk products are widely used in food processing, and they can be classified into various types based on their fat content and processing methods. Some common milk products utilized in food processing include:

    (a) Whole Milk: Contains around 3.5% milk fat and is used in a variety of food products, including dairy-based desserts, sauces, and baked goods.

    (b) Skim Milk: Also known as fat-free or non-fat milk, it contains less than 0.5% milk fat. It is used in the production of low-fat dairy products and as an ingredient in processed foods.

(c) Condensed Milk: This is whole milk with water removed and sugar added, resulting in a thick, sweet product. It is commonly used in the production of confectionery, desserts, and sweetened beverages.

(d) Evaporated Milk: Like condensed milk but without added sugar, evaporated milk is used in the production of creamy sauces, soups, and desserts.

(e) Powdered Milk: Milk that has been evaporated to dryness, leaving behind a milk powder that has a longer shelf life and is used in a wide range of food products, including baked goods, confectionery, and beverages.

Milk is one of the Big 8 food allergens, and food safety concerns associated with milk products primarily center on the risk of microbial contamination, particularly by bacteria such as Salmonella, Escherichia coli, and Listeria monocytogenes. These pathogens can cause foodborne illnesses if present in milk products. Additionally, improper handling and storage of milk products can lead to spoilage and deterioration in quality.

To prevent and regulate these issues, several measures can be taken:

(a) Pasteurization: Most milk products undergo pasteurization, a process of heating the milk to kill harmful bacteria and extend shelf life. Proper pasteurization is crucial for ensuring the safety of milk products.

(b) Hygienic Processing: Strict adherence to good manufacturing practices (GMP) and sanitation procedures in food processing facilities is essential to prevent contamination of milk products.

(c) Quality Control: Regular testing for microbial contamination and quality parameters such as fat content, protein content, and sensory attributes helps ensure the safety and quality of products.

(d) Cold Chain Management: Maintaining proper refrigeration and storage conditions throughout the supply chain is critical to prevent spoilage and microbial growth in milk products.

(e) Regulatory Oversight: Government agencies and regulatory bodies play a key role in setting and enforcing food safety standards for milk products, as well as conducting inspections and audits of food processing facilities.

By implementing these measures, the food industry can mitigate food safety concerns associated with milk products and ensure that they are safe for consumption.

12. **Wheat**: Wheat is a grass widely cultivated for its seed, a cereal grain that is a worldwide staple food. Globally, it is the leading source of vegetable protein in

human food, with a higher protein content than other major cereals, such as maize (corn) and rice. In terms of total production tonnage used for food, it is currently second only to rice among the main human food crops and ahead of maize.

Wheat is one of the Big 8 allergens, and when it comes to other food safety and health concerns associated with wheat, there are a few key points to consider:

(a) Gluten-related Disorders: Wheat contains gluten, a complex mixture of proteins that can trigger adverse reactions in some individuals. Celiac disease is an autoimmune disorder in which the ingestion of gluten leads to damage in the small intestine. Additionally, some individuals may have non-celiac gluten sensitivity, which can cause similar symptoms to those with celiac disease but without the autoimmune response.

(b) Pesticide and Chemical Residues: Like other crops, wheat cultivation may involve the use of pesticides and other chemicals. Residues of these substances on wheat can pose health risks if not properly regulated and monitored.

(c) Mycotoxins: Fungal toxins known as mycotoxins can contaminate wheat and wheat products if the crop is not properly stored. Mycotoxins such as aflatoxins and deoxynivalenol can have detrimental health effects if consumed in high quantities.

In the United States, the regulation of wheat and its related products falls under the purview of several governmental agencies, including the Food and Drug Administration (FDA) and the United States Department of Agriculture (USDA). These agencies oversee various aspects of food safety, including monitoring pesticide residues, mycotoxin levels, and gluten content in food labeling.

Additionally, the FDA enforces regulations related to food labeling and the presence of allergens, including wheat, in food products. For example, the Food Allergen Labeling and Consumer Protection Act (FALCPA) requires that food products containing wheat as an ingredient clearly list this on the label. Furthermore, industry standards and certifications, such as those for organic and non-GMO (genetically modified organism) products, can affect the production and regulation of wheat-based foods. Overall, the regulation of wheat and its related products aims to ensure that they are safe for consumption, that potential allergens are clearly labeled, and that any chemical residues are within safe limits.

(13) Allspice: Allspice is a spice made from the dried, unripe berries of the Pimenta dioicia tree, which is native to the Caribbean, Central, and South America. The name “allspice” is derived from the fact that its flavor is said to resemble a combination of cinnamon, cloves, and nutmeg. It is commonly used in both sweet and savory dishes, particularly in Caribbean and Latin American cuisine, and it is a key ingredient in many spice blends.

Food Safety Concerns:

Allspice itself does not pose significant food safety concerns. However, as with any spice or food product, there are a few potential issues to be aware of:

(a) Contamination: Allspice, like other spices, can be contaminated with pathogens such as Salmonella or E. coli. This can occur during the growing, harvesting, processing, storage, or packaging stages. To prevent contamination, it's important to purchase allspice from reputable sources and store it properly in a cool, dry place.

(b) Allergies: Some individuals may have allergies to allspice. If you or someone you are cooking for has a known allergy, it's important to avoid using allspice in dishes and to carefully check ingredient labels on packaged foods.

(c) Storage: Improper storage of allspice can lead to a loss of flavor and potency over time. To maintain its quality, store allspice in a sealed container in a cool, dark place away from heat and moisture.

Health Issues:
Allspice is generally safe for consumption when used in normal culinary quantities. However, excessive consumption can potentially lead to some health issues:

(a) Cautions for certain health conditions: Allspice, like many spices, contains compounds that may affect certain health conditions. For example, people with gastroesophageal reflux disease (GERD) or irritable bowel syndrome (IBS) may find that allspice exacerbates their symptoms.

(b) Pregnancy: There is some evidence to suggest that large amounts of allspice may have a negative impact during pregnancy, so it's best for pregnant women to consume it in moderation.

Prevention:

(a) Purchase allspice from reputable sources and check for signs of spoilage before use.

(b) Store allspice properly in a cool, dry place to maintain its quality.

(c) Use allspice in moderation, especially if you have known health concerns or are pregnant.

(d) When cooking for others, be mindful of potential allergies and dietary restrictions.

As with any food or ingredient, moderation and mindfulness are key to enjoying allspice safely. If you have specific health concerns, it's always best to consult with a healthcare professional.

(14)Anise: Anise, also known as Pimpinella anisum, is a flowering plant native to the Eastern Mediterranean region and Southwest Asia. It is cultivated for its aromatic seeds, which are used as a spice in cooking and for their medicinal properties. Anise has a sweet, licorice-like flavor and is commonly used in baking, cooking, and herbal medicine.

Food Safety Concerns:

When it comes to food safety concerns associated with anise, the main issue is contamination. Anise seeds can become contaminated with bacteria, mold, or other pathogens during cultivation, harvesting, processing, or storage. Proper handling and storage of anise seeds are essential to prevent contamination and ensure food safety.

Health Issues:

Anise itself is generally safe for consumption when used in moderate amounts as a spice or flavoring. However, some individuals may be allergic to anise or sensitive to its compounds, which can cause allergic reactions or skin irritation. In addition, excessive consumption of anise oil or anise-flavored products may cause adverse effects, especially in individuals with certain medical conditions.

Prevention of Sickness from Anise:

To ensure food safety and prevent health issues associated with anise, consider the following measures:

(a) Purchase from reputable sources: Buy anise seeds from reliable suppliers who follow proper food safety and quality control measures.

(b) Check for freshness: Ensure that the anise seeds are fresh and have not been stored for an extended period. Fresh spices are less likely to be contaminated.

(C)Proper storage: Store anise seeds in airtight containers in a cool, dry place away from direct sunlight to prevent mold growth and maintain freshness.

(d)Allergy awareness: Be mindful of potential allergic reactions to anise and its compounds. If you have a known allergy to similar plants (such as fennel, celery, or dill), use anise with caution or avoid it altogether.

(e)Moderate consumption: Use anise in moderation, especially anise oil, as excessive consumption can lead to adverse effects. If you have any underlying health conditions, consult a healthcare professional before using anise products medicinally. By following these guidelines, you can minimize the risks associated with anise and safely enjoy its flavor and potential health benefits.

(15)Basil: Basil is a popular herb in the mint family, known for its strong, pungent aroma and flavor. It is commonly used in various cuisines around the world, particularly in Italian and Southwest Asian dishes. Basil is also known for its potential health benefits, as it contains a variety of nutrients and has been associated with antioxidant and anti-inflammatory properties.

When it comes to food safety concerns and health issues associated with basil, there are a few key points to consider:

(a) Contamination: Like any fresh produce, basil can be contaminated with bacteria, viruses, or parasites, which can pose a risk of foodborne illness if consumed. Contamination can occur at any point along the production and distribution chain, from the farm to the consumer's kitchen.

(b) Pesticide residues: Basil, if not grown organically, may contain pesticide residues, which can be harmful if consumed in large quantities over time.

To prevent food safety concerns and health issues associated with basil, consider the following measures:

-Source from reputable suppliers: When purchasing basil, whether from a grocery store, market, or directly from a supplier, ensure that it comes from a reputable and trustworthy source with good food safety practices.

-Wash thoroughly: Before using basil in cooking or eating it raw, it's important to wash the leaves thoroughly under running water to remove any dirt, debris, or potential contaminants.

-Proper storage: Store fresh basil properly to maintain its quality and reduce the risk of contamination. Keep it refrigerated and use it within a few days of purchase.

-Consider organic options: Choosing organically grown basil can help reduce the risk of pesticide residues, as organic farming practices typically minimize the use of synthetic pesticides.

-Proper food handling: When using basil in food preparation, follow good food handling practices, such as washing hands, utensils, and surfaces to prevent cross-contamination with other foods.

By following these practices, you can help minimize the food safety concerns and health issues associated with basil and enjoy its flavor and potential health benefits with greater peace of mind.

(16)Sage: Sage (Salvia officinalis) is a perennial herb that is commonly used in cooking, particularly in Mediterranean cuisine. It is known for its distinctive earthy flavor and aromatic fragrance. In addition to its culinary uses, sage has been used in traditional medicine for its potential health benefits. When it comes to food safety concerns and health issues associated with sage, there are a few key points to consider:

(a) Pesticide Residues: Like many agricultural products, sage may be treated with pesticides during cultivation. It's important to purchase sage from reputable sources and, if possible, choose organic varieties to minimize exposure to pesticide residues.

(b) Allergies: Some individuals may be allergic to sage or other members of the mint family. Allergic reactions to sage can range from mild to severe, so it's essential to be aware of potential allergens in the herb and to seek medical attention if an allergic reaction occurs.

(c) Medicinal Use: While sage has been traditionally used for its potential health benefits, it's important to use medicinal herbs with caution. Consuming large quantities of sage or using it in concentrated forms (such as essential oils) may lead to adverse effects or interactions with medications. Pregnant women should also avoid consuming large amounts of sage due to its potential to stimulate uterine contractions.

To prevent these potential issues associated with Sage, here are some tips:

(a) Purchase sage from reputable sources and consider choosing organic varieties to reduce exposure to pesticide residues.

(b) Be mindful of potential allergic reactions. If you know you have a sensitivity to other herbs in the mint family, exercise caution when using sage.

(c) When using sage for medicinal purposes, consult with a healthcare professional or a qualified herbalist to ensure safe and appropriate use.

(d) If you are pregnant, it's best to avoid consuming large amounts of sage or using concentrated sage products.

(e) By being mindful of these considerations and using sage in moderation, you can enjoy its culinary and potential health benefits while minimizing associated risks.

(17)Mustard: Mustard is a condiment made from the seeds of a mustard plant, which is a member of the Brassicaceae family, the same family as cabbage, broccoli, and Brussels sprouts. The seeds can be ground into a powder to make prepared mustard, or mixed with other ingredients such as vinegar, water, and spices to create a paste or sauce.

Food safety concerns and health issues associated with mustard are generally rare, but they can include:

(a) Allergic reactions: Mustard, to some people, is an allergen, and allergic reactions to mustard can range from mild to severe. People with mustard allergies should avoid consuming mustard products and carefully read food labels to identify potential sources of mustard in processed foods.

(b) Contamination: As with any agricultural product, there is a risk of contamination with bacteria, molds, or other pathogens. To prevent contamination, it's important to store mustard and mustard-based products properly, following recommended storage guidelines, and to use clean utensils and containers when handling mustard.

To prevent food safety concerns and health issues associated with mustard, consider the following tips:

-Allergen awareness: If you or someone you are cooking for has a mustard allergy, always read food labels carefully to identify mustard-containing ingredients. Be cautious when dining out and ask about the ingredients in dishes to avoid accidentally consuming mustard.

-Storage: Store prepared mustard, mustard powder, and other mustard-based products according to the manufacturer's instructions. Typically, this involves keeping them in a cool, dry place, away from direct sunlight and heat sources.

-Cross-contamination: When handling mustard and preparing food with mustard, use separate utensils and containers to avoid cross-contamination with other foods. Clean utensils and surfaces thoroughly after using mustard to prevent the spread of allergens or pathogens.

By following these guidelines, you can enjoy mustard safely and minimize the risk of foodborne illness and other health issues associated with this popular condiment.

(18)Peanuts: Peanuts are a type of legume that is widely consumed around the world. They are a good source of protein, healthy fats, and various important nutrients, including vitamin E, niacin, folate, and manganese. Peanuts are used in a variety of food products, including peanut butter, snacks, and cooking oils.

However, there are some food safety concerns and health issues associated with peanuts:

(a) Allergies: Peanut is one of the Big 8 allergens. Peanut allergies are one of the most common food allergies, especially in children. For individuals with peanut allergies, even small amounts of peanuts or peanut products can cause severe allergic reactions, including anaphylaxis, which can be life-threatening.

(b) Aflatoxin contamination: Peanuts are susceptible to contamination by a mold called Aspergillus flavus, which produces a toxin known as aflatoxin. Aflatoxins are potent carcinogens and can have harmful effects on the liver. Proper storage and processing of peanuts are essential to minimize the risk of aflatoxin contamination.

(c) Salmonella contamination: Peanuts can also be contaminated with Salmonella, a type of bacteria that can cause foodborne illness. Improper handling, storage, and processing of peanuts can lead to the growth of Salmonella bacteria.

To prevent these food safety concerns and health issues associated with peanuts, the following measures can be taken:

-Allergen labeling: Food products containing peanuts or peanut derivatives should be clearly labeled to alert consumers who have peanut allergies.

-Allergen separation: In food manufacturing facilities, measures should be taken to prevent cross-contamination between products containing peanuts and those that do not.

-Aflatoxin control: Proper storage conditions, including controlling temperature and humidity, can help reduce the growth of mold and the production of aflatoxins in peanuts. Additionally, good agricultural practices and proper sorting of peanuts can help minimize aflatoxin contamination.

-Salmonella control: Adequate processing, storage, and handling of peanuts are essential to minimize the risk of Salmonella contamination. This includes proper roasting of peanuts and maintaining good hygiene practices during processing.

For individuals with peanut allergies, the best way to prevent adverse reactions is to avoid peanuts and peanut products altogether and to always read food labels carefully. If someone has a known peanut allergy, it's important to carry an epinephrine auto-injector (such as an EpiPen) in case of accidental exposure.

(19)Soy: Soy, or soya, is a species of legume native to East Asia, widely grown for its edible bean, which has numerous uses. Soybeans are an important source of protein and oil, and they are used in a variety of foods, including tofu, tempeh, soy milk, and soy sauce. Soy products are also used in the production of animal feed and as an ingredient in many processed foods.

Food Safety and Health Issues Associated with Soy:

(a) Allergies: Soy is one of the eight most common food allergens, and soy allergies can cause a range of symptoms from mild itching and hives to severe reactions such as anaphylaxis. Individuals with soy allergies need to carefully read food labels and avoid products containing soy.

(b) Phytoestrogens: Soy contains compounds called phytoestrogens, which are plant-derived compounds that can mimic the hormone estrogen in the body. There has been some concern that consuming large amounts of soy-based products could negatively affect hormone balance, particularly in individuals with certain medical conditions. However, the evidence on this topic is still inconclusive.

(c) Genetically Modified Organisms (GMOs): The majority of soybeans grown globally are genetically modified to be herbicide-tolerant. Some people are concerned about the potential health and environmental impacts of consuming GMO soy products. However, regulatory authorities generally consider GMO soy to be safe for consumption.

Preventing Soy-Related Food Safety and Health Issues:

-Allergies: Individuals with soy allergies should carefully read food labels to identify and avoid products containing soy. They should also inform restaurant staff of their allergies when dining out.

-Moderation: As with any food, moderation is key. Eating a varied diet that includes a balanced mix of protein sources can help mitigate potential concerns associated with consuming large amounts of soy.

-Non-GMO or Organic Options: For those concerned about GMOs, choosing organic or non-GMO soy products can be a way to minimize exposure to genetically modified organisms.

It's important to note that for many people, consuming moderate amounts of soy as part of a balanced diet is considered safe and may offer health benefits, such as providing a good source of protein and containing compounds that may positively affect heart health. However, individuals with specific health concerns or conditions should consult with a healthcare professional for personalized dietary advice.

(20)Caraway: Caraway, also known as meridian fennel and Persian cumin, is a biennial plant in the Apiaceae family, which is native to western Asia, Europe, and North Africa. Caraway seeds are used as a spice to add flavor to various dishes, particularly in European and Mediterranean cuisines. The seeds have a pungent anise-like flavor and are often used in rye bread, sauerkraut, and various meat dishes.

When it comes to food safety and health concerns about caraway, there are a few points to consider:

(a) Allergies: Some individuals may be allergic to caraway seeds. Allergic reactions to caraway are rare but can occur in sensitive individuals. Symptoms of an allergic reaction can include hives, itching, swelling, or, in severe cases, anaphylaxis. People with known seed or spice allergies should avoid caraway.

(b) Contamination: As with any spice or dried herb, there is a risk of contamination with pathogens such as Salmonella or foreign matter such as insects, dirt, or other plant materials. To minimize this risk, it's important to purchase caraway seeds from reputable sources and store them in a cool, dry place.

(c) Pesticide Residues: Depending on the source of the caraway seeds and the agricultural practices used, there is a possibility of pesticide residues. To reduce exposure to pesticides, choosing organic caraway seeds may be an option.

To avoid potential health concerns related to caraway, individuals can take the following precautions:

(a) Purchase caraway seeds from reputable suppliers to ensure quality and safety.

(b) Store caraway seeds in a cool, dry place in airtight containers to maintain their flavor and reduce the risk of contamination.

(c) Individuals with known seeds or spice allergies should avoid consuming caraway and products containing caraway.

(d) When using caraway seeds in cooking, it's important to follow proper food safety practices, such as washing hands, utensils, and surfaces to prevent cross-contamination.

By being mindful of these considerations and taking appropriate precautions, individuals can enjoy the flavor of caraway while minimizing potential health and safety concerns.

(21)Cardamom is a highly aromatic "queen of spices." Cardamom has a strong, pungent flavor and is often used to flavor both sweet and savory dishes and beverages. It also has several health benefits.

In terms of food safety concerns and health issues associated with cardamom, there are a few things to consider:

(a) Pesticide Residues: Like many agricultural products, cardamom can be treated with

pesticides to prevent pests and diseases. Residues of these
Pesticides can sometimes remain in the spice, posing potential health risks. To minimize exposure to pesticides, it's best to purchase organic cardamom whenever possible. When using conventionally grown cardamoms, thoroughly washing them before use can help reduce pesticide residues.

(b) Allergies: While rare, some individuals may have allergic reactions to cardamom. Allergic reactions to cardamom can range from mild to severe and may include symptoms such as itching, swelling, hives, and, in severe cases, anaphylaxis. If you or someone you are serving food to has a known allergy to cardamom, it's important to avoid consuming it and to carefully check food labels for its presence.

(c) Contamination: Cardamom, like other spices, can be susceptible to contamination with mold, bacteria, or other pathogens during harvesting, processing, or storage. To minimize the risk of contamination, it's important to purchase cardamom from reputable sources and store it in a cool, dry place to prevent mold and bacterial growth.

To prevent these potential issues, here are some general tips:

-Purchase high-quality cardamom from reputable sources. Look for organic options to minimize exposure to pesticides.

-Thoroughly wash and rinse cardamom before use to remove any potential pesticide residues.

-Store cardamom in a cool, dry place to prevent contamination and the growth of mold and bacteria.

-Be aware of potential allergic reactions and avoid consuming cardamom if you or others have known allergies to it.

-When using cardamom in cooking, follow proper food safety practices, such as cooking foods to the appropriate temperatures and storing them at safe temperatures.

By being mindful of these potential concerns and taking steps to mitigate risks, you can continue to enjoy cardamom in your culinary creations while minimizing any associated health or safety issues.

(22)Cassia: a type of cinnamon commonly used as a spice and for its medicinal properties. It comes from the bark of the Cinnamomum cassia tree, which is native to China and is also found in other parts of East Asia. Cassia is known for its strong, sweet taste and is often used in baking and cooking, as well as in traditional medicine.

Food safety concerns associated with cassia primarily center on coumarin, a natural compound found in many plants, including cassia. High levels of coumarin can be harmful to the liver and kidneys. The European Food Safety Authority (EFSA) has set a tolerable daily intake (TDI) for coumarin, and cassia contains higher levels of coumarin than Ceylon cinnamon.

Consuming large amounts of cassia over a long period of time may lead to health issues such as liver damage, especially for individuals who are sensitive to coumarin. To prevent these potential health issues, it's important to use cassia in moderation and be mindful of the following tips:

(a) Use in moderation: Limit your consumption of cassia and other cinnamon varieties to moderate amounts. If you're using it as a supplement, consult with a healthcare professional to determine a safe dosage.

(b) Consider Ceylon cinnamon: If you're concerned about coumarin levels, consider using Ceylon cinnamon instead of cassia. Ceylon cinnamon contains much lower levels of coumarin and is considered safer for regular consumption.

(c) Check labels: When purchasing cinnamon products, check the labels to see if they specify the type of cinnamon used. Choose products that clearly indicate they contain Ceylon cinnamon to minimize coumarin intake.

(d) Be mindful of cumulative intake: Remember that coumarin is found in various foods and beverages, so be mindful of your overall intake from different sources, especially if you regularly consume cassia.

By being aware of these considerations and using cassia and other cinnamon varieties in moderation, you can enjoy their flavor and potential health benefits while minimizing the associated risks. If you have specific health concerns or dietary restrictions, it's always best to consult with a healthcare professional or a registered dietitian for personalized advice.

(23)Celery Seed: Celery seed is the seed of the Apium graveolens plant, which is related to the celery vegetable. It is commonly used as a spice to flavor various dishes and has been used in traditional medicine for its potential health benefits.

Food Safety Concerns:

(a) Allergies: Celery seed can cause allergic reactions in some individuals, especially those who are allergic to celery or other plants in the Apiaceae family, such as carrots, fennel, and parsley.

(b) Contamination: Like other seeds and spices, celery seed can be susceptible to contamination with bacteria, mold, or other pathogens if not stored or handled properly.

Health Issues:

(a) High blood pressure: Celery seed is often used as a natural remedy to help lower blood pressure, but consuming large amounts may have adverse effects, especially when combined with pressure-lowering medications.

(b) Pregnancy: Pregnant women should avoid consuming large amounts of celery seed, as it may have uterotonic effects and could potentially cause contractions.

(c)Allergies: Individuals with known allergies to celery or related plants should avoid consuming celery seed or products containing celery seed. Always read food labels to check for celery seed as an ingredient.

(d)Contamination: To prevent contamination, store celery seed in a cool, dry place in airtight containers. Purchase celery seed from reputable sources and check for signs of spoilage before using.

(e)Moderation: if using celery seed for its potential health benefits, such as for controlling blood pressure, it's important to use it in moderation and consult with a healthcare professional, especially if you have underlying health conditions or are taking medications.

(24)Chives: Chives are a popular herb that belongs to the same family as onions, garlic, and leeks. They have a mild, onion-like flavor and are commonly used as a garnish or ingredient in various dishes, including soups, salads, and sauces. Chives are known for their nutritional benefits, being rich in vitamins A and C, as well as minerals such as calcium and iron.

Food Safety Concerns: While chives are generally safe to consume, there are some food safety concerns associated with them, particularly in relation to contamination and proper storage:

(a) Contamination: Like other fresh produce, chives can be contaminated with bacteria, such as E. coli or Salmonella, especially if they are not handled and stored properly.

(b) Pesticide Residues: If chives are not organically grown, there is a possibility of pesticide residues, which can pose health risks if consumed in large amounts.

Health Issues:

Health issues associated with chives are relatively minor. However, some individuals may experience allergic reactions to chives, particularly if they are allergic to other members of the allium family, such as onions or garlic. Symptoms of an allergic reaction to chives may include itching, swelling, or a rash.

To prevent food safety concerns and health issues associated with chives, consider the following tips:

(a) Purchase chives from reputable sources, and if possible, choose organic chives to reduce the risk of pesticide residues.

(b) Wash chives thoroughly under running water before removing any potential contaminants.

(c) Store chives in the refrigerator at temperatures below 40 F (4 C) to maintain freshness and reduce the risk of bacterial growth.

(d) Practice good kitchen hygiene, such as washing hands and utensils after handling chives, to prevent cross-contamination with other foods.

(e) If you have known allergies to the allium family, such as onions or garlic, be cautious when consuming chives and seek medical advice if you experience any adverse reactions.

By following these practices, you can minimize the food safety concerns and health risks associated with chives and enjoy their flavor and nutritional benefits with peace of mind.

(25)Cloves: Cloves are the aromatic flower buds of a tree in the family Myrtaceae, Syzygium aromaticum. They are native to the Maluku Islands in Indonesia and are commonly used as spices in cuisines worldwide. Cloves are known for their strong, sweet, and aromatic flavor, and they are often used in both sweet and savory dishes, as well as in the preparation of certain beverages.

When it comes to food safety concerns associated with cloves, there are a few points to consider:

(a) Allergies: Some individuals may be allergic to cloves. Allergic reactions to cloves can range from mild to severe and may include symptoms such as itching, swelling, and difficulty breathing. Those with known allergies to spices should exercise caution when consuming cloves.

(b) Coumarin content: Cloves, like many other spices, contain coumarin, a natural chemical compound that has been linked to potential liver toxicity in high doses. While the levels of coumarin in cloves are generally not high enough to pose a significant risk to most people when used in normal culinary quantities, individuals with existing liver conditions should be mindful of their intake.

(c) Contamination: As with any spice, there is a risk of contamination with microbes, mold, or other impurities. To minimize this risk, it's important to purchase cloves from reputable sources, store them in a cool, dry place, and use them before their expiration date.

To prevent and avoid potential health issues associated with cloves:

-Allergies: If you have known allergies to spices, including cloves, it's important to read food labels carefully, inquire about ingredients when dining out, and consider avoiding dishes that may contain cloves or other potentially problematic spices.

-Coumarin content: For most people, the typical culinary use of cloves is unlikely to pose a health risk related to coumarin. However, individuals with liver conditions or those who are concerned about coumarin intake may choose to limit their consumption of cloves or seek guidance from a healthcare professional.

-Contamination: To minimize the risk of contamination, purchase high-quality cloves from reputable suppliers, store them properly, and practice good food hygiene when using them in cooking.

As with any food or ingredient, moderation, awareness of potential allergens, and attention to quality and storage practices can help mitigate potential health risks associated with cloves.

(26)Stockfish: Stockfish is a type of dried fish popular in several countries, particularly Norway and Nigeria. It is usually made from unsalted fish, typically cod, which is dried by cold air and wind on wooden racks. Stockfish has a long shelf life and is often used in traditional dishes.

Food Safety Concerns and Health Issues Associated with Stockfish:

- Stockfish dried outside is vulnerable to contamination from environmental, biological, and handling factors. Including bacterial pathogens (E. coli, Salmonella, C. botulinum), mold, mycotoxins, heavy metals, microplastics, and insect infestation. Open-air drying exposes fish to windborne dust, animal excrement, and poor sanitation.

-Histamine formation: Improper drying and storage of fish can lead to the formation of histamine, a compound that can cause food poisoning. Symptoms of histamine poisoning include flushing, headache, sweating, and gastrointestinal issues.

-Parasites: Like many types of fish, stockfish can potentially contain parasites. Consuming raw or undercooked stockfish can lead to parasitic infections, which can cause gastrointestinal problems and other health issues.

-Salt content: While stockfish is traditionally not salted, the fish may absorb salt during processing or from the environment, leading to a high salt content. Excessive salt consumption can contribute to high blood pressure and other health problems.

Prevention and Avoidance of Food Safety & Health Issues:

-Proper drying and storage: It is important to ensure that stockfish is properly dried and stored in clean, dry conditions to prevent the growth of harmful bacteria and the formation of histamine.

-Cooking thoroughly: To reduce the risk of parasitic infections, stockfish should be cooked thoroughly before consumption. Cooking fish to an internal temperature of

145 F (63 C) can help kill parasites and ensure food safety.

-Moderation: Given the potential for high salt content in stockfish, it's important to consume it in moderation to avoid excessive sodium intake. This is particularly important for individuals with high blood pressure or other health conditions affected by salt intake.

-Quality control: Purchasing stockfish from reputable sources and ensuring its quality and freshness can help reduce the risk of food safety issues associated with stockfish.

By following these guidelines, individuals can enjoy stockfish while minimizing the associated food safety concerns and health risks.

(27)Egusi: Egusi is a melon seed used to prepare a variety of dishes in West Africa, particularly in countries such as Nigeria, Ghana, and Cameroon. The seeds are ground into a fine powder and used as a major ingredient in soups and stews. Egusi is known for its slightly sweet, nutty flavor and is often used to thicken soups and add texture to dishes.

When it comes to food safety concerns and health issues associated with eating Egusi, there are a few things to consider:

(a) Aflatoxin Contamination: One of the primary concerns with Egusi is the potential for aflatoxin contamination. Aflatoxins are naturally occurring mycotoxins produced by certain molds, and they can contaminate crops such as nuts and seeds if they are not stored properly. Aflatoxins are known to be carcinogenic and can have adverse effects on the liver. To prevent aflatoxin contamination, store Egusi seeds in a cool, dry place and keep them free of moisture, as moisture can promote mold growth.

(b) Hygiene and Sanitation: As with any food product, proper hygiene and sanitation practices are crucial when preparing Egusi-based dishes. It's important to wash hands, properly wash Egusi seeds, thoroughly wash grinding and cooking utensils, and ensure that the Egusi powder is stored in a clean, airtight container to prevent contamination. To minimize the risk of food safety issues and health concerns associated with Egusi, here are some tips:

-Purchase high-quality Egusi seeds or ground Egusi from reputable sources to reduce the risk of aflatoxin contamination.

-Store Egusi seeds in a cool, dry place to prevent mold growth and aflatoxin contamination.

-When preparing Egusi-based dishes, ensure that cooking utensils and surfaces are clean and that good hygiene practices are followed.

-Thoroughly cook Egusi-based dishes to kill any potential bacteria or contaminants.

By following these guidelines, you can help minimize food safety concerns associated with Egusi and enjoy this flavorful ingredient as part of a healthy, diverse diet.

(28)Ogbono: Ogbono, also known as “ogbono seed” or “African mango seed”, is a popular ingredient in African cuisine, particularly in

Nigeria and other West African countries. It is derived from the seeds of the related African mango (Irvingia gabonensis) and is used to make a traditional, delicious soup known as "ogbono soup." The seeds are ground into a powder, which is then used as a thickening agent and flavoring for soups and stews.

In terms of food safety and health concerns associated with ogbono, there are a few things to consider:

(a) Potential for contamination: As with any food product, there is a risk of contamination with bacteria, mold, or other pathogens. Proper handling, storage, and processing of ogbono seeds are important to prevent contamination. Consumers should also ensure that they purchase ogbono powder from reputable sources to minimize the risk of contamination.

(b) Allergies: Some individuals may be allergic to ogbono seeds or other ingredients commonly used in dishes that contain ogbono. Allergic reactions can range from mild to severe and may include symptoms such as itching, swelling, hives, or, in severe cases, anaphylaxis. People with known food allergies should avoid consuming ogbono or dishes containing it.

To prevent and avoid food safety and health concerns associated with ogbono, the following measures can be taken:

-Source from reputable suppliers: When purchasing ogbono or other ogbono-related products, it's important to buy from reputable suppliers who adhere to food safety standards and regulations.

-Proper storage: Store ogbono powder in a cool, dry place to prevent mold growth and preserve its quality. Proper storage helps minimize the risk of contamination and ensures the product's longevity.

-Allergy awareness: Individuals with known food allergies should carefully read food labels and inquire about the ingredients used in dishes before consuming food containing ogbono. It's crucial for those with allergies to avoid foods that could trigger an allergic reaction.

-Hygiene food preparation: When using ogbono powder to prepare dishes, it's important to follow good hygiene practices, such as washing hands, utensils, and cooking surfaces thoroughly to prevent cross-contamination and the spread of harmful bacteria.

By following these measures, consumers can enjoy ogbono-based dishes while minimizing the associated food safety and health risks.

(29)Dried crayfish: These are small crustaceans that have been sun-dried or dried through other methods, which gives them a special flavoring. They are commonly used as a flavoring in various foods around the world, particularly in Asian and African dishes.

When it comes to food safety and health concerns associated with consuming dried crayfish, there are a few things to consider:

a. Quality and Source: The quality of dried fish can vary widely depending on how they were processed and stored. It is important to ensure you are sourcing them from a reputable supplier to minimize the risk of contamination and spoilage.

b. Allergies: Some individuals have allergies to crayfish. It is important to be aware of any potential allergies in yourself and those for whom the food is prepared.

c. Contamination and Hygiene: Like any seafood product, dried crayfish can be susceptible to contamination if they are not handled and stored properly. Crayfish should not be exposed to outside to flies, pests, or outside pollution that can lead to heavy metal and other buildups.

To address these concerns and ensure the safety of consuming dried crayfish, here are some steps you can take: -Buy from trusted sources. In some countries, crayfish may be intentionally contaminated or adulterated to increase size and profit, so buy from reputable sellers with empathy and business acumen.

-Storage: Store dried fish in a cool, dry place to prevent spoilage and the growth of bacteria. They can have a long shelf life when they are properly stored.

-Promote Allergy Awareness: Many people are not aware that dried crayfish could be an allergen. Be mindful of potential crayfish allergy and inform others to stay away from consuming crayfish if there is a risk of an allergic reaction.

-Thorough Cooking: If you are using dried crayfish in cooking, you must ensure that they are thoroughly cooked before consumption. Cooking at high temperatures can kill bacteria and other pathogens.

-Hygiene: Practice good hygiene when handling dried crayfish and other fish products. Wash your crayfish and remove every visible contaminant. Wash every hand and utensil that will come in contact with your crayfish.

By taking these precautions and being mindful of potential risks, you can enjoy dried Crayfish safely and incorporate them into your culinary creations without compromising your health.

(30)Curry Powder: Curry powder is a blend of various spices commonly used in South Asian and Indian cuisines. The specific ingredients in curry powder can vary widely, but they often include spices such as turmeric, coriander, cumin, fenugreek, and chili peppers. Curry powder is known for its rich flavor and vibrant color, and it can be used to season a wide variety of dishes, including meat, vegetables, and sauces.

When it comes to food safety and health issues associated with curry powder, there are a few considerations to keep in mind:

(a) Quality and Contamination: The quality of the spices used in curry powder can vary, and there is a risk of contamination with bacteria, molds, or other pathogens. To minimize this risk, it's important to purchase curry powder from reputable sources and to check the expiration date to ensure freshness.

(b) Allergens: Some curry powder blends may contain allergens such as mustard, gluten, or nuts. It's crucial to read the ingredient labels carefully, especially if you or your guests have food allergies or sensitivities.

(c) Adulteration: Turmeric is a common ingredient in curry powder and is known for its vibrant yellow color. However, turmeric has been adulterated with lead chromate and other harmful substances. To minimize this risk, choose curry powder from trusted brands and suppliers that adhere to food safety standards.

To minimize food safety and health risks associated with curry powder, consider the following tips:

(a) Purchase from Reputable Sources: Buy curry powder from reputable brands or spice shops that adhere to quality and safety standards. Check for certifications or seals of approval from relevant food safety authorities.

(b) Check for Freshness: Check the expiration date on the packaging and look for signs of freshness, such as a strong aroma and vibrant color. Stale or old curry powder may have a diminished flavor and potency.

(c) (c)Store Properly: Keep curry powder in a cool, dry place away from direct sunlight and moisture. Storing it in airtight containers can help maintain its quality and flavor over time.

(d) (d)Read Labels: Always read the ingredient labels to check for allergens and other potential health concerns. If you have specific dietary restrictions or allergies, choose curry powder that meets your needs. By following these guidelines, you can enjoy the flavorful benefits of curry powder while minimizing potential food safety and health risks.

(31)Garlic powder: Provides a convenient way to add garlic flavor without fresh garlic. Garlic powder is a popular seasoning made from dehydrated, ground garlic cloves. It is commonly used in cooking to add flavor to a wide variety of dishes, including meats, vegetables, and sauces. Garlic powder offers the taste of fresh garlic in a convenient, shelf-stable form, making it a versatile ingredient in many kitchens. When it comes to food safety concerns associated with garlic powder, there are several factors to consider:

(a)Microbial Contamination: Like any dried spice, garlic powder can be susceptible to contamination by bacteria, molds, and other microorganisms. To minimize this risk, it's important to store garlic powder in a cool, dry place and to use it within a reasonable period to prevent the growth of harmful microbes.

(b)Allergen Cross-Contamination: Individuals with garlic allergies need to be cautious, as garlic powder could be processed in facilities that also handle other allergens. Cross-contamination with allergens like wheat, soy, or other spices is a concern, so it's essential to check labels and buy from reputable sources if you have allergies.

©Quality of Raw Materials: The quality of the garlic used to produce the powder can impact safety. Contaminated or spoiled garlic used in the production process can lead to food safety issues. Therefore, it's important to purchase garlic powder from trusted suppliers with good quality control measures.

To minimize these food safety concerns associated with garlic powder, consider the following best practices:

-Purchase from Reputable Sources: Buy garlic powder from reputable and trustworthy suppliers to ensure the product's quality and safety standards.

-Store Properly: Store garlic powder in airtight containers in a cool, dry place, away from direct sunlight and moisture. Proper storage can help prevent microbial contamination and maintain product quality.

-Check for Allergen Information: If you have food allergies, carefully read labels to check for potential allergen cross-contamination. Look for products with clear allergen information and consider contacting the manufacturer if you have specific concerns.

-Use Before Expiration Date: Use garlic powder before its expiration date to ensure its freshness and minimize the risk of consuming a product that may have degraded in quality.

By following these guidelines, you can help minimize the food safety concerns associated with garlic powder and enjoy its flavor-enhancing properties in your cooking with confidence.

(32)**Onion powder**: Adds savory depth and flavor to dishes. Onion powder is a popular food ingredient that is derived from dehydrated onions. While it adds flavor to dishes, it also raises some food safety and health concerns. Here are a few concerns and ways to minimize them:

(a)Contamination: Onion powder, like any powdered food product, can be susceptible to contamination with bacteria, mold, or other pathogens. To minimize this risk, store onion powder in a cool, dry place in an airtight container. Additionally, using good manufacturing practices and purchasing from reputable sources can help ensure the product's safety.

(b)Allergies: Some individuals may have allergies to onions. It's important to be aware of this when using onion powder in recipes. Manufacturers should clearly label products containing onion powder to alert consumers with allergies.

(c) Quality and freshness: Over time, onion powder can lose its flavor and potency. To ensure the best quality and freshness, it's important to use onion powder within its shelf life and to store it properly in airtight containers away from light and heat.

(d)Hygiene during production: During the production of onion powder, it's essential to maintain strict hygiene standards to prevent contamination with harmful microorganisms. This includes ensuring that processing equipment is clean and sanitized and that employees follow proper hygiene practices.

(e)Chemical contamination: Onion powder should be free from chemical contaminants such as pesticides, heavy metals, or other harmful substances. Purchasing onion powder from reputable suppliers who adhere to safety and quality standards can help minimize the risk.

By being mindful of these concerns and taking appropriate steps to minimize them, you can safely enjoy the flavor and convenience of onion powder in your cooking. Additionally, following good food safety practices when handling and preparing foods containing onion powder can further minimize associated health risks.

(33)Paprika: Adds color and a mild, sweet flavor to dishes. Paprika is a spice made from the ground, dried fruits of sweet red or chili peppers. It is widely used to add flavor, color, and sometimes a mild heat to a variety of dishes in many different cuisines, including Hungarian, Spanish, American, and Indian.

In terms of food safety and health concerns associated with paprika, here are a few points to consider:

Contamination: Like any spice, paprika can be contaminated with bacteria, molds, and other microorganisms, especially if not stored properly. This can lead to foodborne illnesses if contaminated paprika is used in cooking.

(a)Allergies: Some individuals may be allergic to paprika. Allergic reactions can range from mild to severe, and in some cases, they can be life-threatening.

(b)Pesticide residues: If paprika is not grown and processed using proper agricultural and manufacturing practices, there is a risk of pesticide residues being present in the spice, which can pose health hazards if consumed in large quantities over time.

(c)Irritation: Paprika, particularly hot varieties, can cause irritation to the skin, eyes, and respiratory system, especially in sensitive individuals.

To minimize these concerns, it's important to purchase paprika from reputable sources, store it in a cool, dry place, and be mindful of any allergies or sensitivities when cooking. Additionally, washing hands after handling paprika and avoiding direct contact with the eyes can help prevent irritation.

(34)Chili powder: Chili powder is a spice blend made from dried chili peppers and other ingredients such as cumin, oregano, and garlic. It is commonly used to add heat and flavor to dishes, particularly in Mexican, Indian, and Tex-Mex cuisines.

In terms of food safety and health concerns associated with chili powder, there are a few potential issues to be aware of:

(a)Contamination: Chili powder can be contaminated with bacteria, such as Salmonella, during the production process. This can happen if the peppers are not properly cleaned and dried, or if the equipment used to process the powder is not adequately sanitized.

(b)Allergens: Some chili powder blends may contain added ingredients, such as cumin or garlic, that could be allergens for some individuals. Cross-contamination with allergens during processing or packaging is also a concern.

To prevent these issues and ensure the safety of chili powder, consider the following tips:

-Purchase from reputable sources: Buy chili powder from reputable brands or sources that adhere to food safety standards. Look for products that have been tested for contaminants and are labeled with allergen information.

-Check for freshness: Choose chili powder that is fresh and has not expired. Stale spices can harbor bacteria and may not have the same flavor potency.

-Store properly: Keep chili powder in a cool, dark place to maintain its flavor and quality. Proper storage can help prevent the growth of bacteria and mold.

-Use proper hygiene: When cooking with chili powder, practice good hygiene by washing your hands and utensils thoroughly, and avoiding cross-contamination with other foods, especially those that will be consumed raw.

-Cook thoroughly: When using chili powder in recipes, ensure that the food is cooked thoroughly to kill any potential bacteria that may be present.

By following these guidelines, you can minimize the risks associated with chili powder and enjoy its flavor and heat safely.

(35)Cumin: Cumin is a flowering plant in the family Apiaceae, native to the eastern Mediterranean region and Upper Egypt. The plant's seeds are used in cooking and are known for their distinctive, warm flavor. Cumin is a key ingredient in many cuisines, particularly in Indian, Middle Eastern, and Mexican dishes.

Food Safety Concerns:

(a)Contamination: Cumin, like other spices, can be contaminated with bacteria, mold, or other harmful substances during growing, harvesting, processing, or storage.

(b)Allergies: Some people may have allergies to cumin, which can cause adverse reactions ranging from mild to severe.

Health Concerns:

-Irritant effects: In large amounts, cumin can potentially cause irritation, particularly in individuals with sensitive digestive systems.

-Interaction with medication: Cumin may interact with certain medications, so individuals taking medication should consult a healthcare professional before consuming large quantities of cumin.

-Purchase from reputable sources: Buy cumin from reputable suppliers who adhere to food safety standards and regulations.

-Storage: Store cumin in a cool, dry place in airtight containers to prevent contamination and preserve its flavor.

-Allergen awareness: Be mindful of potential allergens, and if you have known allergies, read food labels carefully and inquire about ingredients when dining out.

-Moderation: Consume cumin in moderation, particularly if you have a sensitive digestive system or are taking medication.

(36)Cinnamon: Often used in both sweet and savory dishes, providing warmth and sweetness. Cinnamon is a spice obtained from the inner bark of several tree species from the genus Cinnamomum. It is widely used in both sweet and savory dishes and in beverages. There are two main types of cinnamon: Ceylon cinnamon (also known as “true” cinnamon) and Cassia cinnamon. Ceylon cinnamon is often considered higher quality and milder in flavor, while Cassia cinnamon is more common and has a stronger, more pungent flavor.

When it comes to food safety concerns and health issues associated with cinnamon, there are a few key points to consider:

(a)Coumarin Content: Cassia cinnamon contains a compound called coumarin, which can be harmful to the liver in high doses. The European Food Safety Authority (EFSA) has set a tolerable daily intake (TDI) for coumarin and recommends limiting intake of Cassia cinnamon, particularly for people at risk of liver damage.

(b)Cinnamon Allergies: Some individuals may be allergic to cinnamon, experiencing symptoms such as skin rashes, swelling, or difficulty breathing. While cinnamon allergies are relatively rare, they can be serious for those affected.

(c)Microbiological Contamination: Like any spice, cinnamon can be susceptible to microbiological contamination if not stored properly. This can lead to foodborne illnesses if contaminated cinnamon is used in food preparation.

To minimize a few concerns, Cinnamon, here are some recommendations:

-Choose Ceylon Cinnamon: If you are concerned about coumarin content, consider using Ceylon cinnamon, which has lower levels of coumarin compared to Cassia cinnamon.

-Moderation: Use cinnamon in moderation, particularly if you are using Cassia cinnamon. This can help reduce the risk of excessive coumarin intake.

-Storage: Store cinnamon in a cool, dry place to prevent microbial contamination. Proper storage can help maintain the quality and safety of the spice.

-Allergy awareness: Be mindful of potential cinnamon allergies, and if you or others have known allergies, be sure to avoid the spice and check food labels for its presence.

By being aware of these concerns and taking appropriate precautions, you can continue to enjoy the flavor and aroma of cinnamon while minimizing potential health risks.

(37)Nutmeg: Adds warmth and depth to dishes, particularly in baked goods and creamy sauces. Nutmeg is a popular spice that is derived from the seed of the Myristica fragrans tree, which is native to Indonesia. It is used in both sweet and savory dishes and is known for its warm, nutty flavor. Nutmeg is used in a variety of cuisines around the world and is often used in baking, desserts, and holiday dishes. However, nutmeg also has some potential food safety and health concerns. When consumed in large quantities, nutmeg can have psychoactive effects due to a compound called myristicin. Ingesting high doses of nutmeg can lead to symptoms such as nausea, dizziness, dry mouth, hallucinations, and a feeling of being “out of touch” with reality. In extreme cases, it can lead to more serious health issues such as a rapid heart rate, agitation, and delirium. To prevent these concerns, it's important to use nutmeg in moderation and avoid consuming it in large quantities. In culinary use, the typical amount of nutmeg in recipes is insufficient to cause these effects. However, individuals should be cautious about consuming very high amounts of nutmeg, as might occur if someone were to intentionally ingest it for its psychoactive effects. In terms of food safety, nutmeg itself is not a common source of foodborne illness. However, with any spice or dried herb, it's important to store nutmeg properly in a cool, dark place to maintain its flavor and prevent mold growth. Additionally, it's important to purchase nutmeg from reputable sources to ensure its quality and safety. Yes, nutmeg is a popular spice with a rich culinary history, but it's important to use it in moderation due to its potential psychoactive effects at high doses. Storing nutmeg properly and purchasing it from reputable sources can help prevent any food safety concerns associated with this spice.

(38)Oregano: a popular culinary herb in the mint family. It is commonly used in Mediterranean,

Italian, Latin American and southwestern cuisines and is known for its strong, savory flavor. Oregano can be used fresh or dried, and both forms are widely used in cooking to add flavor to a variety of dishes, including pasta sauces, pizzas, grilled meats, and salads. When it comes to food safety and health concerns associated with oregano, there are a few important considerations to keep in mind:

(a)Contamination: Like any other spice or herb, oregano can become contaminated with bacteria, molds, or other pathogens. Contamination can occur during various stages of production, processing, or storage.

(b)Pesticide residues: Oregano, if not grown organically, may contain pesticide residues, which can pose health risks if consumed in large quantities.

(c)Allergies: Some individuals may be allergic to oregano or other plants in the mint family. Allergic reactions can range from mild to severe and can pose a significant health concern for sensitive individuals.

To prevent these concerns and ensure the safe consumption of oregano, here are some best practices:

-Purchase from reputable sources: Buy oregano from trusted suppliers who adhere to food safety regulations and quality standards. Look for organic or pesticide-free options if possible.

-Proper storage: Store oregano in a cool, dry place in an airtight container to maintain its quality and reduce the risk of contamination.

-Wash fresh oregano: If using fresh oregano, wash it thoroughly before use to remove any potential contaminants.

-Use safe handling practices: When using oregano in cooking, follow safe food handling practices, such as washing hands, utensils, and surfaces to prevent cross-contamination.

(39)Thyme: Offers a subtle, earthy flavor and is often used in soups, stews, and meat dishes. Thyme is a popular herb that is used in cooking, particularly in Mediterranean and European cuisines. It is a member of the mint family and is known for its aromatic leaves, which are often used to add flavor to a wide variety of dishes, including soups, stews, sauces, and meat dishes.

Like many herbs and spices, thyme can sometimes contain pesticide residues, which can be harmful to human health. Herbs, including thyme, can sometimes be contaminated with harmful microbes such as bacteria, mold, and yeast. Some individuals may be allergic to thyme or other herbs in the mint family. Allergic reactions to thyme can vary in severity and may include skin rashes, itching, and respiratory problems.

To prevent these concerns and ensure the safe use of thyme in cooking, here are some tips:

(a)Purchase organic thyme to minimize exposure to pesticide residues.

(b)When using fresh thyme, wash it thoroughly under running water to remove any dirt or other contaminants. Dry it with a clean paper towel before use.

(c)Store thyme in a cool, dry place to prevent microbial growth. It's best to use fresh thyme within a few days of purchase, but it can also be dried for longer-term storage.

(40)Rosemary: Provides a fragrant, pine-like flavor and is often used with roasted meats and vegetables. Rosemary is an aromatic evergreen herb native to the Mediterranean region. It's commonly used in cooking to add flavor to a variety of dishes, such as meats, soups, and sauces. Rosemary is known for its distinct aroma and earthy, slightly bitter flavor, and it's often used both fresh and dried. When it comes to food safety and health concerns associated with rosemary, there are a few things to consider:

(a)*Allergies*: Some individuals may be allergic to rosemary, which can cause allergic reactions ranging from mild to severe. People with known allergies to other members of the Lamiaceae family, such as mint or oregano, may be at higher risk of an allergic reaction to rosemary. To prevent allergic reactions, it's important to be aware of any known allergies and to carefully read food labels or ask the restaurant about the ingredients.

(a)Contamination: Like all fresh herbs, rosemary can be a source of foodborne illness if it's not handled and stored properly. Fresh rosemary, especially if it's not thoroughly washed, can carry bacteria or other pathogens that may cause illness if ingested. To prevent contamination, it's important to wash fresh rosemary thoroughly before use and to store it properly in the refrigerator.

(b)Pesticide Residues: If rosemary has been treated with pesticides during cultivation, there is a risk of pesticide residues remaining on the herb. While this is not a direct health concern, some individuals may prefer organically grown or pesticide-free rosemary to minimize potential pesticide exposure.

To prevent these concerns, here is some general Advice to keep in mind:

(a)Those Allergic to Rosemary: If you or someone you are serving is allergic to rosemary, it's important to avoid using it in dishes. When dining out, always inquire about the ingredients used in dishes to ensure they are safe to consume.

(b)Avoid Contamination: When using fresh rosemary, wash it thoroughly under running water before adding it to your dishes. Store fresh rosemary in the refrigerator and use it within a few days to ensure freshness and minimize the risk of contamination.

(c)Avoid Pesticide Residues: Consider choosing organic or pesticide-free rosemary to minimize potential exposure to pesticides. When purchasing rosemary, look for products that are certified organic or labeled as pesticide-free.

(41)Turmeric: Adds vibrant color and a warm, slightly bitter flavor, often used in curries and rice dishes. Turmeric is a flowering plant, widely cultivated in tropical regions of Asia, and is best known for its bright yellow-orange rhizome, which is commonly used as a spice in cooking. Turmeric has been used for centuries in traditional medicine and cooking, particularly in Indian and Southeast Asian cuisines. It contains an active compound called curcumin, which is believed to have anti-inflammatory and antioxidant properties. When it comes to food safety and health concerns associated with turmeric, there are a few important points to consider:

(a)Contamination: Turmeric can be contaminated with various substances, including lead, pesticides, and other heavy metals. This is particularly concerning in regions where turmeric is produced under less stringent agricultural and food manufacturing safety practices.

(b)Allergies: Some individuals may be allergic to turmeric, experiencing symptoms such as rashes, hives, and even anaphylaxis in severe cases.

©*Medicinal Interactions*: Turmeric supplements or large amounts of turmeric in food may interact with certain medications, such as blood thinners, and could potentially lead to complications.

To prevent these concerns, the following measures can be taken:(a)*Source from reputable suppliers*: When purchasing turmeric, especially in the form of supplements, it's important to choose products from reputable and trusted suppliers to minimize the risk of contamination.

-*Allergen awareness*: Individuals with known allergies should carefully read food labels and ask about turmeric use in restaurant dishes to avoid potential allergic reactions.

-*Moderation*: While turmeric is generally safe when used in normal culinary amounts, excessive consumption or high-dose supplementation should be avoided, particularly without consulting a healthcare professional, as it may interact with certain medications or cause gastrointestinal discomfort.

-*Quality control*: Regulatory bodies in different countries should implement and enforce quality control measures to ensure the safety and purity of turmeric products, including limits on heavy metal contamination.

(42)Ginger: Provides a pungent, spicy flavor and is commonly used in both sweet and savory dishes. Ginger is a flowering plant whose rhizome, or underground stem, is widely used as a spice and folk medicine. It is commonly used in cooking, especially in Asian and Indian cuisine, and is known for its unique flavor and potential health benefits. Food Safety & Health Concerns:

(a)Contamination: Like all produce, ginger can be contaminated with bacteria, viruses, or parasites, which can cause foodborne illnesses if consumed. Proper handling and washing of ginger can help reduce this risk.

(b)Allergies: Some individuals may have allergies to ginger, which can cause adverse reactions ranging from mild to severe. It's important for people with known allergies to avoid consuming ginger.

(c)Interaction with Medications: Ginger may interact with certain medications, such as blood thinners, and affect their effectiveness. Individuals taking medications should consult with a healthcare professional before consuming large amounts of ginger.

(d)Gastrointestinal Issues: In some cases, consuming large amounts of ginger, particularly in concentrated forms like supplements, may lead to gastrointestinal discomfort such as heartburn or diarrhea.

Food Safety Preventive Measures:

(a)Washing: Thoroughly wash fresh ginger under running water before consumption.

(b)Allergy Awareness: Individuals with known allergies to ginger should carefully read food labels and avoid consuming products that contain ginger.

©*Moderation*: Consume ginger in moderation, particularly when using concentrated forms such as supplements, to avoid potential gastrointestinal issues.

(d)Consultation with Healthcare Professionals: Individuals with underlying health conditions or those taking medications should consult with healthcare professionals before consuming large amounts of ginger, especially in supplemental form.

(43) Bay leaves: Bay leaves are aromatic leaves commonly used in cooking to add flavor to various dishes, particularly in soups, stews, and sauces. They are derived from the bay laurel tree, scientifically known as Laurus nobilis, and are available dried. Bay leaves are often used whole and removed from the dish before serving due to their strong, pungent flavor. In terms of food safety and health issues associated with bay leaves, there are a few considerations to keep in mind:

(a)Choking Hazard: Bay leaves are typically used whole and removed from the dish before serving. However, if a leaf is accidentally ingested, it could pose a choking hazard, especially if it is not thoroughly crushed or chopped.

(b)Potential Allergic Reactions: Some individuals may have allergic reactions to bay leaves. Although relatively rare, allergic reactions can cause symptoms such as itching, swelling, or respiratory issues in sensitive individuals.

©Contamination: Like any dried herb or spice, bay leaves can become contaminated with mold, insects, or other impurities if not stored properly.

Contaminated bay leaves can pose health risks if consumed.

To prevent these issues, consider the following precautions:

(a)Use Bay Leaves Properly: When using bay leaves in cooking, it's important to remove them before serving the dish. This can help prevent the risk of choking. Additionally, consider crushing or chopping the bay leaves before adding them to a dish to release their flavor and make them easier to detect and remove.

(b)Allergic Reactions: If you or someone who has known allergies is preparing food, be sure to check for any potential sensitivity to bay leaves before using them in a dish.

(c)Storage: Store bay leaves in a cool, dry place in an airtight container to prevent contamination. Check the leaves for signs of mold or insect infestation before use.

*(44)***Dill**: Dill is an aromatic herb in the celery family, Apiaceae, and is used both for its leaves (dill) and its seeds (dill seed). It is widely used in cooking for its fresh, grassy flavor and is often

used to flavor pickles, salads, soups, and fish dishes. When it comes to food safety and health issues associated with dill, there are a few considerations to keep in mind:

(a) *Pesticide Residues*: Like with any herb or vegetable, there is a possibility of pesticide residues on dill. To minimize exposure to pesticides, it's best to choose organic dill when possible, or to thoroughly wash conventionally grown dill before use.

(b) *Allergies*: Some individuals may be allergic to dill or other related plants in the Apiaceae family. Allergic reactions to dill can range from mild to severe. So, it's important for individuals with known allergies to avoid dill, and food preparers to be mindful of potential allergens when serving dishes containing dill.

(c) *Contamination*: As with any fresh herb, there is a risk of contamination with bacteria such as Salmonella or E. coli. To prevent contamination, it's important to purchase fresh dill from reputable sources, store it properly, and wash it thoroughly before use. Additionally, following proper food safety practices, such as using clean utensils and surfaces, can help prevent contamination.

To prevent these issues, it's essential to follow these food safety guidelines:

(a) Don't make unnecessary assumptions about Dill sellers by believing that all sellers are reliable in terms of sanitary processing. Purchase dill from reputable sources and consider choosing organic dill to minimize pesticide exposure.

(b) In consideration of how many hands that touched the Dill before it got to you as a consumer, it is necessary that you thoroughly wash the dill before use, especially if it will be consumed raw.

(c)Store dill properly in the refrigerator to maintain its freshness and reduce the risk of bacterial contamination.

(d)Be aware of potential allergens and communicate with guests or diners about the ingredients in dishes containing dill.

(e)Follow standard food safety practices, including using clean utensils and surfaces, and avoiding cross-contamination.

(45)Marjoram: Marjoram is a somewhat cold-sensitive perennial herb or undershrub with sweet pine and citrus flavors. It is often used in Mediterranean and Middle Eastern cuisines. Marjoram is a member of the mint family and is closely related to oregano, with a similar but gentler flavor. When it comes to food safety and health concerns associated with marjoram, there are a few things to consider:

(a)*Contamination*: Like any fresh herb, marjoram can be contaminated with bacteria, such as Salmonella or E. coli, which can cause foodborne illness if consumed.

(b)*Pesticide Residues*: If not organically grown, marjoram may contain pesticide residues, which can be harmful if ingested in large quantities over time.

(c)Allergies: Some individuals may be allergic to marjoram or other herbs in the mint family, which can lead to allergic reactions.

To prevent these concerns, here are a few tips:

(a)Washing: Rinse fresh marjoram thoroughly under running water before use, in order to remove any potential contaminants. If you are concerned about bacteria, you can also use a produce wash.

(b)Cooking: Heat can kill bacteria, so cooking marjoram at the appropriate temperature can help ensure its safety. However, the flavor of marjoram is best preserved when it is added towards the end of the cooking process.

(c)Organic Sources: If you are concerned about pesticide residues, consider purchasing organic marjoram, which is grown without synthetic pesticides.

Allergy Awareness: Be aware of any allergies to marjoram or related herbs and avoid consuming them if you are allergic.

By following these tips, you can enjoy marjoram in your cooking while minimizing any associated food safety and health risks.

(46)Beet Sugar: Beet sugar is a type of sugar that is extracted from sugar beets, which are a root vegetable. It is commonly used as a sweetener in various food products. The process of extracting sugar from beets involves slicing the beets, extracting the sugar with water, and then purifying and crystallizing it. In terms of food safety and health risks associated with beet sugar, there are a few factors to consider:

(a) Pesticide and chemical residues: Sugar beets may be treated with pesticides and other chemicals during cultivation. Residues of these substances could potentially remain in the sugar if not properly removed during processing. Prevention: To mitigate this risk, it's important for sugar beet farmers to follow proper agricultural practices, including the judicious use of pesticides and adherence to safety regulations. Additionally, sugar processing companies should have quality control measures in place to test for and remove any harmful residues.

(b) Genetic modification: Some sugar beets are genetically modified to be resistant to certain herbicides. While the safety of genetically modified organisms (GMOs) is a topic of ongoing debate, some consumers may have concerns about consuming products derived from GMO crops. Prevention: Food labeling regulations in many countries require the disclosure of GMO ingredients, allowing consumers to make informed choices. Additionally, organic certification standards often prohibit the use of GMOs, so choosing organic beet sugar can help avoid this potential risk.

(c) Nutritional concerns: Like other forms of sugar, beet sugar is high in calories and provides empty calories with no significant nutritional value. Excessive consumption of added sugars has been linked to various health issues, including obesity, type 2 diabetes, and dental problems. To minimize health risks associated with beet sugar consumption, individuals should practice moderation and be mindful of their overall sugar intake. Choosing whole foods and minimizing consumption of processed foods and beverages with added sugars can contribute to a healthier diet.

In general, the food safety and health risks associated with beet sugar can be mitigated through responsible agricultural practices, rigorous quality control in processing facilities, transparent labeling, and informed consumer choices. Additionally, promoting a balanced diet that limits added sugar intake is important for overall health and well-being.

(47)Invert Sugar: Invert sugar is a type of sugar that has been broken down into its component parts, glucose and fructose. This process is called inversion, and it is typically achieved by treating sucrose (table sugar) with acid or enzymes. Invert sugar is often used in the food industry for its unique properties that can improve the texture, shelf life, and flavor of various products, such as confectionery, baked goods, and beverages. When it comes to food safety and health risks associated with inverted sugar, there are a few considerations to keep in mind:

(a) *Microbial Contamination*: Like any other sugar or sweetener, invert sugar can potentially support the growth of microorganisms if it becomes contaminated. To prevent this, it's important to store invert sugar in a clean, dry, and airtight container and to follow good manufacturing practices to minimize the risk of contamination during production.

(b) *Allergen Cross-Contamination*: Invert sugar itself does not pose a significant risk of allergen cross-contamination, but if it is used in products that contain allergens (such as nuts, dairy, or gluten), there is a risk that those allergens could be present in the final product. Proper cleaning and sanitation procedures should be followed to prevent allergen cross-contamination during manufacturing.

(c) *Dental Health*: Invert sugar, like other sugars, can contribute to dental cavities if consumed in excess. It's important for consumers to be mindful of their overall sugar intake and to practice good oral hygiene.

To prevent these potential risks, food manufacturers should adhere to strict hygiene and sanitation practices, including regular cleaning and sanitizing of equipment. Additionally, proper labeling and allergen control measures should be implemented to prevent cross-contamination. Consumers should be informed of the sugar content of products containing invert sugar and should consume them in moderation as part of a balanced diet. As with any ingredient, it's important to use invert sugar responsibly and in accordance with food safety guidelines to minimize any associated risks.

(48)Bouillon: Bouillon is a seasoning available in cube, powder, and broth forms, made by simmering meat, vegetables, and aromatic ingredients in water. It is commonly used as a base for soups, stews, sauces, and other dishes to add flavor. Bouillon can be made from various meats, such as beef, chicken, or fish, as well as vegetables, and is often seasoned with herbs and spices. In terms of food safety, bouillon, like any other perishable food product, can pose certain risks if not handled and stored properly. Here are some food safety and health risks associated with bouillon and how to prevent them:

(a) Bacterial Contamination: Bouillon, especially homemade varieties, can be susceptible to contamination by harmful bacteria such as E. coli, Salmonella, or Listeria if not prepared, stored, or handled properly. To prevent bacterial contamination: -Use proper hygiene and sanitation practices when preparing bouillon, including washing hands, utensils, and cooking surfaces. -Ensure that all ingredients used in the bouillon are fresh and of good quality. Cook the bouillon thoroughly to kill any harmful bacteria that may be present. It should be brought to a boil and then simmered for an adequate amount of time.

(b) Storage and Temperature Control: Improper storage of bouillon can lead to bacterial growth and spoilage. To prevent these issues: - Store homemade bouillon broth in the refrigerator and use it within a safe timeframe, typically a few days. -If storing bouillon for a longer period, consider freezing it in airtight containers or ice cube trays for convenient use in smaller portions.

(c) Allergen Concerns: Depending on the ingredients used, bouillon may contain common food allergens such as soy, wheat, fish, or shellfish. To prevent allergic reactions, clearly label bouillon that contains allergens and communicate allergen information to consumers when using bouillon in food service settings.

(d) Sodium Content: Commercial bouillon cubes and powders often contain high amounts of sodium, which can be a concern for individuals on restricted sodium diets. To reduce sodium intake: Look for low-sodium or no-salt-added bouillon. -Consider making homemade bouillon with reduced salt or using alternative flavoring agents like herbs and spices.

By following proper food safety practices and being mindful of potential health risks, bouillon can be enjoyed safely as a flavorful addition to various dishes.

(49)Maggi cube & powder: Maggi cubes and Maggi powder are popular condiments used in many cuisines worldwide, particularly in Africa and Asia. They are produced by Nestle and are used to add flavor to dishes. Maggi cubes are small blocks of concentrated seasoning, while Maggi powder is a powdered form of the same seasoning.

Food safety or health risks associated with Maggi cubes and Maggi powder can include:

(a) *Excessive Sodium*: Maggi cubes and powder often contain a high amount of sodium, which can contribute to health issues such as high blood pressure and heart disease when consumed in large quantities.

(b) *Contamination*: There is a risk of contamination during production, storage, or transportation, which could lead to foodborne illnesses if consumed.

To prevent these risks, the following measures can be taken:

-*Regulation and Labeling*: Governments can regulate the amount of sodium allowed in food products, including Maggi cubes and powder, and require clear labeling of the sodium content to inform consumers of the potential health risks.

-*Quality Control*: Manufacturers should implement strict quality control measures to ensure that their products are free from contaminants and are produced in hygienic conditions.

-*Consumer Education*: Educating consumers about the potential health risks of excessive sodium consumption and the importance of moderation in the use of seasoning products can help prevent overconsumption.

-*Proper Storage*: Consumers should store Maggi cubes and powder in a cool, dry place to prevent contamination and spoilage.

(50)Dawadawa: Dawadawa, also known as "Iru" in some parts of Nigeria, is a traditional fermented condiment popular in Africa. It is made from the fermented seeds of African locust beans (Parkia biglobosa). The fermentation process can take several days to a few weeks, during which the seeds are boiled, dehulled, and allowed to ferment naturally. The resulting product is a strong-smelling, flavorful condiment used to flavor soups, stews, and other dishes in African cuisine. Food Safety and Health Risks Associated with Dawadawa:

(a) *Microbial Contamination*: During the fermentation process, Dawadawa can be susceptible to contamination by harmful microorganisms such as bacteria, molds, and yeasts, which can pose a risk of foodborne illness if consumed.

(b) *Allergies*: Some individuals may be allergic to the proteins present in Dawadawa, especially if they have a known sensitivity to legumes or related food products.

(c) *Histamine Formation*: Improper fermentation and storage of Dawadawa can lead to the accumulation of histamine, a compound that can cause adverse reactions in some people, particularly those with histamine intolerance.

In order to prevent the Risks Associated with Dawadawa:

(a) *Hygienic Practices*: Ensure that the utensils, containers, and surfaces used in the preparation of Dawadawa are clean and sanitized to minimize the risk of contamination.

(b) *Proper Fermentation and Storage*: Follow traditional methods for fermenting Dawadawa carefully, ensuring that the fermentation process occurs under hygienic conditions and that the product is stored in a cool, dry place to prevent spoilage.

(c) *Allergen Information*: If serving Dawadawa-containing dishes, it's important to be aware of potential allergens and inform consumers of the ingredients used.

(d) *Quality Ingredients*: Start with fresh, high-quality locust beans and ensure that they are properly processed and handled before fermentation.

(e) Regulatory Compliance: If producing Dawadawa for commercial purposes, adhere to relevant food safety regulations and standards to ensure the product's safety and quality.

By following these guidelines, you can help minimize the food safety and health risks associated with Dawadawa and enjoy this traditional condiment safely.

(51)Sweet Bay: Sweet Bay, also known as bay laurel or Laurus nobilis, is a popular culinary herb used to add flavor to a wide variety of dishes, particularly in Mediterranean cuisine. The leaves of the sweet bay tree are often used in cooking to add a subtly aromatic and earthy flavor to soups, stews, sauces, and other dishes. In terms of food safety and health risks associated with sweet bay, there are a few considerations to keep in mind:

-*Conta*mination: Like any other herb or spice, sweet bay leaves can become contaminated with bacteria, mold, or other pathogens if they are not stored properly or if they come into contact with contaminated surfaces. This can pose a risk of foodborne illness if the contaminated leaves are consumed.

-*Allergies*: Some individuals may be allergic to sweet bay leaves. Allergic reactions to sweet bay are rare but can occur in sensitive individuals, causing symptoms such as skin rashes, itching, and difficulty breathing. To prevent these safety issues

-*Purchase from reputable sources*: When buying sweet bay leaves, purchase them from reputable suppliers to ensure that they have been handled and stored in a way that minimizes the risk of contamination.

-*Store properly*: Store sweet bay leaves in a cool, dry place in an airtight container to help prevent contamination and preserve their flavor. Proper storage can also help prevent the growth of mold and bacteria.

-*Wash hands and utensils*: When handling sweet bay leaves, wash your hands and any utensils or surfaces that come into contact with the leaves to minimize the risk of contamination.

-*Allergen awareness*: If you or others have known allergies to sweet bay or related plants, be sure to avoid using sweet bay in your dishes or clearly label foods containing sweet bay to alert those with allergies.

-*Use in moderation*: While sweet bay is generally safe when used in small quantities as a culinary herb, excessive consumption can potentially lead to adverse effects. Use sweet bay leaves in moderation to avoid any potential health risks.

(52)Coriander: Coriander, also known as cilantro or Chinese parsley, is a popular herb used in many cuisines worldwide. Both its leaves and seeds are commonly used in cooking and have a citrusy, slightly spicy flavor. It's a rich source of vitamins, minerals, and antioxidants, and it's known for its potential health benefits, including digestive aid, cholesterol-lowering effects, and anti-inflammatory properties. However, like any food product, coriander can pose some food safety and health risks:

(a) *Microbial Contamination*: Fresh coriander, like many other fresh produce items, can be contaminated with bacteria, such as Salmonella, E. coli, or Listeria, which can cause foodborne illnesses.

(b) *Pesticide Residues*: Improperly regulated or excessive use of pesticides in coriander cultivation can lead to the presence of pesticide residues, which can have harmful effects on health if consumed in large quantities.

To prevent these risks and safely enjoy coriander, consider the following measures:

-*Selection and Storage*: When purchasing coriander, choose fresh-looking leaves or seeds that are vibrant in color, have no signs of wilting or yellowing, and are free from mold. Store fresh coriander in the refrigerator and use it within a few days to minimize the risk of microbial contamination.

-*Washing and Preparation*: Thoroughly wash fresh coriander under running water before use to remove any dirt, debris, or potential contaminants. This is especially important if you plan to consume it raw, such as in salads.

-*Source*: Whenever possible, choose organic or locally sourced coriander to reduce the risk of pesticide residues. Organic produce is typically grown and handled without synthetic pesticides or fertilizers.

-Hygiene: Practice good food hygiene when handling coriander and other fresh produce. This includes washing hands, utensils, and cutting boards thoroughly before and after handling the herb to prevent cross-contamination with other foods.

-*Cooking*: When cooking with coriander, the heat will help kill any harmful bacteria that might be present, reducing the risk of foodborne illness.

By following these guidelines, you can minimize the potential health and safety risks associated with coriander and safely enjoy its flavorful addition to your meals.

(53)Capers: Capers are small, green buds commonly used as a seasoning or garnish in many cuisines, particularly Mediterranean dishes. They have a sharp, tangy flavor and are often used in salads, pasta dishes, and various sauces. Capers are typically pickled in vinegar or preserved in salt, which gives them their distinctive flavor. When it comes to food safety and health risks associated with capers, there are a few considerations to keep in mind:

(a) *Allergies*: Some individuals may be allergic to capers, particularly those who are sensitive to other plants in the Capparidaceae family, such as mustard or cabbage. Allergic reactions to capers can range from mild to severe, so it’s essential to be aware of any known allergies before consuming capers.

(b) *Sodium Content*: Capers that are preserved in salt can contribute to the overall sodium content of a dish. Excessive sodium intake can be harmful to individuals with high blood pressure or other health conditions. It's important to consume capers in moderation, particularly if you are watching your sodium intake.

(c) *Foodborne Illness*: As with any pickled or preserved food, there is a risk of foodborne illness if capers are not handled or stored properly. Contamination with bacteria, such as Listeria or Clostridium botulinum, can occur if capers are not processed and stored correctly. To prevent food safety and health risks associated with capers, consider the following tips:

-Purchase capers from reputable sources to ensure quality and proper handling.

-Store capers according to the manufacturer's instructions, typically in a cool, dark place.

Once opened, capers preserved in brine should be refrigerated.

-Use clean utensils when removing capers from the jar to prevent contamination.

-Be mindful of sodium intake if consuming capers that are preserved in salt, especially if you have dietary restrictions related to sodium.

-If you or others consuming the capers have known allergies, be sure to check for any sensitivity before using them in dishes.

By following these guidelines, you can enjoy capers safely and minimize potential health risks associated with their consumption.

(54)Chicken Broth: Chicken broth is a liquid made by simmering chicken bones, meat, and vegetables in water, often with herbs and spices. It's commonly used as a base for soups, sauces, and other dishes, and it's valued for its rich flavor and nutritional benefits.

When it comes to food safety and health issues associated with chicken broth, there are a few potential concerns:

Food safety concerns regarding chicken broth involve risks from bacterial contamination, improper storage, and potential heavy metal leaching in homemade versions. Major concerns include bacterial growth from Improper Cooling: Chicken broth is a prime medium for bacteria (such as Salmonella and E. coli) to grow if left in the “danger zone” (40oF-140oF) for more than 2 hours. Leaving homemade broth in the store overnight is a high-risk practice, as toxins produced by bacteria may not be destroyed by re-boiling. Packaging Failures (Store-bought): A major recall in late 2024-early 2025 at a popular grocery store occurred due to packaging failures in cartons that allowed bacterial contamination and spoilage. Signs of Spoilage: Spoiled broth may appear cloudy, contain mold, or have a slimy texture. It often gives off a sour, rancid, or “wet dog” odor.

Lead Contamination (Homemade Bone Broth): Studies have found that bone broth made from chicken bones can contain high levels of lead (over 10x higher than tap water in some tests), which can cause serious health issues. Storage Timeframes: Once opened, commercial chicken broth should be refrigerated and used within 3—4 days to avoid illness.

To prevent these issues, here are some important steps to take:

-*Use fresh, high-quality ingredients*: Start with fresh chicken and vegetables, and make sure to use them before their expiration dates.

-*Proper cooking temperature*: When making chicken broth, ensure that the ingredients are cooked to the recommended internal temperatures to kill any harmful bacteria. For chickens, the safe internal temperature is 165°F (74°C).

-Safe storage: After making chicken broth, cool it quickly and refrigerate or freeze it promptly. Bacteria can grow rapidly at temperatures between 40 F (4 C) and 140 F (60 C), so it's important to store broth at temperatures below 40 F (4 C) to slow bacterial growth.

-Reheating: When reheating chicken broth, make it sure it reaches a temperature of at least 165 F (74 C) to kill any bacteria that may have developed during storage.

-Proper hygiene: Wash your hands, utensils, and cooking surfaces thoroughly before and after handling raw chicken to prevent cross-contamination.

(55) Beef Broth: Beef broth is also a flavorful liquid, but made by simmering beef bones, vegetables, herbs, and spices. It is commonly used as a base for soups, stews, sauces, and other dishes. Beef broth is rich in nutrients such as protein, vitamins, and minerals, and it can add depth of flavor to a wide variety of recipes. When it comes to food safety and health issues associated with beef broth, there are a few key considerations to keep in mind:

(a) *Bacterial contamination*: Beef broth, like any other food made from animal products, can be a breeding ground for harmful bacteria such as E. coli, Salmonella, and Listeria if it is not handled and stored properly.

(b) *Temperature control*: Maintaining proper temperature is crucial in preventing bacterial growth in beef broth. It's important to keep the broth either hot (above 140 F or 60 C) or cold (below 40 F or 4 C) to minimize the risk of bacterial contamination.

(c) *Cross-contamination*: When preparing beef broth, it's essential to prevent cross-contamination by keeping raw meat separate from other ingredients and using separate cutting boards, utensils, and containers for raw meat and other foods.

(d)Use fresh ingredients: Start with fresh, high-quality beef bones and vegetables. If using store-bought broth or stock, check the expiration date and ensure that the packaging is intact.

(e)Proper cooking and storage: Cook the beef broth thoroughly, bringing it to a rolling boil and then simmering it for an extended period to kill any harmful bacteria. After preparing the broth, cool it quickly and refrigerate it promptly to prevent bacterial growth.

(f)Reheat properly: When reheating beef broth, make sure it reaches a temperature of

at least 165 F (74 C) to kill any bacteria that may have developed during storage.

(g)Hygienic practices: Follow good hygiene practices in the kitchen, including washing your hands thoroughly before and after handling raw meat, using clean utensils and surfaces, and keeping the cooking area clean. By following these food safety practices, you can greatly reduce the risk of contamination and ensure that the beef broth you prepare is safe and healthy to consume.

(56)Vegetable Broth: Vegetable broth, like any perishable food product, can pose certain food safety concerns if not handled and stored properly. Here are some potential concerns and ways to prevent them:

-Bacterial contamination: Vegetables used to make broth can carry bacteria such as E. coli and Salmonella. To prevent bacterial contamination, it's important to handle vegetables in a clean environment, wash them thoroughly before use, and ensure that the broth is cooked to a safe temperature (at least 165 F or 74 C) to kill any harmful bacteria.

-Time and temperature abuse: Vegetable broth, like any perishable food, should be kept out of the "temperature danger zone" (40 F to 140 F or 4 C to 60 C) where bacteria can grow rapidly. After making vegetable broth, it should be cooled quickly and refrigerated within two hours. When reheating, bring it to a boil.

-Cross-contamination: When preparing vegetable broth, it's important to prevent cross-contamination by using separate cutting boards and utensils for raw vegetables and other ingredients. Keep raw vegetables separate from ready-to-eat foods, and wash hands, utensils, and surfaces thoroughly after handling them.

-Storage: Proper storage of vegetable broth is essential for food safety. Store broth in airtight containers in the refrigerator for 3-4 days or freeze it for longer storage. When reheating, make sure the broth reaches a safe temperature throughout.

-Quality of ingredients: The quality of the vegetables used to make the broth is important. Use fresh, high-quality vegetables and check for any signs of spoilage before using them.

By following these food safety practices, I'm sure you can minimize the risks associated with vegetable broth and ensure that it is safe to consume.

(57)Fish Broth: Fish Broth, like any other food product, can pose food safety concerns if not handled and prepared properly. Here are some potential food safety concerns related to fish broth and how to prevent them:

(a) Cross-contamination: Cross-contamination occurs when harmful bacteria from raw fish or other sources are transferred to the broth or other foods. To prevent this, it's important to keep raw fish separate from other ingredients, utensils, and surfaces. Use separate cutting boards and utensils for raw fish, and wash your hands and surfaces thoroughly after handling raw fish.

(b) Bacterial growth: Fish broth, like any perishable food, can be a breeding ground for bacteria if not stored and handled properly. To prevent bacterial growth, store fish broth in the refrigerator at or below 40°F (4°C) and use it within a few days. If you're not going to consume it within that time frame, consider freezing it or using it later.

(c) Proper cooking: When making fish broth, it's important to ensure that the fish is cooked to a safe temperature to kill any bacteria or parasites that may be present. The broth should be brought to a boil, then simmered long enough to ensure the fish is fully cooked.

(d) Quality of fish: Using fresh, high-quality fish is essential for making safe fish broth. Avoid using fish that could be contaminated or have an odor or slimy texture.

(e) Serving temperature: When serving fish broth, make sure it is served hot, at a temperature of 140 F (60 C) or above, to prevent the growth of bacteria.

(f) Thawing fish safely: If you're using frozen fish to make broth, be sure to thaw it safely. The safest way to thaw fish is in the refrigerator or under cold running water. Avoid thawing fish at room temperature, as this can allow bacteria to multiply.

By following these food safety guidelines, I'm sure you can help prevent foodborne illness and ensure that your fish broth is safe to consume.

(58)Bone Broth: Bone broth has become increasingly popular in recent years due to potential health benefits, but there are some food safety concerns to be aware of:

Bacterial Contamination: If bone broth is not prepared and stored properly, it can be a breeding ground for bacteria such as E. coli, Salmonella, and Listeria. These pathogens can cause foodborne illness, especially if the broth is not heated to a sufficiently high temperature during cooking or is not refrigerated promptly.

(a)Heavy metal contamination: Bones used to make broth may contain trace amounts of heavy metals such as lead and cadmium. If the bones are sourced from animals exposed to environmental pollutants, or if the cooking process is not carefully managed, there is a risk that these metals will leach into the broth.

-Histamine production: Bone broth, especially when made from fish or poultry, has the potential to accumulate histamine if not handled and stored properly. Histamine levels can increase during improper storage or if the bones used are not fresh, which can lead to adverse reactions in sensitive individuals.

To minimize these risks, it is important to follow safe food handling practices when preparing and storing bone broth. This includes using fresh, high-quality ingredients, cooking the broth to a safe temperature (165 F or 74 C), and promptly refrigerating or freezing any leftovers. Additionally, individuals with compromised immune systems, pregnant women, young children, and the elderly should take extra precautions when consuming bone broth.

(59)Sichuan Peppercorn: Sichuan peppercorns, also known as Szechuan peppercorns, is a popular spice used in Chinese and other Asian cuisines. While they are generally safe to consume, there are a few food safety concerns associated with them:

(a) Pesticide Residues: Sichuan peppercorns are often imported from various regions, and there have been concerns about the presence of pesticide residues. To address this concern, it's essential to purchase Sichuan peppercorns from reputable sources that adhere to food safety regulations and standards.

(b) Citrus Canker: Sichuan peppercorns can be carriers of citrus canker, a bacterial disease that affects citrus trees. This disease can be a concern in regions where citrus canker is prevalent. To mitigate this risk, it's important to ensure that the Sichuan peppercorns are sourced from areas free of citrus canker or have been appropriately treated to eliminate the risk of contamination.

© Allergen Cross-Contamination: Cross-contamination with allergens is a potential concern, especially if Sichuan peppercorns are processed in facilities that also handle allergenic ingredients. Individuals with known allergies should be cautious and seek out Sichuan peppercorns processed in facilities with stringent allergen-control measures. To prevent these food safety concerns, here are some general guidelines:

-Purchase Sichuan peppercorns from reputable suppliers who adhere to food safety regulations and standards.

-Look for organic or certified pesticide- free Sichuan peppercorns to minimize the risk of pesticide residues.

-Ensure that the Sichuan peppercorns are sourced from regions free of citrus canker or have been appropriately treated to eliminate the risk of contamination.

-For individuals with allergies, seek out Sichuan peppercorns processed in facilities with stringent allergen control measures and always check product labels for allergen information.

By following these guidelines, consumers can enjoy Sichuan peppercorns while minimizing potential food safety concerns.

(60)Ogiri: Ogiri is a traditional ingredient used in African cooking. It is a fermented condiment made from oil bean, castor, or melon seeds. While it adds a unique flavor to dishes, its health benefits are not extensive. It's high in protein and has some nutritional value, but it's also high in salt, which should be consumed in moderation. As with any food or ingredient, moderation and balance are key. If you have specific health concerns or dietary restrictions, it's best to consult with a healthcare professional or a nutritionist for personalized advice. But Ogiri, being a fermented condiment, can present some food safety and health issues. Here are some potential concerns and how to prevent and regulate them:

(a) Microbial Contamination: During fermentation, Ogiri can be contaminated with harmful bacteria. To prevent this, it's important to ensure that the seeds used are high-quality and free of mold and other contaminants. Additionally, maintaining proper hygiene among employees and ensuring that equipment and the environment are cleaned and sanitized during the production and storage process.

(b) High Salt Control: Ogiri is often high in salt, which can contribute to health issues such as high blood pressure. To regulate salt content, producers can consider using less salt during fermentation, or consumers can use it sparingly in their cooking.

(c) Allergen Concerns: Depending on the type of seeds used in its production, Ogiri may pose allergen concerns for some individuals. Producers should clearly label the product to indicate the specific ingredients used, and consumers should be mindful of any allergies they may have.

(d) Storage and Shelf Life: Proper storage is important to prevent spoilage and ensure food safety. Ogiri should be stored in airtight containers in a cool, dry place to extend its shelf life and prevent contamination.

(e) Regulatory Oversight: It is a need for many, so Government agencies responsible for food safety should provide regulations and oversight to ensure that Ogiri producers adhere to food safety standards. This can include inspections, quality control measures, and labeling requirements.

Overall, promoting good manufacturing practices, proper labeling, and consumer education on the safe handling and consumption of Ogiri can help mitigate food safety and health concerns associated with this traditional condiment that enhances the taste of African dishes.

CHAPTER TEN

Cooking Oil & Safety

Cooking oil is often overlooked in food safety discussions, but it can pose several real risks if handled improperly. Here's a clear breakdown of the concerns, how to prevent them, and why people tend to ignore them.

Degradation from Overheating: When oil is heated beyond its smoke point, it begins to break down, forming harmful compounds such as aldehydes and free radicals that can be unsafe if consumed regularly. Use oils appropriate for the cooking method. Avoid letting oil visibly smoke. Every oil will smoke if you keep heating it, so monitor cooking temperatures carefully.

Repeated Use of Oil (Reused Frying Oil). Reusing oil multiple times leads to oxidation and the buildup of toxic substances, increasing health risks. For Prevention, limit how many times oil is reused. Filter oil after use to remove food particles. Discard oil that is dark, thick, or rancid-smelling. Rancidity can occur through oxidation, caused by exposure to oxygen, heat, light, and moisture. Light and heat can cause oil to go rancid, producing off-flavors and potentially harmful compounds. : Store oil in a cool, dark place. Keep containers tightly sealed. Pay attention to expiration dates. Oil can become contaminated by Food particles, which can harbor bacteria, water, which can cause splattering and microbial growth in residues, and improper storage containers. Prevention: Keep oil free from water and debris. Use clean utensils and containers. Avoid mixing fresh oil with old oil. Formation of trans fats, Improper heating, and repeated reuse can increase trans-fat levels, which are linked to cardiovascular disease. In all, avoid prolonged high-heat cooking and do not reuse oil excessively. Store your oil properly in a cool, dark, tightly sealed container.

Why Consumers Pay Less Attention to Cooking Oil Safety

Many people care less about the food safety concerns that are associated with food oils, and there are several reasons for that.

(a) Lack of awareness: Many people may not be fully informed about the potential food safety issues related to different types of oils. They might not be aware of the unsanitary conditions associated with oil processing, the quality of products used, and how some oils are stored.

(b) Conflicting information: There is often a great deal of conflicting information in the media and online about the health and safety aspects of different oils. This can lead to confusion and make it difficult for people to discern accurate safety information.

(c) Cultural or culinary traditions: In some cultures, or in some culinary traditions, certain types of oils are commonly used without much consideration for food safety issues. This could be due to long-standing culinary practices or a lack of awareness about potential risks. In many households and food businesses, reusing oil is common for economic reasons, and the risks are underestimated or normalized.

(d) Cost considerations: Some people may prioritize cost over quality and safety when purchasing oils, opting for cheaper options without considering potential health risks associated with lower-quality products.

(e) Trust in regulation and labeling: Many consumers trust that the oils they purchase meet regulatory standards and assume that the labels accurately reflect the contents of the product. As a result, the government should continue to ensure that the products used in oil processing are safe; that the processing environment is sanitary; and that workers' hygiene practices meet the government's regulatory standards.

(f) Invisible Risk, unlike spoiled food, degraded oil doesn't always look unsafe. People may not notice subtle changes in color or odor.

It's important to note that food safety issues related to oils, such as adulteration, contamination, and rancidity, can have serious health implications. It's crucial for consumers to be well-informed about the oils they use and to prioritize food safety when making purchasing decisions.

1. **Olive oil**: Food safety concerns regarding olive oil can include issues related to quality, authenticity, and potential contamination. Here are some specific concerns:

   (a) Adulteration: One of the primary concerns with olive oil is the potential for adulteration, where lower-quality oils or other vegetable oils are mixed with or sold as olive oil. This can compromise the product's quality and authenticity and raise concerns about allergens for those with specific dietary restrictions.

   (b) Rancidity: Olive oil, like other fats, can become rancid when exposed to oxygen, light, and heat. Rancid oil not only affects the flavor and aroma but can also form potentially harmful compounds. Proper storage and handling are important to minimize the risk of rancidity.

(c) Contamination: Contamination with substances such as pesticides, heavy metals, or other pollutants can be a concern, particularly in regions where environmental regulations may be lax. Additionally, improper handling during production and storage can lead to microbial contamination, potentially causing foodborne illnesses.

(d) Mislabeling: Incorrect or misleading labeling can lead to food safety concerns. For example, if olive oil is labeled "extra virgin" but does not meet the standards for this designation, it could mislead consumers and compromise their expectations of quality and safety.

To mitigate these concerns, consumers should purchase olive oil from reputable sources, look for certifications or seals of authenticity, and store the oil properly in a cool, dark place. Additionally, regulatory agencies and industry standards play a crucial role in ensuring the quality and safety of olive oil products.

2. **Palm oil:** Palm oil is a versatile and commonly used vegetable oil in food products, but there are some food safety concerns associated with its production and consumption. Some of the concerns include:

   (a) Potential for contamination: During the production process, palm oil may be exposed to contaminants such as polycyclic hydrocarbons (PAHs), which can be formed during the processing of the oil at high temperatures. These contaminants can pose health risks if consumed in high quantities.

(b) Trans fat content: Partially hydrogenated palm oil can contain trans fats, which have been linked to an increased risk of heart disease. While many producers have moved away from using partially hydrogenated oils, there may still be concerns about trans-fat content in palm oil-based products. Partially hydrogenated palm oil undergoes a chemical process to alter its texture, creating artificial trans fats linked to heart disease. Conversely, regular unmodified or fully hydrogenated palm oil contains no trans fats but is naturally high is saturated fat.

(c) Impact of processing: The high-temperature processing methods used to extract and refine palm oil can lead to the formation of potentially harmful compounds, such as glycidyl fatty acid esters (GE), which are considered a potential health concern.

(d) Sudan dyes are a group of synthetic industrial colorants (commonly Sudan I,

II, III, and IV) that are sometimes illegally added to foods, especially palm oil, to enhance or restore a deep red color and make the product look fresher or higher quality. What exactly is Sudan dye? It is a chemical dye used for non-food purposes, such as coloring oils, waxes, fuels, and plastics. It is not approved for human consumption. Some types (like Sudan I) are classified as carcinogenic (cancer-causing). Why is it found in palm oil? Palm oil naturally has a reddish color due to carotenoids, but Poor-quality or over-processed palm oil may lose its color. Unscrupulous producers may illegally add chemicals like Sudan dye to improve appearance, increase market value, and mislead consumers into thinking the oil is fresh and high-quality. Consumption of Sudan dye–contaminated food has been linked to Increased cancer risk, Liver and kidney damage, and general toxicity with long-term exposure. How can dye contamination be prevented? At the production level, strict regulation and enforcement by food safety authorities, routine laboratory testing of palm oil and food products, use of approved natural colorants only, and proper processing methods to preserve natural carotenoids. In the supply chain, source palm oil from certified, reputable suppliers, implement traceability systems (know where the oil comes from), and conduct regular quality checks and audits. At the retail and consumer level, buy palm oil from trusted brands or regulated markets. Avoid oils with unnaturally intense or overly bright red color. Look for food safety certifications or regulatory approval labels. Public awareness and education: Educate consumers about the dangers of adulteration; encourage reporting of suspicious products; promote understanding that appearance does not equal quality. In all, Palm oil, like every food, deserves:

(a) Good sanitary and manufacturing practices: Adopting good sanitary and manufacturing practices can help minimize contamination of palm oil with harmful substances. This includes proper handling, storage, and processing techniques to reduce the formation of contaminants.

(b) Quality control and testing: Regular testing and quality control measures can help ensure that palm oil meets safety standards and regulatory requirements. This can help identify and mitigate potential contaminants, ensuring the oil is safe for consumption.

(c) Regulation and labeling: Governments and food safety authorities can implement regulations and standards for palm oil production and use, including limits on harmful compounds and mandatory labeling of trans fat content. This can help consumers make informed choices and incentivize producers to adhere to safety standards.

(d) Alternative processing methods: Research and development of alternative processing methods that minimize the formation of harmful compounds in palm oil can help improve its safety profile.

(e) Consumer awareness: Educating consumers about the potential risks associated with palm oil consumption and encouraging them to make informed choices can also contribute to preventing food safety concerns. By implementing these measures, it is possible to mitigate the food safety concerns associated with palm oil and ensure its safe consumption.

3. **Canola oil:** Canola oil is generally considered to be safe for consumption, but there are a few potential food safety concerns associated with it:

(a) Rancidity: Canola oil can become rancid if stored improperly or for too long. This can happen when the oil is exposed to light, heat, or air, leading to off-flavors and potentially harmful compounds. Prevention: To prevent rancidity, store canola oil in a cool, dark place, such as a pantry or cupboard, and tightly seal the container after each use. It's also important to check the expiration date and use the oil within a reasonable time frame.

(b) Contamination: Like all cooking oils, canola oil can be contaminated with bacteria, such as Salmonella or E. coli, during the production process or through improper handling. To prevent contamination, it's important to purchase canola oil from reputable sources and ensure it is properly processed and handled in accordance with food safety standards. Additionally, practicing good kitchen hygiene, such as washing hands and utensils, can help prevent contamination during cooking and food preparation.

(C) Allergies: While canola oil is not a common allergen, some individuals may have a rare allergy to it. Allergic reactions to canola oil can cause symptoms such as hives, swelling, or difficulty breathing. If you or someone you are cooking for has a known allergy to canola oil, it's important to avoid using it in cooking and food preparation. Always check food labels and be cautious when dining out to avoid potential exposure to canola oil.

Overall, practicing proper storage, handling, and hygiene measures can help prevent food safety concerns associated with canola oil. Additionally, being aware of potential allergies and cross-contamination is important to ensure food safety for all individuals.

4. Vegetable oil: Food safety concerns associated with vegetable oil primarily revolve around its potential for oxidation, contamination, and rancidity. Here are some of the key concerns and how to prevent them:

(a) Oxidation: Vegetable oils can oxidize when exposed to air, light, and heat, leading to the formation of harmful compounds. To prevent oxidation, store vegetable oils in a cool, dark place, tightly sealed to minimize air exposure. Using dark-colored or opaque containers can also help protect the oil from light.

(b) Rancidity: Rancidity occurs when vegetable oils break down and develop off-flavors and odors. To prevent rancidity, use fresh oil and store it properly. Check the "best use" date on the oil and try to use it within that period. Smell and taste the oil before using it in cooking to ensure it hasn't gone rancid.

(c) Contamination: Contamination of vegetable oils can occur during processing, handling, or storage, leading to microbial growth or the presence of harmful substances. To prevent contamination, it's important to choose high-quality oils from reputable sources and to ensure proper hygiene and sanitation during oil handling and storage.

(d) Quality and authenticity: There are concerns regarding the quality and authenticity of vegetable oils, including the possibility of adulteration with lower-quality oils. To address this, purchase oils from trusted suppliers and look for certifications and quality seals on the packaging.

(e) Acrylamide formation: When vegetable oils are heated to high temperatures, they can contribute to the formation of acrylamide, a potentially carcinogenic compound. To minimize acrylamide formation, avoid overheating oils and use cooking methods such as baking, steaming, and sauteing instead of deep-frying.

By being mindful of these concerns and following best practices for handling, storing, and using vegetable oils, you can help ensure their safety and quality in your cooking.

5. Coconut oil: Coconut oil is generally considered safe, but there are some food safety issues to be aware of. Here are a few potential concerns and ways to prevent them:

(a) Microbial contamination: Like any food product, coconut oil can be contaminated with bacteria, molds, and yeasts. To prevent microbial contamination, it's important to store coconut oil in a cool, dry place and to use clean, dry utensils when handling it. Proper hygiene and sanitation practices should be followed during production and packaging to minimize the risk of contamination.

(b) Rancidity: Coconut oil can become rancid due to oxidation, especially when exposed to heat, light, and air. To prevent rancidity, it's important to store coconut oil in a tightly sealed container away from light and heat. Additionally, using fresh, high-quality coconut oil and consuming it within a reasonable time frame can help minimize the risk of rancidity.

(c) Quality and adulteration: There have been instances of coconut oil being adulterated with lower-quality oils or being misrepresented as virgin coconut oil when it is not. To ensure the quality and authenticity of coconut oil, consumers should purchase from reputable sources and look for certifications such as organic or fair trade. Producers should follow proper quality control measures and ensure accurate labeling of their products.

(d) Allergen cross-contamination: Coconut is considered a tree nut, and some individuals may have allergies to coconut or other tree nuts. Cross-contamination with allergens can occur during processing or packaging. To prevent allergen cross-contamination, facilities should have proper cleaning procedures in place, and allergen information should be clearly labeled on packaging. By following good manufacturing practices, proper storage and handling procedures, and ensuring accurate labeling and quality control, many of the food safety issues associated with coconut oil can be effectively prevented.

6. Peanut or groundnut oil: Peanut or groundnut oil, like any other type of oil, can pose certain food safety concerns. Some of the potential issues include:

(a) Allergen cross-contamination: Peanut oil may cause allergic reactions in individuals with peanut allergies. Even highly refined peanut oil can still contain trace amounts of peanut protein, which could potentially trigger an allergic response in sensitive individuals.

(b) Rancidity: Like all types of oil, peanut oil can become rancid over time due to exposure to light, air, and heat. Rancid oil not only affects the taste and odor of food but can also pose health risks if consumed regularly.

(c) Contamination during processing: If the peanut oil is not properly processed, it may become contaminated with bacteria, fungi, or other pathogens. This can happen during production, transportation, or storage if proper hygiene and quality control measures are not followed.

(d) Mislabeling and adulteration: There is a risk of mislabeling or adulteration, where other oils may be mixed with peanut oil without proper labeling. This can pose a risk for individuals with specific dietary requirements or allergies.

To mitigate these concerns, it's important to purchase peanut oil from reputable sources, store it properly in a cool, dark place, and check for any signs of rancidity before use. Additionally, individuals with peanut allergies should always be cautious and read food labels carefully to avoid potential exposure to allergens.

7. Sesame oil: Sesame oil, like any other food product, can have food safety issues if not handled or stored properly. Some potential food safety issues associated with sesame oil include:

(a) Rancidity: Sesame oil can become rancid due to exposure to light, heat, or air. Rancid oil can develop an off-flavor and potentially harmful compounds.

(b) Contamination: Contamination with bacteria, molds, or other pathogens can occur if the oil is not processed, stored, or handled properly.

(c) Allergen cross-contact: Sesame oil is a common allergen, and cross-contact with other allergens during production, such as nuts or other seeds, can pose a risk to allergic individuals.

To prevent these food safety issues, the following measures can be taken:

(a) Proper storage: Store sesame oil in a cool, dark place to prevent light and heat exposure. Airtight containers should be used to minimize air exposure and slow down rancidity.

(b) Quality sourcing: Purchase sesame oil from reputable suppliers to ensure it has been processed and stored under safe conditions.

(c) Good manufacturing practices: Food producers should follow good manufacturing practices to prevent contamination during processing. This includes maintaining cleanliness, proper sanitation, and regular equipment maintenance.

(d) Allergen control: Facilities handling sesame oil should have strict allergen control procedures in place to prevent cross-contact with other allergens. Clear labeling and segregation of allergens can help prevent accidental exposure.

(e) Regular quality checks: Regularly check sesame oil for signs of spoilage, such as off-odor or off-flavor, and discard any product that appears to be rancid. By following these preventive measures, the food safety issues associated with sesame oil can be minimized, ensuring a safe product for consumers.

8. Avocado oil: Avocado oil is generally considered safe and healthy, but there are a few potential food safety issues associated with it. Here are some of the main concerns and ways to prevent them:

(a) Rancidity: Avocado oil, like any other oil, can become rancid if not stored properly. This can happen over time due to exposure to light, heat, or air. To prevent rancidity, store avocado oil in a cool, dark place, preferably in a dark glass bottle to protect it from light. Keep the bottle tightly sealed when not in use.

(b) Contamination: There is a risk of contamination with bacteria, molds, or other microorganisms, especially if the oil is not produced and handled under proper hygiene and safety standards. To prevent contamination, it's important to source avocado oil from reputable producers who follow good manufacturing practices. Consumers should also ensure that the oil is stored and handled properly to prevent contamination.

(c) Adulteration: Adulteration with lower-quality oils is a concern in the avocado oil industry. To prevent this, consumers should purchase avocados from trusted sources and look for reputable brands. Reading labels and understanding where the oil comes from can help ensure its purity.

(d) Chemical residues: Pesticide residues and other chemical contaminants can be a concern in avocado oil. To mitigate this risk, look for organic or certified pesticide-free avocado oil. These products are produced without synthetic chemicals, reducing the risk of chemical residues in the oil.

CHAPTER ELEVEN

PREVENTING FOODBORNE ILLNESSES ANYWHERE & ANYTIME.

1. **Wash your hands & surroundings:** You must regularly wash your hands and food contact surfaces like utensils utilized in preparing the food. It is critical that you wash your hands before preparing any food; after sneezing or coughing; after using the toilet; after handling raw meat; after handling

cleaning supplies like brooms, dust pans, or trash cans; and after handling chemicals or your pets. Ensure that utensils and equipment, such as cutting boards, knives, slicers, dicers, saws, casing peelers, packing materials, tables, conveyors, belts, and equipment controls, are thoroughly cleaned. It is best to scrub with hot soapy water, wash, rinse, and sanitize. Washing your hands regularly when handling food is extremely important for many reasons. (a) Preventing the spread of bacteria and viruses: Hands can be a significant source of contamination, carrying pathogens that can cause foodborne illnesses. Washing hands with soap and running water effectively removes these microorganisms, reducing the risk of transferring them to food or surfaces. (b) Promoting personal hygiene: Wash your hands each time you touch any object that has not been cleaned and sanitized. Also, wash your hands or replace your gloves when you pick something up off the floor. I cannot overemphasize the importance of washing your hands. Handwashing is a fundamental practice of personal hygiene. It helps remove dirt, debris, and other potential contaminants from the hands, ensuring they are clean before or during food handling. (c) Minimizing cross-contamination: Cross-contamination occurs when harmful microorganisms are transferred from one surface or food to another. By washing your hands between tasks, such as handling raw meat and ready-to-eat foods, you can prevent the transfer of bacteria and reduce the risk of cross-contamination. (d) Protecting consumer health: Proper hand hygiene significantly reduces the risk of foodborne illnesses. These illnesses can cause a range of symptoms, from mild discomfort to severe complications and even death, particularly among vulnerable populations such as young children, the elderly, and individuals with weakened immune systems. (e) Meeting regulatory requirements: Handwashing is a standard practice mandated by food safety regulations in many countries. Compliance with these regulations is necessary to maintain food safety standards and ensure the health and well-being of consumers. (f) Demonstrating professionalism and trust: Practicing good safety practices, such as proper handwashing, in a food-handling environment demonstrates professionalism and a commitment to food safety. It instills confidence in customers, employees, and regulatory authorities and enhances the reputation and trustworthiness of food establishments.

How to Effectively Wash Hands

(a) Wet hands with clean, running water. (b) Apply soap and lather well, covering all surfaces of the hands, including the backs, between the fingers, and under the nails. (c) Scrub hands vigorously for at least 20 seconds. (d) Rinse hands thoroughly under running water, ensuring all soap is washed away. (e) Dry hands using a clean towel or air dryer. (f) Use a towel or tissue to turn off the faucet to avoid recontamination.

2. Separate: Regardless of the setting, you must separate raw food from cooked. Raw meat must be stored and processed separately from non-meat food items. When you are at the grocery store, make sure the grocer places your raw meat in a separate bag, as it can carry disease-causing bacteria. Here are some of the best practices for effective separation. (a) Use separate cutting boards: Designate specific cutting boards for raw meat, poultry, and seafood, and for fruits, vegetables, and other ready-to-eat foods. This prevents the transfer of harmful bacteria from raw foods to foods that will be consumed without further cooking.

(b) Use separate utensils: Use separate utensils, such as knives, tongs, and spatulas, for handling raw and cooked foods. This includes using different utensils when flipping raw meat on the grill and removing cooked meat from the grill.

(c) Store raw and cooked foods separately in the refrigerator: Keep raw meats, poultry, seafood, and their juices securely wrapped and stored on the lowest shelf of the refrigerator. This prevents any potential drips or leaks from contaminating cooked or ready-to-eat foods stored above.

(d) Separate storage containers and bags: Use separate storage containers, bags, or wraps for raw and cooked foods. This ensures that raw juices or bacteria from raw foods do not come into contact with cooked foods during storage.

(e) Use different preparation surfaces: When preparing meals, use different surfaces or areas of your kitchen for handling raw and cooked foods. This helps prevent cross-contamination by keeping raw juices and bacteria away from surfaces that come into contact with food.

(f) Maintain a clean and organized workspace: Regularly clean and sanitize countertops, cutting boards, utensils, and other food preparation surfaces to prevent cross-contamination. Keep your workspace organized to minimize the risk of accidentally mixing raw and cooked foods.

(g) Educate and train kitchen staff: If you're working in a professional kitchen, provide proper training for staff members on the importance of separating raw and cooked foods. Ensure that everyone understands and follows the correct procedures for handling and storing different types of foods. By implementing these practices, you can effectively separate raw foods from cooked foods, reducing the risk of cross-contamination and promoting food safety in your kitchen or food-preparation area.

(3)Cook: You must cook your food properly. It is critical to have your thermometer and the USDA, FDA, or World Health

Organization (WHO) recommended chart for the minimum required temperature within reach. The recommended minimum safe cooking temperatures for various foods are important for ensuring that harmful bacteria and other pathogens are killed, reducing the risk of foodborne illness. Here are the general guidelines for minimum safe cooking temperatures:

(a)Poultry (including whole or ground chicken, turkey, duck):

-Chicken and turkey: 165F (74C)-Duck: 165F (74C)
(b) Ground meats (including beef, pork, veal, lamb):
-Ground meat: 160F (71 °C)

(c) Steaks, roasts, and fish:

-Steaks (beef, pork, veal, and lamb):
145 F (63C) for medium-rare, 160F (71C) for medium-rare, 160F (71C) for medium, and 170F (77C f) for well-done.
-Roast (beef, pork, veal, and lamb): 145F (63C) for medium-rare, 160 F (71 C) for medium-rare, 160 F (71 C) for medium, and 170 F (77 C) for well-done.
-Fish 145F (63C) or until the flesh is opaque and separates easily with a fork.

(d) Ground poultry:
-Ground poultry (chicken, turkey): 165F (74 °C)

(e) Pork, ham, and bacon:
-Pork, ham, and bacon: 145F (63 C)

(f) Egg dishes:

-Eggs and dishes containing eggs (such as quiches or casseroles): 160 F (71 °C)

(g) Leftovers and casseroles:
-Leftovers and casseroles: 165F (74 °C)

It's important to note that these temperatures are guidelines and may vary depending on your region's specific food safety regulations. Always refer to local food safety guidelines and use a food thermometer to ensure accurate measurements. Additionally, certain food items, such as ground meats and poultry, may have different temperature recommendations due to safety concerns about bacteria like Salmonella and E. coli.

4. Don't prepare food for others when sick. It is important that you stay out of the food processing area or the kitchen if you are sick and experiencing vomiting, diarrhea, jaundice, a sore throat, fever, a common cold, eye discharge, sneezing, coughing, a runny nose, or mouth discharge. Also, when you have wounds or sores on your hands, wear gloves when preparing food and replace them as needed. Also, replace gloves when you pick things up off the floor or touch objects that could contaminate products.

-Pathogen transmission: When you are sick, especially with an infectious illness like a cold, flu, or gastrointestinal infection, you may have higher levels of pathogens in your body. These pathogens can be transmitted through respiratory droplets, saliva, or fecal matter. If you handle food while you are sick, you can transfer these pathogens to the food, increasing the risk of contamination.

-Weakened immune system: When you are sick, your immune system is already compromised as it fights off the illness. This can make you more susceptible to additional foodborne infections. By preparing more food while sick, you increase the risk of introducing pathogens that your weakened immune system may not handle effectively.

5. Cooling. Never allow perishable foods to sit out for long periods. To reduce the risk of bacterial growth, you should refrigerate any food you're not eating right away. You can also put your food in smaller containers and let it sit in ice water to bring the temperature down faster before refrigeration. Yes, it is important to cool your food properly to prevent the growth of harmful bacteria and ensure food safety. When food is left at room temperature for too long, bacteria can multiply rapidly, increasing the risk of foodborne illness. Cooling food quickly and properly helps minimize the time it spends in the temperature danger zone (between 40 F/4 C and 140

F/60 C), where bacteria can thrive.

Guidelines for cooling food properly

(a) Divide large portions: If you have a large batch of hot food, divide it into smaller portions. Smaller portions cool more quickly than larger ones.

(b) Use shallow containers: Transfer the hot food into shallow, wide containers. This increases the surface area of the food, allowing it to cool down faster.

(c) Rapid cooling methods: There are a few methods you can use to facilitate rapid cooling:

-Ice bath: Place the container of hot food in an ice bath, either in a sink filled with ice and water or in a larger container with ice water. Stir the food occasionally to promote even cooling.

-Blast chiller: If you have access to a blast chiller, use it to quickly cool the food. Blast chillers are designed to rapidly lower the temperature of hot food to safe levels.

Refrigerator or freezer: If you're cooling smaller portions or leftovers, you can place them directly in the refrigerator or freezer. However, it's important to note that placing large containers of hot food directly into the refrigerator can raise its internal temperature, potentially compromising the safety of other food items.

(d) Monitor the temperature: Use a food thermometer to

monitor the temperature of the food as it cools. The goal is to cool the food from 140 F/60 C to below 40 F/4 C within two hours.

(e) Proper storage: Once the food has cooled, transfer it to appropriate storage containers and place it in the refrigerator or freezer. Label the containers with the date to keep track of their freshness.

Again, proper cooling practices are essential to prevent bacterial growth and maintain food safety.

Effective Cooling

Refrigerated Storage. Refrigeration slows bacterial growth. Bacteria live in all places in nature. They exist in soil, air, water, and foods that we consume. When they have nutrients such as food, moisture, and favorable temperatures, they grow swiftly. Bacteria grow most rapidly in temperatures between 40 and 140 degrees Fahrenheit, so set your temperature at 40 degrees or less. Your raw meat, poultry, and seafood must be placed in a sealed container or very well wrapped, so they do not drip raw juices and contaminate other foods.

Proper maintenance of coolers and refrigerated storage is crucial for ensuring food safety and preventing the growth of harmful bacteria. Here are some important guidelines to follow:

(a) Temperature Control: Maintain the internal temperature of your cooler or refrigerator at or below 40 F (4 C). This temperature range helps slow bacterial growth. Use a refrigerator thermometer to regularly monitor the temperature.

(b) Organization: Keep your cooler or refrigerator organized to ensure proper airflow and temperature distribution. Avoid overcrowding and leave space between food items to allow cold air to circulate.

(c) Storage Zones: Assign specific zones for different food categories to prevent cross-contamination. Store raw meats, poultry, and seafood on the bottom shelf or in sealed containers to prevent drips from contaminating other foods. Place ready-to-eat foods, such as cooked meats and leftovers, on higher shelves to avoid any potential contamination from raw foods.

(d) Proper Packaging: Store food in appropriate airtight containers or tightly sealed packaging to maintain freshness and prevent odors from transferring between different foods. This helps to minimize the risk of cross-contamination and extends the shelf life of perishable items.

(e) Regular Cleaning: Clean your cooler or refrigerator regularly to prevent the buildup of spills, food residues, and bacteria. Use mild detergent to clean the interior and shelves, and rinse well. Pay special attention to any areas where drips or spills commonly occur, such as door seals and vegetable crisper drawers.

(f) Hygiene Practices: Practice good personal hygiene when you are handling food. Wash your hands thoroughly with soap and warm water before and after handling food, especially when switching between different types of food. This helps to prevent the transfer of bacteria and other contaminants.

(g) Regular Maintenance: Periodically inspect your cooler or refrigerator for any signs of malfunction, such as temperature fluctuations or unusual noises. Promptly maintain and repair any faulty equipment to ensure consistent temperature control.

(h) Codes: Understand the codes on your foods and regularly check the use or expiration dates on food items stored in your cooler or refrigerator. Discard any expired or spoiled foods promptly to avoid the risk of foodborne illness. By following these guidelines, you can maintain the quality and safety of food stored in your cooler or refrigerator, thereby minimizing the risk of foodborne illnesses and ensuring the freshness of your perishable items.

(6)Proper Taste Testing. When conducting taste testing of foods, it is essential to maintain strict hygiene practices to prevent contamination. Here are some proper ways to conduct taste testing while ensuring food safety:

(a)Avoid dipping your fingers in food; also, avoid licking your utensils when testing your food because a sick person can transfer sickness to the food through a contaminated utensil. It is pertinent to use a new or disposable apparatus when doing your taste testing. More than one person should not use the same spoon, fork, or other utensil in taste testing.

(b) Utensils and serving ware: Use clean utensils, such as spoons or forks.

Individual Portion: Serve individual portions of the food being tasted to prevent multiple people from sampling the same dish. This reduces the risk of cross-contamination between people.

(c)Food Separation: Ensure that each food being tasted is presented separately and not in direct contact with other items. This prevents contamination and maintains the integrity of each sample.

(d)Clean Surfaces: Clean and sanitize the tasting area, including countertops, cutting boards, and utensils, before and after each tasting session. This will minimize the risk of cross-contamination.

(e)Allergen Awareness: If you are conducting taste testing for a group, be mindful of potential allergens. Clearly label the food items and provide information on any common allergens present in each sample, such as nuts, dairy, and gluten.

(f)Temperature Control: If the food samples require

specific temperature control, ensure they are kept at the appropriate temperature before and during the taste testing. Hot foods should be hot (above 140°F or 60°C), and cold foods should be cold (below 40°F or 4°C) to minimize bacterial growth.

(g)Encourage participants to minimize playing with their dogs, pets, and other pets while the test is on. If possible, leave the pets at home.

(h)Minimize Talking and Coughing: Ask participants to avoid talking or coughing directly over the food samples to prevent the spread of respiratory droplets that may contain bacteria or viruses.

(I)Time Limit: Set a specific time limit for tasting sessions. Do not leave food samples at room temperature for extended periods, as this can lead to bacterial growth. Discard any uneaten samples promptly.

(7)Desist from eating while you are shopping. People can eat contaminated foods with such practices. The grocery stores should protect their food against those sample fruits or open packages at the stores. Food Safety is a vital concern. Eating directly from the shelf can contaminate the food with bacteria, viruses, or other pathogens. This can pose a health risk to other customers who might purchase the contaminated food.

(a) Product Integrity: Eating from the shelf can damage the packaging or alter the appearance of the product, making it unappealing or unsellable to other customers. It can also lead to product waste and financial losses for the store. Be empathetic.

(b) Consideration for Others: Grocery stores are communal spaces where many people shop for their daily necessities. Eating from the shelf can be seen as disrespectful to other shoppers who expect to purchase unopened and uncontaminated products.

(c) Legal and Store Policies: Eating from the shelf is typically against store policies, and in some cases, it may even be illegal. Most grocery stores have clear guidelines against consuming products before purchase. Instead of eating directly from the shelf, it's important to follow proper etiquette and hygiene practices when shopping. If you're hungry while at the store, it's advisable to purchase a designated snack or meal from the store's prepared food section, if available, and consume it in the designated eating area or once you have paid for it.

(8)Defrost Your Food Safely. Properly defrosting frozen food is essential to maintain food safety and prevent the growth of harmful bacteria. Here are some best practices to defrost frozen food safely:

(a)Refrigerator Thawing: The safest method for defrosting most foods is to place them in the refrigerator. This method requires advanced planning, as it takes more time. Place the frozen food on a plate or in a container to catch any drips and allow it to thaw in the refrigerator. Make sure to keep raw meat separate from other foods to prevent cross-contamination. It may take several hours, a day, or longer, depending on the food's size and thickness.

(b) Cold Water Thawing: If you need to defrost food more quickly, you can use the cold-water thawing method. Ensure the food is in a leak-proof plastic bag or sealed container to prevent water contamination. Submerge the package in cold water, changing the water every 30 minutes to maintain a safe temperature. Small items may thaw in an hour or less, while larger items may take several hours.

(C)Microwave Thawing: Microwaving can be used for small, uniformly shaped items. Follow the microwave's instructions and use the defrost setting or lower power setting. Be cautious, as microwaving can partially cook the food, creating uneven thawing and potential food safety issues if it is not cooked immediately after defrosting.

It's important to note that I do not recommend room-temperature thawing (leaving food at room temperature), hot-water thawing, or using a microwave's regular cooking setting. These methods can lead to uneven thawing, promote bacterial growth, and compromise food safety.

(9)Don't eat raw eggs, meat, or seafood. There are several reasons why consuming raw eggs, seafood, and meat is unsafe.

(a) Bacterial Contamination: Raw eggs, seafood, and meat can be contaminated with bacteria such as Salmonella, E. coli, Listeria, and Campylobacter. These bacteria can cause foodborne illnesses, resulting in symptoms such as nausea, vomiting, diarrhea, abdominal pain, and, in severe cases, hospitalization.

(b) Salmonella in Eggs: Raw eggs, particularly their shells, can be contaminated with Salmonella bacteria. Consuming raw or undercooked eggs increases the risk of Salmonella infection, which can be especially dangerous for infants, young children, pregnant women, older adults, and individuals with weakened immune systems.

(c) Parasites in Seafood: Raw or undercooked seafood, such as oysters, clams, mussels, and sushi/sashimi, may contain parasites like Vibrio, Hepatitis A, Norovirus, or parasites like tapeworms and roundworms. These parasites can cause various gastrointestinal illnesses, liver problems, or other infections.

(d) Bacterial Infections in Meat: Raw or undercooked meat, including beef, pork, poultry, and game meats, can contain harmful bacteria like Salmonella, E. coli, and Campylobacter. These bacteria can cause severe food poisoning and lead to serious health complications.

(e) Cross-Contamination: Raw eggs, seafood, and meat can also contaminate other surfaces and foods in the kitchen if proper food handling practices are not followed. Cross-contamination can occur when raw juices come into contact with cooked or ready-to-eat foods, utensils, cutting boards, or countertops, increasing the risk of foodborne illnesses.

.

To ensure food safety, it is recommended to cook eggs, seafood, and meat thoroughly to their recommended internal temperatures. This helps to kill bacteria, parasites, and other pathogens, making the food safe to consume. Following proper food-handling practices, such as washing hands, preventing cross-contamination, and storing food at appropriate temperatures, also plays a crucial role in reducing the risk of foodborne illness.

(10) Be Careful at Buffets. When dining at buffets, it's important to practice proper food safety measures to reduce the risk of foodborne illnesses. Here are some essential food safety precautions to keep in mind:

(a)Choose a Reputable Buffet: Select a buffet that has a good reputation for cleanliness and food safety practices. Check online reviews or ask for recommendations from trusted sources. You can even visit there and observe for yourself.

(b)Observe Food Handling Practices: Pay attention to how the buffet staff handles the food. They should wear gloves, use utensils, avoid speaking or coughing on the food, and frequently replenish it to maintain freshness.

(c)Check Food Temperature: Hot food items should be served hot (above 140 F or 60 C), and cold food items should be served cold (below 40 F or 4 C). Avoid consuming any dishes that are at an unsafe temperature or appear to have been sitting out for too long.

(d)Practice Good Hygiene: Wash your hands thoroughly with soap and water before and after eating at the buffet. If handwashing facilities are unavailable, use hand sanitizers containing at least 60% alcohol.

(e)Use Serving Utensils: Avoid using your hands to pick up food. Always use the provided serving utensils to avoid cross-contamination between different food items.

(f)Separate Utensils for Different Dishes: Buffets typically have separate serving

utensils for each food item. Ensure that these utensils are used appropriately and not shared between dishes.
(g) Avoid Cross-Contamination: Be mindful of cross-contamination risks. Do not allow utensils from one dish to touch another, especially when handling raw and cooked foods. Use separate plates or bowls for different food items.

(h)Fresh and Well-Stocked Food: Choose foods that are freshly prepared and well-stocked. Avoid items that appear to have been sitting out for extended periods or that look unappetizing.

(i)Be Cautious with Raw or Undercooked Foods: If there are raw or undercooked food options, such as sushi or rare meats, consider the potential risks and your own tolerance for consuming these items. Ensure they are handled safely and stored at the appropriate temperatures.

(j)Watch Allergen Cross-Contamination: If you have food allergies or sensitivities, be cautious about potential cross-contamination. Buffets often have a wide variety of dishes, and allergenic ingredients may be present or come into contact with other foods.

By following these food safety measures, you can enjoy your buffet dining experience while minimizing the risk of foodborne illnesses.

(11)Pay attention to the high-risk population. When preparing food for infants, elderly individuals, pregnant women, and people with weakened immune systems, extra care should be taken to ensure their safety and well-being. Basic food safety practices are essential because of their vulnerability. Here are some special measures to consider:

(a) Hygiene: Hygiene practices around high-risk populations should even be stricter. Maintain rigorous personal sanitation by washing hands thoroughly with soap and water before handling any food. Help high-risk individuals, as most may not be able to do it themselves. This helps to prevent the spread of bacteria and other pathogens.

(b) Cross-Contamination Prevention: Those who prepare food for pregnant women, the elderly, and the very young must ensure that cross-contamination prevention practices are strictly maintained by keeping raw and cooked foods separate. Use different cutting boards, utensils, and storage containers for raw meats and other ingredients. Clean and sanitize surfaces and utensils between tasks.

(c) Cook Food Thoroughly: Ensure that all foods that are supposed to be eaten by anyone, especially meats, poultry, seafood, and eggs, are cooked thoroughly to the recommended internal temperatures. This helps to kill harmful bacteria and parasites that can cause foodborne illnesses.

(d) Safe Handling of Fresh Produce: Thoroughly wash their fruits and vegetables under running water before consumption to remove dirt, bacteria, and pesticides. Peel or cook produce when necessary to reduce the risk of contamination.

(e) Safe Storage and Temperature Control: Food for the high-risk population must be stored at appropriate temperatures to prevent bacterial growth. Keep refrigerated foods below 40 F (4 C) and hot foods above 140 F (60 C). Discard any perishable foods that have been left at room temperature for more than two hours.

(f) Avoid High-Risk Foods: Certain foods, such as unpasteurized dairy products, soft cheeses, deli meats, and raw sprouts, pose a higher risk of foodborne illnesses. It is advisable for infants, elderly individuals, pregnant women, and those with weakened immune systems to avoid these high-risk foods.

(g) Consider Allergies and Dietary Restrictions: Consider any specific dietary restrictions or allergies when preparing food for these individuals. Avoid cross-contamination with allergens and ensure that their dietary needs are met.

(h) Proper Food Storage: Follow safe food storage practices, including proper labeling, dating, and storage of leftovers. Use refrigerated leftovers within a safe timeframe (typically within 3-4 days) and reheat them thoroughly before serving.

(i) Seek Medical Advice: If you have specific concerns about food preparation for infants, elderly individuals, pregnant women, or those with weakened immune systems, it is advisable to consult with a healthcare professional or registered dietitian for personalized guidance.

(12)Be thoughtful of food recalls. Try to stay informed about food recalls and remove any recalled food from your store, cooler, refrigerator, school, and church Cafeteria. To stay informed about recalled foods and understand how to properly handle them, you can take the following steps:

(a) Sign Up for Notifications: Register for email alerts or subscribe to relevant newsletters from reputable sources such as the U.S. Food and Drug Administration (FDA) or the U.S. Department of

Agriculture (USDA, FSIS). These agencies regularly issue food recall announcements.

(b) Follow News and Media: Stay updated on food safety news through reliable sources, including news websites, newspapers, television, and radio. They often report significant food recalls and provide information on proper handling.

(c) Utilize Online Resources: Visit the websites of regulatory agencies like the USDA, FSIS, and FDA, which maintain dedicated sections for food recalls. These websites provide detailed information about the recalled products, reasons for the recall, and instructions for consumers.

(d) Check Product Labels: Stay vigilant about checking product labels, especially for perishable foods, as they often include information about recalls or instructions for safe disposal.

(e) Keep Receipts or Packaging: Retain receipts or packaging for recently purchased products. In the event of a recall, this information can help you identify the affected products.

(f) Contact Retailers: If you suspect that you have purchased a recalled product, contact the retailer or store where you made the purchase. They can provide guidance on returning or disposing of the product.

(g) Report Issues: If you believe you have encountered a food safety issue or have fallen ill after consuming a product, report it to the appropriate regulatory agency. Your report can help identify potential problems and contribute to consumer protection efforts. Remember, staying informed about food recalls and properly handling recalled foods is crucial for maintaining food safety. By taking these steps, you can protect yourself and others from consuming potentially harmful products.

(13)Be Mindful of food expiration dates. You must check food labels for "use by" or "sell by dates" and don't eat foods that are past their expiration dates. Here are the steps you can follow to check these dates properly:

(a) Locate the date labels: The "use by" and "sell by" dates are typically printed on the packaging of the food item. Look for these labels on the front, back, or side of the packaging. They are usually printed clearly and visibly.

(b) Understand the difference between "use by" and "sell by" dates:
- "Use by" date: This is the date recommended by the manufacturer until which the product is expected to maintain its quality and safety. It indicates the last date by which the food should be consumed to maintain its best quality.
- "Sell by" date: This date is intended for retailers and indicates the last day the

store should display the product for sale.

(c) Check the dates: Once you've located the date labels, check the "use by" and "sell by" dates to determine the freshness of the food item. Ensure that the dates have not passed.

(d) Inspect the packaging: In addition to the dates, examine the packaging for any signs of damage, such as tears, leaks, or bulges. Damaged packaging can indicate that the food may have been compromised, even if the dates haven't expired.

(e) Observe the food's appearance and smell: Open the packaging and assess the food's appearance and smell. If the food looks discolored, moldy, or has an off-putting odor, it may have spoiled and should not be consumed, even if the dates haven't expired.

(f) Follow storage guidelines: Proper storage plays a crucial role in maintaining food freshness. Make sure you have stored the item according to the manufacturer's, USDA, and FDA instructions. Remember, the "use by" and "sell by" dates are guidelines for quality and safety, but they are not absolute indicators of spoilage. Trust your senses and use your judgment when determining if a food item is still safe to consume. When in doubt, it's better to err on the side of caution and discard the food.

(14) **Food preparers' hygiene practices for all**. Hygiene practices are essential for preventing contamination of Ready-to-Eat (RTE) products. Some established food industries have quality control personnel and supervisors trained in food safety management. Some establishments have government inspectors assigned to perform daily tasks for the public and the government. Unfortunately, some small restaurants or processors lack the knowledge or resources to maintain a standardized food safety management system. As a result, this information will help them protect their

health, their employees' health, and their customers' health. I want my readers to know that deadly bacteria could be in the hands of employees or anyone handling the food. It does not matter how well-known their business name is. People's hands are very critical in the transmission of foodborne pathogens. Workers' or cooks' dirty hands or fingernails can contaminate the food. Activities that are bound or likely to contaminate your hands must subsequently be followed by proper washing of your hands. Such activities include leaving the food processing area to use the toilet, picking things up off the floor, touching your nose & other body parts, handling trash cans and boxes, and touching door handles and other hand-contact areas.

Many still do not understand that workers' shoes can carry contamination to the floors where food is processed and prepared. As a USDA Inspector in charge, I oversaw an establishment that the Plant managers designated a 'Changing Room.' No one was allowed to wear street clothes beyond the changing room; street clothes were covered with white coats before entering the processing area, including other government inspectors and me. This endeavor, along with a standard sanitation team and management that are food-safety-conscious, contributed to the absence of pathogens for years, as evidenced by samples from the government and the establishment.

15. At the Store, Place Raw Animal Foods in a Separate Bag: It is important to place raw animal food in a separate bag in the grocery store for several reasons:

(a) Prevent cross-contamination: Raw animal foods, such as meat, poultry, and seafood, can contain harmful bacteria like Salmonella, E. coli, and Campylobacter. These bacteria can contaminate other foods, particularly ready-to-eat foods that won't undergo further cooking. By placing raw animal food in separate bags, you reduce the risk of cross-contamination and keep other foods safe.

(b) Maintain food quality: Raw animal foods can release juices that may contain bacteria. If those juices come into contact with other foods, they can compromise their quality and safety. By separating raw animal foods, you help maintain the quality and freshness of other foods and reduce the potential for spoilage.

(c) Prevent odor transfer: Raw animal foods can have distinct odors that, if not properly contained, can transfer to other foods. Separating raw animal food in a separate bag helps prevent odor transfer, ensuring that other groceries remain unaffected and appetizing.

(d) Comply with food safety regulations: Many grocery stores and food safety regulations require the separation of raw animal foods for safety purposes. These regulations are in place to protect consumers and maintain high food safety standards.

12. **Use Hand Sanitizer after handling Raw animal food, shake many hands, or use the bathroom in the store:** It is advisable to carry your hand sanitizer when you go to stores or restaurants.

13. **Talk to your server about allergies:** Talk to the manager or chef and ascertain ingredients in the food. My wife is allergic to seafood, and we try to ask

questions about it before being served food anywhere. Many people are unaware of the allergens present, and most of these allergens are used in food processing establishments, restaurants, and at home. People can perish for lack of knowledge. Awareness of allergens in food products is important for the safety, well-being, and dietary management of people with allergies or other intolerances. It gives them the opportunity to make good choices, minimize health risks, and maintain a healthy diet. The presence of food allergens in foods can cause a significant danger to people with allergies. Accidental consumption of allergens can lead to allergic reactions, anaphylaxis, and many health conditions. For those with food allergies, methodically seeking ways to avoid specific allergens is an important part of their daily lives. Being informed about allergens helps them select desirable products and meals and avoid contamination during food preparation. No doubt about the fact that allergens can hugely affect an individual's dietary arrangements. For example, a person with a gluten allergy or intolerance should avoid gluten-containing ingredients like wheat, barley, and rye. Consumers with food allergies or intolerances need to ascertain whether a product contains any of the following common allergens: peanuts, tree nuts, eggs, milk, wheat, soy, fish, or shellfish. Amazingly, most processed foods contain most of these known allergens. It is important that you pay attention to labels and the ingredients in the food you buy, particularly if you or a member of your household is allergic to certain foods. The USDA, FDA, and other food regulatory agencies expect the food manufacturers to accurately state the allergens contained in their foods. Consumers want to clearly understand the comprehensive ingredient lists. Furthermore, consumers may want to know whether there is a risk of cross-contamination with allergens. For instance, they may want to know whether a product is made on shared equipment that also processes allergens.

16. Wisely Choose Where You Eat: If a food preparer's outfit, restroom, serving areas, and surroundings are nasty and not adequately maintained, it is likely that the place where the food is prepared is not sanitary and properly maintained.

17Protect Food from Flies & Other Contaminants: Cover your food. To protect your food from flies and other contaminants, here are some effective strategies:

(a) Cover food: Keep all food items covered with lids, plastic wrap, aluminum foil, or food storage containers. This prevents files from landing on the food or laying eggs on it.

(b) Use mesh screens: Install mesh screens on windows and doors to prevent flies from entering your living spaces and kitchen. Make sure there are no gaps or holes in the screens.

(c) Maintain cleanliness: Clean your kitchen regularly, wiping down surfaces and floors to remove any food residue that may attract flies. Pay special attention to spills and crumbs.

(d) Store food properly: Store food items in airtight containers or resealable bags to prevent access by flies and other pests. This is especially important for open packages or items like bread, fruits, and vegetables.

(e) Dispose of garbage properly: Keep your garbage cans tightly sealed and empty them regularly. Flies are attracted to decaying organic matter, so proper disposal of food waste is crucial in preventing infestations.

(f) Don't leave food uncovered during picnics: Avoid leaving food uncovered for extended periods, especially outdoors. If you're having a picnic or outdoor gathering, use food covers or mesh netting to protect the food from flies.

(g) Keep dining areas clean: Whether indoors or outdoors, clean dining areas regularly to remove food particles and spills that may attract flies.

(h) Use natural deterrents: Some natural remedies, such as essential oils like citronella, eucalyptus, or lavender, can help repel flies. Place a few drops on cotton balls and strategically position them around your kitchen or dining area.

(i) Maintain good hygiene: Wash your hands before handling food and encourage others to do the same. Flies are attracted by insanitary odors and can spread bacteria and other contaminants, so proper hygiene is essential.

(j) Consider fly traps or insecticides: If you're dealing with a significant fly problem, you may use fly traps or insecticides specifically designed for fly control. Follow the instructions carefully and keep them out of the reach of children and pets. Remember, prevention is key to protecting your food from flies and contaminants. By implementing these practices, you can significantly reduce the risk of food contamination and maintain a beneficial hygienic environment.

18. Transport Food in Ways to Ensure Food Safety: Transporting food safely is crucial to maintaining its quality and preventing contamination. Here are some guidelines to follow when transporting food:

(a) Use insulated containers: When transporting perishable foods that need to be kept cold, such as meats, dairy products, and prepared dishes, use insulated coolers or containers with ice packs to maintain proper temperatures. This helps prevent bacterial growth and food spoilage.

(b) Separate raw and cooked foods: To avoid cross-contamination, keep raw and cooked foods separate during transport. Use separate containers or bags and ensure they are properly sealed to prevent any leakage.

(c) Secure packaging: Make sure all containers and packaging are tightly sealed to prevent spills, leaks, or contamination from external sources. Use appropriate food-grade containers or wrapping materials that are clean and in good condition.

(d) Maintain proper temperatures: For hot foods, use insulated containers to keep them at a safe temperature above 140 F (60 C) to prevent bacterial growth. For cold foods, keep them below 40 F (4 C) to slow down microbial growth. Use ice packs, dry ice, or frozen gel packs to maintain cold temperatures during transport.

(e) Minimize transport time: Reduce the time between preparing the food and reaching its destination. Plan your transportation route efficiently to minimize travel time and avoid unnecessary delays.

(f) Protect against physical damage: Take precautions to prevent physical damage to the food during transport. Secure containers and packages to prevent them from shifting or falling. Use padding or cushioning materials when needed.

(g) Follow food safety regulations: Familiarize yourself with any local or regional food safety regulations and guidelines related to transporting food. Ensure compliance with any specific requirements, such as permits, licenses, or temperature monitoring devices, if applicable.

(h) Maintain cleanliness: Keep the transport containers, vehicles, and equipment clean and sanitized. Regularly clean and sanitize coolers, bags, and other reusable containers to prevent bacterial buildup.

(i) Monitor temperatures: If possible, use temperature monitoring devices, such as thermometers or data loggers, to monitor the temperature of the food during transport. This helps ensure the food remains within safe temperature ranges.

(j) Handle with care: During transit, handle the food containers and packages with care to minimize the risk of damage or contamination. Avoid rough handling or exposure to extreme temperatures. By following these guidelines, you can help ensure the safety of transported food and maintain its quality until it reaches its intended destination.

19. Manage Your Trash Cans Properly in the Kitchen or Processing Areas: Proper management of trash cans in food processing and kitchen areas is crucial to prevent food contamination and maintain a hygienic environment. Here are some guidelines to help you manage your trash cans effectively:

(a) Separate waste categories: Use different trash cans or bins for different types of waste. For instance, have separate containers for food waste, recyclables, and non-recyclable items. This segregation helps reduce the risk of cross-contamination.

(b) Use color-coded bins: Implement a color-coded system for your trash cans to ensure easy identification of waste categories. For example, use green for food waste, blue for recyclables, and black for non-recyclable waste. Clearly label the bins with appropriate signage.

(c) Secure lids: Ensure that all trash cans have tightly fitting lids to prevent odors, pests, and insects from accessing the waste. This helps maintain a clean and sanitary environment, reducing the risk of contamination.

(d) Regular emptying and cleaning: Schedule regular emptying of trash cans to avoid overflowing and the accumulation of waste. Clean the trash cans regularly using appropriate cleaning agents to eliminate any potential sources of contamination.

(e) Line the trash cans: Always use disposable liners or garbage bags to line the trash cans. This prevents direct contact between the waste and the cans, making it easier to empty and clean. Replace the liners after each use or when they become full.

(f) Proper placement: Position the trash cans away from food preparation areas to minimize the risk of cross-contamination. Ideally, place them in designated waste disposal zones or outside the kitchen area.

(g) Train staff or Family: Educate your staff about proper waste management practices and the importance of maintaining a clean environment. Train them to segregate waste, empty trash cans, and clean them properly.

(h) Pest control: Implement effective pest control measures to prevent insects and rodents from accessing the trash cans. Regular pest inspections and appropriate pest management strategies can help minimize the risk of contamination.

(i) Monitor and enforce guidelines: Regularly monitor the trash can area to ensure compliance with waste management guidelines. Enforce the proper disposal practices among your staff and address any issues promptly.

(j) Stay updated with regulations: Keep yourself updated with government health and safety regulations regarding waste management in food processing and kitchen areas. Adhere to these regulations to maintain compliance and ensure the safety of your facility. By following these practices, you can effectively manage your trash cans and minimize the risk of food contamination in food processing and kitchen areas.

20. **Cover Your Hair Properly if you prepare food**: Covering your hair properly when working around food is necessary for several reasons:

(a) Hygiene: Hair can carry dirt, nonfood oils, and bacteria, which can contaminate food if they come into contact with it. Properly covering your hair helps prevent these contaminants from falling into the food, reducing the risk of foodborne illnesses.

(b) Compliance with food safety regulations: Many food safety regulations and guidelines, such as those set by the United States Department of Agriculture (USDA), Food and Drug Administration (FDA), and local health departments, require food handlers to cover their hair to maintain hygienic practices in food preparation areas.

(c) Prevention of physical contamination: Even if your hair is clean, loose hair can easily fall out and end up in the food. This can be unappetizing to customers and can pose a choking hazard. Proper hair coverings, such as hairnets, caps, or hats, help contain the hair and prevent it from falling into the flood.

(d) Professional appearance: Food service establishments often have guidelines or dress codes that require employees to maintain a professional appearance. Properly covering your hair is part of maintaining your appearance and projecting a clean, well-groomed image to customers. By following these practices and guidelines, you help create a safe, hygienic environment in the food service industry, reducing the risk of food contamination and promoting customer satisfaction.

21. Wear Hand Gloves & Don't Expose long painted Nails: Covering long and painted nails is important for food preparers to prevent contamination for the following reasons:

(a) Bacteria and pathogens: Long nails can harbor dirt, bacteria, and pathogens, even if they are washed regularly. This poses a risk of transferring these contaminants to the food during food preparation. Bacteria can multiply rapidly in the warm, moist conditions found in the kitchen, potentially causing foodborne illnesses.

(b) Hygiene: Maintaining good hygiene is crucial in food preparation to prevent the spread of harmful microorganisms. Even with regular handwashing, debris and bacteria are more likely to accumulate under long nails, making thorough cleaning difficult. By covering the nails, the risk of contaminating food is significantly reduced.

(c) Foreign objects: Painted nails, especially if the paint is chipped, can contribute to foreign objects getting into food. Nail polish or acrylic nails can chip off and mix with food, posing a choking hazard or compromising the food's quality and safety.

(d) Regulatory requirements: In many countries, food safety regulations and standards require food handlers, including food preparers, to maintain proper personal hygiene practices. These practices often include guidelines regarding nail length and covering to minimize the risk of contamination.

To maintain food safety and prevent contamination, it is advisable for food preparers to keep their nails short, clean, and free from nail polish or acrylics. Additionally, wearing disposable gloves or using finger cots can provide an extra layer of protection when handling food.

(22). Speak up and Protect our Health: If food workers or servers are doing inappropriate things related to food contamination, speak up. If someone serving your food touches their face with bare hands and then handles your food, speak up. Speak on behalf of those who cannot read or comprehend food safety matters. Ensuring food safety for all individuals is crucial for our well-being. Here are a few measures that can be taken to help them:

(a) Visual Aids: Use clear and easily understandable visual aids such as pictures and symbols to convey important food safety information. For example, you can use images to illustrate safe handling practices, such as washing your hands or storing food at the correct temperature.

(b) Multilingual Materials: Develop food safety materials in multiple languages to cater to individuals who may have limited literacy skills in a particular language. Translate important food safety information into the community's commonly spoken languages.

(c) Pictorial Menus: If you're providing food services in a setting like a cafeteria or restaurant, consider using pictorial menus. Instead of relying solely on written descriptions, include pictures or illustrations of the dishes to help individuals make informed choices based on visual cues.

(d) Hands-On Demonstrations: Organize practical demonstrations or workshops to teach basic food safety practices. This can involve hands-on activities like showing how to wash hands properly, store food correctly, or identify signs of spoiled food.

(e) Community Education Programs: Collaborate with local community organizations, health centers, or social service agencies to conduct food safety education programs specifically targeted at individuals with limited literacy skills. These programs can include interactive activities, demonstrations, and discussions to enhance understanding.

(f) Audio and Video Resources: Create audio and video resources that explain food safety practices in a clear and engaging manner. These resources can be distributed through community centers, public libraries, or online platforms accessible to everyone.

(g) Peer Education: Train individuals with better literacy skills, such as family members, friends, or community leaders, to act as food safety ambassadors. They can help disseminate important information and provide guidance to those who have difficulty understanding food safety matters.

(h) Clear Labeling: Encourage food manufacturers to use clear and simple labeling practices. This includes using large, easy-to-read fonts, color-coded labels, and providing essential information, such as allergen warnings and storage instructions, in a straightforward manner.

(i) Engage Health Professionals: Involve healthcare providers, nutritionists, nurses, government officials, and dietitians in educating individuals about food safety during routine check-ups or consultations. These professionals can offer personalized guidance and answer any questions individuals may have.

(j) Ongoing Support: Establish mechanisms for ongoing support, such as helplines or community hotlines, where individuals can seek advice or clarification about food safety matters. Having a reliable resource to turn to can be valuable for those who struggle to understand written information. Remember, it's important to tailor these approaches to the specific needs and cultural context of the community you are working with. By implementing these measures, you can help individuals with limited literacy skills better understand and practice food safety.

CHAPTER TWELVE

Udo's Food Safety Songs(UFSS)

1. The Clean Hands Song

Wash your hands before you eat,
Make it a delicious and safe treat.
Soap and water, clean and bright,
Help protect you both day and night.

Tiny germs we cannot see
May live in your food free.
But careful habits every night & day
Can keep food borne disease away.

Separate the cooked food from the raw,
Respect government's safety and its law.
Keep foods cold and cook them right,
Guard your family morning and night.

All parts of your hand must be clean,
Smart safe practice is the way we lean.
Appreciate love, order, truth, and care,
And health is a blessing we all share.

2. **Wisdom in the Kitchen**

The kitchen is a special place
Where love and labor interface.
A meal prepared with love & a clean hand
Brings joy and strength throughout the land.

But food pathogens hide in invisible ways,
In foods left warm for many days,
In dirty surfaces and utensils they roam,
seeking to bring sickness to a happy home.

So, guard the pots and guard the flame,
Cook each food well & label by name.
Check temperatures, inspect each part,
Safety is always the wisdom & the art.

A wise or knowledgeable food worker
understands
Food safety is synonymous with clean hands.
Each food safety decision that we make
Can spare a life from pain and ache.

3. **Always Guard Your Food Wisely**

Always guard your health with wise food safety
care.
For biological, chemical, & Physical hazards are so
unfair.
The food we buy, the meals we share,
Require clean hearts and actions so fair.

From farm to market, dock to store,
Food safety matters more and more.
The farmer, processor, cook, inspector, too,
Must faithfully perform their expected due.

Beware of bacteria, parasites, and viruses
molds, chemical, & physical hazards crisis.
Be careful, for Invisible threats may still arise
Beneath those appealing smells and eyes.

Get wisdom, for it is a shining light
That helps us choose and act aright.
Cook meats well and chill foods fast,
And make good safety habits unsurpassed.

Don’t condone an insanitary condition to please an ego
Food Safety practices are something everyone should know.
Our home, schools, and church teach through standard and grace
To ensure we keep food safety habits in their deserved place.

So let food safety lead the way
In every kitchen and every day.
Protecting health through meaningful food safety practices
Showing love for God and people, & avoiding safety malpractices.

4. Help us, God of Safe Foods

Father God, with thankful hearts we pray,

For every meal before us presented today.

Guide our hands in wisdom's light,

To keep all food wholesome, pure, and right.

Protect our tables from sickness and harm,

Surround our homes with heaven's calm.

Fill us with love that forever stays,

And cover us with grace through all our days.

God, pull us from intentional & unintentional contaminants

Protect our plates from greed's unsafe determinants

Let honesty restore the seeds and grain we trust,

Ahaaa! guide us toward meals that are nutritionally robust.

AUTHOR

Dr. Udo F. Ufomadu worked as a Consumer Food Safety Protection Specialist for the Alabama Department of Agriculture and as a United States Department of Agriculture (USDA) Inspector in Charge (IIC) for 31 years. He is USDA-trained with certifications in Food Safety Regulatory Essentials, Hazard Analysis and Critical Control Points (HACCP), Rules of Practice (ROP), Sanitation Performance Standards (SPS), and Sanitation Standard Operating Procedures (SSOP).

He has a certificate in Science of Well-being from Yale University; A certificate in Business Improving Through Health from Harvard University; and A certificate in Processed Food Inspection from USDA/FSIS (Food Safety Inspection Service) at Texas A&M University, College Station.

He is a Food Safety Consultant with Ufomadu Consulting. He is also a member of the World Health Organization's Food Systems Community of Practice. He is the author of 5 Essential Habits, Light, Shine, How to Become Successful in Business, Ways of Valuable People, Anthology of

Inspiration, and Secrets of Elevation. He is a member of the Institute of Food Technologists and Tabernacle of Praise Church. He is an educational advocate who served as a member of the Selma City Board of Education & TOP school Board.

ACKNOWLEDGEMENTS

I thank Almighty God for granting me the grace to make this possible. I also thank Him for putting food establishments and food regulatory agencies along my path. Food establishments like Nigerian Breweries Limited, Coca Cola Bottling Co., Country's BBQ, Super Brand Frozen Food, & Food safety Regulatory agencies like Imo State Dept. of Trade & Industries, Alabama Department of Agriculture & Industries, and United States Department of Agriculture, Food Safety and Inspection Service (USDA, FSIS). Special thanks to my wife, Rita; my children, Evelyn/Joe, Joy, Ezekiel, and Godson; my grandchildren, Prince & Amias; My brothers, Ike and Iheanyi; my sister Joy Kalu, Umuohuonu family, Ufomadu family, and all other physical and spiritual family members and friends for their ongoing prayer & support.

REFERENCES:

Centers for Disease Control and Prevention (2023).Estimates of Foodborne Illness in the United States. https:// www.cdc.gov

Columbia University Irvin Medical Center (2023). Food Safety: Serving. Columbiadoctors.org

Cornell University (2021). Food Safety & Storage. https:// orleans.cce.cornel.edu

Eatingwell.com(2025).

Food safety Inspection Service (2023). Food Safety Basics. https://www.fsis.usda.gov

Harvard University (2023). Environmental Public Health Food Safety Program: Keeping Safe. https://ehs.harvard.edu

Princeton University (2023). Tips for Safe Food Practices. https:// insidefacilities.princeton.edu

Science Arena Publications (2018). Chemical and Microbial Evaluation of 'Ogiri'. *Specialty Journal of Biological Sciences*. ISSN:2412-7396

WHO Food safety: https://www.who.int/health-topics/food safety

Yale School of Medicine (2012). Connecticut's Food Safety Program strengthened by YSPH. https://medicine.yale.edu

C. Arndt

Information Measures

Information and its Description in Science and Engineering

1st Edition 2001, 2nd Printing

With 64 Figures

Springer

Dr.-Ing. Christoph Arndt
Universität Siegen
Zentrum für Sensorsysteme
Projektbereich II
Paul-Bonatz-Str. 9 - 11
57068 Siegen, Germany

E-mail: arndt@nv.et-inf.uni-siegen.de

ISBN 3-540-40855-X Springer-Verlag Berlin Heidelberg New York

Cataloging-in-Publication Data applied for
Bibliographic information published by Die Deutsche Bibliothek.
Die Deutsche Bibliothek lists this publication in the Deutsche Nationalbibliografie;
detailed bibliographic data is available in the Internet at <http://dnb.ddb.de>

Springer-Verlag Berlin Heidelberg New York
a member of BertelsmannSpringer Science+Business Media GmbH

http://www.springer.de

Typesetting: Digital data supplied by authors
Cover-Design: design & production GmbH, Heidelberg
Printed on acid-free paper 62/3020 hu 5 4 3 2 1 0

TO MY PARENTS

NOTRE ÂME EST JETÉE DANS LE CORPS,

OÙ ELLE TROUVE NOMBRE, TEMPS, DIMENSIONS;

ELLE RAISONNE LÁ-DESSUS ET APPELLE CELA NATURE, NÉCESSITÉ,

ET NE PEUT CROIRE AUTRE CHOSE.

(Pensées, Blaise Pascal)

Preface

This book is intended to be an introduction to the mathematical description of information in science. The necessary mathematical theory of this introduction will be treated in a more vivid way than in the usual theorem–proof structure. This, however, enables us to develop an idea of the connections between different information measures and to understand the trains of thought in their derivation, which is a crucial point for correct applications. It is therefore our intention in the mathematical descriptions to evolve the important ideas of the derivations, so that we obtain the resulting functions as well as the main thoughts and the conditions for the validity of the result.

This simplifies the handling of the information measures, which are sometimes hard to classify without any additional background information. Though the mathematical descriptions are the exact formulations of the measures examined, we do not restrict ourselves to rigorous mathematical considerations, but we will also integrate the different measures into the structure and context of possible information measures. Nevertheless the mathematical approach is unavoidable when we are looking for an objective description and for possible applications in optimization.

Communication theory, the original science of telecommunication (i.e. the processing and transmission of information), and physics, the science that established the concept of entropy, are the foundations, serving to explain the idea of information. As there exist a great number of different possible ways to describe information, we are going to present these measures in a coherent description. Some examples of the information measures examined are: Shannon`s information, applied in coding theory, Akaike`s information criterion, used in system identification to determine auto-regressive models and in neural networks to identify the number of neurons, and Cramér–Rao bound or Fisher`s information, describing the minimal variances achieved by unbiased estimators. This coherent description enables us to obtain a more complete overview of the different approaches and a better understanding of the idea of information and the related mathematical descriptions.

In the context of the great number of information functions, which currently exist, one may ask for reasonable applications of all these information measures. Does it make sense to generate additional degrees of freedom by creating more parameters to the existing information functions? Would it not be more reasonable to use the existing information measures, which we are able to understand and to apply much better than the extensions created by the additional parameters? Of course, additional parameters can be used to match certain information measures to certain kinds of problems, but do we need a separate information measure for each kind of problem?

Here the user has to decide between the different possible information measures, which we are going to introduce. From the users point of view, we should argument with Occams razor not to multiply the entities (degrees of

freedom) in an unnecessary way. The more complex the information measure becomes the more we loose the ability to understand and to interpret the results of our computations. So the interpretation of the results of the computations encourage the use of the least possible number of degrees of freedom, which we should permit in an applicable information measure.
Based on this reflection, we are going to introduce several information measures and examine their connections. But the examination of information transmission and the possible optimizations are demonstrated by using the main important, i. e. the most common and uncomplex information measures, which already allow us to draw a lot of conclusions for signal processing applications.

Table of Contents

Symbols, expressions and abbreviations

$A(k)$ = local state transition matrix

$C(k)$ = observation matrix

$D_1(P\|U)$ = Kullback–Leibler distance

$E\{\ \}$ = expectation function

$\delta(\bullet)$ = Dirac impulse

$\underline{f}[\underline{x}(k),k]$ = nonlinear state transition function

$f_x(\xi)$ = scalar distribution density function

$f_{\underline{x}}(\underline{\xi})$ = n-dimensional distribution density of the multi-dimensional random variable $\underline{x}$

$f_{\underline{y}|\underline{x}}(\underline{\rho}|\underline{\xi})$ = conditional distribution density function (likelihood function)

F = Fisher's information matrix (stochastic observability)

G_α = G-divergence (Rényi)

$G_1 = D_1(P\|U)$ = Kullback's minimum discrimination information (Kullback–Leibler distance)

$h(A)$ = relative frequency of the event A

$H(X)$ = entropy of the discrete random variable X

$H(X,Y)$ = joint entropy of the discrete random variable X and Y

$H(X|Y)$ = equivocation

$H(Y|X)$ = dissipation

$H(X;Y)$ = transinformation of the discrete random variable Kolmogorov's information

$I(\underline{x},\underline{y}) = S(\underline{x};\underline{y})$ = Kolmogorov's information (transinformation)

i.i.d. = independent identically distributed

iff = if and only if

$l = f_{\underline{y}|\underline{x}}(\underline{\rho}|\underline{\xi})$ = likelihood function

$L = \ln l$ = log-likelihood function

ME estimation = maximum entropy estimation

ML = maximum likelihood

MLE = maximum likelihood estimate

$P = \{p_1, p_2, ..., p_n\} \in \Delta_n^0$ = probability distribution, where certain probabilities may be equal to zero

$P = \{p_1, p_2, ..., p_n\} \in \Delta_n$ = probability distribution, where no probability of a single event is equal to zero

$P^-(k)$ = covariance matrix of the prediction error

$P^+(k)$ = covariance matrix of the estimation error

P = realization space of the random variable $y(k)$

ρ = dummy variable of the random variable y

$\Phi(k+1,k)$ = discrete state transition function

Φ_i = entropy function

$\Phi(f_{\underline{x}}(\underline{\xi}))$ = concave function in maximum entropy estimation

$Q(k)$ = driving noise covariance matrix

$R(k)$ = covariance matrix of the measurement error

$\Re$ = set of real numbers

$\Re^n$ = n-dimensional space of the real numbers

S_1= Shannon's information (Shannon's entropy, negentropy)

S_α = Rényi's α-information (α-entropy)

$S(x)$ = entropy of the continuous random variable x

$S(x,y)$ = joint entropy of the continuous random variables x and y

$S(x|y)$ = equivocation

$S(y|x)$ = dissipation

$S(x;y) = I(x,y)$ = Kolmogorov's information, transinformation

σ = standard deviation

σ_A = sigma-algebra

σ_B = Borel field

T = sample time

tr{•} = Sp {•} = trace of a matrix

$\underline{v}(k)$ = measurement noise

$\underline{w}(k)$ = system noise, driving noise

WLS = weighted least-squares

$\underline{x}(k)$ = state vector

Ξ = realization space of the random variable x

ξ = dummy variable of the random variable x

$\underline{y}(k)$ = measurement vector, random vector in the observation space

Z = set of integers

|•| = det(•) = determinant

||•|| = absolute value of the Jacobi determinant

Abstract

The aim of this book is a description of the possible ways we have to measure information with mathematical formulations. In order to understand the idea of information, we begin the intended examination with some thoughts about the concept of information, where our everyday experiences show us that we continually deal with information and transmission of information. One reason for this presence of information is that all living beings secure their life by receiving information about their environment, enabling them to detect food and danger ("information is the very essence of life"; H. Haken). In spite of the importance of information, we have not been able to find a definition of this concept. One reason for this lack of definition is the different levels we need when we examine the information of a message, an object, etc. The mathematical description of information, used for example in communication theory, can only provide a description of a restricted number of levels. Therefore the concept of information, as described by the mathematical measures, can supply us with only a reduced insight into the whole idea of information. Nevertheless these restrictions, which result from leaving out of consideration the subjective part of information, are justified by the common method of science, trying to eliminate the human being as an incalculable source of error from its experiments. Such a subjective aspect of information can be either complicated to implement or very inaccurate to model.

A possible way out could be the use of the effect caused by the received information and the description of the information processing by means of this effect. This could easily be done by sending a message to someone and by judging his reaction afterwards. But this simple information transmission already consists of so many different steps of information processing that one would only describe an amalgamation of these different steps of information processing. (These different steps are: The formulation of the intended reaction = semantic information, transmission of the message to the receiver = syntactic information, receiving, processing and conversion of the message = pragmatic information.) Afterwards we need an evaluation on the basis of certain criteria, where we have to judge the comparison between intention and reaction. This, however, is an easy task because of the objective criteria that can be applied. But the crux of the transmission of information lies instead in the necessary interposed steps and the transformation of information that has to be done in this transmission. Exactly at these interposed steps we require a description of the details, which are the fundamental basis of modeling practical realizations. With the chance to optimize single details in a suitable way, the entire system will likewise profit from this optimization. This, however, requires a comprehension of the concept of information, which at least permits a limited definition of information. When we are able to define at least a part of the concept of information, we are also able to realize applications, based on this partial understanding, and thus to gather additional information about the concept of information via learning from doing. So we are able to extend our comprehension of this abstract concept in special regions by constantly scrutinizing our actions with regard to the information. If we only optimize our systems with respect to the rate of transmission, we do not need to think about the concept of information, because the boundary conditions (bandwidth, time, and energy) impose a limit on the intended optimization, which has to be adjusted with respect to these conditions. There is no need to understand the concept of information to be able to do such an optimization. So this aspect has the advantage that we are able to construct and optimize information transmission systems, though we do not have a complete idea of what information actually is.

The complexity of information evidently causes great difficulties concerning a definition, and this limits the range of applications of this concept. Nevertheless, we are able to comprehend certain parts of the concept of information and to apply them in scientific theories (coding theory, estimation theory, cryptology, ...). Thus it has been possible to develop a theory of information in science, which presumes the comprehension of certain facts and allows us to use this comprehension.

We thus approach the concept of information in an inductive way by benefiting from available, reproducible experiences, which we are able to gain in a limited area of the field of information. As such an inductive strategy is merely based on subareas, which can be verified by experience or experiments, the resulting theory has to be continually developed in order to include all changes gathered by additional experience. Within the scope of those experiences many information measures and other options for a description of information in communication systems have been developed over the past 60 years. They all extend the existing theory and do, because of their great number, meanwhile even provide a complex accumulation of quite different functions. So it becomes hard to tell whether an appointed function serves to describe, understand or even solve a given problem in the desired way. Such a variety is, however, the result of the inductive approach, which – according to Popper [POP94] – cannot lead to a complete characterization of the problem, because of the necessary restriction to special cases (see also chapter 15: Maximum entropy estimation).

A deductive approach, reducing from a description of the complete subject to descriptions of certain, limited areas by adequately adapting the boundary conditions, would require a characterization of information, which is less amenable to mathematical options. But this is not necessarily a negative property of information, because every mathematical system is incomplete, so that we are not able to prove all statements of a mathematical system with its own inherent resources [HOF91]. If one changes the level of description of the system (e.g., from the syntactic information to the pragmatic information), it is possible to obtain additional insight that has not been achievable before. So the different levels of information supplement each other and we use these chances to verify and interpret the results that we obtain in our mathematical treatment of information.

Communication theory, the classical science of transmission and processing of information, composed the following definition of information in the NTG (Nachrichtentechnische Gesellschaft) and in the ITG (Informationstechnische Gesellschaft) in the year 1982 (NTG 0902 Empfehlungen 1982 [DIE86])

Information is:

Pursuant to the common linguistic usage a statement – in the furthest sense of the word – about subjects, facts, procedures and similar.

Remark 1: Information forms the substance of a message.

Remark 2: Information is handled by a communication device as

- continuous in time and value (analog) signals, e.g., music, speech, ...

or as

- discrete in time and value (digital) signals, e.g., data, text, which are taken over at the input and delivered at the output.

Remark 3: In the area of information theory or communication theory the word information is exclusively used in the singular.

In 'IEEE Standard Dictionary of Electrical and Electronic Terms' we find the definition:

Information (general): The meaning assigned to data by convention.

Structure and Structuring

Initially we will discuss several aspects of the concept of information in order to create an information function on the basis of intuitive requirements. In doing so we will derive the character of an information function out of these intuitive guidelines in chapter 2 and subsequently we consider different scales and the related effects on the information functions by using Shannon's information as an example of an information function.

In chapter 3 we will examine some historical milestones in the development of information theory and arrive at Shannon's information, which is the foundation of information theory. The character of this measure of information is more that of an entropy, and measures of entropy have been known in physics since Clausius [CLA96], who was the first to use the concept of entropy (εντρεπειν = to transform, to change). Having introduced Shannon's information, or Shannon's entropy, we will then undertake a short excursion into thermodynamics in chapter 4. Here we take a look at the concept of entropy in the classical, thermodynamic way and at the description used in statistical mechanics.

After this excursion into entropy we turn towards some generalized forms of entropy measures, which partially stem from Shannon's information. Because of the large number of different generalized entropy or information measures, we are able to present only a short enumeration of a few generalized measures, since many scientists have been engaged in finding such measures since 1960. The result of this research is a multitude of entropy measures with different properties that can be applied to specific problems ([TAN89], [KAP89], [KAP94], [COV91], and [CSI81]). We begin this overview with Rényi's information, which is a generalization of Shannon's information, and continue with a look at measures describing an information difference. In chapter 6 we proceed with a short introduction of several other generalized measures of information, which are not direct generalizations of Shannon's information. Next we examine measures of information and measures of information differences.

This is an important difference, as the absolute measures of entropy assign an entropy value to one probability distribution, whereas the relative measures of entropy or information assign one information to two probability distribution densities. This differentiation can be understood as between absolute and relative entropy measures, where the relative measures are known as measures of divergence or measures of distance, because they describe the information difference between two distribution densities. As an example of such measures we start in chapter 7 with Rényi's divergence, which is already a generalized version, consisting of an additional parameter, and proceed in chapter 9 with the derivation of the more familiar Kullback–Leibler distance and its special case, the Kolmogorov information.

A remarkable difference between absolute and relative information measures appears in chapter 8, when we turn towards coordinate transformation and take a look at the transition from discrete to continuous distribution densities. Here we verify the intuitive requirements that we placed on the information function in the introduction.

Furthermore we notice that the information functions always produce scalar values, even though the problem and the corresponding probability distribution function may be multi-dimensional. Thus information evidently is a quantity without a direction. It produces only a scalar value to be assigned to a system or a variable. A direction of information can be evaluated by applying the directed measures of divergence or measures of distance, describing the replacement of a distribution density by another probability distribution density with a scalar measure as well. The most familiar measure of distance is the Kullback–Leibler distance, examined in chapter 9, which describes the difference between two probability distribution densities.

In chapter 10 we find a measure of information, described by a matrix. It can be used to derive Kullback's information with a scalar result as a generalization of Shannon's information. The matrix measuring the information is known as Fisher's information and is closely related to the achievable bounds of estimators, which we are going to derive in later chapters. Here we show some important properties of this matrix, to obtain an impression of this different measure of information. This matrix of information not only appears in Kullback's information, but also is the inverse covariance matrix of the well known Cramér–Rao bound, building the lower bound of the variance of an unbiased estimator. Motivated by this relation between the scalar information measures and the information matrix as a bound of variance, we will present bounds to the uncertainties of estimators, which lead us to other multidimensional measures that can be used to describe the information of systems. Uncertainty and information are closely related, as we will see when we examine the remaining uncertainty of estimators. In this examination of the estimation we find theoretical boundaries, where the estimator cannot fall below these boundaries, and which therefore limit the maximal gain of information. As we can find different boundary conditions for the estimation process, we are also able to find different bounds, depending on the conditions (e.g., biased – unbiased, do the regularity conditions hold?). Furthermore we need not restrict ourselves to second order moments, which are bounded by Fisher's information or the Cramér–Rao bound, which is directly related to measures of the Shannon kind. It is, of course, also possible to find boundaries for higher order moments. Some of them are presented in chapter 10 to complete the overview of these information bounds.

Schweppe [SCH73] calls a part of Kullback's information, which we used to derive Fisher's information and the ambiguity function. The name, however, already indicates the possibility to describe the ambiguities commonly existing in inverse nonlinear mappings, which we often find in estimation scenarios with nonlinear observation models. When we solve such problems with nonlinear observation functions by application of the maximum likelihood estimation, we find several maximum likelihood estimates and thus get the ambiguities, which can be described by the ambiguity function. This function, which we investigate in chapter 11, once more shows a relation to Kullback's information and Fisher's information, describing the width of the maximum likelihood maximums and thus the quality of the maximum likelihood estimate.

This relation of these functions to the maximum likelihood estimation can then be used in chapter 12 to develop the so-called information criteria. These criteria are utilized to determine the necessary parameters in an ARMA (auto regressive moving average) or in an AR (auto regressive) modeling or the number of neurons in neuronal networks. This chapter leads us to the information criteria AIC (Akaike's information criterion), BIC (Bayes's information criterion), HQ (Hannan–Quinn criterion) and others, for example FPE (final prediction error), which we describe in a short overview.

Having introduced the measures of information, we turn to chapter 13, where we examine several relations as well as information measures of second order when we consider the communication channel. The communication channel, the classical description of the transmission of information from a source to a receiver, is used to introduce joint entropy measures and conditional entropies, used to characterize the channel.

With these entropies at hand, the introduced measures of information are explained and applied to the transmission of information from source to receiver. As such a transmission of information is present in all systems, this model, which is completed by a state space model, can be applied to all imaginable arrangements. On the one hand it allows an interpretation of the information measures, which can be evaluated via empirical experiences. On the other hand it provides some insight into the concept of information, resulting from the application of the information measures to such a system. This model thus extends the comprehension of what information is or of what the information measures describe.

Chapter 14 aims at the characterization of state space models with information measures, and we will find that information essentially depends on the surprising part of the message. This surprising part is the new, the informational part in the message.

In chapter 15 we describe estimators, operating on the basis of entropy measures, and this estimation finally results in the option to generate an 'arbitrary' information function. The method that allows us to do this is the maximum entropy estimation developed by Edward Jaynes [JAY83]. It is based on the idea of choosing the optimization function in such a way that no further restrictions be implicated with respect to the possible result. Maximizing the measures of entropy leads to a maximum uncertainty in the result obtained by this optimization. The result, or the uncertainty, can then be diminished by additional boundary conditions, which serve to adapt the solution to certain problems. But the result, obtained via maximum entropy estimation, is the most general result that can be achieved with the given restrictions and without any additional implicated restriction by the optimization process. This then serves as the basis to develop not only distribution densities, which are the results of the optimization of the maximum likelihood estimation. But it also enables us to determine entropy functions for given distribution densities, when we interchange the given boundary conditions.

In the last chapter we briefly summarize the results, discuss some relations, have a look at some areas of information theory and show some current fields of research and topics that are of interest today. As information theory is used in many different sciences, this summary can only be a limited selection of many topics in this interesting field of research.

Information theory is therefore something like cybernetics, which Norbert Wiener presented as the science to build a connection between the existing sciences. Cybernetics was thought to fill the white spots on the map of science to improve communication between different disciplines and to promote synergy. Information seems more and more to be a method to be used by different disciplines and different areas of science, as it is not limited to a certain discipline. The only question that frequently arises is that of suitable definition of this concept information to achieve a standardized utilization.

The classical definition of information, as we use it in communication theory and as we can define in an objective way, is the main point of this book, which tries to give an introduction to and an overview of these information measures.

1 Introduction

Information is an essential component of the transmission of messages or knowledge and it was being used with or without awareness in our everyday life long before computers made their way into our life. Every organism that wants to survive in its environment has to incorporate information concerning food, energy or, more generally, information with respect to environmental conditions, which is necessary for it to exist (independently). This starts with simple monads, becomes more important for crops and animals and finally culminates in human beings, who are the only creatures able to communicate abstract concepts and ideas, because of the development of a complex language.

The transmission of information between human beings (i.e. the transmission of commonly acknowledged characters, signs, symbols, sounds or other patterns, which can be detected by our senses) has developed from simple reactions to signs up to our complex languages. The instinct of herds to flee, because of a certain behavior pattern of single members, is an example for such simple reactions, transmitting information from one member to the whole herd. As soon as one animal shows a certain behavior pattern, or certain reactions, the herd responds accordingly. Similar to this behavior of animals, we are able to orient ourselves in foreign countries without any knowledge of the language by merely observing gestures and reactions of the people. This gives us at least rudimentary information about the mood of these people.

It is possible that we developed our language during our history out of such rudimentary transmissions of signals. Our lack of ability or inability to adapt to the environment possibly required – contrary to the animals, being masters of acclimatization to their environment – an increasing communication to secure our lives [POP97].

Simple sounds became complex phonemes and finally culminated in our languages. During this process abstract concepts arose, which designate no immediately subsumable objects, but abstract meanings or values. One of these abstract concepts is the concept of information.

The spatial and temporal boundaries of the language have been minimized during our history via several different methods, which we use to transmit data from source to receiver. Light, smoke, knots in ropes, etc. and finally script have been used. Because of these methods of transmission of information it was possible to eliminate spatial and temporal barriers and we are thus able to transport or conserve our knowledge over great distances and over a long time (archaeology). In this context we distinguish information and knowledge, which is sometimes called compressed or condensed information. Knowledge is gained by the reduction of some a priori uncertainty or inaccurate knowledge with information. The evaluation of the received information and its transformation into knowledge assumes some a priori knowledge, which is necessary to generate a correlation with the received information. This correlation is fundamental for the analysis and the utility of the information. The resulting question arising from this consideration concerns the basis of our knowledge. But this question leads us to philosophical problems (epistemology, rationalism, Descartes: cogito ergo sum, empiricism, Hume, Locke: tabula rasa, nihil est in intellectu, quid non ante fuerit in sensu), which we are not able to discuss in this volume. Instead we occupy ourselves with the scientific, mathematical aspect of information and thus restrict our attention to those characterizations that are familiar in scientific approaches.

Science and information

Knowledge as the result of science is the essence that we gain through information about our environment. We gain information about nature via empirical investigations (inductive strategy) or via general theories and their verification (deductive strategy). On the basis of these theories the gained information can then be transferred to applications. We thus employ our experiences to gain knowledge. We do this either by searching inductively for the reason for a special experience or by deductively choosing the boundary conditions to apply a general theory to a special problem. Thus experience essentially differs from knowledge, because experience does not search for a reason, but only notices the facts [ARI94]. Nevertheless we connect both concepts with the reception of information. Experience thereby is a collection of impressions and experiences, while knowledge is the conscious combination (processing, fusion) of these experiences. We can thus think of experience as an accumulation without necessary preliminaries, whereas knowledge is based on an a priori knowledge. This a priori knowledge can emerge from experience (empiricism) and consideration (epistemology, rationalism) and it presumes an evaluation and a comprehension of the received information. This comprehension is the main difference from experience, which we simply adopt without further reflection.

Brillouin [BRI56][BRI64] contemplates the information contained in a complex machine from exactly this point of view. The improvement of a theory or the construction of a complex machine requires the design of regularity (order) in the form of a complex structure. This diminishes the confusion (disorder) of the system by application of gained knowledge and we are thus able to construct a complex machine, where the construction 'contradicts' the second law of thermodynamics. This second law says that the entropy cannot decrease in closed spaces. Anyhow, our knowledge enables us to construct complex machines and to generate specific order. The actual function of this machine, of course, obeys the second law of thermodynamics.

Karl Popper [POP97] divides the world, as we observe it, into three areas:

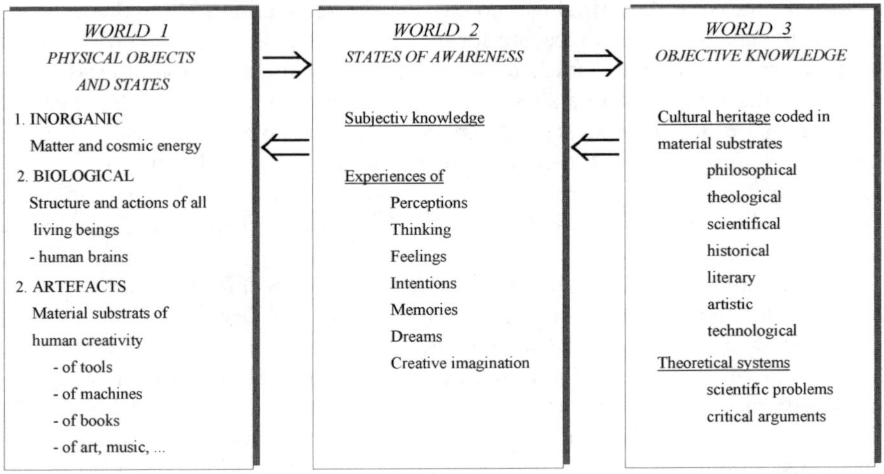

Fig. 1.1: Popper's classification of the perceived world (This figure is taken from: Facing Reality, Springer-Verlag, New York/Heidelberg/Berlin, 1970)

The first, material area (world) contains all physically detectable quantities. the second area applies to our states of awareness and the third area finally gathers the world of our theories, which are the man made abstractions of our real world. This third world is immediately connected to a person, as the second world is, too. Man can surely influence the third world, and as a person one also has access to this world, but it is a purely abstract world consisting of accumulated knowledge. Furthermore there is a constant transition of information between these three worlds. Material events can cause certain sensations on the one hand, but they can also serve to verify theories by empirical experiences. On the other hand one can be so fascinated (world 2) by a theory (world 3), that this fascination results in some application (world 1) of the abstract theoretical idea. So there is a continual interchange of information between these completely different worlds and thus between systems that can and/or cannot be described by physical equations. So, do we have any chance of defining a consistent, uniform concept of information? We certainly find that information cannot be connected with a certain physical variable, but it has to be some measure 'behind the physical process', as it interacts with physical as well as with mental levels of our description of the world. In addition, it is very interesting that we are able to communicate on the level of world 3 (theories) and that it is possible to exchange and gain information between the levels of theory to reduce uncertainty and entropy. This is achieved as the transition of information creates order in the theories and reduces disorder.

Does the world of theories have to obey the second law of thermodynamics at all? Well, there is certainly no direct relation between world 3 and the second law of thermodynamics, because theories are theoretical forms, conceived by our minds, and they therefore are not measurable by physical devices. But if we construct a device, an ordered structure, by applying an abstract theory, then we are able to measure the difference created by the realization of this theory. This gives us the chance to get a measure for the theory, or at least for the utilization of the theory, as we are able to measure the progress in some operation process.

In a similar way we may also measure the learning of robots. Such a machine, which is able to incorporate information, can be described with absolute information measures. But the absolute result achieved will not give us any information as long as we are unable to classify this information without a reference. Therefore we always need some reference to classify the information gained.

We may never be able to find an absolute measure of information, because of the absence of an absolute reference and because of our description of nature by the use of abstract models. Nature would be a reference in modeling, but we are only able to verify our model with measurements and comparison with nature. As the uncertainty relation demonstrates that exact measurements are not feasible for us, we can drop the idea of any exact modeling. If we were able to find such a model, we would have to ask questions about the information contained in this model. How can we measure this information? Do we use a quantitative or a qualitative measure to describe or characterize this model? Where is the information in this model? Is it contained in the dynamical processes or in the parameters or the states? Moreover, nature constantly changes. It continually 'tries' new forms and it therefore cannot be described by a time-invariant modeling. But we cannot even build a time-variant model of nature, because of the enormous number of degrees of freedom that are possible for such a system.

The abstract, complex concept of information is therefore similar to many other complex concepts (knowledge, thinking, education, ...) that we use to describe our everyday experience. We use them in the same way, as we learned them by experience, without thinking about their exact definitions.

Since the early days of computers, the problem of the optimal processing of information (thinking of the computer) has constantly been a matter of research. How should we design a machine, which will at least reproduce rudimentary functions of our brain? The main problem with this question is that, though we use certain concepts to describe the actions of our brain, we are not at all capable of defining these concepts so exactly that we are able to design a device based on these definitions. As an example of this statement we use the question: "What is thinking?" John von Neumann answered: "Define the concept of thinking and I will design a device that will do exactly that."

In this context we meet the classical problem that science presumes a reproducibility of the events that are of relevance to the experiment. It thus eliminates the human being as a source of error from its configurations and experiments as far as possible. However, the human being is always the one who evaluates, for example as a user, the results of his experiments and tries to gather information about the observed events. The presentation of the observations in the form of a transformation onto some measurement device, which can be perceived by our senses, allows us to receive information as a 'last instance' in several different ways (most of the time visual, but also audio-visual, acoustic, via temperature, ...). To be able to reproduce the reactions of human beings to the perceived information (in view of an automatic evaluation), we need a description of the information processing taking place during this process. But how do we incorporate information, how do we evaluate information, how do we react, etc.?

As an example we examine pattern recognition, used in image processing, where the problem of information processing is very much adjusted to the processing of visual data in the human brain [POP97]. The human brain searches the perceived images for edges with different orientations, for circles, for movement and for other patterns predetermined through experience and evolution. The information that is extracted from this pattern recognition is then transported to higher levels for further processing. This processing in the higher levels of our brain becomes more and more beyond our comprehension [HOF91]. But the lower level extraction of information, such as detection of edges, corners and other patterns is used in image processing.

These methods of detection are, however, only very rudimentary steps of processing in our brain, which solely performs many transformations of data in the reception of optical images. These transformations always contain a feedback (the eye locates and focuses on a certain object that has been found interesting), and they also depend on the subject who makes the observation (conditional information).

Our eyes and the subsequent transmission and processing of these data bring about a lot of transformations, which we are still trying to understand, to describe and then to use in information theory. First the optical system of our eyes focuses an image on our retina, which provokes a reaction in 125 Mio. cells [RES87]. The reaction of such an enormous number of sensors cannot be transported to our brain via the optic nerve, because it would simply be swamped with so many data. Therefore an information reduction takes place, to enable the optic nerve to cope with these data. The reduced set of data contains all the important components of the perceived image that we observe with our eyes. So the reduction does not lead to a noticeable loss of information. Of course, we may object that we do not notice the loss of information, because we do not have any chance to circumvent this reduction. We are used to it and we therefore cannot imagine how it would be without this reduction. Afterwards our brain carries out further reduction of data by searching for patterns, which are then processed in higher levels, as we already discussed.

The functionality of the lower levels of our brain, the so-called neuronal connections, is quite well understood. The higher levels of the processing are further and further from our physical comprehension, which is justified by the fact that the physical description of the levels develops into a description of our mind. But a description of our mind is certainly not possible with our physical models. Nevertheless, not only the fact that our mind does not admit a physical description, but also the great number of possible neuronal connections that our brain is able to generate restrict our ability to understand our brain. One may quote the statement of [ORN93]: "The brain of an earthworm consists of only a few cells and we are able to understand its functionality. But the earthworm will never understand its brain." In other words: We would have to have more cells to understand all procedures of our brain and their meanings to understand our brain. This would be a form of the Laplacian demon, knowing the exact position and velocity of all particles in our universe. Such a demon would be able to extrapolate into the future and past without any error or deviation. But it is impossible to understand our brain completely, because this comprehension requires that we know the place and impulse of every single particle of our brain. This means six states for every particle, which consequently demands a brain at least six times greater than ours to describe these states. But this would still not give us any hint about the sense of these states and about their meaning.

The exchange of information in complex natural processes (as for example in our brain) is done between layers of the same level (horizontally) and also vertically, between different levels (between higher and lower levels) [HOF91].

The classical way for science to deal with complex systems is the canonical transformation, which enables us to examine subsystems at first, and then to proceed by combining these subsystems one by one into the whole system. As the examination of the subsystems does not provide information about the coupling and possible feedback, etc. the parameters of the subsystems have to be at least modified or restructured during the combination process. So we are able to consider the mutual effects between the subsystems in the whole system. As soon as the description of the subsystems or the actions between the subsystems become nonlinear, the description of the whole system becomes more complex, as we cannot use the superposition any more. This inevitably leads to more complex connections and a more complex modeling as a consequence of the nonlinearity. Thus the whole system is even more than the sum of its subsystems and equivalently the transmission of information increases in systems where the subsystems are mutually correlated.

In general the gain of information and the accumulation of knowledge are the aim of science (world 3 of Popper) and in doing so the scientist generates subjective knowledge (world 2) as well as objective knowledge. If we want to measure the concrete information, then we discover that information is multi-layered and this is exactly the point that complicates the definition of a measure or a unified concept of information. We could say with John von Neumann: "Define information and I will construct a device, which extracts, processes and applies the information of a given process."

When we, however, are not able to define the whole concept of information in such a way that we are able to utilize it in the form of scientific application, then we may still be able to find scientific formulations for certain aspects of information providing a maximum utility. For that purpose we can use the experiences we already gathered with the idea of information, but the concept of information must still fit into the given scientific systems (mathematics, physics, ...) and obey their formulations and laws.

Man as control loop

(Reception and processing of information of a human being)

Reception of information (not the reception of experience, which is merely a noticing of the environment) by a human being, or, more generally, the reception of information in any feedback system assumes some a priori information or at least an assumption (expectation) with respect to the observed event. Then the reception of information changes the preliminaries in such a way that we accumulate knowledge or a certain know-how. This knowledge qualifies us to step into the event and alter it according to some ideal concept. Information is the parameter that is exchanged between the real event and the knowledge or the know-how of the human being (controller).

But information is again only the pattern or the structure within the physical quantity, which we are able to perceive. This is because we constantly search the incoming data for patterns in order to reduce the large number of data so that we are able to process and finally understand the set of data we observe.

The limit (in scale) of our reception of information can at first be specified by the ability to distinguish two objects. As long as we are able to distinguish two objects, we are also able to find different patterns, making it possible to distinguish the two objects and to identify one object clearly. As soon as we are not able to carry out such a differentiation, we do not gain any additional information from the second object, as they are both equal to us. The 'unhidden' (αληϑεια) or the patterns, which we are able to recognize, are used to gain information about the object and they also provide the information we need to describe the object identically. This kind of concept of information is used in complexity theory.

To get an idea of the information processing of the human being, we describe the human being in a rough abstraction as a control loop.

Sensoric Perception	→ ←	**Human being**	→ ←	**Action Reaction**
	↖		↙	
		Environment		

Fig. 1.2: Man as a control loop

If our environment deviates sufficiently from the ideal concept of man (from reference values, from index values, ...), then we (motivated by the sensation of the pressure to establish the desired ideal imagination) try to eliminate or minimize the difference between the actual state and the desired state of the environment.

So we have an image (an idea) of the environment, which is used as a standard of comparison or as an image to compare with reality. If this image differs from reality, we use it to adapt the environment to this ideal stipulation (idea, theory, and concept). Unfortunately this adaptation is done (or has to be done) without a sufficient consideration of all the consequences of such a manipulation (utility, cost, ethical standards as in genetic engineering, ...). Often the only motivation is the scientific proof, the verification or the demonstration of the possibility or ability of the scientist to carry out the experiment in the intended way.

Information, complexity and typical sequences

In order to reduce the increasing flood of data we constantly look for patterns or structures, giving us the chance to reduce the number of data to be processed. We therefore use existing structures to simplify the representation of the given set of data. A set of data that has no patterns or structures cannot be reduced or simplified and we call such a sequence a 'stochastic sequence'. As we can only call a sequence a stochastic sequence in the strict sense when we know the complete set of data, we are actually not able to assign this attribute to a finite sequence of data.

As a consequence of this train of thought, the random generators (e.g., LCG = linear congruent generator, MLCG = multiplicative LCG [BRA92]) of common computers do not produce random sequences, but contain a periodicity, as a consequence of the deterministic generation of the random numbers. So we are not able to obtain a really random sequence with the deterministic computer.

The patterns we use to reduce the sequences do contain information, but they can be arranged, rearranged and stored with little information. For that the reduction of the set of data is restricted to the description of the patterns and the specifications of where these patterns can be found in the original sequence. The original sequence can thus always be reconstructed with these pieces of information.

An important example of such a reduction of information is DNA, containing the complete design for construction of a living being. On the basis of the DNA the complete human being is constructed. The functionality of our organs as well as their positions and arrangements are determined by this sequence. If we were able to decode this sequence, we would also be able to create artificial forms of life or to reanimate dinosaurs in a Jurassic park à la Spielberg and Crichton. But is it possible to reduce a human being to this sequence? The information needed to 'construct' a human being is certainly contained in the DNA, but it does not define the human being emerging from this sequence. Decisive factors, such as environment and education are not contained in this sequence, so that the DNA only determines a part of the 'complex construction' which we call a human being.

The crucial condition for the reduction of data certainly is the possibility to reconstruct the original set of data. This implies a correlation between the original and the reduced set of data, to ensure the recovery of all the information or at least of the important information. But the correlation must not be too distinct, to preserve the chance to reduce the original data in an appropriate way without decreasing the essential information. Thus the reduction and reconstruction of data require the decrease of the correlation between the original and the reduced data set to obtain a minimum length (complexity) of the compressed data sequence.

However, such a minimization of the correlation reduces the ability to reconstruct the original data in a proper way. So we have to find a trade off between the contradictory requirements of an optimal compression and an optimal reconstruction of all information. Images, nowadays stored in JPEG or GIF format, are compressed by reducing certain parts of the data to reduce the quantity of data. Here we use the properties of our eyes, which do not possess a sufficient sensitivity for the neglected data. So the trade off is made on the basis of the user, providing a subjective measure of the quality of data reduction or information reduction.

Concepts of information

The definition of a scientifically useful concept of information first requires a formulation of the part of information that can be described via scientific measures. Here Waever divides information into the following three dimensions [FLE72],[NIC91]:

Table 1.1: Dimensions of information

THREE DIMENSIONS OF INFORMATION
1. How *precisely* can we represent a message with a given set of characters (semantic dimension, semantic problem)?
2. How *accurately* can we transmit this set of characters to a receiver (syntactic dimension, technical dimension)?
3. How *effectively* does the message generate the desired reaction of the recipient (pragmatic dimension, pragmatic problem)?

In this enumeration the technical dimension is the only one that we are able to describe with our mathematical tools. It allows the comparison of the set of characters before and after the transmission and thus allows us to identify errors in transmission as a loss of information and to correct transmission as a gain of information. This gives us an objective criterion, providing the possibility to measure information in a transmission.

The other two dimensions of information are both ‘contaminated’ with a subjective aspect. The semantic dimension forces us to compare the set of characters with the meaning of the message, which is an impossible task for mathematical theories, because we simply cannot measure subjective meaning with objective physical quantities. The problem of the pragmatic dimension is equally posed, as we need to characterize the meaning of the received message for the recipient to describe the reaction produced. As we see, the latter dimensions of information (semantic and pragmatic) are not transferable to practice. This is the reason we turn to the information that we are able to describe in a mathematical way, as we do not want to deliver another philosophical treatise on the concept of information.

But we have to remark that the pragmatic dimension of information provides at least hints about the desired transmission of information or about the gain of information of the recipient. The learning of a robot can for example be judged by its reactions to a certain message. But this reaction is – as mentioned earlier – only the result of the sum of all steps of the information processing that has taken place in the system, and it is therefore not suitable for an exact judgement of the information gained. This judgement again contains a subjective aspect, being part of all transmissions of information. There is always some subject (human being) to judge the observation, which complicates the objective definition. We therefore reduce our consideration to the objectively describable, technical dimension of information.

Do we need certain preliminaries to do this? Which boundary conditions do we have to consider? Generally information diminishes an ignorance, which requires certain knowledge. When the receiver does not have any knowledge about the signal or about the meaning of the message, the information can be totally useless. But if the recipient possesses a knowledge that can be increased by the information, then the information decreases the uncertainty of the recipient. Thus information is useless when we do not have a basis or a condition that is known a priori to recipient and transmitter of a message.

When, for example, a specialist in atomic physics explains one of his theories, then it is fundamental for the audience to have some a priori knowledge of atomic physics, to be able to catch at least some of the ideas of this theory. The more a priori knowledge the listener has the greater the comprehension will be. But the more a priori knowledge one has the less new information the theory will convey. So previous knowledge is certainly an essential requirement for the utility of the information, but with a growing a priori knowledge the uncertainty that is reduced by the received information also reduces. Pierce [PIE80] puts it this way: "One is only able to understand, appreciate and use the solution of a problem, if one has a suspicion of what the problem is, i.e. when one has understood the problem first."

Information, eliminating ignorance (making decisions possible, ...) therefore always depends on the actual state of the recipient. A message can contain a lot of information, little information or no information at all for different recipients. This again implicates a subjective element in our familiar concept of information, excluding a unifying mathematical formulation of information.

Information, its technical dimension and the meaning of a message

A totally meaningless sentence can contain the same syntactic information as a meaningful statement. This is because the mathematical, technical measures of information only permit the assignment of a measure to a set of characters, occurring with a certain frequency in a given sequence. The meaning of this sequence cannot be measured, at least not with mathematical methods.

An interesting point in this context is the universal library. The question is: How many books would be needed in this library, when we allow all possible permutations of the letters of our alphabet, including some additional symbols. Let us assume that we use 32 symbols (26 letters, space, colon, dot, dash, question mark, exclamation mark) and furthermore that an average book consists of 500 pages with 40 lines, each consisting of 60 symbols. Then a single volume contains $60 \cdot 40 \cdot 500 = 1\,200\,000$ Symbols.

We now search for the number of possible combinations of our 32 symbols in a sequence of 1.2 million symbols. It is easy to deduce that we would get 32 possible combinations if the sequence consisted of only one symbol (We would get 32 sequences, consisting of only one symbol). For two symbols this number increases to 32^2. And in our case the number of possible combination is $32^{1\,200\,000} = 10^{1\,800\,000}$.

Such a library contains all volumes that we are able to create with our alphabet. This means that all books that have ever been and all books that will ever be written are contained in this universal library. There are books, differing only in one single symbol. There are other books filled with one letter. One of these books will have only empty pages; one will be filled with colons, another one with dashes, etc. Though there is certainly a lot of nonsense in this library, it is a very interesting thought that such an universal library will contain a lot of sense and even sense that will only be developed in future times.

Thus a machine that is only able to type a lot of characters at a very high speed (brute force) would be able to produce the same amount of sense that the human race has created, is still creating <u>and</u> will create by the application if its mind. However, as this machine is only able to type the characters and not to create sense, it will never be able to understand the sense of the books that it is writing in this senseless way. It can produce all possible books with a very simple instruction for their construction, but it will not be able to understand them, because it only follows the given instructions in a senseless way. This points out another problem of information. It is very easy to create a lot of information, but the use of this information, and especially the meaningful use of this information, depends on the recipient.

Nevertheless, it is also quite astonishing that the amount of information that we are able to produce with our set of characters is finite. Does this mean that our knowledge is finite, too? The question, we might ask, is the question for the set of symbols that we used to generate this knowledge. Are we able to describe all knowledge with this set of symbols? Certainly not, because there are abstract concepts, e.g., certain values, which we cannot exactly define with our language. For example in religion we try to characterize God with our limited set of symbols by creating artificial superlatives (almighty, omniscient, ...). But the assignment of these properties can certainly not grasp the nature of God. The methods inherent to our world are not sufficient for such a characterization. Thus our means are certainly not sufficient to describe or to grasp all knowledge. They are only capable of describing a limited knowledge and so our systems may even limit our view of the world around us (see also the statement of Pascal in the preface).

It is, however, not an easy task to find a certain book in this universal library. When we assume each book to have a breadth of 3.3 cm, then $3 \cdot 10^4$ volumes already require a length of 1 km. Passing by these books at the speed of light (about $3 \cdot 10^5$ km/s) would take 1 year with $3 \cdot 10^7$ seconds to travel along the number of $3 \cdot 10^5 \cdot 3 \cdot 10^7 \cdot 3 \cdot 10^4 = 9 \cdot 10^{16}$ books. This is, of course, only a negligible part of the whole $10^{1\ 800\ 000}$ volumes of our library. The search for a certain volume, revealing future theories, is therefore a task we may easily leave to mister Sisyphus. Even if we are able to find a certain volume, it will not be at all certain that this volume describes the future developments. There is, however, always a sufficient number of other volumes with equal preliminaries (with equal descriptions of past and present), which describe the future developments. Who will be the one to choose the right volume? Even the Laplacian demon would need help, because no one knows what we have to look for and thus the possibilities for our choice are enormous.

Information as a central concept?

Up to now we have in fact used the concept of information, but we have never given a definition. We always described applications and examples of transmission of information, of processing of information and of the possible gain of information to point out the presence of information in our life. We therefore want to go into the concept of information with the question: What is information, the concept that occupies the scientist in information theory, and what are the characteristics of information? Information theory attained its importance out of the perception that information is the essential preliminary of all life, all development and thus all processes of our world relying on different kinds of communication with each other or the environment.

In his book 'The illusion of reality' Howard Resnikov characterizes information as an abstract quantity, a pattern or an attribute, lying behind the observed physical, chemical or biological processes. Because of this categorical coupling of information to physical quantities, and especially to very different physical quantities, the abstract concept of information defies every attempt to obtain a unified definition. Nevertheless, information is a quantity independent of the actual medium, which can be of benefit to measure or evaluate the processing of our information processing systems. The categorical coupling of information to physical values does not mean that only physical quantities contain information, because ideas also contain information, but they need physical values to be transmitted.

As Carl Friedrich von Weizsäcker points out [vWE85] [vWE92], energy is a quantity similar to information. This is because energy possesses the capability of being transported and stored in several physical quantities. Flechtner [FLE72] confirms this, when he says that we more and more tend to look at information as the third basic quantity beside matter and energy because of the central importance of information. For both energy and matter we only observe the effects they initiate, which is a general resemblance of these two quantities.

Energy is an extensive state variable, i.e. adding the energy of two systems means that the energy of the whole system is equal to the sum of the energies of the subsystems. Information on the other hand is an intensive quantity, which means that the summation of two pieces of information leads to an accumulated information that is less than or equal to the sum of the two pieces of information. The accumulated information is only equal to the sum of the single pieces of information when there is no mutual information between the subsystems that are used to generate the total system (independence of the random variables).

There have been attempts to define theorems of conservation of information analogous to the conservation of energy, as for example in the time–frequency measurements. Here the uncertainty relation of the Fourier transformation

$$\Delta t \cdot \Delta f \geq \frac{1}{4\pi}$$

is used. This relation shows us that any gain of information that we obtain by smaller sample intervals in time leads to a loss of information because of larger frequency intervals in the frequency domain.

Humans constantly receive information in the form of sounds, odors and optical impulses, which are transformed on different levels and finally transmitted to our brain for further processing. Here we evaluate these different pieces of information and produce adequate reactions depending on the evaluation. But we do not only receive information on the conscious level. All our biological, physical and chemical processes have to cooperate to create a functional entity. This cooperation, however, is only possible, when the transmission of information between all the subsystems (organs, etc.) functions. The impulses of our nerves are, for example, transported as electrical signals and at the end of our nerves they are finally transformed into chemical signals, transmitting the same information [ORN93/2].

Now our biological information processing is by no means so fast that it allows us to evaluate all information that we are able to receive with our senses. So the transmission from our senses to our brain and the processing in our brain both require a reduction of the amount of data while the processing preserves the important information. This reduction of data, which needs to identify and evaluate the important information (What is important information?), is not at all well understood [RES87].

We are furthermore able to transform the received information into abstract concepts, which do not implicate any reactions, but a learning or an amalgamation of knowledge. This can be described by the storage of these aspects of information, which are not necessarily describable by physical quantities, as the meaning plays an important role in these aspects. And – as we already discussed – the meaning of information is too subjective to be measured with mathematical definitions. Popper's [POP97] partitioning of information into the three worlds describes the relations between the theories, i.e. the abstractions that man made during history to be able to characterize nature and to handle nature according to his ideas. As we are not able to keep in mind or to handle long sets of data, we create models and systems, allowing us to reproduce these sets of data and thus containing the same amount of information. These systems and models are then manageable by us.

All these abstractions are products of the human mind, allowing us to form a self-confidence in our physically existing brain by defining ourselves in our environment and history. This self-confidence allows us to create abstractions of different events, which we observe in nature as an uninvolved observer. With these abstractions we form our theories and finally a whole description of nature. We are thus able to intervene with specific interferences in the processes of nature.

"Theories are nets we throw out to capture nature." When the theories are not as exact as we want them to be, we have to scale down the meshes. This decreasing of the meshes is done by the reception and evaluation of information (see also chapter 1 and the intuitive derivation of information measures).

Information describes certain attributes of objects, but it does not have any physical unit, and as a consequence it has no immediate physical meaning. It is, however, not possible to assign a certain physical unit to the concept of information, as information is able to change the carrier and therefore the classical units of physics. And this change of the carrier does not necessarily lead to a change of the information content of the transported message.

Because of this lack of a physical unit, it is possible that absolutely different attributes or patterns contain the same information. So a biologist can gain the same amount of information by examining a sequence of DNA as an engineer gets from the observation of a dynamical system with the intention of generating a mathematical model of its dynamics. Both amounts of information can lead to the same numerical value, but the application and the meaning are completely different.

So the ability to describe chemical, biological and other quantities by measuring the information permits us to obtain a consistent description, based on the concept of information. But it also prohibits the definition of a unit of information, which cannot be fixed to one physical, chemical or biological reference system.

Information, as Shannon introduced it, is in general a quantity assigned to an ensemble or to a random variable, describing certain properties of random processes. Thus information does not immediately depend on the physical units and quantities of the chosen reference system, but on the relative frequency of the event occurring in the observed system. These events are subjectively chosen by the observer, who makes his choice concerning relevance or irrelevance (boundary conditions) of all elementary events. He thus combines elementary events to create events according to his choice, and he thus limits the number of relevant objects in the sample space (eventually with conditional events and changes in the sigma-algebra, etc.).

The information assigned to the ensembles has to be referred to some profitable application in the space of possible realizations (physical, chemical, ... reference system). The sole evaluation of the numerical value of the information does not contain any evidence for the observer. Knowledge of the contents of the system is required to be able to apply the numerical value of the measured information. The more similar the knowledge of transmitter and receiver of a message is, the simpler is the transmission, reception and use of the information for the receiver. Thus a correlation between the knowledge of transmitter and receiver is important for the transmission of information. This becomes very evident when we think of coding theory, where both have to know the codebook, i.e. the way the message is coded.

In electrical engineering and especially in communication theory the transmission of information began its development with the invention of telegraphy and the Morse alphabet. Nowadays it forms an important part of communication theory, based on the works of Claude E. Shannon (1948). Applications in coding theory, algorithmic complexity, cryptology (security of data on the internet), and in other areas, e.g., the portfolio analysis, lead to a continually increasing number of and also to a constant improvement of the possible applications of information theory.

Having discussed several aspects of the concept of information, we will now occupy ourselves with measures of information, allowing us to characterize the information with mathematical functions. These permit us to compute, interpret and evaluate the processing of information in scientific systems.

2 Basic considerations

2.1 Formal derivation of information

In order to derive a function describing information, we will follow a train of thought, demonstrated by Howard Resnikoff [RES87], where we first define the preliminaries that we require of the resulting information function.

- We are going to measure an unknown value m, with no a priori information (i.e. a previous knowledge limiting the uncertainty) available before the measurement.
- The measurement merely determines that the value m lies between x and y ($x < m < y$). The measurement thus constitutes the limits of the interval or it delivers an interval.

We now want to consider the information that we gain by the determination of this interval and we ask what a function that measures this information has to look like. We call I the information function and proceed with the following considerations:

- Information cannot depend on the zero point of our chosen scale, because the information regarding m has to be independent of the absolute value of m.
- The information regarding m can only depend on the limits x and y of the interval we obtained by our measurement, which contains the value m.

The first point stems from the consideration that we are able to identify the position of a desired value on some given scale by specifying the interval (x,y) in which the value has to be. We continually do this when we measure the length of an object. During such a measurement we use the maximum accuracy of the available scale and obtain a value describing the length of our object. The accuracy of this value is determined by the available scale of our measurement device and not by the location of zero. Therefore a function of information has to be independent of a shifting of the scale.

If we have two scales, each with 100 divisions, but one with a scale of centimeters and one with a scale of decimeters, then the identification of one of these intervals on each scale provides the same information, independent of the different scales. The accuracy of our measurement differs because on the one hand we identified an interval of one centimeter, while on the other hand we obtained only an interval of a decimeter. But the information gained by this selection of one single interval is the same in each scale. We specified 1 interval out of 100 possible intervals each time.

From these considerations we find that the information function has to be independent of the absolute values x and y when

1. The measurement scale is shifted $x + b, y + b$
2. The measurement scale is dilated $a \cdot x , a \cdot y$

For our information function this means:

$$I(x,y) = I(x+b, y+b) \qquad \text{shifting} \qquad (2.01)$$

$$I(x,y) = I(a \cdot x, a \cdot y) \qquad \text{dilation} \qquad (2.02)$$

We can combine both operations in an affine transformation

$$I(x,y) = I(a \cdot x + b, a \cdot y + b) \qquad \text{affine transformation} \qquad (2.03)$$

which holds for arbitrary values of a and b.

These are the demands that we make on our information function $I(\ )$, and we are going to use these demands to find a function describing the quantity 'information'.

If equation $I(x,y) = I(a \cdot x + b, a \cdot y + b)$ is to hold for arbitrary a and b, we may choose suitable substitutions for a and b to transform our equation:

Substitution 1: $a = 1$; $b = -x$

$$I(x,y) = I(a \cdot x + b, a \cdot y + b) = I(x - x, y - x) = I(0, y - x) \qquad (2.04)$$

Substitution 2: $a = 1/x$; $b = 0$

$$I(x,y) = I(a \cdot x + b, a \cdot y + b) = I\left(\frac{x}{x}, \frac{y}{x}\right) = I\left(1, \frac{y}{x}\right) \qquad (2.05)$$

For the third substitution we first write

$$I(x,y) = I(a \cdot x + b, a \cdot y + b) = I(0,1) \qquad (2.06)$$

and we then ask whether there exists a substitution meeting this demand. We thus write

$$a \cdot x + b = 0 \qquad (2.07)$$

$$a \cdot x + b = 1 \qquad (2.08)$$

and solve the system of equations

$$a = -\frac{b}{x} \tag{2.09}$$

Inserting this in the second equation leads to

$$-\frac{b}{x} \cdot y + b = 1$$

$$b \cdot \left(1 - \frac{y}{x}\right) = 1 \qquad b = \frac{1}{1 - \frac{y}{x}} = \frac{x}{x - y} \tag{2.10}$$

We thus find the resulting value for a

$$a = -\frac{b}{x} = -\frac{1}{x} \cdot \frac{x}{x - y} = \frac{1}{y - x} \tag{2.11}$$

Inserting these results in our equation, finally yields

$$I(x, y) = I(0,1) \tag{2.12}$$

which always takes a constant value, totally independent of the limits of our interval.

Thus the only possibility to obtain an information function on the basis of these preliminaries is to take constant information for all measurements. This, of course, is a very unsatisfactory result, as such a function does not permit any evaluation of our measurement. Obviously a single measurement does not provide any information about the desired value of m when it is observed without any relation to reference values. Why is that? Well, we want to answer this question by adding a third limit of intervals to our considerations and thus examine two intervals containing the desired value m. These two intervals are then the result of two successive measurements.

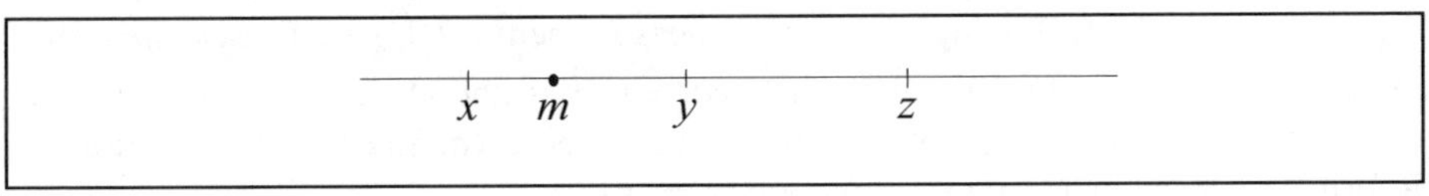

Fig. 2.1: Division of interval

Our basic considerations do not change with the arrival of the second interval. The information function of the three variables is therefore

$$I(x, y, z) = I(a \cdot x + b, a \cdot y + b, a \cdot z + b) \tag{2.13}$$

because again the information function is invariant to translation and dilation.

Here, too, we choose appropriate substitutions by assigning certain values to the free parameters a and b

Substitution 1: $a = 1\ ;\ b = -x$

$$\begin{aligned} I(x,y,z) &= I(a\cdot x+b, a\cdot y+b, a\cdot z+b) \\ &= I(x-x, y-x, z-x) = I(0, y-x, z-x) \end{aligned} \tag{2.14}$$

Substitution 2: $a = 1/x\ ;\ b = 0$

$$I(x,y,z) = I(a\cdot x+b, a\cdot y+b, a\cdot z+b) = I\left(\frac{x}{x},\frac{y}{x},\frac{z}{x}\right) = I\left(1,\frac{y}{x},\frac{z}{x}\right) \tag{2.15}$$

For the third substitution we just calculate the values

$$a = \frac{1}{y-x} \tag{2.16}$$

and

$$b = \frac{1}{1-\dfrac{y}{x}} = \frac{x}{x-y} \tag{2.17}$$

which we are going to insert in the extended form

$$I(x,y,z) = I(a\cdot x+b, a\cdot y+b, a\cdot z+b) = I\left(0,1,\frac{z}{y-x}+\frac{x}{x-y}\right) = I\left(0,1,\frac{z-x}{y-x}\right) \tag{2.18}$$

Now the information function obviously is not constant, as it is not independent of the limitations of the given intervals. So we have found a meaningful information function, depending on the quotient $(z-x)/(y-x)$, i.e. on the relation of the two intervals following from our measurements. $I(x,y,z)$ measures the incremental information gained by the change of our knowledge from the interval $(z-x)$ to the interval $(y-x)$. So we find a reasonable (non-constant) measure of information as soon as we use it in a relative way.

In the beginning we considered the information function to be invariant regarding an affine transformation. When we now scale the observed axis by applying an affine transformation, the shifts cancel out because of the differences and the dilation is canceled because of the quotient. So the measure of information we found is actually invariant regarding an affine transformation of the complete scale.

The above considerations lead us to the following statement:

> Information is a relative measure, which is only reasonable when we are able to build some relation, i.e. when we have more than one measured quantity.

We already discussed this restriction, because information always depends on the source and receiver and their mutual basis of knowledge. It is only possible to diminish an existing uncertainty and to generate knowledge when at least the basis of the knowledge of the source and receiver is correlated, or when they have some a priori information.

As we will examine in detail in later chapters, information is a quantity depending on the probability with which certain events occur in an observation process. As we are most of the time only able to measure the relative frequencies, instead of the probabilities, information always depends on the number of favorable events (hits) and the total number of all possible events. The quotient of these two quantities defines the relative frequency we may use to describe the information. Here we also have a quotient, a relative measure, enabling us to describe information.

In the next step we are going to demonstrate the transitivity of the measure of information. Therefore we again add another limitation to our axis.

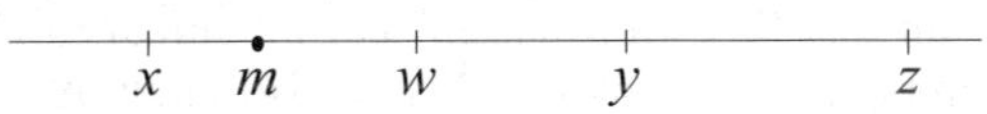

Fig. 2.2: Division of interval

We use our previously acquired knowledge that information depends on the quotient of the width of two intervals, and write

$$I(x, y, z) = I\left(0,1, \frac{z-x}{y-x}\right) \tag{2.19}$$

Now this equation is intended to fulfil transitivity, so we demand:

$$I\left(0,1, \frac{z-x}{w-x}\right) = I\left(0,1, \frac{z-x}{y-x}\right) + I\left(0,1, \frac{y-x}{w-x}\right) \tag{2.20}$$

The information gained by the change from (z–x) to (w–x) must also be gained by first changing the interval (z–x) to (y–x) and then to (w–x). As we do not want to keep the constants 0 and 1 in our information function, we shorten the description by writing

$$I\left(\frac{z-x}{w-x}\right) = I\left(\frac{z-x}{y-x}\right) + I\left(\frac{y-x}{w-x}\right) \tag{2.21}$$

and this equation can also be written as

$$I\left(\frac{t}{v}\right)=I\left(\frac{t}{u}\right)+I\left(\frac{u}{v}\right) \tag{2.22}$$

or

$$I(s\cdot t)=I(t)+I(s) \tag{2.23}$$

The requirements resulting from the transitivity are met by functions containing a logarithm or some multiples of a logarithmic function. As a result of this requirement we thus obtain information functions of the form:

$$I(s)=c\cdot\ln s \tag{2.24}$$

The constant c may include a dependence on the basis of the logarithm. This basis can be varied according to the particular system (dual, octal, decimal, ...). The use of a binary system, which is the common system in computer science, very often implicates the use of the binary logarithm with basis 2. This leads to

$$I_2(s)=\mathrm{ld}(s)=\frac{\ln s}{\ln 2} \qquad c = 1/\ln 2 \tag{2.25}$$

In this case the unit of this information is a bit (binary digit). The use of the natural logarithm on the other hand provides the formulation

$$I(s)=\ln s \tag{2.26}$$

and the unit is a nats (natural units). Both units are pseudo-units, because they do not result from physical deduction, but from definition or convention connecting the units to the measured values of information.

2.1.1 Unit and reference scale

A real unit of information, stemming from a physical derivation (induction or deduction), does not exist. Even a unit quantity (unit element) does not exist, so that in principle we cannot adopt the common descriptions of physical quantities.

The lack of such a unit makes it difficult to find relations between information and real systems and to find appropriate ways to use the measures of information for an evaluation of the trials performed. Common physical quantities can be combined with or connected to other physical quantities by taking care of the units, which provides a much easier understanding and makes them easier to use.

As we already mentioned, the deficiency of a real physical unit for the information functions is due to the fact that information does not depend directly on the actual physical quantity. It depends on the probability with which certain events (an interval of a given scale) occur or on the uncertainty of the measurement (width of the interval). The width of the intervals and the number of the intervals depend on each other, when we observe a predetermined scale.

Information, as Shannon defines it, is defined for ensembles and cannot be applied to one single object. The probabilities of certain events have, however, no deterministic physical values. Furthermore, as the probabilities of events of completely different systems (biological, physical, chemical, ...) may provide the same amount of information, we simply cannot limit the notation and the description of information to a single system. So information is a quantity describing an ensemble, without assigning a real physical meaning to the numerical value resulting from the observed experiment. This ensemble can then be transformed to arbitrary realization spaces, where we are able to observe the realization of the events.

As a consequence of the lack of a unit as well as the lack of a unit element or a unified scale of information, the knowledge of the numerical value of the information alone cannot provide a sufficient statement about our observed experiment. We need a reference value from some previous experiments or from some similar, altered experiments, to get a chance to classify the information of the actual experiment. In principle every scale (time, length, velocity, ...) only serves to provide us with these references, but we need to compare the actual measurement with our experience. So we are able to classify our measurements and to draw the appropriate conclusions from the experiments. As we do not have any standardized scale to measure information, we are not able to compare a single value of information with our experience. We thus need a second value of information to have a chance to compare these two pieces of information with each other.

Another problem preventing a unified scale of information is the great number of different aspects of information currently used to describe physical systems. Depending on the postulates they want the information function to meet, several scientists have found many different information functions (see for example Hartley's information, Shannon's information, Rényi's information and other information measures in chapters 3 and 5). These cannot be reduced to one description of information. So we have many different functions meeting different postulates and assigning different values of information to one single experiment.

Thus an absolute scale of information does not have any sense. It would instead be necessary to find a scale and a unit for every single information function. But this would certainly lead to chaotic and confused conditions, preventing a further development of information measures.

2.1.2 Information and the unit element

Does there exist a single unit of information, an information cell, which cannot be divided into smaller pieces of information (some kind of atom of information)? Looking at the extensive field of computer science, we can easily answer this question, as we use the binary alphabet of '0' and '1' to describe the basic decisions. Thus we measure a unit of information as one of these decisions. Each number, even each object, can be characterized by this binary alphabet, when we determine the rules, the patterns or the coding of these objects in an unambiguous way. So each describable object (random objects or sequences do not have any patterns or recognizable characteristics, so that they cannot be coded with a finite sequence) can be unambiguously coded with a binary number. This binary number can then be used to describe the information of this object. This approach is used in complexity theory, which attempts to assign information (complexity) to a single deterministic object, rather than to an ensemble of random events.

A handicap of this approach is the fact that we have to agree on the way we use to perform the coding. The coding has to be known to the person who performs the coding and to the receiver performing the decoding (previous knowledge, a priori information). So the information contained in the coded message can only be of use if sender and receiver know the code (conditional information). If we describe certain properties of objects (e.g., the length of a line, its starting point and the angle to the ordinate), then sender and receiver have to know the properties of the numbers (length, starting point, angle to the ordinate) being transmitted. Coding theory makes use of codebooks, where sender and receiver can look up and identify the admissible code words. Without this a priori information the transmission of coded data would be completely senseless.

Again the quantity information is only useful when we use it as a conditional quantity with respect to some previous knowledge. We have to know the preliminaries or boundary conditions providing the coding of an original data sequence and thus of the original information.

In digital systems information therefore has the unit bit (binary digit). The reason for this specification is that information (or entropy) measures the average number of binary digits that we need to code a given message with a set of symbols with a given probability.

But again we emphasize that this is not a physical unit, such as we know from familiar physical systems, but it is a pseudo-unit, only assigned to this value, which we call information.

This information measures the average length of the binary sequences. (This indicates the connection with complexity, measuring the number of data necessary to describe a deterministic sequence or a deterministic object.)

Nats (natural units) is also a familiar unit of information, requiring the use of the natural logarithm instead of the binary logarithm to compute the numerical value of the information. This merely is a change in the basis of the logarithms. It does not change the aspects of the calculated value.

2.2 Application of the information measure (here: Shannon's information)

Given: An image, presented with three different scales of pixels [RÖH67].

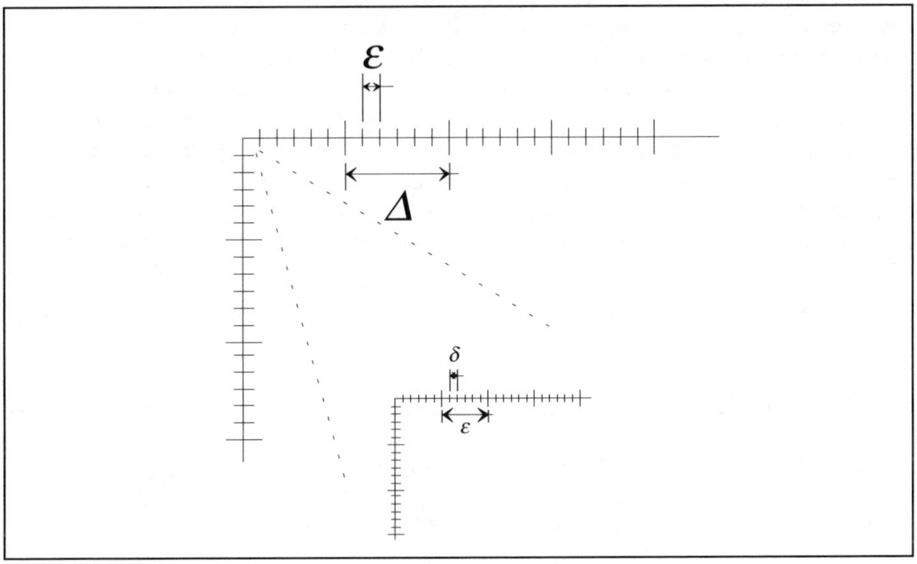

Fig. 2.3: Subdivision in different pixels, two-dimensional intervals

Δ = width of the large interval
ε = width of the small interval
δ = width of the smallest interval

Let the width of the whole object be L, which results in the following number of intervals:

$A_k = L/\varepsilon$ number of small intervals in L (2.27)

and

$A_g = L/\Delta$ number of large intervals in L (2.28)

The probability for a measurement to lie in a certain interval is then given by

$$p_k = \left(\frac{L}{\varepsilon}\right)^{-1} = \frac{\varepsilon}{L} \qquad \text{for the small intervals} \qquad (2.29)$$

and

$$p_g = \left(\frac{L}{\Delta}\right)^{-1} = \frac{\Delta}{L} \qquad \text{for the large intervals} \tag{2.30}$$

We do not prefer any of the intervals, but the probability of all intervals is assumed to be equal and to depend only on the width of the intervals. This means that we may be able to modify this example by the introduction of additional boundary conditions, changing the probability of the subintervals. But in this example we refrain from doing so to keep this model easy to understand.

When we concentrate our attention on only one component of the two-dimensional image, these considerations can immediately be applied to obtain an examination of the information. The information measure that we are going to apply in this case is Shannon's information for discrete events (see also section 3.2.3), which we find in the form of the pixels of our image. The information is:

$$S = -\sum_{i=1}^{N} p_i \cdot \ln p_i \tag{2.31}$$

(or ld instead of ln, leading merely to a change of the whole term by a factor of 1/ld 2)
Inserting the values that we found for the small intervals, we find

$$N = A_k = L/\varepsilon \qquad = \text{number of small intervals in } L \tag{2.32}$$

$$p_i = p_k = \left(\frac{L}{\varepsilon}\right)^{-1} = \frac{\varepsilon}{L} \tag{2.33}$$

and Shannon's information reads

$$S_k = -\sum_{i=1}^{\frac{L}{\varepsilon}} \left(\frac{\varepsilon}{L}\right) \cdot \ln\left(\frac{\varepsilon}{L}\right) \tag{2.34}$$

All terms in this sum are equal, which means that we are summing up L/ε times the same value. This can be written as a product

$$S_k = -\frac{L}{\varepsilon} \cdot \left(\frac{\varepsilon}{L}\right) \cdot \ln\left(\frac{\varepsilon}{L}\right)$$

$$S_k = -\ln\left(\frac{\varepsilon}{L}\right) = \ln\left(\frac{L}{\varepsilon}\right) \tag{2.35}$$

Comparing this information with our previously derived information function

$$I(x, y, z) = I\left(0,1,\frac{z-x}{y-x}\right) \tag{2.36}$$

we notice that in both cases the information arises from the quotient of the larger interval divided by the smaller interval. Thus Shannon's information does not evaluate an absolute measure for the information, caused by the knowledge of the small interval. Shannon's information gives us a relative measure for the information we get by the determination of the small interval, provided that we already know that there exists a large interval (the whole length of the image) divided into a certain number of small intervals. Again information is a relative measure.

We may also put it as follows: The boundary condition L is given and thus we know that the measurement value has to be in this interval. The determination of a certain small interval diminishes the given initial uncertainty by the information $S = -\ln\left(\frac{\varepsilon}{L}\right) = \ln\left(\frac{L}{\varepsilon}\right)$, which simply is the negative logarithm of the probability $p_k = \left(\frac{L}{\varepsilon}\right)^{-1} = \frac{\varepsilon}{L}$ with which the small interval occurs. This is equivalent to Hartley's formulation of information or to the information content, as all p_k are equal (see also section 3.2.1).

In the same way we derive the information that we obtain by determining one of the large intervals. In this case we obtain the formulation

$$S = -\sum_{i=1}^{N} p_i \cdot \ln p_i \tag{2.37}$$

$$N = A_g = L/\Delta \qquad = \text{number of large intervals in } L \tag{2.38}$$

$$p_i = p_g = \left(\frac{L}{\Delta}\right)^{-1} = \frac{\Delta}{L} \tag{2.39}$$

and Shannon's information is equal to

$$S_g = -\sum_{i=1}^{\frac{L}{\Delta}} \left(\frac{\Delta}{L}\right) \cdot \ln\left(\frac{\Delta}{L}\right)$$

$$S_g = -\frac{L}{\Delta} \cdot \left(\frac{\Delta}{L}\right) \cdot \ln\left(\frac{\Delta}{L}\right)$$

$$S_g = -\ln\left(\frac{\Delta}{L}\right) = \ln\left(\frac{L}{\Delta}\right) \tag{2.40}$$

Now this is the information gained by the determination of one of the large intervals. Again we require the knowledge of the whole interval L as an initial condition.

If we already know which of the large intervals contains our desired value and thus only need to determine a certain small interval in the given large interval, then the resulting information has to be a function of the number of the small intervals contained in the large interval. To describe this information, we need the following numbers:

$A_k = L/\varepsilon$ small intervals in L

and

$A_g = L/\Delta$ large intervals in L

which leads to

$A_{k/g} = \Delta/\varepsilon$ length of the large interval / length of the small interval

The probability of occurrence of a certain small interval in the large interval is then

$$p_{k/g} = \left(\frac{\Delta}{\varepsilon}\right)^{-1} = \frac{\varepsilon}{\Delta} \tag{2.41}$$

Given the initial condition that we already know the large interval, we obtain the following information from the determination of a certain small interval:

$$S = -\sum_{i=1}^{N} p_i \cdot \ln p_i \tag{2.42}$$

with

$$N = A_{k/g} = \Delta/\varepsilon \quad = \text{number of small intervals in the large interval} \tag{2.43}$$

$$p_i = p_{k/g} = \left(\frac{\Delta}{\varepsilon}\right)^{-1} = \frac{\varepsilon}{\Delta} \tag{2.44}$$

We thus determine the gain of information by Shannon's information

$$S_{k/g} = -\sum_{i=1}^{\frac{\Delta}{\varepsilon}} \left(\frac{\varepsilon}{\Delta}\right) \cdot \ln\left(\frac{\varepsilon}{\Delta}\right)$$

$$S_{k/g} = -\frac{\Delta}{\varepsilon} \cdot \left(\frac{\varepsilon}{\Delta}\right) \cdot \ln\left(\frac{\varepsilon}{\Delta}\right)$$

$$S_{k/g} = -\ln\left(\frac{\varepsilon}{\Delta}\right) = \ln\left(\frac{\Delta}{\varepsilon}\right) \tag{2.45}$$

As we expected, the information is equal to the logarithm of the quotient of the large interval divided by the small interval. This is equal to the conditional entropy of the small interval under the condition that we know the large interval. This can also be computed for the whole field with width L. Here we find:

$$S = -\sum_{i=1}^{N} p_i \cdot \ln p_i \tag{2.46}$$

$$S_{k/g} = -\sum_{i=1}^{N} p_g \cdot p_{k/g} \cdot \ln p_{k/g} \tag{2.47}$$

with

$$p_{k/g} = \left(\frac{\Delta}{\varepsilon}\right)^{-1} = \frac{\varepsilon}{\Delta} \tag{2.48}$$

$$p_g = \left(\frac{L}{\Delta}\right)^{-1} = \frac{\Delta}{L} \tag{2.49}$$

The summation is done over all small intervals occurring in the whole field with the length L. Thus we find:

$$N = A_k = L/\varepsilon \tag{2.50}$$

It follows that

$$S_{k/g} = -\sum_{i=1}^{N} p_g \cdot p_{k/g} \cdot \ln p_{k/g}$$

$$S_{k/g} = -\sum_{i=1}^{\frac{L}{\varepsilon}} \frac{\Delta}{L} \cdot \frac{\varepsilon}{\Delta} \cdot \ln\left(\frac{\varepsilon}{\Delta}\right) \tag{2.51}$$

Again all components in the summation are equal, enabling us to multiply a single term of this sum by the number of terms to obtain

$$S_{k/g} = -\frac{L}{\varepsilon} \cdot \frac{\Delta}{L} \cdot \frac{\varepsilon}{\Delta} \cdot \ln\left(\frac{\varepsilon}{\Delta}\right)$$

$$S_{k/g} = -\ln\left(\frac{\varepsilon}{\Delta}\right) = \ln\left(\frac{\Delta}{\varepsilon}\right) \tag{2.52}$$

Furthermore information has to be a transitive measure. When we thus reduce the whole set of events to the events occurring in one large interval in a first step and proceed with a second step, reducing the events to those occurring in a small interval, then the gain of information has to be equal to the information gained by the determination of a small interval in a single measurement. This means that

$$S_k = S_g + S_{k/g} \tag{2.53}$$

$$S_{k/g} = -\ln\left(\frac{\varepsilon}{\Delta}\right) = \ln\left(\frac{\Delta}{\varepsilon}\right) \tag{2.54}$$

$$S_g = -\ln\left(\frac{\Delta}{L}\right) = \ln\left(\frac{L}{\Delta}\right) \tag{2.55}$$

$$S_k = -\ln\left(\frac{\varepsilon}{L}\right) = \ln\left(\frac{L}{\varepsilon}\right)$$

$$S_k = \ln\left(\frac{L}{\varepsilon}\right) = \ln\left(\frac{L}{\Delta}\right) + \ln\left(\frac{\Delta}{\varepsilon}\right) = \ln\left(\frac{L}{\Delta}\cdot\frac{\Delta}{\varepsilon}\right) \qquad \text{q.e.d.} \tag{2.56}$$

Thus the conditional entropy is merely an entropy with a changed set (here the reduction of the complete interval L to the large interval Δ). In terms of estimation theory we change the sigma-algebra of the experiment with the condition and thus generate a new sigma-algebra with new assignments of probabilities to the events occurring in the new set. Furthermore we find:

$$I(x,y) = S(x) - S(x \mid y) = S(y) - S(y \mid x) = I(y,x) \tag{2.57}$$

and we use the equalities $x = \varepsilon$ and $y = \Delta$ to obtain

$$I(\varepsilon,\Delta) = S_k - S_{k/g} \tag{2.58}$$

with

$$S_{k/g} = -\ln\left(\frac{\varepsilon}{\Delta}\right) = \ln\left(\frac{\Delta}{\varepsilon}\right);\ S_g = -\ln\left(\frac{\Delta}{L}\right) = \ln\left(\frac{L}{\Delta}\right);\ S_k = -\ln\left(\frac{\varepsilon}{L}\right) = \ln\left(\frac{L}{\varepsilon}\right) \tag{2.59}$$

$$I(\varepsilon,\Delta) = \ln\left(\frac{L}{\varepsilon}\right) - \ln\left(\frac{\Delta}{\varepsilon}\right)$$

$$I(\varepsilon,\Delta) = \ln\left(\frac{L}{\varepsilon}\cdot\frac{\varepsilon}{\Delta}\right) = \ln\left(\frac{L}{\Delta}\right) \tag{2.60}$$

It is, of course, possible to compute the conditional information $S_{g/k}$ from these last connections, which leads us to the equations

$$I(\varepsilon,\Delta) = S_g - S_{g/k} \tag{2.61}$$

$$S_{g/k} = S_g - I(\varepsilon,\Delta)$$

$$S_{g/k} = \ln\left(\frac{L}{\Delta}\right) - \ln\left(\frac{L}{\Delta}\right)$$

$$S_{g/k} = 0 \tag{2.62}$$

Having determined a certain small interval, the subsequent determination of the large interval does not provide any additional gain of information, because the large interval is already known by the determination of the small one. It thus follows that the gain of information we obtain by the determination of the large interval, when we already know the small interval, is equal to zero and therefore the measurement is meaningless. *So we only gain information, when we are able to decrease an existing uncertainty. If there is no existing uncertainty, or if we do not diminish the existing uncertainty, then the gain of information that we obtain with our measurement is equal to zero.*

From this point of view we only take into account the position of the interesting object. As soon as we have determined the minimal interval (atom of the sigma-algebra) that contains the interesting object, the subsequent determination of all larger intervals is completely useless or completely redundant, as the larger intervals contain no additional information. All larger intervals can be generated out of combinations of the smallest intervals (atoms), so that the observer already knows the information of these coarser intervals.

Thus this point of view is restricted to discrete events that may or may not occur and thus eliminate all uncertainty regarding the occurrence of a certain event. If we examine continuous distribution densities instead of discrete ones, then we are only able to eliminate a part of the initial uncertainty that we had before we were able to evaluate the observation. A part of the uncertainty remains, because we are not able to carry out exact measurements. We will return to this point later.

We thus notice that there is no absolute information, but there is always information referring to two ‘semantic levels’. These are the levels of a previous knowledge and the level of the desired knowledge.

C. F. von Weizsäcker calls the previous knowledge a macro state and the n more detailed states of knowledge micro states, which are part of the macro state. The occurrence of a certain realization of one of these micro states can only be determined by an observation.

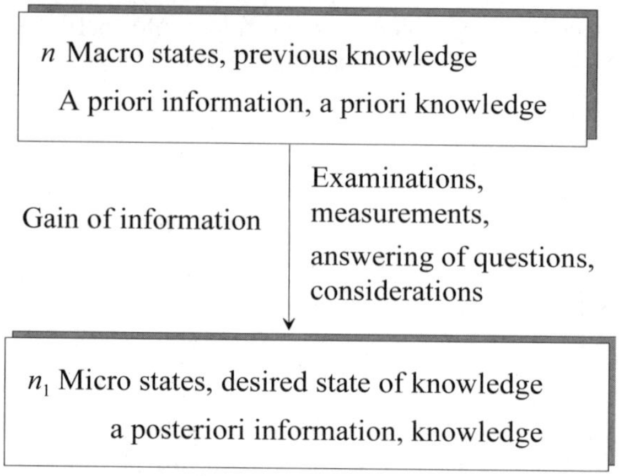

Fig. 2.4: Gain of information

Before we obtain any observation the system has the information content $I = \log n$, because we know one macro state out of n possibilities, and we want to specify this knowledge by determining a certain micro state. After the observation we determine one of n_1 micro states and thus finally have the information $I_1 = \log n_1$, because Shannon's information reduces to the information content for equally probable events. The gain of information that we obtained by this measurement is therefore equal to $\Delta I = \log(n/n_1)$ and is always positive, as $n > n_1$. The observation reduces the possible number of micro states and thus reduces the uncertainty. So the gain of information reduces the information content (in the sense of the a priori uncertainty) in the arranged experiment or in the given problem. The information content is obviously connected with the number of possible states. The reduction of the number reduces the information content and increases the probability for a single event to occur in an observation. We will again find this concept of information when we examine Hartley's information. But for now we want to proceed with our example to develop some additional considerations about information.

Another interesting point of view is, for example, the independence of information from the absolute width of the intervals. To examine further the relativity of information we increase the number of intervals by adding the smallest interval δ to the intervals Δ and ε already examined. This interval satisfies the equation

$$\frac{\delta}{\varepsilon} = \frac{\varepsilon}{\Delta} \tag{2.63}$$

The relation between the small intervals is thus equal to the relation between the larger intervals. With these preliminaries the information gained by the refinement of the scale can on the one hand be expressed by

$$S_{k/g} = -\ln\left(\frac{\varepsilon}{\Delta}\right) = \ln\left(\frac{\Delta}{\varepsilon}\right) \tag{2.64}$$

and on the other hand by

$$S_{\partial/k} = -\ln\left(\frac{\delta}{\varepsilon}\right) = \ln\left(\frac{\varepsilon}{\delta}\right) \tag{2.65}$$

As both quotients are equal, we immediately get

$$S_{\partial/k} = S_{k/g} \tag{2.66}$$

The absolute length of the intervals does not affect the gain of information. As long as we examine quantities with equal units, the units cancel out and the only part remaining in the determination of the gain of information is the quotient of two numbers. Thus information can be applied to arbitrary chemical, biological and physical systems, which can all be described with this measure. This certainly contains the disadvantage that we are not able to determine the origin of the evaluated numerical value for the gain of information, when we have no further information (previous knowledge about the system).However, the great advantage is that we may transfer information from one quantity to another quantity by using the quotients to change the medium of transmission or to change the scale.

What happens to the gain of information when we determine the smallest interval δ knowing only the large interval Δ? Well, we know that

$$S_{k/g} = \ln\left(\frac{\Delta}{\varepsilon}\right) \tag{2.67}$$

we conclude

$$S_{\partial/g} = \ln\left(\frac{\Delta}{\delta}\right) = \ln\left(\frac{\Delta}{\varepsilon}\cdot\frac{\varepsilon}{\delta}\right) = \ln\left(\frac{\Delta}{\varepsilon}\right) + \ln\left(\frac{\varepsilon}{\delta}\right) = 2\cdot\ln\left(\frac{\varepsilon}{\delta}\right) = 2\cdot\ln\left(\frac{\Delta}{\varepsilon}\right) \tag{2.68}$$

This information is twice the information we gain by determining the small interval by knowing the large interval. The number of possible intervals in the large interval increases quadratically by the refinement of the resolution from ε to δ. The doubling of the gain of information thus simply follows from the properties of the logarithm, which doubles when we build the square of the argument.

2.2.1 Summary

Looking back at the formal derivation of a measure of information, we required an information function to be independent regarding an affine transformation. In our example we obtain a change in information resulting from a change in the size of the intervals. How do these two results match? Well, when we use an affine transformation to change the whole axis, the information does not change in our example, because all intervals are dilated and shifted with the same factor. This does not affect the information, which merely depends on the ratio of two such intervals. The information that we need to determine a small interval out of n equal small intervals is not affected by a mapping of the whole scale, because it is the number of these small intervals that determines the information we obtain by its identification.

A similar relation is known in probability theory [LOF90] [LOE63]. There we get a change in the sigma-algebra, which means a change in the observable events or subsets, when we replace a probability distribution by a conditional probability distribution. In probability theory as well as in estimation theory an experiment is described by the specification of the set of possible events Ω (sample space), by the sigma-algebra σ_A defined on this sample space and by the probability assigned to the elements of the sigma-algebra. When we now observe an experiment once without conditions and another time with certain conditions, then the preliminaries of the experiment change if the conditions and the events of our experiment are not independent. For example, if we take a look at the throwing of a die, we have six elementary events. We combine them into the two events: 'even numbers' and 'odd numbers'. These are the events we are interested in and they form the basis elements of our sigma-algebra. Our sigma-algebra thus contains all intersections, all union sets and the complementary events of the interesting events {Ω, $\varnothing$, 'even numbers', and 'odd numbers'}. The probability of the events in a single throwing of our dice are: $P(\Omega) = 0$, $P(\varnothing) = 0$, P('even numbers') = 1/2, P('odd numbers') = 1/2.

If we have two dice (A,B), with one die containing the numbers from 1 to 6 and the second die containing the numbers from 7 to 12, then the probabilities of our events do not change, and the occurrence of one event cannot tell us which of the two dice has been used to produce the event. Each throwing of a die leads to the result 'even number' or 'odd number'. Thus each result consists of two possible sets (in the case of 'even numbers' the possible sets are {2,4,6} and {8,10,12}) that may have occurred, each with probability 1/2.

An additional a priori knowledge (a priori information), defining the die that is used in the actual experiment allows us to determine unambiguously the set that occurred in the observed throw ({2,4,6} or {8,10,12}). Our a priori information thus reduces the number of possible outcomes of our experiment and therefore redefines the sigma-algebra, which now does not consist of the events

'even numbers' = {2,4,6,8,10,12}

and

'odd numbers' = {1,3,5,7,9,11}

but of the events

'even numbers' of die A = {2,4,6}
'even numbers' of die B = {8,10,12}
'odd numbers' of die A = {1,3,5}
'odd numbers' of die B = {7,9,11}

This refines our sigma-algebra, and the information that we get with the additional a priori information is greater than the information we get without any additional information, as the number of possible events in our experiment is 4 instead of 2.

With this change in the sigma-algebra, the probabilities in the subsets of our sample space also change and this leads to an improvement in the gain of information we obtain by the observation of the random variable. Probability theory assigns measures to the possible outcomes of an experiment. If we assign equal probabilities to equal intervals, we will have the same arrangement as in the previous example, where we observed intervals with different widths.

It is, of course, possible to think of an extension of this experiment by assigning different possibilities to the intervals or by using no equidistant partitioning of the intervals, etc. This, however, would require a more detailed introduction to probability theory, which we do not want to carry out here, because we have already obtained a sufficient impression of information and possible measures. Instead we continue with the introduction of functions, which we denote as information functions. These are applied to describe the information in many different systems.

Before we start with the definitions of the information function, we first take another look at the a priori information and at the information functions introduced in this chapter. The application of information, gained by an observation or some kind of transmission, essentially needs a priori information or knowledge to assign a meaning to the received signals and to use the message in a meaningful way. In the example given it is therefore necessary to have some knowledge about the limitations or intervals to be able to choose a certain interval on the basis of a given interval (a priori knowledge). This choice actually contains the information of our observation and it reduces the initial uncertainty. Without an initial uncertainty we would not be able to classify the determination of a certain interval in the right way and we would not obtain a measurable gain of information. As we already said, Pierce formulated this statement this way: "One cannot understand and appreciate the solution of a problem until one has some idea of what the problem is."

In our example we may say: If we know that a desired value is contained in a certain region of a plane, then we are able to refine this region further with measurements and with the information contained in these measurements. If we do not know that we are measuring the location of a desired value, the measurement will not contain any information for the observer.

This statement can also be applied to semantic information. If someone receives information (a message), which he cannot use or understand as his a priori knowledge is not sufficient, then he will classify the message as meaningless, irrelevant or simply as 'not containing information'. But here we use the term information in the sense of knowledge (a priori knowledge) and not in the sense of eliminated uncertainty.

2.3 The law of Weber and Fechner

In experiments relating human perceptions to the physical stimuli, Weber and Fechner found the 'psychophysical law' in the year 1859 [SAL97]. The statement of this 'psychophysical law' is:

"Perceptions are proportional to the logarithm of their stimuli"

We want to examine two examples to explain this law. We start with the perception of sounds of different pitches

Table 2.1: Tone pitches and frequencies

Tone pitch	a_0	a_1	a_2	a_3	a_4
Frequency in Hz	220	440	880	1760	3520

We easily find an exponential or logarithmic relation:

$$f = 220 \cdot 2^n \qquad \text{for } a_n \tag{2.69}$$

The relative frequency per halftone is not altered. For each of the 12 semitones in an octave it takes the value

$$\frac{\Delta f}{f} = 2^{\frac{1}{12}} - 1 \approx 0.05946 \tag{2.70}$$

The change of frequency, caused by a semitone is therefore

$$\begin{aligned} 220\ \text{Hz} \cdot \frac{\Delta f}{f} &= 13.082\ \text{Hz} \qquad \text{for} \quad a_0 \\ 440\ \text{Hz} \cdot \frac{\Delta f}{f} &= 26.164\ \text{Hz} \qquad \text{for} \quad a_1 \\ 880\ \text{Hz} \cdot \frac{\Delta f}{f} &= 52.328\ \text{Hz} \qquad \text{for} \quad a_2 \\ 1760\ \text{Hz} \cdot \frac{\Delta f}{f} &= 104.655\ \text{Hz} \qquad \text{for} \quad a_3 \end{aligned} \tag{2.71}$$

We obviously recognize differences in the frequencies of two tones much better for lower absolute frequencies. However, as the absolute differences between the frequencies (of the semitones) increase with increasing absolute frequencies, we are still able to detect semitones, even for higher frequencies. The relative frequency per halftone $\frac{\Delta f}{f}$ is approximately 5%.

Another example of the logarithmic relation between the physical signal and our perception is the loudness of sounds.

Table 2.2: Volume and the power of sound

L = Volume in decibel (db)	0	20	40	60	80
S = Power of sound (W/m^2)	1	10^2	10^4	10^6	10^8

$$L = 10 \cdot \log_{10} S = 10 \cdot \frac{\ln S}{\ln 10} \tag{2.72}$$

$$S = 10^{\frac{L}{10}}$$

Again smaller values are much easier to distinguish, because we again find a constant quotient $\frac{\Delta S}{S}$ taking values in the range of approximately 5%.

Thus, for a human being the information content of a system immediately corresponds to the number of perceptions that can be distinguished

$$I = \log n \tag{2.73}$$

The more characteristic patterns we are able to distinguish, the more information we are able to gain about a system. But the more we burden our senses, the coarser becomes the scale that we are able to use to distinguish certain characteristics. The louder a sound is, the less is our ability to distinguish absolute differences in the loudness of our perceptions. If we want to talk to a visitor at a rock concert, we necessarily have to adapt the volume of our voice to the music to have a chance of being understood.

The scale of our perception is therefore quantized in the divisions $\frac{\Delta S}{S}$ or $\frac{\Delta f}{f}$ and it depends on the absolute value of the stimuli or on the physical quantities. The greater the absolute value of the physical quantity S or f becomes, the coarser becomes the scale that allows us to distinguish differences in the received physical quantity. Moreover, all separate perceptions contain the same amount of information. Because of the logarithmic relation between the perception and the physical quantity, we establish that our subjective perceptions are appropriate measures for the information content of the physical quantities that we observe.

Nature obviously equipped our senses with a scale adapted to the information content. The logarithmic function, occurring in the following information function, is therefore not only justified by the postulate of the additivity of independent random variables, but also a function established in nature and in our perception of it.

2.4 Information of discrete random variables

When the random variables describing our experiment have discrete realizations, then the number of possible realizations is finite and we may code the realizations with a unique binary sequence. In this case we have a countable amount of information in the observed system. Thus the information is quantized or becomes quantized by the choice of a discrete random variable describing the observation.

This method of description is – as we already know – also used in complexity theory, where the information of a single deterministic object corresponds to the length of the code word used to code the accordant object. In this case statements of yes or no can uniquely identify every element of the given set.

The least information we get from the answering of a question is a statement of yes or no. This least discrete information is thus also used to define the unit of information, which we call a bit (binary digit). Furthermore, each discrete piece of information can be coded into a sequence of answers yes/no, and thus in a sequence of bits.

This division of information into least information is not at all a new idea of information theory. The ancient Greeks had already thought of a scenario to find the least information [REN82]. The starting point of their considerations was a soldier who was sent in a besieged hostile city to search out the information necessary for the conquest of this city. Unfortunately he was detected and his tongue was cut out. But he managed to return to his camp knowing all the information, but without being able to communicate it to his fellows. Only skillful questions, which he was able to answer by shaking his head or by nodding (yes, no = 1,0) could elicit the information. This scenario led the Greeks to the question of how many skillful questions (smart coding) would require the minimum effort (cost) to obtain the desired information.

To be able to work out a smart strategy of such questions, we have to get an idea of what information is, and how we are able to represent information. When we represent information as a sequence of yes/no statements, we get 2^N possible answers for N questions, when the questions are independent of each other. This number of possible answers thus meets the number N_A of events in our sample space and we get

$$N = \text{ld}\, N_A \tag{2.74}$$

And this is the number of questions that we need to obtain the desired information, when the questions are independent of each other.

This quantity occurs in information theory in two central definitions, describing the capacity of a communication channel on the one hand and the information of symbols with equal probability on the other hand. The capacity describes a quality or an attribute of a communication channel or of a storage device. It characterizes the efficiency of such a system by the number of sequences (configurations) that the system can handle, transmit or store per unit of time. Here, however, the capacity is affected not only by the length of the sequence, but also by the number of different symbols (the alphabet) contained in the sequence. We might use binary sequences (0,1) or alphabets with three characters (−1,0,1) or our alphabet with additional characters (A,a,B,b,...,Z,z,!,-,?,...). The set of symbols that we have at our disposal to use in information processing is known as the alphabet.

The channel capacity C is then defined by using the number n of possible different configurations that we are able to generate with the alphabet we apply in the coding process.

$$C = \log n \tag{2.75}$$

The channel capacity can be defined with regard to a unit of time or per symbol.

Example: A communication channel can transmit k symbols per second of an alphabet consisting of m characters. Then we might transmit

$$n' = m^k \tag{2.76}$$

symbols per second over this channel. In two seconds (2s) this leads to

$$n'^2 = m^{2k} \tag{2.77}$$

symbols that can be sent over the channel. Thus the channel capacity per second is equal to

$$C' = 1/\mathrm{s} \log n' = 1/\mathrm{s} \log m^k = k/\mathrm{s} \log m \tag{2.78}$$

$$C' = 1/(2\mathrm{s}) \log n'^2 = 1/(2\mathrm{s}) \log m^{2k} = k/\mathrm{s} \log m \tag{2.79}$$

The channel capacity per symbols is then given by

$$C_e = C'/k = 1/\mathrm{s} \log m \tag{2.80}$$

The definition of this channel capacity is, apart from the preliminaries of the considerations we used in the derivation, identical to Hartley's information, which we are going to introduce in the following chapter.

3 Historic development of information theory

In a short overview – which cannot include all events in information theory – we want to introduce the measures of information familiar to scientists of information theory, communication theory and mathematics. In this overview our main attention is not focused on a philosophic, semantic view of information, but is focused on the information that we are able to describe with mathematical methods and thus to apply in our theories.

3.1 Development of information transmission

3.1.1 Samuel F. B. Morse 1837

The Morse alphabet was developed to transmit messages in an optimal way (a length of time as short as possible and a given alphabet (short, long)) over a communication channel. The assignment of the characters of the alphabet to the symbols, which was carried out by S. Morse, was not based on the relative frequency with which the letters appeared in given texts. It was instead carried out by S. Morse by counting the types in a letter case of the typesetter in a print office. Thus the Morse alphabet is not the optimum coding on the basis of the preliminaries (two characters) and it can be improved by a factor of 15% [PIE80].

With this coding the rate of transmission of a message is bounded by the divergence, leading to a merging of two subsequent pulses on the channel (broadening of the pulses). A line (long character) requires three times the period of a dot (short character) and thus prevents the fast transmission of messages, which was the reason to replace the line by a dot with a negative sign. This leads to an increased speed of the transmission.

3.1.2 Thomas Edison 1874

Thomas Edison introduced the quadruplex coding, using the symbols +3, +1, –1, –3 to increase further the rate of information transmission. The amount of information that can be transmitted over a given channel obviously depends on the following three factors:

1. How fast can we send two successive characters without leading to a merging of both on the channel (divergence, intersymbol interference)? (*frequency, bandwidth*)
2. How many different characters are we able to transmit (number of levels)? (*amplitudes*)
3. How much time do we have for the transmission? (*time*)

From these three quantities we are able to build a cuboid for the message, describing the limitations of the transmission of information.

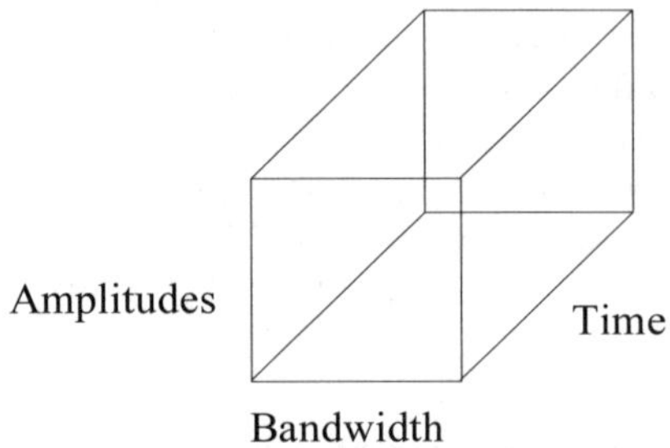

Fig. 3.1: Message cuboid

3.1.3 Nyquist 1924

In his paper 'Certain factors affecting telegraph speed' [PIE80] Nyquist says:

If we send symbols (successive amplitudes of current) at a constant rate, the rate of transmission W (symbols per second) is connected to the number of possible different symbols that can be sent, by the equation:

$$W = B \cdot \ln m \tag{3.01}$$

Where

B is a constant, depending on the number of different amplitudes of current that can be sent per second (bandwidth);

m is the number of different symbols at our disposal for the transmission.

If we change the coding from a sequence of the characters –1,1 to a sequence of the characters –1,0,1 we are able to increase the rate of transmission by 58.5%.

$$W_2 = B \cdot \ln 2 \qquad W_3 = B \cdot \ln 3$$

$$\frac{W_3}{W_2} = \frac{B \cdot \ln 3}{B \cdot \ln 2} = \frac{\ln 3}{\ln 2} = 1.585 \tag{3.02}$$

If we once more change to the sequence given by the quadruplex coding (–3, –1,1,3), as used by Edison, the rate of transmission again increases by further 26.2%.

$$\frac{W_4}{W_3} = \frac{B \cdot \ln 4}{B \cdot \ln 3} = \frac{\ln 4}{\ln 3} = 1.262 \tag{3.03}$$

Now, of course, we might ask how many characters have to be transmitted to optimize the rate of transmission. The answer to this question depends heavily on the boundary conditions we choose. We want to pick an example and determine the optimal number of characters to maximize the rate of transmission.

3.1.4 Optimal number of characters of the alphabet used for the coding

We take:
- a circuit without any noise
- a time τ for the processing of one received symbol

If the circuit is excited by the energy E_n, then the energy of this circuit falls off according to the equation

$$E = E_n \cdot e^{-\frac{t}{\tau}} \tag{3.04}$$

We may now distinguish n equidistant levels of energy

$0, E_0, 2E_0, 3E_0, \ldots, (n-1)E_0 = E_n$

To circumvent the problem of intersymbol interference of two subsequent impulses, we have to wait until the excitation with the energy E_n decreases below a certain level $E_0/2$. If this level is achieved, we set $E = 0$ and are thus able to detect a subsequent impulse.
The time that the system needs to reach this level is

$$\frac{E_0}{2} = E_n \cdot e^{-\frac{t}{\tau}} = (n-1) \cdot E_0 \cdot e^{-\frac{t}{\tau}} \tag{3.05}$$

$$\frac{t}{\tau} = -\ln \frac{1}{2 \cdot (n-1)}$$

$$t = \tau \cdot \ln[2 \cdot (n-1)] \tag{3.06}$$

Thus at least the interval $t = \tau \cdot \ln[2 \cdot (n-1)]$ has to elapse between two consecutive impulses, to allow us to distinguish them. For a transmission time T we are therefore able to detect

$$Z = \frac{T}{t} = \frac{T}{\tau \cdot \ln[2 \cdot (n-1)]} \tag{3.07}$$

pulses with our circuit. If all pulses or all amplitudes are adopted with the same probability, we find

$$I = B \cdot \ln n^Z \qquad B \text{ is again a constant, referring to the bandwidth} \tag{3.08}$$

n = number of energy levels that can be sent $[E_0, 2E_0, \ldots, (n-1)E_0 = E_n]$

Z = number of pulses that are sent per time interval

$$I = B \cdot Z \cdot \ln n = \frac{B \cdot T}{\tau} \cdot \frac{\ln n}{\ln[2 \cdot (n-1)]} \tag{3.09}$$

This is the information content, depending on the constant τ of our circuit and on the number of different levels that we are able to use for the transmission of our message. If we examine the information content, depending on the number n of possible levels, we find

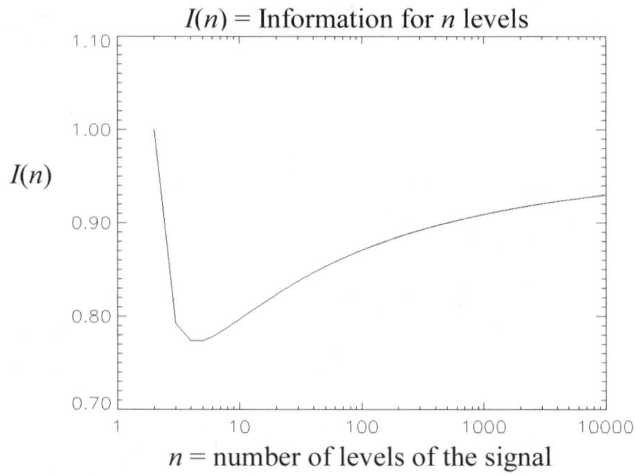

Fig. 3.2: Information content and the number of different levels

This function $I(n)$ assumes a maximum if n = 2 (which is the least number of possible levels) or if $n \to \infty$. When we have only two levels in our system, the energy of the signal rapidly decreases below the level of significance and thus the time between two consecutive signals can be very short. When, on the other hand, we use an infinite number of levels (this could be done for language), then the transmission of information also assumes a maximum, because even a single signal contains maximum information because of the infinite number of possible levels.

3.2 Development of information functions

3.2.1 Hartley 1928

Hartley defined a unit and a measure of information [REN73] [HAK95] by stating the following predication:

> "The answer to a question that can assume the two values 'yes' or 'no' (without taking into account the meaning of the question) contains one unit of information."

The unit of information is called – as we already mentioned – a bit, because we can realize the two answers in a dual system ('yes' = 1 and 'no' = 0). Now we are also able to realize complex systems or procedures by an arbitrary number of these binary digits. The characterization of a set E_N consisting of $N = 2^n$ dual enumerated elements, where the occurrence of the elements can be answered with yes or no, requires the information

$$I(E_N) = n = \text{ld}\, N = \frac{\ln N}{\ln 2}\,\text{bit} \tag{3.10}$$

This definition of information is commonly known as Hartley's formula and it holds under the assumption that all N elements of the set E_N occur with equal probability ($p_{Ni} = \frac{1}{N}$ for $i = 1, ..., N$). As Hartley's formula is a definition, we need not perform any proof. But this definition is not at all an arbitrary selection. It meets the following postulates (of Hartley) summarizing the requirements of an information function.

For the sets E_N, E_{N+1}, E_M, E_{MN}, consisting of N, N+1, M, MN equal probable elements we postulate:

H1: $I(E_{MN}) = I(E_N) + I(E_M)$

The sum of the pieces of information of two independent sets E_N and E_M is equal to the information of the union set E_{MN} (all sets consist of elements occurring with equal probability).

H2: $I(E_N) \le I(E_{N+1})$

Information is a monotonically increasing function.

H3: $I(E_2) = 1$

Determination of a unit quantity to obtain an absolute reference.

Hartleys information (information content)

$$I(E_N) = n = \mathrm{ld}\, N = \frac{\ln N}{\ln 2}\,\mathrm{bit} \tag{3.11}$$

and the channel capacity

$$C = \log n \tag{3.12}$$

are identical, though they are defined with different preliminaries. Hartley examined the probability of the occurring events (symbols), while the channel capacity aims at the number of events (symbols) to define a quantum to measure the quality of a signal processing system or a communication system. Since for equally probable events the reciprocal probability is equal to the number of occurring events, both definitions are identical and we thus get:

Hartley's information = channel capacity

The next scientists to deal with the problem of information theory were Norbert Wiener [WIE63] [WIE50] and Andrej N. Kolmogorov [KOL93]. They studied the problem of predicting a process from contaminated data and developed independent solutions for this problem. Furthermore the evolution of probability theory by Richard von Mises [vMI64/3] [vMI72] and Andrej N. Kolmogorov [KOL77] was an essential prerequisite for the development of the following information measures, which are merely defined on the basis of relative frequency or the probability of the occurring events.

Before we proceed to these definitions of information, we want to take a look at one other possible information function, defined on the basis of patterns of a signal.

3.2.2 Dennis Gabor 1946

This approach is based on the uncertainty relation of the Fourier transformation and tries to find signals with a minimum time–frequency window. The resulting transformation, the Gabor transformation ('Theory of Communication' [GAB46]):

$$S(f, t_0, \alpha) = \int_{-\infty}^{\infty} s(t) \cdot \exp\left[-\alpha^2 \cdot (t - t_0)^2\right] \cdot \exp[-j(2\pi f t + \varphi(t))] \cdot dt \tag{3.13}$$

$$s(t, f_0, \alpha) = \int_{-\infty}^{\infty} S(f) \cdot \exp\left[-\left(\frac{\pi}{\alpha}\right)^2 \cdot (f - f_0)^2\right] \cdot \exp[j(2\pi f t + \varphi(t))] \cdot df \tag{3.14}$$

is a short-time Fourier transformation with a windowing function, selected to minimize the time–frequency windows. Dennis Gabor calls time–frequency windows 'logons'. With this approach we can decompose every signal into such time–frequency windows. By counting the number of these windows, we obtain the information contained in the observed signal. In principle this is pattern recognition and a determination of the information by counting the number of these patterns. This approach, however, is not established as an information function, because it merely looks for deterministic structures and disregards the uncertainty. Pierce thus comments on this approach with the statement: "This ingenious paper, suggestive as it is, missed the inclusion of noise, which is the heart of modern communication theory." Nevertheless, this method is established in image processing and in pattern recognition. But we do not want to go into details, as this method does not include uncertainties in the signal, which form the basis of all other measures of information.

3.2.3 Shannon 1948

Claude E. Shannon enlarged Hartley's concept of information, by permitting sets (and thus also elements), which need not occur with equal probability. The starting point of Shannon's derivation is the mutually disjoint sets $E_1, E_2,.., E_n$ with $E = E_1 + E_2 +...+ E_n$. The number of elements contained in the set E_k is N_k, which means that the union set E consists of $N = \sum_{k=1}^{n} N_k$ elements. As a consequence of these preliminaries, the set E_k occurs with the relative frequency

$$p_k = \frac{N_k}{N} \tag{3.15}$$

The elements of the set E_k all have the same probability of occurrence. Thus, if we know that a certain element is a member of the set E_k, the exact characterization of this element in the set E_k only require (according to Hartley) the information

$$I(N_k) = \text{ld}\, N_k \tag{3.16}$$

The characterization of an element of the set E merely requires the average information I_{binary}, when we already know that the element is a member of the set E_k

$$\begin{aligned} I_{\text{binary}} &= \sum_{k=1}^{n} \frac{N_k}{N} \cdot \text{ld}\, N_k = \sum_{k=1}^{n} p_k \cdot \text{ld}(N \cdot p_k) \\ &= \sum_{k=1}^{n} p_k \cdot (\text{ld}\, N + \text{ld}\, p_k) \\ &= \text{ld}\, N + \sum_{k=1}^{n} p_k \cdot \text{ld}\, p_k \end{aligned} \tag{3.17}$$

thus it follows that

$$\text{ld}\, N = I_{\text{binary}} - \sum_{k=1}^{n} p_k \cdot \text{ld}\, p_k \tag{3.18}$$

The information required to characterize an element of the set E is an amalgamation of two pieces of information

$$\text{ld}\, N = I_{\text{binary}} + I_{\text{Shannon}} \tag{3.19}$$

I_{Shannon} specifies the set E_k that contains the element,

I_{binary} specifies the particular element in the known set E_k.

Shannon's entropy (informationentropy) is thus given by:

$$I_{\text{Shannon}} = \sum_{k=1}^{n} p_k \cdot \text{ld}\left(\frac{1}{p_k}\right) = -\sum_{k=1}^{n} p_k \cdot \text{ld}\, p_k \tag{3.20}$$

Shannon's information is an expectation value of Hartley's formula $E\left\{\text{ld}\left(\frac{1}{p_k}\right)\right\}$, where the probabilities are not required to be equal for all possible events. If the probabilities are equal for all events $p_k = \frac{1}{n}$, Shannon's information reduces to the formulation of the Hartley information, which is

$$I_{\text{Shannon}} = \sum_{k=1}^{n} \frac{1}{n} \cdot \text{ld}\, n = \frac{n}{n} \cdot \text{ld}\, n = \text{ld}\, n \tag{3.21}$$

The character of Shannon's information is more an entropy, which is the reason for several authors to talk about Shannon's entropy. (John von Neumann advised Shannon in a talk to call this quantity entropy. In all discussions this would provide the advantage that nobody actually knows what entropy is [CAM93].) Because of the connection between entropy and information, the negative Shannon entropy is sometimes called information (or negentropy). H. Haken uses both formulations, when he calls it informationentropy [HAK88].

We might, of course, use the binary logarithm instead of the natural logarithm

$$I_{Shannon} = -\sum_{k=1}^{n} p_k \cdot \text{ld}\, p_k = -\sum_{k=1}^{n} p_k \cdot \frac{\ln p_k}{\ln 2} = -\frac{1}{\ln 2} \sum_{k=1}^{n} p_k \cdot \ln p_k \tag{3.22}$$

but this only changes the information by a constant factor.

As we mentioned earlier, there is no unit of information and also no unit element of information that allows us to choose the scale at our will. The natural logarithm is easier to handle, which is the reason we prefer its use. The reasons for the use of the binary logarithm are based on digital systems, dealing with only two possible basic events ('yes' and 'no', or '1' and '0'). In our following examinations there is no reason prohibiting the use of the natural logarithm.

The mathematical motivation for the logarithm is the additivity of the information of independent events. We can imagine such independent events, when we observe four channels, all ensuring the same transmission of information. This configuration is thus able to transmit four times the information of a single channel. In this example we do not change the number of characters of the alphabet that we use for transmission

m = number of characters of the alphabet used

B = number of characters of the alphabet that can be transmitted per second (bandwidth)

but the number of symbols we are able to transmit from the sender to the receiver in the same time. that is, the value of B is multiplied by 4. We thus transmit m^{4B} instead of m^B symbols.

$$W_{4\text{channels}} = 4 \cdot B \cdot \ln m \tag{3.23}$$

(see also the channel information of Nyquist 3.01)

The information is multiplied by the factor 4, which is the reason for the logarithm.

Taking a further look at Shannon's entropy, commonly denoted by $H(X)$, i.e. the entropy of the random variable X, we are usually going to use the notation H or the previously used notation I_{Shannon}.

$$H = H(X) = -\sum_{k=1}^{n} p_k \cdot \ln p_k \tag{3.24}$$

The maximum entropy is achieved when all n symbols occur with the same (equal) probability. We thus find:

$$H = H(X) = -\sum_{k=1}^{n} \ln\left(\frac{1}{n}\right) \cdot \frac{1}{n} = -\ln\left(\frac{1}{n}\right) \tag{3.25}$$

When we have N symbols, as we did in Hartley's information measure, we get:

$$I_{\text{Hartley}} = \ln N = -\ln \frac{1}{p(x)} \tag{3.26}$$

N = number of possible events, occurring with equal probability. Such an event can be a sequence of characters of a certain alphabet with s different characters. When

$$N = s^n$$

and

n = number of characters of the sequence formed by an alphabet consisting of s different characters

$$I_{\text{Hartley}} = \ln s^n = n \cdot \ln s \tag{3.28}$$

Simple transformation leads us to the probability of one event

$$s = \exp\left(n \cdot I_{\text{Hartley}}\right) \tag{3.29}$$

$$p(s) = \frac{1}{s} = \exp\left(-n \cdot I_{\text{Hartley}}\right) \tag{3.30}$$

on the base of the information and the entropy, respectively. In Hartley's definition of information all symbols or sequences of symbols occur with the same probability. As we most of the time do not deal with equally probable sequences in the transmission or storage, the need to extend this concept of information to unequal probabilities is obvious.

The general problem arises when we take a look at the possible results (A_1, A_2, ..., A_n) of a random experiment. This, however, leads us to the basic question: 'How much information do we gain by the observation of one single experiment?'

To obtain an answer to this question, D. K. Faddejew established the following postulates to be met by information functions

P1: Information merely depends on the (discrete) distribution density $P = (p_1, p_2, ..., p_N)$ and is thus denoted as $I(P)$ or $I(p_1, p_2, ..., p_N)$.

P2: $I(p, 1-p)$ is a continuous function of p with $(0 \le p \le 1)$.

P3: $I\left(\frac{1}{2}, \frac{1}{2}\right) = 1$ unit element (see Postulate H3 of Hartley)

P4:

$$I(p_1, p_2, ..., p_N) = I(p_1 + p_2, p_3, ..., p_N) + (p_1 + p_2) \cdot I\left(\frac{p_1}{p_1 + p_2}, \frac{p_2}{p_1 + p_2}\right)$$

(see Postulate H2 of Hartley)

These postulates are only met by the previously described Shannon information, which we will now proof.

3.2.3.1 Validity of the postulates for Shannon's Information

For P1:
$$I_{\text{Shannon}} = \sum_{k=1}^{n} p_k \cdot \text{ld}\left(\frac{1}{p_k}\right) = -\sum_{k=1}^{n} p_k \cdot \text{ld}\, p_k \tag{3.31}$$

is a function, depending only on the probabilities p_k.

For P2:
$$I_{\text{Shannon}}(p,1-p) = -p \cdot \text{ld}\, p - (1-p) \cdot \text{ld}(1-p)$$
$$I_{\text{Shannon}}(p,1-p) = -\text{ld}(1-p) + p \cdot [\text{ld}(1-p) - \text{ld}\, p]$$
$$I_{\text{Shannon}}(p,1-p) = -\text{ld}(1-p) + p \cdot \text{ld}\left(\frac{1-p}{p}\right)$$
$$I_{\text{Shannon}}(p,1-p) = -\text{ld}(1-p) + p \cdot \text{ld}\left(\frac{1}{p} - 1\right) \tag{3.32}$$

In the range $0 < p < 1$ the argument of the first ld() is exclusively positive and the second ld() also leads to a positive value, as $1/p > 1$ for $0 < p < 1$. The logarithm is a continuous function, so that Shannon's information also is a continuous function. At the edges of the interval, the function $I(p,1{-}p)$ assumes the value 0, which is reached continuously from the left side at the right margin and from the right side at the left margin of the interval.

For P3:
$$I_{\text{Shannon}}\left(\frac{1}{2},\frac{1}{2}\right) = -\frac{1}{2} \cdot \text{ld}\left(\frac{1}{2}\right) - \frac{1}{2} \cdot \text{ld}\left(\frac{1}{2}\right) = -\text{ld}\left(\frac{1}{2}\right) = \text{ld}\, 2 = 1 \tag{3.33}$$

For P4:
$$I(p_1, p_2, \dots, p_N) = I(p_1 + p_2, p_3, \dots, p_N) + (p_1 + p_2) \cdot I\left(\frac{p_1}{p_1 + p_2}, \frac{p_2}{p_1 + p_2}\right) \tag{3.34}$$

We examine

$$I(p_1, p_2, p_3) = I(p_1 + p_2, p_3) + (p_1 + p_2) \cdot I\left(\frac{p_1}{p_1 + p_2}, \frac{p_2}{p_1 + p_2}\right) \tag{3.35}$$

and write in an explicit form

$$I(p_1, p_2, p_3) = -p_1 \cdot \text{ld}\, p_1 - p_2 \cdot \text{ld}\, p_2 - p_3 \cdot \text{ld}\, p_3 \tag{3.36}$$

$$I(p_1 + p_2, p_3) = -(p_1 + p_2) \cdot \text{ld}(p_1 + p_2) - p_3 \cdot \text{ld}\, p_3$$

and thus:

$$I(p_1, p_2, p_3) = I(p_1 + p_2, p_3) + (p_1 + p_2) \cdot \text{ld}(p_1 + p_2) - p_1 \cdot \text{ld}\, p_1 - p_2 \cdot \text{ld}\, p_2$$

$$I(p_1, p_2, p_3) = I(p_1 + p_2, p_3) + p_1 \cdot \text{ld}(p_1 + p_2) + p_2 \cdot \text{ld}(p_1 + p_2) - p_1 \cdot \text{ld}\, p_1 - p_2 \cdot \text{ld}\, p_2$$

$$I(p_1, p_2, p_3) = I(p_1 + p_2, p_3) - p_1 \cdot [\text{ld}\, p_1 - \text{ld}(p_1 + p_2)] - p_2 \cdot [\text{ld}\, p_2 - \text{ld}(p_1 + p_2)]$$

$$I(p_1, p_2, p_3) = I(p_1 + p_2, p_3) - p_1 \cdot \text{ld}\left(\frac{p_1}{p_1 + p_2}\right) - p_2 \cdot \text{ld}\left(\frac{p_2}{p_1 + p_2}\right)$$

$$I(p_1, p_2, p_3) = I(p_1 + p_2, p_3) + (p_1 + p_2) \cdot \left[-\frac{p_1}{p_1 + p_2} \cdot \text{ld}\left(\frac{p_1}{p_1 + p_2}\right) - \frac{p_2}{p_1 + p_2} \cdot \text{ld}\left(\frac{p_2}{p_1 + p_2}\right)\right]$$

$$I(p_1, p_2, p_3) = I(p_1 + p_2, p_3) + (p_1 + p_2) \cdot I\left(\frac{p_1}{p_1 + p_2}, \frac{p_2}{p_1 + p_2}\right) \quad \text{q.e.d.} \tag{3.37}$$

With $p_1 = \sum_{i=0}^{n_1} p_i$, $p_2 = \sum_{i=0}^{n_2} p_i$, $p_3 = \sum_{i=0}^{n_3} p_i$ this postulate is also met.

Thus the recursive evaluation of the entropy can always be applied when the event E_{12} occurring with the probability $P(E_{12}) = p_1 + p_2$ can be separated into two subsets E_1 with $P(E_1) = p_1$ and E_2 with $P(E_2) = p_2$.

$$I(p_1, p_2, \ldots, p_N) = I(p_1 + p_2, p_3, \ldots, p_N) + (p_1 + p_2) \cdot I\left(\frac{p_1}{p_1 + p_2}, \frac{p_2}{p_1 + p_2}\right) \tag{3.38}$$

Brillouin demonstrates the recursive computability of Shannon's information with an example:

$$I\left(\frac{1}{2},\frac{1}{3},\frac{1}{6}\right)=I\left(\frac{1}{2},\frac{1}{2}\right)+\frac{1}{2}\cdot I\left(\frac{2}{3},\frac{1}{3}\right)$$

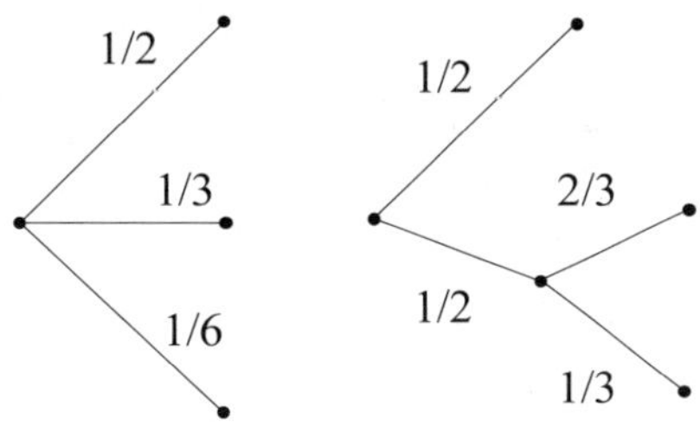

Fig. 3.3: Equivalent trees

3.2.3.2 Shannon's information (another possibility of a derivation)

We now want to show another way to derive Shannon's information and use this way for some additional explanations. Therefore we assume that we have an alphabet $\{x_1, x_2, ..., x_M\}$ and that we are able to assign the relative frequency h_i to the characters of the given alphabet. If we now take a look at a sequence with $N \gg M$ of these M characters, then the numbers $n_1, n_2, ..., n_M$ determine the number of characters $\{x_1, x_2, ..., x_M\}$ occurring in the observed sequence. The following equation must hold

$$n_1+n_2+...+n_M=N \tag{3.39}$$

The sum of all frequencies of occurrence is equal to the number of characters in the sequence. And the relative frequencies can be found by dividing this equation by the number of occurrences N:

$$h_1+h_2+...+h_M=n_1/N+n_2/N+...+n_M/N=1 \tag{3.40}$$

The relative frequency for a sequence of N symbols thus is

$$h_N=\underbrace{h_1\cdot h_1\cdot...\cdot h_1}_{n_1\text{ times}}\cdot\underbrace{h_2\cdot h_2\cdot...\cdot h_2}_{n_2\text{ times}}\cdot.....\cdot\underbrace{h_M\cdot h_M\cdot...\cdot h_M}_{n_M\text{ times}} \tag{3.41}$$

To compute the relative frequency of the sequence, we merely multiply the relative frequencies by the number of occurrences of the related events. This leads us to

$$h_N=h_1^{n_1}\cdot h_2^{n_2}\cdot.....\cdot h_M^{n_M} \tag{3.42}$$

In the next step we use the equivalence:

$$h_i=\frac{n_i}{N}\qquad\text{or}\qquad n_i=h_i\cdot N \tag{3.43}$$

and find

$$h_N = h_1^{h_1 \cdot N} \cdot h_2^{h_2 \cdot N} \cdot \ldots \cdot h_M^{h_M \cdot N}$$

$$h_N = \left[h_1^{h_1} \cdot h_2^{h_2} \cdot \ldots \cdot h_M^{h_M} \right]^N$$

$$h_N^{\frac{1}{N}} = h_1^{h_1} \cdot h_2^{h_2} \cdot \ldots \cdot h_M^{h_M} \tag{3.44}$$

The average of the occurrence of a character x_i in the sequence is equal to the relative frequency h_i. The sequence of the length N itself occurs in the mean with the relative frequency $1/N$. In the mean we may thus calculate with such an average sequence, which is created by characters, occurring with the specified relative frequencies. If we take logarithms on both sides of this equation, we obtain the known formula:

$$\ln h_N = N \cdot \sum_{k=1}^{M} h_k \cdot \ln h_k \tag{3.45}$$

If we furthermore replace the relative frequencies by the probabilities of the characters, we finally find

$$\ln p_N = N \cdot \sum_{k=1}^{M} p_k \cdot \ln p_k$$

and

$$\ln p_N = -N \cdot H$$

$$p_N = e^{-N \cdot H} \tag{3.46}$$

and we are thus able to characterize the possibility with which a sequence of the length N occurs when we know the relative frequencies of the characters. This possibility depends on the entropy of the system and on the length N of the sequence.

Shannon's information

$$-\frac{1}{N} \cdot \ln p_N = -\sum_{k=1}^{M} p_k \cdot \ln p_k = S_1 \tag{3.47}$$

can be obtained by dividing ln p_N, Hartley's information of the sequence, by the number of symbols contained in this sequence.

This approach is used in the definition of typical sequences [CSI81] [vMI64/3]. All sequences having such a probability are said to be typical sequences for the experiment. Sequences producing heavily deviating probabilities are said to be atypical sequences, which can sometimes be classified as outliers. Such outliers might be neglected in the determination of the probabilities out of the observation of sample functions.

The idea of determining the probability of events from the observation of a sequence of realizations is based on the idea that the probability is the limit of the relative frequency or that the probability is the expectation value of the relative frequency. Both concepts fail to provide a sufficient definition of probability. On the one hand the limit of the relative frequency does not exist in a strict mathematical sense, but only as a value to be achieved after the sampling of an infinite sequence. We are, however, not able to carry out the observation of such an infinite sequence, and a scientist has added the remark: "In a long term run we are all dead." The second 'definition' of information, where we use the expectation of the relative frequency, leads us to a circular conclusion (a vicious circle). This is because the determination of the expectation requires the knowledge of the probabilities that we are looking for, when we build the expectation of the relative frequencies (C. F. von Weizsäcker [vWE85]).

In spite of these difficulties in defining the probability with the relative frequencies, the relative frequency is still the only chance to verify the theoretical probabilities with practical measurements. But the verification of the probability can only be achieved with a certain probability, which is the result of the law of large numbers. The more observations we make, the more the relative frequency of a certain event approaches the theoretical probability of the same event with a certain probability.

With this connection between entropy and probability of a sequence of symbols, it is possible to determine the theoretical possibility from the measurement of the entropy.

3.2.3.3 Properties of Shannon's information, entropy

Shannon's information

$$I_{\text{Shannon}} = \sum_{k=1}^{n} p_k \cdot \ln\left(\frac{1}{p_k}\right) = -\sum_{k=1}^{n} p_k \cdot \ln p_k \tag{3.48}$$

achieves its maximum when all n symbols occur with the same probability $p = p_k$. To determine the maximum, we additionally have to take into consideration that the sum of all probabilities is equal to 1

$$\sum_{k=1}^{n} p_k = 1 \tag{3.49}$$

This boundary condition is included in the optimization with a Lagrangian multiplier λ, which provides the equation

$$\frac{d}{dp_k}\left[-\sum_{k=1}^{n} p_k \cdot \ln p_k + \lambda \cdot \left(\sum_{k=1}^{n} p_k - 1\right)\right] \overset{!}{=} 0 \tag{3.50}$$

$$-\ln p_k - p_k \cdot \frac{1}{p_k} + \lambda = 0$$

$$-\ln p_k - 1 + \lambda = 0$$

$$\ln p_k = \lambda - 1$$

$$p_k = e^{\lambda - 1} \tag{3.51}$$

To determine the Lagrangian multiplier, we insert this result, which does not depend on the index k any more, into the boundary condition

$$\sum_{k=1}^{n} p_k = \sum_{k=1}^{n} e^{\lambda-1} = n \cdot e^{\lambda-1} = 1 \tag{3.52}$$

$$e^{\lambda-1} = \frac{1}{n}$$

$$\lambda - 1 = \ln\left(\frac{1}{n}\right) = -\ln n$$

$$\lambda = -\ln n + 1 \tag{3.53}$$

The probability then becomes

$$p_k = e^{\lambda-1}$$

$$p_k = e^{-\ln n + 1 - 1} = e^{-\ln n}$$

$$p_k = e^{\ln\left(\frac{1}{n}\right)}$$

$$p_k = \frac{1}{n} \tag{3.54}$$

and this value has to be inserted into Shannon's information to determine the maximum. It is thus equal to

$$\max\left(I_{\text{Shannon}}\right) = -\sum_{k=1}^{n} \frac{1}{n} \cdot \ln\left(\frac{1}{n}\right) = n \cdot \frac{1}{n} \cdot \ln\left(\frac{1}{n}\right) = \ln\left(\frac{1}{n}\right) \tag{3.55}$$

The maximum of Shannon's information is thus achieved when all events occur with the same probability $1/n$. But this is exactly the formulation that Hartley found to express the information. Therefore Hartley's information maximizes Shannon's information.

Example:

Let us have a look at a binary variable x that can assume the values 0 and 1. When the probability for the occurrence of 0 is equal to $p(x) = 0.1$, then the probability for the occurrence of 1 must be equal to $q(x) = 1 - p(x) = 0.9$. If we plot the entropy of the random variable x as a function of the probability $p(x)$, we see that the maximum occurs when $p(x) = q(x)$. When the two probabilities are equal, i.e. when we have a uniform distribution, then the entropy achieves its maximum, as we see in figure 3.4.

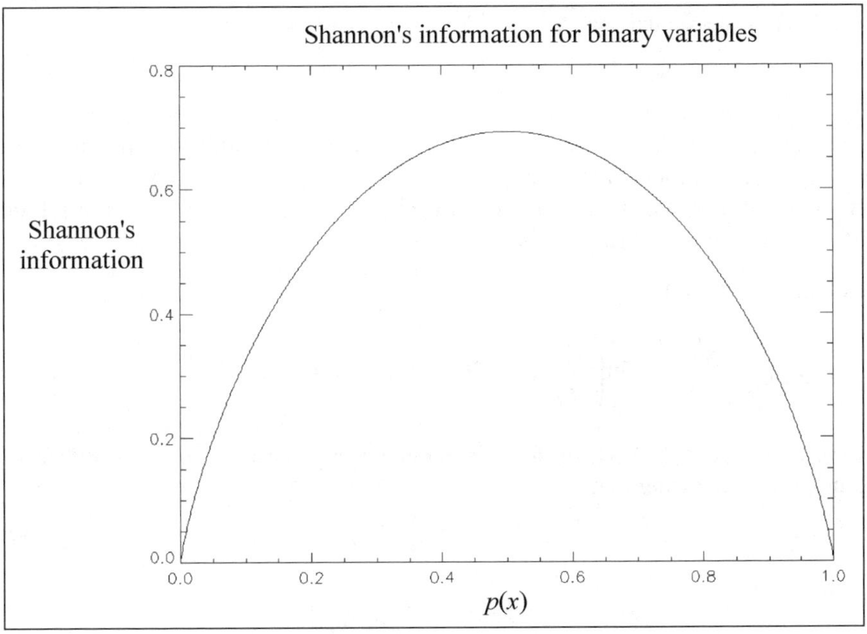

Fig. 3.4: Shannon's information for binary variables

The concave function $-p \ln p$, we find in the summation is shown in figure 3.5.

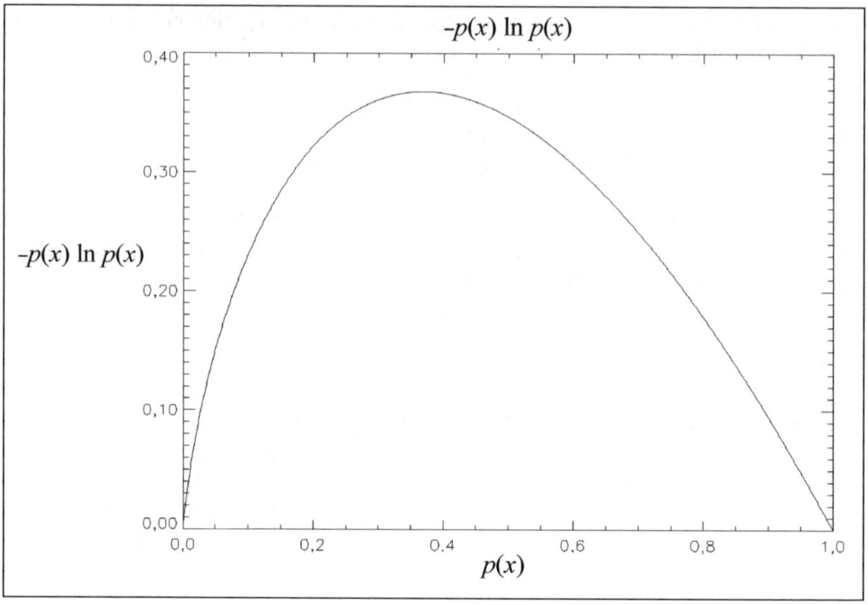

Fig. 3.5: The concave function $-p \ln p$

The maximum of the function $-p \ln p$ lies, as we easily verify, at $1/e$. Probabilities of $p = 1/e$ thus provide the greatest contribution to the sum. If a realization occurs more often with an increase of Δp, then the contribution to the sum does not decrease as fast as it does when the realization occurs more seldom with a decreased probability of Δp.

Furthermore we find

$$I_{\text{Shannon}} = \sum_{k=1}^{n} p_k \cdot \ln\left(\frac{1}{p_k}\right) = -\sum_{k=1}^{n} p_k \cdot \ln(p_k) \geq 0 \tag{3.56}$$

for discrete events. The reason for this nonnegativity is based on the postulates of Kolmogorov, requiring that

$$0 \leq p_k \leq 1 \tag{3.57}$$

and this leads to

$$\ln p_k \leq 0 \tag{3.58}$$

$$p_k \cdot \ln p_k \leq 0 \tag{3.59}$$

As every term of the sum is less than or equal to zero, the whole sum is negative and we get

$$-\sum_{k=1}^{n} p_k \cdot \ln p_k \geq 0 \tag{3.60}$$

Thus the Shannon information or the Shannon entropy of discrete events meets the conditions

$$0 \leq I_{\text{Shannon}} \leq \max\left(I_{\text{Shannon}}\right)$$

$$0 \leq I_{\text{Shannon}} \leq I_{\text{Hartley}} \tag{3.61}$$

Further transformation of the inequalities provides

$$e^0 \leq e^{I_{\text{Shanno}\,n}} \leq e^{I_{\text{Hartley}}}$$

$$1 \leq \exp\left(-\sum_{k=1}^{n} p_k \cdot \ln p_k\right) \leq \exp\left[-\ln\left(p_k = p\right)\right] = \exp\left(\ln n\right)$$

$$1 \leq \exp\left(-\sum_{k=1}^{n} p_k \cdot \ln p_k\right) \leq n \tag{3.62}$$

where n is the number of available symbols.

In Hartley's examination all these symbols are assigned the same probability, while Shannon allows them to have different probabilities. Therefore $\exp\left(-\sum_{k=1}^{n} p_k \cdot \ln p_k\right)$ is the average number of symbols that occur in a sufficiently long sequence. It depends on the probability of the symbols occurring in the sequence. The other two results limit the average value. The average achieves a limit because on the left-hand side we restrict the alphabet to one symbol, which is used to generate the sequence, and on the right-hand side all n symbols occur with the same probability. Thus the average or the number of symbols in a sufficiently long sequence is n on the right-hand side and 1 on the left-hand side, as there is only one symbol.

The reciprocal values

$$1 \geq \exp\left(\sum_{k=1}^{n} p_k \cdot \ln p_k\right) \geq \frac{1}{n} \tag{3.63}$$

can be used to characterize the probability with which the events occur. The single event on the left-hand side of equation (3.63) occurs with probability 1, the average value with $\exp\left(\sum_{k=1}^{n} p_k \cdot \ln p_k\right)$, and the right-hand side values (equally probable) occur all with the same probability $1/n$.

3.2.3.4 Shannon's entropy or Shannon's information?

From the previous calculations and from figure 3.4 we see that Shannon's entropy or Shannon's information achieves its maximum when the events occur with an equal probability. In this case we do not prefer a certain event and thus we get a maximum uncertainty regarding the next realization. In figure 3.4 this can be found for $p(x) = q(x) = 1-p(x) = 0.5$. A subsequent observation eliminates the uncertainty that exists in the initial information or in the initial knowledge about the outcome of the experiment.

As soon as we are able to recognize a tendency, assigning a higher probability to a certain event, the initial uncertainty is reduced before we get any realizations, and we tend to expect the realization with the maximum probability. In the limit we get $p(x) = 1$ or $q(x) = 1$ and the next realization is a certain event, so that the observation does not eliminate any uncertainty and therefore contains no additional information.

Shannon's information, however, describes exactly the uncertainty eliminated by the measurement. Without any observation we can give only some vague description of the possible realizations by assigning probabilities to the possible events. Having obtained an observation, the a priori uncertainty vanishes, as we are able to determine one event as the result of our measurement. Thus Shannon's information measures the information contained in an observation or the eliminated uncertainty.

The original notation of this quantity (as John von Neumann advised Shannon) was entropy. Shannon's measure describes the initial uncertainty that we have before we make the observation. We then only know the distribution density of the experiment, as we do not have a concrete realization. But the distribution density enables us to calculate the uncertainty (the entropy) that we have regarding the desired realization. So Shannon's measure is a measure of entropy, when we look at the experiment on the basis of the initial knowledge.

Shannon's measures is also a measure of information, because by the description of the initial uncertainty it also describes the information that we are able to obtain by the following observation. When we examine Shannon's measure from this point of view, which is more with the tendency to a future gain of information, we may also speak of Shannon's information, as it describes the information contained in the succeeding observations [vWE85].

Here we have to distinguish between discrete probabilities and continuous distribution density functions, because the discrete probabilities are assigned to countable events, which may or may not occur. It is thus possible to find an exact determination of the occurrence of an event by a simple yes/no decision, eliminating all uncertainty regarding the realization of the observed discrete random variable. The description of a random experiment with continuous distribution densities on the other hand does not lead to an exact realization, because we are not able to perform any exact measurement. We still retain an uncertainty after we make the observation. In this case the gain of information that we get from the measurement not only depends on the initial uncertainty (as it is in the case for discrete random variables), but also on the uncertainty remaining after the measurement (which is equal to zero in the discrete case). We will look at the differences between discrete events and continuous distribution densities in a more detailed way in sections 8.1 and 9.5.

As Shannon's measure can be viewed as an entropy as well as an information, we will use both names, always implicating the previous considerations.

3.2.3.5 The Kraft inequality

Now we take a look at a typical application of Shannon's information in a very short example. Such applications, for instance, can be found in coding theory, where Shannon's information is used to construct codes with optimal lengths of the codewords. One of the basic equations in coding theory is the Kraft inequality, which will now derive.

If we are able to arrange messages or codewords in a code tree, then we are able to state something about the number of codewords used to code the given message. Let:

D = number of symbols in the coding alphabet

$n_1, n_2, \ldots, n_M$ = given set of positive integers

Kraft's inequality:

$$\sum_{k=1}^{M} D^{-n_k} \leq 1 \tag{3.64}$$

is a necessary and sufficient condition for the existence of M codewords, corresponding to the end points of a tree. Their length is equal to the given numbers n_k (= length of the codewords).

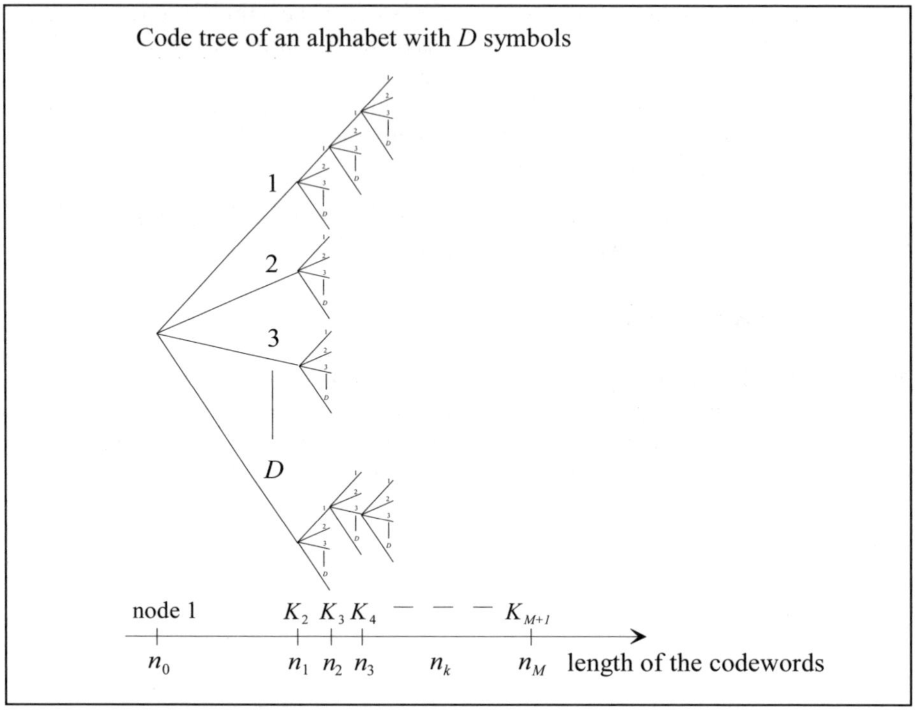

Fig. 3.6: Code tree

Proof of Kraft's inequality:

As the coding alphabet consists of D symbols, there are at most D branches starting at each vertex (node). If we now examine the number of possible vertices for different lengths of the characters (sequences of symbols) generated, we find:

1st vertex D branches,	produce maximally D codewords of length $n_k = 1$
2nd vertex D^2 branches,	produce maximally D^2 codewords of length $n_k = 2$ (each of the D branches of the first node provides D branches in the following node)
3rd vertex D^3 branches,	produce maximally D^3 codewords of length $n_k = 3$
…	
M^{th} vertex D^M branches,	produce maximally D^M codewords of length $n_k = M$

Let $n_{\max}$ be the maximum length of codewords. Such a maximum length of possible codewords produces at most a number of $D^{n_{\max}}$ codewords. Thus $D^{n_{\max}}$

is the maximum number of codewords of an alphabet, consisting of D elements and a maximum length n_{max} of these codewords.

A codeword at the location n_k can only have $D^{n_{max}-n_k}$ branches up to the location n_{max}. As all codewords in our coding tree are disjoint (no codeword is a part of another codeword, which is the characterization of an instantaneous or prefix code), we can sum all codewords starting at the node k. We obtain

$$\sum_{k=1}^{M} D^{n_{max}-n_k} \leq D^{n_{max}} \tag{3.65}$$

because the number of codewords of a part of our tree is always less than the number of codewords of the complete tree. Equality holds when we start at the first node and sum up all conceivable codewords of the tree.

Dividing by $D^{n_{max}}$ leads to

$$\sum_{k=1}^{M} D^{-n_k} \leq 1 \tag{3.66}$$

and this is the well-known Kraft inequality.

It is important that the summation is done only over all disjoint codewords (or all vertices at the end of the tree). There must not be any part of a codeword in the summation or a multiple summation of short codewords.

Kraft's inequality is thus a sufficient condition for the existence of a set of codewords, where the lengths of these codewords are specified in another set. It is possible to generate prefix codes with a minimal average length, when the Kraft inequality holds.

We therefore want to determine the lengths l_1, l_2, ..., l_m of our codewords, meeting the Kraft inequality as a sufficient condition for the construction of a prefix code and furthermore minimizing the average length of the codewords $L = \sum_i p_i \cdot l_i$. Thus our task is the minimization of the term $L = \sum_i p_i \cdot l_i$ under the constraints that $\sum_i D^{-l_i} \leq 1$. This leads us to a Lagrangian optimization

$$J = \sum_i p_i \cdot l_i + \lambda \cdot \left(\sum_i D^{-l_i} - 1 \right) \tag{3.67}$$

As we want to optimize each length l_i of our codewords, we differentiate with respect to this length and find

$$\frac{\partial}{\partial l_i} J = p_i + \lambda \cdot \frac{\partial}{\partial l_i} D^{-l_i} \stackrel{!}{=} 0$$

$$\frac{\partial}{\partial l_i} J = p_i + \lambda \cdot \frac{\partial}{\partial l_i} \exp\left(\ln D^{-l_i}\right)$$

$$= p_i + \lambda \cdot \frac{\partial}{\partial l_i} \exp(-l_i \ln D) = p_i - \lambda \cdot \ln D \cdot \exp(-l_i \ln D)$$

$$\frac{\partial}{\partial l_i} J = p_i - \lambda \cdot \ln D \cdot D^{-l_i} \stackrel{!}{=} 0$$

$$p_i = \lambda \cdot \ln D \cdot D^{-l_i}$$

or

$$D^{-l_i} = \frac{p_i}{\lambda \cdot \ln D} \tag{3.68}$$

We use this result in our boundary condition $\sum_i D^{-l_i} \leq 1$ to determine the still unknown Lagrangian multiplier

$$\sum_i D^{-l_i} = \sum_i \frac{p_i}{\lambda \cdot \ln D} = \frac{1}{\lambda \cdot \ln D} \sum_i p_i = \frac{1}{\lambda \cdot \ln D} \leq 1$$

$$\lambda \geq \frac{1}{\ln D} \tag{3.69}$$

We use this inequality together with $D^{-l_i} = \frac{p_i}{\lambda \cdot \ln D}$ to obtain

$$D^{-l_i} = \frac{p_i}{\lambda \cdot \ln D} \leq \frac{p_i}{\frac{1}{\ln D} \cdot \ln D} = p_i \tag{3.70}$$

Taking the logarithm on both sides, we get

$$\ln D^{-l_i} = -l_i \cdot \ln D \leq \ln p_i$$

$$-l_i \leq \frac{\ln p_i}{\ln D}$$

$$l_i \geq -\frac{\ln p_i}{\ln D} \tag{3.71}$$

and thus the minimum and therefore also optimum length of a codeword is given by $l_i = -\frac{\ln p_i}{\ln D}$. As this equation does not necessarily provide integer results for the length of the codewords, practical applications can only use suboptimal approximations. The average of the optimum length of the codewords is then given by

$$L = \sum_i p_i \cdot l_i = -\sum_i p_i \cdot \frac{\ln p_i}{\ln D} = -\sum_i p_i \cdot \log_D p_i = H_D(X)$$

$$L = H_D(X) = -\frac{1}{\ln D}\sum_i p_i \cdot \ln p_i = \frac{H(X)}{\ln D} \qquad (\ \ln D^L = H(X)\ ) \tag{3.72}$$

The average of the length of the codewords is thus equal to the entropy divided by the natural logarithm of the number of symbols that our alphabets consist of.

As the optimum, minimum length l_{i_opt} can only achieve integer values in real applications, this optimum average length of codewords may not necessarily be attained. The achievable length of codewords is thus always greater than or equal to the optimum length of codewords, which leads us to the following theorem:

Theorem:

> The expected length L of a prefix coding on an alphabet consisting of D elements, which is used to code the random variable $\underline{x}$, is greater than or equal to the entropy $H_D(X)$:
>
> $$L \geq H_D(X) = -\frac{1}{\ln D}\sum_i p_i \cdot \ln p_i = \frac{H(X)}{\ln D} \tag{3.73}$$
>
> Equality holds if all lengths meet the condition $l_i = -\frac{\ln p_i}{\ln D} \in Z$ (Z = integer).

Proof:

Let us examine

$$L - H_D(X) = \sum_i p_i \cdot l_i + \frac{1}{\ln D}\sum_i p_i \cdot \ln p_i \geq 0 \tag{3.74}$$

Again the boundary condition $\sum_i D^{-l_i} \leq 1$ holds, because we want to create a prefix code. Furthermore we write $l_i = l_i \cdot \frac{\ln D}{\ln D} = \frac{\ln D^{l_i}}{\ln D}$ and insert this into the equation

$$L - H_D(X) = \sum_i p_i \cdot \frac{\ln D^{l_i}}{\ln D} + \frac{1}{\ln D} \sum_i p_i \cdot \ln p_i \geq 0$$

$$L - H_D(X) = \frac{1}{\ln D} \left[-\sum_i p_i \cdot \ln D^{-l_i} + \sum_i p_i \cdot \ln p_i \right] \geq 0$$

$$L - H_D(X) = \frac{1}{\ln D} \left[\sum_i p_i \cdot \ln \frac{p_i}{D^{-l_i}} \right] \geq 0 \tag{3.75}$$

Before we proceed with our computation, we want to take a closer look at this equation. D is the number of elements in our alphabet, which is always greater than 1 (otherwise the alphabet is somewhat senseless), so that the logarithm of D always takes positive values. For the verification of nonnegativity this factor can thus be neglected and we proceed with the term in the brackets. Here we find an information function consistent with the Kullback information (see section 9.1). Such an information function is always nonnegative. It furthermore takes the value zero only if the two assignments of probabilities are equal, which would mean that $p_i = D^{-l_i}$. This can only be achieved with minimum length of the codewords, $l_i \geq -\frac{\ln p_i}{\ln D}$, when $l_i = -\frac{\ln p_i}{\ln D} \in Z$, i.e. when all lengths of the codewords attain integer values.

This proves the theorem, but we proceed with some further transformations, which enable us to obtain a further insight into the connections between the information measures.

We use the equation

$$L - H_D(X) = \frac{1}{\ln D} \left[\sum_i p_i \cdot \ln \frac{p_i}{D^{-l_i}} \right] \geq 0 \tag{3.76}$$

to proceed in our consideration.

If we use $r_i = \frac{D^{-l_i}}{\sum_i D^{-l_i}}$ to express the 'probabilities' yielded by integer lengths of the codewords and $c = \sum_i D^{-l_i} \leq 1$ to obtain a term for standardization, we get

$$L - H_D(X) = \frac{1}{\ln D}\left[\sum_i p_i \cdot \ln\left(\frac{p_i}{D^{-l_i}} \cdot \frac{\sum_i D^{-l_i}}{\sum_i D^{-l_i}}\right)\right] \geq 0$$

$$L - H_D(X) = \frac{1}{\ln D}\left[\sum_i p_i \cdot \ln\left(\frac{p_i}{r_i} \cdot \frac{1}{c}\right)\right] \geq 0$$

$$L - H_D(X) = \frac{1}{\ln D}\left[\sum_i p_i \cdot \ln\frac{p_i}{r_i} + \sum_i p_i \cdot \ln\frac{1}{c}\right] \geq 0$$

$$L - H_D(X) = \frac{1}{\ln D}\left[\sum_i p_i \cdot \ln\frac{p_i}{r_i} - \sum_i p_i \cdot \ln c\right] \geq 0 \tag{3.77}$$

The first term in the brackets is again a Kullback information, resulting from the difference between the real probabilities and the 'probabilities' yielded by the necessary integer values of the length of the real codewords. This Kullback information is always nonnegative. When both 'probabilities' are equal, we get $l_i = -\frac{\ln p_i}{\ln D} \in Z$ and $c = \sum_i D^{-l_i} = 1$, and the last sum in the brackets vanishes.

Another transformation of this equation leads to

$$L = H_D(X) + \frac{1}{\ln D}\left[\sum_i p_i \cdot \ln\frac{p_i}{r_i} - \sum_i p_i \cdot \ln c\right]$$

$$L = H_D(X) + \sum_i p_i \cdot \log_D \frac{p_i}{r_i} - \sum_i p_i \cdot \log_D c \tag{3.78}$$

with

$$H_D(X) = -\frac{1}{\ln D}\sum_i p_i \cdot \ln p_i = \frac{H(X)}{\ln D} \tag{3.79}$$

$$r_i = \frac{D^{-l_i}}{\sum_i D^{-l_i}} \quad \text{'probability' of real possible codewords of integer length} \tag{3.80}$$

$$\text{and} \quad c = \sum_i D^{-l_i} \le 1 \quad \text{boundary condition} \tag{3.81}$$

The average length L of the codewords of a real possible coding (integer length of the codewords) is given by the average length of the codewords of an optimal (non-integer lengths) coding and the information difference between the integer and the non-integer coding. This difference is expressed by a Kullback information (Kullback–Leibler distance) in the second sum on the right hand side of equation (3.78). The last sum finally introduces the boundary condition $c = \sum_i D^{-l_i} \le 1$ of an integer length of codewords.

Another useful inequality in this context is the *Gibbs's inequality*

$$\sum_i p_i \cdot \ln p_i \ge \sum_i p_i \cdot \ln r_i \tag{3.82}$$

because of

$$\sum_i p_i \cdot \ln \frac{p_i}{r_i} \ge 0 \tag{3.83}$$

and because of Jensen's inequality. Some authors use this equation, yielded by the Jensen inequality (see Appendix A.1).

3.2.3.6 Limits of the optimal length of codewords

3.2.3.6.1 Shannon's coding theorem

For each source we can find an appropriate binary coding, where the average length lies in a 1 bit limit of the previously derived lower bound. This means that

$$H(X) \le L < H(X) + 1 \tag{3.84}$$

Again we start our proof of this theorem with the optimal length of the codewords

$$l_{i\_opt} = -\frac{\ln p_i}{\ln D} \qquad \text{from} \qquad l_i \ge -\frac{\ln p_i}{\ln D} \tag{3.85}$$

As this expression does not necessarily lead us to a realizable, integer length of our codewords, practical applications replace the value l_{i_opt} with the smallest integer, which is greater than or equal to this optimal value. The mathematical formulation of this statement is

$$l_i = \lceil l_{i\_opt} \rceil = -\left\lceil \frac{\ln p_i}{\ln D} \right\rceil \tag{3.86}$$

$\lceil x \rceil :=$ smallest integer $\ge x$

If we insert this into the Kraft inequality, we obtain

$$\sum_i D^{-l_i} = \sum_i D^{-\lceil l_{i\_opt} \rceil} \le \sum_i D^{-l_{i\_opt}} = 1 \tag{3.87}$$

we notice that these codes meet the Kraft inequality (which can be shown more easily by using $l_i \ge l_{i\_opt} = -\frac{\ln p_i}{\ln D}$) and thus generate prefix codes. For the average length of the codewords we obtain

$$-\frac{\ln p_i}{\ln D} = l_{i\_opt} \le l_i < -\frac{\ln p_i}{\ln D} + 1$$

or equivalently

$$\log_D \frac{1}{p_i} = l_{i\_opt} \le l_i < \log_D \frac{1}{p_i} + 1 \tag{3.88}$$

As l_i is equal to the subsequent integer $\geq \log_D \frac{1}{p_i}$, l_i has to be smaller than the incremented subsequent integer.

Multiplication with p_i

$$-p_i \cdot \frac{\ln p_i}{\ln D} = p_i \cdot l_{i\_\text{opt}} \leq p_i \cdot l_i < -p_i \cdot \frac{\ln p_i}{\ln D} + p_i \tag{3.89}$$

and summation over all i yields

$$-\sum_i p_i \cdot \frac{\ln p_i}{\ln D} = \sum_i p_i \cdot l_{i\_\text{opt}} \leq \sum_i p_i \cdot l_i < -\sum_i p_i \cdot \frac{\ln p_i}{\ln D} + \sum_i p_i$$

$$\frac{H(X)}{\ln D} \leq L < \frac{H(X)}{\ln D} + 1 \tag{3.90}$$

In our special case D is equal to 2, because we wanted to use the binary alphabet. The average length of a binary coding is in fact in the range of 1 bit above the optimal average length of the codewords $H(X)$.

3.2.3.6.2 A sequence of n symbols (elements)

We now send a sequence of n elements of a set X. These elements are drawn according to the probability $p(x)$ and we can thus imagine the sequence of n symbols as a single hypersymbol of an alphabet Ξ^n. With our previously examined equation we get:

$$H_D(X_1, X_2, \dots, X_n) \leq E\{l(X_1, X_2, \dots, X_n)\} < H_D(X_1, X_2, \dots, X_n) + 1 \tag{3.91}$$

with

$$H_D(X) = \frac{H(X)}{\ln D}$$

$$L = E\{l(X_1, X_2, \dots, X_n)\}$$

If we divide by n and define the average length per symbol as

$$L_n = \frac{1}{n} \cdot E\{l(X_1, X_2, \dots, X_n)\} \tag{3.92}$$

we get

$$\frac{1}{n} \cdot H_D(X_1, X_2, \dots, X_n) \leq L_n < \frac{1}{n} \cdot H_D(X_1, X_2, \dots, X_n) + \frac{1}{n} \tag{3.93}$$

If the observed process is stationary, the entropy of the hypersymbol is related to the joint entropy according to the equation

$$\frac{1}{n}\cdot H_D(X_1,X_2,...,X_n)\to H(\Xi) \tag{3.94}$$

and for $n\to\infty$ we obtain

$$L_n=\frac{1}{n}\cdot H_D(X_1,X_2,...,X_n)$$

or

$$L=n\cdot L_n=H_D(X_1,X_2,...,X_n) \tag{3.95}$$

If the X_i are additionally identically distributed and independent (i.i.d., independent identically distributed), we can transformation the joint entropy in

$$H_D(X_1,X_2,...,X_n)=\sum_{i=1}^{n}H_D(X_i)=n\cdot H_D(X) \tag{3.96}$$

Thus

$$L=n\cdot L_n=H_D(X_1,X_2,...,X_n)=n\cdot H_D(X) \tag{3.97}$$

for i.i.d. random variables

To illustrate briefly this equation, we use the relation between the optimal length l_i of the codewords of an alphabet consisting of D elements, where the elements occur with the probabilities p_i. This relation is expressed by the equation $D^{-l_i}=\frac{p_i}{\lambda\cdot\ln D}\leq\frac{p_i}{\frac{1}{\ln D}\cdot\ln D}=p_i$ From this equation we get the relation $D^{-l_i}\leq p_i$ or $D^{-l_i}=p_i$ for an optimum coding. Taking the logarithm on both sides, we find

$$\ln D^{-l_i}=\ln p_i$$

$$-l_i\cdot\ln D=\ln p_i$$

$$\ln D=-\frac{1}{l_i}\cdot\ln p_i$$

$$\ln D=\ln p_i^{-\frac{1}{l_i}} \tag{3.98}$$

We already know this relation from the determination of the length of the codewords $l_i = -\frac{\ln p_i}{\ln D}$ of an optimum coding.

What happens, when we choose a coding on the basis of a wrong distribution density, i. e. with a wrong assignment of probabilities? This can happen, for instance, when we do not know the real distribution density and thus have to use an assumption instead.

The real distribution $p(x)$

The assumed distribution $q(x)$

The optimum real distribution would produce an average length of our codewords

$$L_p = H_{Dp}(X) = -\frac{1}{\ln D}\sum_i p_i \cdot \ln p_i \tag{3.99}$$

D = number of elements of our alphabet

The assumed distribution, which can permit only integer coding, leads us to the following inequality

$$H_{Dq}(X) \le |L_q| < H_{Dq}(X) + 1$$

$$-\frac{1}{\ln D}\sum_i q_i \cdot \ln q_i \le \lceil L_q \rceil < -\frac{1}{\ln D}\sum_i q_i \cdot \ln q_i + 1 \tag{3.100}$$

For the real distribution and the wrong coding we get

$$-\frac{1}{\ln D}\sum_i p_i \cdot \ln q_i \le E_p\{\lceil l(X)\rceil\} < -\frac{1}{\ln D}\sum_i p_i \cdot \ln q_i + 1 \tag{3.101}$$

We now add $H_D(X)$ to all parts of the inequality and simultaneously subtract this term again, so that we only expand the equation

$$\left[H_D(X) + \frac{1}{\ln D}\sum_i p_i \cdot \ln p_i\right] - \frac{1}{\ln D}\sum_i p_i \cdot \ln q_i \le E_p\{\lceil l(X)\rceil\}$$

$$< \left[H_D(X) + \frac{1}{\ln D}\sum_i p_i \cdot \ln p_i\right] - \frac{1}{\ln D}\sum_i p_i \cdot \ln q_i + 1$$

(3.102)

This leads to

$$H_D(X)+\frac{1}{\ln D}\sum_i p_i\cdot\ln\frac{p_i}{q_i}\leq E_p\{l(X)\}< H_D(X)+\frac{1}{\ln D}\sum_i p_i\cdot\ln\frac{p_i}{q_i}+1$$

and finally to

$$H_D(X)+\frac{1}{\ln D}D(p\parallel q)\leq E_p\{l(X)\}< H_D(X)+\frac{1}{\ln D}D(p\parallel q)+1 \quad (3.103)$$

$$D(p\parallel q)=\sum_i p_i\cdot\ln\frac{p_i}{q_i} = \text{Kullback's information}$$

The assumption of a wrong distribution $q(x)$ instead of $p(x)$ increases the average length of the codewords by $D(p\parallel q)=\sum_i p_i\cdot\ln\frac{p_i}{q_i} \geq 0$, which is the Kullback–Leibler distance. We will examine this information function in detail in subsequent chapters.

3.2.3.6.3 Application of the previous results

If n is the number of independently sent symbols of a sequence and s is the number of elements of a given alphabet, then we have

s^{nH} = either the maximum number (if $H = H_{max}$) of messages that can be sent with this configuration, or the most probable number (if $H \neq H_{max}$) of messages.

n = number of elements in the sequence

H = entropy of the elements

Example 1: Let $s = 2$, which means that we send a binary sequence of 1,0 and the probability of both elements is $p(0) = p(1) = ½$

Then we find: $H = H_{max} = 1$ bit/symbol and the number of possible symbols is

$$s^{nH} = 2^n \quad (3.104)$$

For a sequence of eight elements we are able to send $2^8 = 256$ possible messages.

Example 2: The probability of the two elements is now $p(1) = 0.75$ and $p(0) = 0.25$, so that we do not have equally probable elements any more. Thus the entropy does not achieve its maximum any more, but is equal to

$$H = -\frac{3}{4} \cdot \mathrm{ld}\frac{3}{4} - \frac{1}{4} \cdot \mathrm{ld}\frac{1}{4} = 0.311 + 0.5 = 0.811 \tag{3.105}$$

The number of equally probable elements generating such a sequence is

$$2^{n \cdot 0.811} = 2^{8 \cdot 0.811} = 2^{6.488} = 89.76 \text{ for } n = 8 \text{ elements in the sequence} \tag{3.106}$$

There are again 256 possible sequences that can be created with 8 bit, but only 89.76, i.e. 90 of these sequences occur with a sufficient probability. The remaining sequences appear so seldom that we may cancel them in the examination of the experiment, which will immediately lead to shortened sequences, as we will only need 7 bit to code 90 symbols.

With H_{max} we get the maximum number of possible sequences that can be created with the given alphabet. With H, the real entropy, we get the number of the probable messages coded with the elements of our alphabet.

Example 3: Our alphabet consists of 26 letters.
If we assume our language to consist merely of words with a length $n = 5$ and if all letters were to occur with the same probability, we would obtain:

$$26^{5 \cdot H_{max}} = 26^{5 \cdot \ln 26} = 26^{16.29} = 1.123 \cdot 10^{23} \tag{3.107}$$

This number of words could be generated with sequences containing only five letters, when we do not take into consideration such things as meaning, ability to speak the words, etc. But our language surely contains fewer words.

If we are able obtain an entropy that is less than the maximum entropy, because of the application of letters or elements not identically distributed, while we still use all elements of the alphabet to generate the code, then the chosen coding contains a redundancy or a diffuseness.

$$\text{Redundancy} = H_{max} - H \tag{3.108}$$

As we saw in the previous example, only 90 messages of 256 possible messages occurred with a sufficient probability. Thus we obtain a redundancy of

$$\text{Redundancy} = 256 - 90 = 166 \tag{3.109}$$

So there are 166 messages that scarcely occur and can possibly be cancelled in a transmission. This enables us to reduce the number of elements in the sequences from 8 bit to 7 bit, as 7 bit are sufficient to encode 90 different symbols. *The reduction of the number of elements in the data sequences allows us to minimize the redundancy in the coding of the symbols. This minimizes the time of the transmission.* The number of bits needed to transmit a message is equal to the number of symbols (sequences consisting of single bits) multiplied by the entropy.

s^{nH} = either the maximum number (if $H = H_{\max}$) of messages that can be sent with this configuration, or the most probable number (if $H \neq H_{\max}$) of messages.

n = number of elements in the sequence

H = entropy of the elements

$s = 2$ for binary transmission

This merely holds for independent elements! When the elements are not independent, but somehow correlated (Markovian chains, conditional probabilities, ...), the entropy decreases and thus the number of maximal possible sequences also decreases.

<u>*Another Example:*</u>

Given: 28 balls, 27 of them having the same weight, one ball being heavier. The measurements are made on a balance.

We want to know how many measurements we have to make to determine the heavier ball (without allowing random hits).

Each measurement can provide three results: 'as heavy as', 'heavier than', 'not so heavy as'.

Our system (alphabet) thus consists of three elements and we get the following equation for the coding of the 28 balls

$3^n \geq 28$ where n has to be rounded off to get the next integer

A sequence of n (integer) measurements with three results for each measurement allows us to encode the set of 28 balls uniquely and thus also enables us to identify a single element of this set. We therefore find:

$$n \cdot \ln 2 \geq \ln 28$$

$$n \geq \frac{\ln 28}{\ln 3} = 3.033 \tag{3.110}$$

The next greater integer is

$$n = 4$$

So four measurements will be sufficient to determine the heavier ball.

Strategy: We build three subset, each containing seven balls

1. Measurement: set 1 = set 2 (the heavier ball is not contained in set 1 or set 2)
2. Measurement: set 1 < set 3 (equality means that ball no. 28 is the one sought)

Now we divide set 3 into three subsets with two balls each and proceed:

3. Measurement: set 3_1 = set 3_2

The ball sought is in the set 3_3, which merely contains two balls.

4. Measurement: The remaining two balls.

Thus we have determined the heaviest ball with a maximum of four measurements.

This problem is equal to the coding of 28 symbols with an alphabet consisting of three elements. How many elements do we need in a sequence to encode unambiguously the 28 symbols?

$$n \cdot \ln 3 \geq \ln 28$$

$$n \geq \frac{\ln 28}{\ln 3} = 3.033 \tag{3.111}$$

The next greater integer is

$$n = 4$$

3.2.3.7 Information and utility (coding, porfolio analysis)

B. de Finetti [vWE85] created an experiment in which the participants have to determine the probabilities of n disjoint events E_k excluding each other. The probability has to meet the boundary conditions

$$0 \leq p(x_k) \leq 1 \tag{3.112}$$

$$\sum_{k=1}^{n} p(x_k) = 1 \tag{3.113}$$

Every participant thus has to determine a vector consisting of the probabilities of the particular events, excluding each other. Each determination of a probability is connected with a utility function $N_k(x)$, depending only on the probability $p(x_k)$ of the events, which has to be determined by the participants. The amount $N_k(x)$ is then paid as a utility (profit) to the participants. The experiment is repeated many times, but the probabilities $p(x)$ of the random vector determined by the participants are not altered during these repetitions. Thus the choice of the random vector has to be made to optimize the utility, which then leads to an optimized profit paid to the participant.

The experiment and the optimization of the utility function:

For m experiments the event E_k has occurred m_k times. Furthermore we meet the condition

$$\sum_{k=1}^{n} m_k = m \tag{3.114}$$

The relative frequency of the event E_k in the complete experiment is

$$h(E_k) = h_k = \frac{m_k}{m} \tag{3.115}$$

N_i denotes the utility in the ith trial and the complete utility up to the Mth trial is equal to the sum of all utilities up to the Mth trial.

$$\sum_{i=1}^{M} N_{ik}(x) = N \tag{3.116}$$

N_{ik} = utility in the ith trial of the kth component of the random vector.

The utility function has to achieve a maximum when the probabilities $p(x_k)$ determined by the participant are equal to the relative frequencies h_k obtained by the repeated measurements. The value of the utility function $N_k(x)$ depends for each k only on the component $p(x_k)$ of the random vector $p(x)$. Thus we may also write

$$N_k(p(x)) = N_k(p(x_k)) = f(p(x_k)) \tag{3.117}$$

This dependence will be maintained for all alternatives and for all sequences of trials. We thus define the utility function as

$$N = \sum_{k=1}^{n} h_k \cdot f(p(x_k)) \tag{3.118}$$

This utility function is to be maximized, with the additional boundary condition, that $\sum_{k=1}^{n} p(x_k) = 1$. This leads us to the Lagrangian function

$$L = \sum_{k=1}^{n} h_k \cdot f(p(x_k)) + \lambda \cdot \left(\sum_{k=1}^{n} p(x_k) - 1 \right) \tag{3.119}$$

As we are searching for the optimum probabilities, we proceed as follows:

$$\frac{\partial L}{\partial p(x_k)} = h_k \cdot \frac{\partial f(p(x_k))}{\partial p(x_k)} + \lambda \stackrel{!}{=} 0$$

$$\frac{\partial f(p(x_k))}{\partial p(x_k)} = -\frac{\lambda}{h_k} \tag{3.120}$$

Now we put $p(x_k) = h_k$ and obtain

$$\frac{\partial f(p(x_k))}{\partial p(x_k)} = -\frac{\lambda}{p(x_k)}$$

$$f(p(x_k)) = -\lambda \cdot \ln p(x_k) \tag{3.121}$$

for our utility function. The complete utility function is then given by the equation

$$N = -\lambda \cdot \sum_{k=1}^{n} h_k \cdot \ln h_k \tag{3.122}$$

and this exactly meets Shannon's information function (see equation 3.20).

The maximum utility, which arises from the correct determination of the probabilities of the interesting events, is thus equal to Shannon's information. This is used in coding theory, where the utility is connected to the length of the encoded message. The shorter the length of the sequence, the greater is the utility, as we are able to send more signals per second over the channel.

Thomas Cover [COV91] [ERK98] uses this function for portfolio analysis, where the utility in the form of a pecuniary profit of the investor is to be optimized. When we are able to predict tendencies in the development of the prices, then we get the chance to draw a maximum utility from the prognosis, which is proportional to the utility function described.

4 The concept of entropy in physics

The concept of entropy began its development in physics in the 19th century, when scientists began to describe thermodynamic processes. In the progress of these considerations the physicists first assumed some substance able to receive, store and deliver heat. Carnot talked about a power residing in the fire. Boltzmann was the first to develop a statistical description of entropy, which led to a lot of resistance in his time.

After a short description of the laws of thermodynamics, we occupy ourselves with the historical development of the concept of entropy, to characterize the concept of entropy and its comprehension in science parallel to the concept of information, which is our main interest in this book. We do this because we already noticed in the characterization of Shannon's information that this measure of information is an entropy measure. But what is entropy? To answer this question, we take a look at the historical development and at some basic considerations regarding entropy.

The laws of thermodynamics:

1. Law: Preservation of energy (in closed systems, isolated from the environment)

2. Law: $\Delta q = T \cdot \Delta S$

The thermal energy Δq occurring in a process or being exchanged in a process is connected to the entropy with a coefficient describing this exchange. This coefficient is temperature.

3. Law: The theorem of Nernst: The entropy achieves the value zero at the absolute zero temperature.

A basic quantity of thermodynamics is temperature, which is a macroscopic amalgamation of all the elementary motions of the molecules in a substance. We start our examination of entropy therefore with the macroscopic entropy, which is directly connected to the temperature.

4.1 Macroscopic entropy

4.1.1 Sadi Carnot 1824

Sadi Carnot (1796–1832) [CAR95] was the first to write an essay about the power residing in fire ("About the moving power of fire and the machines suitable to produce this power"). In this essay he already describes applications and assumes some kind of thermal substance in the fire. "Everywhere, where a difference in temperature exists, the production of this power can take place." Thus he defines a relative measure, needing at least two reservoirs to be applied. Carnot thinks of the steam engine and he estimates the development of this working machine, which could be used because of the 'moving force', in a very realistic way: "They (these machines) seem to be determined to give rise to a great revolution in culture." They are, however, cheap enough for nearly everyone and they possess the great advantage that they can be applied at every time and at every place without interrupting their work. This leads to the necessity to obtain a scientific comprehension of the processes taking place in such a machine. And Carnot examined different thermodynamic transitions, but he did not develop a concept of entropy.

4.1.2 Clausius's entropy 1850

Rudolph Clausius (1822–1888) [CLA96] presented a statement of the second law of thermodynamics, which reads as follows: "Temperature cannot rise from a cold to a warm body without external help". In his works he for the first time connects the Mayer–Joule principle of the equivalence of heat and work (first law of thermodynamics) in logical way with the principle of the transfer of heat from higher to lower temperatures (the second law of thermodynamics) developed by Carnot. 1865 he defined the entropy function

$$dS = \left(\frac{\partial Q}{T}\right)_{\text{rev}} \tag{4.01}$$

Q = thermal energy
T = absolute temperature
$d(.)$ = exact differential (the integral does not depend on the integration path)
$\partial\,(.)$ = inexact differential (depends on the nature of the process)
rev = this definition only holds for reversible processes

Entropy from εντρεπειν = to transform, to convert

For all reversible processes taking place in closed (isolated) systems, he finds

$$dS \geq 0 \tag{4.02}$$

From this inequality Clausius concludes that the universe has to increase its entropy continuously, which finally results in heat death where no difference in energy exists any more. But not all systems are closed systems, so that there is no need for all systems to increase their entropy constantly and thus to approach heat death.

This entropy has a unit, which is J/K (joules per kelvin).

1g water, for instance, contains an entropy of 1.3069 J/K at 100°C. Steam contains an entropy of 7.3554 J/K at the same temperature. Thus the energy used for the evaporation is about 6 J/K. This quantity, which possesses a unit, allows us to describe transitions between different aggregate states, where the uncertainty increases with the supply of heat, but the temperature of the system remains constant (boiling water has a temperature of 100°C).

If we take a look at the entropy in closed systems, we obtain the results shown in the following sections.

4.1.3 Increase of entropy in a closed system

We have two bodies with temperatures $T_1 > T_2$. An amount of heat Δq is passed from the first body and received by the second body. This leads to the equation

$$\Delta q = T_1 \cdot \Delta S_1 = T_2 \cdot \Delta S_2 \tag{4.03}$$

$$\frac{T_1}{T_2} = \frac{\Delta S_2}{\Delta S_1} > 1 \tag{4.04}$$

Thus the entropy increases because of the transfer of heat from the warm to the cold body. So the entropy increases in a closed system. The change in the heat and the change in entropy are connected with a constant T, measured in kelvins. This is the second law of thermodynamics.

$$\Delta q = T \cdot \Delta S \tag{4.05}$$

When the total process consists of several sub-processes, the total entropy can be computed by the equation

$$S = s_1 + s_2 + \dots + s_n \tag{4.06}$$

when s_i are entropies of reversible processes. If these processes are not reversible (and it is sufficient if only one sub-process is irreversible), we get the inequality

$$S \geq s_1 + s_2 + \dots + s_n \tag{4.07}$$

4.1.4 Prigogine's entropy

Ilya Prigogine [PRI90] [PRI92] extended Clausius's entropy to open systems, being able to exchange matter, energy, information and entropy with their environment. Systems that receive energy and deliver energy to their environment are called 'dissipative structures'. Prigogine demonstrates that such dissipative structures may lead to self-organization. This means that complex structures develop from simpler structures, which contradicts the second law of thermodynamics, as Clausius formulated it, because such systems create order in a world moving towards chaos.

By the delivery of entropy to the environment, open systems, which exchange matter and energy and especially exchange and deliver entropy with and to their surroundings, are able to develop levels of higher complexity and thus a higher degree of order. This ability can be viewed as the basic cause of life on our earth. We receive entropy from the sun in form of heat. At the same time we send out entropy in the form of heat to the universe. This provides the ability to decrease the entropy in certain subsystems and thus to create complex structures and to create life on earth.

For open and closed systems the following equation holds

$$dS_T = dS_I + dS_E \tag{4.08}$$

dS_T = total change of entropy
dS_I = change of entropy in the system
dS_E = change of entropy by an exchange with the environment

$dS_I \geq 0$, i.e. the entropy in a closed system always increases (this is what Clausius said)

$$dS_E \begin{cases} > 0 & \text{entropy enters the system} \\ = 0 & \text{if the system is reversible and isolated} \\ < 0 & \text{entropy leaves the system} \end{cases} \tag{4.09}$$

We conclude that

$dS_T \gtreqless 0$ is possible!

For $dS_T < 0$ the system is freed from chaos, noise or more generally from uncertainty and creates structured forms or complex arrangements increasing the information of the system because of the increasing number of patterns.

Structures with $dS_T < 0$ are called 'self-organizing' [HAK88] [HAK95] [EBE90] [EBE94].

$$dS_I + dS_E = S_T < 0$$
$$dS_I < -dS_E \tag{4.11}$$

This inequality merely occurs in a non-equilibrium state. In the Equilibrium State both changes of entropy vanish. In a non-equilibrium state the system can deliver entropy to its environment and simultaneously increase its inner entropy by a smaller amount. Thus we conclude: In the non-equilibrium state self-organizing structures can develop. We may also put it in another way: Order develops from chaos. And we actually see that all self-organizing structures are far from the Equilibrium State. Each approach to the Equilibrium State results in a disintegration of the organized structures and the attainment of the Equilibrium State finally leads to destruction of the order and for living beings it means their death. The entropy achieves its maximum and the system suffers the heat death predicated by Clausius.

But not all non-equilibrium states lead to a creation of order, some of them become chaotic. But there is also a structure in chaos, as we see in the strange attractors or in the simple, recursive mathematical rules creating chaotic processes. The logistic mapping, for instance, creates a constant quotient between the distances of two subsequent bifurcation points (Feigenbaum constant).

4.1.5 Entropy balance equation

Transitions in the entropy of systems are usually described by change or balance equations. We are not able to store entropy, which is an important difference from energy. Thus we can only use changes in the entropy occurring when entropy flows from one system to another system. Such a flow of entropy is described via an entropy balance equation.

The entropy that the system exchanges with its environment is described with a small volume V (voxel), applied to obtain an appropriate equation. In this equation we consider the following conditions:

$$\left|\begin{matrix}\text{Change of entropy}\\ \text{in the system}\end{matrix}\right| + \left|\begin{matrix}\text{Exchange of entropy}\\ \text{through the border}\end{matrix}\right| = \left|\begin{matrix}\text{Production of entropy}\\ \text{in the system}\end{matrix}\right|$$

$$\frac{\partial}{\partial t}(\rho \cdot s) + \nabla \cdot J_s = \dot{\sigma}_s \tag{4.12}$$

ρ = density
s = entropy per voxel
J_s = density of the flow of entropy, i.e. the entropy crossing the border of the system in a certain time, is a vector. Thus the product with the nabla operator ∇ provides a scalar quantity.
σ_s = rate of entropy production

This equation describes the change of the inner entropy of the system and the exchange of entropy with the environment. Both changes lead to the production rate of the entropy and are therefore another possible formulation of the description that we already discussed in the previous sections. For continuous flows, as for laminar flows, no significant production of entropy is expected, while turbulent flows actually do produce entropy.

4.1.6 Gibbs's free energy and the quality of the energy

Available energy = total energy – unavailable energy

In our cars we only transform a part of the energy contained in the fuel into the movement of the vehicle. In communication theory we may formulate the above equation as

Information = Signal – Noise

But here the concept of information has to be treated with some caution, because we again use information in the sense of the knowledge contained in a message. The common concept of information, as applied in communication theory, depends more on the noise than on the deterministic signal, when we examine conditional random processes, where the variances correspond to the inverse information.

Josiah Willard Gibbs (1839–1903) found, that the 'free energy' in a chemical process can be described by the Gibbs function:

$$F = E - T \cdot S \tag{4.13}$$

F = Gibbs's free energy
E = enthalpy (represents the energy)
T = absolute temperature
S = entropy

In this equation we detect a certain rivalry between energy and entropy. The free energy is increased by an increasing enthalpy, but it is decreased by an increase of entropy. The augmentation of the entropy of a system, the heat death described by Clausius, leads to an increase of the second term on the right-hand side. This reduces all free energy to a minimum (zero), so that the system cannot perform any work by utilizing heat energy. The condition for the generation of work from heat was a difference in temperature, as Clausius already remarked.

The complete energy of a system can be computed by adding the energies of all subsystems, because energy is an extensive quantity, i.e. a quantity we can simply add when we merge subsystems.

$U = f$(physical + chemical energy) = inner energy of a system

For ideal gases U is equal to the kinetic energy of all molecules of this gas:

$$U = W_{\text{kin}} \tag{4.14}$$

With this considerations we return to the laws of thermodynamics

1: The conservation of energy (in closed systems)

2: Increasing entropy in closed systems.

3: The theorem of Nernst ($S = 0$ for 0 K)

to explain an important train of thought regarding the quality of energy.

The first law of thermodynamics does not state anything about the quality of energy. According to Kelvin (W. Thompson and later Lord Kelvin) Carnot's law (Brillouin calls the second law of thermodynamics Carnot's law) describes a degradation of energy. This is because Kelvin divides energy according to different qualities:

High quality energy:	Mechanical, electrical energy
Medium quality energy:	Chemical energy
Low quality energy:	Heat

The total energy of a system is constant (first law). Reversible transformations preserve the quality or grade of the energy in the system. Irreversible transformations lead to a degradation or to a loss of quality of the total energy of the system. The grade of energy can beexactly defined corresponding to the negative entropy ($-S$). This degradation of the energy is the formulation used by Thompson to express the second law of thermodynamics. This expression developed approximately at the same time as Clausius's formulation of the second law. This second law says that the entropy of closed systems increases according to a monotonic function. This means that the negative entropy also decreases according to this monotonic function.

4.1.7 Considerations on the macroscopic entropy

A system receiving thermal energy Δq increases its entropy by the amount of [BRI56]

$$\Delta S = \frac{\Delta q}{T} \qquad T \cdot \Delta S = \Delta q \qquad T = \text{temperature in kelvins} \tag{4.15}$$

Let us now consider two systems A and B, which are in contact with each other but isolated from the environment.

Table 4.1: Preliminaries for the system

System	Temperature	Work (output)	Work (input)	Entropy (input)
A	T_A	W_A	q_A	$\Delta S_A = \frac{q_A}{T_A}$
B	T_B	W_B	q_B	$\Delta S_B = \frac{q_B}{T_B}$

From the first law we find:

$$W_A - q_A + W_B - q_B = 0 \qquad \text{(conservation of energy)} \tag{4.16}$$

The second law leads to:

$$\Delta S_A + \Delta S_B \geq 0 \tag{4.17}$$

4.1.7.1 Irreversible transformations

1.) We now examine the transfer of heat from the warmer to the colder body $T_A > T_B$

There is no work performed during this transfer: $W_A = W_B = 0$

We merely notice a transmission of heat: $q_A = -q_B$ $\quad q_B > 0$

From the second law we find

$$\Delta S_A + \Delta S_B = \frac{q_A}{T_A} + \frac{q_B}{T_B} = q_B\left(\frac{1}{T_B} - \frac{1}{T_A}\right) > 0 \qquad T_A > T_B \tag{4.18}$$

We notice that this transformation is irreversible, because the total entropy of the system increases and therefore the system approaches heat death.

2.) For friction or attenuation the temperature of both systems is equal $T_A = T_B = T$

System *A* produces work and heats system *B*.

$W_A > 0 \qquad q_A = 0$

$W_B = 0 \qquad q_B > 0$

From the conservation of energy we find $W_A - q_B = 0$

And the second law of thermodynamics leads us to:

$$\Delta S_A + \Delta S_B = \frac{q_A}{T_A} + \frac{q_B}{T_B} = \frac{W_A}{T} > 0 \tag{4.19}$$

Again the process is irreversible. In both examples the entropy of the total system increases.

4.1.7.2 Perpetuum mobile and transfer of heat

We have two heat reservoirs *A, B* with the temperatures $T_A > T_B$, which means that the system is able to produce work by heat transfer. The functionality of this machine is:

A: Reservoir *A* transfers the amount q_A of heat to the machine ($q_A > 0$, input to the machine)

B: The machine performs a certain work *W*.

C: The machine delivers an amount of heat q_B to reservoir *B* ($q_B < 0$, output of the machine).

The sum of all energies, resulting from the application of the first law, is

$$W - q_A - q_B = 0 \tag{4.20}$$

When the process is reversible, it has to satisfy the equations

$$\Delta S_A + \Delta S_B = 0 \qquad \Delta S_A = -\Delta S_B = \Delta S$$

$$\frac{q_A}{T_A} + \frac{q_B}{T_B} = 0 \qquad \text{or} \qquad q_B = T_B \cdot \frac{q_A}{T_A} \tag{4.21}$$

the amount of heat $q_A = \Delta S_A \cdot T_A$ transferred from the reservoir *A* must be received from the machine as $q_B = \Delta S_B \cdot T_B$. Thus we get

$$q_A + q_B = (T_A - T_B) \cdot \Delta S > 0 \tag{4.22}$$

The machine, however, does not use all the heat to heat reservoir *B*, but it applies a part of this heat energy to do the work *W*. This explains the lack of the equality in the last equation, because the process is irreversible. The thermal efficiency of this machine can be measured with the quotient

$$\text{thermal efficiency} = \frac{\text{work done by the system}}{\text{heat received by the system}}$$

$$\frac{W}{q_A} = \frac{q_A + q_B}{q_A} = \frac{T_A - T_B}{T_A} < 1 \tag{4.23}$$

The more efficiently such a machine is able to transform heat into work, the closer the thermal efficiency approaches the limit 1. In this limit all heat is transformed into work W.

4.2 Statistical entropy

4.2.1 Boltzmann's entropy

Ludwig Boltzmann (1822–1888)

Starting point:

> Newton's mechanics is reversible → time invariant, while the formulation of the second law of thermodynamics, found by William Thompson (Lord Kelvin) and Clausius, insists on a direction of time and thus depends on time (time variant).

Thus Boltzmann insisted on an asymmetric direction of time on the microscopic level. The verification of this assumption was observed as Brownian motion. For the transport in gases the equation of Boltzmanns is

$$\frac{\partial f}{\partial t} + \underline{v} \cdot \frac{\partial f}{\partial \underline{r}} + \underline{F} \cdot \frac{\partial f}{\partial \underline{v}} = \left(\frac{\partial f}{\partial t}\right)_{\text{coll}} \tag{4.24}$$

$f = f(\underline{v}, \underline{r}, t)$
$\underline{v}$ = velocity
$\underline{r}$ = position
t = time
$\underline{F} = \underline{F}(\underline{r}, t)$
$\left(\frac{\partial f}{\partial t}\right)_{\text{coll}}$ = 'collision term', describing the collision of the particles

Boltzmann developed his famous H(Eta) theorem, describing the tendency of particles to approach a uniform distribution density independent of the initial distribution. The distribution of the velocities of the particles thus approaches a Maxwell–Boltzmann distribution density.

4.2.2 Derivation of Boltzmann's entropy

4.2.2.1 Variation, permutation and the formula of Stirling

If we want to arrange a number of n elements in an arbitrary order we have $n!$ possible arrangements, when we do not consider the order of the elements. Then we take a sample of k elements, out of a set Z, consisting of n elements (with $k \leq n$), where the elements of the set Z occur without any repetition and under consideration of the order. The result of such a trial is called a variation of the class k of the set Z without repetitions [DRE75].

$$V_n^{(k)} = \frac{n!}{(n-k)!} \tag{4.25}$$

= number of variations of the class k of the set Z without repetitions

Variation considers the repetitions and the order of the sequence. If the order does not matter, such a sample is called a permutation and is described by:

$$P_n^{(k)} = \frac{V_n^{(k)}}{k!} = \frac{n!}{k! \cdot (n-k)!} = \binom{n}{k} \tag{4.26}$$

If we do not consider the inequality of the equation $k \leq n$, but focus our attention on the equality, we get

$$P_n^{(n)} = \frac{V_n^{(n)}}{n!} = \frac{n!}{n! \cdot (n-n)!} = 1 \tag{4.27}$$

As soon as we are able to arrange n elements in an arbitrary way with consideration of the order, we have $n!$ possible arrangements. When the order of the elements is not important, we do not distinguish the possible arrangements and thus all arrangements are equal to us. We only have the event 'all n elements are somehow arranged'.

Now we consider different sets E_i with n, n_1, n_2,..., n_m elements. In this case the number of possible arrangements is:

$$P_n = n! \quad \text{for } n \text{ elements} \tag{4.28}$$

$$P_{n_1,n_2} = n_1! \cdot n_2! \quad \text{for } n_1 \text{ elements in the set } E_1 \text{ and } n_2 \text{ in the set } E_2$$

The arrangements of the sets E_1 and E_2 are independent, which allows us to compute the total number of arrangements from the product.

$$P_{n_1,n_2} = P_{n_1} \cdot P_{n_2} = n_1! \cdot n_2!$$

Boltzmann now uses the number n to count the possible microscopic states of the system and the numbers n_i to count the number of molecules of a substance (gas), residing in the ith state. The micro states are also known as the Planck complexions or Planck micro states.

Thus the number of possible arrangements of m states of a substance is given by the equation

$$P_{n_1,n_2,\ldots,n_m} = \prod_{i=1}^{m} P_{n_i} = \prod_{i=1}^{m} n_i! \tag{4.29}$$

The ratio of the possible arrangements of the states of n molecules and the possible allocation of m states with n_i molecules is thus

$$P = \frac{P_n}{P_{n_1,n_2,\ldots,n_m}} = \frac{n!}{\prod_{i=1}^{m} n_i!} \tag{4.30}$$

And Boltzmann uses this number to define his entropy

$$S = k \cdot \ln P \tag{4.31}$$

P is the statistical weight or the thermodynamic probability and it can be calculated from the entropy. The entropy can be observed and thus

$$P = e^{\frac{S}{k}} \tag{4.32}$$

When the entropy increases, the possibility that a state with higher probability is adopted in the system also increases. The Boltzmann constant results from the relation of this entropy of statistical mechanics to the classical, physical entropy of thermodynamics

$$S = \frac{dQ}{dT} \quad = \text{thermal energy per temperature} \tag{4.33}$$

introduced by Clausius.

We are able to measure macroscopic variables. Microscopic variables, on the other hand, are so numerous that we can never observe all desired variables (positions of atoms, their velocities, quantum numbers of the atoms or molecules, ...). All these quantities allow the system to assume a great number of quantized states, the so-called 'Planck complexions' [BRI64]. P is the total number of the Planck complexions, used to define the entropy. The entropy S is therefore a measure for the ratio of all possible arrangements of states to the actual arrangements.

$$k = \frac{R}{N_A} = 1.38 \cdot 10^{-23} \frac{\mathrm{J}}{\mathrm{K}} \qquad = \text{Boltzmann's constant} \tag{4.34}$$

$$R = 8.317 \frac{\mathrm{J}}{\mathrm{mol} \cdot \mathrm{K}} \qquad = \text{gas constant} \tag{4.35}$$

$$N_A = 6.025 \cdot 10^{23} \frac{1}{\mathrm{mol}} \qquad = \text{Avogadro's constant} \tag{4.36}$$

This measure can easily be compared with Hartley's information, because in Hartley's information measure we have

$$I(E_N) = n = \mathrm{ld}\, N = \frac{\ln N}{\ln 2} \mathrm{bit} \qquad \text{(see Hartley's information)} \tag{4.37}$$

and the number N of the equally probable arrangements of the states is used to assign an information to the system. For Boltzmann's entropy measure we have, apart from the additional factor k, exactly the same approach. All states are assigned the same, equal probability. But we have n elements distributed over the events E_i, which in each case can be occupied by n_i elements. Thus we may also use the average of the possibilities of one element and write

$$S = \frac{k}{n} \cdot \ln P \tag{4.38}$$

allowing different probabilities for the sub-events E_i.

Now we are able transform the equation of Boltzmann by applying the formula of Stirling to approximate the factorial. This approximation of the factorial is done with the equation

$$n! = \left(\frac{n}{e}\right)^n \cdot \sqrt{2\pi \cdot n} \cdot \left(1 + \frac{1}{n}\right) \tag{4.39}$$

Usually, for large values of n, we use the form

$$n! \approx \left(\frac{n}{e}\right)^n \cdot \sqrt{2\pi \cdot n} \tag{4.40}$$

and approximate the approximation error with the inequality

$$\left(n + \frac{1}{2}\right) \cdot \ln n - n + \frac{3}{2} \cdot \left(1 - \ln \frac{3}{2}\right) < \ln n! < \left(n + \frac{1}{2}\right) \cdot \ln n - n + 1 \tag{4.41}$$

Here it is sufficient to use

$$n! \approx \left(\frac{n}{e}\right)^n \tag{4.42}$$

Inserting the approximation in the averaged equation

$$S = \frac{k}{n} \cdot \ln P \tag{4.43}$$

with

$$P = \frac{P_n}{P_{n_1,n_2,\ldots,n_m}} = \frac{n!}{\prod_{i=1}^{m} n_i!}$$

we get

$$S = \frac{k}{n} \cdot \ln \frac{P_n}{P_{n_1,n_2,\ldots,n_m}} = \frac{k}{n} \cdot \ln \frac{n!}{\prod_{i=1}^{m} n_i!} = \frac{k}{n} \cdot \left(\ln n! - \ln \prod_{i=1}^{m} n_i! \right) \tag{4.44}$$

and Stirling's equation becomes

$$\ln n! = n \ln n - n \tag{4.45}$$

As we may neglect the additive square in equation (4.40) for sufficiently large n, we get the equation

$$S \approx \frac{k}{n} \cdot \left(\ln n! - \sum_{i=1}^{m} \ln n_i! \right)$$

$$S \approx \frac{k}{n} \cdot \left(n \cdot \ln n - n - \sum_{i=1}^{m} [n_i \cdot \ln n_i - n_i] \right)$$

$$S \approx \frac{k}{n} \cdot \left(n \cdot \ln n - n - \sum_{i=1}^{m} n_i \cdot \ln n_i + \sum_{i=1}^{m} n_i \right) \tag{4.46}$$

with $n = \sum_{i=1}^{m} n_i$ we now get

$$S \approx \frac{k}{n} \cdot \left(n \cdot \ln n - n - \sum_{i=1}^{m} n_i \cdot \ln n_i + n \right)$$

$$S \approx \frac{k}{n} \cdot \left(n \cdot \ln n - \sum_{i=1}^{m} n_i \cdot \ln n_i \right) \tag{4.47}$$

This is Boltzmann's entropy, resulting from the ratio of the number of all possible permutation of n elements to the number of permutations that can be achieved for m states. Further transformation of the equation leads us to

$$S \approx \frac{k}{n} \cdot \left(n \cdot \ln n - \sum_{i=1}^{m} n_i \cdot \ln n_i \right)$$

$$S \approx \frac{k}{n} \cdot \left(\sum_{i=1}^{m} n_i \cdot \ln n - \sum_{i=1}^{m} n_i \cdot \ln n_i \right)$$

$$S \approx \frac{k}{n} \cdot \left(- \sum_{i=1}^{m} n_i \cdot \ln\left(\frac{n_i}{n} \right) \right) \tag{4.48}$$

This term nearly equals Kullback's information, but in Kullback's information we use the probabilities instead of the number of states. Expansion with n gives

$$S \approx \frac{k}{n} \cdot n \left(- \sum_{i=1}^{m} \frac{n_i}{n} \cdot \ln\left(\frac{n_i}{n} \right) \right) \tag{4.49}$$

and now we can use the relative frequency for the quotients, which results in the equation

$$S \approx k \cdot \left(- \sum_{i=1}^{m} h(A_i) \cdot \ln(h(A_i)) \right) \tag{4.50}$$

or, using the probabilities instead of the relative frequencies

$$S \approx k \cdot \left(- \sum_{i=1}^{m} P(A_i) \cdot \ln(P(A_i)) \right) = k \cdot \left(- \sum_{i=1}^{m} p_i \cdot \ln p_i \right)$$

$$S \approx -k \cdot \sum_{i=1}^{m} p_i \cdot \ln p_i \tag{4.51}$$

This is Boltzmann's entropy [BOL81], which is connected to Shannon's entropy by the equation

$$S \approx k \cdot S_{\text{Shannon}} \tag{4.52}$$

If all events have the same probability $p_i = 1/m$ we get

$$S \approx -k \cdot \sum_{i=1}^{m} \frac{1}{m} \cdot \ln\left(\frac{1}{m}\right) = k \cdot \ln m \tag{4.53}$$

This shows the dependence on the number m of available events E_i, each occupied with n_i elements. This, of course, is Hartley's information with respect to the events that all occur with the same probability.

Shannon's information uses different probabilities of sets E_i. The elements of these sets occur with equal probability. For the characterization of a set E_i containing a certain element we thus in the mean need the information, which is described by Shannon's measure.

If we generate a sequence of length n by sequentially adding independent, equally distributed elements, the entropy of the sequence equals n times the entropy of a single element n. This is exactly the case when we fill n states with molecules, and the energies of these molecules are mutually independent. To compute the probability of the levels of energy E_i occupied by the molecules we need the information

$$S \approx k \cdot S_{\text{Shannon}} \tag{4.54}$$

4.2.2.2 Special case: Two states

If we look at the entropy of a system consisting of N states and two groups of molecules, each being able to assume N_1 and N_2 states, respectively, we find:

$$S = k \cdot \ln P \tag{4.55}$$

$$S = k \cdot \ln\left(\frac{N!}{N_1! \cdot N_2!}\right) = k \cdot \left(N \ln N - N_1 \cdot \ln N_1 - N_2 \cdot \ln N_2\right) \tag{4.56}$$

As we obtain the number of total states from the equation $N_1 + N_2 = N$, we proceed:

$$S = k \cdot \left(\left(N_1 + N_2\right) \cdot \ln\left(N_1 + N_2\right) - N_1 \cdot \ln N_1 - N_2 \cdot \ln N_2\right)$$

$$S = k \cdot \left(N_1 \cdot \left[\ln\left(N_1 + N_2\right) - \ln N_1\right] + N_2 \cdot \left[\ln\left(N_1 + N_2\right) - \ln N_2\right]\right)$$

$$S = k \cdot \left(N_1 \cdot \ln\left(\frac{N_1 + N_2}{N_1} \right) + N_2 \cdot \ln\left(\frac{N_1 + N_2}{N_2} \right) \right)$$

$$S = -k \cdot (N_1 + N_2) \cdot \left(\frac{N_1}{N_1 + N_2} \cdot \ln\left(\frac{N_1}{N_1 + N_2} \right) + \frac{N_2}{N_1 + N_2} \cdot \ln\left(\frac{N_2}{N_1 + N_2} \right) \right) \quad (4.57)$$

The quotients are now replaced by the relative frequencies, which leads us to the equation

$$S = -k \cdot N \cdot (h(N_1) \cdot \ln(h(N_1)) + h(N_2) \cdot \ln(h(N_2))) \quad (4.58)$$

Now we divide this equation by the number N of total states and thus find the entropy per element

$$S = -k \cdot (h(N_1) \cdot \ln(h(N_1)) + h(N_2) \cdot \ln(h(N_2)))$$

$$S = -k \cdot \sum_{i=1}^{2} h(N_i) \cdot \ln(h(N_i)) \quad (4.59)$$

If we additionally use $P = e^{\frac{S}{k}}$ to find the number of possible arrangements, we are able to assign the probabilities for a uniform distribution function.

4.2.2.3 Example: Lottery

In our example we want to find the possible arrangements of six out of 49 numbers. This number of arrangements can be computed as

$$\frac{n!}{k! \cdot (n-k)!} = \frac{49!}{6! \cdot 43!} = \frac{6.0828 \cdot 10^{62}}{720 \cdot 6.0415 \cdot 10^{52}} = 13\ 983\ 816 \quad (4.60)$$

Here we merely have two states. One state consisting of the six numbers, which we call hits, and the second state consisting of the remaining 43 numbers, which we summarize as flops. If we calculate the entropy of this lottery, we find

$$S = k \cdot \ln \frac{49!}{6! \cdot 43!} = 1.38 \cdot 10^{-23} \frac{\mathrm{J}}{\mathrm{K}} \cdot \ln 13\ 983\ 816 = 2.2705 \cdot 10^{-22} \frac{\mathrm{J}}{\mathrm{K}} \quad (4.61)$$

and this is really small. When we compare it with thermodynamic entropies

$$S = \frac{dQ}{T} \quad (4.62)$$

we would have to generate thermal energy of $dQ = 6.6528 \cdot 10^{-20}\,\mathrm{J}$ at a temperature of 293 K to get such an entropy with thermodynamic methods. We thus remark that the statistical entropy leads us to very small values, which has also been stated by Wolfowitz [WOL56], who tries to measure the information production of the human race by means of entropy. He therefore assumes that the human race has written approximately 10^8 books, each consisting of 40 000 mutually independent characters, all assuming the information of 5 bit. The change (reduction) in entropy by the generation of this enormous order is $S = 5 \cdot 10^{-9}\,\text{J/K}$. Again this is a very small value for the complete information which has been produced by mankind. This is, of course, caused by the very small value of the Boltzmann constant. But we notice that the measurement of information in thermodynamic units obviously leads to quantities that we have no experience of.

The quantity *H* or *S*, as we called it here, is a measure for the distance between the actual state of the system and the Equilibrium State, because in this Equilibrium State the system achieves a maximum entropy (which we are going to examine in detail in later chapters).

$$H = -\sum_i p_i \cdot \ln p_i \tag{4.63}$$

Max Planck described the relation of this notation to entropy

$$S = k \cdot \log W \tag{4.64}$$

W = number of micro states a system is able to assume [vWE85] [BOL81]

$$S = k \cdot H = -k \cdot \sum_i p_i \cdot \ln p_i \tag{4.65}$$

$$S_{\text{ges}} = S_1 + S_2 \tag{4.66}$$

$$W_{\text{ges}} = W_1\, W_2 \tag{4.67}$$

Boltzmann's entropy is an indication of disorder (measure of chaos) and an announcement of a lack of information.

Before we proceed with an examination of the entropy in the Equilibrium State, we want to explain the Boltzmann factor by taking a closer look at the entropy in a non-equilibrium state.

4.2.3 The Boltzmann factor

We started the derivation of Boltzmann's entropy with the polynomial coefficient

$$P = \frac{P_n}{P_{n_1, n_2, \ldots, n_m}} = \frac{n!}{\prod_{i=1}^{m} n_i!} \tag{4.68}$$

where

$$n = \sum_{i=1}^{m} n_i \tag{4.69}$$

This term describes the number of possible arrangements that can be achieved with respect to the given preliminaries. However, there may always occur statistical fluctuations in the form of deviations from the stable equilibrium and we want to examine them as drifts of a particle from n_i to n_{i+1} and of another particle from n_i to n_{i-1}. We do not assume any preferred direction of these drifts, which leads us to a symmetrical drift. The change in our polynomial coefficient is therefore

$$P_D = \frac{n!}{n_1! \cdot \ldots \cdot (n_{i-1}+1)! \cdot (n_i - 2)! \cdot (n_{i+1}+1)! \cdot \ldots \cdot n_m!} \tag{4.70}$$

The ratio of both coefficients becomes

$$\frac{P_D}{P} = \frac{\dfrac{n!}{n_1! \cdot \ldots \cdot (n_{i-1}+1)! \cdot (n_i - 2)! \cdot (n_{i+1}+1)! \cdot \ldots \cdot n_m!}}{\dfrac{n!}{n_1! \cdot \ldots \cdot (n_{i-1})! \cdot (n_i)! \cdot (n_{i+1})! \cdot \ldots \cdot n_m!}} \tag{4.71}$$

$$\frac{P_D}{P} = \frac{(n_{i-1})! \cdot (n_i)! \cdot (n_{i+1})!}{(n_{i-1}+1)! \cdot (n_i - 2)! \cdot (n_{i+1}+1)!} = \frac{n_i \cdot (n_i - 1)}{(n_{i-1}+1) \cdot (n_{i+1}+1)} \tag{4.72}$$

For $n_j >> 1$ with $j = 1, \ldots, i-1, i, i+1, \ldots, m$ we proceed

$$\frac{P_D}{P} \cong \frac{n_i^2}{n_{i-1} \cdot n_{i+1}} = \frac{n_i}{n_{i+1}} : \frac{n_{i-1}}{n_i} \tag{4.73}$$

Now we have

$$\frac{n_i}{n_{i+1}} = \frac{n_{i-1}}{n_i} = q \tag{4.74}$$

the ratio of the occupation of two subsequent states, which are symmetrical, as we did not presume any preference in the direction of the drifts.
With the boundary conditions

$$n = \sum_{i=1}^{m} n_i \quad \text{as a limited number of } m \text{ states, occupied } n_i \text{ times} \tag{4.75}$$

and

$$E = \sum_{i=1}^{m} E_i \cdot n_i \tag{4.76}$$

= a bounded energy of the total system, with the partial energies E_i

we cannot choose a constant ratio $q = 1$ and thus the solution leads us to

$$n_1 = n_0 \cdot \exp(-\beta \cdot E_1)$$

resp.

$$n_i = n_0 \cdot \exp(-\beta \cdot E_i) \tag{4.77}$$

and therefore

$$\begin{aligned} \frac{n_i}{n_{i+1}} = q = \frac{n_0 \cdot \exp(-\beta \cdot E_i)}{n_0 \cdot \exp(-\beta \cdot E_{i+1})} &= \exp(-\beta \cdot E_i + \beta \cdot E_{i+1}) \\ &= \exp[\beta \cdot (E_{i+1} - E_i)] = \exp(\beta \cdot \varepsilon) \end{aligned} \tag{4.78}$$

Now we have to make an appropriate choice for the remaining constant β, and we start with a consideration of the dependence on temperature. If the temperature of a gas decreases, the number of energy portions ε also decreases, because the molecules are only able to occupy lower energy levels. Thus the number of higher level states approaches zero, and this leads to

$$n_i \to 0 \qquad \text{for} \qquad i > 1 \tag{4.79}$$

Because of

$$n_i = n_0 \cdot \exp(-\beta \cdot E_i) \tag{4.80}$$

this implies a proportionality of the form

$$\beta \sim \frac{1}{T}$$

Then β goes to infinity, when the temperature goes to zero and vanishes

$$n_i = n_0 \cdot \exp(-\beta \cdot E_i) \tag{4.82}$$

Furthermore the supply with an energy portion ε can increase the entropy in a certain area (thus the small letter s) of the system by the amount

$$\Delta s = \frac{\varepsilon}{T} \qquad \text{from the thermodynamic entropy } ds = dQ/T \tag{4.83}$$

This results in a fluctuation, causing a drift, which we describe with the statistical entropy

$$\Delta s = k \cdot \ln \frac{p_D}{p} = k \cdot \ln \frac{n_i}{n_{i+1}} = k \cdot \ln q = k \cdot \ln \exp(\beta \cdot \varepsilon) = k \cdot \beta \cdot \varepsilon \tag{4.84}$$

Equating these two entropies, we find

$$\frac{\varepsilon}{T} = k \cdot \beta \cdot \varepsilon$$

and

$$\beta = \frac{1}{k \cdot T} \tag{4.85}$$

Thus the occupations of the states (Planck's micro states or Planck's complexions) can be described by the equation

$$n_i = n_0 \cdot \exp\left(-\frac{E_i}{k \cdot T}\right) \tag{4.86}$$

consisting of a basic number n_0, the energy of the state E_i, the temperature T and the Boltzmann constant k. The energy E_i determines the number of energy portions ε needed for the occupation of the state. When the temperature rises, the possibility that such a state is occupied increases. With

$$n = \sum_{i=1}^{m} n_i \tag{4.87}$$

and the quotient

$$Z = \frac{n}{n_0} = \frac{1}{n_0} \cdot \sum_{i=1}^{m} n_i = \sum_{i=1}^{m} \exp\left(-\frac{E_i}{k \cdot T}\right) \tag{4.88}$$

we are able to determine the possibility of occurrence of a state with the energy E_i. This possibility is equal to

$$p_i = \frac{n_i}{n} = \frac{\exp\left(-\frac{E_i}{k \cdot T}\right)}{\sum_{i=1}^{m} \exp\left(-\frac{E_i}{k \cdot T}\right)} \tag{4.89}$$

=probability that a molecule occupies the ith state with the energy E_i.

4.2.4 Maximum entropy in equilibrium

We now maximize the statistical entropy under the following constraints

$$P = \frac{P_n}{P_{n_1,n_2,\ldots,n_m}} = \frac{n!}{\prod_{i=1}^{m} n_i!} \tag{4.90}$$

$$n = \sum_{i=1}^{m} n_i \tag{4.91}$$

and

$$E = \sum_{i=1}^{m} n_i \cdot E_i \tag{4.92}$$

This leads us to an optimization with Lagrangian multipliers, allowing us to include these constraints. We get

$$S = k \cdot \ln P \tag{4.93}$$

or

$$S = k \cdot \ln \frac{n!}{\prod_{i=1}^{m} n_i!} + \lambda \cdot \left(n - \sum_{i=1}^{m} n_i \right) + \mu \cdot \left(E - \sum_{i=1}^{m} n_i \cdot E_i \right) \tag{4.94}$$

With the approximation obtained by Stirling's formula, this becomes

$$S = k \cdot \left(n \cdot \ln n - \sum_{i=1}^{m} n_i \cdot \ln n_i \right) + \lambda \cdot \left(n - \sum_{i=1}^{m} n_i \right) + \mu \cdot \left(E - \sum_{i=1}^{m} n_i \cdot E_i \right) \tag{4.95}$$

The parameter for our optimization is n_i = number of the ith states. Thus the derivation becomes:

$$\frac{\partial S}{\partial n_i} = \frac{\partial}{\partial n_i} k \cdot \left(n \cdot \ln n - \sum_{i=1}^{m} n_i \cdot \ln n_i \right) + \frac{\partial}{\partial n_i} \lambda \cdot \left(n - \sum_{i=1}^{m} n_i \right) + \frac{\partial}{\partial n_i} \mu \cdot \left(E - \sum_{i=1}^{m} n_i \cdot E_i \right)$$

$$\frac{\partial S}{\partial n_i} = k \cdot \left(-\ln n_i - n_i \cdot \frac{1}{n_i} \right) + \lambda \cdot (-1) + \mu \cdot (-E_i)$$

$$\frac{\partial S}{\partial n_i} = -k \cdot \ln n_i - k - \lambda - \mu \cdot E_i = 0$$

$$k \cdot \ln n_i = -k - \lambda - \mu \cdot E_i$$

$$n_i = \exp\left(-1 - \frac{\lambda}{k} - \frac{\mu}{k} \cdot E_i\right) \tag{4.96}$$

The remaining Lagrangian multipliers λ and μ are now determined by inserting the present result in our constraints.

1. $\quad n = \sum_{i=1}^{m} n_i$

$$n = \sum_{i=1}^{m} \exp\left(-1 - \frac{\lambda}{k} - \frac{\mu}{k} \cdot E_i\right) = \exp\left(-1 - \frac{\lambda}{k}\right) \cdot \sum_{i=1}^{m} \exp\left(-\frac{\mu}{k} \cdot E_i\right)$$

$$\exp\left(1 + \frac{\lambda}{k}\right) \cdot n = \sum_{i=1}^{m} \exp\left(-\frac{\mu}{k} \cdot E_i\right) \qquad \text{first constraint} \tag{4.97}$$

2. $\quad E = \sum_{i=1}^{m} n_i \cdot E_i$

$$E = \sum_{i=1}^{m} \exp\left(-1 - \frac{\lambda}{k} - \frac{\mu}{k} \cdot E_i\right) \cdot E_i = \exp\left(-1 - \frac{\lambda}{k}\right) \cdot \sum_{i=1}^{m} \exp\left(-\frac{\mu}{k} \cdot E_i\right) \cdot E_i$$

$$\exp\left(1 + \frac{\lambda}{k}\right) \cdot E = \sum_{i=1}^{m} \exp\left(-\frac{\mu}{k} \cdot E_i\right) \cdot E_i \qquad \text{second constraint} \tag{4.98}$$

Now the equation has to be solved to obtain the multipliers. The normal way to solve this equation is a numerical solution, which – depending on the magnitude of m – can entail a remarkable cost. But if we use the equation $\lambda = -k$ in both equations, we obtain

$$\exp\left(1 + \frac{-k}{k}\right) \cdot n = n = \sum_{i=1}^{m} \exp\left(-\frac{\mu}{k} \cdot E_i\right)$$

$$\exp\left(1 + \frac{-k}{k}\right) \cdot E = E = \sum_{i=1}^{m} \exp\left(-\frac{\mu}{k} \cdot E_i\right) \cdot E_i \tag{4.99}$$

and with $\mu = 1/T$ we finally find

$$E = \sum_{i=1}^{m} \exp\left(-\frac{1}{kT} \cdot E_i \right) \cdot E_i \tag{4.100}$$

$$n = \sum_{i=1}^{m} \exp\left(-\frac{1}{kT} \cdot E_i \right) \qquad \text{with} \qquad n_i = \exp\left(-\frac{1}{kT} \cdot E_i \right) \tag{4.101}$$

These are exactly the terms that we already obtained in the examination of the drift caused by fluctuations. With these quantities we have the necessary prerequisites to derive and describe the Boltzmann distribution. If we use

$$p_i = \frac{n_i}{n} = \frac{\exp\left(-\frac{1}{kT} \cdot E_i \right)}{\sum_{i=1}^{m} \exp\left(-\frac{1}{kT} \cdot E_i \right)} \tag{4.102}$$

(probability that a molecule occupies the ith state with the energy E_i) then the distribution density, described by the probabilities p_i is the Boltzmann distribution.

An important physical quantity of the Boltzmann distribution is

$$Z = \sum_{i=1}^{m} \exp\left(-\frac{1}{kT} \cdot E_i \right) \qquad = \text{sum of states of a system} \tag{4.103}$$

Thus we get:

$$p_i = \frac{n_i}{n} = \frac{\exp\left(-\frac{1}{kT} \cdot E_i \right)}{Z} \tag{4.104}$$

and we are able to derive all thermodynamic characteristics of a body (energy, entropy and free energy) on the basis of its sum of states.

For 1 mol of a gas, for instance, we have

$$E = R \cdot T^2 \cdot \frac{d \ln Z}{dT} \tag{4.105}$$

$$S = \frac{E}{T} + R \cdot \ln Z \tag{4.106}$$

The free energy is determined by the equation of Gibbs (4.13)

$$F = E - T \cdot S = -R \cdot T \cdot \ln Z \tag{4.107}$$

and we get

$$Z = \exp\left(-\frac{F}{R \cdot T}\right) \tag{4.108}$$

The sum of states is connected to the free energy F with a simple exponential term. We may also write

$$Z = \exp\left(-\frac{E}{k \cdot T}\right) \tag{4.109}$$

The energy E is composed of the sum of the different energies occurring in the observed system. For example

$$E = E_{\text{translation}} + E_{\text{rotation}} + E_{\text{oscillation}} + E_{\text{electric}} + \ldots \tag{4.110}$$

This additive connection of the energies affects the sum of states in such a way, that we obtain a multiplicative connection

$$Z = \exp\left(-\frac{E_{\text{translation}} + E_{\text{rotation}} + E_{\text{oscillation}} + E_{\text{electric}} + \ldots}{k \cdot T}\right)$$

$$Z = \exp\left(-\frac{E_{\text{translation}}}{k \cdot T}\right) \cdot \exp\left(-\frac{E_{\text{rotation}}}{k \cdot T}\right) \cdot \exp\left(-\frac{E_{\text{oscillation}}}{k \cdot T}\right) \cdot \exp\left(-\frac{E_{\text{electric}}}{k \cdot T}\right) \cdot \ldots$$

$$Z = Z_{\text{translation}} \cdot Z_{\text{rotation}} \cdot Z_{\text{oscillation}} \cdot Z_{\text{electric}} \cdot \ldots \tag{4.112}$$

We need this expression in Boltzmann's distribution density, describing the states and the energies.

If we reduce the energies of the previous equations to the kinetic energy, we obtain the Maxwell–Boltzmann distribution density as a special case of Boltzmann's distribution density [WOL90]. We want to derive the Maxwell–Boltzmann distribution now, because it is a three-dimensional gaussian distribution density. The gaussian distribution density is the distribution density possessing the maximum entropy, when we know only the second order moments (here mean and kinetic energy) as constraints. We thus start with the kinetic energy:

$$E_{\text{kin}} = \frac{1}{2} \cdot m \cdot v^2 \tag{4.113}$$

If we examine the three components of the velocity vector instead of a one-dimensional component of the velocity, we obtain the following equation because of the independence of these three components:

$$E_{\text{kin}} = \frac{1}{2} \cdot m \cdot \left(v_x^2 + v_y^2 + v_z^2\right) \tag{4.114}$$

The number of particles having a velocity in the interval between $\underline{v}$ and $\underline{v} + d\underline{v}$ is given by the formula

$$dN(\underline{v}) = N \cdot f(\underline{v}) \cdot d\underline{v} \tag{4.115}$$

N is the total number of particles and $f(\underline{v})$ is the distribution density of the velocities, which we may write as the 'continuous version' of the already known distribution density

$$p_i = \frac{n_i}{n} = \frac{\exp\left(-\frac{1}{kT} \cdot E_i\right)}{Z} = \frac{1}{Z} \cdot \exp\left(-\frac{m \cdot v_i^2}{2kT}\right) \tag{4.116}$$

This leads to the continuous version

$$f(\underline{v}) = \frac{1}{Z} \cdot \exp\left(-\frac{m}{2kT} \cdot \left(v_x^2 + v_y^2 + v_z^2\right)\right) \tag{4.117}$$

We now determine the sum of states Z with respect to the constraint that the distribution density of the velocities has to meet. This is

$$\int_{\mathrm{Y}} f(\underline{v}) \cdot d\underline{v} = 1$$

$$\frac{1}{Z} \cdot \int_{-\infty}^{\infty} \exp\left(-\frac{m}{2kT} \cdot \left(v_x^2 + v_y^2 + v_z^2\right)\right) \cdot d\underline{v} = 1 \tag{4.118}$$

As we examine a distribution density with three mutually independent components, we may also write:

$$\frac{1}{Z_x} \cdot \int_{-\infty}^{\infty} \exp\left(-\frac{m}{2kT} \cdot v_x^2\right) \cdot dv_x \cdot \frac{1}{Z_y} \cdot \int_{-\infty}^{\infty} \exp\left(-\frac{m}{2kT} \cdot v_y^2\right) \cdot dv_y$$
$$\cdot \frac{1}{Z_z} \cdot \int_{-\infty}^{\infty} \exp\left(-\frac{m}{2kT} \cdot v_z^2\right) \cdot dv_z = 1 \tag{4.119}$$

This is a product of three independent distribution densities and this product only achieves the value 1 if all three factors are equal to 1, as these factors can only achieve values in the range from 0 to 1. This means that

$$\frac{1}{Z_x} \cdot \int_{-\infty}^{\infty} \exp\left(-\frac{m}{2kT} \cdot v_x^2\right) \cdot dv_x = 1 \tag{4.120}$$

From the gaussian distribution we know that

$$\int_{-\infty}^{\infty} \exp\left(-\frac{1}{2} \cdot \frac{\xi^2}{\sigma^2}\right) \cdot d\xi = \sqrt{2\pi} \cdot \sigma \tag{4.121}$$

By comparing the coefficients we find

$$\sigma^2 = \frac{kT}{m} \tag{4.122}$$

and thus

$$\int_{-\infty}^{\infty} \exp\left(-\frac{m}{2kT} \cdot v_x^2\right) \cdot dv_x = Z_x = \sqrt{\frac{2\pi kT}{m}} \tag{4.123}$$

finally

$$Z = Z_x \cdot Z_y \cdot Z_z = \left(\sqrt{\frac{2\pi kT}{m}}\right)^3 \tag{4.124}$$

Thus the number of molecules that have a velocity in the range from $\underline{v}$ to $\underline{v} + d\underline{v}$ is given by the equation

$$dN(\underline{v}) = N \cdot f(\underline{v}) \cdot d\underline{v}$$

$$dN(\underline{v}) = N \cdot \frac{1}{Z} \cdot \exp\left[-\frac{m}{2kT} \cdot \left(v_x^2 + v_y^2 + v_z^2\right)\right] \cdot d\underline{v}$$

$$dN(\underline{v}) = N \cdot \left(\sqrt{\frac{2\pi kT}{m}}\right)^{-3} \cdot \exp\left[-\frac{m}{2kT} \cdot \left(v_x^2 + v_y^2 + v_z^2\right)\right] \cdot d\underline{v} \tag{4.125}$$

The *Maxwell–Boltzmann distribution density* is a three-dimensional zero-mean gaussian distribution density of the velocities with a variance of kT/m. If we are merely interested in the absolute values of the velocities, we obtain the Maxwell distribution.

For limited velocities, we are able to determine the number of molecules occurring in the observed interval. The necessary equation is:

$$N(v_1, v_2) = \int_{v_1}^{v_2} dN(\underline{v}) = N \cdot \int_{v_1}^{v_2} \left(\sqrt{\frac{2\pi kT}{m}} \right)^{-3} \cdot \exp\left[-\frac{m}{2kT} \cdot \left(v_x^2 + v_y^2 + v_z^2 \right) \right] \cdot d\underline{v} \tag{4.126}$$

Now the components of the velocities in this equation can only achieve positive values, because of the squares. Thus all particles with the same velocity are arranged on a surface of a sphere and the number of particles on this surface results from the formulation

$$N(\underline{v})_{\text{sphere}} = N(\underline{v}) \cdot 4\pi \cdot \underline{v}^2 \tag{4.127}$$

$4\pi \cdot \underline{v}^2$ = surface of a sphere with radius $|\underline{v}|$.

The average velocity of all particles located in this system can be found by integrating over all these numbers $N(\underline{v})_{\text{sphere}}$ and we get:

$$\overline{\underline{v}^2} = \int_0^\infty \underline{v}^2 \cdot N(\underline{v})_{\text{sphere}} \cdot d\underline{v} = \int_0^\infty \underline{v}^2 \cdot N(\underline{v}) \cdot 4\pi \cdot \underline{v}^2 \cdot d\underline{v}$$

$$\overline{\underline{v}^2} = 3 \cdot \frac{kT}{m} \tag{4.128}$$

thus the mean of the kinetic energy becomes

$$\overline{E}_{\text{kin}} = \frac{1}{2} \cdot m\overline{\underline{v}^2} = \frac{3}{2} \cdot kT \tag{4.129}$$

We get the same result when we apply the gas law and assume that the inner energy is merely a kinetic energy.

4.2.5 Statistical interpretation of entropy

Entropy, defined in thermodynamics by

$$\Delta S = \frac{\Delta q}{T} \tag{4.130}$$

corresponds, according to the interpretation of Lord Kelvin, to the inverse grade of energy or to the quality of energy. Because of the second law of thermodynamics the entropy monotonically increases in closed systems, leading to a continuous degradation of energy (loss of quality of the energy).

Because of the enormous number of atoms, molecules, etc., which are located in the substances, we are not able to detect or measure each particle. We observe averages, which we also intend to obtain from the theories, describing the microscopic world.

The statistical definition of the entropy is therefore connected to the probabilities and this entropy also has to increase in closed systems. The exact definition of the statistical entropy is the Boltzmann–Planck equation:

$$S = k \cdot \ln P \tag{4.131}$$

P = number of elementary complexions

k = Boltzmann's constant = $1.38 \cdot 10^{-23}$ J/K

see also: $\Delta S = \dfrac{\Delta q}{T}$ in J/K from equation (4.130).

These considerations show us that the number P, i.e. the ratio of all possible arrangements of states to the possible occupations of the states, also has to increase monotonically. The occupation thus tends to the most probable arrangement, which is the arrangement with the maximum number of possibilities. So each system tends to approach the state in which it has the maximum possibilities to occupy the available states.

Let us take a look at some other examples of statistical entropy in physics. We are going to use these examples in our examination of the random (noise) processes disturbing the transmission on a communication channel.

4.2.6 Examples regarding statistical entropy

Given: Two bodies A_1, A_2 in an isolated container V.

$$E_1 + E_2 = E = \text{const.} \tag{4.132}$$

Examining the transport of heat $q>0$ from A_1 to A_2 we get

$$\Delta E_1 = q \qquad \Delta E_2 = -q \tag{4.133}$$

What we are looking for is the most probable distribution of energy between A_1 and A_2.

Approach: The number of complexions P_1 for A_1 is a function of the energy E_1 (an increasing energy leads to an increased number of elementary states accessible to the system). The same holds for P_2 and E_2.

$$P = P_1(E_1) \cdot P_2(E_2) \tag{4.134}$$

because every state of A_1 can occur with every state of A_2.

In the maximum of P with respect to the variable q we get $\Delta P = \frac{dP}{dq} \cdot q = 0$, because the first order deviation does not lead to a change of the function value, as gradient of the tangent at an extreme point is equal to zero [FEY72].

$$\Delta P = \frac{dP}{dq} \cdot q = \left(-\frac{dP_1}{dE_1} \cdot P_2 + P_1 \cdot \frac{dP_2}{dE_2} \right) \cdot q = 0$$

$$\frac{dP_1}{dE_1} \cdot P_2 = P_1 \cdot \frac{dP_2}{dE_2}$$

$$\frac{1}{P_1} \cdot \frac{dP_1}{dE_1} = \frac{1}{P_2} \cdot \frac{dP_2}{dE_2}$$

$$\frac{d}{dE_1} \ln P_1 = \frac{d}{dE_2} \ln P_2 \tag{4.135}$$

We use $S = k \cdot \ln P$ and obtain the formulation:

$$\frac{d}{dE_1} \frac{S_1}{k} = \frac{d}{dE_2} \frac{S_2}{k} \tag{4.136}$$

Now the exchange of energy is a merely an exchange of thermal energy, which means that

$$\frac{dS_1}{dE_1} = \frac{dQ_1}{dE_1} = \frac{dQ_2}{dE_2} \tag{4.137}$$

with $$\Delta S = \frac{\Delta q}{T} \tag{4.138}$$

and we get:

$$\frac{dQ_1}{dE_1} = \frac{1}{T_1} = \frac{dQ_2}{dE_2} = \frac{1}{T_2} \tag{4.139}$$

$$T_1 = T_2 \qquad \text{for } P = \max. \tag{4.140}$$

The most probable distribution of the energy between A_1 and A_2 is achieved for equal temperature of both bodies. We have assumed $P_1(E_1)$ and $P_2(E_2)$ to be continuous functions with regular derivation, which generally only holds for substances consisting of a very large number of particles. (Boltzmann even says that the transition from discrete to continuous systems can never be achieved in a reasonable way [BOL81].)

4.2.6.1 Energy and fluctuation

Given: A very large body A_1, which is very well isolated from its environment (liquid A_1). The temperature T_1 remains constant, as long as the lost thermal energy q is small compared with the total energy E_1.

We get $\dfrac{dS_1}{dE_1} = \dfrac{dQ_1}{dE_1} = \dfrac{1}{T_1}$ and with $S = k \cdot \ln P$ (4.141)

$$\frac{d \ln P_1}{dE_1} = \frac{1}{k \cdot T_1} \tag{4.142}$$

After performing the integration

$$d \ln P_1 = \frac{1}{k \cdot T_1} dE_1$$

$$\ln P_1 = \int \frac{1}{k \cdot T_1} dE_1 \tag{4.143}$$

we obtain:

$$\ln P_1 = \frac{1}{k \cdot T_1} E_1 + C_1$$

$$P_1 = \exp\left(\frac{1}{k \cdot T_1} E_1\right) \cdot \exp(C_1) \tag{4.144}$$

$$P_1 = P_0 \cdot \exp\left(\frac{1}{k \cdot T_1} E_1\right) \tag{4.145}$$

Such a relation between the number of microscopic states P_1 and the energy E_1 is typical for a thermostat (static temperature).

Now we place a very small body A_2 into the liquid A_1. A_2 remains at constant temperature T_1, because of the most probable distribution of temperature (our previous consideration). But A_2 is subject to fluctuations in the energy, even if these fluctuations occur seldom.

From the first law of thermodynamics we have

$$E_1 = E - E_2 \tag{4.146}$$

and the number of possibilities is given by

$$P_1 = P_0 \cdot \exp\left(\frac{1}{k \cdot T_1} E_1\right)$$

$$P_1 = P_0 \cdot \exp\left[\frac{1}{k \cdot T_1} \cdot (E - E_2)\right] = P_0 \cdot \exp\left(\frac{E}{k \cdot T_1}\right) \cdot \exp\left(\frac{-E_2}{k \cdot T_1}\right) \tag{4.147}$$

The number P of states of the total system is:

$$P = P_1(E_1) \cdot P_2(E_2)$$

$$P = P_2(E_2) \cdot P_0 \cdot \exp\left[\frac{1}{k \cdot T_1} \cdot (E - E_2)\right] = P_2(E_2) \cdot P_0 \cdot \exp\left(\frac{E}{k \cdot T_1}\right) \cdot \exp\left(\frac{-E_2}{k \cdot T_1}\right) \tag{4.148}$$

Brillouin describes this as:

$$P = P_2(E_2) \cdot p(T) \tag{4.149}$$

with

$$P = \frac{1}{P_0} \cdot e^{\frac{-E_2}{k \cdot T_1}} = p(T_1) \tag{4.150}$$

In a thermostat the temperature T_1 each of the P_2 states of A_2, belonging to the energy level E_2, has the probability coefficient $p(T)$.

$$\overline{E}_{\text{kin}} = \frac{1}{2} \cdot kT \qquad \text{per degree of freedom} \tag{4.151}$$

4.2.6.2 Quantized oscillator

A quantized oscillator can only achieve discrete values of energy, which restrict it, for a given basic frequency f, to the following levels of energy

$$E = 0,\ hf,\ 2hf,\ 3hf,\ \ldots \tag{4.152}$$

In a thermostat with $P = P_0 \cdot \exp\left(\frac{E}{k \cdot T}\right)$ and T, these levels of energy lead to the probabilities:

$$p = \frac{1}{P} = \frac{1}{P_0} \cdot \exp\left(-\frac{E}{k \cdot T}\right) = p_0 \cdot \exp\left(-\frac{E}{k \cdot T}\right) \tag{4.153}$$

and for the corresponding values of energy we find the expressions:

$$p_1 = p_0 \cdot \exp\left(-\frac{h \cdot f}{k \cdot T}\right) \qquad p_2 = p_0 \cdot \exp\left(-\frac{2 \cdot h \cdot f}{k \cdot T}\right) \dots$$

$$p_n = p_0 \cdot \exp\left(-\frac{n \cdot h \cdot f}{k \cdot T}\right) \tag{4.154}$$

The following equation must hold:

$$\sum_{i=0}^{\infty} p_i = 1 \tag{4.155}$$

and thus

$$p_0 \cdot \left[1 + \exp\left(-\frac{h \cdot f}{k \cdot T}\right) + \exp\left(-\frac{2 \cdot h \cdot f}{k \cdot T}\right) + \dots + \exp\left(-\frac{n \cdot h \cdot f}{k \cdot T}\right) + \dots\right] = 1 \tag{4.156}$$

The sum in the brackets can be interpreted as a geometric series, which we are able to express as:

$$\sum_{i=0}^{\infty} e^{-i \cdot x} = \sum_{i=0}^{\infty} \left(\frac{1}{e^x}\right)^i = \frac{1}{1 - e^{-x}} \tag{4.157}$$

It thus follows:

$$p_0 \cdot \frac{1}{1 - \exp\left(-\frac{h \cdot f}{k \cdot T}\right)} = 1$$

$$p_0 = 1 - \exp\left(-\frac{h \cdot f}{k \cdot T}\right) \tag{4.158}$$

(probability of occurence of the state with basic frequency in the oscillation).

Now we want to calculate the probabilities of the energies above a certain limit

$$E \geq m \cdot h \cdot f = E_m \tag{4.159}$$

These result from:

$$p(E \geq E_m) = \sum_{i=m}^{\infty} p_i = p_m + p_{m+1} + p_{m+2} + \dots\dots$$

$$p(E \geq E_m) = p_m \cdot (1 + p_1 + p_2 + p_3 + \dots + p_n + \dots)$$

$$p(E \geq E_m) = p_m \cdot \sum_{i=0}^{\infty} p_i = p_m$$

$$p(E \geq E_m) = p_m = p_0 \cdot \exp\left(-\frac{m \cdot h \cdot f}{k \cdot T}\right) \tag{4.160}$$

The energies above the limiting energy E_m result from this connection. The average energy of the complete system can be calculated from the number of existing oscillators. It can be computed as the mean of all oscillations, according to the equation

$$\bar{n} = \sum_{n=0}^{\infty} n \cdot p_n$$

$$\bar{n} = p_0 \cdot \sum_{n=0}^{\infty} n \cdot e^{-\frac{n \cdot h \cdot f}{k \cdot T}}$$

$$\bar{n} = p_0 \cdot \left[0 + \exp\left(-\frac{h \cdot f}{k \cdot T}\right) + 2 \cdot \exp\left(-\frac{2 \cdot h \cdot f}{k \cdot T}\right) + ... + n \cdot \exp\left(-\frac{n \cdot h \cdot f}{k \cdot T}\right) + ...\right]$$

For further computation of this term, we use

$$\left(0 + e^{-x} + 2 \cdot e^{-2 \cdot x} + ... + n \cdot e^{-n \cdot x} + ...\right) = -\frac{d}{dx}\left(1 + e^{-x} + e^{-2 \cdot x} + ... + e^{-n \cdot x} + ...\right)$$

$$= -\frac{d}{dx} \sum_{n=0}^{\infty} e^{-n \cdot x} = -\frac{d}{dx} \cdot \frac{1}{1 - e^{-x}}$$

$$= \frac{0 - e^{-x} \cdot (-1)}{\left(1 - e^{-x}\right)^2} = \frac{e^{-x}}{\left(1 - e^{-x}\right)^2} \tag{4.162}$$

and inserting this result, we find with the equation

$$p_0 = 1 - \exp\left(-\frac{h \cdot f}{k \cdot T}\right) \tag{4.163}$$

the average number of quantizing steps

$$\bar{n} = p_0 \cdot \frac{\exp\left(-\frac{h \cdot f}{k \cdot T}\right)}{\left[1 - \exp\left(-\frac{h \cdot f}{k \cdot T}\right)\right]^2} = \frac{\exp\left(-\frac{h \cdot f}{k \cdot T}\right)}{1 - \exp\left(-\frac{h \cdot f}{k \cdot T}\right)} \tag{4.164}$$

and by multiplication with $\exp\left(\frac{h \cdot f}{k \cdot T}\right)$ in nominator and denominator, we are able to evaluate the simpler form

$$\bar{n} = \frac{1}{\exp\left(\frac{h \cdot f}{k \cdot T}\right) - 1} \quad \text{(average number of quantizing steps)} \tag{4.165}$$

Thus the average energy of the resonator is

$$\bar{E} = \bar{n} \cdot h \cdot f = \frac{h \cdot f}{\exp\left(\frac{h \cdot f}{k \cdot T}\right) - 1} \tag{4.166}$$

Taking a closer look at this average energy and the exponential function in the denominator, we may approximate this function by a Taylor approximation

$$\exp\left(\frac{h \cdot f}{k \cdot T}\right) = 1 + \frac{h \cdot f}{k \cdot T} + \text{h.o.t.} \tag{4.167}$$

for $\quad h \cdot f << k \cdot T$

i.e. for frequencies $f << \frac{k \cdot T}{h} = \frac{1.38 \cdot 10^{-23}}{6.625 \cdot 10^{-34}} \frac{\text{J} \cdot \text{K}}{\text{K} \cdot \text{Ws}} = 2.08 \cdot 10^{10} \cdot \frac{1}{\text{s}}$ (4.168)

(For room temperature this limit of approximately $6.24 \cdot 10^{12}$ Hz is very high.) we neglect the higher order terms and obtain

$$\bar{E} = \frac{h \cdot f}{1 + \frac{h \cdot f}{k \cdot T} - 1} = k \cdot T \tag{4.169}$$

and as a resonator has two degrees of freedom (kinetic and potential energy), this equation yields the result

$$\bar{E}_{\text{kin}} = \bar{E} = \frac{1}{2} \cdot kT \quad \text{per degree of freedom} \tag{4.170}$$

we found earlier.

4.2.7 Brillouin–Schrödinger negentropy

Schrödinger wrote the equation of the statistical entropy in the form

$$S = k \cdot \log D \tag{4.171}$$

D = disorder. If we compare this with the number of available states, we find the equation $W = 1/D$ and furthermore

$$-S = k \cdot \log \frac{1}{D} \tag{4.172}$$

This negative entropy is called negentropy. If we equate the disorder with the number of possible Planck complexions, as we know it from Boltzmann's entropy, we exactly obtain Boltzmann's entropy. Nevertheless, we want to look at the concept of negentropy in a more detailed way.

The term negentropy goes back to Erwin Schrödinger [SCH89]. In connection with the decrease of the entropy in closed systems Schrödinger uses the term *negentropy* and gives us the following explanation for this denotation:

> "If we have two bodies with different temperatures and carry out an exchange of heat (by transformation of heat into mechanical, electrical, chemical, ... energy), we simultaneously increase the entropy of this system."

As is suits our feelings more that a quantity (energy) increases when the producing quantity (entropy) decreases, Schrödinger introduced the denotation negentropy. This negentropy decreases when the energy increases, because

$$\text{Negentropy} \quad N = -S \tag{4.173}$$

Max Planck then defined entropy as

$$S = k \cdot \ln N_0 \tag{4.174}$$

N_0 = number of macroscopic states a system is able to adopt.

These are also called the 'Planck complexions'

S = entropy of the physical system.

This is a statistical description of the entropy, allowing an a priori calculation of the entropy of the system. This is also the starting point for the maximum entropy methods. A system always tends to adopt the most probable state. The most probable state has the maximum probability of occurence and the number of variations in this state is also the maximum that can be achieved in the system (see also chapter 15). Each piece of information (gain of information) decreasing the uncertainty of the system diminishes the entropy and thus increases the negentropy.

The negentropy ($N = -S$) corresponds to the quality or the grade of the energy, as Kelvin put it, and it is a monotone decreasing function (reminding us of the sources of energy). The meaning of negative entropy is also examined in the book 'Was ist Leben? (What is life?)' by Erwin Schrödinger. A living organism lives on food. This food merely supplies our body with negentropy, which it needs to compensate for the loss of negentropy by the degradation of energy (transformation of energy into heat). The energy contained in this food is not the crucial factor, because the energy is conserved. It is the negentropy that is the crucial factor, because the decrease of entropy, enabled by this negentropy, leads to an increase of the grade of energy (Thompson, Kelvin) and this is the reason living organisms can create complex structures.

4.2.7.1 Brillouin: Precise definition of information

Information can only be defined in the sense in which we are able to assign mathematical quantities to it. As a consequence we cannot describe semantic or analytical information (Brillouin also calls this sort of information journalistic information), but we are merely able to describe the statistic (syntactic) information with mathematical terms.

Approach:

Let us assume we have a situation with N_0 different possible events with equal probability. Brillouin assigns an a priori information $I_0 = 0$ to this situation, which can be explained by looking at the following relative definition of information with $N_1 = N_0$. But this assignment contradicts our intuition, as we already have a limited number of events N_0 in our set, which reduces the a priori information from an infinite number of possible events to a finite set with only N_0 events. But the transition from an infinite number of events to N_0 events, would contain an infinite amount of information, which is not a meaningful assignment to an a priori information. We thus take the N_0 events as the a priori information of our situation.

Because of these thoughts Brillouin distinguishes between two possible initializations of a system. If a system consists of N_0 different possible events with equal a priori probability, we use the following notations for

the information $I_0 = 0$ for N_0 possibilities (4.175)

and the entropy $S_0 = k \cdot \ln N_0$ (4.176)

Thus we have the a priori information that merely a limited number of states can be assumed in the beginning. Because of the 'principles of insufficient reasoning', stated by Laplace, we assign equal probabilities to the events, as we do not have any additional information that allows us to prefer or reject certain events. Then the observation provides a limitation of the number of events to $N_1 < N_0$ events that may occur. For the information we thus get

$$I_1 = K \cdot \ln\left(\frac{N_0}{N_1}\right) = K \cdot \ln N_0 - K \cdot \ln N_1 \tag{4.177}$$

As the inequality $N_1 < N_0$ always holds, I_1 is positive and thus the limitation of the number of possible states reduces the uncertainty and leads to a gain of information. The probabilities of the N_1 states are also equal and the constant K can be chosen according to the required unit.

For bit $\rightarrow$ $K = 1/\ln 2 \rightarrow$ $$I_1 = \mathrm{ld}\, N_0 - \mathrm{ld}\, N_1 \tag{4.178}$$

For the entropy (information) we get:

$$K = k = 1.38 \cdot 10^{-23}\ \mathrm{J/K} \tag{4.179}$$

The absolute precondition of N_0 events with I_0 and the subsequent measurement of $N_1 < N_0$ leads to $I_1 > 0$, which means that the reduction of the number of states provides a gain of information.

I_0 is thus a gain of information and not absolute information, when we take the point of view of Brillouin. The behavior of the entropy is different. Here we are able to assign an entropy to the initial state

$$S_0 = k \cdot \ln N_0 \tag{4.180}$$

and an analogous entropy to the state after the observation

$$S_1 = k \cdot \ln N_1 \tag{4.181}$$

$$I_1 = S_0 - S_1$$

$$S_1 = S_0 - I_1 = S_0 - N_S \qquad N_S \text{ here} = \text{negentropy} \tag{4.182}$$

Information makes a 'negative contribution to entropy' and exactly this connection between information and entropy leads Brillouin to the 'negentropy principle of information'.

Information, as Brillouin sees it, is not equal to a negative entropy, but to a difference between entropies, which corresponds to our idea that information is a relative quantity. For N_1 states, defined by the measurement, we get with $N_1 < N_0$:

For the information $I_1 > 0$ $$I_1 = K \cdot \ln \frac{N_0}{N_1} \tag{4.183}$$

For the entropy $S_1 = k \cdot \ln N_1$

$$S_1 = S_0 - I_1(K = k)$$

$$S_1 = S_0 - k \cdot \ln \frac{N_0}{N_1} = k \cdot \ln N_0 - k \cdot \ln \frac{N_0}{N_1} = k \cdot \ln N_1 \tag{4.184}$$

Brillouin calls this sort of information '*bound information*' and explains:

bound information leads to: – 'a decrease in entropy S'

– 'an increase in negentropy N'

He distinguishes two classes of information:

1. Free Information I_f

'Free information occurs, when the possible cases are regarded as abstract and have no specified physical significance.'

2. Bound Information I_b

'Bound information occurs, when the possible cases can be interpreted as complexions of a physical system. It is a special case of free information.'

$$S_1 = S_0 - I_{\mathrm{bound}1}(K = k) \tag{4.185}$$

I, as Brillouin uses it, is not absolute information, but it describes a gain of information and thus a relative information. This completely satisfies the first thought we made regarding information. There (chapter 2) we also identified information as a relative quantity.

Brillouin continues:

'In case of free information, we prefer not to think of a connection between information and entropy, since the relation between entropy and the number of cases is defined only if the cases are complexions of a physical system.'

In this case we get $S = k \cdot \ln N$ N = number of complexions (4.186)

generally: $S = \dfrac{Q}{T}$ in J/K (4.187)

This connection determines the unit of the Boltzmann constant.

$$k = 1.38 \cdot 10^{-23} \text{ J/K} \tag{4.188}$$

4.2.7.2 Negentropy as a generalization of Carnot's principle

(Second law of thermodynamics)

In a closed system we have:

$$\Delta S_1 \geq 0$$

$$\Delta(S_0 - I_{b1}) \geq 0 \tag{4.189}$$

with

$$\Delta I_{b1} = k \cdot \ln \frac{N_0}{N_1} \leq 0 \quad \text{when} \quad N_1 \leq N_0$$

N_0 = number of states (complexions)

Thus Brillouin follows the suggestion of Erwin Schrödinger and uses the negentropy N_S instead of the entropy S and gets

$$\Delta N_{S1} \leq 0$$

$$\Delta(N_{S0} + I_{b1}) \leq 0 \tag{4.190}$$

N_S = negentropy

Negentropy and information can be transformed into each other. Negentropy is a measure of the grade or the quality of the energy of the system and

$$\Delta N_{S1} \leq 0 \tag{4.191}$$

is equivalent to Kelvin's 'principle of degradation of energy'. Furthermore the equation $\Delta I_f \leq 0$ holds in general and with $\Delta N_{S1} \leq 0$ we write:

$$\Delta(N_{S1} + I_f) \leq 0$$

$$\Delta(N_{S0} + I) \leq 0 \qquad \text{with } I = I_f + I_b \tag{4.192}$$

with free and bound information. Brillouin calls this denotation a generalization of Carnot's principle, as this notation contains both classes of information.

Every gain of information has to be buyed with an increase of the entropy.

$$\Delta S \geq \Delta I$$

$$-\Delta N_S \geq \Delta I \qquad N_S = \text{negentropy}$$

$$\Delta I + \Delta N_S \geq 0 \tag{4.193}$$

This connection is also used in the Szilard–Brillouin relation

$$\Delta S \geq k \cdot \Delta I_{\text{Shannon}}$$

$$\Delta I_{\text{Shannon}} \leq k^{-1} \cdot \Delta S \tag{4.194}$$

describing the fact that the transported information is always less than the increase in the entropy, which underpins the second law of thermodynamics, which describes the increase of entropy and thus provides an upper limit for the gain of information.

The greater the gain of information, the greater is the increase of information, or the greater is the decrease of negentropy. Now negentropy is intended to be a limited source. Thus an infinite gain of information, requiring an infinite loss of negentropy, becomes impossible and the paradox of an infinite gain of information is thus solved. Dennis Gabor comments on this with the statement: "You cannot get something for nothing."

Maxwell's demon

The gain of information in a closed system has led to the familiar description of Maxwell's demon residing in this closed system, which consists of two rooms separated by a wall. This wall has a door, which is open at the beginning of the considerations. Thus the molecules are equally distributed in both rooms.

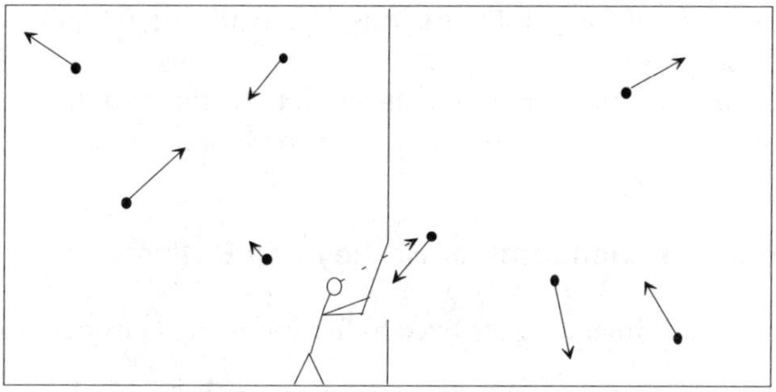

Fig. 4.1: Maxwell's demon

The idea is that there might exist a demon, sitting at the door and opening the door at an appropriate moment so that a molecule from the right room could enter the left room. This demon would be able to order all molecules from the right room into the left room, generating an increased organization out of the initial maximum disorder. So the demon would be able to act against the propositions of the second law of thermodynamics, as it could decrease the entropy in a closed system. We would be able to gain negentropy. But the equations

$$-\Delta N_S \geq \Delta I$$

and

$$\Delta I_{\text{Shannon}} \leq k^{-1} \cdot \Delta S \tag{4.195}$$

tell us that the information gained by Maxwell's demon by creating a certain order in the system can at least achieve the amount of entropy that this demon has to use to perform his actions. Thus the entropy increases in the closed system, even though the demon apparently orders the molecules into the left room.

Some remarks

- A gain of information reduces the entropy.
- A smaller amount of entropy corresponds to an unstable state and a higher grade of energy.
- Evolution in the direction of higher entropy leads to a state of higher stability and thus to a more probable distribution and a lower grade of energy.

The second law of thermodynamics does not state anything about the time required by such an equalization process. This time can be found by a consideration of the kinetic energy of the molecules. Thus entropy is a measure of the amount of disorder in a physical system or a measure of the lack of information regarding such a system. We might use the following statements to 'define' the concept of entropy:

- Entropy is the capability (possibility) to attain the Equilibrium State (heat death).
- Entropy is a measure of the uncertainty (disorder of the system).
- Entropy is an indication of the transmission of information.

4.2.8 Information measures of Hartley and Boltzmann

The definition of information goes back to the following consideration of Hartley:

We want to identify a single element out of a finite number N_0 of elements occurring with equal probability in a given set. The information we gain by the measurement is

$$I_1 = K \cdot \ln N_0 \tag{4.196}$$

If we apply the dual alphabet '0,1', the number $N_0 = 2^n$ of the elements can be uniquely identified by this alphabet and the information of the measurement is described by the equation

$$I = K \cdot \ln N_0 = K \cdot \ln 2^n = K \cdot n \cdot \ln 2 \tag{4.197}$$

This information corresponds to the number n of the required characters in the binary sequences. Therefore we choose

$$K = \frac{1}{\ln 2} \tag{4.198}$$

and thus get

$$I = K \cdot n \cdot \ln 2 = n \tag{4.199}$$

This is the information in bits, or the number of bits required to form the message describing the result of the observation. In general we get

$$I = \frac{1}{\ln 2} \cdot \ln N_0 = \mathrm{ld}\, N_0 \tag{4.200}$$

the information in bits contained in N_0 elements. Comparing this with Boltzmann's entropy

$$S_\mathrm{B} = -k \cdot \sum_{i=1}^{N_0} p_i \cdot \ln p_i \tag{4.201}$$

for equally probable events

$$S_\mathrm{B} = k \cdot \sum_{i=1}^{N_0} \frac{1}{N_0} \cdot \ln N_0 = k \cdot \ln N_0 \qquad k = 1.38 \cdot 10^{-23}\ \mathrm{J/K} \tag{4.202}$$

Now we can build the ratio of entropy and information for this binary measure and obtain

$$\frac{S_\mathrm{B}}{I_\mathrm{bin}} = k \cdot \ln 2 \cdot \frac{\ln N_0}{\ln N_0} = k \cdot \ln 2 = 9.565 \cdot 10^{-24}\ \frac{\mathrm{J}}{\mathrm{K}} \tag{4.203}$$

Wolkenstein (see chapter 4.2.2.3) uses this quantity to define the reduction of entropy achieved by the human race because of the writing of books. We may also calculate the reduction of the entropy caused by the universal library (chapter 1), but this would also lead us to a very small amount of entropy reduction. The reason for these small results is the ratio between the entropy of statistical mechanics on the one hand and the information contained in a binary coding of the events on the other hand, which yields a very small ratio, as we see in equation 4.203.

4.2.8.1 Examples

If an observation does not determine a single element of a set of N_0 equally probable elements, but only a subset consisting of N_1 elements ($N_1 < N_0$), the gain of information of this observation is given by

$$I_1 = K \cdot \ln\left(\frac{N_0}{N_1}\right) = K \cdot \ln N_0 - K \cdot \ln N_1 \tag{4.204}$$

If $N_1 = 1$, i.e. if we identify a single event, the information gain increases to

$$I_1 = K \cdot \ln N_0 \tag{4.205}$$

For an alphabet with m characters a sequence of n elements allows us to encode m^n symbols and the information of a single symbol thus becomes

$$I = K \cdot \ln m^n = K \cdot n \cdot \ln m \tag{4.206}$$

or in binary coding:

$$I = n \cdot ld(m)$$

E. g.: $m = 2 \quad \Rightarrow \quad I_{\text{bin}} = n$

$$m = 10 \quad \Rightarrow \quad I_{\text{bin}} = 3.32\, n \tag{4.207}$$

A decimal number contains 3.32 bit of information. Thus the binary encoding of a decimal number requires 3.32 times the number of elements of the decimal encoding.

Looking at our alphabet of 26 letters, we find

$$m = 26 \quad \Rightarrow \quad I_{\text{bin}} = 4.7\, n \tag{4.208}$$

A letter thus contains 4.7 bit of information.

4.2.9 Shannon's entropy

$$H = -k \cdot \sum_{i=1}^{n} p_i \cdot \text{ld}\, p_i \tag{4.209}$$

H = Shannons uncertainty, entropy or informationentropy.

The relation to Boltzmann's entropy has already been considered in sections 4.2.2.1 and we thus only add some further remarks to this entropy or information, describing the uncertainty or the eliminated uncertainty.

In the description of this entropy Campbell [CAM85] remarks: "In signal processing entropy is quite similar to noise!" This is a justified statement, as we will see in our examination of the communication channel. There we notice that information comprises the noise on the channel on the one hand and on the other hand the uncertainty contained in the signal being sent through the communication channel. This uncertainty has to be created from the emitter by generating and emitting messages containing information in the form on an uncertainty. This uncertainty provides the new or the surprising (not expected a priori) part of the message.

Another example is the information in a translation, where the channel (the translator) performs a transformation (translation). Imagine the following example:

The translation of a sentence from one language into another language and subsequent retranslating back to the initial language

English:	out of sight, out of mind
Language X:	....
English:	invisible idiot

If we translate only the words, then this translation, which is limited to the verbatim connection of two languages, can lead to errors in the allocation leading to a completely changed meaning of the statement. This even happens if we merely use two words or two different meanings to create our statement. We thus cannot restrict the given problem to the sole examination of single words. Instead have to take into account syntax, grammar, etc. to obtain an appropriate translation. Still another problem arises in the judgement of a good or a less succesful translation. Such a differentiation requires the additional use of subjective aspects and thus makes automation difficult.

Similar to this problem is the idea of the Turing Test [HOF91], where questions are put to a human being and a computer. Depending on the answers, the evaluation of this test should be able to determine which is the human being. This again touches the whole context of the different levels of information (semantic, syntactic, pragmatic), where the pragmatic information, i.e. the reaction of the recipient of the questions, produces the observations used for the evaluation of this test.

Shannon's information can merely describe a certain part of the whole concept of information. The other, subjective parts of this concept cannot be described with Shannon's entropy and with other mathematical functions requiring objectivity.

4.3 Dynamic entropy

Dynamic raises the question of a definition of time, or of an objective treatment of time in science.

Aristotle (384–322 B.C.)

"Time is a measure of the change between before and after.“

Our sensing of time varies and we are also able to use various methods to measure time. Some authors state "time, the familiar stranger" or "time is natures way of keeping everything from happening at once". But how can we introduce a single direction of time into physical systems? At the moment the tendency is to involve entropy in physical models, as this quantity is the only one containing a unique direction (arrow) of time.

4.3.1 Eddington and the arrow of time

Lorentz verified that we cannot perform predictions over long periods of time. He realized this in observations of the weather, where he tried to predict the future weather from the actual data. Because of inevitable rounding errors of the measurement devices and of the computers performing the prediction (the rounding error is a nonlinear error), the divergence in the simulations becomes so large after some time that long time predictions are simply impossible because of these initial uncertainties. The statement of this experiment is the same, as the statement of the butterfly effect, describing the fact that even a very small deviation in the initial conditions of a system leads to a large deviation of the trajectory between the original and the predicted process.

The mathematical formulation of this statement is given in the prediction equation. If we have a linear system, describing the evolving state from $\underline{x}(k-1)$ to $\underline{x}(k)$

$$\underline{x}(k) = A(k-1) \cdot \underline{x}(k-1) \tag{4.210}$$

then the error in the prediction results from the error propagation of the existing errors (initial errors) $P^+(k-1)$ and the dynamics of the system, described by the state transition matrix $A(k-1)$.

$$P^-(k) = A(k-1) \cdot P^+(k-1) \cdot A^T(k-1) \tag{4.211}$$

$$P^+(k-1) = E\left\{ (\underline{x}(k-1) - E\{\underline{x}(k-1)\}) \cdot (\underline{x}(k-1) - E\{\underline{x}(k-1)\})^T \right\} \tag{4.212}$$

$$P^-(k) = E\left\{ (\underline{x}(k) - E\{\underline{x}(k)\}) \cdot (\underline{x}(k) - E\{\underline{x}(k)\})^T \right\} \tag{4.213}$$

Even a less uncertain initial condition tends to an increasing error with increasing extrapolation time, because of error propagation. The final resulting error depends on the initial error as well as on the linear mapping $A(k-1)$ and the prediction time. An increasing initial error, an increasing gradient of A and/or a longer period of prediction lead to an increased error at the end of the error propagation.

When we leave the system without making any observations (as we do in prediction), the initial uncertainty will grow in time. This, however, does not hold for the classical models of physics. Brillouin notes that all physical models hold for both possible directions of time, when he says: "The models of physics hold for both directions of time. Only the boundary conditions lead to a unique direction of time."

Example:
The wave equation has two possible solutions, the 'advanced' and the 'retarded' waves. The 'retarded' waves emitted by a source lead us to reasonable solutions, which is the reason for their usage in physics and electrical engineering. 'Advanced' waves approach the source and thus they have to be generated in the past, with knowledge of the actual behavior of the source. They yield unreasonable solutions of the equation and they contradict the causality principle.

We cannot observe the future before it happens. The increase of the entropy leads to a unique arrow of time. Hawking says: There are at least three arrows of time

1. a thermodynamic (Eddington)
2. a psychological (Bergons)
3. a cosmological (Piccard)

which all have to point in the same direction.

Thus entropy is the only quantity that allows us to describe a unique direction of time. All other physical models hold for negative and positive directions of time.

4.3.2 Kolmogorov's entropy

Kolmogorov's entropy is a measure for the loss of information in an N-dimensional state space. Therefore we divide the state space into cells of length ε and we take our samples in time steps of τ. The trajectories proceed from one cell to an adjacent cell. To be able to predict the next cell, we need additional information, which we get from the difference of two succeeding Kolmogorov entropies K_{n+1} - K_n.

$$K = -\lim_{\tau \to 0} \lim_{\varepsilon \to 0} \lim_{N \to \infty} \frac{1}{N \cdot \tau} \cdot \sum_{i=0}^{N-1} p_i \cdot \log p_i \tag{4.214}$$

Possible arrangements:

1. The trajectories do not diverge, $K = 0$.
2. If the trajectories diverge exponentially from the initial conditions, i.e. $N = e^{\lambda}$ λ = Lyapunov exponent, we get chaotic conditions, $K > 0$.
3. If the points are randomly distributed with equal probability, we get a random motion and $K = \infty$.

4.3.3 Rényi's entropy

Rényi generalized the probabilistic entropy, so that the entropy measures also found applications in the examination of fractal dimensions (Hausdorff–Besicovitch), which can be described with this α-entropy

$$S_{\alpha} = \frac{1}{1-\alpha} \log \sum_{i=1}^{N} p_i^{\alpha} \tag{4.215}$$

As fractals do not have integral dimensions, this entropy can be applied to fractals with dim(fractal) = $\alpha \notin Z^{+}$. We are going to describe this entropy in a more detailed way in later chapters.

We want to leave the presentation of the measures of entropy that are used in physics with this quite general entropy and proceed with our introduction of information measures based on the description used by Shannon.

5 Extension of Shannon's information

5.1 Rényi's Information 1960

The starting point of the derivation [REN76/1–3] of Rényi's α-information (α-entropy) S_α and the G-divergence G_α is Hartley's definition of the information

$$I_{\text{Hartley}} = -\ln p \tag{5.01}$$

with known properties, such as additivity of the information for independent events. This definition only holds when we have no a priori knowledge regarding the probabilities p_k of the single events A_k (Laplacian principle of insufficient reasoning), i.e. when all $p_k = p$ have equal probabilities. Shannon extended this description of information by assigning different probabilities to the events and calculating the average of all probabilities occurring in the observed process.

$$I_{\text{Shannon}} = \sum_{k=1}^{n} p_k \cdot \ln\left(\frac{1}{p_k}\right) = -\sum_{k=1}^{n} p_k \cdot \ln p_k \tag{5.02}$$

p_k = weighting factors of the information $I_k = \ln p_k$.

The linear mean value, however, is not the only possible mean value. So Alfred Rényi extended the averaging by applying the theory of the generalized mean, assigning appropriate weights $p_1, p_2, ..., p_n$ (with $0 \le p_k \le 1$ and $\sum_{k=1}^{n} p_k = 1$) to the real numbers $x_1, x_2, ..., x_n$, which results in a generalized mean.

$$m_{\text{g}} = \varphi^{-1}\left[\sum_{k=1}^{n} p_k \cdot \varphi(x_k)\right] \tag{5.03}$$

$\varphi(x)$ is an arbitrary continuous and strictly increasing or decreasing function defined on the set of the real numbers (These conditions ensure the necessary property of the function, which has to have a unique inverse function.) These generalized means are also known as Kolmogorov–Nagumo functions of the mean [KOL30].

If we use the generalized mean instead of the linear mean in the definition of Shannon's information, we obtain the equation

$$I_{\text{Rényi}} = \varphi^{-1}\left[\sum_{k=1}^{n} p_k \cdot \varphi\left(\operatorname{ld}\frac{1}{p_k}\right)\right] \tag{5.04}$$

or with the natural logarithm

$$I_{\text{Rényi}} = \varphi^{-1}\left[\sum_{k=1}^{n} p_k \cdot \varphi\left(\ln\frac{1}{p_k}\right)\right] \tag{5.05}$$

We want to limit our descriptions to the natural logarithm, as all other logarithms only introduce an additional constant factor. If $\varphi(x)$ is a linear function, the generalized mean reduces to the linear mean and we obtain Shannon's information as a special case of this formula.

We may now use $I_{\text{Rényi}}$ as an information measure, but we certainly cannot use an arbitrary function $\varphi(x)$ with the restrictions already stated. The function $\varphi(x)$ has to be selected in such a way that certain postulates are met by the resulting information function. The most important postulate is the requirement that the information of independent events can be added, to obtain the information of the union set of both events. We observe:

The event A consists of two independent events A_1 and A_2.

The event A_1 provides I_h bit of information with the probability p_h.
The event A_2 provides J_k bit of information with the probability q_k.

The requirement of additivity now leads to an information I_h+J_k resulting from the event A with the probability $p_h \cdot q_k$ ($h = 1,2,...,n$; $k = 1,2,...,m$).

In a mathematical description, this equivalence is denoted by the equation:

$$\varphi^{-1}\left[\sum_{h=1}^{n}\sum_{k=1}^{m} p_h \cdot q_k \cdot \varphi(I_h + J_k)\right] = \varphi^{-1}\left[\sum_{h=1}^{n} p_h \cdot \varphi(I_h)\right] + \varphi^{-1}\left[\sum_{k=1}^{m} q_k \cdot \varphi(J_k)\right] \tag{5.06}$$

p_h = probability $\qquad$ $I_h = -\ln p_h$ = information

q_k = probability $\qquad$ $J_k = -\ln p_k$ = information

and this additivity has to hold for arbitrary disjoint events. If we now choose the event A_2 so, that the information $J_k = J$ is independent of k, i.e. that all amounts of information J occur with the same probability $q = 1/m$ (Hartley's information), we get:

$$\varphi^{-1}\left[\sum_{h=1}^{n}\sum_{k=1}^{m} p_h \cdot \frac{1}{m} \cdot \varphi(I_h + J)\right] = \varphi^{-1}\left[\sum_{h=1}^{n} p_h \cdot \varphi(I_h)\right] + \varphi^{-1}\left[\sum_{k=1}^{m} \frac{1}{m} \cdot \varphi(J)\right]$$

$$\varphi^{-1}\left[\sum_{h=1}^{n} p_h \cdot \sum_{k=1}^{m} \frac{1}{m} \cdot \varphi(I_h + J)\right] = \varphi^{-1}\left[\sum_{h=1}^{n} p_h \cdot \varphi(I_h)\right] + \varphi^{-1}\left[m \cdot \frac{1}{m} \cdot \varphi(J)\right]$$

$$\varphi^{-1}\left[\sum_{h=1}^{n} p_h \cdot \varphi(I_h + J)\right] = \varphi^{-1}\left[\sum_{h=1}^{n} p_h \cdot \varphi(I_h)\right] + \varphi^{-1}[\varphi(J)] \tag{5.07}$$

As the function $\varphi(x)$ can be inverted unambiguously, we write

$$\varphi^{-1}\left[\sum_{h=1}^{n} p_h \cdot \varphi(I_h + J)\right] = \varphi^{-1}\left[\sum_{h=1}^{n} p_h \cdot \varphi(I_h)\right] + J \tag{5.08}$$

and this equation merely holds if $\varphi(x)$ is a linear or an exponential function. In the case of a linear function we obtain – as we already noted – Shannon's information. Thus the approach using the exponential function remains. Here Rényi uses the function

$$\varphi(x) = 2^{(1-\alpha)\cdot x} \qquad \alpha \neq 1 \qquad \text{if we use the logarithm with the base 2.} \tag{5.09}$$

We use the natural logarithm with the base e

$$\varphi(x) = e^{(1-\alpha)\cdot x} \qquad \alpha \neq 1 \tag{5.10}$$

The inverse function of this mapping, transforming $\varphi(x)$ unambiguously in x, is

$$\ln[\varphi(x)] = (1-\alpha) \cdot x$$

$$x = \frac{1}{1-\alpha} \cdot \ln[\varphi(x)]$$

$$\varphi^{-1}(y) = \frac{1}{1-\alpha} \cdot \ln[y] \tag{5.11}$$

Now we insert the generalized mean of Rényi and obtain

$$I_{\text{Rényi}} = \varphi^{-1}\left[\sum_{k=1}^{n} p_k \cdot \varphi\left(\ln \frac{1}{p_k}\right)\right]$$

$$I_{\text{Rényi}} = \frac{1}{1-\alpha} \cdot \ln\left\{\sum_{k=1}^{n} p_k \cdot \exp\left[(1-\alpha) \cdot \left(\ln \frac{1}{p_k}\right)\right]\right\}$$

$$I_{\text{Rényi}} = \frac{1}{1-\alpha} \cdot \ln\left[\sum_{k=1}^{n} p_k \cdot \exp\left(\ln \frac{1}{p_k} - \alpha \cdot \ln \frac{1}{p_k}\right)\right]$$

$$I_{\text{Rényi}} = \frac{1}{1-\alpha} \cdot \ln\left[\sum_{k=1}^{n} p_k \cdot \exp(-\ln p_k + \alpha \cdot \ln p_k)\right]$$

$$I_{\text{Rényi}} = \frac{1}{1-\alpha} \cdot \ln\left[\sum_{k=1}^{n} p_k \cdot \frac{\exp(\alpha \cdot \ln p_k)}{\exp(\ln p_k)}\right]$$

$$I_{\text{Rényi}} = \frac{1}{1-\alpha} \cdot \ln\left[\sum_{k=1}^{n} p_k \cdot \frac{\exp\left(\ln p_k^{\alpha}\right)}{\exp(\ln p_k)}\right] \tag{5.12}$$

We have: $\exp\left(\ln p_k^{\alpha}\right) = p_k^{\alpha}$ and $\exp(\ln p_k) = p_k$

$$I_{\text{Rényi}} = \frac{1}{1-\alpha} \cdot \ln\left[\sum_{k=1}^{n} p_k \cdot \frac{p_k^{\alpha}}{p_k}\right]$$

$$I_{\text{Rényi}} = \frac{1}{1-\alpha} \cdot \ln\left[\sum_{k=1}^{n} p_k^{\alpha}\right] \tag{5.13}$$

and this is Rényi's information or Rényi's α-entropy.

If we use the binary logarithm instead of the natural logarithm, we get

$$I_{\text{Rényi}} = \frac{1}{1-\alpha} \cdot \text{ld}\left[\sum_{k=1}^{n} p_k^{\alpha}\right]$$

$$I_{\text{Rényi}} = \frac{1}{1-\alpha} \cdot \frac{\ln\left[\sum_{k=1}^{n} p_k^{\alpha}\right]}{\ln(2)} \tag{5.14}$$

and the change in the basis of the logarithm leads only to an additional constant factor in the information function. We are thus going to use the natural logarithm, which does not require the permanent writing of factors, which do not change the meaning of the function.

5.1.1 Properties of Rényi's entropy

At first Rényi's entropy is a monotonically decreasing function of the additional parameter α, and to prove this we have to show that the derivative of the entropy with respect to this factor is always negative or equal to zero. We therefore write:

$$\frac{d}{d\alpha} I_{\text{Rényi}} = \frac{d}{d\alpha}\left\{\frac{1}{1-\alpha} \cdot \ln\left[\sum_{k=1}^{n} p_k^\alpha\right]\right\} \tag{5.15}$$

$$\frac{d}{d\alpha} I_{\text{Rényi}} = \frac{d}{d\alpha}\frac{1}{1-\alpha} \cdot \ln\left[\sum_{k=1}^{n} p_k^\alpha\right] + \frac{1}{1-\alpha} \cdot \frac{d}{d\alpha}\ln\left[\sum_{k=1}^{n} p_k^\alpha\right]$$

$$\frac{d}{d\alpha} I_{\text{Rényi}} = \frac{1}{(1-\alpha)^2} \cdot \ln\left[\sum_{k=1}^{n} p_k^\alpha\right] + \frac{1}{1-\alpha} \cdot \frac{1}{\sum_{k=1}^{n} p_k^\alpha}\frac{d}{d\alpha}\left[\sum_{k=1}^{n} p_k^\alpha\right]$$

To obtain the derivative of the second addend, we transform it according to

$$\frac{d}{d\alpha} I_{\text{Rényi}} = \frac{1}{(1-\alpha)^2} \cdot \ln\left[\sum_{k=1}^{n} p_k^\alpha\right] + \frac{1}{1-\alpha} \cdot \frac{1}{\sum_{k=1}^{n} p_k^\alpha}\left[\sum_{k=1}^{n} \frac{d}{d\alpha}\exp(\alpha \cdot \ln p_k)\right]$$

$$\frac{d}{d\alpha} I_{\text{Rényi}} = \frac{1}{(1-\alpha)^2} \cdot \ln\left[\sum_{k=1}^{n} p_k^\alpha\right] + \frac{1}{1-\alpha} \cdot \frac{1}{\sum_{k=1}^{n} p_k^\alpha}\left[\sum_{k=1}^{n} \ln p_k \cdot \exp(\alpha \cdot \ln p_k)\right]$$

$$\frac{d}{d\alpha} I_{\text{Rényi}} = \frac{1}{(1-\alpha)^2} \cdot \ln\left[\sum_{k=1}^{n} p_k^\alpha\right] + \frac{1}{1-\alpha} \cdot \frac{1}{\sum_{k=1}^{n} p_k^\alpha}\sum_{k=1}^{n} p_k^\alpha \cdot \ln p_k$$

and obtain

$$\frac{d}{d\alpha} I_{\text{Rényi}} = \frac{1}{(1-\alpha)^2} \cdot \ln\left[\sum_{k=1}^{n} p_k^\alpha\right] + \frac{1-\alpha}{(1-\alpha)^2} \cdot \frac{1}{\sum_{k=1}^{n} p_k^\alpha}\sum_{k=1}^{n} p_k^\alpha \cdot \ln p_k \tag{5.16}$$

The summation of both addends leads to

$$\frac{d}{d\alpha} I_{\text{Renyi}} = \frac{1}{(1-\alpha)^2} \cdot \left\{ \ln\left[\sum_{k=1}^{n} p_k^{\alpha}\right] + \frac{1}{\sum_{k=1}^{n} p_k^{\alpha}} \sum_{k=1}^{n} p_k^{\alpha} \cdot (1-\alpha) \cdot \ln p_k \right\}$$

$$\frac{d}{d\alpha} I_{\text{Renyi}} = \frac{1}{(1-\alpha)^2} \cdot \left\{ \ln\left[\sum_{k=1}^{n} p_k^{\alpha}\right] + \frac{1}{\sum_{k=1}^{n} p_k^{\alpha}} \sum_{k=1}^{n} p_k^{\alpha} \cdot \ln p_k^{1-\alpha} \right\} \tag{5.17}$$

This is the desired derivative, which must not be positive according to our proposition. This means that

$$\frac{1}{(1-\alpha)^2} \cdot \left\{ \ln\left[\sum_{k=1}^{n} p_k^{\alpha}\right] + \frac{1}{\sum_{k=1}^{n} p_k^{\alpha}} \sum_{k=1}^{n} p_k^{\alpha} \cdot \ln p_k^{1-\alpha} \right\} \le 0 \tag{5.18}$$

The quadratic term in the denominator is always positive and need not be regarded in our consideration, as it only causes a dilation, but not a change in the sign of the whole derivative. We thus focus our attention on:

$$\ln\left[\sum_{k=1}^{n} p_k^{\alpha}\right] + \frac{1}{\sum_{k=1}^{n} p_k^{\alpha}} \sum_{k=1}^{n} p_k^{\alpha} \cdot \ln p_k^{1-\alpha} \le 0 \tag{5.19}$$

$$\ln\left[\sum_{k=1}^{n} p_k^{\alpha}\right] \le -\frac{1}{\sum_{k=1}^{n} p_k^{\alpha}} \sum_{k=1}^{n} p_k^{\alpha} \cdot \ln p_k^{1-\alpha}$$

$$\left(\sum_{k=1}^{n} p_k^{\alpha}\right) \cdot \ln\left[\sum_{k=1}^{n} p_k^{\alpha}\right] \le -\sum_{k=1}^{n} p_k^{\alpha} \cdot \ln p_k^{1-\alpha}$$

$$-\left(\sum_{k=1}^{n} p_k^{\alpha}\right) \cdot \ln\left[\frac{1}{\sum_{k=1}^{n} p_k^{\alpha}}\right] \le -\sum_{k=1}^{n} p_k^{\alpha} \cdot \ln p_k^{1-\alpha}$$

If we multiply this by −1, the inequality sign is reversed and we finally find

$$\left(\sum_{k=1}^{n} p_k^{\alpha}\right)\cdot \ln\left[\frac{1}{\sum_{k=1}^{n} p_k^{\alpha}}\right] \geq \sum_{k=1}^{n} p_k^{\alpha}\cdot \ln p_k^{1-\alpha} \tag{5.20}$$

When we are able to verify this equation, then Rényi's entropy is a monotonically decreasing function of the parameter α. This verification can be achieved by application of the log–sum inequality for complex functions (see also Appendix A.2), which is given by:

$$\sum_{k=1}^{n} a_k \cdot \ln\left(\frac{a_k}{b_k}\right) \geq \left(\sum_{k=1}^{n} a_k\right)\cdot \ln\left(\frac{\sum_{k=1}^{n} a_k}{\sum_{k=1}^{n} b_k}\right) \tag{5.21}$$

Our equation is:

$$\sum_{k=1}^{n} p_k^{\alpha}\cdot \ln p_k^{1-\alpha} \leq \left(\sum_{k=1}^{n} p_k^{\alpha}\right)\cdot \ln\left[\frac{1}{\sum_{k=1}^{n} p_k^{\alpha}}\right] \tag{5.22}$$

with $\sum_{k=1}^{n} p_k = 1$ we get

$$\sum_{k=1}^{n} p_k^{\alpha}\cdot \ln\frac{p_k}{p_k^{\alpha}} \leq \left(\sum_{k=1}^{n} p_k^{\alpha}\right)\cdot \ln\left[\frac{\sum_{k=1}^{n} p_k}{\sum_{k=1}^{n} p_k^{\alpha}}\right]$$

Multiplication by −1 provides

$$\sum_{k=1}^{n} p_k^{\alpha}\cdot \ln\frac{p_k^{\alpha}}{p_k} \geq \left(\sum_{k=1}^{n} p_k^{\alpha}\right)\cdot \ln\left[\frac{\sum_{k=1}^{n} p_k^{\alpha}}{\sum_{k=1}^{n} p_k}\right] \tag{5.23}$$

With $a_k = p_k^\alpha$ and $b_k = p_k$ we find exactly the log–sum inequality and thus we have proved that Rényi's entropy is a monotonically decreasing function with respect to the parameter α.

5.1.2 Limits in the interval $0 \leq \alpha < \infty$

As we saw in the previous chapter, Rényi's α-entropy is a monotonically decreasing function of the parameter α. Thus we now want to determine the limits, which can be achieved by variation of the parameter in the interval $[0,\infty)$. Because of the monotonic nature of the function, we restrict our attention to the limits of the interval. The limit achieved for $\alpha = 1$ will be treated later, when we examine the connection between Rényi's and Shannon's information for continuous distribution densities.

$$S_\alpha = \frac{1}{1-\alpha} \cdot \ln\left[\sum_{k=1}^{n} p_k^\alpha\right] \qquad \alpha \neq 1 \text{ and } \alpha > 0 \tag{5.24}$$

For a given distribution density $P = (p_1, p_2, \ldots, p_n)$ S_α is a monotonically decreasing function of the parameter α. In this variation we get only two limits, which we are going to examine now. First

$$\lim_{\alpha \to \infty} S_\alpha = \lim_{\alpha \to \infty} \frac{1}{1-\alpha} \cdot \ln\left[\sum_{k=1}^{n} p_k^\alpha\right] \tag{5.25}$$

This term consists of two factors. The first factor, the quotient $\frac{1}{1-\alpha}$, tends to zero (0^-), when we choose the parameter to approach infinity. The second part, together with the condition that $p_k < 1$, leads to

$$p_k{}^\alpha = 0 \text{ for } \alpha \to \infty \tag{5.26}$$

and thus to a summation of zeroes and finally to the term ln 0. So the second part provides the result $\lim_{\alpha \to \infty} \ln\left[\sum_{k=1}^{n} p_k^\alpha\right] = -\infty$.

But the complete term

$$\lim_{\alpha \to \infty} S_\alpha = \lim_{\alpha \to \infty} \frac{1}{(1-\alpha)} \cdot \ln\left[\sum_{k=1}^{n} p_k^\alpha\right] \tag{5.27}$$

is thus not defined and we therefore use the rule of de l'Hospital by taking the derivative of numerator and denominator. Then we calculate the limit again. The derivatives are:

The denominator (N): $$\frac{\partial}{\partial\alpha}N = \frac{\partial}{\partial\alpha}(1-\alpha) = -1 \tag{5.28}$$

The numerator (Z): $$\frac{\partial}{\partial\alpha}Z = \frac{\partial}{\partial\alpha}\ln\left[\sum_{k=1}^{n} p_k^{\alpha}\right] = \frac{1}{\sum_{k=1}^{n} p_k^{\alpha}} \cdot \frac{\partial}{\partial\alpha}\sum_{k=1}^{n} p_k^{\alpha}$$

$$\frac{\partial}{\partial\alpha}Z = \frac{1}{\sum_{k=1}^{n} p_k^{\alpha}} \cdot \sum_{k=1}^{n} \frac{\partial}{\partial\alpha} p_k^{\alpha} \tag{5.29}$$

We build this derivative

$$\begin{aligned} \frac{\partial}{\partial\alpha} p_k^{\alpha} &= \frac{\partial}{\partial\alpha}\exp\left(\ln p_k^{\alpha}\right) = \frac{\partial}{\partial\alpha}\exp(\alpha\cdot\ln p_k) \\ &= \ln p_k \cdot \exp(\alpha\cdot\ln p_k) = \ln p_k \cdot p_k^{\alpha} \end{aligned} \tag{5.30}$$

and find the following term for the numerator

$$\frac{\partial}{\partial\alpha}Z = \frac{1}{\sum_{k=1}^{n} p_k^{\alpha}} \cdot \sum_{k=1}^{n} p_k^{\alpha} \cdot \ln p_k \tag{5.31}$$

Now we can build the limit again.

$$\lim_{\alpha\to\infty} S_\alpha = \lim_{\alpha\to\infty} \frac{1}{1-\alpha} \cdot \ln\left[\sum_{k=1}^{n} p_k^{\alpha}\right] = \lim_{\alpha\to\infty} \frac{\frac{\partial Z}{\partial\alpha}}{\frac{\partial N}{\partial\alpha}}$$

$$\lim_{\alpha\to\infty} S_\alpha = -\frac{1}{\sum_{k=1}^{n} p_k^{\alpha}} \cdot \sum_{k=1}^{n} p_k^{\alpha} \cdot \ln p_k \tag{5.32}$$

We now use a transformation, which is also used in the calculation of L_∞ norms, and extend numerator and denominator with ${p_{\max}}^{\alpha}$

$$\lim_{\alpha\to\infty} S_\alpha = -\frac{1}{p_{\max}^{\alpha}\cdot\sum_{k=1}^{n}\left(\frac{p_k}{p_{\max}}\right)^{\alpha}}\cdot p_{\max}^{\alpha}\cdot\sum_{k=1}^{n}\left(\frac{p_k}{p_{\max}}\right)^{\alpha}\cdot\ln p_k$$

$$\lim_{\alpha\to\infty} S_\alpha = -\frac{1}{\sum_{k=1}^{n}\left(\frac{p_k}{p_{\max}}\right)^{\alpha}}\cdot\sum_{k=1}^{n}\left(\frac{p_k}{p_{\max}}\right)^{\alpha}\cdot\ln p_k \tag{5.33}$$

and we build the limit. We take a look at the quotient

$$\left(\frac{p_k}{p_{\max}}\right)\begin{cases}=1 & \text{for} \quad p_k = p_{\max}\\ <1 & \text{for} \quad p_k < p_{\max}\end{cases} \tag{5.34}$$

and state that this quotient achieves the value 1 only once. In all other possible cases the quotient is less than 1. Therefore we conclude for the limit

$$\lim_{\alpha\to\infty}\left(\frac{p_k}{p_{\max}}\right)^{\alpha}\begin{cases}=1 & \text{for} \quad p_k = p_{\max}\\ =0 & \text{for} \quad p_k < p_{\max}\end{cases} \tag{5.35}$$

The remaining term is

$$\lim_{\alpha\to\infty} S_\alpha = -\frac{1}{\left(\frac{p_{\max}}{p_{\max}}\right)}\cdot\left(\frac{p_{\max}}{p_{\max}}\right)\cdot\ln p_{\max}$$

and finally we obtain

$$\lim_{\alpha\to\infty} S_\alpha = -\ln p_{\max} \tag{5.36}$$

for the limit, if the parameter α tends to infinity. On the other side we compute the limit for the parameter α, approaching zero, and we find

$$\lim_{\alpha\to 0} S_\alpha = \lim_{\alpha\to 0}\frac{1}{1-\alpha}\cdot\ln\left[\sum_{k=1}^{n} p_k^{\alpha}\right]$$

$$\lim_{\alpha\to 0} S_\alpha = \lim_{\alpha\to 0}\ln\left[\sum_{k=1}^{n} p_k^{0}\right] = \lim_{\alpha\to 0}\ln\left[\sum_{k=1}^{n} 1\right] = \ln n \tag{5.37}$$

We thus summarize: For a given distribution density $P = (p_1, p_2, ..., p_n)$ the term

$$S_\alpha = \frac{1}{1-\alpha} \cdot \ln\left[\sum_{k=1}^{n} p_k^\alpha\right] \qquad \text{with} \qquad \alpha \neq 1 \text{ and } \alpha > 0 \tag{5.38}$$

is a monotonically decreasing function of the parameter α with the limits

$$\ln n \geq \frac{1}{(1-\alpha)} \cdot \ln\left[\sum_{k=1}^{n} p_k^\alpha\right] \geq -\ln p_{\max} \tag{5.39}$$

5.1.3 Nonnegativity for discrete events

For discrete events Rényi's entropy is always nonnegative, which can be verified via the following consideration:

$$I_{\text{Rényi}} = \frac{1}{1-\alpha} \cdot \ln\left[\sum_{k=1}^{n} p_k^\alpha\right]$$

$$I_{\text{Rényi}} = \frac{1}{1-\alpha} \cdot \ln\left[\sum_{k=1}^{n} p_k \cdot p_k^{\alpha-1}\right]$$

$$I_{\text{Rényi}} = \frac{1}{1-\alpha} \cdot \ln\left\{\sum_{k=1}^{n} p_k \cdot \exp\left[-(1-\alpha) \cdot \ln p_k\right]\right\} \tag{5.40}$$

where α is always positive. We exclude the special case $\alpha = 1$, because in this case we obtain Shannon's information, which is a special case of Rényi's information (for the linear mean).

The remaining two intervals are

1. Interval: $0 < \alpha < 1$
2. Interval: $1 < \alpha < \infty$

Let us examine the exponential function $\exp[-(1-\alpha) \cdot \ln p_k]$ for both intervals. The basic conditions that we need for our consideration are the postulates of Kolmogorov, requiring that the probability lies in the interval $0 \leq p_k \leq 1$. Moreover the probability of the occurrence of an event of the sample space is equal to 1, which can be denoted by the equation $\sum_{k=1}^{n} p_k = 1$.

Because of these postulates of Kolmogorov, the term ln p_k is always negative or equal to zero. Thus we get in the

interval 1: $$\underbrace{\exp\left[\underbrace{-(1-\alpha)}_{<0}\cdot\underbrace{\ln p_k}_{\leq 0}\right]}_{\geq 1}$$

and in the

interval 2: $$\underbrace{\exp\left[\underbrace{-(1-\alpha)}_{>0}\cdot\underbrace{\ln p_k}_{\leq 0}\right]}_{\leq 1}$$

The e-function only serves to weight the probabilities in the sum

$$\sum_{k=1}^{n} p_k \cdot \exp[-(1-\alpha)\cdot \ln p_k] \tag{5.41}$$

Here the second boundary condition $\sum_{k=1}^{n} p_k = 1$ is used, because the weights provide the following values for the sum $\sum_{k=1}^{n} p_k \cdot \exp[-(1-\alpha)\cdot \ln p_k]$ and in

interval 1 we find: $$\sum_{k=1}^{n} p_k \cdot \exp[-(1-\alpha)\cdot \ln p_k] \geq 1 \tag{5.42}$$

and in

interval 2 we get: $$\sum_{k=1}^{n} p_k \cdot e - (1-\alpha)\cdot \ln p_k \leq 1 \tag{5.43}$$

We then compute the logarithm of these values and this leads to:

1. interval: $0 < \alpha < 1$

$$I_{\text{Rényi}} = \underbrace{\frac{1}{1-\alpha}}_{>0}\cdot \underbrace{\ln\underbrace{\left\{\sum_{k=1}^{n} p_k \cdot \exp[-(1-\alpha)\cdot \ln p_k]\right\}}_{\geq 1}}_{\geq 0} \geq 0 \tag{5.44}$$

2. interval: $1 < \alpha < \infty$

$$I_{\text{Rényi}} = \underbrace{\frac{1}{1-\alpha}}_{<0} \cdot \underbrace{\ln \underbrace{\left\{ \sum_{k=1}^{n} p_k \cdot \exp\left[-(1-\alpha) \cdot \ln p_k\right] \right\}}_{\leq 1}}_{\leq 0} \geq 0 \tag{5.45}$$

It does not matter in which interval the parameter α lies, the resulting Rényi information is always nonnegative. For the limit $\alpha \to 1$ we obtain Shannon's information, closing the gap in the range that the parameter can achieve. As Shannon's information is also nonnegative for discrete events, we get a completely nonnegative information for all possible values of our parameter.

5.1.4 Additivity and a connection to Minkowski's norm

$$S_\alpha(A) = \frac{1}{1-\alpha} \cdot \ln\left[\sum_{k=1}^{n} p^\alpha(a_k)\right] = \frac{1}{1-\alpha} \cdot \ln\left(\sum_{k=1}^{n} p_k^\alpha\right) \quad \text{with } \alpha \neq 1 \tag{5.46}$$

additionally has the following properties:

$S_\alpha(A)$ is symmetric, normalized, additive and nonnegative (decisive, expandable) [HAK88].

The additivity of the information of two independent random variables can be obtained by computing the Rényi α-information of the joint distribution density of two independent random variables, which means that $f_{z,y} = f_x \cdot f_y$ the joint distribution is equal to the product of the marginal distribution densities. We get

$$S_{\alpha Z} = I_{\alpha Z}(z) = \frac{1}{1-\alpha} \cdot \ln\left(\int_Z f_z^\alpha(\zeta) \cdot d\zeta\right) = \frac{1}{1-\alpha} \cdot \ln\left(\int_\Xi \int_P f_{x,y}^\alpha(\xi, \rho) \cdot d\xi \cdot d\rho\right) \tag{5.47}$$

As the joint distribution density can be created from the product of the marginal distribution densities, we are able to transform the kernel of the integral and thus to separate the integral into two integrals

$$S_{\alpha Z} = \frac{1}{1-\alpha} \cdot \ln\left(\int_\Xi f_x^\alpha(\xi) \cdot d\xi \cdot \int_P f_y^\alpha(\rho) \cdot d\rho\right)$$

$$S_{\alpha Z} = \frac{1}{1-\alpha} \cdot \left[\ln\left(\int_\Xi f_x^\alpha(\xi) \cdot d\xi\right) + \ln\left(\int_P f_y^\alpha(\rho) \cdot d\rho\right)\right]$$

$$S_{\alpha Z} = S_{\alpha X} + S_{\alpha Y} \qquad \text{for independent random variables} \qquad (5.48)$$

Additivity actually means that the entropy of the joint distribution density can be computed from the entropies of a marginal distribution and a conditional distribution (i.e. $f_{x,y} = f_{x|y} \cdot f_y$). If the random variables are independent, then this approach can be simplified and we again obtain the formulation we previously examined.

The entropy $S_\alpha(A)$ is strictly additive. Thus it is not easy to define a concept of a transformation in the form of differences of entropies.

We do not want to prove all these properties in detail, as we do not want to create unnecessary expense, providing exact mathematical derivatives, which possibly also prevent a clear view on the meaning of information.

Another possible way to explain Rényi's information is given by norms. We therefore use the p-norm

$$L_p = \left(\int_\Xi |\phi(\xi)|^p \cdot d\xi \right)^{\frac{1}{p}} \qquad (5.49)$$

also known as Minkowski's norm, which approaches Chebyshev's norm for $p \to \infty$. This norm is very similar to Rényi's measure of information and we therefore want to work out the exact connection. Instead of the absolute value of the function, we directly use the nonnegative distribution density $f_x(\xi)$ and furthermore the substitution $p = \alpha$.

$$-L_\alpha = \left(\int_\Xi f_x^\alpha(\xi) \cdot d\xi \right)^{\frac{1}{\alpha}} = \left(\int_\Xi f_x^\alpha(\xi) \cdot d\xi \right)^{\frac{1}{\alpha} \cdot \frac{1-\alpha}{1-\alpha}} = \left(\int_\Xi f_x^\alpha(\xi) \cdot d\xi \right)^{\frac{1}{1-\alpha} \cdot \frac{1-\alpha}{\alpha}} \qquad (5.50)$$

Taking the logarithm of this norm, we find

$$\ln L_\alpha = \ln \left[\left(\int_\Xi f_x^\alpha(\xi) \cdot d\xi \right)^{\frac{1}{1-\alpha} \cdot \frac{1-\alpha}{\alpha}} \right] = \frac{1-\alpha}{\alpha} \cdot \frac{1}{1-\alpha} \cdot \ln \left(\int_\Xi f_x^\alpha(\xi) \cdot d\xi \right) \qquad (5.51)$$

On the right-hand side we already have got Rényi's α-information, which we only have to write in a more explicit way.

$$\ln L_\alpha = \frac{1-\alpha}{\alpha} \cdot S_\alpha$$

or

$$S_\alpha = \frac{\alpha}{1-\alpha} \cdot \ln L_\alpha \tag{5.52}$$

For $\alpha \to 1$ the quotient tends to infinity while the logarithm tends to zero, because of its argument. Thus the limit for $\alpha \to 1$ requires the application of the rule of de l'Hospital and this leads us to Shannon's information of continuous distribution densities, as we derive in section 8.2.

Rényi's information is thus connected to the p-norm with the substitutions shown and can thus be interpreted as some kind of norm. This measure is actually applied in chaos theory [CAM93], where non-integer dimensions can be handled with the additional parameter occurring in this 'norm'.

5.1.5 The meaning of $S_\alpha(A)$ for $\alpha < 1$ and $\alpha > 1$:

1) $\ln x$ is a concave function and satisfies the Jensen inequality

$$\ln\left(\sum_{k=1}^{n} p_k \cdot x_k\right) \geq \sum_{k=1}^{n} p_k \cdot \ln x_k \tag{5.53}$$

Let $x_k = p_k^{\alpha-1}$ with $\alpha \neq 1$, which provides:

$$\ln\left(\sum_{k=1}^{n} p_k \cdot p_k^{\alpha-1}\right) \geq \sum_{k=1}^{n} p_k \cdot \ln p_k^{\alpha-1}$$

$$\ln\left(\sum_{k=1}^{n} p_k^{\alpha}\right) \geq (\alpha-1) \cdot \sum_{k=1}^{n} p_k \cdot \ln p_k$$

$$\ln\left(\sum_{k=1}^{n} p_k^{\alpha}\right) \geq -(1-\alpha) \cdot \sum_{k=1}^{n} p_k \cdot \ln p_k \tag{5.54}$$

This presents the connection to Shannon's entropy, and we are able to proceed with further transformation with regard to α, leading us to:

a) Let $0 < \alpha < 1$, then the term in the brackets on the right-hand side of equation (5.54) is positive and we write:

$$\frac{1}{1-\alpha} \cdot \ln\left(\sum_{k=1}^{n} p_k^{\alpha}\right) \geq -\sum_{k=1}^{n} p_k \cdot \ln p_k$$

$$S_\alpha(A) = \frac{1}{1-\alpha} \cdot \ln \sum_{k=1}^{n} p_k^\alpha \geq S(A) = -\sum_{k=1}^{n} p_k \cdot \ln p_k$$

$$S_\alpha(A) \geq S(A) \tag{5.55}$$

In this case Rényi's entropy is a measure of uncertainty.

b) Let $\alpha > 1$, then the term in brackets on the right-hand side of equation (5.54) is negative, leading to:

$$\frac{1}{1-\alpha} \cdot \ln\left(\sum_{k=1}^{n} p_k^\alpha\right) \leq -\sum_{k=1}^{n} p_k \cdot \ln p_k \tag{5.56}$$

$$S_\alpha(A) = \frac{1}{1-\alpha} \cdot \ln\left(\sum_{k=1}^{n} p_k^\alpha\right) \leq S(A) = -\sum_{k=1}^{n} p_k \cdot \ln p_k$$

$$S_\alpha(A) \leq S(A) \tag{5.57}$$

Here Rényi's entropy is more a measure of information.

The explanation of these statements of H. Haken can be illustrated in the following way:

The relation of Rényi's entropy and Shannon's entropy depends – as we saw in the previous transformation – on the parameter α.. We may thus define an α-transinformation

$$I_{T\alpha}(A) = S(A) - S_\alpha(A) \tag{5.58}$$

and write:

$$I_{T\alpha}(A) = S(A) - S_\alpha(A) < 0 \qquad \text{for} \qquad 0 < \alpha < 1 \tag{5.59}$$

$$I_{T\alpha}(A) = S(A) - S_\alpha(A) > 0 \qquad \text{for} \qquad \alpha > 1 \tag{5.60}$$

This α-transinformation occurs if we measure the information of a certain event with Shannon's measure and with Rényi's measure of information and calculate the difference between the observed measures of information of the same event A. So this transformation is not a familiar transformation, as we know it from communication theory and the communication channel, where we describe the information of the transmission of information from the input to the output. It actually is a transformation created by the change of the measure, which is used to describe the information of the very same event.

A slight transformation of the equation provides:

$$S_\alpha(A)\cdot(1-\alpha)=\ln\left(\sum_{k=1}^{n}p_k^\alpha\right) \tag{5.61}$$

and Haken states:

α can be denoted as a gain coefficient.

α can describe the efficiency of an observer.

$\alpha > 1$ $S_\alpha(A)$ measures the information

$\alpha < 1$ $S_\alpha(A)$ measures the uncertainty

We can also approximate Rényi's information for $\alpha \cong 1$. We use the equation

2) $$S_\alpha(A)=\frac{1}{1-\alpha}\cdot\ln\left(\sum_{k=1}^{n}p_k^\alpha\right) \tag{5.62}$$

and compute the first two terms of a Taylor series at the point $\alpha = 1$

$$(1-\alpha)\cdot\frac{\partial}{\partial\alpha}S_\alpha(A)-S_\alpha(A)=(1-\alpha)\cdot\frac{\partial}{\partial\alpha}\left[\frac{1}{1-\alpha}\cdot\ln\left(\sum_{k=1}^{n}p_k^\alpha\right)\right]-\frac{1}{1-\alpha}\cdot\ln\left(\sum_{k=1}^{n}p_k^\alpha\right)$$

$$=(1-\alpha)\cdot\left[-\frac{-1}{(1-\alpha)^2}\cdot\ln\left(\sum_{k=1}^{n}p_k^\alpha\right)+\frac{1}{1-\alpha}\cdot\frac{\partial}{\partial\alpha}\ln\left(\sum_{k=1}^{n}p_k^\alpha\right)\right]-\frac{1}{1-\alpha}\cdot\ln\left(\sum_{k=1}^{n}p_k^\alpha\right)$$

$$=\left[\frac{1}{1-\alpha}\cdot\ln\left(\sum_{k=1}^{n}p_k^\alpha\right)+(1-\alpha)\cdot\frac{1}{1-\alpha}\cdot\frac{\partial}{\partial\alpha}\ln\left(\sum_{k=1}^{n}p_k^\alpha\right)\right]-\frac{1}{1-\alpha}\cdot\ln\left(\sum_{k=1}^{n}p_k^\alpha\right)$$

$$=\frac{\partial}{\partial\alpha}\ln\left(\sum_{k=1}^{n}p_k^\alpha\right)+\frac{1}{1-\alpha}\cdot\ln\left(\sum_{k=1}^{n}p_k^\alpha\right)-\frac{1}{1-\alpha}\cdot\ln\left(\sum_{k=1}^{n}p_k^\alpha\right)$$

this leads to the equation

$$(1-\alpha)\cdot\frac{\partial}{\partial\alpha}S_\alpha(A)-S_\alpha(A)=\frac{\partial}{\partial\alpha}\ln\left(\sum_{k=1}^{n}p_k^\alpha\right) \tag{5.63}$$

$$= \frac{1}{\sum_{k=1}^{n} p_k^\alpha} \cdot \frac{\partial}{\partial \alpha} \sum_{k=1}^{n} p_k^\alpha = \frac{1}{\sum_{k=1}^{n} p_k^\alpha} \cdot \sum_{k=1}^{n} \frac{\partial}{\partial \alpha} p_k^\alpha$$

$$= \frac{1}{\sum_{k=1}^{n} p_k^\alpha} \cdot \sum_{k=1}^{n} \frac{\partial}{\partial \alpha} \exp\left(\ln p_k^\alpha\right) = \frac{1}{\sum_{k=1}^{n} p_k^\alpha} \cdot \sum_{k=1}^{n} \frac{\partial}{\partial} \exp(\alpha \cdot \ln p_k)$$

$$= \frac{1}{\sum_{k=1}^{n} p_k^\alpha} \cdot \sum_{k=1}^{n} \ln(p_k) \cdot \exp(\alpha \cdot \ln p_k) = \frac{1}{\sum_{k=1}^{n} p_k^\alpha} \cdot \sum_{k=1}^{n} p_k^\alpha \cdot \ln p_k$$

Finally we get

$$(1-\alpha) \cdot \frac{\partial}{\partial \alpha} S_\alpha(A) - S_\alpha(A) = \frac{1}{\sum_{k=1}^{n} p_k^\alpha} \cdot \sum_{k=1}^{n} p_k^\alpha \cdot \ln p_k \tag{5.64}$$

We obtain the negative value of this term when we create Shannon's entropy on the basis of an incomplete distribution density and additionally magnify the probabilities with the parameter α. We proceed with

$$\lim_{\alpha \to 1} \left[(1-\alpha) \cdot \frac{\partial}{\partial \alpha} S_\alpha(A) - S_\alpha(A) \right] = \lim_{\alpha \to 1} \frac{1}{\sum_{k=1}^{n} p_k^\alpha} \cdot \sum_{k=1}^{n} p_k^\alpha \cdot \ln p_k \tag{5.65}$$

and find

$$S_1(A) = \frac{1}{\sum_{k=1}^{n} p_k} \cdot \sum_{k=1}^{n} p_k \cdot \ln p_k \tag{5.66}$$

which is Shannon's entropy of an incomplete distribution density, providing Shannon's entropy, if

$$\sum_{k=1}^{n} p_k = 1$$

If α is approximately equal to 1, we are able to approximate the probabilities in the Taylor series, which leads us to the equation

$$p_k^\alpha = p_k^1 + \frac{\partial}{\partial\alpha} p_k^\alpha \bigg|_{\alpha=1} \cdot (\alpha-1) + o((\alpha-1))$$

$$p_k^\alpha = p_k^1 + \frac{\partial}{\partial\alpha} \exp\left(\ln p_k^\alpha\right)\bigg|_{\alpha=1} \cdot (\alpha-1) + o((\alpha-1))$$

$$p_k^\alpha = p_k^1 + \frac{\partial}{\partial\alpha} \exp(\alpha \cdot \ln p_k)\bigg|_{\alpha=1} \cdot (\alpha-1) + o((\alpha-1))$$

$$p_k^\alpha = p_k^\alpha + \exp(\alpha \cdot \ln p_k)\big|_{\alpha=1} \cdot (\alpha-1) \cdot \ln p_k + o((\alpha-1))$$

$$p_k^\alpha = p_k^1 + p_k^\alpha \bigg|_{\alpha=1} \cdot (\alpha-1) \cdot \ln p_k + o((\alpha-1))$$

$$p_k^\alpha = p_k^1 + (\alpha-1) \cdot p_k \cdot \ln p_k + o((\alpha-1)) \tag{5.67}$$

Inserting this approximation in the appropriate approximation of Rényi's entropy we get

$$S_\alpha(A) = S_1(A) + \frac{\partial}{\partial\alpha} S_\alpha(A)\bigg|_{\alpha=1} \cdot (\alpha-1) + o((\alpha-1)) \tag{5.68}$$

We already know the derivative from previous transformations

$$(1-\alpha) \cdot \frac{\partial}{\partial\alpha} S_\alpha(A) - S_\alpha(A) = \frac{1}{\sum_{k=1}^{n} p_k^\alpha} \cdot \sum_{k=1}^{n} p_k^\alpha \cdot \ln p_k$$

$$\frac{\partial}{\partial\alpha} S_\alpha(A) = \frac{1}{1-\alpha} \left[S_\alpha(A) + \frac{\sum_{k=1}^{n} p_k^\alpha \cdot \ln p_k}{\sum_{k=1}^{n} p_k^\alpha} \right] \tag{5.69}$$

so that

$$S_\alpha(A)=S_1(A)+\left[\frac{1}{1-\alpha}\cdot\left(S_\alpha(A)+\frac{\sum\limits_{k=1}^{n}p_k^\alpha\cdot\ln p_k}{\sum\limits_{k=1}^{n}p_k^\alpha}\right)\right]_{\alpha=1}\cdot(\alpha-1)+o((\alpha-1))$$

$$S_\alpha(A)=S_1(A)-\left[\left(\frac{1}{1-\alpha}\cdot\ln\sum_{k=1}^{n}p_k^\alpha+\frac{\sum\limits_{k=1}^{n}p_k^\alpha\cdot\ln p_k}{\sum\limits_{k=1}^{n}p_k^\alpha}\right)\right]_{\alpha=1}+o((\alpha-1)) \tag{5.70}$$

We now insert the approximation

$$p_k^\alpha=p_k^1+p_k\cdot\ln p_k\cdot(\alpha-1)+o((\alpha-1)) \tag{5.71}$$

neglecting the higher order terms

$$p_k^\alpha=p_k^1+p_k\cdot\ln p_k\cdot(\alpha-1)+\text{h.o.t.}$$

and thus find

$$S_\alpha(A)=S_1(A)-\left[\frac{1}{1-\alpha}\cdot\ln\left(\sum_{k=1}^{n}\{p_k+(\alpha-1)\cdot p_k\cdot\ln p_k\}\right)\right.$$
$$\left.+\frac{\sum\limits_{k=1}^{n}\{p_k+(\alpha-1)\cdot p_k\cdot\ln p_k\}\cdot\ln p_k}{\sum\limits_{k=1}^{n}\{p_k+(\alpha-1)\cdot p_k\cdot\ln p_k\}}\right]_{\alpha=1}+o((\alpha-1)) \tag{5.72}$$

Performing some transformations leads us to:

$$S_\alpha(A)=S_1(A)-(\alpha-1)\cdot\left[\sum_{k=1}^{n}(\ln p_k)^2\cdot p_k-S^2(A)\right]+o((\alpha-1)) \tag{5.73}$$

which is the approximation of Rényi's entropy on the basis of Shannon entropy if the parameter α is approximately 1.

With the Schwarz inequality, we can also approximate Shannon's entropy by the equation

$$S_1(A) = S(A) = -\sum_{k=1}^{n} p_k \cdot \ln p_k = \sum_{k=1}^{n} p_k \cdot |\ln p_k| \tag{5.74}$$

This approximation holds only, if $0 \le p_k \le 1$. In this case, the logarithm is not positive and can therefore be approximated by its absolute value and by a change of the sign. Thus possible applications of this transformation are restricted to distribution densities, where Kolmogorov's postulates ensure the necessary condition $0 \le p_k \le 1$.

We proceed with the Schwarz inequality

$$S(A) = \sum_{k=1}^{n} |\ln p_k| \cdot \sqrt{p_k} \cdot \sqrt{p_k} \le \left(\sum_{k=1}^{n} \left(\sqrt{p_k}\right)^2 \right)^{\frac{1}{2}} \cdot \left(\sum_{k=1}^{n} \left(|\ln p_k| \cdot \sqrt{p_k}\right)^2 \right)^{\frac{1}{2}}$$

$$S(A) \le \left(\sum_{k=1}^{n} p_k \right)^{\frac{1}{2}} \cdot \left(\sum_{k=1}^{n} (\ln p_k)^2 \cdot p_k \right)^{\frac{1}{2}} \tag{5.75}$$

We now have an estimate of the term $\left(\sum_{k=1}^{n} (\ln p_k)^2 \cdot p_k \right)^{\frac{1}{2}}$, because in case of complete distributions we get

$$\left(\sum_{k=1}^{n} p_k \right)^{\frac{1}{2}} = \sum_{k=1}^{n} p_k = 1 \tag{5.76}$$

and thus

$$S(A) \le \sqrt{\sum_{k=1}^{n} (\ln p_k)^2 \cdot p_k} \tag{5.77}$$

The square is

$$(S(A))^2 \le \sum_{k=1}^{n} (\ln p_k)^2 \cdot p_k \tag{5.78}$$

and we find this expression in the equation

$$S_\alpha(A) = S_1(A) - (\alpha - 1) \cdot \left[\sum_{k=1}^{n} (\ln p_k)^2 \cdot p_k - S^2(A) \right] + o\left((\alpha - 1)^2\right) \tag{5.79}$$

Now we are able to state that the term in the square brackets is always positive. If the parameter $\alpha > 1$, we thus get

$$S_\alpha(A) = S_1(A) - \Delta \qquad \text{with} \qquad \Delta > 0$$

and

$$S_\alpha(A) < S_1(A) \tag{5.80}$$

If the parameter lies in the interval $0 < \alpha < 1$, we get

$$S_\alpha(A) = S_1(A) + \Delta \qquad \text{with} \qquad \Delta > 0$$

and

$$S_\alpha(A) > S_1(A) \tag{5.81}$$

This confirms the previously found result and provides further insight into the relation between the information functions of Shannon and Rényi.
If we additionally use the inequality

$$\ln n \geq \frac{1}{1-\alpha} \cdot \ln\left[\sum_{k=1}^{n} p_k^\alpha\right] = S_\alpha \geq -\ln p_{\max} \tag{5.83}$$

we finally find

$$\ln n \geq S_\alpha(A) > S_1(A) \geq -\ln p_{\max} \qquad \text{for } 0 < \alpha < 1$$

and

$$\ln n \geq S_\alpha(A) = S_1(A) \geq -\ln p_{\max} \qquad \text{for } \alpha = 1 \tag{5.84}$$

$$\ln n \geq S_1(A) > S_\alpha(A) \geq -\ln p_{\max} \qquad \text{for } 1 < \alpha$$

describing the connection between Rényi's entropy and Shannon's entropy for different values of the parameter α.

5.1.6 Graphical presentations of Rényi's information

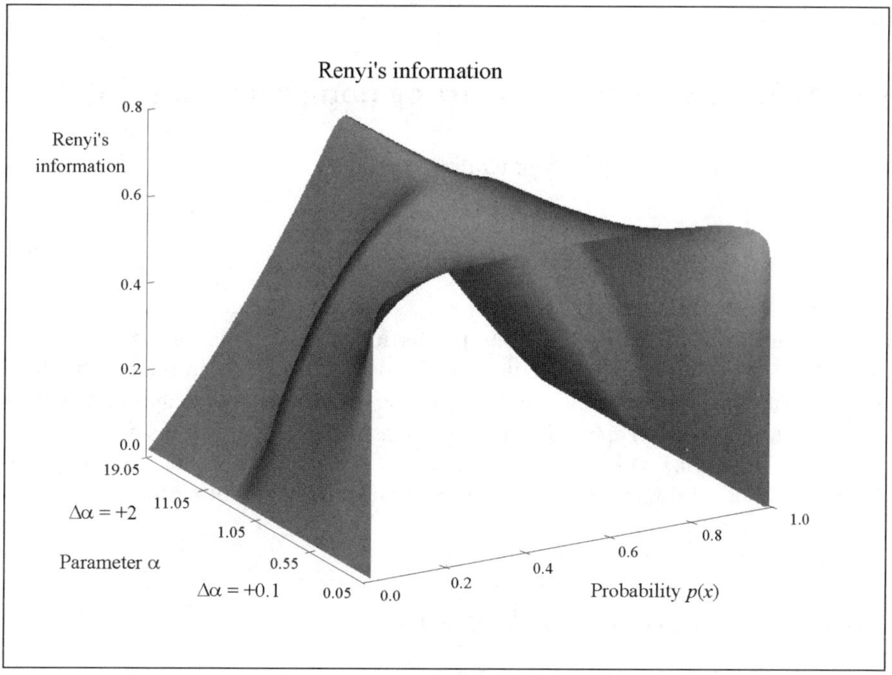

Fig. 5.1: Rényi's information

In this figure we see Rényi's information as a function of two events with the probabilities $p(x)$ and $q(x) = 1-p(x)$, where the parameter α assumes ten values in each of the ranges $\alpha > 1$ and $\alpha < 1$. The difference between two succeeding values of α is 2.0 in the interval $\alpha > 1$ and 0.1 in the interval $\alpha < 1$.

We know from our examinations that a decreasing value of the parameter α leads to a Rényi information tending to $\ln n$ and thus becoming independent of the probabilities of occurrence of the events. Only the number of events occuring has an influence on the value of the information. On the other hand, for $\alpha \to \infty$, Rényi's information approaches $\ln(1/p_{\max})$, which means, that the information is equal to zero at the boundaries of the range of the arguments, i.e. for $p(x) = 1$ or $q(x) = 1$. From $p(x) = 0.5$ to $p(x) = 1.0$ the probability $p(x)$ is the greatest single probability, which also means that $p(x) = p_{\max}$, and we can therefore completely approximate Rényi's information in this range by $\ln(1/p(x))$.

$$p(x) = 0.5 \quad \Rightarrow \quad S_\alpha = \ln(1/0.5) = 0.693$$

$$p(x) = 0.75 \quad \Rightarrow \quad S_\alpha = \ln(1/0.75) = 0.287$$

$$p(x) = 1 \quad \Rightarrow \quad S_\alpha = \ln(1/1.0) = 0.0$$

For $\alpha = 1$ the graph of Rényi's information is equal to the graph of Shannon's information, which we already know from figure 3.4.

5.2 Another generalized entropy (logical expansion)

Rényi's entropy is the result of a generalized mean

$$m_g = \varphi^{-1}\left[\sum_{k=1}^{n} p_k \cdot \varphi(x_k)\right] \tag{5.85}$$

and an additional definition of the properties of the function $\varphi(x)$ (strictly monotonic) in such a way that the necessary inverse function and the additivity of the information of independent events hold. These two demands require a linear or an exponential function $\varphi(x)$. Rényi uses – as we saw in our discussion – an exponential function, which merely consists of linear terms in the exponent. We thus expand the exponent by introducing an additional quadratic term.

$$\varphi(x) = \exp\left(c \cdot x^0 + \alpha \cdot x^1 + \beta \cdot x^2 + \text{h.o.t.}\right)$$

$$\varphi(x) = \exp(c) \cdot \exp\left(\alpha \cdot x + \beta \cdot x^2\right) \cdot \exp(\text{h.o.t.}) \tag{5.86}$$

Neglecting the higher order terms we find

$$\varphi(x) = \gamma \cdot \exp\left(\alpha \cdot x + \beta \cdot x^2\right) \tag{5.87}$$

The parameters are restricted to the range $\alpha,\beta,\gamma > 0$ and the variable is also restricted to $x \geq 0$ (as $x_k = \ln(1/p_k)$ and $0 \leq p_k \leq 1$, meeting the postulates of Kolmogorov), to ensure a unique inverse of the assumed function. The inverse function can be generated in a unique way, if we have a strictly monotonic function, i.e. if the function increases or decreases monotonically. We therefore take a look at the derivative:

$$\frac{d}{dx}\varphi(x) = \gamma \cdot \frac{d}{dx}\exp\left(\alpha \cdot x + \beta \cdot x^2\right) = \gamma \cdot (\alpha + 2 \cdot \beta \cdot x) \cdot \exp\left(\alpha \cdot x + \beta \cdot x^2\right) \tag{5.88}$$

With the preliminaries $\alpha,\beta,\gamma > 0$ and $x \geq 0$ the derivative is always positive and the function is a strictly monotonic increasing function. The inverse function is thus unique.

The second step in the generation of an expanded information function is the calculation of the inverse function, which starts with

$$y = \varphi(x) = \gamma \cdot \exp\left(\alpha \cdot x + \beta \cdot x^2\right) \tag{5.89}$$

and we now transform this equation until we have the inverse mapping.

$$\frac{y}{\gamma} = \exp\left(\alpha \cdot x + \beta \cdot x^2\right)$$

$$\ln\frac{y}{\gamma} = \alpha \cdot x + \beta \cdot x^2$$

$$x^2 + \frac{\alpha}{\beta} \cdot x = \frac{1}{\beta} \cdot \ln\frac{y}{\gamma}$$

$$x^2 + \frac{\alpha}{\beta} \cdot x + \left(\frac{\alpha}{2 \cdot \beta}\right)^2 = \frac{1}{\beta} \cdot \ln\left(\frac{y}{\gamma}\right) + \left(\frac{\alpha}{2 \cdot \beta}\right)^2$$

$$\left(x + \frac{\alpha}{2 \cdot \beta}\right)^2 = \frac{1}{\beta} \cdot \ln\left(\frac{y}{\gamma}\right) + \left(\frac{\alpha}{2 \cdot \beta}\right)^2$$

$$x = \varphi^{-1}(y) = -\frac{\alpha}{2 \cdot \beta} \pm \sqrt{\frac{1}{\beta} \cdot \ln\left(\frac{y}{\gamma}\right) + \left(\frac{\alpha}{2 \cdot \beta}\right)^2} \qquad (5.90)$$

Because of the different signs of the square root, we have to take a closer look at this result, as we desire a unique inverse function. The term in the square root has to be nonnegative. Any negative term in the square root would provide a complex x, which must not occur in the inversion of the given exponential function. Therefore we start with the term in the square root:

$$\frac{1}{\beta} \cdot \ln\left(\frac{y}{\gamma}\right) + \left(\frac{\alpha}{2 \cdot \beta}\right)^2 \qquad (5.91)$$

The second addend is 'per definitionem' always positive. The first addend varies with y. As the natural logarithm is a monotonically increasing function, the minimum of the logarithm function is exactly where the minimum of y is. We get:

$$\min(y) = \gamma \cdot \exp\left(\alpha \cdot 0 + \beta \cdot 0^2\right) \qquad \text{as } \min(x) = 0$$

$$\min(y) = \gamma \qquad (5.92)$$

Inserting the minimum in the term in the square root provides:

$$\min\left[\frac{1}{\beta} \cdot \ln\left(\frac{y}{\gamma}\right) + \left(\frac{\alpha}{2 \cdot \beta}\right)^2\right] = \frac{1}{\beta} \cdot \ln\left(\frac{\gamma}{\gamma}\right) + \left(\frac{\alpha}{2 \cdot \beta}\right)^2 = \left(\frac{\alpha}{2 \cdot \beta}\right)^2 \qquad (5.93)$$

and expression is positive, which means, that the expression of the square root always takes a value which is at least equal to $\left(\frac{\alpha}{2\cdot\beta}\right)^2$. With this consideration we return to

$$x=\varphi^{-1}(y)=-\frac{\alpha}{2\cdot\beta}\pm\sqrt{\frac{1}{\beta}\cdot\ln\left(\frac{y}{\gamma}\right)+\left(\frac{\alpha}{2\cdot\beta}\right)^2} \tag{5.94}$$

and now we can eliminate the negative sign in front of the square root, because this would generate a strictly decreasing function. We obtain

$$x=\varphi^{-1}(y)=-\frac{\alpha}{2\cdot\beta}+\sqrt{\frac{1}{\beta}\cdot\ln\left(\frac{y}{\gamma}\right)+\left(\frac{\alpha}{2\cdot\beta}\right)^2} \tag{5.95}$$

$$\begin{aligned}\min(x)=\varphi^{-1}(\min(y))&=-\frac{\alpha}{2\cdot\beta}+\sqrt{\frac{1}{\beta}\cdot\ln\left(\frac{\gamma}{\gamma}\right)+\left(\frac{\alpha}{2\cdot\beta}\right)^2}\\&=-\frac{\alpha}{2\cdot\beta}+\sqrt{\left(\frac{\alpha}{2\cdot\beta}\right)^2}=0\end{aligned} \tag{5.96}$$

We thus have found a unique inverse function, which allows us to proceed with the generalized mean, and build on the basis of these considerations. We get

$$m_{\mathrm{g}}=\varphi^{-1}\left[\sum_{k=1}^{n}p_k\cdot\varphi(x_k)\right] \tag{5.97}$$

$$I_{\mathrm{avg}}=-\frac{\alpha}{2\cdot\beta}+\sqrt{\frac{1}{\beta}\cdot\ln\left(\frac{1}{\gamma}\cdot\left[\sum_{k=1}^{n}p_k\cdot\gamma\cdot\exp\left(\alpha\cdot x_k+\beta\cdot x_k^2\right)\right]\right)+\left(\frac{\alpha}{2\cdot\beta}\right)^2} \tag{5.98}$$

The constant factor γ vanishes and we find:

$$I_{\mathrm{avg}}=-\frac{\alpha}{2\cdot\beta}+\sqrt{\frac{1}{\beta}\cdot\ln\left[\sum_{k=1}^{n}p_k\cdot\exp\left(\alpha\cdot x_k+\beta\cdot x_k^2\right)\right]+\left(\frac{\alpha}{2\cdot\beta}\right)^2} \tag{5.99}$$

Now we insert the information $x_k=\ln(1/p_k)$ in this equation and thus obtain

$$I_{\mathrm{avg}}=-\frac{\alpha}{2\cdot\beta}+\sqrt{\frac{1}{\beta}\cdot\ln\left[\sum_{k=1}^{n}p_k\cdot\exp\left[-\alpha\cdot\ln p_k+\beta\cdot(\ln p_k)^2\right]\right]+\left(\frac{\alpha}{2\cdot\beta}\right)^2}$$

$$I_{\text{avg}} = -\frac{\alpha}{2\cdot\beta} + \sqrt{\frac{1}{\beta}\cdot\ln\left[\sum_{k=1}^{n} p_k \cdot \exp\left(\ln p_k^{-\alpha}\right)\cdot\exp\left[\beta\cdot(\ln p_k)^2\right]\right] + \left(\frac{\alpha}{2\cdot\beta}\right)^2}$$

$$I_{\text{avg}} = -\frac{\alpha}{2\cdot\beta} + \sqrt{\frac{1}{\beta}\cdot\ln\left[\sum_{k=1}^{n} p_k \cdot p_k^{-\alpha} \cdot\exp\left[\beta\cdot(\ln p_k)^2\right]\right] + \left(\frac{\alpha}{2\cdot\beta}\right)^2}$$

$$I_{\text{avg}} = -\frac{\alpha}{2\cdot\beta} + \sqrt{\frac{1}{\beta}\cdot\ln\left[\sum_{k=1}^{n} p_k^{1-\alpha} \cdot\exp\left[\beta\cdot(\ln p_k)^2\right]\right] + \left(\frac{\alpha}{2\cdot\beta}\right)^2}$$

$$I_{\text{avg}} = -\frac{\alpha}{2\cdot\beta} + \frac{\alpha}{2\cdot\beta}\cdot\sqrt{1+\frac{4\cdot\beta}{\alpha^2}\cdot\ln\left[\sum_{k=1}^{n} p_k^{1-\alpha} \cdot\exp\left[\beta\cdot(\ln p_k)^2\right]\right]}$$

$$I_{\text{avg}} = \frac{\alpha}{2\cdot\beta}\cdot\left\{\sqrt{1+\frac{4\cdot\beta}{\alpha^2}\cdot\ln\left[\sum_{k=1}^{n} p_k^{1-\alpha} \cdot\exp\left[\beta\cdot(\ln p_k)^2\right]\right]} - 1\right\} \tag{5.100}$$

This is an extended information function with

$\alpha,\beta,\gamma > 0$ (γ does not occur in this function any more)

$x \geq 0$ because $x_k = \ln(1/p_k)$ and $0 \leq p_k \leq 1$

We do not want to carry out extensive calculation on this function, as we do not want to extend the mathematical part of this volume unnecessarily. To get a rough idea of the parameters α and β and their effect on this information function, we want to take a look at some values of the function, obtained by the calculation of the information of two events occurring with the probabilities $p(x)$ and $q(x) = 1 - p(x)$.

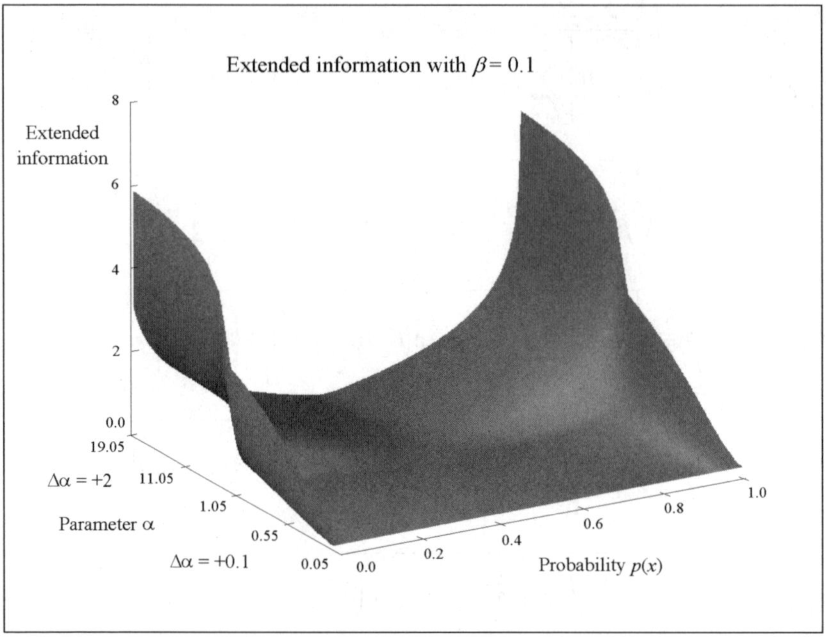

Fig. 5.2: Generalization of Rényi's information I

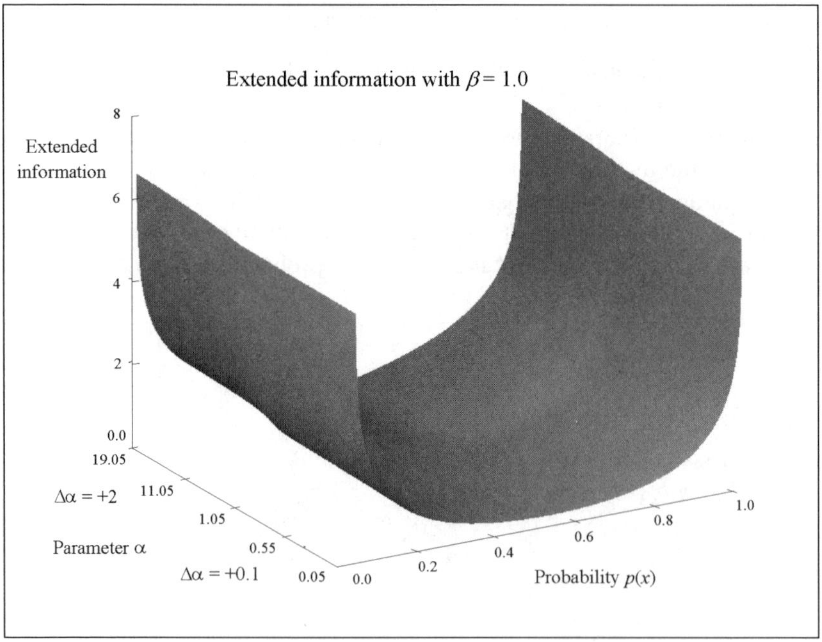

Fig. 5.3: Generalization of Rényi's information II

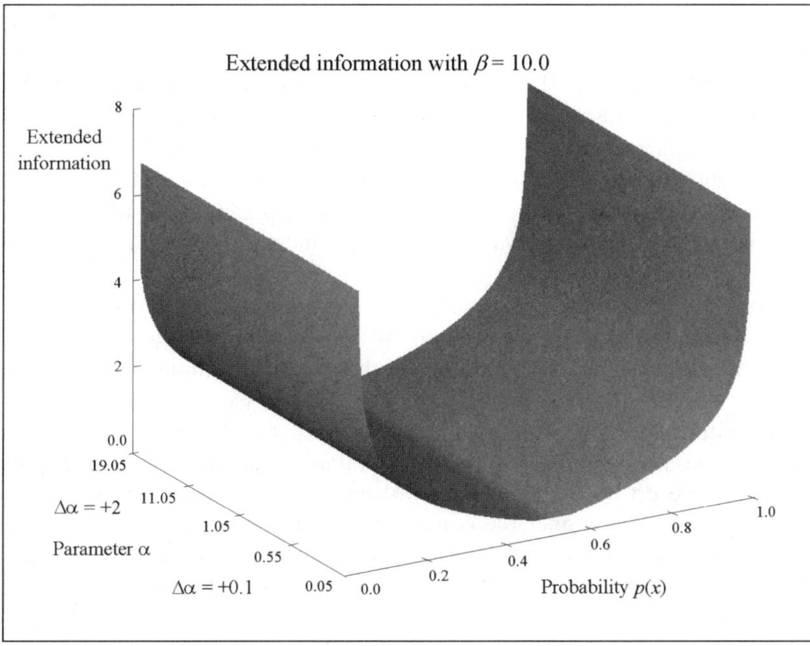

Fig. 5.4: Generalization of Rényi's information III

The three figures 5.2, 5.3 and 5.4 differ in the choice of the parameter β, while the parameter α is varied from 0.1 to 1.0 in steps of 0.1 and from 1.0 to 10.0 in steps of 2.0.

The first observation we make is that this is not a concave function, as we found in the figures for Shannon's and Rényi's information. This is a convex function, which means that the reciprocal function would be concave. The form of the reciprocal function would then be equal to the form we used for the examination of Shannon's and Rényi's information (convexity and symmetry regarding $p(x) = 0.5$).

Furthermore this extended information function yields large values if the probability of one of the events is large or small. Thus the extended information function increases when the probability of one event (with at least the probability 0.5) increases, i.e. when the uncertainty about the occurrence of this event decreases. So we have a measure that does not increase with an increasing number of possible different realizations, as Shannon's and Rényi's measures of information do, but which increases with a decreasing number of possible different realizations. It thus increases with an increasing certainty about the expected realizations. It therefore could be used to describe the a priori information of the number of the expected states. With a decreasing number of possible states this information measure also increases and it thus describes the information that we already have about the system.

Looking at the dependence of this information measure on the parameters, we find that the dependence on the parameter α is similar to the dependence in Rényi's information, but the two measures provide 'reciprocal' values of information. If we choose $\beta < 1$, as in figure 5.2, the sensitivity with respect to a variation of the parameter α increases. For small values of α the values of the information function do not differ when the probabilities are changed. We also saw this in Rényi's information. With an increasing parameter α the differences between the values of the information function increase when the probabilities vary.

The larger we choose the parameter β, the less the variation of the parameter α changes the function (see figures 5.3 and 5.4). Thus the parameter β allows us to determine the sensitivity of the function with respect to variations of the parameter α.

As this information measure (we use this term though it is not a concave function of the probabilities) is not a familiar measure, applied in present computations, we do not want to intensify our considerations. As we are more interested in an overview over the considerations in the derivatives and aim at a presentation of the connections between the information measures, we cannot discuss each measure in detail. We can only present and explain some of the information measures.

If we do not want to determine the information of a random variable on the basis of its distribution density or its probability distribution, but intend to calculate the difference in the information of two random variables, then we somehow have to calculate the information difference, distinguishing both random variables from each other. This can be achieved by calculating the gain of information that is obtained by the replacement of one random variable by another random variable. Rényi pursues this approach in deriving information differences, which we will now inspect.

5.3 Gain of information via conditional probabilities

The basic consideration is connected with the two terms: 'gain of information' and 'gain of uncertainty', which differ only by a negative sign. We can imagine this by considering the following train of thought, where we observe an event A_k with the probability $P(A_k) = p_k$. Then the event B_j occurs and changes the sigma-algebra, so that we are able to determine a conditional probability $P(A_k|B_j) = q_k$ for our event A_k. Thus the measurement, respectively the occurrence of the event B_j, provides a gain of information, which we are able to describe with the equation

$$\Delta I_k = \ln\left(\frac{1}{p_k}\right) - \ln\left(\frac{1}{q_k}\right) = \ln\left(\frac{1}{p_k}\right) + \ln q_k = \ln\left(\frac{q_k}{p_k}\right) \tag{5.101}$$

When we insert the original probabilities from our train of thought, we get the following formulation

$$\Delta I_k = \ln\left(\frac{q_k}{p_k}\right) = \ln\left(\frac{P(A_k \mid B_j)}{P(A_k)}\right) = \ln\left(\frac{P(A_k, B_j)}{P(A_k)\cdot P(B_j)}\right) = \ln\left(\frac{P(B_j \mid A_k)}{P(B_j)}\right) \quad (5.102)$$

and we directly recognize the symmetry of the gain of information. The event A_k contains as much information about the event B_j, as the event B_j contains about the event A_k. This special case will be discussed as Kolmogorov's information function in a later chapter. If we do not limit our attention to Hartley's information, but extend the examination to Shannon's information, we find the information difference

$$\Delta I_{\text{Shannon}} = \sum_{k=1}^{n} q_k \cdot \ln\left(\frac{q_k}{p_k}\right) \quad (5.103)$$

and this is already a discrete formulation of Kullback's information, which actually determines the information difference between two assignments of information (for instance the unconditioned and the conditional probability). It is nothing other than the linear mean of the information differences of all events occurring. The following condition must hold

$$\sum_{k=1}^{n} q_k = \sum_{k=1}^{n} p_k = 1 \quad (5.104)$$

but the single probabilities q_k and p_k may differ from each other, so that we currently sum up positive and negative addends in the linear mean.

We arbitrarily used the formulation $\Delta I_k = \ln\left(\frac{1}{p_k}\right) - \ln\left(\frac{1}{q_k}\right)$ and called it an information difference. If we were to use the reverse formulation

$$\Delta U_k = \ln\left(\frac{1}{q_k}\right) - \ln\left(\frac{1}{p_k}\right) = -\Delta I_k \quad (5.105)$$

we would describe an increase of the uncertainty, having the same absolute value as the information difference, but differing in the sign. Gain of information and an increasing uncertainty thus differ merely in the factor (−1) and can therefore be used as equivalent descriptions.

If we compute Shannon's information difference with the uncertainty, we find

$$\Delta I_{\text{Shannon}} = -\sum_{k=1}^{n} q_k \cdot \ln\left(\frac{p_k}{q_k}\right) \quad (5.106)$$

and thus this formulation generates a certain symmetry to equation (5.104), at least so long as we do not exchange the weights q_k and p_k in the summation, but only the terms in the logarithm function.

In principle we have two possible ways to define the gain of information. The first way is to create the mean of all single gains of information (which can achieve positive as well as negative values, depending on the probabilities of the single events, as we saw in our consideration of Shannon's information difference). The second possibility is to create the mean of all single uncertainty differences and multiply the result by –1. In case of Shannon's information both approaches lead to the same result. First we generated the mean value of a function of the changed probability $\frac{q_k}{p_k} = \frac{P(A_k | B_j)}{P(A_k)}$ and called the result a gain of information. In the second formulation we calculated the mean value of the same function, but we used the reciprocal argument $\frac{p_k}{q_k} = \frac{P(A_k)}{P(A_k | B_j)}$. This is equivalent to a calculation of the mean of an uncertainty and we obtain the gain of information simply by multiplying the calculated mean value by the factor –1. The result is the same in both approaches, when we use Shannon's information.

This identity of both approaches does not hold if we apply Rényi's information measure of the order α. There we obtain the information of one measurement by the equation

$$I_{\text{Rényi}} = \frac{1}{1-\alpha} \cdot \ln\left[\sum_{k=1}^{n} p_k^{\alpha}\right] \tag{5.107}$$

We may also use another formulation of this equation, where the expectation is more obvious

$$I_{\text{Rényi}} = \frac{1}{1-\alpha} \cdot \ln\left[\sum_{k=1}^{n} p_k \cdot p_k^{\alpha-1}\right] \tag{5.108}$$

if we only obtain an observation of the event A_k.

Here we compute the expectation value of $p_k^{\alpha-1}$ or the function of an expectation value of the function ($(\bullet)^{\alpha-1}$) of the probability. Replacing the single probabilities by the quotient $\frac{q_k}{p_k} = \frac{P(A_k | B_j)}{P(A_k)}$ – as we already did in Shannon's information – leads us to the gain of information in the formulation of Rényi's information

$$
\begin{aligned}
I_\alpha(Q \parallel P) &= (-1) \cdot \frac{1}{1-\alpha} \cdot \ln\left[\sum_{k=1}^{n} q_k \cdot \left(\frac{P(A_k | B_j)}{P(A_k)}\right)^{\alpha-1}\right] \\
&= \frac{1}{\alpha - 1} \cdot \ln\left[\sum_{k=1}^{n} q_k \cdot \left(\frac{q_k}{p_k}\right)^{\alpha-1}\right]
\end{aligned}
\tag{5.109}
$$

if we carry out the computation of the mean over the set $Q = (q_1, q_2,..., q_n)$. We thus obtain

$$
I_\alpha(Q \parallel P) = \frac{1}{\alpha - 1} \cdot \ln\left[\sum_{k=1}^{n} \frac{q_k^{\alpha}}{p_k^{\alpha-1}}\right] \tag{5.110}
$$

as Rényi's gain of information for replacing the given discrete probability distribution $P = (p_1, p_2,..., p_n)$ by the probability distribution Q.

The second possibility is to replace the quotient $\frac{q_k}{p_k} = \frac{P(A_k | B_j)}{P(A_k)}$ by its reciprocal quotient $\frac{p_k}{q_k} = \frac{P(A_k)}{P(A_k | B_j)}$, which leads to a measure of uncertainty U_α. This provides the equation

$$
\begin{aligned}
U_\alpha(Q \parallel P) &= \frac{1}{1-\alpha} \cdot \ln\left[\sum_{k=1}^{n} q_k \cdot \left(\frac{P(A_k)}{P(A_k | B_j)}\right)^{\alpha-1}\right] \\
&= \frac{1}{1-\alpha} \cdot \ln\left[\sum_{k=1}^{n} q_k \cdot \left(\frac{p_k}{q_k}\right)^{\alpha-1}\right]
\end{aligned}
$$

$$
U_\alpha(Q \parallel P) = \frac{1}{1-\alpha} \cdot \ln\left[\sum_{k=1}^{n} q_k \cdot \left(\frac{q_k}{p_k}\right)^{1-\alpha}\right]
$$

$$
U_\alpha(Q \parallel P) = \frac{1}{1-\alpha} \cdot \ln\left[\sum_{k=1}^{n} \frac{q_k^{2-\alpha}}{p_k^{1-\alpha}}\right] \tag{5.111}
$$

and this second possibility leads to the following gain of information:

$$\bar{I}_\alpha(Q \parallel P) = \frac{1}{1-\alpha} \cdot \ln\left[\sum_{k=1}^{n} \frac{q_k^{2-\alpha}}{p_k^{1-\alpha}}\right] \tag{5.112}$$

This formulation generates the problem that we have not only the value $\alpha = 1$, requiring a special treatment, but also the value $\alpha = 2$. For $\alpha = 2$ we obtain

$$\bar{I}_{\alpha=2}(Q \parallel P) = \frac{1}{-1} \cdot \ln\left[\sum_{k=1}^{n} \frac{q_k^{0}}{p_k^{-1}}\right]$$

$$\bar{I}_{\alpha=2}(Q \parallel P) = -\ln\left[\sum_{k=1}^{n} p_k\right] = 0 \tag{5.113}$$

The information difference always vanishes for $\alpha = 2$ and therefore this formulation is not suitable for denoting an information difference. The other formulation of the information difference merely consists of the special case $\alpha = 1$, which we do have to examine anyway.

If we apply Jensen's inequality to the gains of information $I_\alpha(Q \parallel P), I_1(Q \parallel P)$ with the concave logarithmic function and another function x^α, which is concave in the interval $0<\alpha<1$, but convex for $\alpha>1$, we find

$$I_\alpha(Q \parallel P) \geq 0$$
$$I_1(Q \parallel P) \geq 0 \qquad \text{(see also Kullback's information in chapter 9)} \tag{5.114}$$

The gain of information is thus always positive and it only equals zero when the two probabilities $p_k = q_k$ are equal for all k.

Rényi's gain of information, however, is not a symmetric function of the two arguments p_k and q_k, because the resulting gains of information

$$I_\alpha(Q \parallel P) = \frac{1}{\alpha - 1} \cdot \ln\left[\sum_{k=1}^{n} \frac{q_k^{\alpha}}{p_k^{\alpha-1}}\right]$$

$$I_\alpha(P \parallel Q) = \frac{1}{\alpha - 1} \cdot \ln\left[\sum_{k=1}^{n} \frac{p_k^{\alpha}}{q_k^{\alpha-1}}\right] \tag{5.115}$$

are in fact nonnegative, but in general they do not provide the same gains of information. To obtain a symmetric gain of information, independent of the order of replacement, it is possible to define a *J*-divergence

$$J_\alpha(P,Q) = I_\alpha(Q \parallel P) + I_\alpha(P \parallel Q)$$

$$= \frac{1}{\alpha - 1} \cdot \ln\left[\sum_{k=1}^{n} \frac{q_k^\alpha}{p_k^{\alpha-1}}\right] + \frac{1}{\alpha - 1} \cdot \ln\left[\sum_{k=1}^{n} \frac{p_k^\alpha}{q_k^{\alpha-1}}\right] \tag{5.116}$$

$$J_\alpha(P,Q) = \frac{1}{\alpha - 1} \cdot \ln\left[\sum_{k=1}^{n} \frac{q_k^\alpha}{p_k^{\alpha-1}} \cdot \sum_{k=1}^{n} \frac{p_k^\alpha}{q_k^{\alpha-1}}\right]$$

$$J_\alpha(P,Q) = \frac{1}{\alpha - 1} \cdot \ln\left[\sum_{k=1}^{n}\sum_{i=1}^{n} \frac{(q_k \cdot p_i)^\alpha}{(p_k \cdot q_i)^{\alpha-1}}\right] \tag{5.117}$$

or for $\alpha = 1$

$$J(P,Q) = J_1(P,Q) = I_1(Q \parallel P) + I_1(P \parallel Q)$$

$$= \sum_{k=1}^{n} q_k \cdot \ln\left(\frac{q_k}{p_k}\right) + \sum_{k=1}^{n} p_k \cdot \ln\left(\frac{p_k}{q_k}\right)$$

$$J(P,Q) = \sum_{k=1}^{n} (p_k - q_k) \cdot \ln\left(\frac{p_k}{q_k}\right) \tag{5.118}$$

These two *J*-divergences are now symmetric measures of the gain of information resulting from the replacement of one probability distribution by another probability distribution.

For the special case that $Q = (q_1, q_2, \dots, q_n) = (1/n, 1/n, \dots, 1/n)$ is a uniform probability distribution and that we replace it by another probability distribution P, we can proceed:

$$I_\alpha(P \parallel Q) = \frac{1}{\alpha - 1} \cdot \ln\left[\sum_{k=1}^{n} \frac{p_k^\alpha}{q_k^{\alpha-1}}\right] = \frac{1}{\alpha - 1} \cdot \ln\left[\sum_{k=1}^{n} n^{\alpha-1} \cdot p_k^\alpha\right]$$

$$I_\alpha(P \parallel Q) = \frac{1}{\alpha - 1} \cdot \ln\left[n^{\alpha-1} \cdot \sum_{k=1}^{n} p_k^\alpha\right]$$

$$I_\alpha(P \parallel Q) = \frac{1}{\alpha - 1} \cdot \ln n^{\alpha-1} + \frac{1}{\alpha - 1} \cdot \ln\left[\sum_{k=1}^{n} p_k^\alpha\right]$$

$$I_\alpha(P \parallel Q) = \ln n - \frac{1}{\alpha - 1} \cdot \ln\left[\sum_{k=1}^{n} p_k^\alpha\right]$$

$$I_\alpha(P \parallel Q) = I_{\text{Hartley}}(Q) - I_\alpha(P)$$

$$I_\alpha(P \parallel Q) = I_{\text{Shannon}}(Q) - I_\alpha(P) = I_1(Q) - I_\alpha(P) \tag{5.119}$$

Analogous to this derivation, we find the divergence based on Shannon's information as

$$\Delta I_{\text{Shannon}} = I(P \parallel Q) = \sum_{k=1}^{n} p_k \cdot \ln\left(\frac{p_k}{q_k}\right) = \sum_{k=1}^{n} p_k \cdot \ln(n \cdot p_k)$$

$$I(P \parallel Q) = \sum_{k=1}^{n} p_k \cdot \ln n + \sum_{k=1}^{n} p_k \cdot \ln p_k$$

$$I(P \parallel Q) = \ln n + \sum_{k=1}^{n} p_k \cdot \ln p_k$$

$$I(P \parallel Q) = I_{\text{Hartley}}(Q) - I_{\text{Shannon}}(P)$$

or

$$I(P \parallel Q) = I_{\text{Shannon}}(Q) - I_{\text{Shannon}}(P) \tag{5.120}$$

Here the information difference is in both cases equal to the difference of two information values. In both cases we subtract an information from Hartley's information. The uniform distribution density maximizes the information function and thus the gain of information is always positive. Thus the entropy described by the information function decreases when we replace the uniform distribution density by any other distribution density.

This approach also allows us to define a gain of information on the basis of completely missing a priori information, because the distribution density describing the complete lack of any previous information is the uniform distribution density. This is the connection to Laplace's principle of insufficient reasoning, expressing the fact that we must not prefer any event, if we do not have any a priori information. We thus have to assign equal probabilities to all events. But the general validity of this principle is limited, because any uniform distribution that is transformed into another arbitrary coordinate system does not necessarily generate another uniform distribution density, but other distribution densities preferring certain events. Which is the coordinate system that forms the basis for the uniform distribution density to describe our lack of knowledge? If there is only one coordinate system, our previous examination allows us to define an absolute gain of information and thus a specific scaling of the gain of information, depending on the coordinate system.

The approach that Alfred Rényi uses in his book [REN73] to introduce his information function and the divergence is somewhat different from this derivation, which started with the expansion of Shannon's information. Rényi starts with information differences respectively with average gains of information. These gains of information occur when we replace a (incomplete) distribution density $P=(p_1,p_2,...,p_n)$ of a random variable x with $p_k>0$ for k=1,...,n by another (incomplete) distribution density $Q=(q_1,q_2,...,q_n)$. The resulting gain of information is denoted by $I(Q||P)$.

An incomplete distribution density, according to Rényi, is a distribution with

$$\sum_{k=1}^{n} p_k < 1 \tag{5.121}$$

This requires certain standardizations in the calculations, which are not required for complete distribution densities with

$$\sum_{k=1}^{n} p_k = 1 \tag{5.122}$$

Before we proceed to the presentation of the postulates that are satisfied by this information function, we have to note two basic conditions for the function. The first is:

> $I(Q||P)$ is only defined if P and Q have an equal number of elements, which are connected to each other in a unique way.

And the second:

> $p_k>0$ has been required. Some of the q_k may satisfy the equation $q_k=0$; they merely should not all be equal to zero.

With these conditions Rényi formulates the following six postulates:

P I: (Generalized formulation of the additivity)

If $P=P_1\cdot P_2$ and $Q=Q_1\cdot Q_2$ (factorable distributions), we get

$I(Q\,||\,P)=I(Q_1\,||\,P_1)+I(Q_2\,||\,P_2)$

P II: If $p_k \le q_k$ for (k=1,...,n), we get $I(Q\,||\,P)\ge 0$

if $p_k \ge q_k$ for (k=1,...,n), we get $I(Q\,||\,P)\le 0$

($p_k = q_k$ for (k=1,...,n), we get $I(Q\,||\,P)=0$)

P III: (Determination of the unit {unit element} of the gain of information)

$$I\left(E_{\frac{1}{2}} \parallel E_{\frac{1}{2}}\right) = 1$$ (see also P3 of Shannon's information)

P IV: $I(Q\|P)$ depends only on the function $F(Q,P,x)$

$$I(Q \parallel P) = I[F(Q,P,x)] = I[F(x)]$$

$F(Q,P,x)$ is a conditional function of the gain of information

$$F(Q,P,x) = F(x) = \sum_{k=1}^{n} \left[\mathrm{ld}\left(\frac{q_k}{p_k}\right) < x \right] \cdot q'_k$$

$$F(Q,P,x) = F(x) = \sum_{k=1}^{n} \left[\mathrm{ld}\left(\frac{q_k}{p_k}\right) < x \right] \cdot \left(\frac{q_k}{\sum_{j=1}^{n} q_j} \right)$$

(Standardization factor $\sum_{j=1}^{n} q_j < 1$ for incomplete distributions)
(see also Shannon's postulate P1)

P V: If $F \in \Theta$, $G \in \Theta$, $F \not\equiv G$ and $G(x) \geq F(x)$ for $x \in (-\infty, \infty)$, then:

$$I[G(x)] \leq I[F(x)]$$

Θ is the set of all finite discrete distribution density functions

P VI: If $F_i \in \Theta$ $(i = 1,2,3)$ and $I[F_2] = I[F_3]$

Then we get for every t with $0 \leq t \leq 1$:

$$I[t \cdot F_1 + (1-t) \cdot F_2] = I[t \cdot F_1 + (1-t) \cdot F_3]$$

furthermore $I[t \cdot F_1 + (1-t) \cdot F_2]$ is a continuous function of t.
(see also Shannon's postulate P2)

Based on these six postulates we are able to develop the two information functions and Rényi formulates the following theorem:

Theorem 1:

If $I(Q\|P)$ satisfies the postulates I to VI, we may either find a real number $\alpha \neq 1$, which meets the equation $I(Q \| P) = I_\alpha(Q \| P)$, where $I_\alpha(Q \| P)$ is defined by

$$I_\alpha(Q \| P) = \frac{1}{\alpha - 1} \operatorname{ld}\left(\frac{1}{\sum_{j=1}^{n} q_j} \cdot \sum_{k=1}^{n} \frac{q_k^\alpha}{p_k^{\alpha-1}} \right) \tag{5.123}$$

or we get $I(Q \| P) = I_1(Q \| P)$, where $I_1(Q \| P)$ is defined by

$$I_1(Q \| P) = \frac{1}{\sum_{j=1}^{n} q_j} \cdot \sum_{k=1}^{n} q_k \cdot \operatorname{ld}\left(\frac{q_k}{p_k} \right) \tag{5.124}$$

with

$$I_1(Q \| P) = \lim_{\alpha \to 1} I_\alpha(Q \| P) \tag{5.125}$$

If P and Q are complete distributions ($\sum_{j=1}^{n} q_j = 1$), then $I(Q\|P)$ is obviously equal to the difference of two Shannon information functions. We know this connection already and we will explain it again in chapter 9.

$I_\alpha(Q\|P)$ = gain of information of order α.

For complete distributions with $n \to \infty$ Rényi finds the gain of information of order α

$$I_\alpha(Q \| P) = \frac{1}{\alpha - 1} \cdot \operatorname{ld}\left(\sum_{k=1}^{\infty} \frac{q_k^\alpha}{p_k^{\alpha-1}} \right) \qquad \text{for } \alpha \neq 1 \quad 0 < \alpha < \infty \tag{5.126}$$

$$I_1(Q \| P) = \sum_{k=1}^{\infty} q_k \cdot \operatorname{ld}\left(\frac{q_k}{p_k} \right) \tag{5.127}$$

if the series converge, which is always true for $0 < \alpha < 1$. The information of order α is

$$I_\alpha(x) = \frac{1}{1-\alpha} \cdot \mathrm{ld}\left(\sum_{k=1}^{\infty} p_k^\alpha\right) \tag{5.128}$$

$$I_1(x) = \sum_{k=1}^{\infty} p_k \cdot \mathrm{ld}\left(\frac{1}{p_k}\right) \tag{5.129}$$

x is a random variable, assuming a countable finite number of possible realization $x = x_k$ with the probabilities $p_k = P(x = x_k)$ $(k = 1,2,..)$.

The extension to continuous distribution densities $f_x^\alpha(\xi)$ provides the following two formulations

$$S_\alpha = I_\alpha(x) = \frac{1}{1-\alpha} \cdot \mathrm{ld}\left(\int_{-\infty}^{\infty} f_x^\alpha(\xi) \cdot d\xi\right) \tag{5.130}$$

$$G_\alpha = I_\alpha(Q \,\|\, P) = \frac{1}{\alpha-1} \cdot \mathrm{ld}\left(\int_{-\infty}^{\infty} \left(\frac{f_q^\alpha(\xi)}{f_p^{\alpha-1}(\xi)}\right) \cdot d\xi\right) \tag{5.131}$$

respectively

$$S_\alpha = I_\alpha(x) = \frac{1}{1-\alpha} \cdot \ln\left(\int_{-\infty}^{\infty} f_x^\alpha(\xi) \cdot d\xi\right) \tag{5.132}$$

$$G_\alpha = I_\alpha(Q \,\|\, P) = \frac{1}{\alpha-1} \cdot \ln\left(\int_{-\infty}^{\infty} \left(\frac{f_q^\alpha(\xi)}{f_p^{\alpha-1}(\xi)}\right) \cdot d\xi\right) \tag{5.133}$$

with $\alpha \in \Re$ and $\alpha \neq 1$

which are known as Rényi's α-information (α-entropy) S_α and as G-divergence G_α, S_α = 'the one-dimensional information of the order α, assigned to the random variable x'.

Rényi himself calls his measure an information measure, and in the same way he calls Shannon's measure an information measure. Nevertheless, both measures are instead measures of entropy, but this should not cause any confusion as we know that entropy and information depend on the point of view (section 3.2.3.4).

The entropy

$$I_\alpha(x,y)=\frac{1}{\alpha-1}\cdot\ln\left(\int_{-\infty}^{\infty}\int_{-\infty}^{\infty}\left(\frac{f_{\underline{y},\underline{x}}^{\alpha}(\underline{\rho},\underline{\xi})}{f_{\underline{x}}^{\alpha-1}(\underline{\xi})\cdot f_{\underline{y}}^{\alpha-1}(\underline{\rho})}\right)\cdot d\underline{\rho}\cdot d\underline{\xi}\right) \tag{5.133a}$$

is a special case of the G-divergence, because we compute the divergence between the joint distribution density and the marginal distribution densities. This special information measure is also known as GMI = Generalized Mutual Information and is used to find relations between random variables.

5.4 Other entropy or information measures

We want to take a short look at some other entropy measures, directly related to Rényi's α-entropy, and we demonstrate these relations very briefly.

5.4.1 Daroczy's entropy

$$S^\alpha(A)=\frac{\sum_{k=1}^{n}p_k^\alpha-1}{2^{1-\alpha}-1} \qquad \text{with} \qquad \alpha\neq 1 \tag{5.134}$$

We get the connection between Rényi's and Daroczy's entropy, if we examine the sums

$$\sum_{k=1}^{n}p_k^\alpha=\exp[(1-\alpha)\cdot S_\alpha(A)] \qquad \text{for Rényi} \tag{5.135}$$

$$\sum_{k=1}^{n}p_k^\alpha=\left(2^{1-\alpha}-1\right)\cdot S^\alpha(A)+1 \qquad \text{for Daroczy} \tag{5.136}$$

Equating both terms leads us to the formula

$$\exp[(1-\alpha)\cdot S_\alpha(A)]=\left(2^{1-\alpha}-1\right)\cdot S^\alpha(A)+1 \tag{5.137}$$

and we find

$$S_\alpha(A)=\frac{1}{1-\alpha}\cdot\ln\left[\left(2^{1-\alpha}-1\right)\cdot S^\alpha(A)+1\right] \qquad \text{Rényi from Daroczy} \tag{5.138}$$

and

$$S^{\alpha}(A)=\frac{1}{2^{1-\alpha}-1}\cdot\exp[(1-\alpha)\cdot S_{\alpha}(A)]-\frac{1}{2^{1-\alpha}-1} \quad \text{Daroczy from Rényi} \tag{5.139}$$

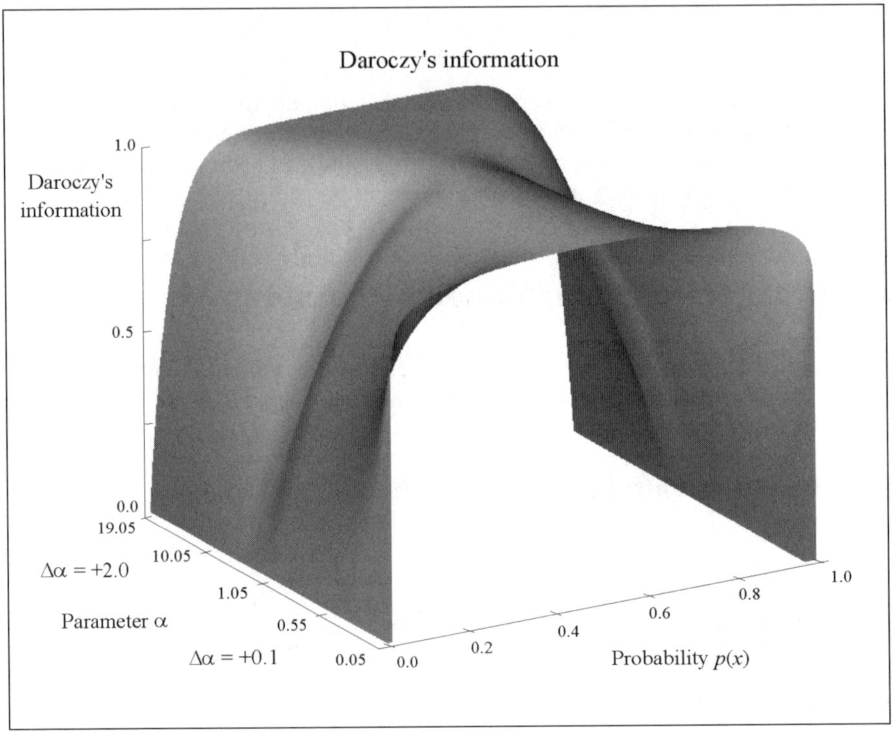

Fig. 5.5: Daroczy's information

5.4.2 Quadratic entropy

$$S_Q(A)=\sum_{k=1}^{n} p_k\cdot(1-p_k) \tag{5.140}$$

We choose a similar approach

$$S_Q(A)=\sum_{k=1}^{n} p_k-\sum_{k=1}^{n} p_k^2=1-\sum_{k=1}^{n} p_k^2 \tag{5.141}$$

$$\sum_{k=1}^{n} p_k^2 = 1 - S_Q(A) \qquad \text{for the quadratic entropy} \qquad (5.142)$$

$$\sum_{i=1}^{m} p^c(a_i) = \exp[(1-c)\cdot H_c(A)] \qquad \text{for Rényi's entropy} \qquad (5.143)$$

Taking $\alpha = 2$, we proceed:

$$\exp[(1-2)\cdot S_2(A)] = 1 - S_Q(A)$$

$$\exp[-S_2(A)] = 1 - S_Q(A)$$

$$-S_2(A) = \ln(1 - S_Q(A))$$

$$S_2(A) = \ln\left(\frac{1}{1 - S_Q(A)}\right) \qquad (5.144)$$

or

$$S_Q(A) = 1 - \exp[(1-2)\cdot S_2(A)] = 1 - \exp[-S_2(A)] \qquad (5.145)$$

Thus the quadratic entropy can also be created from Rényi's entropy.

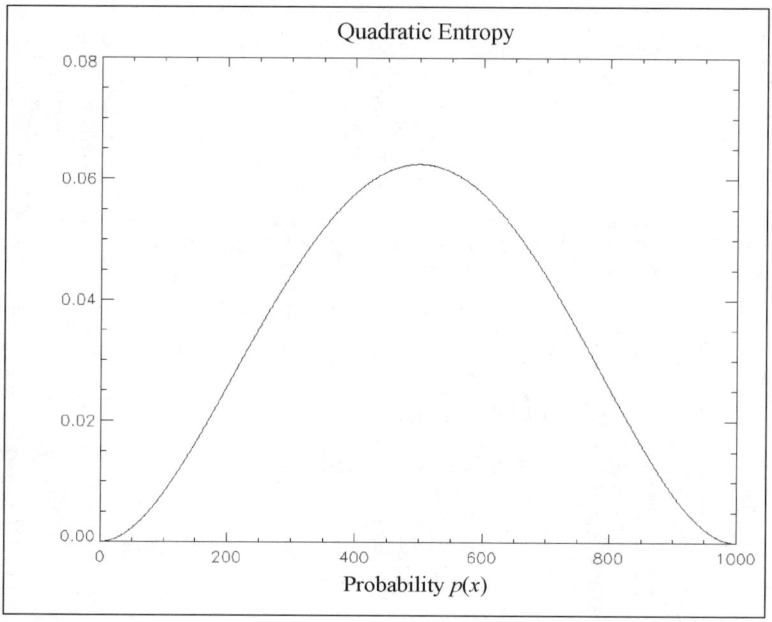

Fig. 5.6: Quadratic entropy

Figure 5.6 shows us that the form of the quadratic entropy is very similar to the form of the Shannon entropy. Only the limits of the range differ from the form that we know from Shannon's entropy. While the slope of the Shannon entropy increases at the limits of the interval, the slope of the quadratic entropy decreases.

5.4.3 *R*-norm entropy

$$S_R(A) = \frac{R}{R-1} \cdot \left[1 - \left(\sum_{k=1}^{n} p_k^R \right)^{\frac{1}{R}} \right] \qquad \text{with} \qquad R \neq 1 \tag{5.146}$$

The connection between the *R*-norm entropy and Rényi's entropy

$$S_R(A) = \frac{R}{R-1} + \frac{R}{1-R} \cdot \left(\sum_{k=1}^{n} p_k^R \right)^{\frac{1}{R}} \tag{5.147}$$

$$S_\alpha(A) = \frac{1}{1-\alpha} \cdot \ln \sum_{k=1}^{n} p_k^\alpha$$

$$\ln \sum_{k=1}^{n} p_k^\alpha = S_\alpha(A) \cdot (1-\alpha)$$

$$\sum_{k=1}^{n} p_k^\alpha = \exp\{S_\alpha(A) \cdot (1-\alpha)\} \tag{5.148}$$

Inserting this sum of probabilities into the *R*-norm entropy, we obtain the formulation

$$S_R(A) = \frac{R}{R-1} + \frac{R}{1-R} \cdot \left(\sum_{k=1}^{n} p_k^R \right)^{\frac{1}{R}}$$

and if we choose with $R = \alpha$ in the sum we obtain

$$S_R(A) = \frac{R}{R-1} + \frac{R}{1-R} \cdot \left(\exp\{S_{\alpha=R}(A) \cdot (1-R)\}\right)^{\frac{1}{R}}$$

$$S_R(A) = \frac{R}{R-1} + \frac{R}{1-R} \cdot \exp\left\{ \frac{1}{R} \cdot S_{\alpha=R}(A) \cdot (1-R) \right\}$$

$$S_R(A)=\frac{R}{R-1}+\frac{R}{1-R}\cdot\exp\left\{\frac{1}{R}\cdot S_{\alpha=R}(A)\cdot(1-R)\right\} \tag{5.149}$$

and we again have a connection between this entropy and the entropy that was introduced by Alfred Rényi.

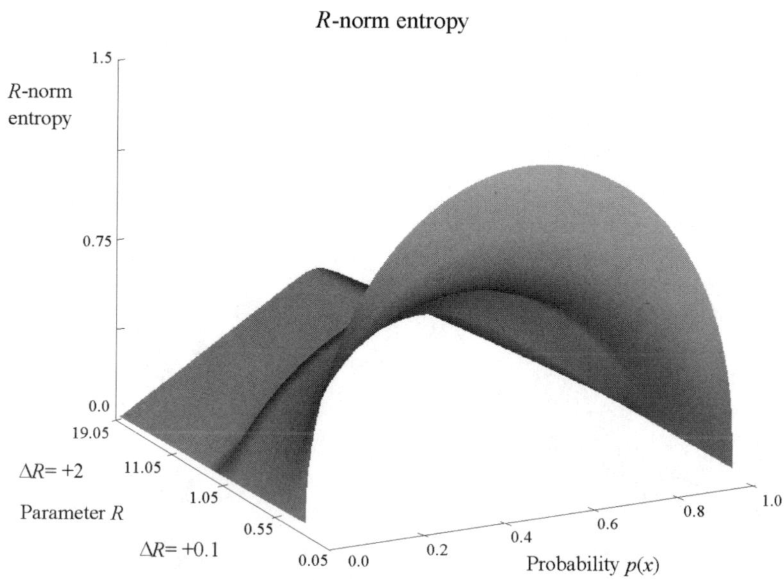

Fig. 5.7: R-norm entropy

Figure 5.7 starts with $\alpha = 0.45$, because the values of the R-norm entropy increase rapidly when we choose smaller values of the parameter α. As the form of the function is not changed for $\alpha < 0.5$, we do not include this interval in figure 5.7. Thus we are able to show the form of the function in the range $1 < \alpha < 20$, which is similar to Rényi's information in this range.

We will now examine some other measures of information to demonstrate the great variety of possible measures, which we are not able to present completely in this volume.

6 Generalized entropy measures

Kapur presents several different generalized measures of entropy in his publications [KAP89] [KAP92] [KAP94]. We will present some of them in this chapter. We do not want to give detailed mathematical derivations, but short presentations in the form of an enumeration of the measures and a short description. Most of the generalized entropy measures have additional parameters, allowing variations of these measures. We present only some of them, just to give an impression of further entropy measures, and we begin with the already known Rényi information

Rényi: $$I_\alpha = \frac{1}{1-\alpha} \cdot \ln \sum_{k=1}^{n} p_k^\alpha \qquad \alpha > 0 \text{ and } \alpha \neq 1 \qquad (6.01)$$

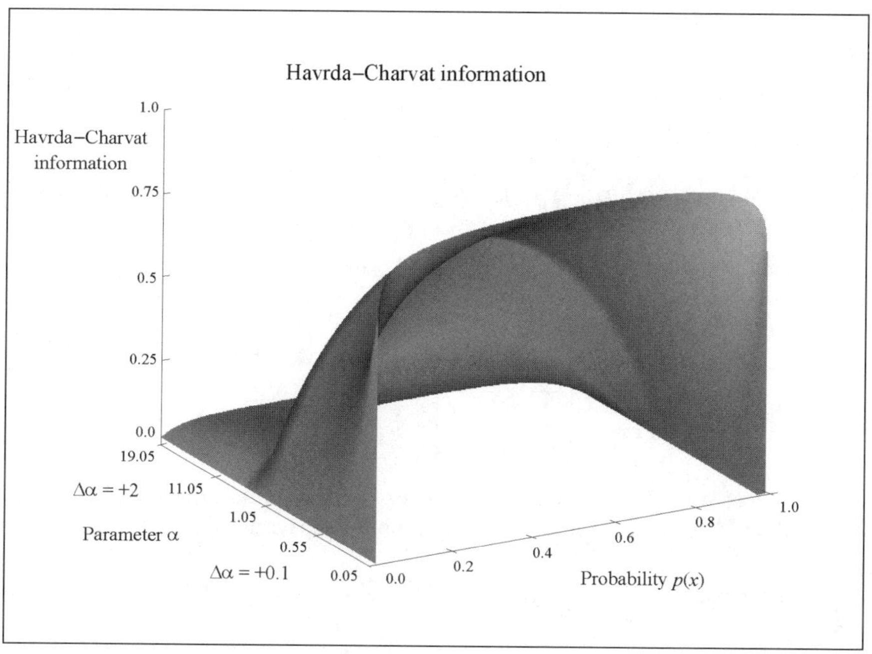

Fig. 6.1: Havrda–Charvat information

A similar information measure, using the parameter α, is given by

$$\text{Havrda–Charvat: } I_{\mathrm{HC},\alpha} = \frac{1}{1-\alpha} \cdot \left(\sum_{k=1}^{n} p_k^{\alpha} - 1 \right) \qquad \alpha > 0 \text{ and } \alpha \neq 1 \tag{6.02}$$

This information measure is very similar to Rényi's information. In the range $0 < \alpha \leq 1$ the similarity is apparent, while in the range $\alpha > 1$ the information of Havrda–Charvat flatter than that of Rényi (see also figure 5.1).

The connection between Rényi's information and the information of Havrda and Charvat:

$$I_{\mathrm{HC},\alpha} = \frac{1}{1-\alpha} \cdot \left(\sum_{k=1}^{n} p_k^{\alpha} - 1 \right) = \frac{1}{1-\alpha} \cdot \sum_{k=1}^{n} p_k^{\alpha} - \frac{1}{1-\alpha} \tag{6.03}$$

$$\sum_{k=1}^{n} p_k^{\alpha} = (1-\alpha) \cdot I_{\mathrm{HC},\alpha} + 1 \tag{6.04}$$

With this transformation we are able to insert the sum into Rényi's information, which leads us to

$$I_{\alpha} = \frac{1}{1-\alpha} \cdot \ln \sum_{k=1}^{n} p_k^{\alpha} \tag{6.05}$$

$$I_{\alpha} = \frac{1}{1-\alpha} \cdot \ln\left((1-\alpha) \cdot I_{\mathrm{HC},\alpha} + 1 \right) \tag{6.06}$$

Do we get Shannon's information, if we choose $\alpha = 1$? Figure 6.1 suggests that we may obtain Shannon's information and thus we examine this case. We build the limit

$$\lim_{\alpha \to 1} I_{\mathrm{HC},\alpha} = \lim_{\alpha \to 1} \frac{1}{1-\alpha} \cdot \left(\sum_{k=1}^{n} p_k^{\alpha} - 1 \right) \tag{6.07}$$

and obtain zero in numerator and denominator of the resulting quotient, so that we use the rule of de l´Hospital again and examine the derivatives of both terms

$$\frac{\partial}{\partial \alpha} \text{denominator} = \frac{\partial}{\partial \alpha} (1-\alpha) = -1 \tag{6.08}$$

$$\frac{\partial}{\partial \alpha} \text{numerator} = \frac{\partial}{\partial \alpha} \left(\sum_{k=1}^{n} p_k^{\alpha} - 1 \right) = \sum_{k=1}^{n} \frac{\partial}{\partial \alpha} p_k^{\alpha} \tag{6.09}$$

with

$$\frac{\partial}{\partial\alpha} p_k^\alpha = \frac{\partial}{\partial\alpha} \exp\left(\ln p_k^\alpha\right) = \frac{\partial}{\partial\alpha} \exp(\alpha \cdot \ln p_k)$$
$$= \ln p_k \cdot \exp(\alpha \cdot \ln p_k) = p_k^\alpha \cdot \ln p_k \tag{6.10}$$

we find

$$\frac{\partial}{\partial\alpha} \text{numerator} = \sum_{k=1}^{n} \ln p_k \cdot p_k^\alpha \tag{6.11}$$

and in the limit we get

$$\lim_{\alpha\to 1} I_{\text{HC},\alpha} = \lim_{\alpha\to 1} -\sum_{k=1}^{n} \ln p_k \cdot p_k^\alpha = -\sum_{k=1}^{n} p_k \cdot \ln p_k = I_{\text{Shannon}}(P) \tag{6.12}$$

As we suggested, the information function of Havrda and Charvat is equal to Shannon's information if we ch0ose $\alpha = 1$.

Another information function of this type is the following function given by Kapur:

$$K_{\alpha,\beta}(P) = I_{\text{Kapur}} = \frac{1}{\beta - \alpha} \cdot \ln \frac{\sum_{k=1}^{n} p_k^\alpha}{\sum_{k=1}^{n} p_k^\beta} \qquad \alpha > 0\,,\ \beta > 0 \text{ and } \alpha \neq \beta \tag{6.13}$$

This information is again similar to Rényi's information. The main difference is the additional parameter β, which we choose equal to 0.5 first and then equal to 5 to present the corresponding graphs.

In the range $\alpha > 1$ figure 6.2 does not show any apparent differences from Rényi's information and for $\alpha = 1$ we get a clear similarity to Shannon's information. But in the range $\alpha < 1$ we do not get a constant function of the probabilities, as we obtained in Rényi's information, but the function changes from a concave function to a symmetrical convex function of the probabilities, achieving its maximum for the extreme values $p(x) = 0$ and $p(x) = 1$.

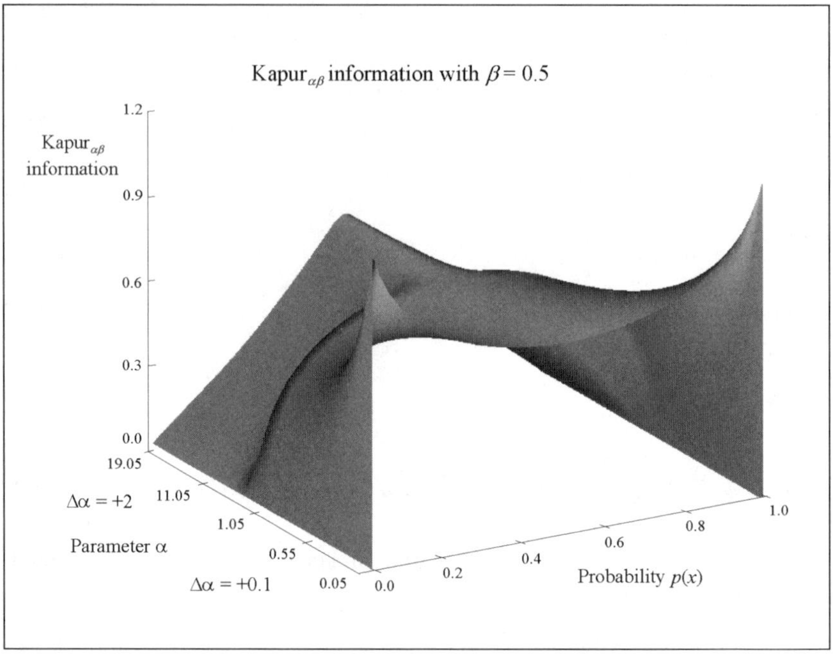

Fig. 6.2: Kapur$_{\alpha\beta}$ information

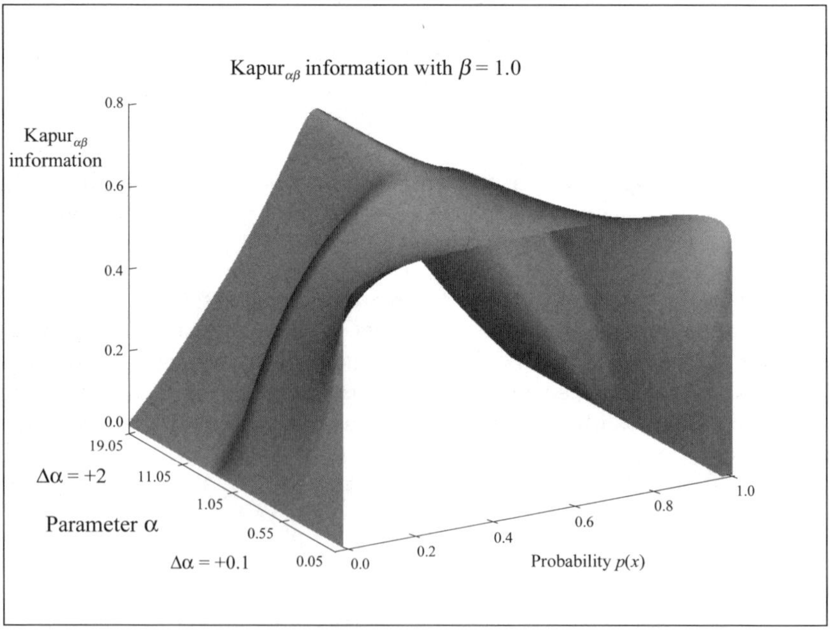

Fig. 6.3: Kapur$_{\alpha\beta}$ for β = 1 is equal to Rényi

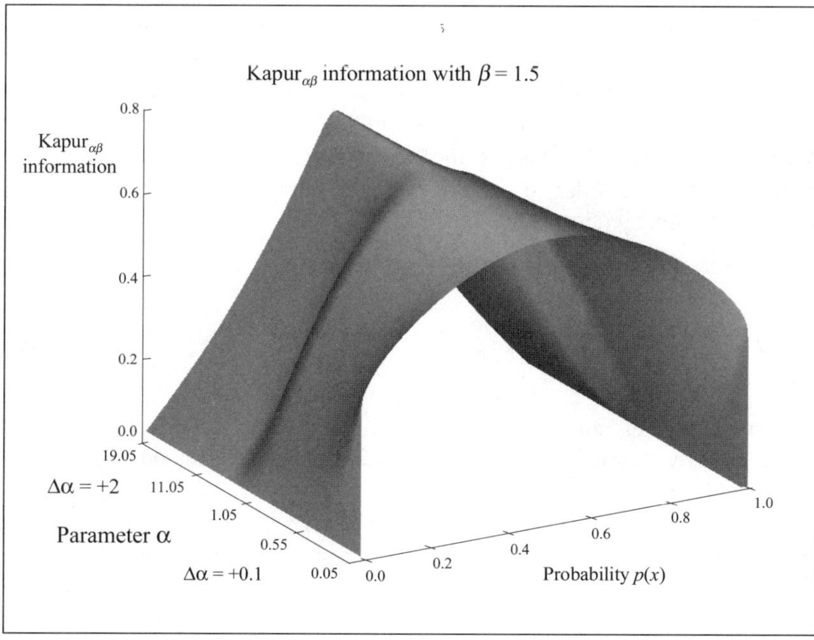

Fig. 6.4: Kapur$_{\alpha\beta}$ for β = 1.5

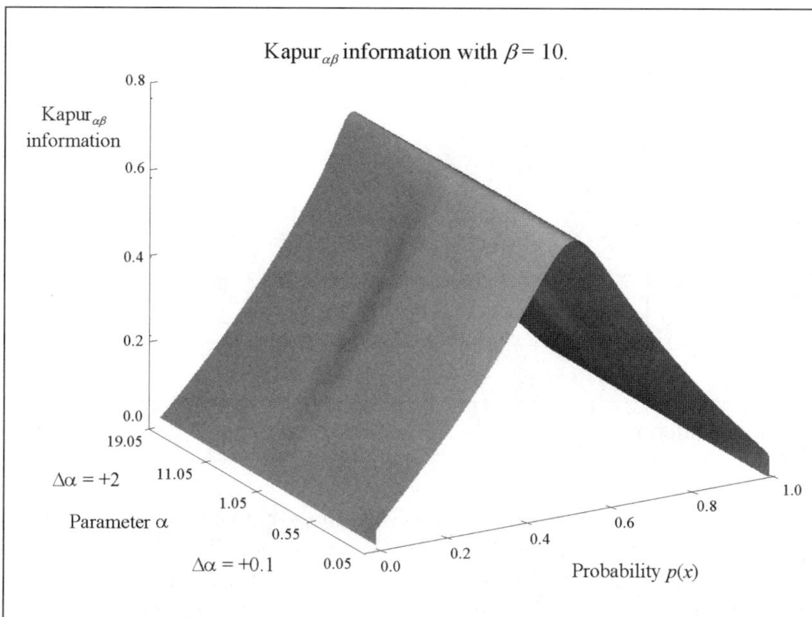

Fig. 6.5: Kapur$_{\alpha\beta}$ for β = 10

If we increase the parameter β to 10, as we see in figure 6.5, we discover that this information measure approaches more and more the information measure of Rényi for large values of α. In this case the variation of the parameter α in $K_{\alpha,\beta}$ does not lead to any further changes in the function.

Thus the parameter β changes the sensitivity of the function $K_{\alpha,\beta}$ to the parameter α. This is similar to the effect of the parameter β in the extension of Rényi's information (section 5.2).

Burg introduced a rather simple information measure:

$$\text{Burg:} \quad I_{\text{Burg}} = \sum_{k=1}^{n} \ln p_k \tag{6.14}$$

This information does not contain any parameters. It can be developed from the maximum entropy estimation [KAP92] [KAP94], where the boundary condition is a set of variables with one missing element.

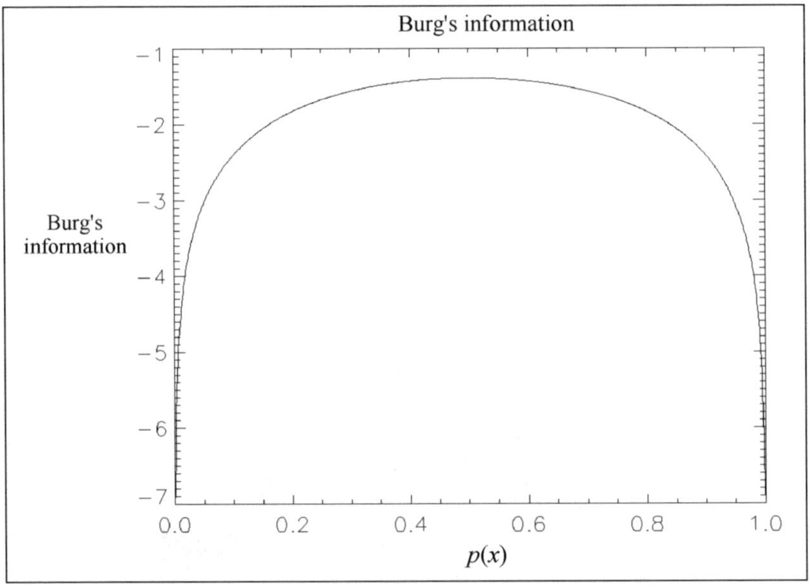

Fig. 6.6: Burg's information

This information function, similar to Shannon's entropy or to the quadratic entropy, is a concave function. It has the disadvantage that the values of this information function are exclusively negative, so that we cannot establish the connection to the uncertainty, expressed by the variances of the random variables, because the values of the variances are always nonnegative. Thus a direct interpretation of this measure as a measure of uncertainty is not possible [KAP94].

Another information measure of Kapur

$$I_{\text{Kapur}} = -\sum_{k=1}^{n} p_k \cdot \ln p_k + \frac{1}{a} \cdot \sum_{k=1}^{n} (1 + a \cdot p_k) \cdot \ln(1 + a \cdot p_k) - \frac{1}{a} \cdot (1 + a) \cdot \ln(1 + a) \tag{6.15}$$

consists of the sum of two information measures. The second will be subject to further considerations and the first is, as we have easily seen, Shannon's information. For the parameter we choose $a > -1$, to meet the range in which the logarithm is defined. As a also appears in the denominator of the quotient, we additionally require $a > 0$.

We now want to maximize this entropy

$$K_a(P) = -\sum_{k=1}^{n} p_k \cdot \ln p_k + \frac{1}{a} \cdot \sum_{k=1}^{n} (1 + a \cdot p_k) \cdot \ln(1 + a \cdot p_k) - \frac{1}{a} \cdot (1 + a) \cdot \ln(1 + a) \tag{6.16}$$

$$a > 0$$

with the conditions that

$$\sum_{k=1}^{n} p_k = 1 \tag{6.17}$$

We have a complete distribution density and that we know the average energy

$$\sum_{k=1}^{n} p_k \cdot E_k = \overline{E} \tag{6.18}$$

where $E_1, E_2, \ldots, E_n$ are n levels of energy and $\overline{E}$ is the expected energy of the system. This leads to the probabilities

$$p_k = \frac{1}{\exp(\lambda + \mu \cdot E_k) + c} \tag{6.19}$$

The Lagrangian multipliers λ and μ can be computed from the initial conditions. c is a constant, which will be at least equal to -1 and different choices of this constant lead to the following entropies:

Bose–Einstein entropy for $c = 1$

$$K_1(P) = -\sum_{k=1}^{n} p_k \cdot \ln p_k + \sum_{k=1}^{n} (1 + p_k) \cdot \ln(1 + p_k) - 2 \cdot \ln 2 \tag{6.20}$$

and Fermi–Dirac entropy for $c = -1$

$$K_{-1}(P) = -\sum_{k=1}^{n} p_k \cdot \ln p_k - \sum_{k=1}^{n} (1 - p_k) \cdot \ln(1 - p_k) \qquad (6.21)$$

Both entropies are assigned to the corresponding distributions in statistical mechanics.

Thus the specified entropy can be used for optimization in this field. The entropy is therefore a measure that is specified for a given problem, as we found, for instance, in this example of statistical mechanics, where either the conditions of Bose–Einstein or the conditions of Fermi–Dirac are applied to obtain the desired entropy. Obviously the entropy measures cannot be applied to arbitrary problems, but only to certain problems that can then be optimized via maximum entropy estimation. We return to a more detailed description of this optimization in chapter 15.

Another extended information measure given by Kapur is:

$$K_a(P) = \sum_{k=1}^{n} \ln(1 + a \cdot p_k) - \ln(1 + a) \qquad a > 0 \qquad (6.22)$$

and it has the following properties:

$K_a(P)$ is a continuous function of p_k, i.e. small variations in the values of p_k do only generate small variations of this information measure.

$K_a(P)$ is symmetric with respect to possible permutations of $p_1, p_2, \ldots, p_n$. Thus arbitrary permutations do not change this measure.

The maximization of $K_a(P)$ under the condition $\sum_{k=1}^{n} p_k = 1$ leads to a uniform distribution density with $p_1 = p_2 = \ldots = p_n$. This maximized function increases monotonically with an increasing number of equally probable events.

$$K_a(p_1 = p_2 = \ldots = p_n) = \sum_{k=1}^{n} \ln\left(1 + \frac{a}{n}\right) - \ln(1 + a)$$

$$K_a(p_1 = p_2 = \ldots = p_n) = n \cdot \ln\left(\frac{n + a}{n}\right) - \ln(1 + a) \qquad (6.23)$$

The first derivative is

$$\frac{\partial}{\partial n}K_a(p_1 = p_2 = ... = p_n) = \ln\left(\frac{n+a}{n}\right) + \frac{n^2}{n+a} \cdot \frac{n-(n+a)}{n^2}$$

$$\frac{\partial}{\partial n}K_a(p_1 = p_2 = ... = p_n) = \ln\left(\frac{n+a}{n}\right) - \frac{a}{n+a} \tag{6.24}$$

and the second derivative is equal to

$$\frac{\partial^2}{\partial n^2}K_a(p_1 = p_2 = ... = p_n) = \frac{n}{n+a} \cdot \frac{n-(n+a)}{n^2} - \frac{-a}{(n+a)^2}$$

$$\frac{\partial^2}{\partial n^2}K_a(p_1 = p_2 = ... = p_n) = -\frac{1}{n+a} \cdot \frac{a}{n} + \frac{a}{(n+a)^2} = \frac{n \cdot a - (n+a) \cdot a}{n \cdot (n+a)^2}$$

$$\frac{\partial^2}{\partial n^2}K_a(p_1 = p_2 = ... = p_n) = -\frac{a^2}{n \cdot (n+a)^2} < 0 \tag{6.25}$$

The second derivative is always less than zero, so that the first derivative is a monotonically decreasing function. Thus $K_a(p_1 = p_2 = ... = p_n)$ is a concave function of the number of equally probable events.

$K_a(P)$ is a concave function of $p_1, p_2, ..., p_n$.

$K_a(P)$ achieves its maximum if only one event occurs with probability 1 while all other events are weighted by a probability of zero. This probability distribution is then described by $\Delta_k = \left(0,0,...0,\ \underset{k^{th}\text{ position}}{\underset{\downarrow}{1}}\ ,0,...,0\right)$.

$K_a(P)$ does not meet the requirements of additivity and the recursiveness, but these two properties are less important properties of an entropy function, when we focus our attention on the maximization.

In figure 6.7 we vary the parameter a from 0.05 to 20.05 and the maximum changes, while the convex form is preserved. When we choose a larger value of a, the difference of the information of the certain event $p(x) = 0$ and the most uncertain event $p(x) = 0.5$ increases, which means that the parameter a enhances the possible discrimination between the uncertainty of the events.

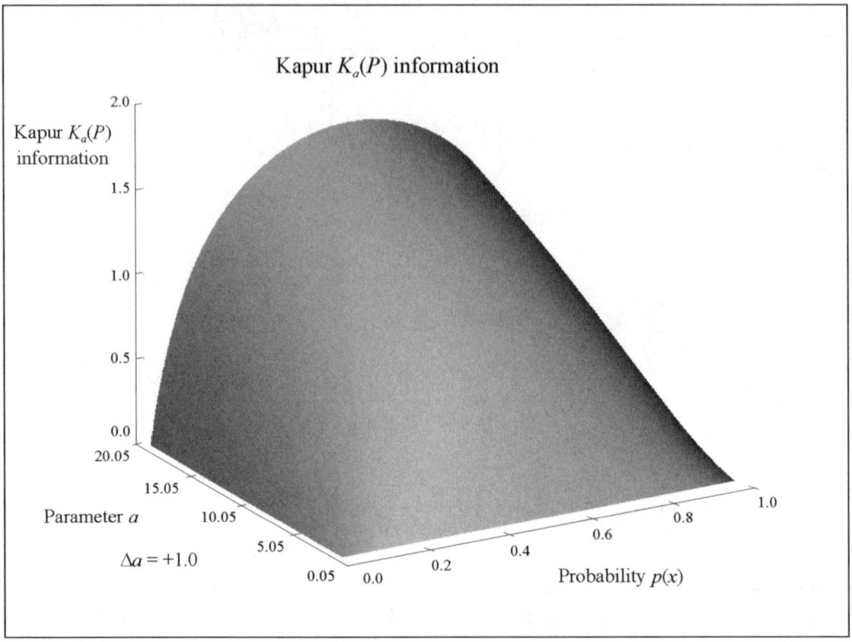

Fig. 6.7: Kapur$_a$ information

The connection between Kapur's entropy

$$K_a(P)=\sum_{k=1}^{n}\ln(1+a\cdot p_k)-\ln(1+a) \qquad a>0 \tag{6.26}$$

and Burg's entropy

$$B(P)=\sum_{k=1}^{n}\ln p_k \tag{6.27}$$

can be found by transforming

$$K_a(P)=\sum_{k=1}^{n}\ln(1+a\cdot p_k)-\ln(1+a) \tag{6.28}$$

with

$$K_a(P)=\sum_{k=1}^{n}\ln\left[a\cdot\left(p_k+\frac{1}{a}\right)\right]-\ln\left[a\cdot\left(1+\frac{1}{a}\right)\right]$$

$$K_a(P) = \sum_{k=1}^{n} \ln\left(p_k + \frac{1}{a}\right) + \sum_{k=1}^{n} \ln a - \ln\left(1 + \frac{1}{a}\right) - \ln a$$

$$K_a(P) = \sum_{k=1}^{n} \ln\left(p_k + \frac{1}{a}\right) + n \cdot \ln a - \ln\left(1 + \frac{1}{a}\right) - \ln a$$

$$K_a(P) = \sum_{k=1}^{n} \ln\left(p_k + \frac{1}{a}\right) - \ln\left(1 + \frac{1}{a}\right) - (n-1) \cdot \ln a \tag{6.29}$$

If we compare this equation with Burg's entropy

$$B(P) = \sum_{k=1}^{n} \ln p_k \tag{6.30}$$

we realize that Burg's entropy and $K_a(P)$ differ only by an additive constant $(n-1)\ln a$ for large values of a. This means that the maximization of both entropies leads to the same result. In the limit we get

$$\lim_{a \to \infty} \left(K_a(P) - B(P)\right) = -\infty \tag{6.31}$$

and the difference becomes infinitely large.

6.1 The corresponding measures of divergence

There are also measures of divergence corresponding to the previously introduced entropy measures. These can be derived by application of the approach that we presented in the derivation of Rényi's divergence. We thus obtain the following measures of divergence for the extended entropies

1. Possibility

$$D_{a,1}(P \parallel Q) = \sum_{k=1}^{n} (1 + a \cdot p_k) \cdot \ln\left(\frac{1 + a \cdot p_k}{1 + a \cdot q_k}\right) \tag{6.32}$$

We find:

$$D_{a,1}(P \| Q) \geq 0 \tag{6.33}$$

$$D_{a,1}(P \| Q) = 0 \ \text{ if } P = Q \tag{6.34}$$

This measure also provides a displacement between the two probability distributions P and Q, but it is not a distance, because the symmetry does not hold. If we assume a uniform distribution density for P, we find

$$D_{a,1}(U \parallel Q) = \sum_{k=1}^{n} \left(1 + a \cdot \frac{1}{n}\right) \cdot \ln\left(\frac{1 + a \cdot \frac{1}{n}}{1 + a \cdot q_k}\right)$$

$$D_{a,1}(U \parallel Q) = \sum_{k=1}^{n} \left(\frac{n+a}{n}\right) \cdot \ln\left(\frac{\frac{n+a}{n}}{1 + a \cdot q_k}\right)$$

$$D_{a,1}(U \parallel Q) = \left(\frac{n+a}{n}\right) \cdot \sum_{k=1}^{n} \ln\left(\frac{n+a}{n}\right) - \sum_{k=1}^{n} \left(\frac{n+a}{n}\right) \cdot \ln(1 + a \cdot q_k)$$

$$D_{a,1}(U \parallel Q) = (n+a) \cdot \ln\left(\frac{n+a}{n}\right) - \left(\frac{n+a}{n}\right) \cdot \sum_{k=1}^{n} \ln(1 + a \cdot q_k) \tag{6.35}$$

With

$$K_a(P) = \sum_{k=1}^{n} \ln(1 + a \cdot p_k) - \ln(1 + a) \tag{6.36}$$

we then obtain

$$D_{a,1}(U \parallel Q) = (n+a) \cdot \ln\left(\frac{n+a}{n}\right) - \left(\frac{n+a}{n}\right) \cdot (K_a(P) + \ln(1+a)) \tag{6.37}$$

and we see, that the divergence is a direct function of the entropy measure if we use a uniformly distributed random variable.

2. Possibility

$$D_{a,2}(P \parallel Q) = \sum_{k=1}^{n} \left(\frac{1 + a \cdot p_k}{1 + a \cdot q_k} - \ln\left(\frac{1 + a \cdot p_k}{1 + a \cdot q_k}\right) - 1\right) \tag{6.38}$$

We get:

$$D_{a,2}(P \| Q) \geq 0 \tag{6.39a}$$

$$D_{a,2}(P \| Q) = 0 \ \text{ if } P = Q \tag{6.39b}$$

Furthermore $D_{a,2}(P\|Q)$ is a convex function of both arguments P and Q and, of course, also a measure of the displacement. For a uniform distribution $Q = U$ we get

$$D_{a,2}(P \| Q) = \sum_{k=1}^{n} \left[\frac{1 + a \cdot p_k}{1 + \frac{a}{n}} - \ln\left(\frac{1 + a \cdot p_k}{1 + \frac{a}{n}} \right) - 1 \right]$$

$$D_{a,2}(P \| Q) = \sum_{k=1}^{n} \left[\frac{n}{n+a} \cdot (1 + a \cdot p_k) - \ln\left(\frac{n}{n+a} \cdot (1 + a \cdot p_k) \right) - 1 \right]$$

$$D_{a,2}(P \| Q) = \frac{n}{n+a} \cdot \sum_{k=1}^{n} (1 + a \cdot p_k) - \sum_{k=1}^{n} \ln\left(\frac{n}{n+a} \cdot (1 + a \cdot p_k) \right) - n$$

$$D_{a,2}(P \| Q) = \frac{n}{n+a} \cdot \left(n + a \cdot \sum_{k=1}^{n} p_k \right) - \sum_{k=1}^{n} \ln\left(\frac{n}{n+a} \right) - \sum_{k=1}^{n} \ln(1 + a \cdot p_k) - n$$

$$D_{a,2}(P \| Q) = n - n \cdot \ln\left(\frac{n}{n+a} \right) - \sum_{k=1}^{n} \ln(1 + a \cdot p_k) - n$$

$$D_{a,2}(P \| Q) = -[K_a(P) + \ln(1 + a)] - n \cdot \ln\left(\frac{n}{n+a} \right) \tag{6.40}$$

and thus we have found a connection to $K_a(P)$.

<u>*3. Furthermore*</u>

$$\frac{1}{1+a} \cdot D_{a,1}(P \| Q) = \frac{1}{1+a} \cdot \sum_{k=1}^{n} (1 + a \cdot p_k) \cdot \ln\left(\frac{1 + a \cdot p_k}{1 + a \cdot q_k} \right) \tag{6.41}$$

$$\lim_{a \to \infty} \frac{1}{1+a} \cdot D_{a,1}(P \| Q) = \lim_{a \to \infty} \sum_{k=1}^{n} \frac{1 + a \cdot p_k}{1+a} \cdot \ln\left(\frac{1 + a \cdot p_k}{1 + a \cdot q_k} \right)$$

$$\lim_{a \to \infty} \frac{1}{1+a} \cdot D_{a,1}(P \| Q) = \sum_{k=1}^{n} p_k \cdot \ln\left(\frac{p_k}{q_k} \right)$$

$$\lim_{a \to \infty} \frac{1}{1+a} \cdot D_{a,1}(P \| Q) = D_1(P \| Q) \tag{6.42}$$

The limit of the extended divergence multiplied with the quotient 1/(1+a) leads us to the familiar Kullback–Leibler distance.

If we compute the limit for the second measure of divergence, we find

$$\lim_{a\to\infty} D_{a,2}(P\,\|\,Q) = \lim_{a\to\infty} \sum_{k=1}^{n}\left[\frac{1+a\cdot p_k}{1+a\cdot q_k} - \ln\left(\frac{1+a\cdot p_k}{1+a\cdot q_k}\right) - 1\right]$$

$$\lim_{a\to\infty} D_{a,2}(P\,\|\,Q) = \lim_{a\to\infty} \sum_{k=1}^{n}\left[\frac{p_k}{q_k} - \ln\left(\frac{p_k}{q_k}\right) - 1\right] = D_2(P:Q) \tag{6.43}$$

and this is also a measure of weighted divergences.

Other measures of divergence are for example:

Havrda–Charvat measure of directed divergence

$$D_\alpha(P\,\|\,Q) = \frac{1}{\alpha\cdot(1-\alpha)}\cdot\left(\sum_{k=1}^{n} p_k^{\alpha}\cdot q_k^{1-\alpha} - 1\right) \tag{6.44}$$

Csiszár's family of divergence measures

$$D(P\,\|\,Q) = -\left(\sum_{k=1}^{n} p_k\cdot\Phi\left(\frac{p_k}{q_k}\right)\right) \tag{6.45}$$

where $\Phi(\,)$ has to be an appropriate (concave) function. If we chose $\Phi(\,) = \ln(\,)$ we obtain the Kullback–Leibler distance.

The generalized transinformation (GMI) also fits into this enumeration. Generalized mutual information (GMI)

$$I_\alpha(P,Q) = \frac{1}{\alpha-1}\cdot\ln\left(\sum_{k=1}^{n}\sum_{j=1}^{n}\frac{pq_{jk}^{\alpha}}{p_k^{\alpha-1}\cdot q_j^{\alpha-1}} - 1\right) \tag{6.46}$$

measures the similarity between random variables, where p_k, q_j are the marginal probabilities and pq_{jk} is the joint probability.

6.2 Weighted entropies and expectation values of entropies

Additional implementation of weighting factors allows us to generalize the previously presented measures and we obtain weighted entropies. We start with the familiar information measures of

Shannon: $$H(P) = -\sum_{k=1}^{n} p_k \cdot \ln p_k \tag{6.47}$$

Kullback $$D(P \parallel Q) = \sum_{k=1}^{n} p_k \cdot \ln\left(\frac{p_k}{q_k}\right) \tag{6.48}$$

Csiszár $$D_{\text{Csiszár}}(P \parallel Q) = -\sum_{k=1}^{n} p_k \cdot \Phi\left(\frac{p_k}{q_k}\right) \tag{6.49}$$

In principle these entropy measures can be generalized by additional weighting factors, but we may not chose arbitrary weighting factors. These factors must not change the 'character' of the entropy functions. This means, that the function has to remain a measure of the entropy resp. a measure of the information. So these considerations are to be met by these weighted entropies and the simplest of these possible extensions are given by

Guiasu: $$H(P : w) = -\sum_{k=1}^{n} w_k \cdot p_k \cdot \ln p_k \tag{6.50}$$

Guiasu, Belis: $$H(P : u) = -\sum_{k=1}^{n} u_k \cdot p_k \cdot \ln p_k \tag{6.51}$$

Both measures are equal, but in one case the weighting factor w_k is called a weight and in the other case it is called utility u_k.

Taneja, Tuteja: $$D(P \parallel Q, w) = -\sum_{k=1}^{n} w_k \cdot p_k \cdot \ln\left(\frac{p_k}{q_k}\right) \tag{6.52}$$

and

Csiszár: $$D_{\text{Csiszár}}(P \parallel Q, w) = -\sum_{k=1}^{n} w_k \cdot p_k \cdot \Phi\left(\frac{p_k}{q_k}\right) \tag{6.53}$$

We may also generate a generalized form of

Havrda–Charvat: $I_{\text{HC},\alpha} = H^{\alpha} = \frac{1}{1-\alpha} \cdot \left(\sum_{k=1}^{n} p_k^{\alpha} - 1 \right) \quad \alpha > 0 \text{ and } \alpha \neq 1$ (6.54)

This becomes

$$H_a^{\alpha} = \frac{1}{1-\alpha} \cdot \left(\sum_{k=1}^{n} p_k^{\alpha} - \sum_{k=1}^{n} p_k \right) - \frac{1}{a} \cdot \frac{1}{1-\alpha} \cdot \left(\sum_{k=1}^{n} (1 + a \cdot p_k)^{\alpha} - \sum_{k=1}^{n} (1 + a \cdot p_k) \right)$$
$$+ \frac{1}{a} \cdot \frac{1}{1-\alpha} \cdot \left((1+a)^{\alpha} - (1+a) \right) \tag{6.55}$$

for $\alpha > 0$ and $\alpha \neq 1$

This generalization contains the following special cases::

$\lim_{\alpha \to 1} H_a^{\alpha}(P) = H_a(P)$ Kapur entropy (6.56)

$\lim_{\alpha \to 0} H_a^{\alpha}(P) = H^{\alpha}(P)$ Havrda–Charvat entropy (6.57)

$\lim_{\alpha \to 1, a \to 1} H_a^{\alpha}(P) = K_1(P)$ Bose–Einstein entropy (6.58)

$\lim_{\alpha \to 1, a \to -1} H_a^{\alpha}(P) = K_{-1}(P)$ Fermi–Dirac entropy (6.59)

$\lim_{\alpha \to 1, a \to 0} H_a^{\alpha}(P) = H(P)$ Shannon's entropy (6.60)

There are, of course, several other entropy measures, which can be found in the books [KAP89] [KAP92] and in the publication [TAN89] in a clear presentation. Many other publications describe generalized measures of entropy and point out their special features. We are not able to present all these descriptions, but there is a short enumeration in Appendix A.3, where some additional entropy measures are presented. As we intend to keep this volume handy, we abandon detailed introductions or descriptions of the information measures, as the generalizations presented by Taneja in [TAN89] are already extensive.

We find an elegant possibility to construct entropy measures for specified problems in chapter 15. There we extend the maximum entropy method, motivated by Edward Jaynes [JAY83], and use this extension to derive other entropy measures, including additional a priori information in the derivation.

But first we want to examine the familiar information measures in a more detailed way by pointing out their connections to each other and by calculating the value of the information measures for gaussian random variables. This allows us to obtain an improved understanding and a better feeling for the relation between the information functions and the quantities that they describe.

7 Information functions and gaussian distributions

Starting with Rényi's α-information S_α

$$S_\alpha = I_\alpha(x) = \frac{1}{1-\alpha} \cdot \mathrm{ld}\left(\int_{-\infty}^{\infty} f_x^\alpha(\xi) \cdot d\xi \right) \tag{7.01}$$

and with the G-divergence

$$G_\alpha = I_\alpha(Q \parallel P) = \frac{1}{\alpha - 1} \cdot \mathrm{ld}\left(\int_{-\infty}^{\infty} \left(\frac{f_q^\alpha(\xi)}{f_p^{\alpha-1}(\xi)} \right) \cdot d\xi \right) \tag{7.02}$$

we use this chapter to derive the connections between the most familiar information functions and to point out some problems in applications. In all of these presentations we replace the binary logarithm (ld) by the natural logarithm (ln) to reduce the necessary computational effort. We start our examinations with the two information measures introduced by Alfred Rényi, and we calculate the values of these information functions for gaussian distribution densities.

7.1 Rényi's information of a gaussian distributed random variable

We have a gaussian distribution density $N(\mu, \sigma^2)$ with the expectation value μ and with the variance σ^2:

$$f_x(\xi) = \frac{1}{\sqrt{2\pi} \cdot \sigma} \cdot \exp\left\{ -\frac{1}{2} \cdot \frac{(\xi - \mu)^2}{\sigma^2} \right\} \tag{7.03}$$

An extension to a multi-dimensional gaussian random variable of dimension n provides the distribution density

$$f_{\underline{x}}(\underline{\xi}) = \frac{1}{(2\pi)^{\frac{n}{2}} \cdot |P|^{\frac{1}{2}}} \cdot \exp\left\{ -\frac{1}{2} \cdot (\underline{\xi} - \underline{\mu})^T \cdot P^{-1} \cdot (\underline{\xi} - \underline{\mu}) \right\} \tag{7.04}$$

with the expectation vector

$$\underline{\mu} = E\{\underline{x}\} \tag{7.05}$$

and the covariance matrix

$$P = E\{(\underline{x} - \underline{\mu})(\underline{x} - \underline{\mu})^{T}\} \tag{7.06}$$

7.1.1 Rényi's α-information

If we look at S_α, we realize that this information is equal to the expectation value of a distribution density

$$S_\alpha = \frac{1}{1-\alpha} \cdot \ln\left[E\left\{f_{\underline{x}}^{\alpha-1}(\underline{\xi})\right\}\right] \tag{7.07}$$

which is a typical form of an information function. Such functions often contain an expectation of a quantity, defined as an information, which consists of a logarithm of a distribution density. Another familiar way to generate such functions is to build the logarithm of the expectation value of a mean of a function of the distribution density. Shannon's information has, for instance, such a form as $S = -E\{\ln(p(x))\}$, were the expectation is taken over Hartley's information measure.

Now we insert the multi-dimensional gaussian distribution into this expectation function and we get

$$S_\alpha = \frac{1}{1-\alpha} \cdot \ln\left[\int_{\Xi}\left(\frac{1}{(2\pi)^{\frac{n}{2}} \cdot |P|^{\frac{1}{2}}} \cdot \exp\left\{-\frac{1}{2} \cdot (\underline{\xi}-\underline{\mu})^T \cdot P^{-1} \cdot (\underline{\xi}-\underline{\mu})\right\}\right)^{\alpha} \cdot d\underline{\xi}\right] \tag{7.08}$$

$$S_\alpha = \frac{1}{1-\alpha} \cdot \ln\left[\left(\frac{1}{(2\pi)^{\frac{n}{2}} \cdot |P|^{\frac{1}{2}}}\right)^{\alpha-1} \cdot \int_{\Xi} \frac{1}{(2\pi)^{\frac{n}{2}} \cdot |P|^{\frac{1}{2}}} \cdot \exp\left\{-\frac{1}{2} \cdot (\underline{\xi}-\underline{\mu})^T \cdot \alpha \cdot P^{-1} \cdot (\underline{\xi}-\underline{\mu})\right\} \cdot d\underline{\xi}\right] \tag{7.09}$$

and we use a new covariance matrix $P_{\text{new}} = \left(\alpha \cdot P^{-1}\right)^{-1}$ in the exponent, so that

$$\int_{\Xi} \frac{1}{(2\pi)^{\frac{n}{2}} \cdot |P_{\text{new}}|^{\frac{1}{2}}} \cdot \exp\left\{-\frac{1}{2} \cdot \left(\underline{\xi} - \underline{\mu}\right)^T \cdot P_{\text{new}}^{-1} \cdot \left(\underline{\xi} - \underline{\mu}\right)\right\} \cdot d\underline{\xi} = 1 \tag{7.10}$$

To obtain the determinant $|P_{\text{new}}|$ we have to calculate the matrix P_{new}

$$P_{\text{new}} = \left(\alpha \cdot P^{-1}\right)^{-1} = \frac{1}{\alpha} \cdot P \tag{7.11}$$

and thus we are able to determine the determinant of the $[n \times n]$-dimensional matrix.

$$\left|P_{\text{new}}\right| = \left|\frac{1}{\alpha} \cdot P\right| = \frac{1}{\alpha^n}|P| \tag{7.12}$$

The integral of the new distribution density is given by

$$\int_{\Xi} \frac{1}{(2\pi)^{\frac{n}{2}} \cdot \left(\frac{1}{\alpha^{\frac{n}{2}}}\right) |P|^{\frac{1}{2}}} \cdot \exp\left\{-\frac{1}{2} \cdot \left(\underline{\xi} - \underline{\mu}\right)^T \cdot \alpha \cdot P^{-1} \cdot \left(\underline{\xi} - \underline{\mu}\right)\right\} \cdot d\underline{\xi} = 1 \tag{7.13}$$

Multiplying both sides by $\alpha^{-\frac{n}{2}}$ leads to

$$\int_{\Xi} \frac{1}{(2\pi)^{\frac{n}{2}} \cdot |P|^{\frac{1}{2}}} \cdot \exp\left\{-\frac{1}{2} \cdot \left(\underline{\xi} - \underline{\mu}\right)^T \cdot \alpha \cdot P^{-1} \cdot \left(\underline{\xi} - \underline{\mu}\right)\right\} \cdot d\underline{\xi} = \alpha^{-\frac{n}{2}} \tag{7.14}$$

and thus the integral in equation (7.09) is solved. We therefore write

$$S_\alpha = \frac{1}{1-\alpha} \cdot \ln\left[\left(\frac{1}{(2\pi)^{\frac{n}{2}} \cdot |P|^{\frac{1}{2}}}\right)^{\alpha-1} \cdot \alpha^{-\frac{n}{2}}\right] \tag{7.15}$$

$$S_\alpha = \ln\left[(2\pi)^{\frac{n}{2}} \cdot |P|^{\frac{1}{2}}\right] - \frac{1}{1-\alpha} \cdot \frac{n}{2} \cdot \ln\alpha \tag{7.16}$$

with $\alpha > 0$ and $\alpha \neq 1$

Rényi's information depends on the logarithm of the determinant of the covariance matrix, which describes the uncertainty of the expectation value of the gaussian distribution density. Thus Rényi's information is proportional to the variance and has the character of an entropy rather than that of an information. This is also confirmed in the additional dimension n, where the increasing dimension enhances the uncertainty and thus the entropy.

For $\alpha > 1$ the logarithm of the second term on the right-hand side of equation (7.16) is positive, but the quotient is negative, so that the whole second term is positive. For $\alpha < 1$ the logarithm is negative and the quotient positive, which again leads to a positive second term. From this calculation we additionally realize that we have to choose $\alpha > 0$, which is required by the range of the logarithm. Independent of the choice of α the Rényi information [REN82] S_α always behaves like an entropy and it is therefore called an entropy by most of the authors who apply Rényi's α-entropy.

7.1.2 Rényi's *G*-divergence

In the computation of the G-divergence we have to assume two gaussian distribution densities, both with scalar random variables, to reduce the computational effort.

$$f_x(\xi) = \frac{1}{\sqrt{2\pi} \cdot \sigma} \cdot \exp\left\{-\frac{1}{2} \cdot \frac{(\xi - \mu)^2}{\sigma^2}\right\} \tag{7.17}$$

$$f_{x_s}(\xi) = \frac{1}{\sqrt{2\pi} \cdot \sigma_s} \cdot \exp\left\{-\frac{1}{2} \cdot \frac{(\xi - \mu_s)^2}{\sigma_s^2}\right\} \tag{7.18}$$

The dummy variable ξ is the same for both densities. Otherwise one of the two distribution densities would be independent of the integration variable and could be extracted out of the integral.

We get

$$G_\alpha = \frac{1}{\alpha - 1} \cdot \ln\left(\int_\Xi \left(\frac{f_x^\alpha(\xi)}{f_{x_s}^{\alpha-1}(\xi)}\right) \cdot d\xi\right)$$

$$= \frac{1}{\alpha - 1} \cdot \ln\left(\int_{\Xi} \left(\frac{\left[\frac{1}{\sqrt{2\pi} \cdot \sigma} \cdot \exp\left\{ -\frac{1}{2} \cdot \frac{(\xi - \mu)^2}{\sigma^2} \right\} \right]^{\alpha}}{\left[\frac{1}{\sqrt{2\pi} \cdot \sigma_s} \cdot \exp\left\{ -\frac{1}{2} \cdot \frac{(\xi - \mu_s)^2}{\sigma_s^2} \right\} \right]^{\alpha - 1}} \right) \cdot d\xi \right)$$

$$= \frac{1}{\alpha - 1} \cdot \ln\left(\int_{\Xi} \frac{\left(\sqrt{2\pi} \cdot \sigma_s \right)^{\alpha - 1}}{\left(\sqrt{2\pi} \cdot \sigma \right)^{\alpha}} \cdot \exp\left\{ -\frac{1}{2} \cdot \alpha \cdot \frac{(\xi - \mu)^2}{\sigma^2} + \frac{1}{2} \cdot (\alpha - 1) \cdot \frac{(\xi - \mu_s)^2}{\sigma_s^2} \right\} \cdot d\xi \right)$$

(7.19)

and in this equation we first focus our attention on the term in the exponent, which we abbreviate as E:

$$E = -\frac{1}{2} \cdot \alpha \cdot \frac{(\xi - \mu)^2}{\sigma^2} + \frac{1}{2} \cdot (\alpha - 1) \cdot \frac{(\xi - \mu_s)^2}{\sigma_s^2}$$

$$= -\frac{1}{2} \cdot \left[\frac{\alpha}{\sigma^2} \cdot \left(\xi^2 - 2\xi\mu + \mu^2 \right) + \frac{1 - \alpha}{\sigma_s^2} \cdot \left(\xi^2 - 2\xi\mu_s + \mu_s^2 \right) \right]$$

$$= -\frac{1}{2} \cdot \left[\xi^2 \left(\frac{\alpha}{\sigma^2} + \frac{1 - \alpha}{\sigma_s^2} \right) - 2\xi \left(\mu \frac{\alpha}{\sigma^2} + \mu_s \frac{1 - \alpha}{\sigma_s^2} \right) + \mu^2 \frac{\alpha}{\sigma^2} + \mu_s^2 \frac{1 - \alpha}{\sigma_s^2} \right] \quad (7.20)$$

We introduce the abbreviations

$$a = \frac{\alpha}{\sigma^2} + \frac{1 - \alpha}{\sigma_s^2} \quad (7.21a)$$

$$b = \mu \frac{\alpha}{\sigma^2} + \mu_s \frac{1 - \alpha}{\sigma_s^2} \quad (7.21b)$$

$$c = \mu^2 \frac{\alpha}{\sigma^2} + \mu_s^2 \frac{1 - \alpha}{\sigma_s^2} \quad (7.21c)$$

and thus obtain the simpler formulation

$$E = -\frac{1}{2} \cdot \left[a \cdot \xi^2 - b \cdot 2 \cdot \xi + c \right] \quad (7.22)$$

and

$$E = -\frac{1}{2} \cdot a \cdot \left[\xi^2 - \frac{b}{a} \cdot 2 \cdot \xi + \frac{c}{a} \right] \tag{7.23}$$

$$E = -\frac{1}{2} \cdot a \cdot \left[\left(\xi - \frac{b}{a} \right)^2 + \frac{c}{a} - \left(\frac{b}{a} \right)^2 \right] \tag{7.24}$$

Now we insert this term into equation (7.19) to obtain

$$\begin{aligned}
G_\alpha &= \frac{1}{\alpha - 1} \cdot \ln \left(\int_\Xi \frac{\left(\sqrt{2\pi} \cdot \sigma_s \right)^{\alpha - 1}}{\left(\sqrt{2\pi} \cdot \sigma \right)^\alpha} \cdot \exp\left\{ -\frac{1}{2} \cdot a \cdot \left[\left(\xi - \frac{b}{a} \right)^2 + \frac{c}{a} - \left(\frac{b}{a} \right)^2 \right] \right\} \cdot d\xi \right) \\
&= \frac{1}{\alpha - 1} \cdot \ln \left(\frac{\sigma_s{}^{\alpha - 1}}{\sigma^\alpha} \int_\Xi \frac{1}{\sqrt{2\pi}} \cdot \exp\left\{ -\frac{1}{2} \cdot a \cdot \left[\left(\xi - \frac{b}{a} \right)^2 + \frac{c}{a} - \left(\frac{b}{a} \right)^2 \right] \right\} \cdot d\xi \right) \\
&= \frac{1}{\alpha - 1} \cdot \ln \left(\frac{\sigma_s{}^{\alpha - 1}}{\sigma^\alpha} \right) + \frac{1}{\alpha - 1} \cdot \ln \left(\int_\Xi \frac{1}{\sqrt{2\pi}} \cdot \exp\left\{ -\frac{1}{2} \cdot a \cdot \left(\xi - \frac{b}{a} \right)^2 \right\} \cdot d\xi \right. \\
&\qquad \left. \cdot \exp\left\{ -\frac{1}{2} \cdot a \cdot \left[\frac{c}{a} - \left(\frac{b}{a} \right)^2 \right] \right\} \right)
\end{aligned} \tag{7.25}$$

We solve the integral

$$\int_\Xi \frac{1}{\sqrt{2\pi}} \cdot \exp\left\{ -\frac{1}{2} \cdot a \cdot \left(\xi - \frac{b}{a} \right)^2 \right\} \cdot d\xi \tag{7.26}$$

by using the identity

$$\int_\Xi \frac{1}{\sqrt{2\pi} \cdot \frac{1}{\sqrt{a}}} \cdot \exp\left\{ -\frac{1}{2} \cdot a \cdot \left(\xi - \frac{b}{a} \right)^2 \right\} \cdot d\xi = 1 \tag{7.27}$$

multiply both sides by $\frac{1}{\sqrt{a}}$ and get the solution to our integral

$$\frac{1}{\sqrt{a}} \cdot \int_{\Xi} \frac{1}{\sqrt{2\pi} \cdot \frac{1}{\sqrt{a}}} \cdot \exp\left\{-\frac{1}{2} \cdot a \cdot \left(\xi - \frac{b}{a}\right)^2\right\} \cdot d\xi$$
$$= \int_{\Xi} \frac{1}{\sqrt{2\pi}} \cdot \exp\left\{-\frac{1}{2} \cdot a \cdot \left(\xi - \frac{b}{a}\right)^2\right\} \cdot d\xi = \frac{1}{\sqrt{a}} \tag{7.28}$$

With this identity we find

$$G_\alpha = \frac{1}{\alpha - 1} \cdot \ln\left(\frac{{\sigma_s}^{\alpha-1}}{\sigma^\alpha}\right) + \frac{1}{\alpha - 1} \cdot \ln\left(\frac{1}{\sqrt{a}} \cdot \exp\left\{-\frac{1}{2} \cdot a \cdot \left[\frac{c}{a} - \left(\frac{b}{a}\right)^2\right]\right\}\right) \tag{7.29}$$

or

$$G_\alpha = \frac{1}{\alpha - 1} \cdot \ln\left(\frac{{\sigma_s}^{\alpha-1}}{\sigma \cdot \sigma^{\alpha-1}}\right) + \frac{1}{\alpha - 1} \cdot \ln\left(\frac{1}{\sqrt{a}}\right) - \frac{1}{\alpha - 1} \cdot \frac{1}{2} \cdot a \cdot \left[\frac{c}{a} - \left(\frac{b}{a}\right)^2\right] \tag{7.30}$$

$$G_\alpha = \ln\left(\frac{\sigma_s}{\sigma}\right) - \frac{1}{\alpha - 1} \cdot \left\{\ln(\sigma) + \frac{1}{2} \cdot \ln(a) + \frac{1}{2} \cdot a \cdot \left[\frac{c}{a} - \left(\frac{b}{a}\right)^2\right]\right\} \tag{7.31}$$

Inserting the terms for a, b and c from equation (7.21) finally leads to

$$G_\alpha = \ln\left(\frac{\sigma_s}{\sigma}\right) - \frac{1}{\alpha - 1} \cdot \left\{\ln(\sigma) + \frac{1}{2} \cdot \ln\left(\frac{\alpha}{\sigma^2} + \frac{1-\alpha}{\sigma_s^2}\right)\right.$$
$$\left. + \frac{1}{2} \cdot \left(\frac{\alpha}{\sigma^2} + \frac{1-\alpha}{\sigma_s^2}\right) \cdot \left[\frac{\left(\mu^2 \frac{\alpha}{\sigma^2} + \mu_s^2 \frac{1-\alpha}{\sigma_s^2}\right)}{\left(\frac{\alpha}{\sigma^2} + \frac{1-\alpha}{\sigma_s^2}\right)} - \left(\frac{\left(\mu \frac{\alpha}{\sigma^2} + \mu_s \frac{1-\alpha}{\sigma_s^2}\right)}{\left(\frac{\alpha}{\sigma^2} + \frac{1-\alpha}{\sigma_s^2}\right)}\right)^2\right]\right\} \tag{7.32}$$

G_α-divergence for two gaussian distributions.

If both expectation values are equal, the last addend on the right-hand side vanishes and we merely get

$$G_\alpha = \ln\left(\frac{\sigma_s}{\sigma}\right) - \frac{1}{\alpha - 1} \cdot \left\{\ln(\sigma) + \frac{1}{2} \cdot \ln\left(\frac{\alpha}{\sigma^2} + \frac{1-\alpha}{\sigma_s^2}\right)\right\} \tag{7.33}$$

The first term on the right-hand side describes the information difference between the different uncertainties of both gaussian distributions, while the term depending on the parameter introduces an additional weighting that cannot be interpreted directly.

$$G_\alpha = \ln\left(\frac{\sigma_s}{\sigma}\right) - \frac{1}{\alpha - 1} \cdot \left\{\ln(\sigma) + \frac{1}{2} \cdot \ln\left[\frac{1}{\sigma_s^2} + \alpha\left(\frac{1}{\sigma^2} - \frac{1}{\sigma_s^2}\right)\right]\right\} \tag{7.34}$$

$$G_\alpha = \ln\left(\frac{\sigma_s}{\sigma}\right) - \frac{1}{\alpha - 1} \cdot \left\{\ln(\sigma) + \frac{1}{2} \cdot \ln\left[\frac{1}{\sigma_s^2} + \alpha\left(\frac{\sigma_s^2 - \sigma^2}{\sigma^2 \cdot \sigma_s^2}\right)\right]\right\} \tag{7.35}$$

Nevertheless, this transformation shows that the argument of the logarithm is built by the reciprocal variance σ_s^2 and by an additional reciprocal variance weighted with the parameter α.

7.2 Shannon's information

We obtain Shannon's information, when we compute the limit of the more general Rényi α-information for $\alpha \to 1$.

$$S_1(\underline{x}) = \lim_{\alpha \to 1} S_\alpha(\underline{x})$$

$$= \lim_{\alpha \to 1} \frac{1}{1-\alpha} \cdot \ln\left(\int_{\underline{\Xi}} f_{\underline{x}}^\alpha(\underline{\xi}) \cdot d\underline{\xi}\right) \tag{7.36}$$

As we easily see, the quotient $1/(1-\alpha)$ tends to infinitely large values for $\alpha \to 1$ and the logarithm also tends to infinitely large values, as the integral is taken from a distribution function, which means that

$$\int_{\underline{\Xi}} f_{\underline{x}}(\underline{\xi}) \cdot d\underline{\xi} = 1 \tag{7.37}$$

Thus we have to apply the rule of de l'Hospital, which provides for the denominator

$$\frac{d}{d\alpha}(1-\alpha)=-1 \tag{7.38}$$

and in the numerator we find

$$dZ=\frac{d}{d\alpha}\ln\left(\int_{\underline{\Xi}} f_{\underline{x}}^{\alpha}(\underline{\xi})\cdot d\underline{\xi}\right) \tag{7.39}$$

$$dZ=\frac{1}{\int_{\underline{\Xi}} f_{\underline{x}}^{\alpha}(\underline{\xi})\cdot d\underline{\xi}}\frac{d}{d\alpha}\left(\int_{\underline{\Xi}} f_{\underline{x}}^{\alpha}(\underline{\xi})\cdot d\underline{\xi}\right) \tag{7.40}$$

We now apply Leibnitz's rule to differentiate the integral:

$$\frac{d}{dx}\int_{a(t)}^{b(t)} f(x,t)\cdot dx=\int_{a(t)}^{b(t)}\frac{d}{dx}f(x,t)\cdot dx+f(b(t),t)\cdot\frac{d}{dx}b(t)-f(a(t),t)\cdot\frac{d}{dx}a(t) \tag{7.41}$$

Because of the constant limits of the interval (which are independent of the parameter α) the last two terms vanish, and we get

$$\int_{\underline{\Xi}}\frac{d}{d\alpha}f_{\underline{x}}^{\alpha}(\underline{\xi})\cdot d\underline{\xi}=\int_{\underline{\Xi}}\frac{d}{d\alpha}\exp\left\{\ln\left[f_{\underline{x}}^{\alpha}(\underline{\xi})\right]\right\}\cdot d\underline{\xi} \tag{7.42}$$

$$=\int_{\underline{\Xi}}\frac{d}{d\alpha}\exp\left\{\alpha\cdot\ln\left[f_{\underline{x}}(\underline{\xi})\right]\right\}\cdot d\underline{\xi}$$

$$=\int_{\underline{\Xi}}\ln\left[f_{\underline{x}}(\underline{\xi})\right]\cdot\exp\left\{\alpha\cdot\ln\left[f_{\underline{x}}(\underline{\xi})\right]\right\}\cdot d\underline{\xi}$$

$$=\int_{\underline{\Xi}}\ln\left[f_{\underline{x}}(\underline{\xi})\right]\cdot f_{\underline{x}}^{\alpha}(\underline{\xi})\cdot d\underline{\xi}$$

Thus the numerator becomes

$$dZ=\frac{1}{\int_{\underline{\Xi}} f_{\underline{x}}^{\alpha}(\underline{\xi})\cdot d\underline{\xi}}\cdot\int_{\underline{\Xi}}\ln\left[f_{\underline{x}}(\underline{\xi})\right]\cdot f_{\underline{x}}^{\alpha}(\underline{\xi})\cdot d\underline{\xi} \tag{7.43}$$

in the last step of this derivation we have to build the limit in the derivatives of numerator and denominator for $\alpha\to 1$. This results in

$$S_1(\underline{x}) = \lim_{\alpha \to 1} \left[(-1) \cdot \frac{1}{\int\limits_{\underline{\Xi}} f_{\underline{x}}^{\alpha}(\underline{\xi}) \cdot d\underline{\xi}} \cdot \int\limits_{\underline{\Xi}} \ln\left[f_{\underline{x}}(\underline{\xi})\right] \cdot f_{\underline{x}}^{\alpha}(\underline{\xi}) \cdot d\underline{\xi} \right] \tag{7.44}$$

The integral in the denominator is equal to 1, and we therefore obtain Shannon's information of a continuous multi-dimensional random variable $\underline{x}$

$$S(\underline{x}) = S_1(\underline{x}) = -\int\limits_{\underline{\Xi}} \ln\left[f_{\underline{x}}(\underline{\xi})\right] \cdot f_{\underline{x}}(\underline{\xi}) \cdot d\underline{\xi} \tag{7.45}$$

The order of the multiplications in the integral of the Shannon information is of no importance, because even multi-dimensional distribution densities provide only scalar results.

Equation (7.45) describes Shannon's information of a stochastic process, which is described by a continuous distribution density. So far we introduced only Shannon's information of discrete probability distributions

$$I_{\text{Shannon}} = -\sum_{k=1}^{n} p_k \cdot \text{ld}\, p_k \tag{7.46}$$

Thus the questions arises whether it is possible to deduce the description of Shannon's information for continuous distribution densities from the description of the discrete Shannon information function. We will examine this question in the next chapter, but we are already able to cite Boltzmann, who describes such a transition from a discrete to a continuous system in his lecture of the theory of gases [BOL81].

"...Denn ein wirkliches Continuum muss aus mathematisch unendlich vielen Theilen bestehen; eine im mathematischen Sinne wirklich unendliche Zahl aber ist undefinirbar. Ferner muss die Annahme eines Continuums die partiellen Differentialgleichungen für das Verhalten desselben als das ursprünglich gegebene auffassen...."

"... A real continuum has to consist of an infinite number of mathematical parts; but a really infinite number in the mathematical sense cannot be defined. Moreover the assumption of a continuum has to interpret the partial differential equations for its behavior as the originally given conditions...."

With this statement Boltzmann decides that the direct transition from a discrete to a continuous system is not possible. Nevertheless, we will try to find a connection between the discrete and the continuous formulation of Shannon's entropy.

8 Shannon's information of discrete probability distributions

We begin our considerations with the discrete formulation of Shannon's entropy, where we use the familiar notation with the binary logarithm

$$I_{Shannon} = -\sum_{k=1}^{n} p_k \cdot \mathrm{ld}\, p_k \tag{8.01}$$

To derive the continuous formulation, we replace the probabilities p_k by the distribution functions $p_k = P\{x : x(\omega) \in (\xi_k, \xi_k + d\xi]\} = f_x(\xi_k) \cdot d\xi$, and build the limit for $N \to \infty$, respectively $d\xi \to 0$, which leads us to the integral

$$\begin{aligned} I_{\text{Shannon}} &= -\int_{\Xi} \mathrm{ld}(f_x(\xi) \cdot d\xi) \cdot f_x(\xi) \cdot d\xi \\ &= -\int_{\Xi} [\mathrm{ld}(f_x(\xi)) + \mathrm{ld}(d\xi)] \cdot f_x(\xi) \cdot d\xi \\ &= -\int_{\Xi} \mathrm{ld}(f_x(\xi)) \cdot f_x(\xi) \cdot d\xi - \int_{\Xi} \mathrm{ld}(d\xi) \cdot f_x(\xi) \cdot d\xi \end{aligned} \tag{8.02}$$

$$I_{\text{Shannon}} = H_0 + K \tag{8.03}$$

H_0 is known as Shannon's differential entropy, while the additional term K is neglected in all practical applications of this entropy measure. For $d\xi \to 0$ this additional term K tends to $+\infty$, so that information – independent of H_0 – would always assume infinitely large values. Every continuous information would be meaningless in this case, because it would always have the same infinite negative value. We therefore might use the lower limit of the resolution of two succeeding values in our system to limit $d\xi$, or – as suggested by Middleton [MID87/1] [MID87/2] – we might directly use Planck's constant $h = 6.625 \cdot 10^{-34}$ Ws^2 as the lower limit of our systems. Both would be justified. In the first case, we would have the limited resolution (uncertainty) of the actual measurements, which is directly related to the information that we are able to extract from these measurements. In the second case Planck's constant would generate an absolute reference, which could be used independently of the system and which would always provide the same value of K.

Anyhow the term K is constant in a test series and it therefore does not produce any changes in the information during such a series. Such changes are solely caused by the term H_0. The information, described with the term K, would thus merely describe a constant shift of the whole information, which can be neglected without any limitations of the applicability (here we may additionally mention the lack of a unit element and a scale for information measurements). K vanishes anyway if we calculate the difference between two Shannon information functions. These differences lead us to the equations

$$I_{\text{Shannon1}} - I_{\text{Shannon2}} = H_{01} + K - H_{02} - K = H_{01} - H_{02}$$
$$= -\int_{\Xi} \operatorname{ld}\left(f_{x_1}(\xi)\right) \cdot f_{x_1}(\xi) \cdot d\xi + \int_{\Xi} \operatorname{ld}\left(f_{x_2}(\xi)\right) \cdot f_{x_2}(\xi) \cdot d\xi \tag{8.04}$$

In these equations the coordinate system and the limitation $d\xi$ are preserved. This enables us to compute the difference between two Shannon information measures and it eliminates the term K. This difference measures the discrepancy in the uncertainty of the random variables x_1 and x_2 occurring in the same coordinate system.
If we compare the difference of two discrete random variables

$$I_{\text{Shannon1}} - I_{\text{Shannon2}} = -\sum_{k=1}^{n} p_k \cdot \operatorname{ld} p_k + \sum_{i=1}^{m} q_i \cdot \operatorname{ld} q_i \tag{8.05}$$

then we find that this computation also calculates the difference in the uncertainty.

The continuous difference as well as the discrete difference may assume positive and negative values and these differences allow us to compare the discrete Shannon information with the continuous Shannon information, which is impossible for the absolute values. The absolute values cannot be compared, because of the additional constant K and because the information of a continuous distribution density can also assume negative values, which is impossible for discrete distribution densities.

In both representations the description of the certain event can additionally be used to demonstrate the difference between discrete and continuous measures. In the discrete formulation we find with equation (8.01) and $p_k = 1$ ($\operatorname{ld} 1 = 0$) the information of the certain event $I_{\text{Shannon}} = 0$. In equation (7.45) this limit is not so obvious. We first have to describe the distribution density of the certain event by the distribution density

$$f_x(\xi) = \delta(\xi - x_0) \tag{8.06}$$

which is a Dirac impulse, as we do not have any uncertainty about the realization of the event $x = x_0$. We approximate this Dirac impulse, according to Middleton [MID87/1], by a gaussian distribution density, where the variance tends to zero. From equation (8.02) we calculate the terms of equation (8.03) and find $H_0 = -\infty$.

If we add $K = +\infty$, we might also obtain the result $S = 0$ for the certain event, but we have to remark that we cannot add two infinite values like this. One of the reasons is that term K cannot be specified with a sufficient accuracy.

Thus this chapter clearly demonstrates that the information of discrete random variables, measured in bits, cannot be transferred to the information of continuous random variables in a simple way.

8.1 Continuous and discrete random variables

Random variables $x(.)$ are merely mappings from sample space Ω into the realization space $\mathfrak{R}^n$. In sample space we only have events A (specified by the sigma-algebra), which we weight with probabilities $P(A)$. So discrete events A receive a measure, which we call probability, if they meet the three axioms of A. N. Kolmogorov

$$P(A) \geq 0 \tag{8.07a}$$

$$P(\Omega) = 1 \tag{8.07b}$$

and

$$P(A) + P(B) = P(A \cup B) \qquad \text{iff (if and only if)} \quad A \cap B = \varnothing \tag{8.07c}$$

Thus the probability is first of all an abstract measure, which assigns a probability to some event A to describe or weight this event.

The random variable maps the event A into the realization space, where the probability of the realization $x(A)$ in the realization space is the same as the probability of the event A, namely $P(A)$. The realization can be generated by a single event in the realization space or by a set of single events in realization space. The probability of the realization $x(A)$ is not changed! The probability of the realization is always equal to the probability of the event A. Thus the probability is not directly related to the realization. It is instead a measure for the occurrence of the abstract event A, which is connected with the realization by the mapping of the random variable.

Discrete random variables (e.g., dice experiments), their realizations and their probabilities are merely connected to each other via a concrete assignment. If we do not know this assignment and we are able to measure only the probabilities without the connected realization, then we are simply not able to find the realization, providing the probabilities for the observed experiment.

The probability 1/6 can arise from the throwing of a die, but also from each other arbitrary choice from six sets, occurring with equal probability but consisting of arbitrary elements.

The computation of an entropy of such discrete probabilities provides

$$I_{\text{Shannon}} = -\sum_{k=1}^{N} p_k \cdot \text{ld}\, p_k$$

and we may assign this entropy to arbitrary experiments, where we choose one event out of six equally probable events.

If we examine, for instance, a two-dimensional array consisting of nine fields and arrange the occurring values in five intervals, then we are able to determine the relative frequencies of the events occurring in the intervals. We use this knowledge to compute the entropy of this arrangement. To develop these statements further, we take a look at two of these two-dimensional fields

Table 8.1: 3×3 fields with different values

10	24	15
33	7	41
14	48	28

If we arrange the intervals [0,9], [10,19], [20,29], [30,39], [40,49] we obtain the following events in the given intervals (7), (10,14,15), (24, 28), (33) and (41, 44). So we compute the following relative frequencies of the intervals: 1/9, 3/9, 2/9, 1/9, 2/9 for the five intervals (i.e. the relative frequencies of five events).

1	2	1
3	0	4
1	4	2

Analogously we again assume five intervals [0,0.9], [1,1.9], [2,2.9], [3,3.9] and [4,4.9] and arrange the events (0), (1,1,1), (2,2), (3), (4,4) into these intervals. This arrangement also leads to the relative frequencies 1/9, 3/9, 2/9, 1/9, 2/9 for the five intervals.

With these probabilities (relative frequencies) of the intervals in the two different realization spaces, we are now able to compute the entropy of this arrangement. In both cases the entropy is

$$I_{\text{Shannon}} = -\sum_{k=1}^{n} p_k \cdot \text{ld}\, p_k$$

$$I_{\text{Shannon}} = -\left(\frac{1}{9}\cdot\text{ld}\frac{1}{9}+\frac{3}{9}\cdot\text{ld}\frac{3}{9}+\frac{2}{9}\cdot\text{ld}\frac{2}{9}+\frac{1}{9}\cdot\text{ld}\frac{1}{9}+\frac{2}{9}\cdot\text{ld}\frac{2}{9}\right)$$

$$I_{\text{Shannon}} = 2.19716 \tag{8.08}$$

This entropy does not allow us to make any statements about maximum values or minimum values or the resulting width of the intervals. The same distribution of the relative frequencies in five fields can occur for arbitrary maxima or minima and the resulting width of the fields. Thus a partitioning in very small intervals

would provide the same entropy as a partitioning in very large intervals, extending over a wider range.
Comparison of the mean values and variances

$$m_1 = \sum_{k=1}^{n} x_k \cdot p_k = \frac{1}{9} \cdot (7+10+14+15+24+28+33+41+44) = \frac{219}{9} = 24$$

$$m_2 = \sum_{k=1}^{n} x_k \cdot p_k = \frac{1}{9} \cdot (0+1+1+1+2+2+3+4+4) = \frac{18}{9} = 2 \tag{8.09}$$

$$\sigma_1^2 = \sum_{k=1}^{n} (x_k - m_1)^2 \cdot p_k = \sum_{k=1}^{n} x_k^2 \cdot p_k - m_1^2 = 737.33 - (24)^2 = 161.33$$

$$\sigma_2^2 = \sum_{k=1}^{n} x_k^2 \cdot p_k - m_2^2 = 5.78 - 4 = 1.78 \tag{8.10}$$

shows us the differences between the two arrangements very clearly. Usually the variance is connected to the uncertainty, as it describes the range in which the random variable normally varies. The greater this range is, the greater is the uncertainty or the fluctuation in the experiment. In the first arrangement the variance σ_1^2 is nearly seven times greater than the mean value m_1 and about 80 times greater than the variance σ_2^2 of the second arrangement, assuming a value in the range of the mean m_2. Therefore the fluctuations of the realizations are substantially greater than in the second arrangement. The entropy, neglecting the realizations and concentrating merely on the probabilities of the events, cannot consider these differences. Obviously we need the realizations to get a complete image of the experiment. The entropy does not give us enough hints.

The discrete probabilities therefore do not allow a complete description of the experiment! They are merely measures, weighting the realizations of an experiment and allowing us to calculate several typical values, such as mean, variance and other moments. But the probabilities of the discrete events do not contain the information that we need to describe a random experiment completely.

Continuous random variables are completely different. Continuous distribution densities, resulting from the distribution functions, describe the distribution of the probabilities. These distribution densities are concrete connections between the probability space and the realization space (Borel fields).

$$F_x(\xi) = P\{\omega : x(\omega) \le \xi\} \tag{8.11}$$

The distribution function of the random variable $x(.)$ for the argument ξ is equal to the probability of all elementary events ω, having a realization $x(\omega)$, which is less than or equal to the limit ξ. Thus the probability is directly connected to semi-open intervals in the realization space. The abscissa of the distribution function is the range of possible events, while the ordinate indicates the probability that one of the possible realizations, assuming a value less than or equal to the actual value on the abscissa, has occurred. The distribution function increases monotonically and has the limits

$$0 = F_x(-\infty)\ ,\ 1 = F_x(\infty) \tag{8.12}$$

The distribution density

$$f_x(\xi) = \frac{\partial}{\partial \xi} F_x(\xi) \tag{8.13}$$

is the derivative of the distribution function and it also contains the complete information that we need to describe the random variable (the experiment). Both functions allow the computation of the probabilities of certain realizations and the calculation of all typical values, such as mean, mode, variance and so on.

In section 8.2 we will establish the connection between entropy and variance and discuss this relation between the measures of uncertainty.

8.1.1 Summary

An entropy calculated from discrete probabilities provides a measure, but it is meaningless if we do not get further information about the realizations connected to the probabilities. Only the additional knowledge of the realizations allows the complete description of the random experiment and the determination of the statistical values. This is, of course, not only a problem of information theory, but it stems from the definition of the discrete probabilities, which are not directly related to specified events.

An entropy calculated from a continuous distribution density has no need for further explanation as the distribution density already contains all the information that is necessary to describe the complete random variable or the random experiment.

Some concluding remarks regarding information

- Limitations (systematic boundaries, natural limits, ...) provide a certain gain of information I_c regarding the expected measurement.

  $$I_{\text{a posteriori}} = I_{\text{a priori}} - I_c$$

  Boundaries decrease the uncertainty, which is the reason for the negative sign.

- Information is the result of a choice, of a reduction of the uncertainty. Such a choice can be described with mathematical objective measures, assigned to the events, but it requires an interpretation, because without the assignment of a meaning to the experiment, the experiment is simply meaningless.

- Information is not knowledge. A play from Shakespeare and a theorem from Einstein can contain the same amount of information. The interpretation assigns different meanings to both messages (Hofstadter, Goedel) and the knowledge of these ideas is to be judged very different regarding applications, background, areas of interest of the person, etc.

- The formulation of the information measures does not distinguish between useful and useless information (relevance and irrelevance).

- Information $\rightarrow$ scarcity $\rightarrow$ value: In our definition of information rare events contain more information, because with a decreasing probability the information content (Hartley's information) increases. We often connect scarcity with the value of the event or the element. Elements that occur more seldom than other elements are assigned a higher price than the elements that everybody can afford.

- The observation of discrete events eliminates all uncertainty, contained in the experiment. The observation of continuous realizations cannot eliminate all uncertainty, as we simply cannot provide exact measurements. There always remains an uncertainty caused by the inexact measurement.

8.2 Shannon's information of a gaussian distribution

To increase our insight into the meaning of information, we now calculate the S_1-information of a multi-dimensional gaussian random variable of dimension n. This provides some further statements about the connection between information and continuous distributions.

The n-dimensional gaussian distribution density of a random vector $\underline{x}$

$$f_{\underline{x}}(\underline{\xi}) = \frac{1}{(2\pi)^{\frac{n}{2}} \cdot |P|^{\frac{1}{2}}} \cdot \exp\left\{-\frac{1}{2} \cdot (\underline{\xi} - \underline{\mu})^T \cdot P^{-1} \cdot (\underline{\xi} - \underline{\mu})\right\} \tag{8.14}$$

leads to the n-dimensional integral

$$S(\underline{x}) = -\int_{\underline{\Xi}} \ln\left[\frac{1}{(2\pi)^{\frac{n}{2}} \cdot |P|^{\frac{1}{2}}} \cdot \exp\left\{-\frac{1}{2} \cdot (\underline{\xi} - \underline{\mu})^T \cdot P^{-1} \cdot (\underline{\xi} - \underline{\mu})\right\}\right] \cdot f_{\underline{x}}(\underline{\xi}) \cdot d\underline{\xi} \tag{8.15}$$

where $f_{\underline{x}}(\underline{\xi})$ is only written in explicit form in the logarithm, as we have to transform these terms. Because of the nonnegative values of the distribution densities, we do not have any problems with the possible range of arguments of the logarithm.

$$S(\underline{x}) = -\int_{\Xi} \left\{ -\ln\left[(2\pi)^{\frac{n}{2}} \cdot |P|^{\frac{1}{2}} \right] - \frac{1}{2} \cdot (\underline{\xi} - \underline{\mu})^T \cdot P^{-1} \cdot (\underline{\xi} - \underline{\mu}) \right\} \cdot f_{\underline{x}}(\underline{\xi}) \cdot d\underline{\xi} \tag{8.16}$$

We obtain two integrals

$$S(\underline{x}) = \frac{1}{2} \cdot \int_{\Xi} \ln\left[(2\pi)^n \cdot |P| \right] \cdot f_{\underline{x}}(\underline{\xi}) \cdot d\underline{\xi} + \frac{1}{2} \cdot \int_{\Xi} (\underline{\xi} - \underline{\mu})^T \cdot P^{-1} \cdot (\underline{\xi} - \underline{\mu}) \cdot f_{\underline{x}}(\underline{\xi}) \cdot d\underline{\xi} \tag{8.17}$$

and we extract the constant factor $\ln\left[(2\pi)^n \cdot |P|\right]$ from the first integral, which leaves an integral of a distribution density. The value of such an integral is 1. So only the second integral remains

$$S(\underline{x}) = \frac{1}{2} \cdot \ln\left[(2\pi)^n \cdot |P| \right] + \frac{1}{2} \cdot \int_{\Xi} (\underline{\xi} - \underline{\mu})^T \cdot P^{-1} \cdot (\underline{\xi} - \underline{\mu}) \cdot f_{\underline{x}}(\underline{\xi}) \cdot d\underline{\xi} \tag{8.18}$$

and there we use the identity [BAS83]

$$\underline{v}^T \cdot M \cdot \underline{v} = \mathrm{tr}\left\{ M \cdot \underline{v} \cdot \underline{v}^T \right\} \tag{8.19}$$

to transform the quadratic form. We get

$$S(\underline{x}) = \frac{1}{2} \cdot \ln\left[(2\pi)^n \cdot |P| \right] + \frac{1}{2} \cdot \int_{\Xi} \mathrm{tr}\left\{ P^{-1} \cdot (\underline{\xi} - \underline{\mu}) \cdot (\underline{\xi} - \underline{\mu})^T \right\} \cdot f_{\underline{x}}(\underline{\xi}) \cdot d\underline{\xi} \tag{8.20}$$

Only the vectors in the trace operator depend on the variable of the integral. This allows us to extract the matrix from the integral and to write

$$S(\underline{x}) = \frac{1}{2} \cdot \ln\left[(2\pi)^n \cdot |P| \right] + \frac{1}{2} \cdot \mathrm{tr}\left\{ P^{-1} \cdot \int_{\Xi} (\underline{\xi} - \underline{\mu}) \cdot (\underline{\xi} - \underline{\mu})^T \cdot f_{\underline{x}}(\underline{\xi}) \cdot d\underline{\xi} \right\} \tag{8.21}$$

We thus get a familiar integral describing the covariance matrix

$$\int_{\Xi} (\underline{\xi} - \underline{\mu}) \cdot (\underline{\xi} - \underline{\mu})^T \cdot f_{\underline{x}}(\underline{\xi}) \cdot d\underline{\xi} = P \tag{8.22}$$

and therefore

$$S(\underline{x}) = \frac{1}{2} \cdot \ln\left[(2\pi)^n \cdot |P|\right] + \frac{1}{2} \cdot \mathrm{tr}\left\{P^{-1} \cdot P\right\} \tag{8.23}$$

$$S(\underline{x}) = \frac{1}{2} \cdot \ln\left[(2\pi)^n \cdot |P|\right] + \frac{1}{2} \cdot \mathrm{tr}\{I\}$$

and we finally obtain Shannon's information of a n-dimensional gaussian distribution

$$S(\underline{x}) = \frac{1}{2} \cdot \ln\left[(2\pi)^n \cdot |P|\right] + \frac{n}{2} \tag{8.24}$$

We notice that this information depends on the dimension n as well as on the covariance matrix P. It does not depend on the mean value. We get the same result when we calculate the limit for $\alpha \to 1$ in equation (7.16).

Let us examine this result: We interpreted information as certainty regarding an event. We already mentioned that Shannon's information (Rényi calls it information) has the character of an entropy, i.e. it describes the uncertainty, the inaccuracy. Our result confirms this assessment. If we increase the dimension n, we also increase the value of S. Since we get more dimensions, we also get more possible states that can be assumed by our system and so the uncertainty regarding a certain state increases.

The mean value, a deterministic state vector, does not have any uncertainty. The uncertainty is contained in the covariance matrix. Thus the mean value has no effect on the information, which can be noticed in equations (7.16) and (8.24). The assigned uncertainty is contained in the covariance matrix P, which is a part of the S-information in equation (8.24). The greater the determinant $|P|$ becomes, i.e. the greater the elements on the main diagonal of this matrix become (variances), the greater S becomes. This connection also leads us to the designation entropy S.

Moreover the S-information not only has positive values, but for

$$-\ln|P| \le (n) \cdot (1 + \ln(2\pi)) \tag{8.25}$$

or

$$-2 \cdot \ln(\sigma) \le (1 + \ln(2\pi)) \qquad \text{for a scalar random variable} \tag{8.26}$$

it takes negative values. Shannon's informationentropy is only equal to zero when the equality holds in both equations. Then we find

$$\sigma^2 = \exp\{-(1 + \ln(2\pi))\} = \exp\{-1\} \cdot \exp\{-\ln(2\pi)\} = \frac{1}{2\pi \cdot e} \tag{8.27}$$

How does this special case change when we increase the dimension of the random variable? To answer this question, we examine equation (8.25) in its most general form and we get

$$|P| = \exp\{-(n)\cdot(1+\ln(2\pi))\} = \exp\{-n - n\cdot\ln(2\pi)\}$$
$$= \frac{1}{e^n}\cdot\frac{1}{(2\pi)^n} = \frac{1}{(2\pi\cdot e)^n} \tag{8.28}$$

For independent random variables the covariance is a diagonal matrix and the resulting determinant is computed as the product of the main diagonal elements, so that

$$|P| = \prod_{i=1}^{n}\sigma_i^2 = \frac{1}{(2\pi\cdot e)^n} \tag{8.29}$$

and for $n = 2$ we find

$$\sigma_1^2\cdot\sigma_2^2 = \frac{1}{(2\pi\cdot e)^2} \quad \text{or} \quad \sigma_1\cdot\sigma_2 = \frac{1}{2\pi\cdot e} \tag{8.30}$$

and this is equal to the area created by the standard deviations of the two random variables. If we choose the first variance as in equation (8.27)

$$\sigma_1^2 = \frac{1}{2\pi\cdot e} \quad \text{we immediately get} \quad \sigma_2^2 = \frac{1}{2\pi\cdot e} \tag{8.31}$$

and the second variance is equal to the first.

This can be interpreted in such a way that the new component has no information about the already available components (uncorrelated, independent) and thus the second component has to have the same uncertainty as the first component to obtain the equation $S = 0$. The value of S is therefore an amalgamation of all single amounts of information of the marginal distribution densities, which build the joint distribution density function. This, of course, only holds iff P is a diagonal matrix.

Every covariance matrix (symmetric, nonnegative) can be transformed into a diagonal matrix via familiar transformations (PCA = principle component analysis, Karhunen–Loève transformation, LDU decomposition, ...). All these transformations add supplementary amounts of information (see section 9.4.3), which cannot be split into the different components with simple procedures.

We obtain further insight into this entropy measure by rearranging the covariance matrix with the correlation coefficient. We have the continuous entropy of a gaussian distribution, which is equal to

$$S = \frac{1}{2}\cdot\ln\left[(2\pi)^n\cdot|P|\right] + \frac{n}{2} \tag{8.32}$$

And for a two-dimensional random variable the covariance matrix is

$$P = \begin{pmatrix} \sigma_1^2 & \rho \cdot \sigma_1 \cdot \sigma_2 \\ \rho \cdot \sigma_1 \cdot \sigma_2 & \sigma_2^2 \end{pmatrix} \tag{8.33}$$

and the determinant becomes

$$|P| = \det(P) = \sigma_1^2 \cdot \sigma_2^2 \cdot \left(1 - \rho^2\right) \tag{8.34}$$

Thus

$$S = \frac{1}{2} \cdot \ln\left[(2\pi)^2 \cdot \sigma_1^2 \cdot \sigma_2^2 \cdot \left(1 - \rho^2\right)\right] + 1 \tag{8.35}$$

$$S = \frac{1}{2} \cdot \ln\left[(2\pi)^2 \cdot \sigma_1^2 \cdot \sigma_2^2\right] + 1 + \frac{1}{2} \cdot \ln\left(1 - \rho^2\right) \tag{8.36}$$

and the entropy is a function of the variances of both random variables. It further depends on the standardized correlation coefficient ρ. This standardized correlation coefficient assumes values in the interval $-1 \leq \rho \leq 1$ and thus

$$0 \leq \left(1 - \rho^2\right) \leq 1 \tag{8.37}$$

as a consequence we find

$$-\infty < \ln\left(1 - \rho^2\right) \leq 0 \tag{8.38}$$

So this term merely diminishes the entropy, which leads us to the conclusion, that the entropy achieves its maximum if the correlation vanishes, i.e. $\rho = 0$. Then both random variables are uncorrelated and because of the gaussian distribution are also independent. We thus notice that we obtain a maximum of the entropy if we observe independent random variables and that each correlation decreases the entropy of the joint distribution density.

8.3 Shannon's information as the possible gain of information in an observation

During the examination of the entropy of discrete events with discrete probabilities we found that Shannon's information describes the uncertainty that we eliminate by the observation of the realizations. It describes the uncertainty that we have before we reduce the possible number of realizations to a smaller number by an observation. The gain of information therefore reduces or eliminates the uncertainty that we have before we observe the measurement. This means that we interpret this measure as the description of the potential information that we may gain by the observation,

In the description of continuous distribution densities of continuous random variables, we also get an uncertainty, which is expressed by the variance of the distribution density. The variance is thus the uncertainty that we have before we measure.

$$S(\underline{x}) = \frac{1}{2} \cdot \ln\left[(2\pi)^n \cdot | P_{\text{a priori}} |\right] + \frac{n}{2} \tag{8.39}$$

As we cannot eliminate the uncertainty by the measurement, we obtain a remaining uncertainty, based on the fact that we are not able to provide exact measurements. Thus we also get a variance (uncertainty) after the observation, leading to the following entropy

$$S(\underline{x}) = \frac{1}{2} \cdot \ln\left[(2\pi)^n \cdot | P_{\text{a posteriori}} |\right] + \frac{n}{2} \tag{8.40}$$

In spite of the still existing uncertainty, we obtain a gain of information by the measurement, as the measurement in fact reduces the variance. Obviously the examination of continuous realizations requires the evaluation of the uncertainties before and after the observation, to be able to determine the gain of information obtained by the measurement. It is not possible to eliminate the uncertainty completely, because the uncertainty relation and the measurement facilities [BRI53] [BRI64] form natural limits, which we cannot reduce.

Here we have the decisive difference from the discrete entropies, which we already described in section 8.1. These discrete entropies do not have a direct connection to the realizations or to descriptive values of these realizations, such as the moments. Discrete entropies are solely based on the probabilities of the events and do not have any reference to the concrete observations.

The continuous entropies, on the other hand, do have this reference to the concrete observations, as they are functions of the distribution densities describing the distribution of the probability in the realization space. So the gaussian distribution is directly connected to the variance, which describes the width of the variation. Thus the entropy is a direct function of the uncertainty and that is exactly what Shannon's entropy represents: an uncertainty, contained in the next event. But we still retain an uncertainty, because of the limitations of our measurement devices.

For discrete entropies the knowledge of the initial uncertainty is sufficient to determine the information that we may gain by the measurement. For continuous measurements, we have to pay attention to the uncertainty of the measured realization, too. As this uncertainty is limited by natural bounds, which we cannot affect, we have a look at these limitations in the following section.

8.4 Limits of the information, limitations of the resolution

In the introduction we realized that we meet the limits of information when we are no longer able to distinguish two objects. In such a case both objects contain the same amount of information for the observer. We obtain this effect when the observer cannot distinguish the structures of the objects (when they are equal). As we perceive information on the basis of certain patterns, we are only able to find differences in different patterns. If these patterns or structures do not differ, we cannot observe an information difference and so the objects have the same information to us.

We may thus use the uncertainty relation to define the least possible information difference. The uncertainty relation defines relations between conjugated variables (energy–time, impulse–location, time–frequency), including the uncertainty relation of the Fourier transformation and even the demands of the sample theorem.

8.4.1 The resolution or the precision of the measurements

Each measurement affects the observed configuration!

This may not be important for observations where the effects of quanta can be neglected, but each measurement is an interference with the system, which is thus altered. We are thus merely able to get limited accuracy in our measurements (uncertainty relation). Moreover, we are not able to isolate a system completely from its environment (except for an experiment carried out in our minds) and to obtain undisturbed observations.

Borel, E. (1912):

> 'Introduction géométrique á la physique', Ganthier-Villars, Paris:
> A mass of 1 g, moving 1 cm on a star, which is not too far away from us (e.g., Sirius), changes the gravitational field of our Earth by a factor 10^{-100}. Such a deviation from 10^{-100} in the initial values of the positions and the velocities of gas molecules makes it impossible to compute the movement of these molecules for more than a millionth of a second.
>
> $\Rightarrow$ An exact measurement is impossible.

Nowadays precise measurements are made with light waves using interferences. But the more the observed structures reduce, the more the wavelength of the light has to be reduced, to be able to produce interferences with the structures. For an observed structure of

$$\Delta = 10^{-50}\ \text{cm} = 10^{-52}\ \text{m} \tag{8.41}$$

we need a wavelength lying in this range, i.e.

$$\lambda = 10^{-52}\ \text{m} \tag{8.42}$$

The energy that we have to expend to carry out such a measurement is given by the equation

$$E = h \cdot f = h \cdot \frac{c}{\lambda} \tag{8.43}$$

$$E = 6.625 \cdot 10^{-34}\,\text{Js} \cdot \frac{3 \cdot 10^{8}\,\frac{\text{m}}{\text{s}}}{10^{-52}\,\text{m}}$$

and

$$E = 1.987 \cdot 10^{27}\,\text{J} \tag{8.44}$$

If we now use Einstein's formula to calculate the mass that we would be able to destroy with this energy, we find

$$E = m \cdot c^2$$

$$m = \frac{E}{c^2} = \frac{1.987 \cdot 10^{27}\,\text{J}}{\left(3 \cdot 10^{8}\,\frac{\text{m}}{\text{s}}\right)^2} = 2.208 \cdot 10^{10}\,\text{kg} = 2.2 \cdot 10^{7}\ \text{t} \tag{8.45}$$

The energy needed for the measurement of structures in the range of $\Delta = 10^{-52}$ m would be so large that we would be able to destroy $2.2 \cdot 10^{7}$ t at once.

$\Rightarrow$ Such a measurement is impossible.

Realistically we are able to measure in areas of $\Delta = 10^{-18}$ m

$$E = 6.625 \cdot 10^{-34}\,\text{Js} \cdot \frac{3 \cdot 10^{8}\,\frac{\text{m}}{\text{s}}}{10^{-18}\,\text{m}} = 1.987 \cdot 10^{-7}\,\text{J} \tag{8.46}$$

$$m = \frac{E}{c^2} = \frac{1.987 \cdot 10^{-7}\,\text{J}}{\left(3 \cdot 10^{8}\,\frac{\text{m}}{\text{s}}\right)^2} = 2.208 \cdot 10^{-24}\,\text{kg} \tag{8.47}$$

In this range the relation between the measured mass and the mass that is destroyed by the measurement is in a correct range. But thus the initial (a priori) information in continuous systems is also limited and can never be infinitely large!

The paradox of $N_0 = \infty$: If the number of possible initial states of a system is equal to $N_0 = \infty$, the information gained by the first, limiting measurement, reducing this number to the number N_1, is infinitely large

$$I_1 = K \cdot \ln\left(\frac{N_0}{N_1}\right) = K \cdot \ln N_0 - K \cdot \ln N_1 = \infty$$

K = constant, e.g., 1/ld 2

And this result does not depend on N_1. But – as we already know – we are only able to make measurements with limited accuracy and the transition from single points to a real continuum cannot be defined if we rely on Boltzmann [BOL81].

8.4.2 The uncertainty relation of the Fourier transformation

The uncertainty relation describes the relation between the width of the autocorrelation function in time space to the spreading of the power density spectrum in frequency space. A decreasing width of the autocorrelation in time space allows us a better separation of two succeeding signals in time, but it decreases the resolution in frequency space. To derive this relation, we use the autocorrelation function

$$\varphi_{xx}(\tau) = \int_{-\infty}^{\infty} x(t) \cdot x^*(t+\tau) \cdot dt \qquad (8.48)$$

and the related Fourier transform, the power density spectrum

$$\Phi_{xx}(f) = \int_{-\infty}^{\infty} \varphi_{xx}(\tau) \cdot e^{-j(2\pi\ f\tau)} \cdot d\tau \qquad (8.49)$$

After some computation we find the formula that connects the resolution in time space with the resolution in frequency space:

$$\frac{1}{(4\pi)^2} \leq (\Delta t)^2 \cdot (\Delta f)^2 \qquad (8.50)$$

$$\Delta t \cdot \Delta f \geq \frac{1}{4\pi} \quad \underline{\text{Uncertainty relation of the Fourier transformation}} \qquad (8.51)$$

Such uncertainty relations occur between so-called ‘conjugated variables’. If we apply the previous equation to quantum physics, we use the identities

$$\Delta E = h \cdot \Delta f \qquad \Delta f = \frac{\Delta E}{h} \qquad (8.52)$$

and find

$$\Delta t \cdot \Delta E \geq \frac{h}{4\pi} \quad \underline{\text{Heisenberg's uncertainty relation of time and energy}} \qquad (8.53)$$

which can easily be derived from the properties of the Fourier transformation.

Conjugated variables are for example:

$$\begin{array}{lll} t = \text{time} & \text{and} & f = \text{frequency} \\ t = \text{time} & \text{and} & E = \text{energy} \\ x = \text{location} & \text{and} & p = \text{impulse} \end{array} \tag{8.54}$$

Gabor [GAB46] uses the uncertainty relation for the definition of the Gabor transformation, which is a short-time Fourier transformation, where the windowing functions are chosen to achieve the minimum uncertainty. Gabor uses gaussian functions in his transformation.

8.5 Maximization of the entropy of a continuous random variable

Shannon's information of a continuous random variable x is given by the equation

$$S = -\int_{\Xi} \ln(f_x(\xi)) \cdot f_x(\xi) \cdot d\xi \tag{8.55}$$

We add the following two boundary conditions:

$$\int_{\Xi} \xi^2 \cdot f_x(\xi) \cdot d\xi = \sigma^2 \qquad \text{limited power of the error} \tag{8.56}$$

$$\int_{\Xi} f_x(\xi) \cdot d\xi = 1 \tag{8.57}$$

and try to maximize the information. $f_x(\xi)$ is a distribution density, so that we additionally have to meet the condition $f_x(\xi) \geq 0$. What we are looking for is a distribution function, which produces a maximum Shannon entropy or Shannon information if only the limited power of the error is known. The optimization with the Lagrangian multipliers leads to the equation:

$$\frac{d}{df_x(\xi)} \left\{ -\int_{\Xi} \ln(f_x(\xi)) \cdot f_x(\xi) \cdot d\xi - \lambda_1 \cdot \left(\int_{\Xi} f_x(\xi) \cdot d\xi - 1 \right) - \lambda_2 \cdot \left(\int_{\Xi} \xi^2 \cdot f_x(\xi) \cdot d\xi - \sigma^2 \right) \right\} \stackrel{!}{=} 0 \tag{8.58}$$

As all Lagrangian multipliers are constant, we contract the integrals and write:

$$\frac{d}{df_x(\xi)}\left\{-\int_{\Xi}\left[\ln(f_x(\xi))+\lambda_1+\lambda_2\cdot\xi^2\right]\cdot f_x(\xi)\cdot d\xi+\lambda_1+\lambda_2\cdot\sigma^2\right\}\overset{!}{=}0 \tag{8.59}$$

$$-\int_{\Xi}\frac{d}{df_x(\xi)}\left\{\left[\ln(f_x(\xi))+\lambda_1+\lambda_2\cdot\xi^2\right]\cdot f_x(\xi)\right\}\cdot d\xi\overset{!}{=}0$$

$$-\int_{\Xi}\left[\ln(f_x(\xi))+\lambda_1+\lambda_2\cdot\xi^2+\frac{1}{f_x(\xi)}\cdot f_x(\xi)\right]\cdot d\xi\overset{!}{=}0$$

$$-\int_{\Xi}\left[\ln(f_x(\xi))+\lambda_1+\lambda_2\cdot\xi^2+1\right]\cdot d\xi\overset{!}{=}0 \tag{8.60}$$

This integral has to be equal to zero, which only holds iff the kernel is equal to zero. We thus write

$$\ln(f_x(\xi))+\lambda_1+\lambda_2\cdot\xi^2+1=0 \tag{8.61}$$

$$\ln(f_x(\xi))=-\lambda_1-\lambda_2\cdot\xi^2-1$$

$$f_x(\xi)=\exp\left(-1-\lambda_1-\lambda_2\cdot\xi^2\right) \tag{8.62}$$

and the constants are calculated by inserting this solution into the boundary conditions. There we find:

$$\int_{\Xi}f_x(\xi)\cdot d\xi=\int_{\Xi}\exp\left(-1-\lambda_1-\lambda_2\cdot\xi^2\right)\cdot d\xi=1 \tag{8.63}$$

$$\exp(-1-\lambda_1)\cdot\int_{\Xi}\exp\left(-\lambda_2\cdot\xi^2\right)\cdot d\xi=1$$

$$\int_{\Xi}\exp\left(-\lambda_2\cdot\xi^2\right)\cdot d\xi=\exp(1+\lambda_1) \tag{8.64}$$

We have:

$$\int_{\Xi}\exp\left(-\frac{1}{2\cdot\sigma^2}\cdot\xi^2\right)\cdot d\xi=\sqrt{2\pi}\cdot\sigma \tag{8.65}$$

and with

$$\frac{1}{2\cdot\sigma^2}=\lambda_2 \qquad \text{or} \qquad \sigma=\frac{1}{\sqrt{2\cdot\lambda_2}} \tag{8.66}$$

we get:

$$\int_{\Xi}\exp\left(-\lambda_2\cdot\xi^2\right)\cdot d\xi=\sqrt{2\pi}\cdot\frac{1}{\sqrt{2\cdot\lambda_2}}=\sqrt{\frac{\pi}{\lambda_2}}=\exp(1+\lambda_1) \tag{8.67}$$

$$\sqrt{\frac{\pi}{\lambda_2}}=\exp(1+\lambda_1)$$

$$\ln\left(\sqrt{\frac{\pi}{\lambda_2}}\right)-1=\lambda_1 \tag{8.68}$$

From the remaining boundary conditions we calculate:

$$\int_{\Xi}\xi^2\cdot f_x(\xi)\cdot d\xi=\int_{\Xi}\xi^2\cdot\exp\left(-1-\lambda_1-\lambda_2\cdot\xi^2\right)\cdot d\xi=\sigma^2 \tag{8.68}$$

$$\int_{\Xi}\xi^2\cdot\exp\left(-\lambda_2\cdot\xi^2\right)\cdot d\xi=\sigma^2\cdot\exp(1+\lambda_1) \tag{8.70}$$

and we get

$$\int_{\Xi}\xi^2\cdot\exp\left(-\frac{1}{2\cdot\sigma^2}\cdot\xi^2\right)\cdot d\xi=\sigma^2\cdot\sqrt{2\pi}\cdot\sigma \tag{8.71}$$

with

$$\frac{1}{2\cdot\sigma^2}=\lambda_2 \qquad \text{or} \qquad \sigma=\frac{1}{\sqrt{2\cdot\lambda_2}} \tag{8.72}$$

we find

$$\int_{\Xi}\xi^2\cdot\exp\left(-\lambda_2\cdot\xi^2\right)\cdot d\xi=\frac{1}{2\cdot\lambda_2}\cdot\sqrt{2\pi}\cdot\frac{1}{\sqrt{2\cdot\lambda_2}}=\sigma^2\cdot\exp(1+\lambda_1) \tag{8.73}$$

$$\frac{1}{2\cdot\lambda_2}\cdot\sqrt{\frac{\pi}{\lambda_2}}=\sigma^2\cdot\exp(1+\lambda_1) \tag{8.74}$$

With the first solution $\sqrt{\frac{\pi}{\lambda_2}}=\exp(1+\lambda_1)$ we get

$$\frac{1}{2 \cdot \lambda_2} = \sigma^2 \qquad \text{resp.} \qquad \frac{1}{2 \cdot \sigma^2} = \lambda_2 \tag{8.75}$$

and then

$$\ln\left(\sqrt{\frac{\pi}{\lambda_2}}\right) - 1 = \ln\left(\sqrt{2\pi \cdot \sigma^2}\right) - 1 = \ln\left(\frac{\sqrt{2\pi \cdot \sigma^2}}{e}\right) = \lambda_1 \tag{8.76}$$

If we insert both solutions into the computed distribution density, we finally obtain the solution:

$$f_x(\xi) = \exp\left(-1 - \lambda_1 - \lambda_2 \cdot \xi^2\right) \tag{8.77}$$

$$f_x(\xi) = \exp\left\{-1 - \left[\ln\left(\sqrt{2\pi \cdot \sigma^2}\right) - 1\right] - \frac{1}{2 \cdot \sigma^2} \cdot \xi^2\right\}$$

$$f_x(\xi) = \exp\left[-\ln\left(\sqrt{2\pi \cdot \sigma^2}\right) - \frac{1}{2 \cdot \sigma^2} \cdot \xi^2\right]$$

$$f_x(\xi) = \frac{1}{\sqrt{2\pi} \cdot \sigma} \exp\left(-\frac{1}{2 \cdot \sigma^2} \cdot \xi^2\right) \tag{8.78}$$

and this is a gaussian distribution density!

We have found that the gaussian distribution density has the greatest Shannon entropy of all distribution densities if we restrict the power of the error. All the other distribution densities, having the same second order moments as the gaussian distribution density, have a smaller Shannon entropy than the gaussian distribution density. This justifies the assumption of a gaussian distribution density, if we restrict our attention to second order moments. In this case the gaussian distribution density builds the worst case and it contains the most uncertainty of all distribution densities on the basis of the moments, which we assume to be the given boundary conditions. All the other distribution densities require further restrictions and therefore further reduction of the uncertainty (of the entropy).

Having discussed the absolute measures of information from Shannon and Rényi, we now face the measures of distance, which have also been derived by Alfred Rényi in a generalized formulation. But they can also be transformed it to simpler forms. One of those simpler forms is the familiar Kullback information, which we use to start the next discussions.

9 Information functions for gaussian distributions part II

9.1 Kullback's information

The information that Kullback himself [KUL87] calls 'discrimination information' is generated from the difference of two Shannon information functions. As a difference of these information functions it is not affected by the constant factor K (see chapter 8), and it is – contrary to Shannon's information – invariant with respect to a transformation of the coordinate system. We start the derivation with Rényi's G-divergence

$$G_\alpha = \frac{1}{\alpha - 1} \cdot \ln\left(\int_{\underline{\Xi}} \frac{f_{\underline{x}}^{\alpha}\left(\underline{\xi}\right)}{f_{\underline{x}_s}^{\alpha-1}\left(\underline{\xi}\right)} \cdot d\underline{\xi} \right) \tag{9.01}$$

$$\alpha \in \Re \text{ and } \alpha \neq 1$$

and we build the limit for $\alpha \to 1$

$$\begin{aligned} G_1 &= \lim_{\alpha \to 1} G_\alpha \\ &= \lim_{\alpha \to 1} \left[\frac{1}{\alpha - 1} \cdot \ln\left(\int_{\underline{\Xi}} \frac{f_{\underline{x}}^{\alpha}\left(\underline{\xi}\right)}{f_{\underline{x}_s}^{\alpha-1}\left(\underline{\xi}\right)} \cdot d\underline{\xi} \right) \right] \end{aligned} \tag{9.02}$$

Analogously to the derivation in the previous chapter, the numerator with the integral over a distribution density, as well as the denominator tend to zero. We again apply de l'Hospital's rule to find the derivative in the denominator

$$N = \frac{d}{d\alpha}(\alpha - 1) = 1 \tag{9.03}$$

and in the numerator

$$Z = \frac{d}{d\alpha}\left[\ln\left(\int_{\underline{\Xi}} \frac{f_{\underline{x}}^{\alpha}(\underline{\xi})}{f_{\underline{x}_s}^{\alpha-1}(\underline{\xi})} \cdot d\underline{\xi}\right)\right] \tag{9.04}$$

$$Z = \frac{1}{\int_{\underline{\Xi}} \frac{f_{\underline{x}}^{\alpha}(\underline{\xi})}{f_{\underline{x}_s}^{\alpha-1}(\underline{\xi})} \cdot d\underline{\xi}} \cdot \frac{d}{d\alpha}\left(\int_{\underline{\Xi}} \frac{f_{\underline{x}}^{\alpha}(\underline{\xi})}{f_{\underline{x}_s}^{\alpha-1}(\underline{\xi})} \cdot d\underline{\xi}\right) \tag{9.05}$$

which requires Leibnitz's rule, given in equation (7.41). The derivative of the integral is equal to the derivative of the kernel, because of the constant limits of the integral

$$\frac{d}{d\alpha}\left(\int_{\underline{\Xi}} \frac{f_{\underline{x}}^{\alpha}(\underline{\xi})}{f_{\underline{x}_s}^{\alpha-1}(\underline{\xi})} \cdot d\underline{\xi}\right) = \int_{\underline{\Xi}} \frac{d}{d\alpha}\left(\frac{f_{\underline{x}}^{\alpha}(\underline{\xi})}{f_{\underline{x}_s}^{\alpha-1}(\underline{\xi})}\right) \cdot d\underline{\xi} \tag{9.06}$$

We proceed with the calculation in equation (9.06) and write:

$$\frac{d}{d\alpha}\left(\frac{f_{\underline{x}}^{\alpha}(\underline{\xi})}{f_{\underline{x}_s}^{\alpha-1}(\underline{\xi})}\right) = \frac{d}{d\alpha}\left(\frac{f_{\underline{x}}(\underline{\xi})}{f_{\underline{x}_s}(\underline{\xi})}\right)^{\alpha} \cdot f_{\underline{x}_s}(\underline{\xi}) \tag{9.07}$$

$$= f_{\underline{x}_s}(\underline{\xi}) \cdot \frac{d}{d\alpha} \exp\left\{\ln\left(\frac{f_{\underline{x}}(\underline{\xi})}{f_{\underline{x}_s}(\underline{\xi})}\right)^{\alpha}\right\}$$

$$= f_{\underline{x}_s}(\underline{\xi}) \cdot \frac{d}{d\alpha} \exp\left\{\alpha \cdot \ln\left(\frac{f_{\underline{x}}(\underline{\xi})}{f_{\underline{x}_s}(\underline{\xi})}\right)\right\}$$

$$= f_{\underline{x}_s}(\underline{\xi}) \cdot \ln\left(\frac{f_{\underline{x}}(\underline{\xi})}{f_{\underline{x}_s}(\underline{\xi})}\right) \cdot \exp\left\{\alpha \cdot \ln\left(\frac{f_{\underline{x}}(\underline{\xi})}{f_{\underline{x}_s}(\underline{\xi})}\right)\right\}$$

$$= f_{\underline{x}_s}(\underline{\xi}) \cdot \ln\left(\frac{f_{\underline{x}}(\underline{\xi})}{f_{\underline{x}_s}(\underline{\xi})}\right) \cdot \left(\frac{f_{\underline{x}}(\underline{\xi})}{f_{\underline{x}_s}(\underline{\xi})}\right)^{\alpha} \tag{9.08}$$

We insert this term in the integral and thus get the final formulation for the numerator

$$Z = \frac{1}{\int\limits_{\underline{\Xi}} \frac{f_{\underline{x}}^{\alpha}(\underline{\xi})}{f_{\underline{x}_s}^{\alpha-1}(\underline{\xi})} \cdot d\underline{\xi}} \cdot \left(\int\limits_{\underline{\Xi}} f_{\underline{x}_s}(\underline{\xi}) \cdot \ln\left(\frac{f_{\underline{x}}(\underline{\xi})}{f_{\underline{x}_s}(\underline{\xi})} \right) \cdot \left(\frac{f_{\underline{x}}(\underline{\xi})}{f_{\underline{x}_s}(\underline{\xi})} \right)^{\alpha} \cdot d\underline{\xi} \right) \tag{9.09}$$

Then we calculate

$$\lim_{\alpha \to 1} \frac{Z}{N} = \lim_{\alpha \to 1} Z \tag{9.10}$$

$$G_1 = \lim_{\alpha \to 1} \left\{ \frac{1}{\int\limits_{\underline{\Xi}} \frac{f_{\underline{x}}^{\alpha}(\underline{\xi})}{f_{\underline{x}_s}^{\alpha-1}(\underline{\xi})} \cdot d\underline{\xi}} \cdot \left[\int\limits_{\underline{\Xi}} f_{\underline{x}_s}(\underline{\xi}) \cdot \ln\left(\frac{f_{\underline{x}}(\underline{\xi})}{f_{\underline{x}_s}(\underline{\xi})} \right) \cdot \left(\frac{f_{\underline{x}}(\underline{\xi})}{f_{\underline{x}_s}(\underline{\xi})} \right)^{\alpha} \cdot d\underline{\xi} \right] \right\} \tag{9.11}$$

where the integral is taken over a distribution density $f_{\underline{x}}(\underline{\xi})$.

The result

$$G_1 = \int\limits_{\underline{\Xi}} \ln\left(\frac{f_{\underline{x}}(\underline{\xi})}{f_{\underline{x}_s}(\underline{\xi})} \right) \cdot f_{\underline{x}}(\underline{\xi}) \cdot d\underline{\xi} \tag{9.12}$$

is Kullback's 'discrimination information'. Some easy rearrangements lead to

$$G_1 = \int\limits_{\underline{\Xi}} \ln\left(f_{\underline{x}}(\underline{\xi})\right) \cdot f_{\underline{x}}(\underline{\xi}) \cdot d\underline{\xi} - \int\limits_{\underline{\Xi}} \ln\left(f_{\underline{x}_s}(\underline{\xi})\right) \cdot f_{\underline{x}}(\underline{\xi}) \cdot d\underline{\xi} \tag{9.13}$$

$$G_1 = -S\left(f_{\underline{x}}(\underline{\xi})\right) + S\left(f_{\underline{x}_s}(\underline{\xi}), f_{\underline{x}}(\underline{\xi}) \right) \tag{9.14}$$

and we immediately notice the relation to Shannon's information S.

But the second addend is not a 'pure' Shannon information, it is more a cross-information (according to the cross-covariances). Such cross-information measures are applied in the description of channel entropies, where the dissipation, the transinformation and equivocation are determined for the communication channel.

We already know Kullback's information from the examination of Rényi's entropy in chapter 5.3, where we derived the information difference between two different events. The mean of the information difference together with the application of Shannon's information led us to Kullback's information. This information arises when we replace a given probability distribution by another probability distribution, or a coding by another coding (see also section 3.2.3.6).

The derivation from the G-divergence demonstrates that the discrimination information describes the information difference between two distribution densities. It thus provides a tool to calculate the parameters of probability distribution densities (e.g., mean and covariance matrix of a gaussian distribution) according to a given distribution density (e.g., a uniform distribution). Minimizing the discrimination information, i.e. the difference between the two distribution densities, can give the desired approximation.

$$G_1 = \int_{\underline{P}} \ln\left(\frac{f_{\underline{y}}(\underline{\rho})}{f_{\underline{y}_{\text{gauss}}}(\underline{\rho})}\right) \cdot f_{\underline{y}}(\underline{\rho}) \cdot d\underline{\rho} \overset{!}{=} \min \tag{9.15}$$

An explicit calculation is given in chapter 15.

Furthermore Kullback's information allows us to describe the information of the random variable $\underline{x}$ contained in the random variable $\underline{y}$

$$G_1 = \int_{\underline{P}} \ln\left(\frac{f_{\underline{y},\underline{x}}(\underline{\rho},\underline{\xi})}{f_{\underline{x}}(\underline{\xi}) \cdot f_{\underline{y}}(\underline{\rho})}\right) \cdot f_{\underline{y},\underline{x}}(\underline{\rho},\underline{\xi}) \cdot d\underline{\rho} \cdot d\underline{\xi} \tag{9.16}$$

This information is also known as Kolmogorov's information (transinformation or mutual information), and it is a special, symmetric case of Kullback's information.

9.1.1 G_1 for gaussian distribution densites

Equation (9.12) enables us to calculate the discrimination information of two gaussian distribution densities having the same multi-dimensional dummy variable $\underline{\xi}$.

Distribution densities with different dummy variables

$$G_1 = \int_{\underline{\Xi}} \ln\left(\frac{f_{\underline{x}}(\underline{\xi})}{f_{\underline{\Gamma}_s}(\underline{\gamma})}\right) \cdot f_{\underline{x}}(\underline{\xi}) \cdot d\underline{\xi} \tag{9.17}$$

lead to a degeneration of Kullback's information, i.e. to Shannon's information with an additional additive term $-\ln\left(f_{\underline{\Gamma}_s}(\underline{\gamma})\right)$ (Hartley's 'differential' information), as this term is independent of the integration variable $\underline{\xi}$.

Returning to the case where the dummy variables are equal we calculate the discrimination information from the equation

$$G_1 = G(x, x_s) = \int_{\Xi} \ln\left(\frac{f_{\underline{x}}(\underline{\xi})}{f_{\underline{x}_s}(\underline{\xi})}\right) \cdot f_{\underline{x}}(\underline{\xi}) \cdot d\underline{\xi} = -S\left(f_{\underline{x}}(\underline{\xi})\right) + S\left(f_{\underline{x}_s}(\underline{\xi}), f_{\underline{x}}(\underline{\xi})\right)$$

$$\begin{aligned} &= \int_{\Xi} \left(-\frac{1}{2} \cdot \ln\left[(2\pi)^n \cdot |P|\right] - \frac{1}{2} \cdot (\underline{\xi} - \underline{\mu})^T \cdot P^{-1} \cdot (\underline{\xi} - \underline{\mu})\right) \cdot f_{\underline{x}}(\underline{\xi}) \cdot d\underline{\xi} - \\ &\quad - \int_{\Xi} \left(-\frac{1}{2} \cdot \ln\left[(2\pi)^n \cdot |P_s|\right] - \frac{1}{2} \cdot (\underline{\xi} - \underline{\mu}_s)^T \cdot P_s^{-1} \cdot (\underline{\xi} - \underline{\mu}_s)\right) \cdot f_{\underline{x}}(\underline{\xi}) \cdot d\underline{\xi} \end{aligned} \tag{9.18}$$

We already know the first addend of this equation from section 7.2, and this leads to

$$\begin{aligned} G_1 &= -\frac{1}{2} \cdot \ln\left[(2\pi)^n \cdot |P|\right] - \frac{n}{2} \\ &\quad - \int_{\Xi} \left(-\frac{1}{2} \cdot \ln\left[(2\pi)^n \cdot |P_s|\right] - \frac{1}{2} \cdot (\underline{\xi} - \underline{\mu}_s)^T \cdot P_s^{-1} \cdot (\underline{\xi} - \underline{\mu}_s)\right) \cdot f_{\underline{x}}(\underline{\xi}) \cdot d\underline{\xi} \end{aligned} \tag{9.19}$$

and we only have to solve the remaining integral. We get two integrals

$$\begin{aligned} I &= \int_{\Xi} \left(-\frac{1}{2} \cdot \ln\left[(2\pi)^n \cdot |P_s|\right] - \frac{1}{2} \cdot (\underline{\xi} - \underline{\mu}_s)^T \cdot P_s^{-1} \cdot (\underline{\xi} - \underline{\mu}_s)\right) \cdot f_{\underline{x}}(\underline{\xi}) \cdot d\underline{\xi} = \\ &= -\int_{\Xi} \frac{1}{2} \cdot \ln\left[(2\pi)^n \cdot |P_s|\right] \cdot f_{\underline{x}}(\underline{\xi}) \cdot d\underline{\xi} - \int_{\Xi} \frac{1}{2} \cdot (\underline{\xi} - \underline{\mu}_s)^T \cdot P_s^{-1} \cdot (\underline{\xi} - \underline{\mu}_s) \cdot f_{\underline{x}}(\underline{\xi}) \cdot d\underline{\xi} \end{aligned} \tag{9.20}$$

As the term $-\frac{1}{2} \cdot \ln\left[(2\pi)^n \cdot |P_s|\right]$ is independent of the integration variable, and on applying the trace operator with $\underline{v}^T \cdot M \cdot \underline{v} = \mathrm{tr}\left\{M \cdot \underline{v} \cdot \underline{v}^T\right\}$ to the second integral we obtain

$$\begin{aligned} I &= -\frac{1}{2} \cdot \ln\left[(2\pi)^n \cdot |P_s|\right] - \frac{1}{2} \cdot \int_{\Xi} \mathrm{tr}\left\{P_s^{-1} \cdot (\underline{\xi} - \underline{\mu}_s) \cdot (\underline{\xi} - \underline{\mu}_s)^T\right\} \cdot f_{\underline{x}}(\underline{\xi}) \cdot d\underline{\xi} \\ &= -\frac{1}{2} \cdot \ln\left[(2\pi)^n \cdot |P_s|\right] - \frac{1}{2} \cdot \mathrm{tr}\left\{P_s^{-1} \cdot \int_{\Xi} (\underline{\xi} - \underline{\mu}_s) \cdot (\underline{\xi} - \underline{\mu}_s)^T \cdot f_{\underline{x}}(\underline{\xi}) \cdot d\underline{\xi}\right\} \end{aligned} \tag{9.21}$$

The dyadic product must be solved

$$
\begin{aligned}
I_2 &= \int_{\underline{\Xi}} \left(\underline{\xi} - \underline{\mu}_s\right) \cdot \left(\underline{\xi} - \underline{\mu}_s\right)^T \cdot f_{\underline{x}}\left(\underline{\xi}\right) \cdot d\underline{\xi} \\
&= \int_{\underline{\Xi}} \left(\underline{\xi} \cdot \underline{\xi}^T - \underline{\xi} \cdot \underline{\mu}_s^{\ T} - \underline{\mu}_s \cdot \underline{\xi}^T + \underline{\mu}_s \cdot \underline{\mu}_s^{\ T}\right) \cdot f_{\underline{x}}\left(\underline{\xi}\right) \cdot d\underline{\xi} \\
&= \int_{\underline{\Xi}} \underline{\xi} \cdot \underline{\xi}^T \cdot f_{\underline{x}}\left(\underline{\xi}\right) \cdot d\underline{\xi} - \int_{\underline{\Xi}} \underline{\xi} \cdot \underline{\mu}_s^{\ T} \cdot f_{\underline{x}}\left(\underline{\xi}\right) \cdot d\underline{\xi} \\
&\quad - \int_{\underline{\Xi}} \underline{\mu}_s \cdot \underline{\xi}^T \cdot f_{\underline{x}}\left(\underline{\xi}\right) \cdot d\underline{\xi} + \int_{\underline{\Xi}} \underline{\mu}_s \cdot \underline{\mu}_s^{\ T} \cdot f_{\underline{x}}\left(\underline{\xi}\right) \cdot d\underline{\xi}
\end{aligned}
$$

We extract the constants out of the integral and get

$$
\begin{aligned}
I_2 &= \int_{\underline{\Xi}} \underline{\xi} \cdot \underline{\xi}^T \cdot f_{\underline{x}}\left(\underline{\xi}\right) \cdot d\underline{\xi} - \int_{\underline{\Xi}} \underline{\xi} \cdot f_{\underline{x}}\left(\underline{\xi}\right) \cdot d\underline{\xi} \cdot \underline{\mu}_s^T \\
&\quad - \underline{\mu}_s \cdot \int_{\underline{\Xi}} \underline{\xi}^T \cdot f_{\underline{x}}\left(\underline{\xi}\right) \cdot d\underline{\xi} + \underline{\mu}_s \cdot \underline{\mu}_s^{\ T} \cdot \int_{\underline{\Xi}} f_{\underline{x}}\left(\underline{\xi}\right) \cdot d\underline{\xi}
\end{aligned}
\tag{9.22}
$$

Now we write

$$
\begin{aligned}
I_2 &= P + \underline{\mu} \cdot \underline{\mu}^T - \underline{\mu} \cdot \underline{\mu}_s^{\ T} - \underline{\mu}_s \cdot \underline{\mu}^T + \underline{\mu}_s \cdot \underline{\mu}_s^{\ T} \\
&= P + \left(\underline{\mu} - \underline{\mu}_s\right) \cdot \left(\underline{\mu} - \underline{\mu}_s\right)^T
\end{aligned}
\tag{9.23}
$$

Thus Kullback's information from the equations (9.21), (9.22) and (9.23) is

$$
\begin{aligned}
G_1 &= -\frac{1}{2} \cdot \ln\left[(2\pi)^n \cdot |P|\right] - \frac{n}{2} + \frac{1}{2} \cdot \ln\left[(2\pi)^n \cdot |P_s|\right] \\
&\quad + \frac{1}{2} \cdot \mathrm{tr}\left\{P_s^{-1} \cdot \left[P + \left(\underline{\mu} - \underline{\mu}_s\right) \cdot \left(\underline{\mu} - \underline{\mu}_s\right)^T\right]\right\}
\end{aligned}
\tag{9.24}
$$

with the extension $n = \mathrm{tr}\left\{P^{-1} \cdot P\right\}$ and combinations of the logarithms we get

$$
G_1 = -\frac{1}{2} \cdot \ln\left[\frac{|P_s|}{|P|}\right] - \frac{1}{2} \cdot \mathrm{tr}\left\{P^{-1} \cdot P\right\} + \frac{1}{2} \cdot \mathrm{tr}\left\{P_s^{-1} \cdot \left[P + \left(\underline{\mu} - \underline{\mu}_s\right) \cdot \left(\underline{\mu} - \underline{\mu}_s\right)^T\right]\right\} \tag{9.25}
$$

and finally

$$
G_1 = -\frac{1}{2} \cdot \ln\left[\frac{|P_s|}{|P|}\right] + \frac{1}{2} \cdot \mathrm{tr}\left\{\left(P_s^{-1} - P^{-1}\right) \cdot P\right\} + \frac{1}{2} \cdot \mathrm{tr}\left\{P_s^{-1} \cdot \left(\underline{\mu} - \underline{\mu}_s\right) \cdot \left(\underline{\mu} - \underline{\mu}_s\right)^T\right\} \tag{9.26}
$$

or

$$G_1 = -\frac{1}{2} \cdot \ln\left[\frac{|P_S|}{|P|}\right] + \frac{1}{2} \cdot \mathrm{tr}\left\{\left(P_S^{-1} - P^{-1}\right) \cdot P\right\} + \frac{1}{2} \cdot \left(\underline{\mu} - \underline{\mu}_S\right)^T \cdot P_S^{-1} \cdot \left(\underline{\mu} - \underline{\mu}_S\right) \quad (9.27)$$

From this result we notice that G_1 is not symmetric, because the exchange of the distribution densities in equation (9.18) leads to a change in the covariance matrices of the formula, and this causes another information. The exchange does not affect the mean values.

It is an interesting fact, that we find mean values in this information function. Up to now we have interpreted information as the reciprocal value of a variance, i.e. the reciprocal value of an uncertainty. But here we do have additional differences of mean values and thus error variations between the distribution densities. If we look at the term

$$\left(\underline{\mu} - \underline{\mu}_S\right)^T \cdot P_S^{-1} \cdot \left(\underline{\mu} - \underline{\mu}_S\right) = \left\|\underline{\mu} - \underline{\mu}_S\right\|_{P_S^{-1}} \quad (9.28)$$

from equation (9.27) we immediately see that the differences of the mean values create a quadratic error, weighted with the inverse of the covariance matrix P_S^{-1}.

Contrary to the quadratic errors (variances) we observed up to now, which resulted from the calculation of the mean of some moment generating function, we now get a quadratic error that is computed after the calculation of the expectation values. It is created from deterministic mean values and such a formulation is also known as the weighted norm of the vector $\left(\underline{\mu} - \underline{\mu}_S\right)$. As Kullback's information is a distance (apart from the symmetry), the Kullback–Leibler distance, the weighted difference of the mean values is nothing unusual, because it determines the shift between the centers (means) of the two distribution density functions.

Interpretation of Kullback's information (for gaussian distribution densities):

$$\frac{1}{2} \cdot \left\|\underline{\mu} - \underline{\mu}_S\right\|_{P_S^{-1}} \quad : \quad (9.29)$$

This term describes the distance between the centers of the distribution densities f and f_S relative to the contours of the constant density f_S.

$$\frac{1}{2} \cdot \ln\left(\frac{|P_S|}{|P|}\right) : \quad (9.30)$$

This term compares the volumes of the ellipses [LOF90] that can be created from both distribution densities. It vanishes if the volumes of both ellipses are equal.

$$\frac{1}{2}\cdot \operatorname{tr}\left\{\left(P_S^{-1}-P^{-1}\right)\cdot P\right\} \text{ or } \frac{1}{2}\cdot \operatorname{tr}\left\{P_S^{-1}\cdot P\right\} \text{ (see equation (9.24)):} \tag{9.31}$$

This residual term compares the eigenstructures of the matrices P and P_S. If both matrices have the same eigenvectors, we are able to create matrices with these eigenvectors $U = U_S$, and applying the Karhunen–Loève transformation we find the relation:

$$\operatorname{tr}\left\{P_S^{-1}\cdot P\right\}=\operatorname{tr}\left\{\left(U_S\cdot\Lambda_S\cdot U_S^{-1}\right)^{-1}\cdot U\cdot\Lambda\cdot U^{-1}\right\} \tag{9.32}$$

with

$$\begin{aligned}&\left(U_S\cdot\Lambda_S\cdot U_S^{-1}\right)^{-1}\cdot\left(U_S\cdot\Lambda_S\cdot U_S^{-1}\right)=I\\ \Rightarrow\quad&\left(U_S\cdot\Lambda_S\cdot U_S^{-1}\right)^{-1}=U_S\cdot\Lambda_S^{-1}\cdot U_S^{-1}\end{aligned} \tag{9.33}$$

sing the identity of the matrices $U = U_S$ of the eigenvectors we get

$$\begin{aligned}&\operatorname{tr}\left\{P_S^{-1}\cdot P\right\}=\operatorname{tr}\left\{U_S\cdot\Lambda_S^{-1}\cdot U_S^{-1}\cdot U\cdot\Lambda\cdot U^{-1}\right\}=tr\left\{\Lambda_S^{-1}\cdot\Lambda\right\}\\ &\operatorname{tr}\left\{P_S^{-1}\cdot P\right\}=\sum_{i=1}^{n}\frac{\lambda_i}{\lambda_{s_i}}\end{aligned} \tag{9.34}$$

in this case this term is equal to the sum of the quotient of the eigenvalues.

We notice that Kullback's discrimination information not only compares the volumes of the ellipses that can be constructed from both distribution densities. It also compares the positions of the distributions relative to each other and the eigenstructures of the covariance matrices. The minimization of Kullback's information minimizes the sum of all three properties, so that the approximation with Kullback's information optimizes the position as well as the form of the distribution density. We are will use this in chapter 15.

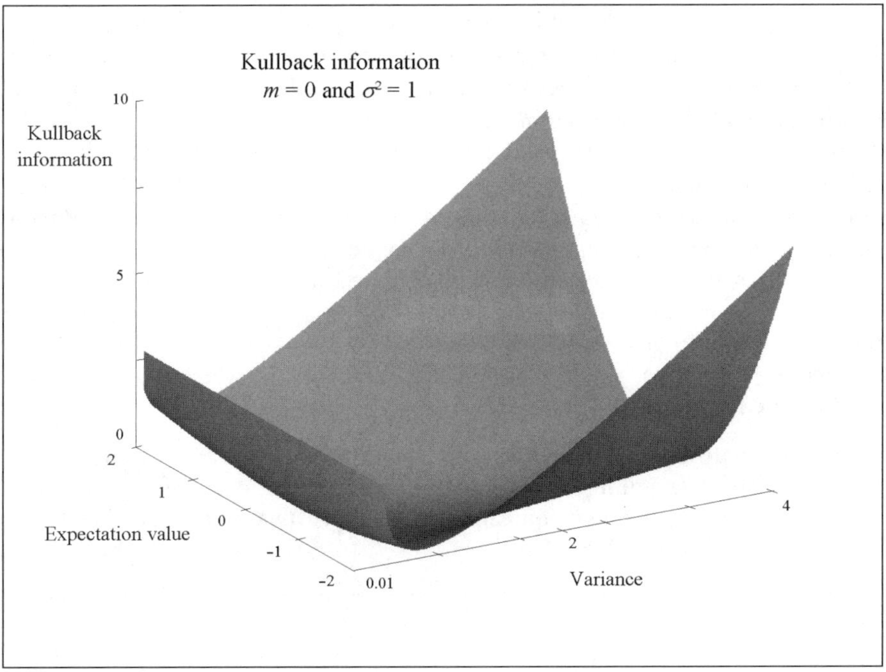

Fig. 9.1: Kullback's information

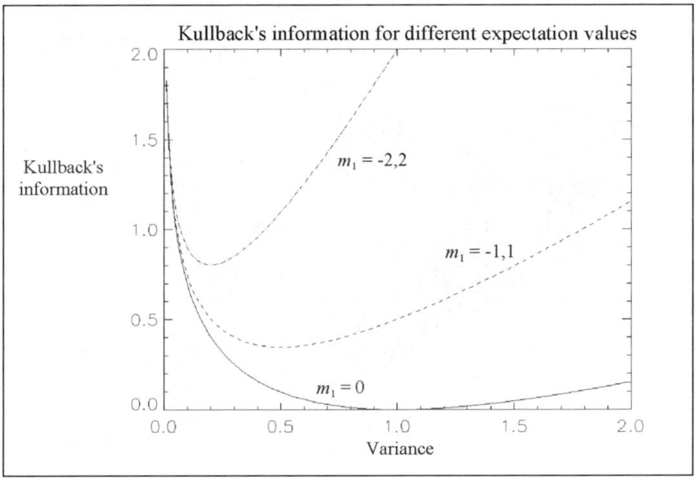

Fig. 9.2: Kullback's information

Kullback's discrimination information as a function of the variance with varying mean value m_1. ($m = 0$, $\sigma^2 = 1$ and σ_1 is the argument)

Figure 9.1 shows Kullback's information of two gaussian distribution densities. The first one has fixed mean $m = 0$ and variance $\sigma^2 = 1$, while both values vary in the second distribution. The mean value m_1 varies in the range from –2 to 2 and the variance σ_1^2 is altered from 0.01 to 4. We clearly notice the parabolic structure, caused by the variation of the means. The minimum of this parabola is achieved at the point $m_1 = 0$, where the second distribution achieves the mean of the reference distribution. The variation of the variance also shows a minimum, which depends on the varying mean m_1. We notice this in figure 9.2.

We see that the absolute minimum is achieved at the point where the mean values $m_1 = m = 0$ and the variances $\sigma^2 = \sigma_1^2 = 1$ are equal to those of the reference density. In this case the two distribution densities do not differ and thus Kullback's information, which measures the difference between two distribution densities, is equal to zero.

If we look only at small variations (–0.25 up to 0.25) in the mean value, we can identify the absolute minimum of the Kullback information, which is achieved when the moments of the two compared gaussian distribution densities are equal. This relation is shown in figure 9.3.

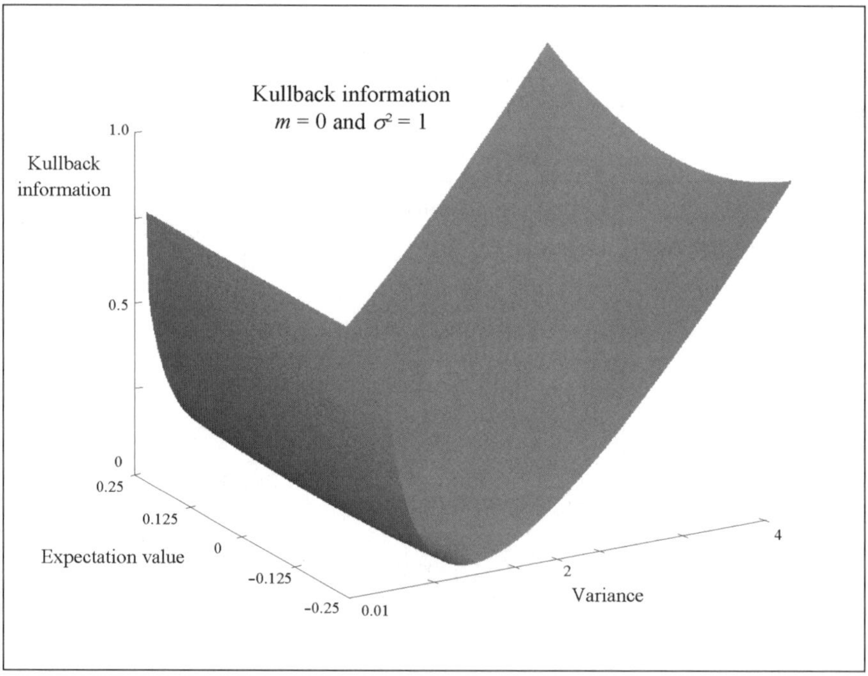

Fig. 9.3: Kullback's information with small variations of the mean

As we see from these figures (especially from figures 9.1 and 9.2), variations in the mean values also lead to variations in the variances, which we need to minimize the Kullback–Leibler distance. This reflects the connections that we already found in the equation, where the difference of the means is weighted with a reciprocal covariance matrix.

9.2 Kullback's divergence

In [ACZ75] [KUL87] the sum of two Kullback discrimination information functions is called divergence and it is defined according to

$$D(\underline{x},\underline{x}_S) = G(\underline{x},\underline{x}_S) + G(\underline{x}_S,\underline{x}) = \int_{\Xi} \ln\left(\frac{f_{\underline{x}}(\underline{\xi})}{f_{\underline{x}_S}(\underline{\xi})}\right)\cdot\left(f_{\underline{x}}(\underline{\xi}) - f_{\underline{x}_S}(\underline{\xi})\right)\cdot d\underline{\xi} \tag{9.35}$$

If we calculate this divergence for two gaussian distribution densities, we get – by applying the results from the previous chapter – the equation

$$D = D(\underline{x},\underline{x}_S) = -\frac{1}{2}\cdot\ln\left[\frac{|P_S|}{|P|}\right] + \frac{1}{2}\cdot\mathrm{tr}\left\{\left(P_S^{-1} - P^{-1}\right)\cdot P\right\} + \frac{1}{2}\cdot\left(\underline{\mu} - \underline{\mu}_S\right)^T\cdot P_S^{-1}\cdot\left(\underline{\mu} - \underline{\mu}_S\right)$$

$$-\frac{1}{2}\cdot\ln\left[\frac{|P|}{|P_S|}\right] + \frac{1}{2}\cdot\mathrm{tr}\left\{\left(P^{-1} - P_S^{-1}\right)\cdot P_S\right\} + \frac{1}{2}\cdot\left(\underline{\mu}_S - \underline{\mu}\right)^T\cdot P^{-1}\cdot\left(\underline{\mu}_S - \underline{\mu}\right) \tag{9.36}$$

$$D = \frac{1}{2}\cdot\mathrm{tr}\left\{P_S^{-1}\cdot P + P^{-1}\cdot P_S\right\} - \frac{1}{2}\cdot\mathrm{tr}\{I + I\}$$
$$+\frac{1}{2}\cdot\left(\underline{\mu} - \underline{\mu}_S\right)^T\cdot P_S^{-1}\cdot\left(\underline{\mu} - \underline{\mu}_S\right) + \frac{1}{2}\cdot\left(\underline{\mu}_S - \underline{\mu}\right)^T\cdot P^{-1}\cdot\left(\underline{\mu}_S - \underline{\mu}\right)$$

For $\dim[I] = n$ we can simplify this equation

$$D = \frac{1}{2}\cdot\mathrm{tr}\left\{P_S^{-1}\cdot P + P^{-1}\cdot P_S\right\} - n + \frac{1}{2}\cdot\left(\underline{\mu} - \underline{\mu}_S\right)^T\cdot\left(P_S^{-1} + P^{-1}\right)\cdot\left(\underline{\mu} - \underline{\mu}_S\right) \tag{9.37}$$

Now we have to anticipate Fisher's information matrix, corresponding to an inverse covariance matrix (section 10.1.5), because in the first addend we find the multiplication of a covariance matrix with an inverse covariance matrix. This is a multiplication of an information with a covariance matrix or the weighting of an information with this covariance matrix. The evaluation uses only the main diagonal elements of this weighted information matrix, which might be called the 'autoinformation elements' in relation to the autocorrelation.

A clear description of this result arises when we examine two independent random noise processes $\underline{v}$ and $\underline{v}_s$, providing diagonal covariance matrices P_s and P. In this special case the matrix equation can be decomposed into the sum of the divergences of the single elements

$$D_{ges} = \frac{1}{2} \cdot \mathrm{tr}\left\{P_s^{-1} \cdot P + P^{-1} \cdot P_s\right\} - n + \frac{1}{2} \cdot \left(\underline{\mu} - \underline{\mu}_s\right)^T \cdot \left(P_s^{-1} + P^{-1}\right) \cdot \left(\underline{\mu} - \underline{\mu}_s\right)$$

$$D_{ges} = D(x_1) + D(x_2) + \ldots + D(x_n) \tag{9.38}$$

$$D(x_i) = \frac{1}{2} \cdot \left(\frac{p_i}{p_{s_i}} + \frac{p_{s_i}}{p_i}\right) - 1 + \frac{1}{2} \cdot (\mu - \mu_s)^2 \cdot \left(\frac{1}{p_i} + \frac{1}{p_{s_i}}\right) \tag{9.39}$$

Contrary to Kullback's information, Kullback's divergence is a symmetrical function of the distribution densities, as it does not change if we exchange the densities.

9.2.1 Jensen's inequality for G_1

In contrast to Shannon's information, providing positive as well as negative values for the information, Kullback's information G_1 provide only nonnegative information values, similar to a distance measure. This can be verified with Jensens inequality

$$\ln x \le x - 1 \qquad \text{(equality holds for } x = 1) \tag{9.40}$$

by setting

$$\ln\left(\frac{f_{\underline{x}_s}(\underline{\xi})}{f_{\underline{x}}(\underline{\xi})}\right) \le \frac{f_{\underline{x}_s}(\underline{\xi})}{f_{\underline{x}}(\underline{\xi})} - 1 \tag{9.41}$$

and multiplying by $f_{\underline{x}}(\underline{\xi})$

$$f_{\underline{x}}(\underline{\xi}) \cdot \ln\left(\frac{f_{\underline{x}_s}(\underline{\xi})}{f_{\underline{x}}(\underline{\xi})}\right) \le f_{\underline{x}_s}(\underline{\xi}) - f_{\underline{x}}(\underline{\xi}) \tag{9.42}$$

We can do this without any restrictions, because even multi-dimensional distribution densities produce scalar values and thus the commutativity holds for multiplication. The following integration of both terms over the range $\underline{\Xi}$ (realization space of the random variable $\underline{x}$) provides the result

$$\int_{\Xi} f_{\underline{x}}(\underline{\xi}) \cdot \ln\left(\frac{f_{\underline{x}_s}(\underline{\xi})}{f_{\underline{x}}(\underline{\xi})}\right) \cdot d\underline{\xi} \leq \int_{\Xi} \left(f_{\underline{x}_s}(\underline{\xi}) - f_{\underline{x}}(\underline{\xi})\right) \cdot d\underline{\xi} \tag{9.43}$$

$$-\int_{\Xi} f_{\underline{x}}(\underline{\xi}) \cdot \ln\left(\frac{f_{\underline{x}}(\underline{\xi})}{f_{\underline{x}_s}(\underline{\xi})}\right) \cdot d\underline{\xi} \leq \int_{\Xi} f_{\underline{x}_s}(\underline{\xi}) \cdot d\underline{\xi} - \int_{\Xi} f_{\underline{x}}(\underline{\xi}) \cdot d\underline{\xi}$$

and then

$$-G_1 \leq 1 - 1 = 0$$

$$G_1 \geq 0 \tag{9.44}$$

This validates the statement that we made at the beginning of this chapter. The discrimination information leads only to nonnegative values. Therefore Kullback's information measures the absolute difference in the information between two distribution. The lack of symmetry in this measure leads to two different positive values for the discrimination information of two densities. On the one hand the result refers to the distribution density f_s and on the other hand to the distribution density f.

9.3 Kolmogorov's information

Andrej N. Kolmogorov presented another information in the year 1920 [GIC71] [KOL93], which is connected to the Kullback information. In order to circumvent another formal derivation, we present an intuitive approach and proceed with the description of the connections to other information measures. Therefore we use two multi-dimensional random vectors $\underline{x}$ and $\underline{y}$. The aim of our consideration is to find the information that is contained in the random variable $\underline{x}$ about the random variable $\underline{y}$. The best possible introduction is to look at the Venn diagram, presenting the amounts of information of $\underline{x}$ as I_x and of $\underline{y}$ as I_y.

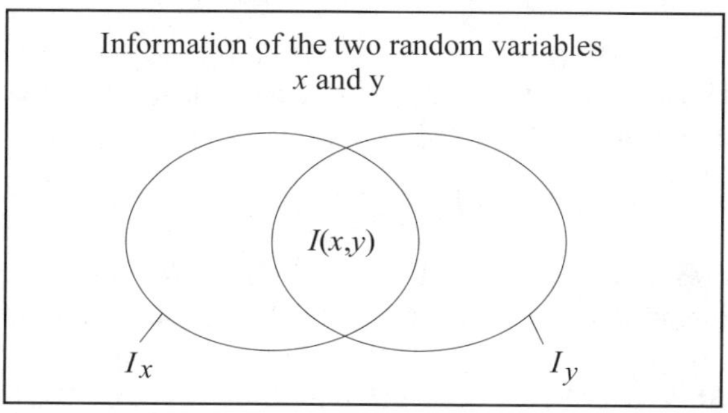

Fig. 9.4: Information diagram

Both random variables are independent of each other if the sets do not have an intersection, because in this case there is no mutual information and the whole information $I_{x,y}$ from both random variables is equal to the sum of the two single amounts of information I_x, I_y. If there is an intersection, then the two random variables are no longer independent and the whole information is smaller than the sum of the single amounts of information. From this consideration we get the inequality

$$I_{x,y} \leq I_x + I_y \tag{9.45}$$

The difference between the sum of the single amounts of information and the whole information, contained in both random variables, is – according to Kolmogorov – equal to the mutual information $I(\underline{x},\underline{y})$,

$$I(\underline{x},\underline{y}) = I_x + I_y - I_{x,y} \tag{9.46}$$

Kolmogorov describes the information that the random variable $\underline{x}$ contains about the random variable $\underline{y}$ by the equation

$$I(\underline{x},\underline{y}) = I(\underline{y},\underline{x}) = \int_{\underline{\Xi}}\int_{\underline{\mathrm{P}}} f_{\underline{x},\underline{y}}(\underline{\xi},\underline{\rho}) \cdot \ln\left[\frac{f_{x,y}(\xi,\rho)}{f_{\underline{x}}(\underline{\xi}) \cdot f_{\underline{y}}(\underline{\rho})}\right] \cdot d\underline{\rho} \cdot d\underline{\xi} \tag{9.47}$$

We obtain this equation when we use the following formulas

$$I_{x,y} = -\int_{\underline{\Xi}}\int_{\underline{\mathrm{P}}} f_{\underline{x},\underline{y}}(\underline{\xi},\underline{\rho}) \cdot \ln\left[f_{\underline{x},\underline{y}}(\underline{\xi},\underline{\rho})\right] \cdot d\underline{\rho} \cdot d\underline{\xi} \tag{9.48}$$

$$I_x = -\int_{\underline{\Xi}} f_{\underline{x}}(\underline{\xi}) \cdot \ln\left[f_{\underline{x}}(\underline{\xi})\right] \cdot d\underline{\xi} = -\int_{\underline{\Xi}}\int_{\underline{\mathrm{P}}} f_{\underline{x},\underline{y}}(\underline{\xi},\underline{\rho}) \cdot \ln\left[f_{\underline{x}}(\underline{\xi})\right] \cdot d\underline{\rho} \cdot d\underline{\xi} \tag{9.49}$$

and

$$I_y = -\int_{\underline{\mathrm{P}}} f_{\underline{y}}(\underline{\rho}) \cdot \ln\left[f_{\underline{y}}(\underline{\rho})\right] \cdot d\underline{\rho} = -\int_{\underline{\Xi}}\int_{\underline{\mathrm{P}}} f_{\underline{x},\underline{y}}(\underline{\xi},\underline{\rho}) \cdot \ln\left[f_{\underline{y}}(\underline{\rho})\right] \cdot d\underline{\rho} \cdot d\underline{\xi} \tag{9.50}$$

in equation (9.46). Thus Kolmogorov's information is created by three Shannon information functions (see equation (7.45))

$$I(\underline{x},\underline{y}) = I_x + I_y - I_{ges} = S(\underline{x}) + S(\underline{y}) - S(\underline{x},\underline{y}) \tag{9.51}$$

This definition is a special case of Kullback's information. It is, contrary to Kullback's information, symmetric and it describes the mutual information of the random variables, while Kullback's information generally describes the information difference between two different distribution densities of a random variable. With Kolmogorovs information we compare the joint distribution density with the marginal distribution densities and evaluate the mutual information contained in the joint distribution density. This mutual information cannot be expressed in the marginal densities. If the random variables are independent, the joint distribution density can be decomposed into the product of the marginal distributions

$$f_{\underline{x},\underline{y}}(\underline{\xi},\underline{\rho}) = f_{\underline{x}}(\underline{\xi}) \cdot f_{\underline{y}}(\underline{\rho}) \tag{9.52}$$

and the mutual information $I(\underline{x},\underline{y})$ vanishes. So two independent random variables do not possess any information about each other. In the Venn diagram we notice this by the lack of an intersection between the two information sets. Then the whole information is equal to the sum of the single amounts of information.

In gaussian distribution densities it is sufficient that the cross-covariances are equal to zero. Then the random variables are uncorrelated and therefore also independent. Thus the cross-covariances in the gaussian distributions are indicators for the mutual dependence of the random variables.

Applying the Bayesian equation

$$f_{\underline{x}|\underline{y}}(\underline{\xi} \mid \underline{\rho}) = \frac{f_{\underline{x},\underline{y}}(\underline{\xi},\underline{\rho})}{f_{\underline{y}}(\underline{\rho})} \tag{9.53}$$

to equation (9.47) we get

$$I(\underline{x},\underline{y}) = \int_{\underline{\Xi}}\int_{\underline{\mathrm{P}}} f_{\underline{x}|\underline{y}}(\underline{\xi} \mid \underline{\rho}) \cdot f_{\underline{y}}(\underline{\rho}) \cdot \ln\left[\frac{f_{\underline{x}|\underline{y}}(\underline{\xi} \mid \underline{\rho}) \cdot f_{\underline{y}}(\underline{\rho})}{f_{\underline{x}}(\underline{\xi}) \cdot f_{\underline{y}}(\underline{\rho})}\right] \cdot d\underline{\rho} \cdot d\underline{\xi} \tag{9.54}$$

$$I(\underline{x},\underline{y}) = \int_{\underline{\Xi}}\int_{\underline{\mathrm{P}}} f_{\underline{x}|\underline{y}}(\underline{\xi} \mid \underline{\rho}) \cdot f_{\underline{y}}(\underline{\rho}) \cdot \left[\ln f_{\underline{x}|\underline{y}}(\underline{\xi} \mid \underline{\rho}) - \ln f_{\underline{x}}(\underline{\xi})\right] \cdot d\underline{\rho} \cdot d\underline{\xi} \tag{9.55}$$

which leaves two integrals

$$\begin{aligned} I(\underline{x},\underline{y}) = & \int_{\underline{\Xi}}\int_{\underline{\mathrm{P}}} f_{\underline{x}|\underline{y}}(\underline{\xi} \mid \underline{\rho}) \cdot f_{\underline{y}}(\underline{\rho}) \cdot \ln f_{\underline{x}|\underline{y}}(\underline{\xi} \mid \underline{\rho}) \cdot d\underline{\rho} \cdot d\underline{\xi} \\ & - \int_{\underline{\Xi}}\int_{\underline{\mathrm{P}}} f_{\underline{x}|\underline{y}}(\underline{\xi} \mid \underline{\rho}) \cdot f_{\underline{y}}(\underline{\rho}) \cdot \ln f_{\underline{x}}(\underline{\xi}) \cdot d\underline{\rho} \cdot d\underline{\xi} \end{aligned} \tag{9.56}$$

If we rearrange these integrals

$$I(\underline{x},\underline{y}) = \int\limits_{\underline{\Xi}}\int\limits_{\underline{\mathrm{P}}} f_{\underline{x}|\underline{y}}(\underline{\xi} \mid \underline{\rho}) \cdot \ln f_{\underline{x}|\underline{y}}(\underline{\xi} \mid \underline{\rho}) \cdot d\underline{\xi} \cdot f_{\underline{y}}(\underline{\rho}) \cdot d\underline{\rho} - \int\limits_{\underline{\Xi}}\int\limits_{\underline{\mathrm{P}}} f_{\underline{x},\underline{y}}(\underline{\xi},\underline{\rho}) \cdot \ln f_{\underline{x}}(\underline{\xi}) \cdot d\underline{\xi} \cdot d\underline{\rho} \tag{9.57}$$

we are able to further simplify the terms

$$I(\underline{x},\underline{y}) = -\int\limits_{\underline{\mathrm{P}}} S_1(\underline{x} \mid \underline{\rho}) \cdot f_{\underline{y}}(\underline{\rho}) \cdot d\underline{\rho} - \int\limits_{\underline{\Xi}} f_{\underline{x}}(\underline{\xi}) \cdot \ln f_{\underline{x}}(\underline{\xi}) \cdot d\underline{\xi} \tag{9.58}$$

to get

$$I(\underline{x},\underline{y}) = -\int\limits_{\underline{\mathrm{P}}} S_1(\underline{x} \mid \underline{\rho}) \cdot f_{\underline{y}}(\underline{\rho}) \cdot d\underline{\rho} + S_1(\underline{x}) \tag{9.59}$$

or

$$I(\underline{x},\underline{y}) = -E_{\underline{y}}\{S_1(\underline{x} \mid \underline{y})\} + S_1[\underline{x}]$$

$$I(\underline{x},\underline{y}) = -S_1(\underline{x} \mid \underline{y}) + S_1[\underline{x}] \tag{9.60}$$

We now have found a connection between Kolmogorov's information $I(\underline{x},\underline{y})$ and Shannon's information S_1.

We proceed by calculating the information of gaussian random variables by using two gaussian marginal distribution densities

$$f_{\underline{x}}(\underline{\xi}) = \frac{1}{(2\pi)^{\frac{n}{2}} \cdot |P_{xx}|^{\frac{1}{2}}} \cdot \exp\left\{-\frac{1}{2} \cdot \left(\underline{\xi} - \underline{\mu}_{\underline{x}}\right)^T \cdot P_{xx}^{-1} \cdot \left(\underline{\xi} - \underline{\mu}_{\underline{x}}\right)\right\} \tag{9.61}$$

$$\dim[\underline{x}] = n$$

$$f_{\underline{y}}(\underline{\rho}) = \frac{1}{(2\pi)^{\frac{n}{2}} \cdot |P_{yy}|^{\frac{1}{2}}} \cdot \exp\left\{-\frac{1}{2} \cdot \left(\underline{\rho} - \underline{\mu}_{\underline{y}}\right)^T \cdot P_{yy}^{-1} \cdot \left(\underline{\rho} - \underline{\mu}_{\underline{y}}\right)\right\} \tag{9.62}$$

$$\dim[\underline{y}] = m$$

and a gaussian joint distribution

$$f_{\underline{z}}(\underline{\zeta}) = \frac{1}{(2\pi)^{\frac{m+n}{2}} \cdot |P_{zz}|^{\frac{1}{2}}} \cdot \exp\left\{-\frac{1}{2} \cdot \left(\underline{\zeta} - \underline{\mu}_{\underline{z}}\right)^T \cdot P_{zz}^{-1} \cdot \left(\underline{\zeta} - \underline{\mu}_{\underline{z}}\right)\right\} \tag{9.63}$$

$$\dim[\underline{z}] = m + n$$

with the partitioned random vector $\underline{z}^T = \left[\underline{x}^T \mid \underline{y}^T\right]$. We have

$$\underline{\mu}_{\underline{z}}^T = \left[\underline{\mu}_{\underline{x}}^T \mid \underline{\mu}_{\underline{y}}^T\right] \tag{9.64}$$

and

$$P_{zz} = \begin{pmatrix} P_{xx} & P_{xy} \\ P_{yx} & P_{yy} \end{pmatrix} \tag{9.65}$$

If we insert this into equation (9.47), we obtain the information

$$I(\underline{x}, \underline{y}) = \int_{\underline{\Xi}}\int_{\underline{P}} f_{\underline{z}}(\underline{\zeta}) \cdot \ln\left[\frac{f_{\underline{z}}(\underline{\zeta})}{f_{\underline{x}}(\underline{\xi}) \cdot f_{\underline{y}}(\underline{\rho})}\right] \cdot d\underline{\rho} \cdot d\underline{\xi} \tag{9.66}$$

$$I(\underline{x}, \underline{y}) = \int_{\underline{\Xi}}\int_{\underline{P}} f_{\underline{z}}(\underline{\zeta}) \cdot \left[\ln f_{\underline{z}}(\underline{\zeta}) - \ln f_{\underline{x}}(\underline{\xi}) - \ln f_{\underline{y}}(\underline{\rho})\right] \cdot d\underline{\rho} \cdot d\underline{\xi} \tag{9.67}$$

and decompose this into three integrals.

<u>Integral 1</u> (with $d\underline{\xi} \cdot d\underline{\rho} = d\underline{\zeta}$) leads to

$$I_1(\underline{x}, \underline{y}) = \int_{\underline{\Xi}}\int_{\underline{P}} f_{\underline{z}}(\underline{\zeta}) \cdot \ln f_{\underline{z}}(\underline{\zeta}) \cdot d\underline{\zeta} \tag{9.68}$$

which is Shannon's information of the random vector $\underline{z}^T = \left[\underline{x}^T \mid \underline{y}^T\right]$, already calculated as

$$I_1(\underline{x}, \underline{y}) = -S(\underline{x}, \underline{y}) = -S(\underline{z}) = -\frac{1}{2} \cdot \left[(m+n) \cdot \ln(2\pi) + \ln(|P_{zz}|) + (m+n)\right] \tag{9.69}$$

in section 8.2.

<u>Integral 2</u>

$$I_2(\underline{x}, \underline{y}) = -\int_{\underline{\Xi}}\int_{\underline{P}} f_{\underline{z}}(\underline{\zeta}) \cdot \ln f_{\underline{x}}(\underline{\xi}) \cdot d\underline{\xi} \cdot d\underline{\rho} \tag{9.70}$$

$$I_2(\underline{x}, \underline{y}) = -\int_{\underline{\Xi}}\int_{\underline{P}} f_{\underline{x},\underline{y}}(\underline{\xi}, \underline{\rho}) \cdot d\underline{\rho} \cdot \ln f_{\underline{x}}(\underline{\xi}) \cdot d\underline{\xi}$$

$$I_2(\underline{x},\underline{y}) = -\int_{\underline{\Xi}} \ln f_{\underline{x}}(\underline{\xi}) \cdot f_{\underline{x}}(\underline{\xi}) \cdot d\underline{\xi} \tag{9.71}$$

is also a Shannon information

$$I_2(\underline{x},\underline{y}) = S(\underline{x}) = \frac{1}{2} \cdot \{n \cdot \ln(2\pi) + \ln(|P_{xx}|) + n\} \tag{9.72}$$

and *Integral 3*

$$I_3(\underline{x},\underline{y}) = -\int_{\underline{\Xi}}\int_{\underline{P}} f_{\underline{z}}(\underline{\zeta}) \cdot \ln f_{\underline{y}}(\underline{\rho}) \cdot d\underline{\xi} \cdot d\underline{\rho} \tag{9.73}$$

$$I_3(\underline{x},\underline{y}) = -\int_{\underline{\Xi}}\int_{\underline{P}} f_{\underline{x},\underline{y}}(\underline{\xi},\underline{\rho}) \cdot d\underline{\xi} \cdot \ln f_{\underline{y}}(\underline{\rho}) \cdot d\underline{\rho}$$

$$I_3(\underline{x},\underline{y}) = -\int_{\underline{P}} \ln f_{\underline{y}}(\underline{\rho}) \cdot f_{\underline{y}}(\underline{\rho}) \cdot d\underline{\rho} \tag{9.74}$$

$$I_3(\underline{x},\underline{y}) = S(\underline{y}) = \frac{1}{2} \cdot \{m \cdot \ln(2\pi) + \ln(|P_{yy}|) + m\} \tag{9.75}$$

together with the equations (9.67), (9.69), (9.72) and (9.75) provides the result

$$I(\underline{x},\underline{y}) = S(\underline{x}) + S(\underline{y}) - S(\underline{x},\underline{y}) \tag{9.76}$$

which is the familiar connection between Shannon's and Kolmogorov's information.

Explicitly we get the formulation

$$\begin{aligned} I(\underline{x},\underline{y}) = {} & \frac{1}{2} \cdot \{m \cdot \ln(2\pi) + \ln(|P_{yy}|) + m\} + \frac{1}{2} \cdot \{n \cdot \ln(2\pi) + \ln(|P_{xx}|) + n\} \\ & - \frac{1}{2} \cdot \{(m+n) \cdot \ln(2\pi) + \ln(|P_{zz}|) + (m+n)\} \end{aligned} \quad 9.77)$$

$$I(\underline{x},\underline{y}) = \frac{1}{2} \cdot [\ln(|P_{yy}|) + \ln(|P_{xx}|) - \ln(|P_{zz}|)] \tag{9.78}$$

Kolmogorov's information (interim result)

$$I(\underline{x},\underline{y}) = \frac{1}{2} \cdot \ln\left(\frac{|P_{yy}| \cdot |P_{xx}|}{|P_{zz}|}\right) \tag{9.79}$$

It is now possible calculate the determinant of the partitioned matrix

$$P_{zz} = \begin{pmatrix} P_{xx} & P_{xy} \\ P_{yx} & P_{yy} \end{pmatrix} \tag{9.80}$$

via the LDU decomposition [LOF90], leading to

$$|P_{zz}| = |P_{xx}| \cdot \left| P_{yy} - P_{yx} \cdot P_{xx}^{-1} \cdot P_{xy} \right| \tag{9.81}$$

or, because of the symmetry, to

$$|P_{zz}| = |P_{yy}| \cdot \left| P_{xx} - P_{xy} \cdot P_{yy}^{-1} \cdot P_{yx} \right| \tag{9.82}$$

This changes Kolmogorov`+s information into

$$\begin{aligned} I(\underline{x}, \underline{y}) &= \frac{1}{2} \cdot \ln\left(\frac{|P_{yy}| \cdot |P_{xx}|}{|P_{xx}| \cdot \left| P_{yy} - P_{yx} \cdot P_{xx}^{-1} \cdot P_{xy} \right|} \right) \\ &= \frac{1}{2} \cdot \ln\left(\frac{|P_{yy}|}{\left| P_{yy} - P_{yx} \cdot P_{xx}^{-1} \cdot P_{xy} \right|} \right) \end{aligned} \tag{9.83}$$

or

$$I(\underline{x}, \underline{y}) = \frac{1}{2} \cdot \ln\left(\frac{|P_{xx}|}{\left| P_{xx} - P_{xy} \cdot P_{yy}^{-1} \cdot P_{yx} \right|} \right) \tag{9.84}$$

The term $P_{xx} - P_{xy} \cdot P_{yy}^{-1} \cdot P_{yx}$ occurs in the maximum a posteriori (MAP) estimation when we determine the error covariance matrix, calculated with $P_{xx} = P_{xx}^{-}$ from the equation

$$P_{xx}^{+} = P_{xx}^{-} - P_{xx}^{-} \cdot C^T \cdot \left[C \cdot \left(P_{xx}^{-}\right)^{-1} \cdot C^T + R \right]^{-1} \cdot C \cdot P_{xx}^{-} \tag{9.85}$$

$$P_{xx}^{+} = P_{xx}^{-} - P_{xy} \cdot \left(P_{yy}^{-}\right)^{-1} \cdot P_{yx} \tag{9.86}$$

We obtain this equation, when we calculate the covariances of the MAP estimation, which starts with the equation

$$\frac{\partial}{\partial \underline{\xi}} \ln f_{\underline{x}|\underline{y}}\left(\underline{\xi} \mid \underline{\rho}\right) \Bigg|_{\underline{\xi} = \underline{\hat{x}}_{\text{MAP}}} \overset{!}{=} 0 \tag{9.87}$$

and application of the Bayesian rule provides

$$\left.\frac{\partial}{\partial \underline{\xi}} \ln \frac{f_{\underline{y}|\underline{x}}(\underline{\rho} \mid \underline{\xi}) \cdot f_{\underline{x}}(\underline{\xi})}{f_{\underline{y}}(\underline{\rho})}\right|_{\underline{\xi}=\hat{\underline{x}}_{\mathrm{MAP}}} \overset{!}{=} 0 \tag{9.88}$$

or

$$\left.\frac{\partial}{\partial \underline{\xi}}\left[\ln f_{\underline{y}|\underline{x}}(\underline{\rho} \mid \underline{\xi}) + \ln f_{\underline{x}}(\underline{\xi})\right]\right|_{\underline{\xi}=\hat{\underline{x}}_{\mathrm{MAP}}} \overset{!}{=} 0 \tag{9.89}$$

We thus calculate an optimal a posteriori estimate from the measurement information, contained in the conditional distribution density, and the a priori information, contained in the marginal distribution of the random variable $\underline{x}$ with mean value $\hat{\underline{x}}^-$. The covariance of this estimate can be calculated with equation (9.86).

Now P_{xx}^- is the covariance matrix of the distribution density function $f_{\underline{x}}(\underline{\xi})$, i.e. it is describes the uncertainty of the a priori knowledge in the MAP estimation. P_{xx}^+ on the other side describes the uncertainty contained in the MAP estimate, consisting of the a priori information as well as the information that is contained in the observed random variable $\underline{y}$. Thus Kolmogorov's information is described by the logarithm of the ratio of the uncertainties of the random variable $\underline{x}$ before and after the observation of the measurement value $\underline{y}$. So this measure of information enables us to describe the information that is contained in the measurement variable $\underline{y}$ about the variable $\underline{x}$.

As Komogorov's information is a special case of Kullback's information, we can also use Jensen's inequality to demonstrate the nonnegativity of this information.

9.4 Transformation of the coordinate system and the effects on the information

In chapter 2 we presented a derivation of a general information function based on an intuitive approach. In the subsequent calculation of the information in different scales in section 2.2, we found that a change in the scale also affects the information that we are able to obtain about some interesting object. Now we derived several different information functions and examined their connections. Consequently we ask how these information measure behave, if we change the scale or the coordinate system.

As we found in section 2.2, the information of a certain event changes when we change the number of possible events. Thus the reduction of the uncertainty caused by the choice of one special event is also changed. Such a change in the number of the available events can be achieved either by changing the scale and maintaining the functions or by dilation of the function while maintaining the scale. In both cases the information of the measurement is changed (see also chapter 2). Thus our idea is to change the scale and to examine the changes in the information functions, resulting from this procedure.

If we change the coordinate system by the transformation

$$\underline{y} = \underline{g}(\underline{x}) = T \cdot \underline{x} \tag{9.90}$$

with $T = \dfrac{\partial \underline{y}}{\partial \underline{x}}$, how does the information behave ?

9.4.1 S_α-information

$$S_\alpha(\underline{x}) = \frac{1}{1-\alpha} \cdot \ln\left\{ \int_{\Xi} f_{\underline{x}}^{\alpha}(\underline{\xi}) \cdot d\underline{\xi} \right\} \tag{9.91}$$

with (equally probable events for different probability distributions)

$$f_{\underline{y}}(\underline{\rho}) \cdot d\underline{\rho} = f_{\underline{x}}(\underline{\xi}) \cdot d\underline{\xi} \tag{9.92}$$

we find by using the general transformation from equation (9.90)

$$f_{\underline{x}}(\underline{\xi}) = f_{\underline{y}}(\underline{g}(\underline{\xi})) \cdot \left\| \frac{\partial \underline{g}(\underline{\xi})}{\partial \underline{\xi}} \right\| \tag{9.93}$$

$\left\| \dfrac{\partial \underline{g}(\underline{\xi})}{\partial \underline{\xi}} \right\|$ = absolute value of the Jacobi determinant

(The computation of the absolute value is made because the areas described by $\partial \underline{\xi}$ do not have a direction. Additionally the distribution densities do not have negative values.)

Analogously we obtain

$$\left[f_{\underline{x}}(\underline{\xi})\right]^{\alpha} = \left[f_{\underline{y}}(\underline{g}(\underline{\xi})) \cdot \left\| \frac{\partial \underline{g}(\underline{\xi})}{\partial \underline{\xi}} \right\| \right]^{\alpha} \tag{9.94}$$

To be able to use this equivalence, we change our denotation of S_α in

$$S_\alpha(\underline{x}) = \frac{1}{1-\alpha} \cdot \ln\left\{ \int_{\underline{\Xi}} f_{\underline{x}}^{\alpha-1}(\underline{\xi}) \cdot f_{\underline{x}}(\underline{\xi}) \cdot d\underline{\xi} \right\} \tag{9.95}$$

and thus find

$$S_\alpha(\underline{x}) = \frac{1}{1-\alpha} \cdot \ln\left\{ \int_{\underline{P}} \left[f_{\underline{y}}(\underline{g}(\underline{\xi})) \cdot \left\| \frac{\partial \underline{g}(\underline{\xi})}{\partial \underline{\xi}} \right\| \right]^{\alpha-1} \cdot f_{\underline{y}}(\underline{\rho}) \cdot d\underline{\rho} \right\} \tag{9.96}$$

[The exact mathematical way would be to replace $f_{\underline{x}}(\underline{\xi})$ by $f_{\underline{y}}(\underline{\rho}) \cdot \left\| \frac{\partial \underline{g}(\underline{\xi})}{\partial \underline{\xi}} \right\|$ and $d\underline{\xi}$ by $\left\| \frac{\partial \underline{\xi}}{\partial \underline{\rho}} \right\| \cdot d\underline{\rho}$, which immediately reveals another demand of this transformation. It is the unique invertibility, which ensures that the product of the Jacobi determinants $\left\| \frac{\partial \underline{\xi}}{\partial \underline{\rho}} \right\|$ and $\left\| \frac{\partial \underline{\rho}}{\partial \underline{\xi}} \right\| = \left\| \frac{\partial \underline{g}(\underline{\xi})}{\partial \underline{\xi}} \right\|$ is equal to 1. Furthermore, even the range of the integration changes from $\underline{\Xi}$ to $\underline{P}$ with this substitution.]

We proceed in our transformation and obtain the equation

$$S_\alpha(\underline{x}) = \frac{1}{1-\alpha} \cdot \ln\left\{ \int_{\underline{P}} \left\| \frac{\partial \underline{g}(\underline{\xi})}{\partial \underline{\xi}} \right\|^{\alpha-1} \cdot \left[f_{\underline{y}}(\underline{\rho}) \right]^{\alpha} \cdot d\underline{\rho} \right\} \tag{9.97}$$

If the transformation is uniquely invertible, we can use $\left\| \frac{\partial \underline{g}(\underline{\xi})}{\partial \underline{\xi}} \right\|^{\alpha-1}$ instead of $\left\| \frac{\partial \underline{g}^{-1}(\underline{\rho})}{\partial \underline{\rho}} \right\|^{1-\alpha}$. This leads to the result

$$S_\alpha(\underline{x}) = \frac{1}{1-\alpha} \cdot \ln\left\{ \int_{\underline{P}} \left\| \frac{\partial \underline{g}^{-1}(\underline{\rho})}{\partial \underline{\rho}} \right\|^{1-\alpha} \cdot \left[f_{\underline{y}}(\underline{\rho}) \right]^{\alpha} \cdot d\underline{\rho} \right\} \tag{9.98}$$

or, in case of the linear mapping of equation (9.90)

$$S_\alpha(\underline{x}) = \frac{1}{1-\alpha} \cdot \ln\left\{ \int_{\underline{\mathrm{P}}} \left\| T^{-1} \right\|^{1-\alpha} \cdot \left[f_{\underline{y}}(\underline{\rho}) \right]^\alpha \cdot d\underline{\rho} \right\} \tag{9.99}$$

$$S_\alpha(\underline{x}) = \frac{1}{1-\alpha} \cdot \ln\left\{ \int_{\underline{\mathrm{P}}} \cdot \left[f_{\underline{y}}(\underline{\rho}) \right]^\alpha \cdot d\underline{\rho} \right\} + \frac{1}{1-\alpha} \cdot \ln\left\| T^{-1} \right\|^{1-\alpha} \tag{9.100}$$

and finally

$$S_\alpha(\underline{x}) = S_\alpha(\underline{y}) + \ln\left\| T^{-1} \right\| = S_\alpha(\underline{y}) - \ln\left\| T \right\| \tag{9.101}$$

Rényi's α-information thus depends on the choice of the coordinate system and it changes in the case of a transformation of the coordinate system according to the equation we derived from this calculation.

9.4.2 *G*-divergence

$$G_\alpha(\underline{x}) = \frac{1}{\alpha-1} \cdot \ln\left\{ \int_{\underline{\Xi}} \left(\frac{f_{\underline{x}}(\underline{\xi})}{f_{\underline{x}_s}(\underline{\xi})} \right)^{\alpha-1} \cdot f_{\underline{x}}(\underline{\xi}) \cdot d\underline{\xi} \right\} \tag{9.102}$$

with

$$f_{\underline{y}}(\underline{\rho}) \cdot d\underline{\rho} = f_{\underline{x}}(\underline{\xi}) \cdot d\underline{\xi} \tag{9.103}$$

we get

$$f_{\underline{x}}(\underline{\xi}) = f_{\underline{y}}(\underline{g}(\underline{\xi})) \cdot \left\| \frac{\partial \underline{g}(\underline{\xi})}{\partial \underline{\xi}} \right\| \tag{9.104}$$

and

$$\left[f_{\underline{x}}(\underline{\xi}) \right]^{\alpha-1} = \left[f_{\underline{y}}(\underline{g}(\underline{\xi})) \cdot \left\| \frac{\partial \underline{g}(\underline{\xi})}{\partial \underline{\xi}} \right\| \right]^{\alpha-1} \tag{9.105}$$

as well as

$$\left[f_{\underline{x}_s}(\underline{\xi}) \right]^{\alpha-1} = \left[f_{\underline{y}_s}(\underline{g}(\underline{\xi})) \cdot \left\| \frac{\partial \underline{g}(\underline{\xi})}{\partial \underline{\xi}} \right\| \right]^{\alpha-1} \tag{9.106}$$

Inserting this into equation (9.102), we get

$$G_\alpha(\underline{x}) = \frac{1}{\alpha-1}\cdot\ln\left\{\int_{\underline{\mathrm{P}}}\left(\frac{f_{\underline{y}}(\underline{g}(\underline{\xi}))\cdot\left\|\frac{\partial\,\underline{g}(\underline{\xi})}{\partial\underline{\xi}}\right\|}{f_{\underline{y}_s}(\underline{g}(\underline{\xi}))\cdot\left\|\frac{\partial\,\underline{g}(\underline{\xi})}{\partial\underline{\xi}}\right\|}\right)^{\alpha-1}\cdot f_{\underline{y}}(\underline{\rho})\cdot d\underline{\rho}\right\} \tag{9.107}$$

Both Jacobi determinants in the quotient are equal so that we can cancel both. With $\underline{\rho} = \underline{g}(\underline{\xi})$ we obtain:

$$G_\alpha(\underline{x}) = \frac{1}{\alpha-1}\cdot\ln\left\{\int_{\underline{\mathrm{P}}}\left(\frac{f_{\underline{y}}(\underline{\rho})}{f_{\underline{y}_s}(\underline{\rho})}\right)^{\alpha-1}\cdot f_{\underline{y}}(\underline{\rho})\cdot d\underline{\rho}\right\} \tag{9.108}$$

and this is the final result

$$G_\alpha(\underline{x}) = G_\alpha(\underline{y}) \tag{9.109}$$

denoting that the *G*-divergence does not vary with a change of the coordinate system. It is therefore transformation invariant.

9.4.3 *S*-information

$$S_1(\underline{x}) = -\int_{\underline{\Xi}} \ln\left[f_{\underline{x}}(\underline{\xi})\right]\cdot f_{\underline{x}}(\underline{\xi})\cdot d\underline{\xi} \tag{9.110}$$

with

$$f_{\underline{y}}(\underline{\rho})\cdot d\underline{\rho} = f_{\underline{x}}(\underline{\xi})\cdot d\underline{\xi} \tag{9.111}$$

we build the entropy in the vector space *Y* by additionally replacing $f_{\underline{x}}(\underline{\xi})$ with

$$f_{\underline{x}}(\underline{\xi}) = f_{\underline{y}}(\underline{g}(\underline{\xi}))\cdot\left\|\frac{\partial\,\underline{g}(\underline{\xi})}{\partial\underline{\xi}}\right\| \tag{9.112}$$

This leads to

$$S_1(\underline{x}) = -\int_{\underline{\mathrm{P}}} \ln\left[f_{\underline{y}}(\underline{g}(\underline{\xi}))\cdot\left\|\frac{\partial\,\underline{g}(\underline{\xi})}{\partial\underline{\xi}}\right\|\right]\cdot f_{\underline{y}}(\underline{\rho})\cdot d\underline{\rho} \tag{9.113}$$

$$S_1(\underline{x}) = -\int_{\underline{P}} \left\{ \ln\left[f_{\underline{y}}\left(\underline{g}(\underline{\xi})\right)\right] + \ln\left[\left\| \frac{\partial \underline{g}(\underline{\xi})}{\partial \underline{\xi}} \right\|\right] \right\} \cdot f_{\underline{y}}(\underline{\rho}) \cdot d\underline{\rho}$$

and we find the result

$$S_1(\underline{x}) = -\int_{\underline{P}} \ln\left[f_{\underline{y}}(\underline{\rho})\right] \cdot f_{\underline{y}}(\underline{\rho}) \cdot d\underline{\rho} - \int_{\underline{\Xi}} \ln\left[\left\| \frac{\partial \underline{g}(\underline{\xi})}{\partial \underline{\xi}} \right\|\right] \cdot f_{\underline{x}}(\underline{\xi}) \cdot d\underline{\xi}$$

$$S_1(\underline{x}) = S_1(\underline{y}) - \int_{\underline{\Xi}} \ln\left[\left\| \frac{\partial \underline{g}(\underline{\xi})}{\partial \underline{\xi}} \right\|\right] \cdot f_{\underline{x}}(\underline{\xi}) \cdot d\underline{\xi} \tag{9.114}$$

and for the linear mapping $\underline{y} = T \cdot \underline{x}$ from equation (9.90):

$$S_1(\underline{x}) = S_1(\underline{y}) - \ln\|T\| \tag{9.115}$$

Thus Shannon's informationentropy varies in a transformation. The same result can be obtained when we calculate the limit $\alpha \to 1$ in equation (9.100) or in equation (9.101). We can use this property to demonstrate the advantages of Data fusion over Data integration [ARN97/1].

9.4.3.1 Example

Let us examine a simple extension

$$y = 3 \cdot x \tag{9.116}$$

and we calculate the information according to

$$S_1(x) = S_1(y) - \ln\|T\| \tag{9.117}$$

for a gaussian distribution. We then find

$$\frac{1}{2}\ln(2\pi \cdot e) + \ln\sigma_x = \frac{1}{2}\ln(2\pi \cdot e) + \ln\sigma_y - \ln 3 \tag{9.118}$$

$$\ln\sigma_x = \ln\left(\frac{\sigma_y}{3}\right)$$

$$\sigma_x = \frac{\sigma_y}{3} \tag{9.119}$$

which confirms the statement that the uncertainty is increased when we stretch a system and maintain the scale of the measurement (the units of the scale are not stretched with the system).

This result verifies the improvement that we got in section 2.2, where we refined the scale for the measurement. There we also obtained an improvement by y factor of ln 3 in the Shannon information by refining the scale by a factor of 3. This confirms the considerations that we made about the information functions in the beginning.

9.4.4 Discrimination information

$$G_1(\underline{x}) = \int_{\underline{\Xi}} \ln\left[\frac{f_{\underline{x}}(\underline{\xi})}{f_{\underline{x}_s}(\underline{\xi})}\right] \cdot f_{\underline{x}}(\underline{\xi}) \cdot d\underline{\xi} \tag{9.120}$$

with

$$f_{\underline{y}}(\underline{\rho}) \cdot d\underline{\rho} = f_{\underline{x}}(\underline{\xi}) \cdot d\underline{\xi} \tag{9.121}$$

and

$$f_{\underline{y}_s}(\underline{\rho}) \cdot d\underline{\rho} = f_{\underline{x}_s}(\underline{\xi}) \cdot d\underline{\xi} \tag{9.122}$$

we find

$$G_1(\underline{x}) = \int_{\underline{\mathrm{P}}} \ln\left[\frac{f_{\underline{y}}(\underline{g}(\underline{\xi})) \cdot \left\| \frac{\partial \underline{g}(\underline{\xi})}{\partial \underline{\xi}} \right\|}{f_{\underline{y}_s}(\underline{g}(\underline{\xi})) \cdot \left\| \frac{\partial \underline{g}(\underline{\xi})}{\partial \underline{\xi}} \right\|}\right] \cdot f_{\underline{y}}(\underline{\rho}) \cdot d\underline{\rho} \tag{9.123}$$

and again the two Jacobi determinants can be cancelled in the quotient. Together with the backward substitution $\underline{\rho} = \underline{g}(\underline{\xi})$ we find

$$G_1(\underline{x}) = \int_{\underline{\mathrm{P}}} \ln\left[\frac{f_{\underline{y}}(\underline{\rho})}{f_{\underline{y}_s}(\underline{\rho})}\right] \cdot f_{\underline{y}}(\underline{\rho}) \cdot d\underline{\rho} \tag{9.124}$$

and thus

$$G_1(\underline{x}) = G_1(\underline{y}) \tag{9.125}$$

the discrimination information does not depend on the choice of the coordinate system. It is invariant with respect to coordinate transformation. Kullback's information is a measure of the distance (apart from the symmetry) between two distribution densities and so we might have expected this result, as distances are invariant with respect to transformations of the coordinate system.

9.4.5 Kolmogorov's information

$$I(\underline{x},\underline{y})=\int\limits_{\underline{\Xi}}\int\limits_{\underline{\mathrm{P}}}\ln\left[\frac{f_{\underline{x},\underline{y}}(\underline{\xi},\underline{\rho})}{f_{\underline{x}}(\underline{\xi})\cdot f_{\underline{y}}(\underline{\rho})}\right]\cdot f_{\underline{x},\underline{y}}(\underline{\xi},\underline{\rho})\cdot d\underline{\xi}\cdot d\underline{\rho} \tag{9.126}$$

This measure provides different possibilities for coordinate transformations, because of the two coordinate systems *X* and *Y* in the given equation. For the new coordinate systems *X'*, *Y'* we get

$$f_{\underline{y}'}(\underline{\rho}')\cdot d\underline{\rho}'=f_{\underline{y}}(\underline{\rho})\cdot d\underline{\rho} \qquad \text{with} \qquad \underline{y}'=\underline{g}_1(\underline{y}) \tag{9.127a}$$

$$f_{\underline{x}'}(\underline{\xi}')\cdot d\underline{\xi}'=f_{\underline{x}}(\underline{\xi})\cdot d\underline{\xi} \qquad \text{with} \qquad \underline{x}'=\underline{g}_2(\underline{x}) \tag{9.127b}$$

and

$$f_{\underline{x}',\underline{y}'}(\underline{\xi}',\underline{\rho}')\cdot d\underline{\xi}'\cdot d\underline{\rho}'=f_{\underline{x},\underline{y}}(\underline{\xi},\underline{\rho})\cdot d\underline{\xi}\cdot d\underline{\rho} \tag{9.127c}$$

with

$$\underline{z}'=\underline{g}_3(\underline{z}) \qquad \text{where} \qquad \underline{z}^T=\left(\underline{x}^T,\underline{y}^T\right)^T$$

resp.

$$f_{\underline{y}}(\underline{\rho})=f_{\underline{y}'}(\underline{g}_1(\underline{\rho}))\cdot\left\|\frac{\partial\underline{g}_1(\underline{\rho})}{\partial\underline{\rho}}\right\| \quad \text{and} \quad f_{\underline{x}}(\underline{\xi})=f_{\underline{x}'}(\underline{g}_2(\underline{\xi}))\cdot\left\|\frac{\partial\underline{g}_2(\underline{\xi})}{\partial\underline{\xi}}\right\| \tag{9.128a}$$

and

$$f_{\underline{x},\underline{y}}(\underline{\xi},\underline{\rho})=f_{\underline{x}',\underline{y}'}(\underline{g}_3(\underline{\zeta}))\cdot\left\|\frac{\partial\underline{g}_3(\underline{\zeta})}{\partial\underline{\zeta}}\right\| \quad \text{with} \quad \underline{\zeta}^T=\left(\underline{\xi}^T,\underline{\rho}^T\right) \tag{9.128b}$$

We substitute this into equation (9.126) and obtain

$$I(\underline{x},\underline{y})=\int\limits_{\underline{\Xi}}\int\limits_{\underline{\mathrm{P}}}\ln\left[\frac{f_{\underline{x}',\underline{y}'}(\underline{g}_3(\underline{\zeta}))\cdot\left\|\frac{\partial\underline{g}_3(\underline{\zeta})}{\partial\underline{\zeta}}\right\|}{f_{\underline{x}'}(\underline{g}_2(\underline{\xi}))\cdot\left\|\frac{\partial\underline{g}_2(\underline{\xi})}{\partial\underline{\xi}}\right\|\cdot f_{\underline{y}'}(\underline{g}_1(\underline{\rho}))\cdot\left\|\frac{\partial\underline{g}_1(\underline{\rho})}{\partial\underline{\rho}}\right\|}\right]\cdot f_{\underline{x}',\underline{y}'}(\underline{\xi}',\underline{\rho}')\cdot d\underline{\xi}'\cdot d\underline{\rho}' \tag{9.129}$$

Further rearrangement provides

$$I(\underline{x},\underline{y}) = \int_{\underline{\Xi}}\int_{\underline{P}} \ln\left[\frac{f_{\underline{x}',\underline{y}'}(\underline{\xi}',\underline{\rho}')}{f_{\underline{x}'}(\underline{\xi}')\cdot f_{\underline{y}'}(\underline{\rho}')}\right]\cdot f_{\underline{x}',\underline{y}'}(\underline{\xi}',\underline{\rho}')\cdot d\underline{\xi}'\cdot d\underline{\rho}'$$

$$+\int_{\underline{\Xi}}\int_{\underline{P}}\left[\ln\left\|\frac{\partial \underline{g}_3(\underline{\zeta})}{\partial \underline{\zeta}}\right\| - \ln\left\|\frac{\partial \underline{g}_2(\underline{\xi})}{\partial \underline{\xi}}\right\| - \ln\left\|\frac{\partial \underline{g}_1(\underline{\rho})}{\partial \underline{\rho}}\right\|\right]\cdot f_{\underline{x}',\underline{y}'}(\underline{\xi}',\underline{\rho}')\cdot d\underline{\xi}'\cdot d\underline{\rho}'$$

and finally

$$I(\underline{x},\underline{y}) = I(\underline{x}',\underline{y}')$$

$$+\int_{\underline{\Xi}}\int_{\underline{P}}\left[\ln\left\|\frac{\partial \underline{g}_3(\underline{\zeta})}{\partial \underline{\zeta}}\right\| - \ln\left\|\frac{\partial \underline{g}_2(\underline{\xi})}{\partial \underline{\xi}}\right\| - \ln\left\|\frac{\partial \underline{g}_1(\underline{\rho})}{\partial \underline{\rho}}\right\|\right]\cdot f_{\underline{x}',\underline{y}'}(\underline{\xi}',\underline{\rho}')\cdot d\underline{\xi}'\cdot d\underline{\rho}' \tag{9.130}$$

Thus Kolmogorov's information does not depend on the coordinate system when the kernel of the remaining integral vanishes. This means that the two compared coordinate systems have to be connected according to the equation

$$\ln\left\|\frac{\partial \underline{g}_3(\underline{\zeta})}{\partial \underline{\zeta}}\right\| = \ln\left\|\frac{\partial \underline{g}_2(\underline{\xi})}{\partial \underline{\xi}}\right\| + \ln\left\|\frac{\partial \underline{g}_1(\underline{\rho})}{\partial \underline{\rho}}\right\| \tag{9.131}$$

This condition is met when the mapping $g_3(\bullet)$ consists of two independent mappings $g_1(\bullet)$ and $g_2(\bullet)$ such that

$$\ln\left\|\frac{\partial \underline{g}_3(\underline{\zeta})}{\partial \underline{\zeta}}\right\| = \ln\left\|\begin{pmatrix}\frac{\partial \underline{g}_2(\underline{\xi})}{\partial \underline{\xi}} & 0 \\ 0 & \frac{\partial \underline{g}_1(\underline{\rho})}{\partial \underline{\rho}}\end{pmatrix}\right\| = \ln\left\|\frac{\partial \underline{g}_2(\underline{\xi})}{\partial \underline{\xi}}\right\| + \ln\left\|\frac{\partial \underline{g}_1(\underline{\rho})}{\partial \underline{\rho}}\right\| \tag{9.132}$$

In this case Kolmogorov's information is invariant with respect to a transformation of the coordinate system. As Kolmogorov's information is merely a special case of Kullback's information, we might also argue that because of this relation we conclude that Kolmogorov's information is invariant with respect to a coordinate transformation. It is merely an extended vector that we use in Kolmogorov's information.

9.4.6 Prerequisites for the transformations

In the previous chapters we noticed that the 'absolute' information measures S_α and S_1 do change with a change of the coordinate system, while the relative information measures (Kullback, G-divergence) are not affected by a change in the coordinate system.

An important requirement for the previous results is bijective mappings, i.e. there has to be a one-to-one relation between the arguments and their images. If this condition is not met, the product of two reciprocal Jacobi determinants is not equal to 1 and can thus not be cancelled in the relative information measures. We would obtain other results, which can be derived from the given derivations without any difficulties.

Another condition that has to be met by the vector spaces X and Y is the condition that both vectors $\underline{x}$ and $\underline{y}$ have the same dimension. Only quadratic matrices $\left\| \frac{\partial \underline{g}(\underline{\xi})}{\partial \underline{\xi}} \right\|$ allow us to compute the determinant. Moreover it is clear that the reduction of the dimension leads to a loss of information. If we map three-dimensional data onto a plane, the mapping is always connected with a loss of information (e.g., geographical maps, where the height information is reduced by the mapping in the two-dimensional plane, 3-D cameras, ...).

Kolmogorov's information is a certain exception to the results obtained from the other examinations, as we need to transform two vector spaces. But this merely requires an analogous extension of the described relations to these two transformations.

9.5 Transformation, discrete and continuous measures of entropy and relative entropies

Kullback's information measure is a relative information, which means that it describes the difference between two continuous distribution densities or between two probability distributions (in the discrete case). This relative definition of the information measure provides three advantages:

At first the value of the calculated information is directly a measure, which describes the difference between the amounts of information in the two random variables. We therefore do not require any additional generation of a reference value in the form of a second information. On the other hand it is a disadvantage of this information that we cannot directly determine which random variable contains more information. This is caused by the nonnegativity of this measure.

The great advantage of this measure is that it does not change with a change of the coordinate system. Thus we can use this measure to determine the information differences before and after some arbitrary transformations. Furthermore this measure allows us to compare discrete and continuous entropies. For absolute information measures, such as Shannon's or Rényi's information, this comparison cannot be carried out, because of the transition to infinitesimal intervals, leading to an infinite term in the continuous information. But here we obtain

$$K = \sum_{k=1}^{n} p_k \ln \frac{p_k}{q_k} \tag{9.133}$$

The transition to the continuous entropies uses the equation

$$p_k = P(A_k) = P(\omega : x(\omega) \in A_k) = \int_{A_k} f_x(\xi) \cdot d\xi \rightarrow f_x(\xi) \cdot d\xi \tag{9.134}$$

$$q_k = P(B_k) = P(\omega : x_2(\omega) \in B_k) = \int_{B_k} f_{x2}(\xi) \cdot d\xi \rightarrow f_{x_2}(\xi) \cdot d\xi \tag{9.135}$$

If the intervals A_k, B_k are small, we are able to approximate the probability by the product of the distribution density and the infinitesimal small interval. This allows the transition to the continuous description of the entropy and we find

$$K = \int_{\Xi} \ln \frac{f_x(\xi) \cdot d\xi}{f_{x_2}(\xi) \cdot d\xi} \cdot f_x(\xi) \cdot d\xi$$

$$K = \int_{\Xi} \ln \frac{f_x(\xi)}{f_{x_2}(\xi)} \cdot f_x(\xi) \cdot d\xi \tag{9.136}$$

This is the continuous formulation of Kullback's information and it allows us to compare discrete and continuous information differences (or entropy differences). We already know from Shannon's information that the transformation from discrete to continuous systems leads to an additional term in the information. We again find this relation for Rényi's α-entropy, where we start with the discrete formulation

$$S_{\alpha\_\text{discrete}} = \frac{1}{1-\alpha} \cdot \ln \sum_{k=1}^{n} p_k^{\alpha} \tag{9.137}$$

and generate the transition into a continuous formulation

$$S_{\alpha\_\text{discrete}} = \frac{1}{1-\alpha} \cdot \ln \sum_{k=1}^{n} (f_x(\xi) \cdot d\xi)^{\alpha}$$

$$S_{\alpha\_\text{discrete}} = \frac{1}{1-\alpha} \cdot \ln \left[\sum_{k=1}^{n} (f_x(\xi))^{\alpha} \cdot (d\xi)^{\alpha} \right]$$

$$S_{\alpha\_\text{discrete}} = \frac{1}{1-\alpha} \cdot \ln\left[(d\xi)^{\alpha} \cdot \sum_{k=1}^{n} (f_x(\xi))^{\alpha} \right]$$

$$S_{\alpha\_\text{discrete}} = \frac{1}{1-\alpha} \cdot \ln(d\xi)^{\alpha} + \frac{1}{1-\alpha} \cdot \ln \sum_{k=1}^{n} (f_x(\xi))^{\alpha}$$

$$S_{\alpha\_\text{discrete}} = \frac{1}{1-\alpha} \cdot \ln(d\xi)^{\alpha} + S_{\alpha\_\text{continuous}} \tag{9.138}$$

The discrete and the continuous formulations are not the same for these absolute information measures, as we again (see Shannon's information) obtain an additional term

$$\lim_{d\xi \to 0} \frac{1}{1-\alpha} \cdot \ln(d\xi)^{\alpha} = -\infty \tag{9.139}$$

approaching infinity. However, differences between two Rényi entropies can be compared, because this additional term vanishes as soon as we build differences.

Kullback's information offers the advantages that we directly obtain a difference information (absolute information measures do not directly provide any meaning because we have neither a reference scale nor a unit element), and that we may compare discrete and continuous entropies. Additionally this information measure is invariant regarding transformations. This advantage is a general property of the relative information measures, calculating a comparison between two distribution densities or between two probability distributions. Absolute information measures on the other hand always provide an additional term, when we compute the transition from discrete to continuous representations, and thus they do not allow any comparison between these two representations.

As we noticed in section 9.4.3 (Transformation of Shannon's information), differences between discrete and continuous information values can be compared with each other. Thus possible comparisons are not restricted to relative information measures. It is also possible to use absolute information measures to determine the differences between the information in continuous and discrete representations, when we use the differences of the information measures obtained.

9.6 Summary of the information functions

We have now presented several information measures and we have described their properties either more explicitly or more from the descriptive point of view. These information functions are but merely the most familiar ones, which are commonly used in applications. It is impossible to enumerate all information functions, because – as we will see in chapter 15 (Maximum entropy estimation) – every stated problem leads to the generation of another information function. The information functions introduced demonstrate already the most important properties and possible variations that are typical for such information functions. Further variations or different generalizations can certainly be created by additional weights, where the properties of these functions would be subject to further examinations. But we might ask the sense of further generalizations, which might not have direct connections to any useful application and thus would not produce any pragmatic information, because their application would have no meaning to us. Further measures of information and their relation can be found in appendix A.3.

Measures of information and measures of entropy are closely related to each other, which caused our excursion into thermodynamics. There the concept of entropy arose in the description of thermal energy and it is nowadays used in the statistical description of physical states, which results for instance in the Maxwell–Boltzmann distribution and the Boltzmann distribution. The entropy, as it is used in physics, describes the probability of the occupation of certain states. It is therefore a measure of uncertainty, which is exactly the concept we used in the documentation of Shannon's information. This uncertainty refers to the actual moment, because an observation provides a gain of information and this diminishes the entropy or, in the special case of discrete systems, eliminates the entropy. Information therefore is a concept referring to the future and describing the possibility of a gain of information by an observation. So we can also denote the entropy functions as information functions, because the only difference is the reference we use in the interpretation.

The comments on the information function proceeded with the computation of the information for gaussian random variables. As the gaussian distribution density is completely described by the mean and the covariance matrix, describing the uncertainty of the random variable, the information could only be a function of these two moments. On the basis of this approach we found different properties of the information functions, depending on the variances and the cross-covariances. Most of the information functions do not depend on the mean, because this is a deterministic value, which does not contain any hint regarding the uncertainty. The only information measure of our enumeration that depends on the means is Kullback's information, which determines the difference between two distribution densities, and there we do also need the shift between the means.

A disadvantage of all these information functions is certainly the scalar value, summarizing even multi-dimensional distribution densities in only one information value. This, of course, reduces the range of unambiguous function values, but there is also a certain sense in this scalar result. Information is – as we explained in the examination of Shannon's information – some kind of amalgamation of the uncertainty caused by the dimension, and of stochastic uncertainty, described by the variances or similar stochastic parameters. With this one-dimensional value the information does not have a direction, which it would have in the case of multi-dimensional values. The direction of the flow of information has to be determined from the boundary conditions, because the information functions do not provide this information per se.

A transformation of the information measures to different dimensions can be achieved by decomposing the observed random variables via transformations (PCA = Principle component analysis, Karhunen–Loève transformation, Givens rotation, ...) in independent random variables and calculating their information, which simply can be added to obtain the information of the whole system. Application of the corresponding backward transformation yields the relevant vector in the real coordinate system and the partitioning of the information can be determined from the position and the direction of the vector.

The examination of information cannot be restricted to one optimal information function initially, because we cannot always foresee which information is the best for the calculations. Cambel [CAM93] states this fact this way:

> "There is no one best entropy definition. Use the one best suited for your purposes. Also, please do not overlook the possibility that describing certain complex systems may require evaluating several of its entropy functions."

We will see this clearly in the examination of the maximum entropy principle, where we are able to generate entropy functions for arbitrary arrangements.

We considered several different descriptions of information, but we did not get a detailed definition for the concept of information. On the basis of certain demands that we required the information to meet, we found some measures that behave according our intuition. Tables 9.1 and 9.2 present the most familiar information functions and show their connections.

Table 9.1: Connections between the information functions

		POSTULATES		VARIED POSTULATES
	⇗	⇩		⇩
S_α Rènyi's α-information	⇨ ΔS_α	Rènyi's G_α-divergence		Daroczy's entropy
⇩ $\alpha \to 1$		⇩ $\alpha \to 1$	⇘ f_{yx} / $f_x f_y$	Generalized mutual information GMI
Shannon's S-information	⇨ ΔS	Kullback's G_1-information	⇨ ΔG_1	Kullback's D-divergence
		⇩ *Series expansion* *Differential quotient*	⇘ f_{yx} / $f_x f_y$	
		Fisher's information matrix		Kolmogorov's Transinformation

Table 9.2: Formulations of the information functions

		POSTULATES		VARIED POSTULATES
	⇖	⇩		⇩
$S_\alpha = \frac{1}{1-\alpha} \cdot \ln \int\limits_{\underline{\Xi}} f_{\underline{x}}^{\alpha}(\underline{\xi}) \cdot d\underline{\xi}$	⇨ ΔS_α	$G_\alpha = \frac{1}{\alpha - 1} \cdot \ln \int\limits_{\underline{\Xi}} \frac{f_{\underline{x}}^{\alpha}(\underline{\xi})}{f_{\underline{xs}}^{\alpha-1}(\underline{\xi})} \cdot d\underline{\xi}$		$S^\alpha = \frac{\sum_{k=1}^{N} p_k^{\alpha-1}}{2^{1-\alpha} - 1}$
			⇘ f_{yx} / $f_x f_y$	
⇩ $\alpha \to 1$		⇩ $\alpha \to 1$		$I_\alpha(\underline{y}, \underline{x}) = \ln \int\limits_{\underline{\Xi}} \int\limits_{\underline{\mathrm{P}}} \frac{\left[f_{\underline{x},\underline{y}}(\underline{\xi}, \underline{\rho})\right]^{\alpha}}{\left[f_{\underline{x}}(\underline{\xi}) \cdot f_{\underline{y}}(\underline{\rho})\right]^{\alpha-1}} \cdot d\underline{\rho} \cdot d\underline{\xi}$
$S = -\int\limits_{\underline{\Xi}} \left(\ln f_{\underline{x}}(\underline{\xi})\right) \cdot f_{\underline{x}}(\underline{\xi}) \cdot d\underline{\xi}$	⇨ ΔS	$G = \int\limits_{\underline{\Xi}} \left(\ln \frac{f_{\underline{x}}(\underline{\xi})}{f_{\underline{xs}}(\underline{\xi})}\right) \cdot f_{\underline{x}}(\underline{\xi}) \cdot d\underline{\xi}$	⇨ ΣG_1	$D = \int\limits_{\underline{\Xi}} \left(\ln \frac{f_{\underline{x}}(\underline{\xi})}{f_{\underline{xs}}(\underline{\xi})}\right) \cdot \left(f_{\underline{x}}(\underline{\xi}) - f_{\underline{xs}}(\underline{\xi})\right) \cdot d\underline{\xi}$
		⇩ *Series expansion* *Differential quotient*	⇘ f_{yx} / $f_x f_y$	
		$G = \frac{1}{2} \cdot \Delta\underline{x}^T \cdot F \cdot \Delta\underline{x}$ (fundamental tensor)		$I(\underline{y}, \underline{x}) = \int\limits_{\underline{\Xi}} \int\limits_{\underline{\mathrm{P}}} f_{\underline{x},\underline{y}}(\underline{\xi}, \underline{\rho}) \cdot \ln\left[\frac{f_{\underline{x},\underline{y}}(\underline{\xi}, \underline{\rho})}{f_{\underline{x}}(\underline{\xi}) \cdot f_{\underline{y}}(\underline{\rho})}\right] \cdot d\underline{\rho} \cdot d\underline{\xi}$

Having described these information functions, we now focus our attention om some other descriptions of the uncertainty in systems, which is expressed by the moments of the distribution densities. The bounds of these moments are the main point of the next chapter and we will again find different possible measures, depending on the boundary conditions and the prerequisites. One of these boundaries is Fisher's information, which we will find in the familiar inequality of Cramér and Rao, describing the least possible variance of an estimator.

10 Bounds of the variance

Information depends on the uncertainty of the signals that we observe. The uncertainty in these signals is commonly described by the variance, representing the power of the error. The variance is not a function of a probability distribution function, but a moment, i.e. a direct component of this distribution density. These moments are used to determine optimal estimators or to find limitations that these estimators cannot exceed. In the literature such limitations are known as bounds and we will adopt this denotation, which describes the main point of this chapter.

Before we start with the derivation of different bounds for variance and other higher order moments, we want to discuss the meaning of these limitations. We start our examination with the general problem of estimation theory:

We have a system that is described in state space and observation space. We want to find the states $\underline{x}(k)$ of our state space, which are mapped via the transformation $\underline{h}[\underline{x}(k), k, \underline{v}(k)]$ into the observation space. Such a mapping can be caused by some states, which we are not able to observe directly, or by measurement devices, performing a transformation of the desired variables. The error occurring in such a mapping is commonly described by measurement noise $\underline{v}(k)$ and modeled as a gaussian random variable with zero mean.

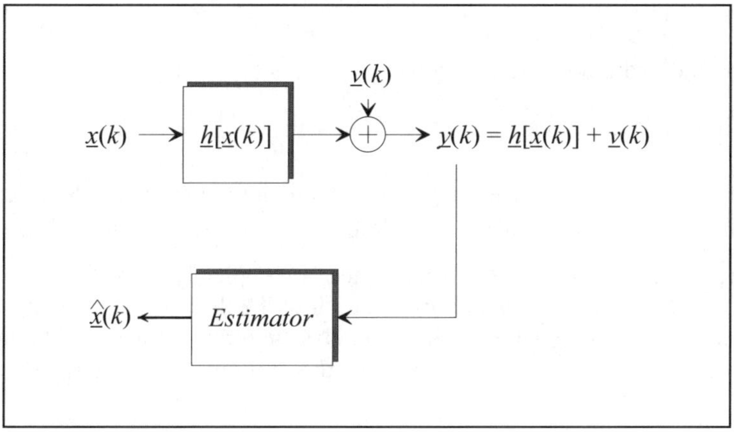

Fig. 10.1: Basic idea of an estimator

An estimator is a device, a function or an algorithm, estimating an optimal state vector $\underline{x}(k)$ on the basis of the given measurements $\underline{y}(k)$ with additional knowledge of the conditional distribution density function $f_{\underline{y}|\underline{x}}(\underline{\rho} \mid \underline{x}(k))$. The variance of such estimators is used to measure the quality of the estimation.

The basic requirements that such an estimator has to meet are the unbiased estimate, which means that the desired parameter or state is estimated without any additional bias, and a minimum variance in the estimation error. Involuntarily the question arises how good can such an estimate be, or how small can the variance of the estimation be. This question is answered by the bounds, leading to different limitations of the variances, which depend on the prerequisites of the estimators.

We already noticed in the introduction that a gain of information from disturbed measurements is connected with an increase in the entropy (second law of thermodynamics), so that we are only able to determine the desired state estimates with a limited accuracy. Thus the accuracy of the evaluation is limited by certain bounds, depending on the accuracy of the measurement and on the given observation equation. We want to derive the most familiar of these bounds on the basis of the conditional distribution density of the measurements.

10.1 Cramér–Rao bound

We want to examine the maximum accuracy with which we are able to estimate the state vector $\underline{x}$ from the measurement vector $\underline{y}$, and the first step is the description of the mapping of this state vector $\underline{x}$ into the observation space Y. This mapping is described by the observation equation

$$\underline{y}(k) = \underline{h}[\underline{x}(k)] + \underline{v}(k) \tag{10.01}$$

or more generally with non-additive noise:

$$\underline{y}(k) = \underline{h}[\underline{x}(k), \underline{v}(k)] \tag{10.02}$$

With this functional dependence we generate the conditional distribution density $f_{\underline{y}|\underline{x}}(\underline{\rho} \mid \underline{x}(k))$ of the measurements, conditioned on the state vector, where the mapping from state space to observation space is completely included. And now we are looking for a function that allows us to describe the reverse mapping from the contaminated measurements to an optimal state vector. Such a function is called an estimator $\underline{t}_{\underline{y}}(\underline{\rho})$ and we use the following equation for the determination:

$$\int_{\underline{P}} \underline{t}_{\underline{y}}(\underline{\rho}) \cdot f_{\underline{y}|\underline{x}}(\underline{\rho} \mid \underline{x}(k)) \cdot d\underline{\rho} = \underline{x}(k) \tag{10.03}$$

The expectation value of this estimator usually provides the desired parameter or state $\underline{x}(k)$. In this case the estimator is also known as an 'unbiased estimator', and the determination of the maximum accuracy requires some further calculations.

Because of the independence of the parameter $\underline{x}(k)$ and the integration variable $\underline{\rho}$ we get

$$\underline{x}(k) = \underline{x}(k) \cdot 1 = \underline{x}(k) \cdot \int_{\underline{P}} f_{\underline{y}|\underline{x}}\left(\underline{\rho} \mid \underline{x}(k)\right) \cdot d\underline{\rho} = \int_{\underline{P}} \underline{x}(k) \cdot f_{\underline{y}|\underline{x}}\left(\underline{\rho} \mid \underline{x}(k)\right) \cdot d\underline{\rho} \tag{10.04}$$

and this is inserted into equation (10.03) to obtain

$$\int_{\underline{P}} \underline{t}_{\underline{y}}(\underline{\rho}) \cdot f_{\underline{y}|\underline{x}}\left(\underline{\rho} \mid \underline{x}(k)\right) \cdot d\underline{\rho} = \int_{\underline{P}} \underline{x}(k) \cdot f_{\underline{y}|\underline{x}}\left(\underline{\rho} \mid \underline{x}(k)\right) \cdot d\underline{\rho}$$

$$\int_{\underline{P}} \left[\underline{t}_{\underline{y}}(\underline{\rho}) - \underline{x}(k)\right] \cdot f_{\underline{y}|\underline{x}}\left(\underline{\rho} \mid \underline{x}(k)\right) \cdot d\underline{\rho} = \underline{0} \tag{10.05}$$

As the estimator is intended to estimate the state $\underline{x}(k)$ optimally, we are looking for an extremum of the created function and such extreme values are determined via derivation

$$\frac{\partial}{\partial \underline{x}(k)} \cdot \int_{\underline{P}} \left[\underline{t}_{\underline{y}}(\underline{\rho}) - \underline{x}(k)\right] \cdot f_{\underline{y}|\underline{x}}\left(\underline{\rho} \mid \underline{x}(k)\right) \cdot d\underline{\rho} = \underline{0} \tag{10.06}$$

If the 'regularity conditions' (Appendix A.5) hold, we can interchange differentiation and integration and we find

$$\int_{\underline{P}} \frac{\partial}{\partial \underline{x}(k)} \left\{ \left[\underline{t}_{\underline{y}}(\underline{\rho}) - \underline{x}(k)\right] \cdot f_{\underline{y}|\underline{x}}\left(\underline{\rho} \mid \underline{x}(k)\right) \right\} \cdot d\underline{\rho} = \underline{0} \tag{10.07}$$

The partial differentiation of this product provides

$$\int_{\underline{P}} \left(-I \cdot f_{\underline{y}|\underline{x}}\left(\underline{\rho} \mid \underline{x}(k)\right) + \left[\underline{t}_{\underline{y}}(\underline{\rho}) - \underline{x}(k)\right] \cdot \frac{\partial}{\partial \underline{x}(k)} \left\{ f_{\underline{y}|\underline{x}}\left(\underline{\rho} \mid \underline{x}(k)\right) \right\} \right) \cdot d\underline{\rho} = \underline{0} \tag{10.08}$$

or, after decomposition into two integrals

$$-\int_{\underline{P}} I \cdot f_{\underline{y}|\underline{x}}\left(\underline{\rho} \mid \underline{x}(k)\right) \cdot d\underline{\rho} + \int_{\underline{P}} \left[\underline{t}_{\underline{y}}(\underline{\rho}) - \underline{x}(k)\right] \cdot \frac{\partial}{\partial \underline{x}(k)} \left\{ f_{\underline{y}|\underline{x}}\left(\underline{\rho} \mid \underline{x}(k)\right) \right\} \cdot d\underline{\rho} = \underline{0} \tag{10.09}$$

and

$$\int_{\underline{P}} \left[\underline{t}_{\underline{y}}(\underline{\rho}) - \underline{x}(k)\right] \cdot \frac{\partial}{\partial \underline{x}(k)} \left\{f_{\underline{y}|\underline{x}}(\underline{\rho} \mid \underline{x}(k))\right\} \cdot d\underline{\rho} = I \tag{10.10}$$

To obtain an expectation value again, we substitute

$$\frac{\partial}{\partial \underline{x}(k)} \left\{f_{\underline{y}|\underline{x}}(\underline{\rho} \mid \underline{x}(k))\right\} = \frac{\partial}{\partial \underline{x}(k)} \ln\left\{f_{\underline{y}|\underline{x}}(\underline{\rho} \mid \underline{x}(k))\right\} \cdot f_{\underline{y}|\underline{x}}(\underline{\rho} \mid \underline{x}(k)) \tag{10.11}$$

and obtain

$$\int_{\underline{P}} \left[\underline{t}_{\underline{y}}(\underline{\rho}) - \underline{x}(k)\right] \cdot \frac{\partial}{\partial \underline{x}(k)} \ln\left\{f_{\underline{y}|\underline{x}}(\underline{\rho} \mid \underline{x}(k))\right\} \cdot f_{\underline{y}|\underline{x}}(\underline{\rho} \mid \underline{x}(k)) \cdot d\underline{\rho} = I \tag{10.12}$$

and this expression can be transformed with the Schwartz inequality

$$(E\{A \cdot B\})^2 \le E\{A^2\} \cdot E\{B^2\} \tag{10.13}$$

to get the equation

$$\int_{\underline{P}} \left[\underline{t}_{\underline{y}}(\underline{\rho}) - \underline{x}(k)\right] \cdot \left[\underline{t}_{\underline{y}}(\underline{\rho}) - \underline{x}(k)\right]^T \cdot f_{\underline{y}|\underline{x}}(\underline{\rho} \mid \underline{x}(k)) \cdot d\underline{\rho} \cdot$$

$$\cdot \int_{\underline{P}} \left[\frac{\partial}{\partial \underline{x}(k)} \ln\left\{f_{\underline{y}|\underline{x}}(\underline{\rho} \mid \underline{x}(k))\right\}\right] \cdot \left[\frac{\partial}{\partial \underline{\xi}} \ln\left\{f_{\underline{y}|\underline{x}}(\underline{\rho} \mid \underline{x}(k))\right\}\right]^T \cdot f_{\underline{y}|\underline{x}}(\underline{\rho} \mid \underline{x}(k)) \cdot d\underline{\rho} \ge I$$

or

$$\int_{\underline{P}} \left[\underline{t}_{\underline{y}}(\underline{\rho}) - \underline{x}(k)\right] \cdot \left[\underline{t}_{\underline{y}}(\underline{\rho}) - \underline{x}(k)\right]^T \cdot f_{\underline{y}|\underline{x}}(\underline{\rho} \mid \underline{x}(k)) \cdot d\underline{\rho}$$

$$\ge I \cdot \left[\int_{\underline{P}} \left[\frac{\partial}{\partial \underline{x}(k)} \ln\left\{f_{\underline{y}|\underline{x}}(\underline{\rho} \mid \underline{x}(k))\right\}\right] \cdot \left[\frac{\partial}{\partial \underline{x}(k)} \ln\left\{f_{\underline{y}|\underline{x}}(\underline{\rho} \mid \underline{x}(k))\right\}\right]^T \cdot f_{\underline{y}|\underline{x}}(\underline{\rho} \mid \underline{x}(k)) \cdot d\underline{\rho}\right]^{-1} \tag{10.14}$$

The left-hand side of this expression is the variance of the unbiased estimator, because the parameter $\underline{x}(k)$ is equal to the expectation value (equation (10.3)). Thus we have found the lowest bound of the variance of an unbiased estimator:

$$\operatorname{var}\left(\underline{t}_{\underline{y}}(\underline{\rho})\right) \geq \left[\int_{\underline{\mathrm{P}}} \left[\frac{\partial}{\partial \underline{x}(k)} \ln\left\{ f_{\underline{y}|\underline{x}}(\underline{\rho} \mid \underline{x}(k)) \right\} \right] \cdot \left[\frac{\partial}{\partial \underline{x}(k)} \ln\left\{ f_{\underline{y}|\underline{x}}(\underline{\rho} \mid \underline{x}(k)) \right\} \right]^T \cdot f_{\underline{y}|\underline{x}}(\underline{\rho} \mid \underline{x}(k)) \cdot d\underline{\rho} \right]^{-1} \tag{10.15}$$

and this inequality is commonly known as the Cramér–Rao inequality. The resulting limit is called the Cramér–Rao bound, and on the right-hand side of the equation we find the inverse of Fisher's information matrix

$$F = \int_{\underline{\mathrm{P}}} \left[\frac{\partial}{\partial \underline{x}(k)} \ln\left\{ f_{\underline{y}|\underline{x}}(\underline{\rho} \mid \underline{x}(k)) \right\} \right] \cdot \left[\frac{\partial}{\partial \underline{x}(k)} \ln\left\{ f_{\underline{y}|\underline{x}}(\underline{\rho} \mid \underline{x}(k)) \right\} \right]^T \cdot f_{\underline{y}|\underline{x}}(\underline{\rho} \mid \underline{x}(k)) \cdot d\underline{\rho} \tag{10.16}$$

which will be subject to further examinations. We now calculate the matrix F for gaussian distribution densities, to obtain some further insight into this limitation of the variance.

10.1.1 Fisher's information for gaussian distribution densities

Let:

$\underline{x}$ = $[n\times1]$-dimensional state vector
$\underline{y}$ = $[m\times1]$-dimensional measurement vector
$\underline{v}$ = $[m\times1]$-dimensional noise vector
R = $[m\times m]$-dimensional measurement covariance matrix

The observation equation is given by

$$\underline{y}(k) = \underline{h}[\underline{x}(k)] + \underline{v}(k) \tag{10.17}$$

and we get the conditional distribution density

$$f_{\underline{y}|\underline{x}}(\underline{\rho} \mid \underline{x}(k))$$

$$= \frac{1}{(2\pi)^{\frac{m}{2}} \cdot |R|^{\frac{1}{2}}} \cdot \exp\left\{ -\frac{1}{2} \cdot \left(\underline{\rho} - E_{\underline{y}|\underline{x}}\{\underline{\rho} \mid \underline{x}(k)\}\right)^T \cdot R^{-1} \cdot \left(\underline{\rho} - E_{\underline{y}|\underline{x}}\{\underline{\rho} \mid \underline{x}(k)\}\right) \right\} \tag{10.18}$$

This conditional distribution density can be inserted into the equation

$$F = \int_{\underline{\mathrm{P}}} \left[\frac{\partial}{\partial \underline{x}(k)} \ln\left\{ f_{\underline{y}|\underline{x}}(\underline{\rho} \mid \underline{x}(k)) \right\} \right] \cdot \left[\frac{\partial}{\partial \underline{x}(k)} \ln\left\{ f_{\underline{y}|\underline{x}}(\underline{\rho} \mid \underline{x}(k)) \right\} \right]^T \cdot f_{\underline{y}|\underline{x}}(\underline{\rho} \mid \underline{x}(k)) \cdot d\underline{\rho} \tag{10.19}$$

where we first calculate the logarithm, to be able to insert the complete term, which is contained in the brackets.

$$\frac{\partial}{\partial \underline{x}(k)} \ln\left[f_{\underline{y}|\underline{x}}\left(\underline{\rho} \mid \underline{x}(k)\right)\right] = \frac{1}{f_{\underline{y}|\underline{x}}\left(\underline{\rho} \mid \underline{x}(k)\right)} \cdot \frac{\partial}{\partial \underline{x}(k)} \left[f_{\underline{y}|\underline{x}}\left(\underline{\rho} \mid \underline{x}(k)\right)\right] \tag{10.20}$$

With the derivative

$$\frac{\partial}{\partial \underline{x}(k)} \left[f_{\underline{y}|\underline{x}}\left(\underline{\rho} \mid \underline{x}(k)\right)\right] = -\frac{1}{2} \cdot 2 \cdot \left[\frac{\partial}{\partial \underline{x}(k)} E_{\underline{y}|\underline{x}}\left\{\underline{\rho} \mid \underline{x}(k)\right\}\right]^T \cdot R^{-1} \cdot (-1) \cdot \left(\underline{\rho} - E_{\underline{y}|\underline{x}}\left\{\underline{\rho} \mid \underline{x}(k)\right\}\right) \cdot f_{\underline{y}|\underline{x}}\left(\underline{\rho} \mid \underline{x}(k)\right) \tag{10.21}$$

(consisting of the derivative of the quadratic form in the exponent) [ATH65] we are able to cancel the conditional distribution density, which leads to the term

$$\frac{\partial}{\partial \underline{x}(k)} \ln\left[f_{\underline{y}|\underline{x}}\left(\underline{\rho} \mid \underline{x}(k)\right)\right] = \left[\frac{\partial}{\partial \underline{x}(k)} E_{\underline{y}|\underline{x}}\left\{\underline{\rho} \mid \underline{x}(k)\right\}\right]^T \cdot R^{-1} \cdot \left(\underline{\rho} - E_{\underline{y}|\underline{x}}\left\{\underline{\rho} \mid\right\}\right) \tag{10.22}$$

And this term can be inserted into the integral in equation (10.19), where we neglect the dependence of the parameter $\underline{x}$ on the time index k

$$F = \int_{\underline{\mathrm{P}}} \left[\left[\frac{\partial}{\partial \underline{x}} E_{\underline{y}|\underline{x}}\left\{\underline{\rho} \mid \underline{x}\right\}\right]^T \cdot R^{-1} \cdot \left(\underline{\rho} - E_{\underline{y}|\underline{x}}\left\{\underline{\rho} \mid \underline{x}\right\}\right)\right] \cdot \left[\left[\frac{\partial}{\partial \underline{x}} E_{\underline{y}|\underline{x}}\left\{\underline{\rho} \mid \underline{x}\right\}\right]^T \cdot R^{-1} \cdot \left(\underline{\rho} - E_{\underline{y}|\underline{x}}\left\{\underline{\rho} \mid \underline{x}\right\}\right)\right]^T \cdot f_{\underline{y}|\underline{x}}\left(\underline{\rho} \mid \underline{x}\right) \cdot d\underline{\rho} \tag{10.23}$$

and with $(A \cdot B)^T = B^T \cdot A^T$, as well as $R^T = R$ we can rearrange the equation

$$F = \int_{\underline{\mathrm{P}}} \left[\left[\frac{\partial}{\partial \underline{x}} E_{\underline{y}|\underline{x}}\left\{\underline{\rho} \mid \underline{x}\right\}\right]^T \cdot R^{-1} \cdot \left(\underline{\rho} - E_{\underline{y}|\underline{x}}\left\{\underline{\rho} \mid \underline{x}\right\}\right) \cdot \left(\underline{\rho} - E_{\underline{y}|\underline{x}}\left\{\underline{\rho} \mid \underline{x}\right\}\right)^T \cdot R^{-1} \cdot \left[\frac{\partial}{\partial \underline{x}} E_{\underline{y}|\underline{x}}\left\{\underline{\rho} \mid \underline{x}\right\}\right]\right] \cdot f_{\underline{y}|\underline{x}}\left(\underline{\rho} \mid \underline{x}\right) \cdot d\underline{\rho} \tag{10.24}$$

and now we extract the factors out of the integral, which do not depend on the integration variable

$$F = \left[\frac{\partial}{\partial \underline{x}} E_{\underline{y}|\underline{x}} \{ \underline{\rho} \mid \underline{x} \} \right]^T \cdot R^{-1} \cdot \int_{\underline{\mathrm{P}}} \left(\underline{\rho} - E_{\underline{y}|\underline{x}} \{ \underline{\rho} \mid \underline{x} \} \right) \cdot \left(\underline{\rho} - E_{\underline{y}|\underline{x}} \{ \underline{\rho} \mid \underline{x} \} \right)^T \cdot f_{\underline{y}|\underline{x}} \left(\underline{\rho} \mid \underline{x} \right) \cdot d\underline{\rho}$$

$$\cdot R^{-1} \cdot \left[\frac{\partial}{\partial \underline{x}} E_{\underline{y}|\underline{x}} \{ \underline{\rho} \mid \underline{x} \} \right] \tag{10.25}$$

The remaining integral is equal to the measurement error covariance matrix R, and thus

$$F = \left[\frac{\partial}{\partial \underline{x}} E_{\underline{y}|\underline{x}} \{ \underline{\rho} \mid \underline{x} \} \right]^T \cdot R^{-1} \cdot R \cdot R^{-1} \cdot \left[\frac{\partial}{\partial \underline{x}} E_{\underline{y}|\underline{x}} \{ \underline{\rho} \mid \underline{x} \} \right] \tag{10.26}$$

finally

$$F = \left[\frac{\partial}{\partial \underline{x}} E_{\underline{y}|\underline{x}} \{ \underline{\rho} \mid \underline{x} \} \right]^T \cdot R^{-1} \cdot \left[\frac{\partial}{\partial \underline{x}} E_{\underline{y}|\underline{x}} \{ \underline{\rho} \mid \underline{x} \} \right] \tag{10.27}$$

This is Fisher's information about the state (or parameter) $\underline{x}$ on the basis of the observation $\underline{y}$ for a gaussian conditional distribution density. The derivative of the expectation value occurring in this matrix depends on the mapping $\underline{h}[\underline{x}(k)]$. In case the of a linear mapping we get

$$\underline{h}[\underline{x}(k)] = C(k) \cdot \underline{x}(k) \tag{10.28}$$

and we obtain

$$\frac{\partial}{\partial \underline{x}} E_{\underline{y}|\underline{x}} \{ \underline{\rho} \mid \underline{x} \} = \frac{\partial}{\partial \underline{x}} C \cdot \underline{x} = C \tag{10.29}$$

Thus we find:

$$F(k) = C^T(k) \cdot R(k)^{-1} \cdot C(k) \tag{10.30}$$

In case the of a nonlinear mapping, the expectation value is given by

$$E_{\underline{y}|\underline{x}} \{ \underline{\rho} \mid \underline{x} \} = E_{\underline{y}|\underline{x}} \{ \underline{h}[\underline{x}(k)] + \underline{v}(k) | \underline{x} \} \tag{10.31}$$

The measurement noise has zero mean in the linear and in the nonlinear case, but the nonlinear mapping has to be linearized, which is achieved by a Taylor series expansion at the state $\underline{x}^-(k)$.

$$\underline{h}[\underline{x}(k)] = \underline{h}[\underline{x}(k)]_{\underline{x}(k)=\underline{x}^-(k)} + \left[\frac{\partial}{\partial \underline{x}}\underline{h}[\underline{x}(k)]\right]\Bigg|_{\underline{x}(k)=\underline{x}^-(k)} \cdot \left(\underline{x}(k) - \underline{x}^-(k)\right) + \text{h.o.t.} \tag{10.32}$$

If we neglect the higher order terms, we find

$$E_{\underline{y}|\underline{x}^-}\left\{\underline{\rho} \mid \underline{x}^-\right\} = E_{\underline{y}|\underline{x}^-}\left\{\underline{h}\left[\underline{x}^-(k)\right] + \left[\frac{\partial}{\partial \underline{x}}\underline{h}[\underline{x}(k)]\right]\Bigg|_{\underline{x}(k)=\underline{x}^-(k)} \cdot \left(\underline{x}(k) - \underline{x}^-(k)\right) + \underline{v}(k) \Bigg| \underline{x}^-\right\} \tag{10.33}$$

and Fisher's information matrix becomes

$$F = \left(\left[\frac{\partial}{\partial \underline{x}}\underline{h}[\underline{x}(k)]\right]\Bigg|_{\underline{x}(k)=\underline{x}^-(k)}\right)^T \cdot R(k)^{-1} \cdot \left(\left[\frac{\partial}{\partial \underline{x}}\underline{h}[\underline{x}(k)]\right]\Bigg|_{\underline{x}(k)=\underline{x}^-(k)}\right) \tag{10.34}$$

10.1.2 Fisher's information and Kullback's information

Now we derive a first connection to the already known information functions, which only provide scalar values for the information. These values are amalgamations of the amounts of information occurring in the observed system or process.

We examine Kullback's discrimination information G_1

$$G_1 = \int_{\Xi} \ln\left[\frac{f_{\underline{x}}(\underline{\xi})}{f_{\underline{x}_s}(\underline{\xi})}\right] \cdot f_{\underline{x}}(\underline{\xi}) \cdot d\underline{\xi} \tag{10.35}$$

and substitute

$$f_{\underline{x}}(\underline{\xi}) = f_{\underline{x}|\underline{\Theta}}(\underline{\xi} \mid \underline{\theta})$$

and

$$f_{\underline{x}_s}(\underline{\xi}) = f_{\underline{x}|\underline{\Theta}}(\underline{\xi} \mid \underline{\theta} + \Delta\underline{\theta}) \tag{10.36}$$

Inserting this into Kullback's information provides

$$G_1 = \int_{\Xi} \ln\left[\frac{f_{\underline{x}|\underline{\Theta}}(\underline{\xi} \mid \underline{\theta})}{f_{\underline{x}|\underline{\Theta}}(\underline{\xi} \mid \underline{\theta} + \Delta\underline{\theta})}\right] \cdot f_{\underline{x}|\underline{\Theta}}(\underline{\xi} \mid \underline{\theta}) \cdot d\underline{\xi}$$

or

$$G_1 = \int_{\underline{\Xi}} \left\{ \ln\left[f_{\underline{x}|\underline{\Theta}}\left(\underline{\xi} \mid \underline{\theta}\right)\right] - \ln\left[f_{\underline{x}|\underline{\Theta}}\left(\underline{\xi} \mid \underline{\theta} + \Delta\underline{\theta}\right)\right] \right\} \cdot f_{\underline{x}|\underline{\Theta}}\left(\underline{\xi} \mid \underline{\theta}\right) \cdot d\underline{\xi} \tag{10.37}$$

Now we expand the logarithm of the distribution density $f_{\underline{x}|\underline{\Theta}}\left(\underline{\xi} \mid \underline{\theta} + \Delta\underline{\theta}\right)$ in a Taylor series at the parameter $\underline{\theta}$

$$\ln\left[f_{\underline{x}|\underline{\Theta}}\left(\underline{\xi} \mid \underline{\theta} + \Delta\underline{\theta}\right)\right] = \ln\left[f_{\underline{x}|\underline{\Theta}}\left(\underline{\xi}\right)\right] + (\Delta\underline{\theta})^T \cdot \frac{\partial}{\partial\underline{\theta}} \ln\left[f_{\underline{x}|\underline{\Theta}}\left(\underline{\xi} \mid \underline{\theta} + \Delta\underline{\theta}\right)\right]\Big|_{\Delta\underline{\theta}=\underline{0}}$$

$$+ \frac{1}{2} \cdot (\Delta\underline{\theta})^T \cdot \frac{\partial^2}{\partial\underline{\theta} \cdot \partial\underline{\theta}^T} \ln\left[f_{\underline{x}|\underline{\Theta}}\left(\underline{\xi} \mid \underline{\theta} + \Delta\underline{\theta}\right)\right]\Big|_{\Delta\underline{\theta}=\underline{0}} \cdot (\Delta\underline{\theta}) + \text{h.o.t.} \tag{10.38}$$

This leads to the formula

$$\ln\left[f_{\underline{x}|\underline{\Theta}}\left(\underline{\xi}\right)\right] - \ln\left[f_{\underline{x}|\underline{\Theta}}\left(\underline{\xi} \mid \underline{\theta} + \Delta\underline{\theta}\right)\right] = -(\Delta\underline{\theta})^T \cdot \frac{\partial}{\partial\underline{\theta}} \ln\left[f_{\underline{x}|\underline{\Theta}}\left(\underline{\xi} \mid \underline{\theta} + \Delta\underline{\theta}\right)\right]\Big|_{\Delta\underline{Q}=\underline{0}}$$

$$- \frac{1}{2} \cdot (\Delta\underline{\theta})^T \cdot \frac{\partial^2}{\partial\underline{\theta} \cdot \partial\underline{\theta}^T} \ln\left[f_{\underline{x}|\underline{\Theta}}\left(\underline{\xi} \mid \underline{\theta} + \Delta\underline{\theta}\right)\right]\Big|_{\Delta\underline{\theta}=\underline{0}} \cdot (\Delta\underline{\theta}) + \text{h.o.t.} \tag{10.39}$$

and the G_1-information

$$G_1 = -\int_{\underline{\Xi}} \left\{ (\Delta\underline{\theta})^T \cdot \frac{\partial}{\partial\underline{\theta}} \ln\left[f_{\underline{x}|\underline{\Theta}}\left(\underline{\xi} \mid \underline{\theta} + \Delta\underline{\theta}\right)\right]\Big|_{\Delta\underline{Q}=\underline{0}} \right\} \cdot f_{\underline{x}|\underline{\Theta}}\left(\underline{\xi} \mid \underline{\theta}\right) \cdot d\underline{\xi}$$

$$- \frac{1}{2} \cdot \int_{\underline{\Xi}} \left\{ (\Delta\underline{\theta})^T \cdot \frac{\partial^2}{\partial\underline{\theta} \cdot \partial\underline{\theta}^T} \ln\left[f_{\underline{x}|\underline{\Theta}}\left(\underline{\xi} \mid \underline{\theta} + \Delta\underline{\theta}\right)\right]\Big|_{\Delta\underline{\theta}=\underline{0}} \cdot (\Delta\underline{\theta}) \right\} \cdot f_{\underline{x}|\underline{\Theta}}\left(\underline{\xi} \mid \underline{\theta}\right) \cdot d\underline{\xi} \tag{10.40}$$

consists of two integrals, which are to be solved now.

$$G_{11} = -\int_{\underline{\Xi}} \left\{ (\Delta\underline{\theta})^T \cdot \frac{\partial}{\partial\underline{\theta}} \ln\left[f_{\underline{x}|\underline{\Theta}}\left(\underline{\xi} \mid \underline{\theta} + \Delta\underline{\theta}\right)\right]\Big|_{\Delta\underline{Q}=\underline{0}} \right\} \cdot f_{\underline{x}|\underline{\Theta}}\left(\underline{\xi} \mid \underline{\theta}\right) \cdot d\underline{\xi}$$

$$G_{11} = -(\Delta\underline{\theta})^T \cdot \int_{\underline{\Xi}} \left\{ \left(\frac{1}{f_{\underline{x}|\underline{\Theta}}\left(\underline{\xi} \mid \underline{\theta} + \Delta\underline{\theta}\right)} \cdot \frac{\partial}{\partial\underline{\theta}} \left[f_{\underline{x}|\underline{\Theta}}\left(\underline{\xi} \mid \underline{\theta} + \Delta\underline{\theta}\right)\right] \right)\Bigg|_{\Delta\underline{Q}=\underline{0}} \right\} \cdot f_{\underline{x}|\underline{\Theta}}\left(\underline{\xi} \mid \underline{\theta}\right) \cdot d\underline{\xi}$$

$$G_{11} = -(\Delta \underline{\theta})^T \cdot \int_{\underline{\Xi}} \left\{ \frac{1}{f_{\underline{x}|\underline{\Theta}}(\underline{\xi} \mid \underline{\theta})} \cdot \left(\frac{\partial}{\partial \underline{\theta}} \left[f_{\underline{x}|\underline{\Theta}}(\underline{\xi} \mid \underline{\theta} + \Delta \underline{\theta}) \right] \right) \Bigg|_{\Delta \underline{Q}=0} \right\} \cdot f_{\underline{x}|\underline{\Theta}}(\underline{\xi} \mid \underline{\theta}) \cdot d\underline{\xi}$$

$$G_{11} = -(\Delta \underline{\theta})^T \cdot \int_{\underline{\Xi}} \left(\frac{\partial}{\partial \underline{\theta}} \left[f_{\underline{x}|\underline{\Theta}}(\underline{\xi} \mid \underline{\theta} + \Delta \underline{\theta}) \right] \right) \Bigg|_{\Delta \underline{Q}=0} \cdot d\underline{\xi} \tag{10.41}$$

Under the condition that we are able to interchange differentiation and integration (regularity conditions), we find

$$G_{11} = -(\Delta \underline{\theta})^T \cdot \frac{\partial}{\partial \underline{\theta}} \int_{\underline{\Xi}} \left(\left[f_{\underline{x}|\underline{\Theta}}(\underline{\xi} \mid \underline{\theta} + \Delta \underline{\theta}) \right] \right) \Big|_{\Delta \underline{Q}=0} \cdot d\underline{\xi} \tag{10.42}$$

and the integral is built with the conditional distribution density, which directly leads to

$$\int_{\underline{\Xi}} \left(\left[f_{\underline{x}|\underline{\Theta}}(\underline{\xi} \mid \underline{\theta} + \Delta \underline{\theta}) \right] \right) \Big|_{\Delta \underline{Q}=0} \cdot d\underline{\xi} = 1 \tag{10.43}$$

independent of the parameter $\underline{\theta}$. The derivative of this constant is equal to zero and thus the first integral G_{11} vanishes.

$$G_{11} = 0 \tag{10.44}$$

So the discrimination information is determined by the second integral

$$G_1 = -\frac{1}{2} \cdot \int_{\underline{\Xi}} \left\{ (\Delta \underline{\theta})^T \cdot \frac{\partial^2}{\partial \underline{\theta} \cdot \partial \underline{\theta}^T} \ln \left[f_{\underline{x}|\underline{\Theta}}(\underline{\xi} \mid \underline{\theta} + \Delta \underline{\theta}) \right] \Bigg|_{\Delta \underline{\theta}=0} \cdot (\Delta \underline{\theta}) \right\} \cdot f_{\underline{x}|\underline{\Theta}}(\underline{\xi} \mid \underline{\theta}) \cdot d\underline{\xi} \tag{10.45}$$

which we rearrange slightly.

$$G_1 = -\frac{1}{2} \cdot (\Delta \underline{\theta})^T \cdot \int_{\underline{\Xi}} \left\{ \frac{\partial^2}{\partial \underline{\theta} \cdot \partial \underline{\theta}^T} \ln \left[f_{\underline{x}|\underline{\Theta}}(\underline{\xi} \mid \underline{\theta} + \Delta \underline{\theta}) \right] \Bigg|_{\Delta \underline{\theta}=0} \right\} \cdot f_{\underline{x}|\underline{\Theta}}(\underline{\xi} \mid \underline{\theta}) \cdot d\underline{\xi} \cdot (\Delta \underline{\theta}) \tag{10.46}$$

(<u>*The negative integral is a second formulation of Fisher's information*</u>)

We examine the kernel

$$\frac{\partial^2}{\partial \underline{\theta} \cdot \partial \underline{\theta}^T} \ln \left[f_{\underline{x}|\underline{\Theta}}(\underline{\xi} \mid \underline{\theta} + \Delta \underline{\theta}) \right] = \frac{\partial}{\partial \underline{\theta}} \left\{ \frac{\partial}{\partial \underline{\theta}^T} \ln \left[f_{\underline{x}|\underline{\Theta}}(\underline{\xi} \mid \underline{\theta} + \Delta \underline{\theta}) \right] \right\}$$

$$
\begin{aligned}
&= \frac{\partial}{\partial \underline{\theta}} \left\{ \frac{1}{f_{\underline{x}|\underline{\Theta}}\left(\underline{\xi} \mid \underline{\theta} + \Delta\underline{\theta}\right)} \cdot \frac{\partial}{\partial \underline{\theta}^T} \left[f_{\underline{x}|\underline{\Theta}}\left(\underline{\xi} \mid \underline{\theta} + \Delta\underline{\theta}\right) \right] \right\} \\
&= -\frac{1}{\left(f_{\underline{x}|\underline{\Theta}}\left(\underline{\xi} \mid \underline{\theta} + \Delta\underline{\theta}\right)\right)^2} \cdot \frac{\partial}{\partial \underline{\theta}} \left[f_{\underline{x}|\underline{\Theta}}\left(\underline{\xi} \mid \underline{\theta} + \Delta\underline{\theta}\right) \right] \cdot \frac{\partial}{\partial \underline{\theta}^T} \left[f_{\underline{x}|\underline{\Theta}}\left(\underline{\xi} \mid \underline{\theta} + \Delta\underline{\theta}\right) \right] \\
&\quad + \frac{1}{f_{\underline{x}|\underline{\Theta}}\left(\underline{\xi} \mid \underline{\theta} + \Delta\underline{\theta}\right)} \cdot \frac{\partial^2}{\partial \underline{\theta} \cdot \partial \underline{\theta}^T} \left[f_{\underline{x}|\underline{\Theta}}\left(\underline{\xi} \mid \underline{\theta} + \Delta\underline{\theta}\right) \right]
\end{aligned} \tag{10.47}
$$

Again the integral is decomposed into two integrals, and we will first examine the second integral

$$
G_{12} = -\frac{1}{2} \cdot (\Delta\underline{\theta})^T \cdot \int_{\underline{\Xi}} \left\{ \frac{1}{f_{\underline{x}|\underline{\Theta}}\left(\underline{\xi} \mid \underline{\theta} + \Delta\underline{\theta}\right)} \cdot \frac{\partial^2}{\partial \underline{\theta} \cdot \partial \underline{\theta}^T} \left[f_{\underline{x}|\underline{\Theta}}\left(\underline{\xi} \mid \underline{\theta} + \Delta\underline{\theta}\right) \right] \Bigg|_{\Delta\underline{\theta}=\underline{0}} \right\} \cdot f_{\underline{x}|\underline{\Theta}}\left(\underline{\xi} \mid \underline{\theta}\right) \cdot d\underline{\xi} \cdot (\Delta\underline{\theta}) \tag{10.48}
$$

Analogously to the previous calculation we find

$$
G_{12} = -\frac{1}{2} \cdot (\Delta\underline{\theta})^T \cdot \int_{\underline{\Xi}} \left\{ \frac{\partial^2}{\partial \underline{\theta} \cdot \partial \underline{\theta}^T} \left[f_{\underline{x}|\underline{\Theta}}\left(\underline{\xi} \mid \underline{\theta} + \Delta\underline{\theta}\right) \right] \Bigg|_{\Delta\underline{\theta}=\underline{0}} \right\} \cdot d\underline{\xi} \cdot (\Delta\underline{\theta}) \tag{10.49}
$$

and after interchanging integration and differentiation we obtain

$$
G_{12} = 0 \tag{10.50}
$$

Thus only the integral of the first part of the second derivative in equation (10.47) remains

$$
G_1 = \frac{1}{2} \cdot (\Delta\underline{\theta})^T \cdot \int_{\underline{\Xi}} \left\{ \left(\frac{\frac{\partial}{\partial \underline{\theta}} \left[f_{\underline{x}|\underline{\Theta}}\left(\underline{\xi} \mid \underline{\theta} + \Delta\underline{\theta}\right) \right] \cdot \frac{\partial}{\partial \underline{\theta}^T} \left[f_{\underline{x}|\underline{\Theta}}\left(\underline{\xi} \mid \underline{\theta} + \Delta\underline{\theta}\right) \right]}{\left(f_{\underline{x}|\underline{\Theta}}\left(\underline{\xi} \mid \underline{\theta} + \Delta\underline{\theta}\right)\right)^2} \right) \Bigg|_{\Delta\underline{\theta}=\underline{0}} \right\} \cdot f_{\underline{x}|\underline{\Theta}}\left(\underline{\xi} \mid \underline{\theta}\right) \cdot d\underline{\xi} \cdot (\Delta\underline{\theta}) \tag{10.51}
$$

(*This integral is a third notation for Fisher's information*)

Because of

$$\frac{1}{\left[f_{\underline{x}|\Theta}\left(\underline{\xi} \mid \underline{\theta}+\Delta\underline{\theta}\right)\right]} \cdot \frac{\partial}{\partial \underline{\theta}}\left[f_{\underline{x}|\Theta}\left(\underline{\xi} \mid \underline{\theta}+\Delta\underline{\theta}\right)\right]=\frac{\partial}{\partial \underline{\theta}} \ln\left[f_{\underline{x}|\Theta}\left(\underline{\xi} \mid \underline{\theta}+\Delta\underline{\theta}\right)\right] \tag{10.52}$$

and

$$\frac{\partial}{\partial \underline{\theta}^{T}}\left[f_{\underline{x}|\Theta}\left(\underline{\xi} \mid \underline{\theta}+\Delta\underline{\theta}\right)\right]=\left(\frac{\partial}{\partial \underline{\theta}}\left[f_{\underline{x}|\Theta}\left(\underline{\xi} \mid \underline{\theta}+\Delta\underline{\theta}\right)\right]\right)^{T} \tag{10.53}$$

we can rearrange the integral once more

$$G_1=\frac{1}{2} \cdot(\Delta\underline{\theta})^{T} \cdot \int_{\underline{\Xi}}\left(\frac{\partial}{\partial \underline{\theta}} \ln\left[f_{\underline{x}|\Theta}\left(\underline{\xi} \mid \underline{\theta}+\Delta\underline{\theta}\right)\right]\right) \cdot\left(\frac{\partial}{\partial \underline{\theta}} \ln\left[f_{\underline{x}|\Theta}\left(\underline{\xi} \mid \underline{\theta}+\Delta\underline{\theta}\right)\right]\right)^{T}\Bigg|_{\Delta\underline{\theta}=0} \cdot f_{\underline{x}|\Theta}\left(\underline{\xi} \mid \underline{\theta}\right) \cdot d\underline{\xi} \cdot(\Delta\underline{\theta}) \tag{10.54}$$

and we already know this integral from the Cramér–Rao inequality as Fisher's information, so that we have finally found the connection between Kullback's discrimination information and Fisher's information matrix

$$G_1=\frac{1}{2} \cdot(\Delta\underline{\theta})^{T} \cdot F \cdot(\Delta\underline{\theta}) \tag{10.55}$$

If we use vector matrix notation, the quadratic form can be expressed by the following norm

$$G_1=\frac{1}{2} \cdot\|\Delta\underline{\theta}\|_{F} \tag{10.56}$$

From both notations we see that Kullback's information consists of the sum of single amounts of information, which occur separately in Fisher's information matrix. Thus Fisher's information is more appropriate to examine the partitioning of the information in multi-dimensional systems. But up to now we do not have any possibility to interpret this information measure, as we merely know F^{-1} as a lower limit of the variance of an estimator. We therefore use the following sections to explain different points of view, which provide further insight into this matrix information.

10.1.3 Fisher's information and the metric tensor

The arc length ds (distance between two points) is described in cartesian coordinates by the formula of Pythagoras

$$ds^2 = dx^2 + dy^2 + dz^2 \tag{10.57}$$

In general, curvilinear coordinates the arc length is calculated from the vector product

$$ds^2 = dr \cdot dr \tag{10.58}$$

with

$$dr = \frac{\partial r}{\partial u_1} \cdot du_1 + \frac{\partial r}{\partial u_2} \cdot du_2 + \frac{\partial r}{\partial u_3} \cdot du_3 = a_1 \cdot du_1 + a_2 \cdot du_2 + a_3 \cdot du_3 \tag{10.59}$$

We thus get the product

$$\begin{aligned} dr \cdot dr = {} & a_1 \cdot a_1 \cdot du_1 \cdot du_1 + a_1 \cdot a_2 \cdot du_1 \cdot du_2 + a_1 \cdot a_3 \cdot du_1 \cdot du_3 + \\ & + a_2 \cdot a_1 \cdot du_2 \cdot du_1 + a_2 \cdot a_2 \cdot du_2 \cdot du_2 + a_2 \cdot a_3 \cdot du_2 \cdot du_3 + \\ & + a_3 \cdot a_1 \cdot du_3 \cdot du_1 + a_3 \cdot a_2 \cdot du_3 \cdot du_2 + a_3 \cdot a_3 \cdot du_3 \cdot du_3 \end{aligned} \tag{10.60}$$

With Einstein's summation convention we can rearrange this to

$$dr \cdot dr = g_{ij} \cdot du_i \cdot du_j \tag{10.61}$$

g_{ij} is called a fundamental tensor or metric tensor. The coefficients $g_{ij} = g_{ji}$ are symmetric. This tensor is applied to describe relations between different vector spaces [SOK64].

In general we get

$$ds_k \cdot ds_l = g_{ij} \cdot du_i \cdot du_j = \frac{\partial s_k}{\partial u_i} \cdot \frac{\partial s_l}{\partial u_j} \cdot du_i \cdot du_j \tag{10.62}$$

$$g_{ij} = \frac{\partial s_k}{\partial u_i} \cdot \frac{\partial s_l}{\partial u_j} \tag{10.63}$$

The metric tensor consists of the curvature of both coordinate systems with respect to each other.

The fundamental quadratic form from equation (10.61) can also be found in the relation between Kullback's and Fisher's information in equations (10.55) and (10.56). As this quadratic form describes a length, i.e. a parameter, which is independent of the chosen coordinate system, we can derive two conclusions for Kullback's information:

1. Kullback's information is invariant with respect to transformations of the coordinate system (a bijective mapping is required).

2. Kullback's information is always greater than or equal to zero, as a length can never be negative.

During the change of the coordinate system Fisher's information occurs as a fundamental tensor. Thus the least resolution of one coordinate system is transformed into a least resolution of another coordinate system, because Fisher's information matrix is equal to the reciprocal Cramér–Rao bound.

Thus these considerations confirm the properties we found in the sections 9.1 and 9.1.1 for Kullback's information, but we are still not able to classify the metric F. What kind of measure is F? To answer this question, we examine state space models and the observability.

10.1.4 Fisher's information and the stochastic observability

In this chapter we investigate the development of Fisher's information when we have a state space model without any driving noise at the input of the system. Then the only information about the state space variables is gained via the measurements. We assume the driving noise to be equal to zero to circumvent any loss of information by uncertainties of the state space model. The only reason for the application of the driving noise is that we need some model parameter describing the lack of information regarding the model.

The state space model is described by the formula

$$\underline{x}(k+i)=\Phi(k+i,k)\cdot\underline{x}(k) \tag{10.64}$$

The observation model is assumed to be linear, to obtain an easier notation and calculation. (The extension to nonlinear observation models is analogous to section 10.1.1)

$$\underline{y}(k)=C(k)\cdot\underline{x}(k)+\underline{v}(k) \qquad \text{with} \qquad \underline{v}(k)=N(\underline{0},R) \tag{10.65}$$

We now generate the extended measurement vector

$$\underline{Y}(k)=\left[\underline{y}(k-N+1)^T,\underline{y}(k-N+2)^T,\ldots,\underline{y}(k-1)^T,\underline{y}(k)^T\right]^T \tag{10.66}$$

and an extended vector of the errors in the measurements

$$\underline{V}(k)=\left[\underline{v}(k-N+1)^T,\underline{v}(k-N+2)^T,\ldots,\underline{v}(k-1)^T,\underline{v}(k)^T\right]^T \tag{10.67}$$

with the covariance matrix $R_a(k)$, and finally the extended observation equation

$$\underline{Y}(k) = C_a(k) \cdot \underline{x}(k) + \underline{V}(k) \tag{10.68}$$

with the observation vector

$$C_a(k) = [C(k-N+1) \cdot \Phi(k-N+1,k), \ldots, C(k-1) \cdot \Phi(k-1,k), C(k) \cdot I] \tag{10.69}$$

If the measurement errors are gaussian random variables, we get the distribution density

$$f_{\underline{y}|\underline{x}}(\underline{\rho} \mid \underline{x}(k))$$
$$= \frac{1}{(2\pi)^{\frac{m \cdot N}{2}} \cdot |R_a|^{\frac{1}{2}}} \cdot \exp\left\{-\frac{1}{2} \cdot (\underline{\rho} - C_a(k) \cdot \underline{x}(k))^T \cdot R_a^{-1} \cdot (\underline{\rho} - C_a(k) \cdot \underline{x}(k))\right\} \tag{10.70}$$

and analogously to the calculation in section 10.1.1 we obtain Fisher's information matrix of the extended observation

$$F_a = C_a^T \cdot R_a^{-1} \cdot C_a \tag{10.71}$$

If the measurement errors are mutually independent

$$R_a = \begin{pmatrix} R(k-N+1) & 0 & \cdots & 0 & 0 \\ 0 & R(k-N+2) & \cdots & 0 & 0 \\ \vdots & \vdots & \ddots & \vdots & \vdots \\ 0 & 0 & \cdots & R(k-1) & 0 \\ 0 & 0 & \cdots & 0 & R(k) \end{pmatrix} \tag{10.72}$$

the cross-covariances vanish and we obtain the following sum from the quadratic form

$$F = \sum_{i=k-N+1}^{k} \Phi(i,k)^T \cdot C(i)^T \cdot R(i)^{-1} \cdot C(i) \cdot \Phi(i,k) = F(k,k-N+1) \tag{10.73}$$

This is Fisher's information matrix for N measurements. The sum can be transformed by using the identity $\Phi(k,k) = I$

$$F = C(k)^T \cdot R(k)^{-1} \cdot C(k) + \sum_{i=k-N+1}^{k-1} \Phi(i,k)^T \cdot C(i)^T \cdot R(i)^{-1} \cdot C(i) \cdot \Phi(i,k) \tag{10.74}$$

$$F = F(k) + F(k-1,k-N+1) \tag{10.75}$$

This is the recursive formulation, providing the gain of information we get from a single measurement $F(k)$, and the information $F(k-1,k-N+1)$ that was available before we made the observation. $F(k)$ is equal to the Fisher information (already derived in chapter 10.1.1) of a gaussian distribution density of a multi-dimensional measurement vector $\underline{y}(k)$ and the related covariance matrix $R(k)$.

Because of the steady gain of information $F(k)$ the matrix F always tends to infinity. As we neglected the driving noise in the derivation of this result, we can explain this increase of information. The neglect of the driving noise means that we know exactly the system and its dynamical behavior and thus the state variables have to be optimized by succeeding measurements. So the system accumulates more and more information, which it extracts from the measurements, and tends more and more to the exact states, as it does not lose any information.

Real state space models do contain a driving noise component – contrary to this optimal system we used in this derivation – to describe errors in the model and effects that have not been modeled. This driving noise leads to a loss of information in each prediction step of the system and this loss of information is at least partially compensated by the information gained by the observations. In stationary states [ARN97/1] [LOF90] the loss of information via the extrapolation and the gain of information via the measurement compensate each other.

Thus Fisher's information matrix is an absolute maximum of the information, which will not be achieved in real systems. This is not a surprising fact, as we already know F^{-1} from the Cramér–Rao bound to be the lowest variance that can be achieved by an unbiased estimator.

In a linear Kalman filter this sum can be examined without the need of any measurements [ARN97/1]. It enables us to compare the two concepts of Data fusion and Data integration on the basis of the model used. So we can compare the components of the measurement vector with regard to their accuracy and their need in the estimation process. Measurements that are not required for the algorithm to meet certain limitations can be neglected, whilie other measurements, which do not always contain information, can be used according to the information they are able to provide for the estimation. So the examination of the information gives us an additional tool to optimize estimation algorithms with regard to minimum errors and to optimal gain of information.

If the mapping from state space to observation space is nonlinear, the filter equations in the covariance cycle have to be linearized at the current optimal estimate and thus the filter equations depend on the actual estimates. In this case we are not able to compute Fisher's information a priori without any measurements, and the optimization as well as the examination of the information processing of the filter require Monte Carlo simulations.

10.1.4.1 Fisher's information and the Matrix-Riccati equation

The Matrix-Riccati equation describes the development of the covariance of a filter algorithm, which is formulated in state space, and it is derived in [GEL90] [JAZ70] [LOF90]. If we consider the system model (Appendix A.6),

$$\frac{d}{dt}\underline{x} = F \cdot \underline{x} + G \cdot \underline{w} \tag{10.76}$$

and the observation model

$$\underline{y} = H \cdot \underline{x} + \underline{v} \tag{10.77}$$

and the resulting change in the covariance matrices, then we obtain the Matrix-Riccati equation

$$\frac{d}{dt}P = F \cdot P + P \cdot F^T + G \cdot Q \cdot G^T - P \cdot H^T \cdot R^{-1} \cdot H \cdot P \tag{10.78}$$

which allows us to derive the stochastic observability under the conditions, that

1. we neglect the driving noise $\rightarrow Q = 0$
2. we do not have any initial information $\rightarrow P^+(0) \rightarrow \infty$ or $(P^+(0))^{-1} = 0$

(F here denotes the state transition matrix!)

We notice from the Matrix-Riccati equation that the last, negative term leads to a decreasing covariance matrix and it thus describes the gain of information, while the other three addends describe the loss of information by the extrapolation $\left(F \cdot P + P \cdot F^T\right)$ and the driving noise $\left(G \cdot Q \cdot G^T\right)$ (see section 10.1.4). The extrapolation is carried out on the basis of the state space model without any measurements. After the extrapolation is done, the measurements are optimally combined with the extrapolated state space predictions, to minimize the covariance matrix of the estimation error. The reduction of the covariance matrix is caused by the measurement information, described by the stochastic observability.

The first condition has already been discussed in section 10.1.4, while the second ensures that all information gained by the filter results from the evaluation of the observations, as the filter does not have any a priori information. With this conditions we find a linear differential equation

$$\frac{d}{dt}P = F \cdot P + P \cdot F^T - P \cdot H^T \cdot R^{-1} \cdot H \cdot P \tag{10.79}$$

which we multiply by P^{-1}

$$P^{-1} \cdot \frac{d}{dt}P \cdot P^{-1} = P^{-1} \cdot F + F^T \cdot P^{-1} - H^T \cdot R^{-1} \cdot H \tag{10.80}$$

With the identity:

$$P^{-1} \cdot \frac{d}{dt}P \cdot P^{-1} = -\left[\frac{d}{dt}P\right]^{-1} \tag{10.81}$$

With [ATH65] we get

$$\left[\frac{d}{dt}P\right]^{-1} = -P^{-1} \cdot F - F^T \cdot P^{-1} + H^T \cdot R^{-1} \cdot H \tag{10.82}$$

and the solution of this equation is

$$P^{-1} = \int_0^t \Phi(\tau,t)^T \cdot H(\tau)^T \cdot R(\tau)^{-1} \cdot H(\tau) \cdot \Phi(\tau,t) \cdot d\tau \tag{10.83}$$

$$\Phi(\tau,t) = \Phi(t,\tau)^{-1} \tag{10.84}$$

and

$$\dot{\Phi}(t,t_0) = F \cdot \Phi(t,t_0) \tag{10.85}$$

this solution enables us to carry out some considerations regarding the inverted covariance matrix P^{-1}. In the first step P^{-1} is – with the initial value $(P^+(0))^{-1}$ - 'proportional' to R^{-1}, and thus the covariance matrix of the estimation error P increases (in one step) when the accuracy of the measurement decreases, as the measurements provide less information. In this case the information is equal to the inverse covariance matrix P^{-1} of the state space vector.

In the discrete case we obtain the stochastic observability [LOF90]

$$\alpha_1 \le \sum_{i=k-N+1}^{k} \Phi(i,k)^T \cdot H(i)^T \cdot R(i)^{-1} \cdot H(i) \cdot \Phi(i,k) \le \alpha_2 \tag{10.86}$$

where we again find Fisher's information matrix.

The limits α_1 and α_2 are measures for the observability. They indicate how well (α_2) or how poorly (α_1) the state space vector can be observed from the measurements. An increasing α_2 indicates increasing information in the measurements, while a decreasing α_1 indicates a decreasing amount of information in the measurements.

Contrary to the deterministic observability, which only distinguishes between 'observable', and 'not observable', the stochastic observability has a much wider range to determine the quality of the measurement. The two deterministic choices $\alpha_1 = 0$ = not observable, and $\alpha_2 \to \infty$ = observable are merely the extreme special cases of the stochastic observability, which therefore is the extension to contaminated measurements $\underline{y}(k) = \underline{h}[\underline{x}(k), k] + \underline{v}(k)$, where the special case of the exact measurement $\underline{v}(k) = \underline{0}$ is included.

Thus Fisher's information is a measure of the quality of the observability.

10.1.5 Fisher's information and maximum likelihood estimation

In the maximum likelihood estimation we search for the value of the parameter that occurs with the maximum likelihood. This could be achieved by deriving the conditional distribution density, conditioned on this parameter, with regard to this parameter. To enable us to handle derivatives of products with several factors or distribution densities of the exponential family (see section 10.1.7) the approach is modified.

At first we build the logarithm of the conditional distribution density

$$L = \ln\left\{f_{\underline{y}|\underline{x}}(\underline{\rho} \mid \underline{x})\right\} \tag{10.87}$$

which is commonly known as the log-likelihood function. The logarithm is a strict monotonically increasing function, so that the likelihood function has the same extreme values as the conditional distribution density. Moreover the requirement that the argument of the logarithm has to be nonnegative is met by the conditional distribution function, which stems from a monotonically increasing distribution function. To determine the maximum, we generate the derivative of the log-likelihood function with regard to the desired parameter

$$\left.\frac{\partial}{\partial \underline{x}} L\right|_{\underline{x}=\underline{x}_{\text{MLE}}} = \left.\frac{\partial}{\partial \underline{x}} \ln\left\{f_{\underline{y}|\underline{x}}(\underline{\rho} \mid \underline{x})\right\}\right|_{\underline{x}=\underline{x}_{\text{MLE}}} \overset{!}{=} 0 \tag{10.88}$$

and call this estimate $\underline{x}_{\text{MLE}}$ the maximum likelihood estimate.

The expectation value of this derivative, which is not the expectation value of the maximum likelihood estimate, is given by

$$E_{\underline{y}|\underline{x}}\left\{\frac{\partial}{\partial \underline{x}} L\right\} = E_{\underline{y}|\underline{x}}\left\{\frac{\partial}{\partial \underline{x}} \ln\left\{f_{\underline{y}|\underline{x}}(\underline{\rho} \mid \underline{x})\right\}\right\} = \int_{\underline{\mathrm{P}}} \frac{\partial}{\partial \underline{x}} \ln\left\{f_{\underline{y}|\underline{x}}(\underline{\rho} \mid \underline{x})\right\} \cdot f_{\underline{y}|\underline{x}}(\underline{\rho} \mid \underline{x}) \cdot d\underline{\rho} \tag{10.89}$$

$$= \int_{\underline{\mathrm{P}}} \frac{1}{f_{\underline{y}|\underline{x}}(\underline{\rho} \mid \underline{x})} \cdot \frac{\partial}{\partial \underline{x}} \left\{f_{\underline{y}|\underline{x}}(\underline{\rho} \mid \underline{x})\right\} \cdot f_{\underline{y}|\underline{x}}(\underline{\rho} \mid \underline{x}) \cdot d\underline{\rho}$$

$$= \int_{\underline{\mathrm{P}}} \frac{\partial}{\partial \underline{x}} \left\{f_{\underline{y}|\underline{x}}(\underline{\rho} \mid \underline{x})\right\} \cdot d\underline{\rho} = \frac{\partial}{\partial \underline{x}} \int_{\underline{\mathrm{P}}} f_{\underline{y}|\underline{x}}(\underline{\rho} \mid \underline{x}) \cdot d\underline{\rho} = \frac{\partial}{\partial \underline{x}} 1 = 0 \tag{10.90}$$

and it always vanishes (when we are able to interchange integration and differentiation).

The expectation value of the maximum likelihood estimate on the other hand is

$$E_{\underline{y}|\underline{x}}\left\{\frac{\partial}{\partial \underline{x}} L\Big|_{\underline{x}=\underline{x}_{\mathrm{MLE}}} \overset{!}{=} 0\right\} = \int_{\underline{\mathrm{P}}} \left\{\left[\frac{\partial}{\partial \underline{x}} \ln\left\{f_{\underline{y}|\underline{x}}(\underline{\rho} \mid \underline{x})\right\}\right]\Big|_{\underline{x}=\underline{x}_{\mathrm{MLE}}} \overset{!}{=} 0\right\} \cdot f_{\underline{y}|\underline{x}}(\underline{\rho} \mid \underline{x}) \cdot d\underline{\rho}$$

$$E_{\underline{y}|\underline{x}}\{\underline{x}_{\mathrm{MLE}}\} = \int_{\underline{\mathrm{P}}} \underline{x}_{\mathrm{MLE}} \cdot f_{\underline{y}|\underline{x}}(\underline{\rho} \mid \underline{x}) \cdot d\underline{\rho} = \underline{x}_{\mathrm{MLE}} \tag{10.91}$$

The necessary condition for the determination of the maximum is that we find a zero point in the derivation and furthermore that the second derivative at this point $\underline{x} = \underline{x}_{\mathrm{MLE}}$ assumes a negative value.

$$\frac{\partial^2}{\partial \underline{x} \cdot \partial \underline{x}^T} \ln\left\{f_{\underline{y}|\underline{x}}(\underline{\rho} \mid \underline{x})\right\}\Big|_{\underline{x}=\underline{x}_{\mathrm{MLE}}} < 0 \tag{10.92}$$

or

$$-\frac{\partial^2}{\partial \underline{x} \cdot \partial \underline{x}^T} \ln\left\{f_{\underline{y}|\underline{x}}(\underline{\rho} \mid \underline{x})\right\}\Big|_{\underline{x}=\underline{x}_{\mathrm{MLE}}} > 0 \tag{10.93}$$

For multi-dimensional arguments $\underline{x}$ this condition is restricted to the elements on the main diagonal of the resulting matrix. If we again build the expectation value of this derivative, we obtain Fisher's information matrix (the second formulation)

$$F = -\int_{\underline{\mathrm{P}}} \frac{\partial^2}{\partial \underline{x} \cdot \partial \underline{x}^T} \ln\left\{f_{\underline{y}|\underline{x}}(\underline{\rho} \mid \underline{x})\right\}\Big|_{\underline{x}=\underline{x}_{\mathrm{MLE}}} \cdot f_{\underline{y}|\underline{x}}(\underline{\rho} \mid \underline{x}) \cdot d\underline{\rho} \tag{10.94}$$

$$F = \int_{\underline{\mathrm{P}}} \left[\frac{\partial}{\partial \underline{x}} \ln\left\{f_{\underline{y}|\underline{x}}(\underline{\rho} \mid \underline{x})\right\}\Big|_{\underline{x}=\underline{x}_{\mathrm{MLE}}}\right] \cdot \left[\frac{\partial}{\partial \underline{x}} \ln\left\{f_{\underline{y}|\underline{x}}(\underline{\rho} \mid \underline{x})\right\}\Big|_{\underline{x}=\underline{x}_{\mathrm{MLE}}}\right]^T \cdot f_{\underline{y}|\underline{x}}(\underline{\rho} \mid \underline{x}) \cdot d\underline{\rho} \tag{10.95}$$

which is greater than zero at the maximum likelihood estimate. Thus Fisher's information matrix is equal to the expectation value of the negative second derivative of the likelihood function at the point where the maximum likelihood estimate is determined.

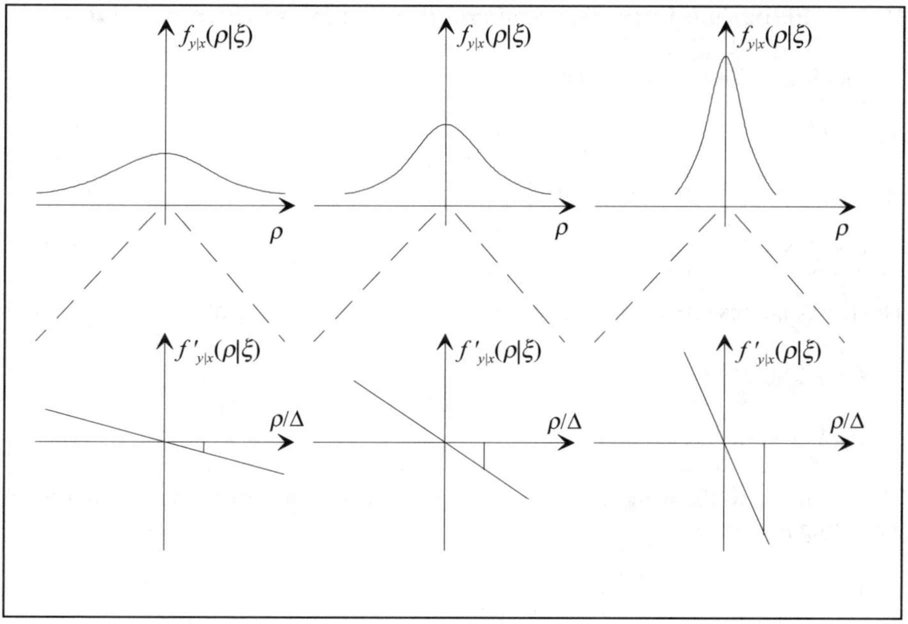

Fig. 10.2: Likelihood functions and their derivatives

In such a maximum the second derivative has to assume negative values, which leads to positive elements on the main diagonal of Fisher's matrix. The more negative the second derivative becomes, the greater Fisher's information becomes and the smaller this maximum becomes. Thus Fisher's information describes the width or the spreading of the maximum of the ML estimation. It is therefore a quantity that measures the quality of the ML estimate. Because of this relation Middleton calls Fisher's information the 'intrinsic curvature'.

This is equivalent to the considerations we made with the metric tensor g_{ij} in the previous chapter. This tensor is a measure for the curvature of a vector space with respect to another vector space and Kullback's information is the sum of all curvatures, weighted with the vectors $\Delta\theta$ (see section 10.1.3).

For orthogonal spaces we get $g_{ij} = 0$ for $i \neq j$. For Fisher's information matrix f_{ij} this means that the components of the random vector are not correlated (e.g., features in pattern recognition). On the main diagonal of Fisher's information we find the reciprocal bounds of the variances of the components. Via comparison of these reciprocal bounds of the variances we may determine the quality of the estimator and the quality of the estimates, and this would be another application for Fisher's information. As the expectation value of the maximum likelihood estimation is always equal to zero, the maximum likelihood estimate is not necessarily an unbiased estimator, as it may also generate biased estimates (see also section 10.7).

10.1.6 Fisher's information and weighted least-squares estimation

The nonlinear observation equation is

$$\underline{y} = \underline{h}[\underline{x}] + \underline{v} \tag{10.96}$$

or with a linearized functional $\underline{h}[\underline{x}]$

$$\underline{y} = C \cdot \underline{x} + \underline{v} \tag{10.97}$$

The least-squares criterion searches for the minimum of the term

$$J = \underline{v}^T \cdot \underline{v}$$

$$J = (\underline{y} - C \cdot \underline{x})^T \cdot (\underline{y} - C \cdot \underline{x}) \tag{10.98}$$

The extension to the weighted least-squares criterion is achieved via an additional weighting matrix G

$$J = \underline{v}^T \cdot G \cdot \underline{v}$$

$$J = (\underline{y} - C \cdot \underline{x})^T \cdot G \cdot (\underline{y} - C \cdot \underline{x}) \tag{10.99}$$

We are looking for the minimum of this expression, which we determine with

$$\frac{\partial J}{\partial \underline{x}} \overset{!}{=} 0 \tag{10.100}$$

We get

$$\frac{\partial J}{\partial \underline{x}} = \frac{\partial}{\partial \underline{x}} \left[(\underline{y} - C \cdot \underline{x})^T \cdot G \cdot (\underline{y} - C \cdot \underline{x}) \right] = -2 \cdot C^T \cdot G \cdot (\underline{y} - C \cdot \underline{x}) \overset{!}{=} 0$$

$$2 \cdot C^T \cdot G \cdot \underline{y} = 2 \cdot C^T \cdot G \cdot C \cdot \underline{x}$$

$$\underline{x}_{\mathrm{WLS}} = \left(C^T \cdot G \cdot C\right)^{-1} \cdot C^T \cdot G \cdot \underline{y} \tag{10.101}$$

This is the weighted least-squares estimate of the state space vector, where the weighting matrix G is not specified up to now. If we examine the conditional gaussian distribution

$$l = f_{\underline{y}|\underline{x}}\left(\underline{\rho} \mid \underline{x}(k)\right)$$

$$= \frac{1}{(2\pi)^{\frac{m}{2}} \cdot |R|^{\frac{1}{2}}} \cdot \exp\left\{-\frac{1}{2} \cdot \left(\underline{\rho} - E_{\underline{y}|\underline{x}}\left\{\underline{\rho} \mid \underline{x}(k)\right\}\right)^T \cdot R^{-1} \cdot \left(\underline{\rho} - E_{\underline{y}|\underline{x}}\left\{\underline{\rho} \mid \underline{x}(k)\right\}\right)\right\}$$

(10.102)

which is the likelihood function l, and we build the log-likelihood function

$$\ln(l) = L = -\frac{1}{2} \cdot \ln\left[(2\pi)^m \cdot |R|\right] - \frac{1}{2} \cdot \left(\underline{\rho} - E_{\underline{y}|\underline{x}}\left\{\underline{\rho} \mid \underline{x}(k)\right\}\right)^T \cdot R^{-1} \cdot \left(\underline{\rho} - E_{\underline{y}|\underline{x}}\left\{\underline{\rho} \mid \underline{x}(k)\right\}\right)$$

$$\ln(l) = L = -\frac{1}{2} \cdot \ln\left[(2\pi)^m \cdot |R|\right] - \frac{1}{2} \cdot J \qquad (10.103)$$

then we immediately notice, by comparing this result with equation (10.101), that the maximum of the log-likelihood function is equal to the minimum of the weighted least-squares estimation. To achieve this equality, we have to chose the weighting matrix equal to $G = R^{-1}$ the inverse covariance matrix of the measurement errors. For gaussian distributions we thus find

Maximum likelihood (ML) estimate = Weighted least-squares (WLS) estimate

because the maximum of the MLE is exactly at that point, where we also find the minimum of the WLS estimation. For the calculation of the second order moments of the weighted least-squares estimation we transform equation (10.101) once more

$$\underline{x}_{\text{WLS}} = \left(C^T \cdot G \cdot C\right)^{-1} \cdot C^T \cdot G \cdot \underline{y}$$

$$\underline{x}_{\text{WLS}} = \left(C^T \cdot G \cdot C\right)^{-1} \cdot C^T \cdot G \cdot (C \cdot \underline{x} + \underline{v})$$

$$C^T \cdot G \cdot C \cdot \underline{x}_{\text{WLS}} = C^T \cdot G \cdot (C \cdot \underline{x} + \underline{v})$$

$$C^T \cdot G \cdot C \cdot \left(\underline{x}_{\text{WLS}} - \underline{x}\right) = C^T \cdot G \cdot \underline{v} \qquad (10.104)$$

and now we are able to determine the expectation value

$$E\left\{C^T \cdot G \cdot C \cdot \left(\underline{x}_{\text{WLS}} - \underline{x}\right)\right\} = E\left\{C^T \cdot G \cdot \underline{v}\right\} \qquad (10.105)$$

$$C^T \cdot G \cdot C \cdot E\left\{\left(\underline{x}_{\text{WLS}} - \underline{x}\right)\right\} = C^T \cdot G \cdot E\{\underline{v}\} = 0 \qquad \text{because of } E\{\underline{v}\} = 0$$

$$E\{\underline{x}\} = \underline{x}_{\text{WLS}} \qquad (10.106)$$

and the covariance matrix with a symmetric weighting matrix G:

$$E\left\{\left[C^T \cdot G \cdot C \cdot \left(\underline{x}_{\text{WLS}} - \underline{x}\right)\right] \cdot \left[C^T \cdot G \cdot C \cdot \left(\underline{x}_{\text{WLS}} - \underline{x}\right)\right]^T\right\}$$

$$= E\left\{\left[C^T \cdot G \cdot \underline{v}\right] \cdot \left[C^T \cdot G \cdot \underline{v}\right]^T\right\}$$

$$C^T \cdot G \cdot C \cdot E\left\{\left(\underline{x}_{\text{WLS}} - \underline{x}\right) \cdot \left(\underline{x}_{\text{WLS}} - \underline{x}\right)^T\right\} \cdot C^T \cdot G \cdot C = C^T \cdot G \cdot E\left\{\underline{v} \cdot \underline{v}^T\right\} \cdot G \cdot C \tag{10.107}$$

Using the equality $G = R^{-1}$ we find

$$C^T \cdot R^{-1} \cdot C \cdot E\left\{\left(\underline{x}_{\text{WLS}} - \underline{x}\right) \cdot \left(\underline{x}_{\text{WLS}} - \underline{x}\right)^T\right\} \cdot C^T \cdot R^{-1} \cdot C = C^T \cdot R^{-1} \cdot R \cdot R^{-1} \cdot C$$

$$C^T \cdot R^{-1} \cdot C \cdot E\left\{\left(\underline{x}_{\text{WLS}} - \underline{x}\right) \cdot \left(\underline{x}_{\text{WLS}} - \underline{x}\right)^T\right\} \cdot C^T \cdot R^{-1} \cdot C = C^T \cdot R^{-1} \cdot C \tag{10.108}$$

and multiplication by $\left(C^T \cdot R^{-1} \cdot C\right)^{-1}$ from the right side (or from the left side) provides

$$C^T \cdot R^{-1} \cdot C \cdot E\left\{\left(\underline{x}_{\text{WLS}} - \underline{x}\right) \cdot \left(\underline{x}_{\text{WLS}} - \underline{x}\right)^T\right\} = I$$

$$E\left\{\left(\underline{x}_{\text{WLS}} - \underline{x}\right) \cdot \left(\underline{x}_{\text{WLS}} - \underline{x}\right)^T\right\} = \left(C^T \cdot R^{-1} \cdot C\right)^{-1} = F^{-1} \tag{10.109}$$

Thus the covariance matrix of the weighted least-squares estimation is equal to the inverse of Fisher's matrix, which uniquely identifies Fisher's matrix as an inverse covariance matrix.

But the inverse of Fisher's information matrix does not generally exist, because there may exist state vector components that cannot be observed by the available measurements. It is Fisher's matrix for one observation that we are currently talking about. But it is possible to determine the information matrix of a recursive algorithm, which combines the estimates (i.e. the components of the estimated vector) of $\underline{x}_{\text{WLS}}$, and this information matrix can be computed recursively and it can be inverted [LOF90]. This is because the connections between the components even provide information about non-observable components of the estimated vector.

10.1.7 The availability of the Cramér–Rao bound

As we explained the matrix F with geometrical formulations and also integrated it into the ML estimation, we at last ask what a distribution density has to look like, to enable an estimator to achieve this lowest limit of the variance, the Cramér–Rao bound. We thus go back to the derivation of the Cramér–Rao bound and take a closer look at the Schwartz inequality and its relation to the correlation coefficient.

From the correlation coefficient we know that it assumes values in the range from −1 to 1 and that the extreme values are only achieved if one of the two random variables is a linear deterministic function of the other random variable. We applied the Schwartz inequality to the equation

$$\int_{\underline{P}} \left[\underline{t}_y(\underline{\rho}) - \underline{x} \right] \cdot \frac{\partial}{\partial \underline{x}} \ln\left\{ f_{\underline{y}|\underline{x}}(\underline{\rho} \mid \underline{x}) \right\} \cdot f_{\underline{y}|\underline{x}}(\underline{\rho} \mid \underline{x}) \cdot d\underline{\rho} = I \tag{10.110}$$

As both estimators are unbiased, we get

$$\int_{\underline{P}} \left[\underline{t}_y(\underline{\rho}) - \underline{x} \right] \cdot f_{\underline{y}|\underline{x}}(\underline{\rho} \mid \underline{x}) \cdot d\underline{\rho} = \underline{0} \tag{10.111}$$

and

$$\int_{\underline{P}} \frac{\partial}{\partial \underline{x}} \ln\left\{ f_{\underline{y}|\underline{x}}(\underline{\rho} \mid \underline{x}) \right\} \cdot f_{\underline{y}|\underline{x}}(\underline{\rho} \mid \underline{x}) \cdot d\underline{\rho} = \underline{0} \tag{10.112}$$

We thus write

$$\begin{aligned} &\int_{\underline{P}} \left[\underline{t}_y(\underline{\rho}) - \underline{x} \right] \cdot \frac{\partial}{\partial \underline{x}} \ln\left\{ f_{\underline{y}|\underline{x}}(\underline{\rho} \mid \underline{x}) \right\} \cdot f_{\underline{y}|\underline{x}}(\underline{\rho} \mid \underline{x}) \cdot d\underline{\rho} \\ &= \operatorname{cov}\left(\left[\underline{t}_y(\underline{\rho}) - \underline{x} \right], \frac{\partial}{\partial \underline{x}} \ln\left\{ f_{\underline{y}|\underline{x}}(\underline{\rho} \mid \underline{x}) \right\} \right) \end{aligned} \tag{10.112}$$

and with the cross-correlation coefficient we obtain

$$\left[\operatorname{cov}\left(\left[\underline{t}_y(\underline{\rho}) - \underline{x} \right], \frac{\partial}{\partial \underline{x}} \ln\left\{ f_{\underline{y}|\underline{x}}(\underline{\rho} \mid \underline{x}) \right\} \right) \right]^2 \leq \operatorname{var}\left(\underline{t}_y(\underline{\rho}) - \underline{x} \right) \cdot \operatorname{var}\left(\frac{\partial}{\partial \underline{x}} \ln\left\{ f_{\underline{y}|\underline{x}}(\underline{\rho} \mid \underline{x}) \right\} \right) \tag{10.114}$$

with

$$\operatorname{var}\left(\underline{t}_y(\underline{\rho}) - \underline{x} \right) = E\left\{ \left(\underline{t}_y(\underline{\rho}) - \underline{x} - E\left\{ \left(\underline{t}_y(\underline{\rho}) - \underline{x} \right) \right\} \right)^2 \right\} \tag{10.115}$$

and because of

$$E\left\{ \left(\underline{t}_y(\underline{\rho}) - \underline{x} \right) \right\} = 0 \qquad \text{(unbiased estimator)} \tag{10.116}$$

we find

$$\operatorname{var}\left(\underline{t}_y(\underline{\rho}) - \underline{x} \right) = \operatorname{var}\left(\underline{t}_y(\underline{\rho}) \right) \tag{10.117}$$

and equality holds, if the estimators are connected by some linear function or if one of the estimators provides constant values (trivial case).

We assume a linear dependence, and with

$$A'(\underline{x}) = \frac{\partial}{\partial \underline{x}} \cdot A(\underline{x}) \qquad \text{and} \qquad B'(\underline{x}) = \frac{\partial}{\partial \underline{x}} \cdot B(\underline{x}) \tag{10.118}$$

we find

$$\frac{\partial}{\partial \underline{x}} \ln\left\{f_{\underline{y}|\underline{x}}(\underline{\rho} \mid \underline{x})\right\} = A'(\underline{x}) \cdot t_{\underline{y}}(\underline{\rho}) + B'(\underline{x}) \tag{10.119}$$

Integration and insertion of a scalar integration constant $K_{\underline{y}}(\underline{\rho})$ provides

$$\ln\left\{f_{\underline{y}|\underline{x}}(\underline{\rho} \mid \underline{x})\right\} = \int_{\underline{\Xi}} \left(A'(\underline{x}) \cdot t_{\underline{y}}(\underline{\rho}) + B'(\underline{x})\right) \cdot d\underline{x} + K_{\underline{y}}(\underline{\rho}) \tag{10.120}$$

and then

$$\ln\left\{f_{\underline{y}|\underline{x}}(\underline{\rho} \mid \underline{x})\right\} = A(\underline{x}) \cdot t_{\underline{y}}(\underline{\rho}) + B(\underline{x}) + K_{\underline{y}}(\underline{\rho}) \tag{10.121}$$

We thus find the conditional distribution density

$$f_{\underline{y}|\underline{x}}(\underline{\rho} \mid \underline{x}) = \exp\left\{A(\underline{x}) \cdot t_{\underline{y}}(\underline{\rho}) + B(\underline{x}) + K_{\underline{y}}(\underline{\rho})\right\}$$

$$f_{\underline{y}|\underline{x}}(\underline{\rho} \mid \underline{x}) = \exp\left\{A(\underline{x}) \cdot t_{\underline{y}}(\underline{\rho})\right\} \cdot \exp\left\{B(\underline{x})\right\} \cdot \exp\left\{K_{\underline{y}}(\underline{\rho})\right\}$$

$$f_{\underline{y}|\underline{x}}(\underline{\rho} \mid \underline{x}) = C(\underline{x}) \cdot \exp\left\{A(\underline{x}) \cdot t_{\underline{y}}(\underline{\rho})\right\} \cdot H_{\underline{y}}(\underline{\rho}) \tag{10.122}$$

and this is a distribution from the ‘exponential family’ [PAZ93], where the logarithmic dimension

$$\log \dim = \ln\left(\frac{f_{\underline{y}|\underline{x}}(\underline{\rho} \mid \underline{x})}{f_{\underline{y}|\underline{x}}(\underline{\rho} \mid \underline{x} + \Delta\underline{x})}\right) \tag{10.123}$$

is finite. This logarithmic dimension occurs in Kullback’s information G_1, where we compute the expectation of this logarithmic dimension. If G_1 leads to a finite discrimination information, the Cramér–Rao bound can theoretically be achieved from an estimator. Such an estimator is called an efficient estimator. This term will be the subject of further considerations in the following chapters.

10.1.8 Efficiency, asymptotic efficiency, consistency, bias

In the process of our considerations of variance bounds and of limitations of higher order moments, we have to make some remarks regarding the notions and the vocabulary that we are going to use in this connection. We therefore use this chapter to introduce and to explain several notions that are commonly used in estimation and information theory to describe estimators, which are the main point in our considerations.

10.1.8.1 Unbiased estimator

An estimator (which is sometimes also called a statistic) $t_{\underline{y}}(\underline{\rho})$ is called unbiased when the expectation value of the loss function $loss[t_{\underline{y}}(\underline{\rho}),\underline{x}]$ meets the following equation:

$$E_{\underline{x}}\{loss[t_{\underline{y}}(\underline{\rho}),\underline{x}_1]\}\geq E_{\underline{x}}\{loss[t_{\underline{y}}(\underline{\rho}),\underline{x}]\} \qquad \text{for all } \underline{x}_1,\ \underline{x} \tag{10.124}$$

If we, for instance, use a quadratic loss function $(t_{\underline{y}}(\underline{\rho})-\underline{x})\cdot(t_{\underline{y}}(\underline{\rho})-\underline{x})^T$ for the estimator $t_{\underline{y}}(\underline{\rho})$, estimation $\underline{x}$, the upper equation becomes

$$E_{\underline{x}}\{(t_{\underline{y}}(\underline{\rho})-\underline{x}_1)\cdot(t_{\underline{y}}(\underline{\rho})-\underline{x}_1)^T\}\geq E_{\underline{x}}\{(t_{\underline{y}}(\underline{\rho})-\underline{x})\cdot(t_{\underline{y}}(\underline{\rho})-\underline{x})^T\} \text{ for all } \underline{x}_1,\ \underline{x} \tag{10.125}$$

and this condition is only met by the estimator that has an expectation value equal to $\underline{x}$, because in this case the right-hand side of this equation is equal to the variance of the estimator (= least central second order moment). If we use the loss function $|t_{\underline{y}}(\underline{\rho})-\underline{x}|$ instead, then the condition for an unbiased estimator is equivalent to requiring that the median of the estimator $t_{\underline{y}}(\underline{\rho})$ is equal to $\underline{x}$ [LOF90]. The familiar definition of an unbiased estimator

$$E\{t_{\underline{y}}(\underline{\rho})\}=\underline{x} \tag{10.126}$$

is consistent with the general definition shown if we choose the quadratic loss function.

10.1.8.2 Consistency

We call a procedure consistent if the expected deviation from the lowest achievable bound (deviation = regret) for every state that has been computed by the estimator approaches zero, if n = number of samples tends to infinity.

'regret': $regret\left[\underline{t}_{\underline{y}}(\underline{\rho}),\underline{x}\right]= loss\left[\underline{t}_{\underline{y}}(\underline{\rho}),\underline{x}\right]-\min_{\underline{t}_{\underline{y}}(\underline{\rho})}\left\{loss\left[\underline{t}_{\underline{y}}(\underline{\rho}),\underline{x}\right]\right\}$ (10.127)

Such a deviation between the optimal and the real bound occurs if we choose a wrong estimator, which does not provide the minimum loss function (e.g., a biased estimator).

If we choose again the quadratic loss function $\left(\underline{t}_{\underline{y}}(\underline{\rho})_n-\underline{x}\right)\cdot\left(\underline{t}_{\underline{y}}(\underline{\rho})_n-\underline{x}\right)^T$, we find

$$\lim_{n\to\infty} E\left\{\left(\underline{t}_{\underline{y}}(\underline{\rho})_n-\underline{x}\right)\cdot\left(\underline{t}_{\underline{y}}(\underline{\rho})_n-\underline{x}\right)^T\right\}=0 \tag{10.128}$$

which means that $\underline{t}_{\underline{y}}(\underline{\rho})_n$ tends to $\underline{x}$ in the quadratic mean (convergence in the quadratic mean), if the number of samples tends to infinity. If an estimator converges in the quadratic mean to $\underline{x}$, this implies that the estimator also converges in probability to $\underline{x}$ [JAZ70].

10.1.8.3 Efficiency

An estimator $\underline{t}_{\underline{y}}(\underline{\rho})$ is more efficient as another estimator $\underline{t}_{2\underline{y}}(\underline{\rho})$ if

$$regret\left[\underline{t}_{\underline{y}}(\underline{\rho}),\underline{x}\right]\le regret\left[\underline{t}_{2\underline{y}}(\underline{\rho}),\underline{x}\right] \tag{10.129}$$

where the initial conditions are the same for both estimators. The quadratic loss function provides in this case

$$E\left\{\left(\underline{t}_{\underline{y}}(\underline{\rho})-\underline{x}\right)\cdot\left(\underline{t}_{\underline{y}}(\underline{\rho})-\underline{x}\right)^T\right\}\le E\left\{\left(\underline{t}_{2\underline{y}}(\underline{\rho})-\underline{x}\right)\cdot\left(\underline{t}_{2\underline{y}}(\underline{\rho})-\underline{x}\right)^T\right\} \tag{10.130}$$

and the relative efficiency

$$e\left(\underline{t}_{2\underline{y}}(\underline{\rho}),\underline{t}_{\underline{y}}(\underline{\rho})\right)=\frac{E\left\{\left(\underline{t}_{\underline{y}}(\underline{\rho})-\underline{x}\right)\cdot\left(\underline{t}_{\underline{y}}(\underline{\rho})-\underline{x}\right)^T\right\}}{E\left\{\left(\underline{t}_{2\underline{y}}(\underline{\rho})-\underline{x}\right)\cdot\left(\underline{t}_{2\underline{y}}(\underline{\rho})-\underline{x}\right)^T\right\}} \tag{10.131}$$

(This is only a formal notation, because both terms in the quotient are matrices.)

$E\left\{\left(\underline{t}_{\underline{y}}(\underline{\rho})-\underline{x}\right)\cdot\left(\underline{t}_{\underline{y}}(\underline{\rho})-\underline{x}\right)^T\right\}=\mathrm{var}\left(\underline{t}_{\underline{y}}(\underline{\rho})\right)$ for an efficient unbiased estimator and thus the minimum can be calculated with the Cramér–Rao inequality.

Inserting this minimum, we obtain efficiencies in the range from 0 to 1, and we thus have the relation between the quality of the real estimator and the optimum estimator, which achieves the Cramér–Rao bound.

$$e\left(t_{2_{\underline{y}}}(\underline{\rho}), t_{\underline{y}}(\underline{\rho})\right) = \frac{F^{-1}}{E\left\{\left(t_{2_{\underline{y}}}(\underline{\rho}) - \underline{x}\right) \cdot \left(t_{2_{\underline{y}}}(\underline{\rho}) - \underline{x}\right)^T\right\}} \tag{10.132}$$

Asymptotic efficiency denotes the limit of the efficiency if n = the number of samples in the observation tends to infinity.

$$\lim_{n\to\infty} e\left(t_{2_{\underline{y}}}(\underline{\rho}), t_{\underline{y}}(\underline{\rho})\right) = \lim_{n\to\infty} \frac{F^{-1}}{E\left\{\left(t_{2_{\underline{y}}}(\underline{\rho}) - \underline{x}\right) \cdot \left(t_{2_{\underline{y}}}(\underline{\rho}) - \underline{x}\right)^T\right\}} \tag{10.133}$$

If the asymptotic efficiency is equal to 1 (equal to the unit matrix I for multi-dimensional estimators), the estimator is called asymptotically efficient, which means that this estimator achieves the Cramér–Rao bound for an infinite number of samples.

10.1.9 Summary

We introduced the Cramér–Rao bound as a lowest limit of the variance and we stated some concepts as well as applications of this bound. The derivation was based on the regularity conditions (see Appendix A.5), which guarantee the existence of a (best asymptotically normal) estimator and thus secure the existence of the variance of such an estimator. If these conditions do not hold for the examined estimator, we are first only able to discover the trivial result that the variance of the applied estimator has to be positive or equal to zero.

In the next section we thus occupy ourselves with a bound of the variance of such an estimator, where the regularity conditions are not met. We will compare the result with the Cramér–Rao bound.

10.2 Chapman–Robbins bound

If the 'regularity conditions' do not hold, Chapman and Robbins [CHA51] suggest another bound of the variance, which also starts with the assumption of an unbiased estimator:

$$\int_{\underline{\mathrm{P}}} t_{\underline{y}}(\underline{\rho}) \cdot f_{\underline{y}|\underline{x}}(\underline{\rho} \mid \underline{x}(k)) \cdot d\underline{\rho} = \underline{x}(k) \tag{10.134}$$

Additionally they use the equation

$$\int_{\underline{P}} \underline{t}_{\underline{y}}(\underline{\rho}) \cdot f_{\underline{y}|\underline{x}}(\underline{\rho} \mid \underline{x}(k) + \underline{h}) \cdot d\underline{\rho} = \underline{x}(k) + \underline{h} \tag{10.135}$$

which is an unbiased estimator, where the desired parameter $\underline{x}$ is varied by the amount $\underline{h}$. We thus may write:

$$\int_{\underline{P}} f_{\underline{y}|\underline{x}}(\underline{\rho} \mid \underline{x}(k)) \cdot d\underline{\rho} = 1 \tag{10.136}$$

and

$$\int_{\underline{P}} f_{\underline{y}|\underline{x}}(\underline{\rho} \mid \underline{x}(k) + \underline{h}) \cdot d\underline{\rho} = 1 \tag{10.137}$$

Using the independence of the parameters $\underline{x}$ and $\underline{h}$ from the integration variable $\underline{\rho}$ we get

$$\int_{\underline{P}} \underline{x} \cdot f_{\underline{y}|\underline{x}}(\underline{\rho} \mid \underline{x}(k)) \cdot d\underline{\rho} = \underline{x} \tag{10.138}$$

and

$$\int_{\underline{P}} \underline{x} \cdot f_{\underline{y}|\underline{x}}(\underline{\rho} \mid \underline{x}(k) + \underline{h}) \cdot d\underline{\rho} = \underline{x} \tag{10.139}$$

We use these equations to transform equations (10.134) and (10.135) to

$$\int_{\underline{P}} \left(\underline{t}_{\underline{y}}(\underline{\rho}) - \underline{x}\right) \cdot f_{\underline{y}|\underline{x}}(\underline{\rho} \mid \underline{x}(k)) \cdot d\underline{\rho} = \underline{0} \tag{10.140}$$

and

$$\int_{\underline{P}} \left(\underline{t}_{\underline{y}}(\underline{\rho}) - \underline{x}\right) \cdot f_{\underline{y}|\underline{x}}(\underline{\rho} \mid \underline{x}(k) + \underline{h}) \cdot d\underline{\rho} = \underline{h} \tag{10.141}$$

The difference of the two equations is

$$\int_{\underline{P}} \left(\underline{t}_{\underline{y}}(\underline{\rho}) - \underline{x}\right) \cdot f_{\underline{y}|\underline{x}}(\underline{\rho} \mid \underline{x}(k) + \underline{h}) \cdot d\underline{\rho} - \int_{\underline{P}} \left(\underline{t}_{\underline{y}}(\underline{\rho}) - \underline{x}\right) \cdot f_{\underline{y}|\underline{x}}(\underline{\rho} \mid \underline{x}(k)) \cdot d\underline{\rho} = \underline{h} \tag{10.142}$$

If we build one integral out of these two integrals we obtain

$$\int_{\underline{\mathrm{P}}} \left(t_{\underline{y}}(\underline{\rho})-\underline{x}\right)\cdot\left[f_{\underline{y}|\underline{x}}(\underline{\rho}\mid\underline{x}(k)+\underline{h})-f_{\underline{y}|\underline{x}}(\underline{\rho}\mid\underline{x}(k))\right]\cdot d\underline{\rho}=\underline{h} \tag{10.143}$$

and expand the kernel with $f_{\underline{y}|\underline{x}}(\underline{\rho}\mid\underline{x}(k))$

$$\int_{\underline{\mathrm{P}}} \left(t_{\underline{y}}(\underline{\rho})-\underline{x}\right)\cdot\frac{\left[f_{\underline{y}|\underline{x}}(\underline{\rho}\mid\underline{x}(k)+\underline{h})-f_{\underline{y}|\underline{x}}(\underline{\rho}\mid\underline{x}(k))\right]}{f_{\underline{y}|\underline{x}}(\underline{\rho}\mid\underline{x}(k))}\cdot f_{\underline{y}|\underline{x}}(\underline{\rho}\mid\underline{x}(k))\cdot d\underline{\rho}=\underline{h} \tag{10.144}$$

which is equal to the expectation value of the term

$$S=\frac{\left[f_{\underline{y}|\underline{x}}(\underline{\rho}\mid\underline{x}(k)+\underline{h})-f_{\underline{y}|\underline{x}}(\underline{\rho}\mid\underline{x}(k))\right]}{f_{\underline{y}|\underline{x}}(\underline{\rho}\mid\underline{x}(k))} \tag{10.145}$$

We now apply the Schwartz inequality

$$\left(E\{A\cdot B\}\right)^2\le E\left\{A^2\right\}\cdot E\left\{B^2\right\} \tag{10.146}$$

and find

$$\begin{aligned}\underline{h}\cdot\underline{h}^T\le&\int_{\underline{\mathrm{P}}}\left(t_{\underline{y}}(\underline{\rho})-\underline{x}\right)\cdot\left(t_{\underline{y}}(\underline{\rho})-\underline{x}\right)^T\cdot f_{\underline{y}|\underline{x}}(\underline{\rho}\mid\underline{x}(k))\cdot d\underline{\rho}\cdot\\&\cdot\int_{\underline{\mathrm{P}}}\left[\frac{\left[f_{\underline{y}|\underline{x}}(\underline{\rho}\mid\underline{x}(k)+\underline{h})-f_{\underline{y}|\underline{x}}(\underline{\rho}\mid\underline{x}(k))\right]}{f_{\underline{y}|\underline{x}}(\underline{\rho}\mid\underline{x}(k))}\right]\cdot\left[\frac{\left[f_{\underline{y}|\underline{x}}(\underline{\rho}\mid\underline{x}(k)+\underline{h})-f_{\underline{y}|\underline{x}}(\underline{\rho}\mid\underline{x}(k))\right]}{f_{\underline{y}|\underline{x}}(\underline{\rho}\mid\underline{x}(k))}\right]^T\\&\cdot f_{\underline{y}|\underline{x}}(\underline{\rho}\mid\underline{x}(k))\cdot d\underline{\rho}\end{aligned} \tag{10.147}$$

and

$$\begin{aligned}&\mathrm{var}\left(t_{\underline{y}}(\underline{\rho})\right)\ge\underline{h}\cdot\underline{h}^T\\&\cdot\left(\int_{\underline{\mathrm{P}}}\left[\frac{\left[f_{\underline{y}|\underline{x}}(\underline{\rho}\mid\underline{x}(k)+\underline{h})-f_{\underline{y}|\underline{x}}(\underline{\rho}\mid\underline{x}(k))\right]}{f_{\underline{y}|\underline{x}}(\underline{\rho}\mid\underline{x}(k))}\right]\cdot\left[\frac{\left[f_{\underline{y}|\underline{x}}(\underline{\rho}\mid\underline{x}(k)+\underline{h})-f_{\underline{y}|\underline{x}}(\underline{\rho}\mid\underline{x}(k))\right]}{f_{\underline{y}|\underline{x}}(\underline{\rho}\mid\underline{x}(k))}\right]^T\cdot f_{\underline{y}|\underline{x}}(\underline{\rho}\mid\underline{x}(k))\cdot d\underline{\rho}\right)^{-1}\end{aligned} \tag{10.148}$$

The Chapman–Robbins bound is achieved by variation of the parameter vector of the denominator on the right-hand side of this inequality.

$$\operatorname{var}\left(\underline{t}_y(\underline{\rho})\right) \geq$$

$$\min_{\underline{h}}\left\{\underline{h}\cdot\underline{h}^T\cdot\left(\int_{\underline{P}}\left[\frac{\left[f_{\underline{y}|\underline{x}}(\underline{\rho}\mid\underline{x}(k)+\underline{h})-f_{\underline{y}|\underline{x}}(\underline{\rho}\mid\underline{x}(k))\right]}{f_{\underline{y}|\underline{x}}(\underline{\rho}\mid\underline{x}(k))}\right]\cdot\left[\frac{\left[f_{\underline{y}|\underline{x}}(\underline{\rho}\mid\underline{x}(k)+\underline{h})-f_{\underline{y}|\underline{x}}(\underline{\rho}\mid\underline{x}(k))\right]}{f_{\underline{y}|\underline{x}}(\underline{\rho}\mid\underline{x}(k))}\right]^T\right.\right.$$

$$\left.\left.\cdot f_{\underline{y}|\underline{x}}(\underline{\rho}\mid\underline{x}(k))\cdot d\underline{\rho}\right)^{-1}\right\} \tag{10.149}$$

A minimum of the denominator leads to a maximum limitation of the variance $\operatorname{var}\left(\underline{t}_y(\underline{\rho})\right)$ because of the reciprocal connection. Thus the Chapman–Robbins bound is not necessarily the lowest limit of the variance. This restriction is caused by the lack of the 'regularity conditions'.

How is this bound connected to the Cramér–Rao bound? To answer this question we have to rearrange Fisher's information matrix in the Cramér–Rao bound. In Fisher's information matrix we find the derivative of the conditional distribution density, which is not found in equation (10.149). In order to generate a connection between the differential quotient and the derivative, we use the Taylor approximation

$$f_{\underline{y}|\underline{x}}(\underline{\rho}\mid\underline{x}(k)+\underline{h})=f_{\underline{y}|\underline{x}}(\underline{\rho}\mid\underline{x}(k))+\frac{\partial}{\partial\underline{x}}f_{\underline{y}|\underline{x}}(\underline{\rho}\mid\underline{x}(k)+\underline{h})\Big|_{\underline{h}=\underline{0}}\cdot\underline{h}+\text{h.o.t.}$$

neglect the higher order terms and find

$$f_{\underline{y}|\underline{x}}(\underline{\rho}\mid\underline{x}(k)+\underline{h})-f_{\underline{y}|\underline{x}}(\underline{\rho}\mid\underline{x}(k))=\frac{\partial}{\partial\underline{x}}f_{\underline{y}|\underline{x}}(\underline{\rho}\mid\underline{x}(k)+\underline{h})\Big|_{\underline{h}=\underline{0}}\cdot\underline{h}$$

$$\left(f_{\underline{y}|\underline{x}}(\underline{\rho}\mid\underline{x}(k)+\underline{h})-f_{\underline{y}|\underline{x}}(\underline{\rho}\mid\underline{x}(k))\right)\cdot\underline{h}^T=\frac{\partial}{\partial\underline{x}}f_{\underline{y}|\underline{x}}(\underline{\rho}\mid\underline{x}(k)+\underline{h})\Big|_{\underline{h}=\underline{0}}\cdot\underline{h}\cdot\underline{h}^T$$

$$\frac{\partial}{\partial\underline{x}}f_{\underline{y}|\underline{x}}(\underline{\rho}\mid\underline{x}(k)+\underline{h})\Big|_{\underline{h}=\underline{0}}=\left(f_{\underline{y}|\underline{x}}(\underline{\rho}\mid\underline{x}(k)+\underline{h})-f_{\underline{y}|\underline{x}}(\underline{\rho}\mid\underline{x}(k))\right)\cdot\underline{h}^T\cdot\left(\underline{h}\cdot\underline{h}^T\right)^{-1}$$

In the following formulation of Fisher's information matrix we use

$$\begin{aligned}\frac{\partial}{\partial\underline{x}}f_{\underline{y}|\underline{x}}(\underline{\rho}\mid\underline{x}(k))&=\frac{\partial}{\partial\underline{x}}f_{\underline{y}|\underline{x}}(\underline{\rho}\mid\underline{x}(k)+\underline{h})\Big|_{\underline{h}=\underline{0}}\\&=\left(f_{\underline{y}|\underline{x}}(\underline{\rho}\mid\underline{x}(k)+\underline{h})-f_{\underline{y}|\underline{x}}(\underline{\rho}\mid\underline{x}(k))\right)\cdot\underline{h}^T\cdot\left(\underline{h}\cdot\underline{h}^T\right)^{-1}\end{aligned} \tag{10.149a}$$

and then we consider the limit for $\underline{h}\rightarrow\underline{0}$.

10.2.1 Cramér–Rao bound versus Chapman–Robbins bound

In order to compare directly the two introduced bounds, we take a look at Fisher's information matrix given by the equation

$$F = \int_{\underline{P}} \left[\frac{\partial}{\partial \underline{x}} \ln\left\{ f_{\underline{y}|\underline{x}}(\underline{\rho} \mid \underline{x}) \right\} \right] \cdot \left[\frac{\partial}{\partial \underline{x}} \ln\left\{ f_{\underline{y}|\underline{x}}(\underline{\rho} \mid \underline{x}) \right\} \right]^T \cdot f_{\underline{y}|\underline{x}}(\underline{\rho} \mid \underline{x}) \cdot d\underline{\rho} \tag{10.150}$$

and write the derivatives as

$$\frac{\partial}{\partial \underline{x}} \ln\left\{ f_{\underline{y}|\underline{x}}(\underline{\rho} \mid \underline{x}) \right\} = \frac{1}{f_{\underline{y}|\underline{x}}(\underline{\rho} \mid \underline{x})} \cdot \frac{\partial}{\partial \underline{x}} f_{\underline{y}|\underline{x}}(\underline{\rho} \mid \underline{x}) \tag{10.151}$$

which leads to

$$F = \int_{\underline{P}} \left[\frac{\partial}{\partial \underline{x}} f_{\underline{y}|\underline{x}}(\underline{\rho} \mid \underline{x}) \right] \cdot \left[\frac{\partial}{\partial \underline{x}} f_{\underline{y}|\underline{x}}(\underline{\rho} \mid \underline{x}) \right]^T \cdot \frac{1}{f_{\underline{y}|\underline{x}}(\underline{\rho} \mid \underline{x})} \cdot d\underline{\rho} \tag{10.152}$$

Now we use the result of equation (10.194*a*) and obtain

$$F \cong \int_{\underline{P}} \left[\left(f_{\underline{y}|\underline{x}}(\underline{\rho} \mid \underline{x} + \underline{h}) - f_{\underline{y}|\underline{x}}(\underline{\rho} \mid \underline{x}) \right) \cdot \underline{h}^T \cdot \left(\underline{h} \cdot \underline{h}^T \right)^{-1} \right]$$

$$\cdot \left[\left(f_{\underline{y}|\underline{x}}(\underline{\rho} \mid \underline{x} + \underline{h}) - f_{\underline{y}|\underline{x}}(\underline{\rho} \mid \underline{x}) \right) \cdot \underline{h}^T \cdot \left(\underline{h} \cdot \underline{h}^T \right)^{-1} \right]^T \cdot \frac{1}{f_{\underline{y}|\underline{x}}(\underline{\rho} \mid \underline{x})} \cdot d\underline{\rho}$$

$$F \cong \underline{h}^T \cdot \left(\underline{h} \cdot \underline{h}^T \right)^{-1} \cdot \left(\underline{h} \cdot \underline{h}^T \right)^{-1} \cdot \underline{h}$$

$$\cdot \int_{\underline{P}} \left[\frac{f_{\underline{y}|\underline{x}}(\underline{\rho} \mid \underline{x} + \underline{h}) - f_{\underline{y}|\underline{x}}(\underline{\rho} \mid \underline{x})}{f_{\underline{y}|\underline{x}}(\underline{\rho} \mid \underline{x})} \right] \cdot \left[\frac{f_{\underline{y}|\underline{x}}(\underline{\rho} \mid \underline{x} + \underline{h}) - f_{\underline{y}|\underline{x}}(\underline{\rho} \mid \underline{x})}{f_{\underline{y}|\underline{x}}(\underline{\rho} \mid \underline{x})} \right]^T \cdot f_{\underline{y}|\underline{x}}(\underline{\rho} \mid \underline{x}) \cdot d\underline{\rho} \tag{10.153}$$

The first term on the right-hand side of this equation can be transformed

$$\underline{h}^T \cdot \left(\underline{h} \cdot \underline{h}^T \right)^{-1} \cdot \left(\underline{h} \cdot \underline{h}^T \right)^{-1} \cdot \underline{h} = tr\left(\left(\underline{h} \cdot \underline{h}^T \right)^{-1} \cdot \left(\underline{h} \cdot \underline{h}^T \right)^{-1} \cdot \underline{h} \cdot \underline{h}^T \right) = tr\left(\left(\underline{h} \cdot \underline{h}^T \right)^{-1} \right)$$

and with this formulation we build the limit for $\underline{h} \rightarrow \underline{0}$ leading to

$$F = \lim_{\underline{h} \to \underline{0}} \left\{ tr\left(\left(\underline{h} \cdot \underline{h}^T \right)^{-1} \right) \right.$$

$$\left. \cdot \int_{\underline{P}} \left[\frac{f_{\underline{y}|\underline{x}}\left(\underline{\rho} \mid \underline{x}+\underline{h}\right) - f_{\underline{y}|\underline{x}}\left(\underline{\rho} \mid \underline{x}\right)}{f_{\underline{y}|\underline{x}}\left(\underline{\rho} \mid \underline{x}\right)} \right] \cdot \left[\frac{f_{\underline{y}|\underline{x}}\left(\underline{\rho} \mid \underline{x}+\underline{h}\right) - f_{\underline{y}|\underline{x}}\left(\underline{\rho} \mid \underline{x}\right)}{f_{\underline{y}|\underline{x}}\left(\underline{\rho} \mid \underline{x}\right)} \right]^T \cdot f_{\underline{y}|\underline{x}}\left(\underline{\rho} \mid \underline{x}\right) \cdot d\underline{\rho} \right\} \tag{10.154}$$

in the denominator of the Cramér–Rao bound. For direct comparison we examine

$$N_{CR-bound} = \min_{\underline{h}} \left\{ \int_{\underline{P}} \left[\frac{f_{\underline{y}|\underline{x}}\left(\underline{\rho} \mid \underline{x}(k)+\underline{h}\right) - f_{\underline{y}|\underline{x}}\left(\underline{\rho} \mid \underline{x}(k)\right)}{f_{\underline{y}|\underline{x}}\left(\underline{\rho} \mid \underline{x}(k)\right)} \right] \right.$$

$$\left. \cdot \left[\frac{f_{\underline{y}|\underline{x}}\left(\underline{\rho} \mid \underline{x}(k)+\underline{h}\right) - f_{\underline{y}|\underline{x}}\left(\underline{\rho} \mid \underline{x}(k)\right)}{f_{\underline{y}|\underline{x}}\left(\underline{\rho} \mid \underline{x}(k)\right)} \right]^T \cdot f_{\underline{y}|\underline{x}}\left(\underline{\rho} \mid \underline{x}(k)\right) \cdot d\underline{\rho} \cdot \left(\underline{h} \cdot \underline{h}^T \right)^{-1} \right\} \tag{10.155}$$

the denominator of the Chapman–Robbins bound from equation (10.149).

In F we always calculate the limit for $\underline{h} \to \underline{0}$, while in the denominator $N_{CR\text{-}bound}$ we determine the minimum via variation of $\underline{h}$. This means that the denominator of the Chapman–Robbins bound $N_{CR\text{-}bound}$ is always less or equal the Fisher information F.

$$F \geq N_{CR-bound} \tag{10.156}$$

If the denominator $N_{CR\text{-}bound}$ is always less or equal to the denominator F, we conclude for the whole quotient that the quotient $1/F$ is always less or equal to the quotient $1/N_{CR\text{-}bound}$. Thus we have show, that the Cramér–Rao bound is always less or equal to the Chapman–Robbins bound and thus the Cramér–Rao bound is the least limitation for the variances of unbiased estimators.

Another bound of the variance, the Bhattacharrya bound, is now examined and its connections to Fisher's information and the Cramér–Rao bound are derived. This requires another approach, enabling us to derive this new bound.

10.3 Bhattacharrya bound

Bhattacharrya starts with the consideration that in such cases, where the Cramér–Rao bound cannot be achieved, there must be another (greater) lowest limit for the variance of an estimator. We want to sketch the derivation for scalar state space and observation space variables in order to keep the mathematical formulations to a handy size.

The essential condition for the availability of the Cramér–Rao bound is the existence of an estimator $t_y(\rho)$, where $t_y(\rho)-b(x)$ is a linear function of

$$\frac{\partial}{\partial x}\ln\{f_{y|x}(\rho\mid x)\}=\frac{1}{f_{y|x}(\rho\mid x)}\cdot\frac{\partial}{\partial x}f_{y|x}(\rho\mid x) \tag{10.157}$$

$b(x)$ denotes the bias with respect to the desired parameter x.

Even if such an estimator does not exist, there will always be another estimator, where $t_y(\rho)-b(x)$ is a linear function of

$$\frac{1}{f_{y|x}(\rho\mid x)}\cdot\frac{\partial}{\partial x}f_{y|x}(\rho\mid x) \qquad \underline{\text{and}} \qquad \frac{1}{f_{y|x}(\rho\mid x)}\cdot\frac{\partial^2}{\partial x^2}f_{y|x}(\rho\mid x) \tag{10.158}$$

or, more generally, of the higher order derivatives of the likelihood function $l=f_{y|x}(\rho\mid x)$.

On the basis of this considerations Bhattacharrya [KEN69/73] formulates an estimator

$$D_s=t_y(\rho)-b(x)-\sum_{r=1}^{s}a_r\cdot\frac{l^{(r)}}{l} \tag{10.159}$$

with

$$l^{(r)}=\frac{\partial^r}{\partial x^r}l \tag{10.160}$$

and

a_r = weights of the derivatives of the likelihood function $l=f_{y|x}(\rho\mid x)$, which have to be determined in the derivation.

Analogously to the derivation of the Cramér–Rao bound the first step is the calculation of the mean of this estimator

$$E_{y|x}\{D_s\}=\int_{\mathrm{P}}D_s\cdot f_{y|x}(\rho\mid x)\cdot d\rho \tag{10.161}$$

$$E_{y|x}\{D_s\}=\int_{\mathrm{P}}\left(t_y(\rho)-b(x)-\sum_{r=1}^{s}a_r\cdot\frac{l^{(r)}}{l}\right)\cdot f_{y|x}(\rho\mid x)\cdot d\rho \tag{10.162}$$

This integral can be decomposed into three integrals. In section 10.7 we use the following equation

$$E\{t_y(\rho)\} = b(x) \tag{10.163}$$

and thus the first two integrals vanish. The remaining integral is

$$E_{y|x}\left\{\sum_{r=1}^{s} a_r \cdot \frac{l^{(r)}}{l}\right\} \tag{10.164}$$

Because of the relation $l = f_{y|x}(\rho \mid x)$ this is an integral of the weighted derivatives of the conditional distribution density

$$E_{y|x}\left\{\sum_{r=1}^{s} a_r \cdot \frac{l^{(r)}}{l}\right\} = \int_{\mathrm{P}} \sum_{r=1}^{s} a_r \cdot f_{y|x}^{(r)}(\rho \mid x) \cdot d\rho \tag{10.165}$$

Interchanging differentiation and integration and further interchanging of summation and integration provides

$$E_{y|x}\left\{\sum_{r=1}^{s} a_r \cdot \frac{l^{(r)}}{l}\right\} = \sum_{r=1}^{s} a_r \cdot 0 = 0 \tag{10.166}$$

So we find the expectation value of the estimator

$$E\{D_s\} = 0 \tag{10.167}$$

We are now looking for the minimum variance of such an estimator. Because of the zero mean this variance can be calculated from the equation

$$E_{y|x}\{D_s^2\} = \int_{\mathrm{P}} \left(t_y(\rho) - b(x) - \sum_{r=1}^{s} a_r \cdot \frac{l^{(r)}}{l} \right)^2 \cdot l \cdot d\rho \tag{10.168}$$

For this calculation we need the weights a_r, which can be found via

$$\int_{\mathrm{P}} \left(t_y(\rho) - b(x) - \sum_{r=1}^{s} a_r \cdot \frac{l^{(r)}}{l} \right) \cdot \frac{l^{(p)}}{l} \cdot l \cdot d\rho = 0 \qquad \text{with the derivative } p = 1,\dots,s \tag{10.169}$$

(analogously to equation (10.10)). Decomposition into two integrals

$$\int_{\mathrm{P}} \left(t_y(\rho) - b(x)\right) \cdot \frac{l^{(p)}}{l} \cdot l \cdot d\rho = \int_{\mathrm{P}} \left(\sum_{r=1}^{s} a_r \cdot l \right) \cdot \frac{l^{(p)}}{l} \cdot l \cdot d\rho \tag{10.170}$$

and interchanging of integration and summation (which can be done for these nonnegative addends) leads to the equation

$$\int_P \left(t_y(\rho)-b(x)\right)\cdot\frac{l^{(p)}}{l}\cdot l\cdot d\rho=\sum_{r=1}^{s}a_r\cdot\int_P\frac{l^{(r)}}{l}\cdot\frac{l^{(p)}}{l}\cdot l\cdot d\rho \tag{10.171}$$

and on the right-hand side of this equation we notice an integral that for $r = p = 1$ provides Fisher's information. To determine the weights a_r, we have to solve the integrals. The integral on the left-hand side can be decomposed into two integrals

$$\int_P \left(t_y(\rho)-b(x)\right)\cdot\frac{l^{(p)}}{l}\cdot l\cdot d\rho=\int_P t_y(\rho)\cdot l^{(p)}\cdot d\rho-\int_P b(x)\cdot l^{(p)}\cdot d\rho \tag{10.172}$$

where the latter integral

$$\int_P b(x)\cdot l^{(p)}\cdot d\rho=b(x)\cdot\int_P l^{(p)}\cdot d\rho=b(x)\cdot\frac{d^p}{dx^p}\int_P l\cdot d\rho=b(x)\cdot 0=0 \tag{10.173}$$

vanishes and the remaining integral

$$\int_P t_y(\rho)\cdot l^{(p)}\cdot d\rho=b^{(p)}(x) \tag{10.174}$$

because of

$$\int_P t_y(\rho)\cdot l\cdot d\rho=b(x) \tag{10.175}$$

leads to

$$\frac{d}{dx}\int_P t_y(\rho)\cdot l\cdot d\rho=\frac{d}{dx}b(x) \tag{10.176}$$

As the estimator $t_y(\rho)$ does not depend on the parameter x, and the 'regularity conditions' hold (see appendix A.5), we get

$$\int_P t_y(\rho)\cdot\frac{d}{dx}l\cdot d\rho=\frac{d}{dx}b(x)$$

or

$$\int_P t_y(\rho)\cdot\frac{\frac{d}{dx}l}{l}\cdot l\cdot d\rho=\frac{d}{dx}b(x) \tag{10.177}$$

The extension to arbitrary derivatives $\frac{d^p}{dx^p}$ is obvious and verifies equation (10.174). Starting with equation (10.171) we arrived at the equation

$$\frac{d^p}{dx^p} b(x) = b^{(p)}(x) = \sum_{r=1}^{s} a_r \cdot \int_{\mathrm{P}} \frac{l^{(r)}}{l} \cdot \frac{l^{(p)}}{l} \cdot l \cdot d\rho \qquad \text{for } p = 1,\ldots,s \tag{10.178}$$

and under the condition that the matrix

$$J_{rp} = E\left\{\frac{l^{(r)}}{l} \cdot \frac{l^{(p)}}{l}\right\} \tag{10.179}$$

is not singular we determine the weights a_r.

$$a_r = \sum_{p=1}^{s} b^{(p)}(x) \cdot J_{rp}^{-1} \tag{10.180}$$

Remark:

It is not always possible to invert this matrix. For $j = p = 1$ we get Fisher's information matrix, which cannot necessarily be inverted, because the rows and columns need not be filled with nonzero quantities. Non-observable components lead to rows and columns that are filled with zeroes and this prevents an inversion of the matrix. This only holds for Fisher's information matrix $C^T R^{-1} C$ (for only one observation), because, when we observe a sequence of samples, the system model connects the observations (Data fusion), and the result is the summation equation of the stochastic observability (see section 10.1.4.1). Only this connection provides an observation matrix, which can be inverted.

Thus all terms of the estimator D_s are calculated and we can therefore proceed with the computation of the variance

$$E_{y|x}\left\{D_s^2\right\} = \int_{\mathrm{P}} \left(t_y(\rho) - b(x) - \sum_{r=1}^{s} a_r \cdot \frac{l^{(r)}}{l}\right)^2 \cdot l \cdot d\rho \tag{10.181}$$

which has already been formulated in equation (10.168).

$$E_{y|x}\left\{D_s^2\right\} = \int_{\mathrm{P}} \left(t_y(\rho) - b(x) - \sum_{r=1}^{s} \sum_{p=1}^{s} b^{(p)}(x) \cdot J_{rp}^{-1} \cdot \frac{l^{(r)}}{l}\right)^2 \cdot l \cdot d\rho \tag{10.182}$$

Multiplication of the terms leads to three integrals

$$E_{y|x}\left\{D_s^2\right\}_1 = \int_{\mathrm{P}} \left(t_y(\rho) - b(x)\right)^2 \cdot l \cdot d\rho \tag{10.183}$$

$$E_{y|x}\left\{D_s^2\right\}_2 = -2 \cdot \int_{\mathrm{P}} \left(t_y(\rho) - b(x)\right) \cdot \left(\sum_{r=1}^{s} \sum_{p=1}^{s} b^{(p)}(x) \cdot J_{rp}^{-1} \cdot \frac{l^{(r)}}{l} \right) \cdot l \cdot d\rho \tag{10.184}$$

$$E_{y|x}\left\{D_s^2\right\}_3 = \int_{\mathrm{P}} \left(\sum_{r=1}^{s} \sum_{p=1}^{s} b^{(p)}(x) \cdot J_{rp}^{-1} \cdot \frac{l^{(r)}}{l} \right)^2 \cdot l \cdot d\rho \tag{10.185}$$

$$E_{y|x}\left\{D_s^2\right\} = E_{y|x}\left\{D_s^2\right\}_1 + E_{y|x}\left\{D_s^2\right\}_2 + E_{y|x}\left\{D_s^2\right\}_3 \tag{10.186}$$

which are solved separately. Because of

$$\int_{\mathrm{P}} t_y(\rho) \cdot l \cdot d\rho = b(x) \tag{10.187}$$

the first integral is equal to the variance of the biased estimator

$$E_{y|x}\left\{D_s^2\right\}_1 = \int_{\mathrm{P}} \left(t_y(\rho) - b(x)\right)^2 \cdot l \cdot d\rho = \mathrm{var}\left(t_y(\rho)\right) \tag{10.188}$$

In the second integral

$$E_{y|x}\left\{D_s^2\right\}_2 = -2 \cdot \int_{\mathrm{P}} \left(t_y(\rho) - b(x)\right) \cdot \left(\sum_{r=1}^{s} \sum_{p=1}^{s} b^{(p)}(x) \cdot J_{rp}^{-1} \cdot \frac{l^{(r)}}{l} \right) \cdot l \cdot d\rho \tag{10.189}$$

we interchange summation and integration to obtain

$$\begin{aligned} E_{y|x}\left\{D_s^2\right\}_2 &= -2 \cdot \sum_{r=1}^{s} \sum_{p=1}^{s} \int_{\mathrm{P}} \left(t_y(\rho) - b(x)\right) \cdot b^{(p)}(x) \cdot J_{rp}^{-1} \cdot \frac{l^{(r)}}{l} \cdot l \cdot d\rho \\ &= -2 \cdot \sum_{r=1}^{s} \sum_{p=1}^{s} \int_{\mathrm{P}} t_y(\rho) \cdot b^{(p)}(x) \cdot J_{rp}^{-1} \cdot \frac{l^{(r)}}{l} \cdot l \cdot d\rho \\ &\quad + 2 \cdot \sum_{r=1}^{s} \sum_{p=1}^{s} \int_{\mathrm{P}} b(x) \cdot b^{(p)}(x) \cdot J_{rp}^{-1} \cdot \frac{l^{(r)}}{l} \cdot l \cdot d\rho \end{aligned} \tag{10.190}$$

$$
\begin{aligned}
&= -2 \cdot \sum_{r=1}^{s} \sum_{p=1}^{s} b^{(p)}(x) \cdot J_{rp}^{-1} \cdot \int_{P} t_y(\rho) \cdot l^{(r)} \cdot d\rho \\
&+ 2 \cdot \sum_{r=1}^{s} \sum_{p=1}^{s} b(x) \cdot b^{(p)}(x) \cdot J_{rp}^{-1} \cdot \int_{P} l^{(r)} \cdot d\rho
\end{aligned}
\tag{10.191}
$$

In the last addend we immediately notice that this term vanishes because of $\int_{P} l^{(r)} \cdot d\rho = 0$, so that only the first addend remains. We already know the solution of this integral from equation (10.177) and thus

$$E_{y|x}\left\{D_s^2\right\}_2 = -2 \cdot \sum_{r=1}^{s} \sum_{p=1}^{s} b^{(p)}(x) \cdot J_{rp}^{-1} \cdot b^{(r)}(x) \tag{10.192}$$

because

$$\int_{P} t_y(\rho) \cdot l \cdot d\rho = b(x) \tag{10.193}$$

$$\frac{d^r}{dx^r} \int_{P} t_y(\rho) \cdot l \cdot d\rho = \int_{P} t_y(\rho) \cdot \frac{d^r}{dx^r} l \cdot d\rho = \int_{P} t_y(\rho) \cdot l^{(r)} \cdot d\rho = \frac{d^r}{dx^r} b(x) = b^{(r)}(x) \tag{10.194}$$

We thus examine the third integral from equation (10.182)

$$E_{y|x}\left\{D_s^2\right\}_3 = \int_{P} \left(\sum_{r=1}^{s} \sum_{p=1}^{s} b^{(p)}(x) \cdot J_{rp}^{-1} \cdot \frac{l^{(r)}}{l} \right)^2 \cdot l \cdot d\rho \tag{10.195}$$

$$E_{y|x}\left\{D_s^2\right\}_3 = \int_{P} \sum_{r=1}^{s} \sum_{p=1}^{s} \sum_{m=1}^{s} \sum_{n=1}^{s} b^{(p)}(x) \cdot J_{rp}^{-1} \cdot \frac{l^{(r)}}{l} \cdot b^{(n)}(x) \cdot J_{mn}^{-1} \cdot \frac{l^{(m)}}{l} \cdot l \cdot d\rho \tag{10.196}$$

Interchanging of summation and integration leads to

$$E_{y|x}\left\{D_s^2\right\}_3 = \sum_{r=1}^{s} \sum_{p=1}^{s} \sum_{m=1}^{s} \sum_{n=1}^{s} b^{(p)}(x) \cdot J_{rp}^{-1} \cdot b^{(n)}(x) \cdot J_{mn}^{-1} \cdot \int_{P} \frac{l^{(r)}}{l} \cdot \frac{l^{(m)}}{l} \cdot l \cdot d\rho \tag{10.197}$$

and the remaining integral is known as

$$J_{rm} = \int_{\mathrm{P}} \frac{l^{(r)}}{l} \cdot \frac{l^{(m)}}{l} \cdot l \cdot d\rho \tag{10.198}$$

Thus the third integral is equal to

$$E_{y|x}\left\{D_s^2\right\}_3 = \sum_{r=1}^{s}\sum_{p=1}^{s}\sum_{m=1}^{s}\sum_{n=1}^{s} b^{(p)}(x) \cdot b^{(n)}(x) \cdot J_{rp}^{-1} \cdot J_{mn}^{-1} \cdot J_{rm} \tag{10.199}$$

As all indices run from 1 to s we rearrange the summation

$$E_{y|x}\left\{D_s^2\right\}_3 = \sum_{p=1}^{s}\sum_{n=1}^{s} b^{(p)}(x) \cdot b^{(n)}(x) \cdot J_{pn}^{-1} \tag{10.200}$$

and thus find the result for all three integral

$$E_{y|x}\left\{D_s^2\right\} = E_{y|x}\left\{D_s^2\right\}_1 + E_{y|x}\left\{D_s^2\right\}_2 + E_{y|x}\left\{D_s^2\right\}_3 \tag{10.201}$$

with

$$E_{y|x}\left\{D_s^2\right\}_1 = \int_{\mathrm{P}} \left(t_y(\rho) - b(x)\right)^2 \cdot l \cdot d\rho = \mathrm{var}\left(t_y(\rho)\right) \tag{10.202}$$

$$E_{y|x}\left\{D_s^2\right\}_2 = -2 \cdot \sum_{r=1}^{s}\sum_{p=1}^{s} b^{(p)}(x) \cdot J_{rp}^{-1} \cdot b^{(r)}(x) \tag{10.203}$$

$$E_{y|x}\left\{D_s^2\right\}_3 = \sum_{p=1}^{s}\sum_{n=1}^{s} b^{(p)}(x) \cdot b^{(n)}(x) \cdot J_{pn}^{-1} \tag{10.204}$$

The last two terms differ only by a factor of 2, which leads to

$$E_{y|x}\left\{D_s^2\right\} = \mathrm{var}\left(t_y(\rho)\right) - \sum_{r=1}^{s}\sum_{p=1}^{s} b^{(p)}(x) \cdot J_{rp}^{-1} \cdot b^{(r)}(x) \tag{10.205}$$

On the left-hand side of this equation we have a variance, which is always nonnegative. Consequently the right-hand side has the same property.

Thus the Bhattacharrya bound is given by:

$$\mathrm{var}\left(t_y(\rho)\right) \geq \sum_{r=1}^{s}\sum_{p=1}^{s} b^{(p)}(x) \cdot J_{rp}^{-1} \cdot b^{(r)}(x) \tag{10.206}$$

We always have to remember that the calculation of the weights a_r requires that the terms J_{rs} can be inverted. If we cannot determine the terms J_{rs}^{-1} explicitly, but have to rely on the inversion, the validity of this bound is doubtful.

Furthermore we notice that the derivation with multi-dimensional bias inevitably leads to tensors, because the higher order derivatives of vectors with respect to other vectors cannot be denoted in vector matrix notation. In the special case $s=1$ this bound is equal to the Cramér–Rao bound

$$\mathrm{var}(t_y(\rho)) \geq b^{(1)}(x) \cdot J_{11}^{-1} \cdot b^{(1)}(x) \tag{10.207}$$

J_{11}^{-1} is Fisher's information matrix and is to be computed from the observation matrix and the inverse of the covariance matrix of the measurement error.

Remark

Here we denote the biased estimator with the equation

$$\int_{\mathrm{P}} t_y(\rho) \cdot f_{y|x}(\rho \mid x) \cdot d\rho = b(x) \tag{10.208}$$

In section 10.7 we use the formulation

$$\int_{\mathrm{P}} t_y(\rho) \cdot f_{y|x}(\rho \mid x) \cdot d\rho = x + b(x) \tag{10.209}$$

which is merely a shift $b(x) \to x + b(x)$, which has to be considered in the result. The approach of section 10.7 leads to a Bhattacharrya bound

$$\mathrm{var}(t_y(\rho)) \geq \sum_{r=1}^{s} \sum_{p=1}^{s} (x + b(x))^{(p)} \cdot J_{rp}^{-1} \cdot (x + b(x))^{(r)} \tag{10.210}$$

In the next section we want to verify the statement, which we made at the beginning of this chapter, that this bound of the variance is greater than the Cramér–Rao bound.

10.3.1 Bhattacharrya bound and Cramér–Rao bound

To be able to compare these two bound, we use another formulation of the Bhattacharrya bound for $s = 2$. We find this equation in [KEN69/73]

$$\begin{vmatrix} \mathrm{var}(t_y(\rho)) & b'(x) & b''(x) \\ b'(x) & J_{11} & J_{21} \\ b''(x) & J_{12} & J_{22} \end{vmatrix} \geq 0 \tag{10.211}$$

and explicitly

$$\begin{aligned}&\operatorname{var}(t_y(\rho))\cdot[J_{11}\cdot J_{22}-J_{12}\cdot J_{21}]-b'(x)\cdot[b'(x)\cdot J_{22}-b''(x)\cdot J_{12}]\\&\quad+b''(x)\cdot[b'(x)\cdot J_{21}-b''(x)\cdot J_{11}]\ \geq 0\end{aligned} \tag{10.212}$$

and in the familiar formulation of the Cramér–Rao inequality we get

$$\operatorname{var}(t_y(\rho))\geq\frac{b'(x)\cdot[b'(x)\cdot J_{22}-b''(x)\cdot J_{12}]-b''(x)\cdot[b'(x)\cdot J_{21}-b''(x)\cdot J_{11}]}{[J_{11}\cdot J_{22}-J_{12}\cdot J_{21}]} \tag{10.213}$$

which holds if the denominator is not equal to zero, but this is true for the given denominator (because of Hadamard's inequality, see section 13.3).

$$\operatorname{var}(t_y(\rho))\geq\frac{(b'(x))^2\cdot J_{22}-b'(x)\cdot b''(x)\cdot J_{12}-b''(x)\cdot b'(x)\cdot J_{21}+(b''(x))^2\cdot J_{11}}{[J_{11}\cdot J_{22}-J_{12}\cdot J_{21}]} \tag{10.214}$$

In tensor notation we are able to combine the covariance matrices $J_{12}=J_{12}{}^T=J_{21}$ which leads to

$$\operatorname{var}(t_y(\rho))\geq\frac{(b'(x))^2\cdot J_{22}-2\cdot b'(x)\cdot b''(x)\cdot J_{12}+(b''(x))^2\cdot J_{11}}{[J_{11}\cdot J_{22}-J_{12}\cdot J_{21}]} \tag{10.215}$$

After the expansion of the quotient with J_{11} and addition of the term $J_{12}^2\cdot(b'(x))^2-J_{12}^2\cdot(b'(x))^2$ in the numerator we obtain

$$\operatorname{var}(t_y(\rho))\geq\frac{(b'(x))^2}{J_{11}}+\frac{\{b'(x)\cdot J_{12}-b''(x)\cdot J_{11}\}^2}{J_{11}\cdot\left[J_{11}\cdot J_{22}-J_{12}^2\right]} \tag{10.216}$$

While the first term on the right-hand side is known from the Cramér–Rao inequality, the second term denotes the improvement of the Cramér–Rao bound. If the estimator is not to estimate a function $b(x)$ of the parameter x, but directly the parameter x, we get

$$\operatorname{var}(t_y(\rho))\geq\frac{1}{J_{11}}+\frac{J_{12}{}^2}{J_{11}\cdot\left[J_{11}\cdot J_{22}-J_{12}^2\right]} \tag{10.217}$$

If this lower limit (here for $s=2$) has the same value as the Cramér–Rao bound (for $s=1$), we cannot conclude that the Cramér–Rao bound can be achieved by our estimator. But we notice that the improvement of the Cramér–Rao bound, denoted by the second addend on the right-hand side, is always positive or equal to zero, which means that this bound is greater than or equal to the Cramér–Rao bound.

Thus Bhattacharrya's bound is greater or equal to the Cramér–Rao bound.

10.3.2 Bhattacharrya's bound for gaussian distribution densities

The computation of the covariance matrix J_{11}, and the tensors J_{12}, J_{22} can only be achieved by applying tensor calculus, requiring another, tensorial formulation of the gaussian distribution densities [MCC87]. In this notation $\kappa^{i,j}$ is the known covariance matrix and the inverse of this matrix is $\kappa_{i,j}$. κ^i is the tensor of the expectation values. Furthermore the dummy variables ρ^i and ξ^l as well as the random variables y^i and x^l are denoted as contravariant tensors. This leads to the conditional distribution density

$$f_{y^i|x^l}\left(\rho^i \mid x^l\right)=\frac{1}{(2\pi)^{\frac{m}{2}}\cdot\left|\kappa^{i,j}\right|^{\frac{1}{2}}}\cdot\exp\left\{-\frac{1}{2}\cdot\left(\rho^i-\kappa^i\right)\cdot\left(\rho^j-\kappa^j\right)\cdot\kappa_{i,j}\right\} \tag{10.218}$$

and we first use this density to determine Fisher's information J_{11} by solving the following integral

$$J_{11}=\int_{\mathrm{P}}\left[\frac{\partial}{\partial x^p}\ln\left\{f_{y^i|x^l}\left(\rho^i \mid x^l\right)\right\}\right]\cdot\left[\frac{\partial}{\partial x^s}\ln\left\{f_{y^i|x^l}\left(\rho^i \mid x^l\right)\right\}\right]\cdot f_{y^i|x^l}\left(\rho^i \mid x^l\right)\cdot d\rho^i \tag{10.219}$$

The derivative of the logarithm

$$\frac{\partial}{\partial x^p}\ln\left\{f_{y^i|x^l}\left(\rho^i \mid x^l\right)\right\}=-\frac{1}{2}\cdot 2\cdot\left(\rho^i-\kappa^i\right)\cdot\kappa_{i,j}\cdot\left(-\frac{\partial}{\partial x^p}\right)\cdot\kappa^j \tag{10.220}$$

leads to the equation

$$\begin{aligned}J_{11}=&\int_{\mathrm{P}}\left[\left(\rho^t-\kappa^t\right)\cdot\kappa_{t,j}\cdot\left(\frac{\partial}{\partial x^p}\right)\cdot\kappa^j\right]\cdot\left[\left(\rho^j-\kappa^j\right)\cdot\kappa_{i,j}\cdot\left(\frac{\partial}{\partial x^p}\right)\cdot\kappa^t\right]\\&\cdot f_{y^i|x^l}\left(\rho^i \mid x^l\right)\cdot d\rho^i\end{aligned}$$

$$\begin{aligned}J_{11}=&\,\kappa_{t,j}\cdot\left(\frac{\partial}{\partial x^p}\right)\cdot\kappa^j\cdot\kappa_{i,j}\cdot\left(\frac{\partial}{\partial x^p}\right)\cdot\kappa^t\\&\cdot\int_{\mathrm{P}}\left(\rho^t-\kappa^t\right)\cdot\left(\rho^j-\kappa^j\right)\cdot f_{y^i|x^l}\left(\rho^i \mid x^l\right)\cdot d\rho^i\end{aligned} \tag{10.221}$$

where the remaining integral is equal to the covariance matrix $\kappa^{t,j}$. We thus get

$$J_{11} = \kappa_{t,j} \cdot \left(\frac{\partial}{\partial x^p} \cdot \kappa^j \right) \cdot \kappa_{i,j} \cdot \left(\frac{\partial}{\partial x^p} \cdot \kappa^t \right) \cdot \kappa^{t,j} \tag{10.222}$$

or as Fisher's information in tensor notation

$$J_{11} = \kappa_{t,j} \cdot \left(\frac{\partial}{\partial x^p} \cdot \kappa^j \right) \cdot \left(\frac{\partial}{\partial x^p} \cdot \kappa^t \right) \tag{10.223}$$

and this is equivalent to the result in section 10.1.1 which we calculated with vector matrix notation.

Let us now examine J_{12}: Here we have to calculate the terms $\frac{l^{(1)}}{l}$ and $\frac{l^{(2)}}{l}$, to be able to formulate the integral.

$$\frac{l^{(1)}}{l} = \left(\rho^i - \kappa^i\right) \cdot \kappa_{i,j} \cdot \frac{\partial}{\partial x^p} \kappa^j \tag{10.224}$$

and

$$l^{(2)} = \frac{\partial}{\partial x^r} \left(\left(\rho^i - \kappa^i\right) \cdot \kappa_{i,j} \cdot \frac{\partial}{\partial x^p} \kappa^j \cdot l \right) \tag{10.225}$$

If we assume that the covariance matrix of the measurement error does not depend on the state, we have $\frac{\partial}{\partial x^p} \kappa_{i,j} = 0$. And thus we get

$$\begin{aligned} l^{(2)} = & -\frac{\partial}{\partial x^r} \kappa^i \cdot \kappa_{i,j} \cdot \frac{\partial}{\partial x^p} \kappa^j \cdot l + \left(\rho^i - \kappa^i\right) \cdot \kappa_{i,j} \cdot \frac{\partial^2}{\partial x^r \cdot \partial x^p} \kappa^j \cdot l \\ & + \left[\left(\rho^i - \kappa^i\right) \cdot \kappa_{i,j} \cdot \frac{\partial}{\partial x^p} \kappa^j \right] \cdot \left[\left(\rho^s - \kappa^s\right) \cdot \kappa_{s,t} \cdot \frac{\partial}{\partial x^r} \kappa^t \right] \cdot l \end{aligned} \tag{10.226}$$

and then

$$\begin{aligned} \frac{l^{(2)}}{l} = & -\frac{\partial}{\partial x^r} \kappa^i \cdot \kappa_{i,j} \cdot \frac{\partial}{\partial x^p} \kappa^j + \left(\rho^i - \kappa^i\right) \cdot \kappa_{i,j} \cdot \frac{\partial^2}{\partial x^r \cdot \partial x^p} \kappa^j \\ & + \left[\left(\rho^i - \kappa^i\right) \cdot \kappa_{i,j} \cdot \frac{\partial}{\partial x^p} \kappa^j \right] \cdot \left[\left(\rho^s - \kappa^s\right) \cdot \kappa_{s,t} \cdot \frac{\partial}{\partial x^r} \kappa^t \right] \end{aligned} \tag{10.227}$$

We now calculate the expectation value

$$J_{12} = E\left\{\left[\left(\rho^a - \kappa^a\right)\cdot \kappa_{a,b} \cdot \frac{\partial}{\partial x^c}\kappa^b\right]\cdot\left[-\frac{\partial}{\partial x^r}\kappa^i \cdot \kappa_{i,j} \cdot \frac{\partial}{\partial x^p}\kappa^j + \left(\rho^i - \kappa^i\right)\right.\right.$$

$$\left.\left.\cdot \kappa_{i,j} \cdot \frac{\partial^2}{\partial x^r \cdot \partial x^p}\kappa^j + \left[\left(\rho^i - \kappa^i\right)\cdot \kappa_{i,j} \cdot \frac{\partial}{\partial x^p}\kappa^j\right]\cdot\left[\left(\rho^s - \kappa^s\right)\cdot \kappa_{s,t} \cdot \frac{\partial}{\partial x^r}\kappa^t\right]\right]\right\} \tag{10.228}$$

As all odd central moments of the gaussian distribution vanish, we obtain

$$J_{12}^{crp} = E\left\{\left(\rho^a - \kappa^a\right)\cdot\left(\rho^i - \kappa^i\right)\right\}\cdot \kappa_{a,b} \cdot \frac{\partial}{\partial x^c}\kappa^b \cdot \kappa_{i,j} \cdot \frac{\partial^2}{\partial x^r \cdot \partial x^p}\kappa^j \tag{10.229}$$

If we insert the covariance matrix, resulting from the expectation value, we obtain the following tensor

$$J_{12}^{crp} = \kappa^{a,i} \cdot \kappa_{a,b} \cdot \kappa_{i,j} \cdot \frac{\partial}{\partial x^c}\kappa^b \cdot \frac{\partial^2}{\partial x^r \cdot \partial x^p}\kappa^j \tag{10.230}$$

With

$$\kappa^{a,i} \cdot \kappa_{a,b} \cdot \kappa_{i,j} = \kappa_{b,j} \tag{10.231}$$

we find

$$J_{12}^{crp} = \kappa_{b,j} \cdot \frac{\partial}{\partial x^c}\kappa^b \cdot \frac{\partial^2}{\partial x^r \cdot \partial x^p}\kappa^j \tag{10.232}$$

With Einstein's summation convention we notice that the free indices *c, r, p* determine the dimension of J_{12}. If we examine the dimension of J_{12} we find that this term is not a covariance matrix. Only the multiplication with the derivative of the bias $b'(\xi)$ leads again to a covariance matrix.

At last we examine the computation of J_{22}: As we already determined the term l^2/l, we are able to start the computation with the integral. But we only want to state the result, because the computation of the integral is 'straightforward'. Using the property of the gaussian distribution density that all odd higher order moments vanish, we find the following result:

$$J_{22}^{\alpha\beta rp} = \kappa_{a,b} \cdot \kappa_{j,l} \cdot \frac{\partial \kappa^a}{\partial x^\alpha} \cdot \frac{\partial \kappa^b}{\partial x^\beta} \cdot \frac{\partial \kappa^j}{\partial x^r} \cdot \frac{\partial \kappa^l}{\partial x^p} + \frac{\partial^2 \kappa^b}{\partial x^\alpha \cdot \partial x^\beta} \cdot \kappa_{b,l} \cdot \frac{\partial^2 \kappa^l}{\partial x^r \cdot \partial x^p}$$
$$+ \frac{\partial \kappa^b}{\partial x^\beta} \cdot \frac{\partial \kappa^d}{\partial x^\alpha} \cdot \frac{\partial \kappa^l}{\partial x^p} \cdot \frac{\partial \kappa^t}{\partial x^r} \cdot \left(\kappa_{b,d} \cdot \kappa_{l,t} + \kappa_{b,l} \cdot \kappa_{d,t} + \kappa_{b,t} \cdot \kappa_{l,d}\right)$$
$$- \frac{\partial \kappa^a}{\partial x^\alpha} \cdot \frac{\partial \kappa^b}{\partial x^\beta} \cdot \frac{\partial \kappa^l}{\partial x^p} \cdot \frac{\partial \kappa^t}{\partial x^r} \cdot \kappa_{a,d} \cdot \kappa_{l,t} - \frac{\partial \kappa^j}{\partial x^r} \cdot \frac{\partial \kappa^l}{\partial x^{p\alpha}} \cdot \frac{\partial \kappa^b}{\partial x^\beta} \cdot \frac{\partial \kappa^d}{\partial x^\alpha} \cdot \kappa_{j,l} \cdot \kappa_{b,d} \tag{10.233}$$

With

$$\kappa^{a,i} \cdot \kappa_{a,b} \cdot \kappa_{i,j} = \kappa_{b,j} \tag{10.234}$$

we find

$$J_{22}^{\alpha\beta rp} = \frac{\partial^2 \kappa^b}{\partial x^\alpha \cdot \partial x^\beta} \cdot \kappa_{b,l} \cdot \frac{\partial^2 \kappa^l}{\partial x^r \cdot \partial x^p} + \frac{\partial \kappa^b}{\partial x^\beta} \cdot \frac{\partial \kappa^d}{\partial x^\alpha} \cdot \frac{\partial \kappa^l}{\partial x^p} \cdot \frac{\partial \kappa^t}{\partial x^r} \cdot \left(\kappa_{b,l} \cdot \kappa_{d,t} + \kappa_{b,t} \cdot \kappa_{l,d}\right) \tag{10.235}$$

It is certainly not wise to carry out such extensive computations if we do not see any necessity for it. If Fisher's information of the Cramér–Rao bound allows us to make sufficiently precise statements, it is unnecessary to worry about an improvement of this information measure.

10.4 Barankin bound

A short and easy derivation of this bound can be found in [SCH73], which also starts with an unbiased estimator and thus applies the approach that we already used in the derivation of the Cramér–Rao bound:

$$\int_{\mathrm{P}} t_y(\rho) \cdot f_{y|x}(\rho \mid x) \cdot d\rho = x \tag{10.236}$$

To obtain an improved bound with respect to the Cramér–Rao bound, Barankin weights the desired parameters $x_1, x_2, \ldots, x_n$ with the weights a_k where $k = 1,2,..,n$. Each estimator is therefore described by the equation

$$a_k \cdot \int_{\mathrm{P}} t_y(\rho) \cdot f_{y|x}(\rho \mid x_k) \cdot d\rho = a_k \cdot x_k \tag{10.237}$$

and for all estimators we get

$$\sum_{k=1}^{n}\left(a_k \cdot \int_{\mathrm{P}} t_y(\rho)\cdot f_{y|x}(\rho \mid x_k)\cdot d\rho\right) = \sum_{k=1}^{n}(a_k \cdot x_k) \tag{10.238}$$

Interchanging of summation and integration provides

$$\int_{\mathrm{P}} t_y(\rho)\cdot \sum_{k=1}^{n} a_k \cdot f_{y|x}(\rho \mid x_k)\cdot d\rho = \sum_{k=1}^{n}(a_k \cdot x_k) \tag{10.239}$$

Now we subtract the 'true value' x_t from both sides of this equation to obtain a form that allows us to apply the Schwartz inequality.

$$\int_{\mathrm{P}} \left[t_y(\rho) - x_t\right]\cdot \sum_{k=1}^{n} a_k \cdot f_{y|x}(\rho \mid x_k)\cdot d\rho = \sum_{k=1}^{n}\left(a_k \cdot [x_k - x_t]\right) \tag{10.240}$$

Expanding the integral with the conditional distribution density, conditioned on the true value, we now find

$$\int_{\mathrm{P}} \left[t_y(\rho) - x_t\right]\cdot \frac{\sum_{k=1}^{n} a_k \cdot f_{y|x}(\rho \mid x_k)}{f_{y|x}(\rho \mid x_t)}\cdot f_{y|x}(\rho \mid x_t)\cdot d\rho = \sum_{k=1}^{n}\left(a_k \cdot [x_k - x_t]\right) \tag{10.241}$$

and with the Schwartz inequality we get

$$\int_{\mathrm{P}} \left[t_y(\rho) - x_t\right]^2 \cdot f_{y|x}(\rho \mid x_t)\cdot d\rho \cdot \int_{\mathrm{P}} \left[\frac{\sum_{k=1}^{n} a_k \cdot f_{y|x}(\rho \mid x_k)}{f_{y|x}(\rho \mid x_t)}\right]^2 \cdot f_{y|x}(\rho \mid x_t)\cdot d\rho$$

$$\geq \left[\sum_{k=1}^{n}\left(a_k \cdot [x_k - x_t]\right)\right]^2 \tag{10.242}$$

and finally

$$\operatorname{var}(t_y(\rho) \mid x_t) \geq \frac{\left[\sum_{k=1}^{n} (a_k \cdot [x_k - x_t])\right]^2}{\int\limits_{\mathrm{P}} \left[\frac{\sum_{k=1}^{n} a_k \cdot f_{y|x}(\rho \mid x_k)}{f_{y|x}(\rho \mid x_t)}\right]^2 \cdot f_{y|x}(\rho \mid x_t) \cdot d\rho} \tag{10.243}$$

If we maximize the right-hand side of this equation with respect to x_k and a_k, we obtain the greatest lower bound ('tightest' bound), which Schweppe calls the Barankin bound.

The determination of this bound is not a 'straightforward' task, because the optimization with respect to two sets of parameters requires some expense. Therefore practical calculations of an improved bound fall back on the computation of the Bhattacharrya bound. The introduced weights remind us of the weights in Bhattacharrya's bound, where these factors are determined in the course of the derivation.

This derivation, however, does not lead to the complete Barankin bound. We may notice the basic proceeding of the derivation, but this derivation merely provides a limitation for quadratic functions. But Barankin's bound [BAR49] provides limitations of higher order moments. We therefore examine a more detailed formulation of the derivation and we start with an estimator, estimating a function of the parameter x

$$\int\limits_{\mathrm{P}} t_y(\rho) \cdot f_{y|x}(\rho \mid x) \cdot d\rho = g(x) \tag{10.244}$$

Subtracting the true value on both sides of the equation we get

$$\int\limits_{\mathrm{P}} (t_y(\rho) - g(x_t)) \cdot f_{y|x}(\rho \mid x) \cdot d\rho = g(x) - g(x_t) = h(x) \tag{10.245}$$

For non-constant $h(x)$ we introduce the parameters x_1, x_2, ..., x_n with the real weights a_1, a_2,..., a_n and obtain

$$\sum_{k=1}^{n} a_k \cdot \int\limits_{\mathrm{P}} [t_y(\rho) - g(x_t)] \cdot f_{y|x}(\rho \mid x_k) \cdot d\rho = \sum_{k=1}^{n} a_k \cdot h(x_k)$$

or

$$\int\limits_{\mathrm{P}} [t_y(\rho) - g(x_t)] \cdot \sum_{k=1}^{n} a_k \cdot f_{y|x}(\rho \mid x_k) \cdot d\rho = \sum_{k=1}^{n} a_k \cdot h(x_k) \tag{10.246}$$

$$\int_{\mathrm{P}} \left[t_y(\rho) - g(x_t)\right] \cdot \sum_{k=1}^{n} a_k \cdot \frac{f_{y|x}(\rho \mid x_k)}{f_{y|x}(\rho \mid x_t)} \cdot f_{y|x}(\rho \mid x_t) \cdot d\rho = \sum_{k=1}^{n} a_k \cdot h(x_k) \quad (10.247)$$

as $\left[t_y(\rho) - g(x_t)\right]$ does not depend on the parameter x_k. We now apply Hölder's inequality. To do this, we first define two normed spaces L_{1r} and L_{1s} with

$$r = \frac{s}{s-1} \qquad \text{(see Hölder's inequality (10.250))} \qquad (10.248)$$

and the norms (Minkowski distance) on the whole sample space Ω

$$\|x\|_r = \left(\int_{\Omega} |x|^r \cdot dv \right)^{\frac{1}{r}}$$

$$\|\eta\|_r = \left(\int_{\Omega} |\eta|^s \cdot dv \right)^{\frac{1}{s}} \qquad (10.249)$$

to be able to use Hölder's inequality as a more general formulation of the Schwartz inequality.
We get Hölder's inequality

$$\sum_j \left| x_j \cdot y_j \right| \le \left(\sum_j \left| x_j \right|^s \right)^{\frac{1}{s}} \cdot \left(\sum_j \left| y_j \right|^r \right)^{\frac{1}{r}} \qquad (10.250)$$

For continuous random variables this is

$$\int_{\Omega} \left| x_j \cdot y_j \right| \cdot dv \le \left(\int_{\Omega} \left| x_j \right|^s \cdot dv \right)^{\frac{1}{s}} \cdot \left(\int_{\Omega} \left| y_j \right|^r \cdot dv \right)^{\frac{1}{r}} \qquad (10.251)$$

and together with the Minkowski distances in realization space P we get

$$\int_{\mathrm{P}} \left| \left[t_y(\rho) - g(x_t)\right] \cdot \sum_{k=1}^{n} a_k \cdot \frac{f_{y|x}(\rho \mid x_k)}{f_{y|x}(\rho \mid x_t)} \right| \cdot f_{y|x}(\rho \mid x_t) \cdot d\rho$$

$$\le \left\| \left[t_y(\rho) - g(x_t)\right] \right\|_s \cdot \left\| \sum_{k=1}^{n} a_k \cdot f_{y|x}(\rho \mid x_k) \right\|_r \qquad (10.252)$$

and with equation (10.247) we find

$$\left\| \left[t_y(\rho) - g(x_t) \right] \right\|_s \cdot \left\| \sum_{k=1}^{n} a_k \cdot \frac{f_{y|x}(\rho \mid x_k)}{f_{y|x}(\rho \mid x_t)} \right\|_r \geq \sum_{k=1}^{n} a_k \cdot h(x_k) \tag{10.253}$$

We have thus derived the limitation of the sth absolute central moment (s.a.c.m.), the Barankin bound

$$\left\| \left[t_y(\rho) - g(x_t) \right] \right\|_s \geq \frac{\sum_{k=1}^{n} a_k \cdot h(x_k)}{\left\| \sum_{k=1}^{n} a_k \cdot \frac{f_{y|x}(\rho \mid x_k)}{f_{y|x}(\rho \mid x_t)} \right\|_r} \tag{10.254}$$

Explicitly

$$\left[\int_{\mathrm{P}} \left[t_y(\rho) - g(x_t) \right]^s \cdot f_{y|x}(\rho \mid x_t) \cdot d\rho \right]^{\frac{1}{s}} \geq \frac{\sum_{k=1}^{n} a_k \cdot h(x_k)}{\left[\int_{\mathrm{P}} \left[\sum_{k=1}^{n} a_k \cdot \frac{f_{y|x}(\rho \mid x_k)}{f_{y|x}(\rho \mid x_t)} \right]^r \cdot f_{y|x}(\rho \mid x_t) \cdot d\rho \right]^{\frac{1}{r}}} \tag{10.255}$$

with $\quad h(x_k) = g(x_k) - g(x_t)$

This bound is not only a limitation of the variance, but a limitation of all absolute central higher order moments and thus in any case a limitation of the even central higher order moments.

For $s = r = 1$ we obtain the Barankin bound as it was derived in [SCH73], and the additional assumption of $h(x) = x$ and $a_k = 1$ leads to the Cramér–Rao bound

$$\left\| \left[t_y(\rho) - x_t \right] \right\|_2 \geq \frac{\sum_{k=1}^{n} a_x \cdot (x_k - x_t)}{\left\| \sum_{k=1}^{n} a_k \cdot \frac{f_{y|x}(\rho \mid x_k)}{f_{y|x}(\rho \mid x_t)} \right\|_2} \tag{10.256}$$

$$\left[\int_{\mathrm{P}}\left[t_y(\rho)-x_t\right]^2\cdot f_{y|x}(\rho\mid x_t)\cdot d\rho\right]^{\frac{1}{2}}\geq\frac{\sum_{k=1}^{n}a_k\cdot(x_k-x_t)}{\left[\int_{\mathrm{P}}\left[\sum_{k=1}^{n}a_k\cdot\frac{f_{y|x}(\rho\mid x_k)}{f_{y|x}(\rho\mid x_t)}\right]^2\cdot f_{y|x}(\rho\mid x_t)\cdot d\rho\right]^{\frac{1}{2}}}\tag{10.257}$$

and squared

$$\operatorname{var}(t_y(\rho))=\int_{\mathrm{P}}\left[t_y(\rho)-x_t\right]^2\cdot f_{y|x}(\rho\mid x_t)\cdot d\rho$$

$$\geq\frac{\left(\sum_{k=1}^{n}a_k\cdot(x_k-x_t)\right)^2}{\int_{\mathrm{P}}\left[\frac{\sum_{k=1}^{n}a_k\cdot f_{y|x}(\rho\mid x_k)}{f_{y|x}(\rho\mid x_t)}\right]^2\cdot f_{y|x}(\rho\mid x_t)\cdot d\rho}\tag{10.258}$$

the Cramér–Rao bound.

Barankin's derivation is in principle equal to the derivation of Cramér and Rao, apart from the step in the derivation where the inequality is introduced. While Cramér and Rao apply the Schwartz inequality and thus restrict their bound to quadratic central moments, Barankin applies Hölder's inequality, which allows the determination of limitations for all absolute central higher order moments (*s*. a. c. m.: s^{th} absolute central moment). Thus Barankin obtains a more general formulation for the lower limits of the variance and additionally for the lower limits of all absolute central moments. But even this bound of the variance is not smaller than the Cramér–Rao bound!

10.5 Other bounds

As we do not want to present several other extensive derivations, we just enumerate some other bounds by describing the starting point of their approach and the resulting limitation of the moments. The application of such bounds is certainly limited to special cases, so that anyone who is interested in a more detailed description of these bounds is referred to [BLI69], where these bounds are presented in a more detailed derivation.

Fraser–Guttman bound

Similar to the bound of Chapman and Robbins, this bound is based on an extension of the distribution density by the parameter $\underline{h}$. This extension allows the generation of a difference between two distribution densities, which are standardized with the original distribution density and weighted with additional parameters c_j.

$$S = \sum_{j=1}^{r} c_j \cdot \frac{\left[f_{\underline{y}|\underline{x}}\left(\underline{\rho} \mid \underline{x}(k) + j \cdot \underline{h}\right) - f_{\underline{y}|\underline{x}}\left(\underline{\rho} \mid \underline{x}(k) + (j-1) \cdot \underline{h}\right)\right]}{f_{\underline{y}|\underline{x}}\left(\underline{\rho} \mid \underline{x}(k)\right)} \tag{10.259}$$

On the basis of this weighted distribution density (see also equation (10.145)) we find the lower limit of the variance

$$\operatorname{var}\left(t_{\underline{y}}(\underline{\rho})\right) \geq \frac{1}{\inf\limits_{\left[c_1,..,c_r;\underline{h}\in H, j=1,..,r\right]}\left\{\frac{1}{\underline{h}^2} \cdot \int\limits_{P}..\int\limits_{P} \sum_{j=1}^{r} c_j \cdot \frac{\left[\prod f_{\underline{y}|\underline{x}}\left(\underline{\rho} \mid \underline{x}(k) + j \cdot \underline{h}\right) - \prod f_{\underline{y}|\underline{x}}\left(\underline{\rho} \mid \underline{x}(k) + (j-1) \cdot \underline{h}\right)\right]^2}{\prod f_{\underline{y}|\underline{x}}\left(\underline{\rho} \mid \underline{x}(k)\right)} \cdot \prod d\underline{\rho}\right\}} \tag{10.260}$$

Kiefer bound

For different measures $\mu_1(\underline{h})$ and $\mu_2(\underline{h})$ and different weights E_1, E_2 Kiefer formulates the following approach (see equation (10.145))

$$S = \frac{\int\limits_{H} f_{\underline{y}|\underline{x}}\left(\underline{\rho} \mid \underline{x}(k) + \underline{h}\right) \cdot d\left[\mu_1(h) - \mu_2(h)\right]}{E_1 \cdot h - E_2 \cdot h} \tag{10.261}$$

and we obtain the limitation for the variance

$$\operatorname{var}\left(t_{\underline{y}}(\underline{\rho})\right) \geq \sup_{\left[\mu_1,\mu_2\right]} \frac{\left(E_1 \cdot h - E_2 \cdot h\right)^2}{\int\limits_{P}\cdots\int\limits_{P} \frac{\left\{\left[\int\limits_{H} f_{\underline{y}|\underline{x}}\left(\underline{\rho} \mid \underline{x}(k) + \underline{h}\right) \cdot d\left[\mu_1(h) - \mu_2(h)\right]\right]^2\right\}}{f_{\underline{y}|\underline{x}}\left(\underline{\rho} \mid \underline{x}(k) + \underline{h}\right)} \cdot d\underline{\rho} - 1} \tag{10.262}$$

If the distribution densities are not least-square integrable, we may find other, more general limits for the variance, for example the bound resulting from the approach

$$S = k^n(\vartheta, p) \cdot \frac{\left[f^p_{\underline{y}|\underline{x}}\left(\underline{\rho} \mid \underline{x}(k) + \underline{h}\right) - f^p_{\underline{y}|\underline{x}}\left(\underline{\rho} \mid \underline{x}(k)\right) \right]}{h \cdot f_{\underline{y}|\underline{x}}\left(\underline{\rho} \mid \underline{x}(k)\right)} \tag{10.263}$$

where we again find the weighted difference. The result is the

Extended Fraser–Guttman bound

$$\operatorname{var}\left(t_{\underline{y}}\left(\underline{\rho}\right)\right) \geq \sup_{\underline{h}, p \in H, P} \frac{\dfrac{h^2}{k^{2n}(\vartheta, p)}}{\int\limits_{P} \cdots \int\limits_{P} \dfrac{\left[f^p_{\underline{y}|\underline{x}}\left(\underline{\rho} \mid \underline{x}(k) + \underline{h}\right) - f^p_{\underline{y}|\underline{x}}\left(\underline{\rho} \mid \underline{x}(k)\right) \right]^2}{f_{\underline{y}|\underline{x}}\left(\underline{\rho} \mid \underline{x}(k)\right)} \cdot d\underline{\rho}} \tag{10.264}$$

If the maximum occurs at the point $p = 1$, this bound is reduced to the Chapman–Robbins bound.

Bhattacharrya's bound for densities that are not square integrable

This approach starts with a covariant tensor of second order H_{r0} (mixed $H_r^{\ 0}$)

$$H_{r0Ps} = \left[\left(\underline{h}, p\right) \middle| k \cdot f^p_{\underline{y}|\underline{x}}\left(\underline{\rho} \mid \underline{x}(k) + j \cdot \underline{h}\right) \in \sigma_{\text{Algebra}} \quad \text{for } j = 0, \ldots, r \right] \tag{10.265}$$

If the minimum in equation (10.266) occurs at the point $p = 1$, this bound reduces to the Fraser–Guttman bound, and for $p = 1$, $\underline{h} = 0$ we obtain the Bhattacharrya bound.

$$\operatorname{var}\left(t_{\underline{y}}\left(\underline{\rho}\right)\right) \geq \inf_{\left[c_1, .., c_r; \left(\underline{h}, p\right) \in H_r^0 Ps\right]} \left\{ \frac{k^{2n}\left(\underline{h}, p\right)}{\underline{h}^2 \cdot \int\limits_{P} \cdots \int\limits_{P} \sum_{j=1}^{r} c_j \cdot \dfrac{\left[f^p_{\underline{y}|\underline{x}}\left(\underline{\rho} \mid \underline{x}(k) + j \cdot \underline{h}\right) - f^p_{\underline{y}|\underline{x}}\left(\underline{\rho} \mid \underline{x}(k) + (j-1) \cdot \underline{h}\right) \right]}{f^p_{\underline{y}|\underline{x}}\left(\underline{\rho} \mid \underline{x}(k)\right)} \cdot d\underline{\rho}} \right\} \tag{10.266}$$

This bound shows a strong similarity to the Fraser–Guttman bound.

We now end this enumeration and summarize the main point of the description of the error bounds in the next chapter.

10.6 Summary

In the examination of the different bounds we noticed that the Cramér–Rao bound provides the least lower bound of the introduced bounds. Thus the optimum estimation of an estimator that is restricted to second order moments (Kalman filter), is limited by this bound. All other bounds contain this bound in some form, but they cannot achieve a lower bound, because of additional assumptions (Bhattacharrya) or because of additional conditions regarding the interchanging of differentiation and integration (Chapman–Robbins).

Because of the presence of the Cramér–Rao bound in all other variance bounds, and because of the fact that it is in fact the lowest limit of the variance and the most familiar bound in practical applications, the Cramér–Rao bound certainly is the most important and best-known bound. Extensions to other bounds (e.g., Bhattacharrya) require additional computational effort, which may not be necessary in the analysis of the information.

Nevertheless, the calculation of the limitations of higher order moments may provide further insight into the behavior of the examined estimator, as Fisher's information merely describes a maximum information, which cannot be achieved by a real system with a finite number of measurements. Especially not if we have nonlinear mappings of the random variable, which alter the gaussian distribution and introduce higher order moments into the description of the complete distribution density.

Examining the information on the basis of these bounds requires an additional consideration: whether it makes any sense to carry out extra computation in order to improve the variance bound or whether the results of Fisher's information matrix are sufficient for the current purposes.

In Table no. 10.1 we find a list of the most important and most familiar bounds, which we introduced in this chapter.

Table 10.1: Limitations of the moments of a distribution density function

Unbiased estimates		Unbiased estimates		Biased estimates
No regularity conditions		*Regularity conditions*		
Chapman–Robbins bound	$\geq$	Cramér–Rao bound =1/(Fisher's information)	$\leq$ $\geq$	Cramér–Rao bound
		Hölder's inequality instead of Schwartz's inequality		
Fraser–Guttman, Kiefer		Barankin bounds (s^{th} absolute central moments)		Barankin bounds
Non square integrable distribution densities		*Derivation with an extended estimator*		
Bhattacharrya bounds		Bhattacharrya bounds		Bhattacharrya bounds

10.7 Biased estimator

In the derivation of the Cramér–Rao bound and of all other bounds, we demanded an unbiased estimator. If we examine nonlinear systems, howver, it is quite possible to obtain a bias in the estimation, which causes the estimator $t_{b\underline{y}}(\underline{\rho})$ to estimate the parameter together with an additional shift or bias, instead of estimating the sole parameter. Such a bias can always occur in the inverse mapping of nonlinear functions and we may not even notice it or have any chance to eliminate it. Equivalent to equation (10.03) we derive a lower bound for a biased estimator

$$\int_{\mathrm{P}} t_{b\underline{y}}(\underline{\rho}) \cdot f_{\underline{y}|\underline{x}}(\underline{\rho} \mid \underline{x}) \cdot d\underline{\rho} = \underline{x} + \underline{b}(\underline{x}) \tag{10.267}$$

As the following steps are analogous to the derivation in section 10.1, we neglect any explanation of the transformations. We get

$$\int_{\mathrm{P}} \left(t_{b\underline{y}}(\underline{\rho}) - \underline{x}\right) \cdot f_{\underline{y}|\underline{x}}(\underline{\rho} \mid \underline{x}) \cdot d\underline{\rho} = \underline{b}(\underline{x}) \tag{10.268}$$

$$\frac{\partial}{\partial \underline{x}} \int_{\mathrm{P}} \left(t_{b\underline{y}}(\underline{\rho}) - \underline{x}\right) \cdot f_{\underline{y}|\underline{x}}(\underline{\rho} \mid \underline{x}) \cdot d\underline{\rho} = \frac{\partial}{\partial \underline{x}} \underline{b}(\underline{x}) \tag{10.269}$$

interchange integration and differentiation

$$\int_{\mathrm{P}} \frac{\partial}{\partial \underline{x}} \left[\left(t_{b\underline{y}}(\underline{\rho}) - \underline{x}\right) \cdot f_{\underline{y}|\underline{x}}(\underline{\rho} \mid \underline{x})\right] \cdot d\underline{\rho} = \frac{\partial}{\partial \underline{x}} \underline{b}(\underline{x}) \tag{10.270}$$

Decomposition of the partial derivative into two integrals provides

$$-\int_{\mathrm{P}} I \cdot f_{\underline{y}|\underline{x}}(\underline{\rho} \mid \underline{x}) \cdot d\underline{\rho} + \int_{\mathrm{P}} \left(t_{b\underline{y}}(\underline{\rho}) - \underline{x}\right) \cdot \frac{\partial}{\partial \underline{x}} f_{\underline{y}|\underline{x}}(\underline{\rho} \mid \underline{x}) \cdot d\underline{\rho} = \frac{\partial}{\partial \underline{x}} \underline{b}(\underline{x}) \tag{10.271}$$

and

$$\int_{\mathrm{P}} \left(t_{b\underline{y}}(\underline{\rho}) - \underline{x}\right) \cdot \frac{\partial}{\partial \underline{x}} f_{\underline{y}|\underline{x}}(\underline{\rho} \mid \underline{x}) \cdot d\underline{\rho} = I + \frac{\partial}{\partial \underline{x}} \underline{b}(\underline{x}) \tag{10.272}$$

To obtain an expectation value, we use the substitution

$$\frac{\partial}{\partial \underline{x}} f_{\underline{y}|\underline{x}}(\underline{\rho} \mid \underline{x}) = \frac{\partial}{\partial \underline{x}} \ln\left[f_{\underline{y}|\underline{x}}(\underline{\rho} \mid \underline{x})\right] \cdot f_{\underline{y}|\underline{x}}(\underline{\rho} \mid \underline{x}) \tag{10.273}$$

and transform this via the Schwartz inequality

$$\left(E\{A \cdot B\}\right)^2 \le E\{A^2\} \cdot E\{B^2\} \tag{10.274}$$

to obtain the equation

$$\begin{aligned} &\int_{\mathrm{P}} \left(t_{b\underline{y}}(\underline{\rho}) - \underline{x}\right) \cdot \left(t_{b\underline{y}}(\underline{\rho}) - \underline{x}\right)^T \cdot f_{\underline{y}|\underline{x}}(\underline{\rho} \mid \underline{x}) \cdot d\underline{\rho} \\ &\cdot \int_{\mathrm{P}} \left(\frac{\partial}{\partial \underline{x}} \ln\left[f_{\underline{y}|\underline{x}}(\underline{\rho} \mid \underline{x})\right]\right) \cdot \left(\frac{\partial}{\partial \underline{x}} \ln\left[f_{\underline{y}|\underline{x}}(\underline{\rho} \mid \underline{x})\right]\right)^T \cdot f_{\underline{y}|\underline{x}}(\underline{\rho} \mid \underline{x}) \cdot d\underline{\rho} \\ &\ge \left(I + \frac{\partial}{\partial \underline{x}} \underline{b}(\underline{x})\right) \cdot \left(I + \frac{\partial}{\partial \underline{x}} \underline{b}(\underline{x})\right)^T \end{aligned} \tag{10.275}$$

We thus find the lower bound of biased estimators

$$\int_{P} \left(t_{\underline{by}}(\underline{\rho}) - \underline{x}\right) \cdot \left(t_{\underline{by}}(\underline{\rho}) - \underline{x}\right)^T \cdot f_{\underline{y}|\underline{x}}(\underline{\rho} \mid \underline{x}) \cdot d\underline{\rho}$$

$$\geq \left(I + \frac{\partial}{\partial \underline{x}} \underline{b}(\underline{x})\right) \cdot \left(I + \frac{\partial}{\partial \underline{x}} \underline{b}(\underline{x})\right)^T \tag{10.276}$$

$$\cdot \left\{ \int_{P} \left(\frac{\partial}{\partial \underline{x}} \ln\left[f_{\underline{y}|\underline{x}}(\underline{\rho} \mid \underline{x})\right]\right) \cdot \left(\frac{\partial}{\partial \underline{x}} \ln\left[f_{\underline{y}|\underline{x}}(\underline{\rho} \mid \underline{x})\right]\right)^T \cdot f_{\underline{y}|\underline{x}}(\underline{\rho} \mid \underline{x}) \cdot d\underline{\rho} \right\}^{-1}$$

Notice that the left-hand side of this inequality is not the variance of the biased estimator. To find a more general formulation of the Cramér–Rao bound, we have to rearrange the inequality so that we obtain the variance of the biased estimator $t_{b\underline{y}}(\underline{\rho})$ on the left-hand side of this inequality. This biased variance is

$$\mathrm{var}\left(t_{\underline{by}}(\underline{\rho})\right) = \int_{P} \left(t_{\underline{by}}(\underline{\rho}) - (\underline{x} + \underline{b}(\underline{x}))\right) \cdot \left(t_{\underline{by}}(\underline{\rho}) - (\underline{x} + \underline{b}(\underline{x}))\right)^T \cdot f_{\underline{y}|\underline{x}}(\underline{\rho} \mid \underline{x}) \cdot d\underline{\rho} \tag{10.279}$$

To achieve the desired result, we use two transformations

1.) First we transform the left-hand side of inequality (10.275)

$$\int_{P} \left(t_{\underline{by}}(\underline{\rho}) - \underline{x}\right) \cdot \left(t_{\underline{by}}(\underline{\rho}) - \underline{x}\right)^T \cdot f_{\underline{y}|\underline{x}}(\underline{\rho} \mid \underline{x}) \cdot d\underline{\rho} = E_{\underline{y}|\underline{x}}\left\{\left(t_{\underline{by}}(\underline{\rho}) - \underline{x}\right) \cdot \left(t_{\underline{by}}(\underline{\rho}) - \underline{x}\right)^T \middle| \underline{x}\right\}$$

$$= E_{\underline{y}|\underline{x}}\left\{t_{\underline{by}}(\underline{\rho}) \cdot t_{\underline{by}}(\underline{\rho})^T - t_{\underline{by}}(\underline{\rho}) \cdot \underline{x}^T - \underline{x} \cdot t_{\underline{by}}(\underline{\rho})^T + \underline{x} \cdot \underline{x}^T \middle| \underline{x}\right\}$$

$$= E_{\underline{y}|\underline{x}}\left\{t_{\underline{by}}(\underline{\rho}) \cdot t_{\underline{by}}(\underline{\rho})^T \middle| \underline{x}\right\} - E_{\underline{y}|\underline{x}}\left\{t_{\underline{by}}(\underline{\rho}) \cdot \underline{x}^T \middle| \underline{x}\right\} - E_{\underline{y}|\underline{x}}\left\{\underline{x} \cdot t_{\underline{by}}(\underline{\rho})^T \middle| \underline{x}\right\} + E_{\underline{y}|\underline{x}}\left\{\underline{x} \cdot \underline{x}^T \middle| \underline{x}\right\}$$

$$= E_{\underline{y}|\underline{x}}\left\{t_{\underline{by}}(\underline{\rho}) \cdot t_{\underline{by}}(\underline{\rho})^T \middle| \underline{x}\right\} - E_{\underline{y}|\underline{x}}\left\{t_{\underline{by}}(\underline{\rho}) \middle| \underline{x}\right\} \cdot \underline{x}^T - \underline{x} \cdot E_{\underline{y}|\underline{x}}\left\{t_{\underline{by}}(\underline{\rho})^T \middle| \underline{x}\right\} + \underline{x} \cdot \underline{x}^T \tag{10.280}$$

and the expectation values

$$E_{\underline{y}|\underline{x}}\left\{t_{\underline{by}}(\underline{\rho}) \middle| \underline{x}\right\} = \underline{x} + \underline{b}(\underline{x}) \tag{10.281}$$

are known from the approach in section 10.7. So we find

$$= E_{\underline{y}|\underline{x}}\left\{t_{\underline{by}}(\underline{\rho}) \cdot t_{\underline{by}}(\underline{\rho})^T \middle| \underline{x}\right\} - (\underline{x} + \underline{b}(\underline{x})) \cdot \underline{x}^T - \underline{x} \cdot (\underline{x} + \underline{b}(\underline{x}))^T + \underline{x} \cdot \underline{x}^T$$

which we rearrange to

$$= E_{\underline{y}|\underline{x}}\left\{ t_{b\underline{y}}(\underline{\rho}) \cdot t_{b\underline{y}}(\underline{\rho})^T \middle| \underline{x} \right\} - \underline{x} \cdot \underline{x}^T - \underline{b}(\underline{x}) \cdot \underline{x}^T - \underline{x} \cdot \underline{x}^T - \underline{x} \cdot \underline{b}(\underline{x})^T + \underline{x} \cdot \underline{x}^T$$

and

$$= E_{\underline{y}|\underline{x}}\left\{ t_{b\underline{y}}(\underline{\rho}) \cdot t_{b\underline{y}}(\underline{\rho})^T \middle| \underline{x} \right\} - \underline{x} \cdot \underline{x}^T - \underline{b}(\underline{x}) \cdot \underline{x}^T - \underline{x} \cdot \underline{b}(\underline{x})^T - \underline{b}(\underline{x}) \cdot \underline{b}(\underline{x})^T + \underline{b}(\underline{x}) \cdot \underline{b}(\underline{x})^T$$

we obtain with

$$\begin{aligned} &\int_{\mathrm{P}} \left(t_{b\underline{y}}(\underline{\rho}) - \underline{x} \right) \cdot \left(t_{b\underline{y}}(\underline{\rho}) - \underline{x} \right)^T \cdot f_{\underline{y}|\underline{x}}(\underline{\rho} \mid \underline{x}) \cdot d\underline{\rho} \\ &= E_{\underline{y}|\underline{x}}\left\{ t_{b\underline{y}}(\underline{\rho}) \cdot t_{b\underline{y}}(\underline{\rho})^T \middle| \underline{x} \right\} - (\underline{x} + \underline{b}(\underline{x})) \cdot (\underline{x} + \underline{b}(\underline{x}))^T + \underline{b}(\underline{x}) \cdot \underline{b}(\underline{x})^T \end{aligned} \tag{10.282}$$

a first interim result from the transformation of the left-hand side of inequality (10.275).

In the second step we transform the variance of the biased estimator in a similar way.

2.) The variance of a biased estimator is

$$\operatorname{var}\left(t_{b\underline{y}}(\underline{\rho})\right) = \int_{\mathrm{P}} \left(t_{b\underline{y}}(\underline{\rho}) - E_{\underline{y}|\underline{x}}\left\{ t_{b\underline{y}}(\underline{\rho}) \middle| \underline{x} \right\} \right) \cdot \left(t_{b\underline{y}}(\underline{\rho}) - E_{\underline{y}|\underline{x}}\left\{ t_{b\underline{y}}(\underline{\rho}) \middle| \underline{x} \right\} \right)^T \cdot f_{\underline{y}|\underline{x}}(\underline{\rho} \mid \underline{x}) \cdot d\underline{\rho} \tag{10.283}$$

$$\begin{aligned} &= E_{\underline{y}|\underline{x}}\left\{ \left[t_{b\underline{y}}(\underline{\rho}) - (\underline{x} + \underline{b}(\underline{x})) \right] \cdot \left[t_{b\underline{y}}(\underline{\rho}) - (\underline{x} + \underline{b}(\underline{x})) \right]^T \middle| \underline{x} \right\} \\ &= E_{\underline{y}|\underline{x}}\Big\{ t_{b\underline{y}}(\underline{\rho}) \cdot t_{b\underline{y}}(\underline{\rho})^T - t_{b\underline{y}}(\underline{\rho}) \cdot (\underline{x} + \underline{b}(\underline{x}))^T \\ &\qquad - (\underline{x} + \underline{b}(\underline{x})) \cdot t_{b\underline{y}}(\underline{\rho})^T + (\underline{x} + \underline{b}(\underline{x})) \cdot (\underline{x} + \underline{b}(\underline{x}))^T \Big| \underline{x} \Big\} \end{aligned}$$

We rearrange the expectation of the sum into the sum of the expectation, and get

$$\begin{aligned} &= E_{\underline{y}|\underline{x}}\left\{ t_{b\underline{y}}(\underline{\rho}) \cdot t_{b\underline{y}}(\underline{\rho})^T \middle| \underline{x} \right\} - E_{\underline{y}|\underline{x}}\left\{ t_{b\underline{y}}(\underline{\rho}) \cdot (\underline{x} + \underline{b}(\underline{x}))^T \middle| \underline{x} \right\} \\ &\quad - E_{\underline{y}|\underline{x}}\left\{ (\underline{x} + \underline{b}(\underline{x})) \cdot t_{b\underline{y}}(\underline{\rho})^T \middle| \underline{x} \right\} + E_{\underline{y}|\underline{x}}\left\{ (\underline{x} + \underline{b}(\underline{x})) \cdot (\underline{x} + \underline{b}(\underline{x}))^T \middle| \underline{x} \right\} \end{aligned}$$

and now we extract the factor ($\underline{x}+\underline{b}(\underline{x})$), which is independent of the integration variable, from the integral

$$= E_{\underline{y}|\underline{x}}\left\{t_{b\underline{y}}(\underline{\rho})\cdot t_{b\underline{y}}(\underline{\rho})^T \middle| \underline{x}\right\} - E_{\underline{y}|\underline{x}}\left\{t_{b\underline{y}}(\underline{\rho}) \middle| \underline{x}\right\}\cdot(\underline{x}+\underline{b}(\underline{x}))^T$$

$$-(\underline{x}+\underline{b}(\underline{x}))\cdot E_{\underline{y}|\underline{x}}\left\{t_{b\underline{y}}(\underline{\rho})^T \middle| \underline{x}\right\} + (\underline{x}+\underline{b}(\underline{x}))\cdot(\underline{x}+\underline{b}(\underline{x}))^T$$

with the expectation value of the biased estimator we find

$$= E_{\underline{y}|\underline{x}}\left\{t_{b\underline{y}}(\underline{\rho})\cdot t_{b\underline{y}}(\underline{\rho})^T \middle| \underline{x}\right\} - (\underline{x}+\underline{b}(\underline{x}))\cdot(\underline{x}+\underline{b}(\underline{x}))^T$$

$$-(\underline{x}+\underline{b}(\underline{x}))\cdot(\underline{x}+\underline{b}(\underline{x}))^T + (\underline{x}+\underline{b}(\underline{x}))\cdot(\underline{x}+\underline{b}(\underline{x}))^T$$

and finally

$$\operatorname{var}\left(t_{b\underline{y}}(\underline{\rho})\right) = E_{\underline{y}|\underline{x}}\left\{t_{b\underline{y}}(\underline{\rho})\cdot t_{b\underline{y}}(\underline{\rho})^T \middle| \underline{x}\right\} - (\underline{x}+\underline{b}(\underline{x}))\cdot(\underline{x}+\underline{b}(\underline{x}))^T \tag{10.284}$$

which is the second interim result. If we insert this second interim result into the first interim result from equation (10.282), the left-hand side of inequality (10.275) becomes

$$\int_{\mathrm{P}} \left(t_{b\underline{y}}(\underline{\rho})-\underline{x}\right)\cdot\left(t_{b\underline{y}}(\underline{\rho})-\underline{x}\right)^T \cdot f_{\underline{y}|\underline{x}}(\underline{\rho}\,|\,\underline{x})\cdot d\underline{\rho}$$

$$= E_{\underline{y}|\underline{x}}\left\{t_{b\underline{y}}(\underline{\rho})\cdot t_{b\underline{y}}(\underline{\rho})^T \middle| \underline{x}\right\} - (\underline{x}+\underline{b}(\underline{x}))\cdot(\underline{x}+\underline{b}(\underline{x}))^T + \underline{b}(\underline{x})\cdot\underline{b}(\underline{x})^T \tag{10.285}$$

$$= \operatorname{var}\left(t_{b\underline{y}}(\underline{\rho})\right) + \underline{b}(\underline{x})\cdot\underline{b}(\underline{x})^T$$

and we finally obtain the desired lowest limitation of the variance of a biased estimator

$$\int_{\mathrm{P}} \left(t_{b\underline{y}}(\underline{\rho})-\underline{x}\right)\cdot\left(t_{b\underline{y}}(\underline{\rho})-\underline{x}\right)^T \cdot f_{\underline{y}|\underline{x}}(\underline{\rho}\,|\,\underline{x})\cdot d\underline{\rho} = \operatorname{var}\left(t_{b\underline{y}}(\underline{\rho})\right) + \underline{b}(\underline{x})\cdot\underline{b}(\underline{x})^T$$

$$\geq \left(I + \frac{\partial}{\partial \underline{x}}\underline{b}(\underline{x})\right)\cdot\left(I + \frac{\partial}{\partial \underline{x}}\underline{b}(\underline{x})\right)^T \tag{10.286}$$

$$\cdot\left\{\int_{\mathrm{P}} \left(\frac{\partial}{\partial \underline{x}}\ln\left[f_{\underline{y}|\underline{x}}(\underline{\rho}\,|\,\underline{x})\right]\right)\cdot\left(\frac{\partial}{\partial \underline{x}}\ln\left[f_{\underline{y}|\underline{x}}(\underline{\rho}\,|\,\underline{x})\right]\right)^T \cdot f_{\underline{y}|\underline{x}}(\underline{\rho}\,|\,\underline{x})\cdot d\underline{\rho}\right\}^{-1}$$

or with Fisher's information matrix

$$\operatorname{var}\left(t_{b\underline{y}}(\underline{\rho})\right) \geq \left(I+\frac{\partial}{\partial \underline{x}}\underline{b}(\underline{x})\right)\cdot\left(I+\frac{\partial}{\partial \underline{x}}\underline{b}(\underline{x})\right)^T \cdot F^{-1} - \underline{b}(\underline{x})\cdot\underline{b}(\underline{x})^T \tag{10.287}$$

This is the desired Cramér–Rao bound for biased estimators.

10.7.1 Biased estimator versus unbiased estimator

Now we consider the relation between biased and unbiased estimators. We thus examine the Cramér–Rao bounds for unbiased

$$\operatorname{var}\left(t_{\underline{y}}(\underline{\rho})\right) \geq F^{-1} \tag{10.288}$$

and for biased estimators

$$\operatorname{var}\left(t_{b\underline{y}}(\underline{\rho})\right) \geq \left(I+\frac{\partial}{\partial \underline{x}}\underline{b}(\underline{x})\right)\cdot\left(I+\frac{\partial}{\partial \underline{x}}\underline{b}(\underline{x})\right)^T \cdot F^{-1} - \underline{b}(\underline{x})\cdot\underline{b}(\underline{x})^T \tag{10.289}$$

and try to find the least limitation of these two bounds.

It only makes sense to ask if the biased estimator might generate a smaller variance than the unbiased estimator, because a greater variance would be synonymous with a worse estimator. But we are not interested in such worse estimators, because we can always generate arbitrary poor estimates. Thus our approach is clear and we start the calculation with the proposition

$$\operatorname{var}\left(t_{b\underline{y}}(\underline{\rho})\right) \leq \operatorname{var}\left(t_{\underline{y}}(\underline{\rho})\right) \tag{10.290}$$

that the variance of the biased estimators may be less than or equal to the variance of the unbiased estimator. If we insert the results that we found in the evaluation of the information inequality (Cramér–Rao bound), we get

$$\left(I+\frac{\partial}{\partial \underline{x}}\underline{b}(\underline{x})\right)\cdot\left(I+\frac{\partial}{\partial \underline{x}}\underline{b}(\underline{x})\right)^T \cdot F^{-1} - \underline{b}(\underline{x})\cdot\underline{b}(\underline{x})^T \leq F^{-1} \tag{10.291}$$

and we thus have found a multi-dimensional differential inequality, which we are going to solve as a differential equation. An appropriate extension of this solution to the solution area of the inequality can be achieved by a proper choice of the integration constant. Because of the solution area of the inequality it is obvious that we do not get a unique solution and that we are therefore not able to identify a unique bias $\underline{b}(\underline{x})$. Nevertheless, we want to solve the differential equation, to demonstrate that the statements can be verified.

We reduce the multi-dimensional problem to a scalar problem, where $b(x)$, x, F are scalar quantities. The solution of the multi-dimensional differential equation would lead to the computation of n^2 differential inequalities for an n-dimensional vector $\underline{x}$. As we do not expect a unique solution because of the inequality, we show only the possibility of such a solution, and this can be exemplarily done for the scalar inequality.

The scalar differential equation is thus given by

$$\frac{\left(1+\frac{\partial}{\partial x}b(x)\right)^2}{F}-b(x)^2=\frac{1}{F} \tag{10.292}$$

$$\left(1+\frac{\partial}{\partial x}b(x)\right)^2-F\cdot b(x)^2-1=1+(b'(x))^2+2\cdot b'(x)-F\cdot b(x)^2-1$$
$$=(b'(x))^2+2\cdot b'(x)-F\cdot b(x)^2 \tag{10.293}$$

We now search for the solution of the quadratic equation

$$(b'(x))^2+2\cdot b'(x)+1=F\cdot b(x)^2+1$$

$$(b'(x)+1)^2=F\cdot b(x)^2+1 \tag{10.294}$$

and

$$b'(x)=\frac{\partial}{\partial x}b(x)=-1\pm\sqrt{1+F\cdot b(x)^2} \tag{10.295}$$

We separate the variables

$$db(x)=dx\cdot\left\{-1\pm\sqrt{1+F\cdot b(x)^2}\right\} \tag{10.296}$$

and obtain

$$\frac{db(x)}{\left\{-1\pm\sqrt{1+F\cdot b(x)^2}\right\}}=dx=x+C_x \tag{10.297}$$

where C_x is the integration constant.

The solution of the differential on the left-hand side remains, and we introduce a case discrimination, caused by the signs of the square root.

Case 1

$$\frac{db(x)}{\left\{-1-\sqrt{1+F\cdot b(x)^2}\right\}}=-\frac{db(x)}{\left\{1+\sqrt{1+F\cdot b(x)^2}\right\}}$$
$$=-\frac{db(x)}{\left\{1+\sqrt{F}\cdot\sqrt{\frac{1}{F}+b(x)^2}\right\}} \tag{10.298}$$

In order to eliminate the square root in the denominator, we expand the quotient

$$=-\frac{db(x)}{\left\{1+\sqrt{F}\cdot\sqrt{\frac{1}{F}+b(x)^2}\right\}}\cdot\frac{\left\{1-\sqrt{F}\cdot\sqrt{\frac{1}{F}+b(x)^2}\right\}}{\left\{1-\sqrt{F}\cdot\sqrt{\frac{1}{F}+b(x)^2}\right\}}$$

$$=-db(x)\cdot\frac{\left\{1-\sqrt{F}\cdot\sqrt{\frac{1}{F}+b(x)^2}\right\}}{\left\{1-F\cdot\left(\frac{1}{F}+b(x)^2\right)\right\}}$$

$$=-db(x)\cdot\frac{\left\{1-\sqrt{F}\cdot\sqrt{\frac{1}{F}+b(x)^2}\right\}}{\left\{-F\cdot b(x)^2\right\}}=db(x)\cdot\frac{\left\{1-\sqrt{F}\cdot\sqrt{\frac{1}{F}+b(x)^2}\right\}}{\left\{F\cdot b(x)^2\right\}}$$

and we get two quotients

$$=\frac{db(x)}{\left\{F\cdot b(x)^2\right\}}-\frac{\sqrt{F}\cdot\sqrt{\frac{1}{F}+b(x)^2}}{\left\{F\cdot b(x)^2\right\}}\cdot db(x) \tag{10.299}$$

which we solve separately. With

$$\frac{db(x)}{\left\{F\cdot b(x)^2\right\}}=-\frac{1}{F}\cdot\frac{1}{b(x)}+C_{b1} \tag{10.300}$$

and

$$-\frac{\sqrt{F}\cdot\sqrt{\frac{1}{F}+b(x)^2}}{\left\{F\cdot b(x)^2\right\}}\cdot db(x)=\frac{1}{\sqrt{F}}\cdot\left\{-\frac{\sqrt{\frac{1}{F}+b(x)^2}}{b(x)}+\operatorname{arcsinh}\left(b(x)\cdot\sqrt{F}\right)\right\}+C_{b2} \tag{10.301}$$

from ([BRO87], page 46, no. 190) we find the first result

$$\begin{aligned}&\frac{db(x)}{\left\{F\cdot b(x)^2\right\}}-\frac{\sqrt{F}\cdot\sqrt{\frac{1}{F}+b(x)^2}}{\left\{F\cdot b(x)^2\right\}}\cdot db(x)\\ &=-\frac{1}{F}\cdot\frac{1}{b(x)}+\frac{1}{\sqrt{F}}\cdot\left\{-\frac{\sqrt{\frac{1}{F}+b(x)^2}}{b(x)}+\operatorname{arcsinh}\left(b(x)\cdot\sqrt{F}\right)\right\}+C_b\end{aligned} \tag{10.302}$$

where the two integration constants $C_{b1}+C_{b2}$ have already been summarized to C_b.

Case2

$$\frac{db(x)}{\left\{-1+\sqrt{1+F\cdot b(x)^2}\right\}}=\frac{db(x)}{\left\{-1+\sqrt{F}\cdot\sqrt{\frac{1}{F}+b(x)^2}\right\}} \tag{10.303}$$

and we again expand the quotient, to circumvent the square root in the denominator

$$=\frac{db(x)}{\left\{-1+\sqrt{F}\cdot\sqrt{\frac{1}{F}+b(x)^2}\right\}}\cdot\frac{\left\{1+\sqrt{F}\cdot\sqrt{\frac{1}{F}+b(x)^2}\right\}}{\left\{1+\sqrt{F}\cdot\sqrt{\frac{1}{F}+b(x)^2}\right\}}$$

leading to the term

$$=db(x)\cdot\frac{\left\{1+\sqrt{F}\cdot\sqrt{\frac{1}{F}+b(x)^2}\right\}}{\left\{-1+F\cdot\left(\frac{1}{F}+b(x)^2\right)\right\}}=db(x)\cdot\frac{\left\{1+\sqrt{F}\cdot\sqrt{\frac{1}{F}+b(x)^2}\right\}}{\left\{F\cdot b(x)^2\right\}}$$

which we decompose into two addends

$$= \frac{db(x)}{\left\{F \cdot b(x)^2\right\}} + \frac{\sqrt{F} \cdot \sqrt{\frac{1}{F} + b(x)^2}}{\left\{F \cdot b(x)^2\right\}} \cdot db(x) \tag{10.304}$$

that are solved separately.

With the previously specified solutions, we obtain the second result

$$\frac{db(x)}{\left\{F \cdot b(x)^2\right\}} + \frac{\sqrt{F} \cdot \sqrt{\frac{1}{F} + b(x)^2}}{\left\{F \cdot b(x)^2\right\}} \cdot db(x)$$

$$= -\frac{1}{F} \cdot \frac{1}{b(x)} - \frac{1}{\sqrt{F}} \cdot \left\{ -\frac{\sqrt{\frac{1}{F} + b(x)^2}}{b(x)} + \operatorname{arcsinh}\left(b(x) \cdot \sqrt{F}\right) \right\} + C_b \tag{10.305}$$

The complete result on the left-hand side of the differential equation (10.297) is thus

$$\frac{db(x)}{\left\{-1 \pm \sqrt{1 + F \cdot b(x)^2}\right\}} = -\frac{1}{F} \cdot \frac{1}{b(x)} \mp \frac{1}{\sqrt{F}} \cdot \left\{ -\frac{\sqrt{\frac{1}{F} + b(x)^2}}{b(x)} + \operatorname{arcsinh}\left(b(x) \cdot \sqrt{F}\right) \right\} + C_b \tag{10.306}$$

(we might additionally us $\operatorname{arcsinh}\left(\sqrt{F} \cdot b(x)\right) = \ln\left[\sqrt{F} \cdot b(x) + \sqrt{1 + F \cdot b(x)^2}\right]$)

and so we have the final solutions

1.Result

$$-\frac{1}{F} \cdot \frac{1}{b(x)} + \frac{1}{\sqrt{F}} \cdot \left\{ -\frac{\sqrt{\frac{1}{F} + b(x)^2}}{b(x)} + \operatorname{arcsinh}\left(b(x) \cdot \sqrt{F}\right) \right\} + C_b = x + C_x \tag{10.307}$$

$$-1 - \sqrt{1 + F \cdot b(x)^2} + \sqrt{F} \cdot b(x) \cdot \operatorname{arcsinh}\left(b(x) \cdot \sqrt{F}\right) + F \cdot b(x) \cdot C_b = F \cdot b(x) \cdot (x + C_x)$$

and

2.Result

$$-\frac{1}{F}\cdot\frac{1}{b(x)}-\frac{1}{\sqrt{F}}\cdot\left\{-\frac{\sqrt{\frac{1}{F}+b(x)^2}}{b(x)}+\operatorname{arcsinh}\left(b(x)\cdot\sqrt{F}\right)\right\}+C_b=x+C_x \quad (10.308)$$

$$-1+\sqrt{1+F\cdot b(x)^2}-\sqrt{F}\cdot b(x)\cdot\operatorname{arcsinh}\left(b(x)\cdot\sqrt{F}\right)+F\cdot b(x)\cdot C_b=F\cdot b(x)\cdot(x+C_x)$$

Now we may find appropriate integration constants (ranges for the integration constants for the inequality), which cannot be identified uniquely, for this transcendent equation. Thus we have verified the statement that a biased estimator may generate a smaller variance than an unbiased estimator.

So the Cramér–Rao bound is not necessarily the lowest limitation of the variance when we deal with nonlinear mappings from state space to observation space. According to the choice of the estimator, the variance may fall below the Cramér–Rao bound, which indeed indicates that the chosen estimator is biased. This bias cannot be uniquely identified, which leads to ambiguities. Such ambiguities are familiar in inverse nonlinear mappings and they do not present a special problem of the estimation theory, but a general problem of the inverse mapping of nonlinear functions.

In the following we present a function that allows us to describe these ambiguities. This ambiguity function has a direct connection to the information functions we already examined, and we will range this ambiguity function in the arrangement of the information functions.

11 Ambiguity function

We now introduce and study another interesting function, which is connected to the previous information measures, especially to Kullback's discrimination information. It enables us to derive certain information criteria, applied in system identification, which is the main reason for us to include this function in our presentation. In [SCH73] the author – referring to radar engineering – presents the ambiguity function, which allows a conclusion about the uniqueness of the expected estimation.

$$\gamma(\underline{x}, \underline{x}_t) = \int_{\mathrm{P}} \ln\left[f_{\underline{y}|\underline{x}}(\underline{\rho} \mid \underline{x})\right] \cdot f_{\underline{y}|\underline{x}}(\underline{\rho} \mid \underline{x}_t) \cdot d\underline{\rho} \tag{11.01}$$

This function is not the ambiguity function that is used in radar engineering, where the ambiguity function is directly related to the Wigner–Ville spectrum and represents a second order Fourier transformation. It is instead the expectation value of log-likelihood function conditioned on the true value $\underline{x}_t$.

So we do not obtain a scalar result from the calculation of this integral, but we get another function. The form of this function – especially the positions of the maximums – decides whether we find a unique estimate or not.

With the example of a gaussian distribution, resulting from the measurement of real and imaginary parts of a complex phasor with additive gaussian measurement errors, we examine the ambiguity function in detail. We start with

$$f_{\underline{y}|\underline{x}}(\underline{\rho} \mid \underline{x}) = \frac{1}{(2\pi) \cdot |R|^{\frac{1}{2}}} \cdot \exp\left\{-\frac{1}{2} \cdot \left(\underline{\rho} - \underline{h}(\underline{x})\right)^T \cdot R^{-1} \cdot \left(\underline{\rho} - \underline{h}(\underline{x})\right)\right\} \tag{11.02}$$

where

$$R = \begin{pmatrix} r_1 & 0 \\ 0 & r_2 \end{pmatrix} \qquad \text{and} \qquad \underline{h}(x) = \begin{pmatrix} a \cdot \cos x \\ a \cdot \sin x \end{pmatrix} \tag{11.03}$$

$$\underline{y}(x) = \begin{pmatrix} a \cdot \cos x \\ a \cdot \sin x \end{pmatrix} + \begin{pmatrix} v_1 \\ v_2 \end{pmatrix} \tag{11.04}$$

Inserting the distribution density into the integral provides two integrals, when we calculate the logarithm of the density

$$\gamma_1\left(\underline{x},\underline{x}_t\right) = -\int_P \ln\left[(2\pi)\cdot|R|^{\frac{1}{2}}\right]\cdot f_{\underline{y}|\underline{x}}\left(\underline{\rho}\mid\underline{x}_t\right)\cdot d\underline{\rho} \tag{11.06}$$

$$\gamma_1\left(\underline{x},\underline{x}_t\right) = -\ln\left[(2\pi)\cdot|R|^{\frac{1}{2}}\right] \tag{11.07}$$

and

$$\gamma_2\left(\underline{x},\underline{x}_t\right) = -\frac{1}{2}\cdot\int_P \left(\underline{\rho}-\underline{h}(\underline{x})\right)^T\cdot R^{-1}\cdot\left(\underline{\rho}-\underline{h}(\underline{x})\right)\cdot f_{\underline{y}|\underline{x}}\left(\underline{\rho}\mid\underline{x}_t\right)\cdot d\underline{\rho} \tag{11.08}$$

As the argument x is a scalar value, we find

$$\gamma_2(x,x_t) = -\frac{1}{2}\cdot\int_P \left(\underline{h}(x_t)+\underline{v}-\underline{h}(x)\right)^T\cdot R^{-1}\cdot\left(\underline{h}(x_t)+\underline{v}-\underline{h}(x)\right)\cdot f_{\underline{y}|\underline{x}}\left(\underline{\rho}\mid x_t\right)\cdot d\underline{\rho}$$

$$\gamma_2(x,x_t) = -\frac{1}{2}\cdot\int_P \left(\left[\underline{h}(x_t)-\underline{h}(x)\right]+\underline{v}\right)^T\cdot R^{-1}\cdot\left(\left[\underline{h}(x_t)-\underline{h}(x)\right]+\underline{v}\right)\cdot f_{\underline{y}|\underline{x}}\left(\underline{\rho}\mid x_t\right)\cdot d\underline{\rho} \tag{11.09}$$

The mixed terms of the multiplication vanish because of the independence of measurement noise and the mapped state $\underline{h}(x)$, and we get the integral

$$\gamma_2(x,x_t) = -\frac{1}{2}\cdot\int_P \left(\left[\underline{h}(x_t)-\underline{h}(x)\right]^T\cdot R^{-1}\cdot\left[\underline{h}(x_t)-\underline{h}(x)\right]+\underline{v}^T\cdot R^{-1}\cdot\underline{v}\right)\cdot f_{\underline{y}|\underline{x}}\left(\underline{\rho}\mid x_t\right)\cdot d\underline{\rho} \tag{11.10}$$

which we again decompose into the sum of two integrals

$$\begin{aligned}\gamma_2(x,x_t) = &-\frac{1}{2}\cdot\int_P \left[\underline{h}(x_t)-\underline{h}(x)\right]^T\cdot R^{-1}\cdot\left[\underline{h}(x_t)-\underline{h}(x)\right]\cdot f_{\underline{y}|\underline{x}}\left(\underline{\rho}\mid x_t\right)\cdot d\underline{\rho}\\ &-\frac{1}{2}\cdot\int_P \underline{v}^T\cdot R^{-1}\cdot\underline{v}\cdot f_{\underline{y}|\underline{x}}\left(\underline{\rho}\mid x_t\right)\cdot d\underline{\rho}\end{aligned} \tag{11.11}$$

The quadratic form in the first integral does not depend on the integration variable, so that we proceed

$$\begin{aligned}\gamma_2(x,x_t) = &-\frac{1}{2}\cdot\left[\underline{h}(x_t)-\underline{h}(x)\right]^T\cdot R^{-1}\cdot\left[\underline{h}(x_t)-\underline{h}(x)\right]\\ &-\frac{1}{2}\cdot\int_P \operatorname{tr}\left\{R^{-1}\cdot\underline{v}\cdot\underline{v}^T\right\}\cdot f_{\underline{y}|\underline{x}}\left(\underline{\rho}\mid x_t\right)\cdot d\underline{\rho}\end{aligned}$$

and obtain finally

$$\gamma_2(x,x_t) = -\frac{1}{2}\cdot\left[\underline{h}(x_t)-\underline{h}(x)\right]^T\cdot R^{-1}\cdot\left[\underline{h}(x_t)-\underline{h}(x)\right] - \frac{1}{2}\cdot \mathrm{tr}\left\{R^{-1}\cdot R\right\} \tag{11.12}$$

with

$$\mathrm{tr}\left\{R^{-1}\cdot R\right\} = \mathrm{tr}\{I\} = m$$

which leads to the complete second term.

$$\gamma_2(x,x_t) = -\frac{1}{2}\cdot\left[\underline{h}(x_t)-\underline{h}(x)\right]^T\cdot R^{-1}\cdot\left[\underline{h}(x_t)-\underline{h}(x)\right] - \frac{m}{2} \tag{11.13}$$

Thus we have the ambiguity function for arbitrary gaussian densities

$$\gamma(x,x_t) = -\ln\left[(2\pi)\cdot|R|^{\frac{1}{2}}\right] - \frac{m}{2} - \frac{1}{2}\cdot\left[\underline{h}(x_t)-\underline{h}(x)\right]^T\cdot R^{-1}\cdot\left[\underline{h}(x_t)-\underline{h}(x)\right] \tag{11.14}$$

The first two terms are independent of the variable state x and thus do not add any contribution to the form of the ambiguity function. They only cause a parallel shift of the function. The third part of this equation describes the form of the function, and thus this part contains the information about the uniqueness of the estimates.

If we first examine this third term for the measurement $y_1 = a\cdot\cos(x)+v_1$, we get the expression

$$T_{\mathrm{Form}} = -\frac{1}{2}\cdot\frac{\left[a\cdot\cos(x_t)-a\cdot\cos(x)\right]^2}{r_1} = -\frac{a^2}{2}\cdot\frac{\left[c_t-\cos(x)\right]^2}{r_1} \tag{11.15}$$

with c_t = const., as x_t (true value) has been chosen to be constant.

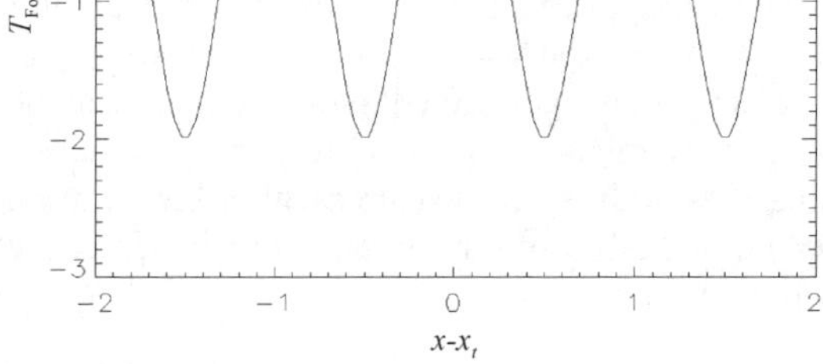

Figure: T_{Form}

with the values

a = 1

$x_t = 0 \Rightarrow c_t = 1$

$r_1 = 1$

Fig. 11.1: T_{Form} of the real part

We obtain a periodic function with an arbitrary number of relative maxima, which all provide equal estimates of the state, because of the ambiguities of the inverse mapping. We cannot find a special best estimate.

Thus the ambiguity function clearly states the drawback of the inverse trigonometric mappings, where the estimates may have a bias $n \cdot (2\pi)$ with $n \in \Re$ with respect to the true value. If we do not have any additional information about the desired state x, we are not able to identify the right value from the infinite number of possible optimum estimates.

An additional measurement of the imaginary part $y_2 = a \cdot \sin(x) + v_2$ does not provide any improvement, because it also maintains an infinite number of relative maxima, preventing a unique determination of the best estimate. The form of the function is described by the term

$$T_{\text{Form2}} = -\frac{a^2}{2} \cdot \left(\frac{[\cos(x_t) - \cos(x)]^2}{r_1} + \frac{[\sin(x_t) - \sin(x)]^2}{r_2} \right) \tag{11.16}$$

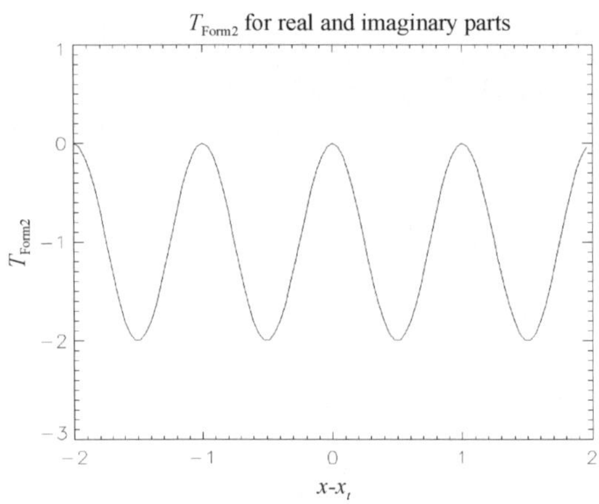

Figure: T_{Form2}

with the values

a = 1

$x_t = 0$ =>

$\cos x_t = 1$; $\sin x_t = 0$

$r_1 = r_2 = 1$

Fig. 11.2: T_{Form2} with real and imaginary part

We only obtain a unique relation between the measurements and the estimates when an additional measurement generates one absolute maximum out of the infinite number of relative maxima. The value of this absolute maximum is greater than the values of all other relative maxima, so that we get one best estimate. But such an absolute maximum only occurs if the additional measurement can be mapped uniquely to one estimate, which means that this measurement alone already provides a unique maximum. This additional measurement, which can be inverted uniquely, is then added to the available measurements and it bends the function so that we obtain an absolute maximum.

If we take for instance $y_3 = k \cdot x + v_3$ with $v_3 \propto N(0,r_3)$ independent from v_1, v_2 we get a third term $T_3 = -\frac{1}{2} \cdot k^2 \cdot \frac{(x_t - x)^2}{r_3}$, describing an upside-down parabola, which is added to the already known term. We get

$$T_{\text{Form3}} = -\frac{a^2}{2} \cdot \left(\frac{[\cos(x_t) - \cos(x)]^2}{r_1} + \frac{[\sin(x_t) - \sin(x)]^2}{r_2} \right) - \frac{1}{2} \cdot k^2 \cdot \frac{(x_t - x)^2}{r_3} \tag{11.17}$$

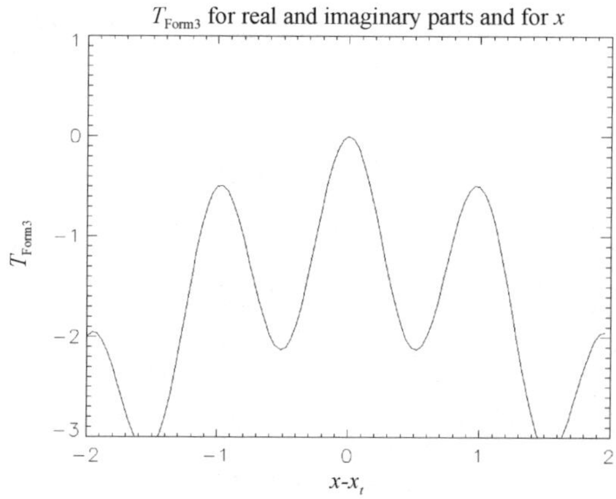

Figure: T_{Form3}

with the values

a = k = 1

$x_t = 0$ =>

$\cos x_t = 1$; $\sin x_t = 0$

$r_1 = r_2 = 1$; $r_3 = 10$

Fig. 11.3: T_{Form3} with a linear term

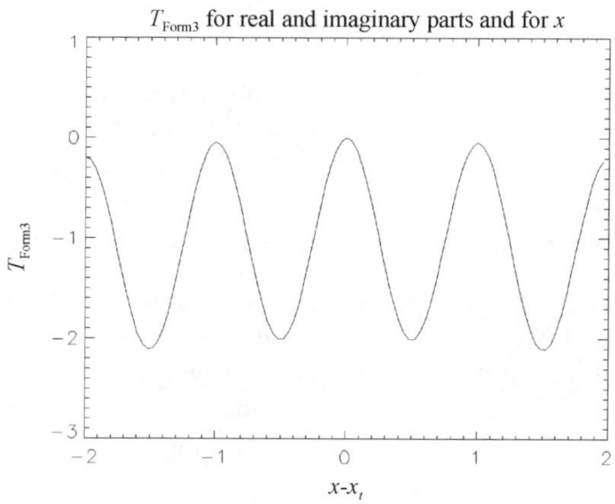

Figure: T_{Form3}

with the values

a = k =1

$x_t = 0$ =>

$\cos x_t = 1$; $\sin x_t = 0$

$r_1 = r_2 = 1$; $r_3 = 100$

Fig. 11.4: T_{Form3} with a linear term II

This function has a unique maximum at the point $x = x_t$.

We meet the problem of a unique identification and an optimal estimation from nonlinear mappings in many applications and it is treated in many different ways. But unique solutions are not achievable without additional a priori knowledge (e.g., a mapping, which can be inverted uniquely, a given restriction of the range or a bounded divergence of the process, ...).

11.1 The ambiguity function and Kullback's information

The ambiguity function is nothing other than the second part of Kullback's information

$$G_1 = \int_{\mathrm{P}} \ln\left[\frac{f_{\underline{y}|\underline{x}}(\underline{\rho} \mid \underline{x}_t)}{f_{\underline{y}|\underline{x}}(\underline{\rho} \mid \underline{x})}\right] \cdot f_{\underline{y}|\underline{x}}(\underline{\rho} \mid \underline{x}_t) \cdot d\underline{\rho} \tag{11.18}$$

$$G_1 = \int_{\mathrm{P}} \ln\left[f_{\underline{y}|\underline{x}}(\underline{\rho} \mid \underline{x}_t)\right] \cdot f_{\underline{y}|\underline{x}}(\underline{\rho} \mid \underline{x}_t) \cdot d\underline{\rho} - \int_{\mathrm{P}} \ln\left[f_{\underline{y}|\underline{x}}(\underline{\rho} \mid \underline{x})\right] \cdot f_{\underline{y}|\underline{x}}(\underline{\rho} \mid \underline{x}_t) \cdot d\underline{\rho} \tag{11.19}$$

$$G_1 = \int_{\mathrm{P}} \ln\left[f_{\underline{y}|\underline{x}}(\underline{\rho} \mid \underline{x}_t)\right] \cdot f_{\underline{y}|\underline{x}}(\underline{\rho} \mid \underline{x}_t) \cdot d\underline{\rho} - \gamma(\underline{x} \mid \underline{x}_t) \tag{11.20}$$

with

$$\gamma(\underline{x}, \underline{x}_t) = \int_{\mathrm{P}} \ln\left[f_{\underline{y}|\underline{x}}(\underline{\rho} \mid \underline{x})\right] \cdot f_{\underline{y}|\underline{x}}(\underline{\rho} \mid \underline{x}_t) \cdot d\underline{\rho} \tag{11.21}$$

The first addend in Kullback's information is a Shannon information, while the second addend builds the ambiguity function. Shannon's information is computed from the distribution density conditioned on the true value of the desired state and it thus depends only on the variance. The actual, directed difference of the two distribution densities $f_{\underline{y}|\underline{x}}(\underline{\rho} \mid \underline{x}_t)$ and $f_{\underline{y}|\underline{x}}(\underline{\rho} \mid \underline{x})$, depending on the position of the random variable $\underline{x}$, as the Kullback–Leibler distance describes it, is generated by the ambiguity function. Thus the ambiguity function contains the complete information regarding the shift in the position between the two distribution densities. Shannon's information consists merely of a term, which is independent of the absolute position of the distribution densities, and which merely describes the uncertainty, contained in the distribution.

Thus the ambiguity function is a part of the Kullback–Leibler distance and it is exactly the part that describes the information about the position. We find this connection again in the next section where we examine the connection between Fisher's information and the ambiguity function.

11.2 Connection between ambiguity function and Fisher's information

The ambiguity function does not lead to a scalar value. It provides a function where the desired parameter is the argument. It thus shows the ambiguities of mappings, which cannot be inverted uniquely. Automatically we may ask for the applications of such a function.

We want to discuss this question by first comparing the ambiguity function with Fisher's information

$$F = -\int_{\mathrm{P}} \left(\frac{\partial^2}{\partial \underline{x} \cdot \partial \underline{x}^T} \ln\left[f_{\underline{y}|\underline{x}}(\underline{\rho} \mid \underline{x}) \right] \right) \cdot f_{\underline{y}|\underline{x}}(\underline{\rho} \mid \underline{x}) \cdot d\underline{\rho} \tag{11.22}$$

$$\gamma(\underline{x}, \underline{x}_t) = \int_{\mathrm{P}} \ln\left[f_{\underline{y}|\underline{x}}(\underline{\rho} \mid \underline{x}) \right] \cdot f_{\underline{y}|\underline{x}}(\underline{\rho} \mid \underline{x}_t) \cdot d\underline{\rho} \tag{11.23}$$

If we build the second derivative of the ambiguity function with respect to the parameter $\underline{x}$, we obtain, if we require that we can interchange integration and differentiation:

$$\frac{\partial^2}{\partial \underline{x} \cdot \partial \underline{x}} \gamma(\underline{x}, \underline{x}_t) = \int_{\mathrm{P}} \left(\frac{\partial^2}{\partial \underline{x} \cdot \partial \underline{x}} \ln\left[f_{\underline{y}|\underline{x}}(\underline{\rho} \mid \underline{x}) \right] \right) \cdot f_{\underline{y}|\underline{x}}(\underline{\rho} \mid \underline{x}_t) \cdot d\underline{\rho} \tag{11.24}$$

and thus

$$F = -\frac{\partial^2}{\partial \underline{x} \cdot \partial \underline{x}} \gamma(\underline{x}, \underline{x}_t) \Bigg|_{\underline{x} = \underline{x}_t} \tag{11.25}$$

i.e. the negative second derivative of the ambiguity function at the point $\underline{x} = \underline{x}_t$ is equal to Fisher's information.

Fisher's information describes – as we already know – the spreading of the maximum of a distribution density. This maximum provides the maximum likelihood estimate (ML estimate), where Fisher's matrix describes the quality. An optimal ML estimator estimates $\underline{x}_{\mathrm{MLE}} = \underline{x}_t$, and thus the maximum of the distribution density is located in the point $\underline{x}_t$. If we now calculate the second derivative of the ambiguity function, then this derivative is exactly the stochastic observability. Moreover the first derivative vanishes at the point $\underline{x} = \underline{x}_t$, which means that we have an extremum at this point. We are going to carry out more detailed examinations of these connections in the next sections.

11.3 Maximum likelihood estimation and the ambiguity function

We already know from section 10.1.5 that Fisher's information is a measure for the quality of an ML estimate. If we calculate the expectation value of the likelihood function at the point where the true value $\underline{x}_t$ lies

$$E_{\underline{y}|\underline{x}}\{L \mid \underline{x}_t\} = \int_{\mathrm{P}} \ln\left[f_{\underline{y}|\underline{x}}(\underline{\rho} \mid \underline{x})\right] \cdot f_{\underline{y}|\underline{x}}(\underline{\rho} \mid \underline{x}_t) \cdot d\underline{\rho} = \gamma(\underline{x}, \underline{x}_t) \tag{11.26}$$

we obtain the ambiguity function. From the equality we notice that the maximum likelihood estimation does not necessarily provide a unique estimate and the resulting ambiguities are described by the ambiguity function. We may ask whether we might be able to find a unique estimator. We therefore take a look at several approaches applied to measurements with additive gaussian noise.

11.3.1 Maximum likelihood estimation = minimum Kullback estimation = maximum Ambiguity estimation = minimum variance estimation

The maximum likelihood estimation calculates the maximum of the likelihood function or of the log-likelihood function (see section 10.1.5)

$$l = f_{\underline{y}|\underline{x}}(\underline{\rho} \mid \underline{x}) \text{ or } \qquad L = \ln l = \ln\left[f_{\underline{y}|\underline{x}}(\underline{\rho} \mid \underline{x})\right] \tag{11.27}$$

The log-likelihood function of a gaussian distribution is

$$L = \ln l = \ln\left[f_{\underline{y}|\underline{x}}(\underline{\rho} \mid \underline{x})\right] = -\frac{1}{2} \cdot \ln\left[(2\pi)^m \cdot |R|\right] - \frac{1}{2} \cdot \left(\underline{\rho} - \underline{h}(\underline{x})\right)^T \cdot R^{-1} \cdot \left(\underline{\rho} - \underline{h}(\underline{x})\right) \tag{11.28}$$

and the maximum is determined via

$$\frac{\partial}{\partial \underline{x}} \ln l = -\frac{1}{2} \cdot \frac{\partial}{\partial \underline{x}} \left[\left(\underline{\rho} - \underline{h}(\underline{x})\right)^T \cdot R^{-1} \cdot \left(\underline{\rho} - \underline{h}(\underline{x})\right)\right] \stackrel{!}{=} 0 \tag{11.29}$$

which is equal to the approach of the weighted least-squares estimation, where the inverse of the covariance matrix is used as the weighting matrix. In the least-squares estimation we determine the minimum of the quadratic form

$$LSE = \min\left[\left(\underline{\rho} - \underline{h}(\underline{x})\right)^T \cdot R^{-1} \cdot \left(\underline{\rho} - \underline{h}(\underline{x})\right)\right] \tag{11.30}$$

where the second derivative has the opposite sign to the second derivative of the ML estimation, which leads to a minimum. Thus maximum likelihood estimation and weighted least-squares estimation provide the same estimate for gaussian distribution densities.

The ambiguity function of a gaussian distribution is

$$\gamma(x,x_t) = -\ln\left[(2\pi)\cdot|R|^{\frac{1}{2}}\right] - \frac{m}{2} - \frac{1}{2}\cdot[\underline{h}(x_t) - \underline{h}(x)]^T \cdot R^{-1}\cdot[\underline{h}(x_t) - \underline{h}(x)] \quad (11.31)$$

and the determination of the maximum leads to the same quadratic form

$$\frac{\partial}{\partial \underline{x}}\left\{[\underline{h}(x_t) - \underline{h}(x)]^T \cdot R^{-1}\cdot[\underline{h}(x_t) - \underline{h}(x)]\right\} \overset{!}{=} 0 \quad (11.32)$$

which is equal to the previous quadratic forms if $\underline{\rho} = \underline{h}(\underline{x}_t)$.

Thus the ambiguity function estimate is equal to the maximum likelihood estimate under the condition that we know the true value $\underline{x}_t$. The discrimination estimation can also be used for estimation, because for gaussian densities we get (see section 9.1.1)

$$G_1 = \int_P \ln\left[\frac{f_{\underline{y}|\underline{x}}(\underline{\rho}\,|\,\underline{x}_t)}{f_{\underline{y}|\underline{x}}(\underline{\rho}\,|\,\underline{x})}\right]\cdot f_{\underline{y}|\underline{x}}(\underline{\rho}\,|\,\underline{x}_t)\cdot d\underline{\rho} \quad (11.33)$$

$$G_1 = \int_P \ln\left[f_{\underline{y}|\underline{x}}(\underline{\rho}\,|\,\underline{x}_t)\right]\cdot f_{\underline{y}|\underline{x}}(\underline{\rho}\,|\,\underline{x}_t)\cdot d\underline{\rho} - \int_P \ln\left[f_{\underline{y}|\underline{x}}(\underline{\rho}\,|\,\underline{x})\right]\cdot f_{\underline{y}|\underline{x}}(\underline{\rho}\,|\,\underline{x}_t)\cdot d\underline{\rho}$$

$$G_1 = S_1\left(f_{\underline{y}|\underline{x}}(\underline{\rho}\,|\,\underline{x}_t)\right) - \gamma(\underline{x},\underline{x}_t) \quad (11.34)$$

with

$$\gamma(x,x_t) = -\ln\left[(2\pi)\cdot|R|^{\frac{1}{2}}\right] - \frac{m}{2} - \frac{1}{2}\cdot[\underline{h}(x_t) - \underline{h}(x)]^T \cdot R^{-1}\cdot[\underline{h}(x_t) - \underline{h}(x)] \quad (11.35)$$

As Shannon's information does not depend on the desired parameter, but only on the covariance of this parameter, the derivative of this equation leads to the term

$$\frac{\partial}{\partial \underline{x}}\left\{[\underline{h}(x_t) - \underline{h}(x)]^T \cdot R^{-1}\cdot[\underline{h}(x_t) - \underline{h}(x)]\right\} \overset{!}{=} 0 \quad (11.36)$$

Because of the positive sign the minimum discrimination estimate is equal to a maximum likelihood estimate, or a minimum least-squares estimate as well as a maximum ambiguity estimate.

None of these criteria is able to find a unique estimate if the inverse mapping of the transformation $\underline{h}(\underline{x})$ is not unique. We thus conclude that the ambiguities cannot be eliminated without any additional a priori information or a priori knowledge.

To return to the connection between the ambiguity function and Fisher's information matrix, we only need to take a closer look at these criteria. There a maximum ambiguity estimate equals a maximum likelihood estimate. The maxima of the ambiguity function do only occur at the positions $\underline{x} = \underline{x}_t + n \cdot$ (time of one period), so that the second derivative is zero at this point. This negative second derivative is equal to Fisher's information matrix, where the elements on the main diagonal have only nonnegative values. Obviously, because of this connection, the ambiguity function is directly related to the maximum likelihood estimation.

11.3.2 Maximum likelihood estimation

Here we are looking for the maximum of a conditional distribution density. But we additionally take the logarithm of this function, which does not change the result, because the logarithm is a strictly monotonically increasing function, which alters the scale of the presentation but does not alter the positions of the extreme values.

11.3.2.1 Application: Discriminator (Demodulation)

The classical form of a discriminator is generated by creating the derivative of the (desired) transition function *H(f)* and then by multiplying the given function with the generated derivative. A subsequent integration over this product provides a value that is a measure for the deviation of the desired and the available function.

Example of a discriminator:

The likelihood function or the more familiar log-likelihood function is

$$L = \ln l = \ln f_{\underline{y}|\underline{x}}\left(\underline{\rho} \mid \underline{\xi}\right) \tag{11.37}$$

The logarithm does not alter the positions of the maximum and it also does not alter the symmetry. If we calculate the derivative of this function and multiply the result by the original distribution, conditioned on $\underline{\xi}_t$, we obtain the integral, which equals the first derivative of the ambiguity function.

$$\int_Y \frac{\partial}{\partial \underline{\xi}} \left[\ln f_{\underline{y}|\underline{x}}\left(\underline{\rho} \mid \underline{\xi}\right)\right] \cdot f_{\underline{y}|\underline{x}}\left(\underline{\rho} \mid \underline{\xi}_t\right) \cdot d\underline{\rho} \tag{11.38}$$

If $\underline{\xi} = \underline{\xi}_t$, we obtain

$$\int_{Y} \frac{\partial}{\partial \underline{\xi}} \left[\ln f_{\underline{y}|\underline{x}}(\underline{\rho} \mid \underline{\xi}) \right] \cdot f_{\underline{y}|\underline{x}}(\underline{\rho} \mid \underline{\xi}) \cdot d\underline{\rho}$$

$$= \int_{Y} \frac{1}{f_{\underline{y}|\underline{x}}(\underline{\rho} \mid \underline{\xi})} \cdot \frac{\partial}{\partial \underline{\xi}} \left[f_{\underline{y}|\underline{x}}(\underline{\rho} \mid \underline{\xi}) \right] \cdot f_{\underline{y}|\underline{x}}(\underline{\rho} \mid \underline{\xi}) \cdot d\underline{\rho} \tag{11.39}$$

$$= \int_{Y} \frac{\partial}{\partial \underline{\xi}} \left[f_{\underline{y}|\underline{x}}(\underline{\rho} \mid \underline{\xi}) \right] \cdot d\underline{\rho} = \frac{\partial}{\partial \underline{\xi}} \int_{Y} \left[f_{\underline{y}|\underline{x}}(\underline{\rho} \mid \underline{\xi}) \right] \cdot d\underline{\rho} = \frac{\partial}{\partial \underline{\xi}} 1 = 0$$

If both functions are equal, the result is equal to zero. If the existing function $f_{\underline{y}|\underline{x}}(\underline{\rho} \mid \underline{\xi}_t) = f_{\underline{y}|\underline{x}}(\underline{\rho} \mid \underline{\xi} + \delta\underline{\xi})$ is shifted with respect to its derivative, then the result of the integral varies (with respect to the shift) positively or negatively from zero. The variation can be use to determine the deviation $\delta\underline{\xi}$ of the parameter. This requires the description of the connection between the resulting deviation and $\delta\underline{\xi}$, which is the next step of our derivation. We use the already known integral

$$\int_{Y} \frac{\partial}{\partial \underline{\xi}} \left[\ln f_{\underline{y}|\underline{x}}(\underline{\rho} \mid \underline{\xi}) \right] \cdot f_{\underline{y}|\underline{x}}(\underline{\rho} \mid \underline{\xi}_t) \cdot d\underline{\rho} \tag{11.40}$$

substitute $f_{\underline{y}|\underline{x}}(\underline{\rho} \mid \underline{\xi}_t) = f_{\underline{y}|\underline{x}}(\underline{\rho} \mid \underline{\xi} + \delta\underline{\xi})$ and insert the function into the integral

$$\int_{Y} \frac{\partial}{\partial \underline{\xi}} \left[\ln f_{\underline{y}|\underline{x}}(\underline{\rho} \mid \underline{\xi}) \right] \cdot f_{\underline{y}|\underline{x}}(\underline{\rho} \mid \underline{\xi} + \delta\underline{\xi}) \cdot d\underline{\rho} \tag{11.41}$$

We again substitute $\underline{\xi} + \delta\underline{\xi} = \underline{\vartheta}$ and obtain the integral

$$\int_{Y} \frac{\partial}{\partial \underline{\xi}} \left[\ln f_{\underline{y}|\underline{x}}(\underline{\rho} \mid \underline{\vartheta} + \delta\underline{\xi}) \right] \cdot f_{\underline{y}|\underline{x}}(\underline{\rho} \mid \underline{\vartheta}) \cdot d\underline{\rho} \tag{11.42}$$

Now we calculate a Taylor series of the log-likelihood function

$$\ln f_{\underline{y}|\underline{x}}(\underline{\rho} \mid \underline{\vartheta} + \delta\underline{\xi}) = \ln f_{\underline{y}|\underline{x}}(\underline{\rho} \mid \underline{\vartheta}) + \frac{\partial}{\partial \underline{\vartheta}} \ln f_{\underline{y}|\underline{x}}(\underline{\rho} \mid \underline{\vartheta} + \delta\underline{\xi}) \Big|_{\delta\underline{\xi}=0} \cdot \delta\underline{\xi} + \text{h.o.t.} \tag{11.43}$$

If we insert this into the integral and neglect the higher order terms, we get

$$\int_{Y} \frac{\partial}{\partial \underline{\xi}} \left[\ln f_{\underline{y}|\underline{x}}(\underline{\rho} \mid \underline{\vartheta}) + \frac{\partial}{\partial \underline{\vartheta}} \ln f_{\underline{y}|\underline{x}}(\underline{\rho} \mid \underline{\vartheta} + \delta\underline{\xi}) \Big|_{\delta\underline{\xi}=0} \cdot \delta\underline{\xi} \right] \cdot f_{\underline{y}|\underline{x}}(\underline{\rho} \mid \underline{\vartheta}) \cdot d\underline{\rho}$$

$$= \int_{Y} \frac{\partial}{\partial \underline{\xi}} \left[\ln f_{\underline{y}|\underline{x}}(\underline{\rho} \mid \underline{\vartheta}) \right] \cdot f_{\underline{y}|\underline{x}}(\underline{\rho} \mid \underline{\vartheta}) \cdot d\underline{\rho} \tag{11.44}$$

$$+ \int_{Y} \frac{\partial}{\partial \underline{\xi}} \left[\frac{\partial}{\partial \underline{\vartheta}} \ln f_{\underline{y}|\underline{x}}(\underline{\rho} \mid \underline{\vartheta} + \delta\underline{\xi}) \Big|_{\delta\underline{\xi}=0} \cdot \delta\underline{\xi} \right] \cdot f_{\underline{y}|\underline{x}}(\underline{\rho} \mid \underline{\vartheta}) \cdot d\underline{\rho}$$

We already know that the first integral vanishes. Thus the computation of the second integrals remains

$$\int_{Y} \frac{\partial}{\partial \underline{\xi}} \left[\frac{\partial}{\partial \underline{\vartheta}} \ln f_{\underline{y}|\underline{x}}(\underline{\rho} \mid \underline{\vartheta} + \delta\underline{\xi}) \Big|_{\delta\underline{\xi}=0} \cdot \delta\underline{\xi} \right] \cdot f_{\underline{y}|\underline{x}}(\underline{\rho} \mid \underline{\vartheta}) \cdot d\underline{\rho} =$$

$$= \underbrace{\int_{Y} \left[\frac{\partial^2}{\partial \underline{\vartheta} \cdot \partial \underline{\vartheta}^T} \ln f_{\underline{y}|\underline{x}}(\underline{\rho} \mid \underline{\vartheta} + \delta\underline{\xi}) \Big|_{\delta\underline{\xi}=0} \right] \cdot f_{\underline{y}|\underline{x}}(\underline{\rho} \mid \underline{\vartheta}) \cdot d\underline{\rho}}_{-F} \cdot \delta\underline{\xi} \tag{11.45}$$

We finally get

$$\int_{Y} \frac{\partial}{\partial \underline{\xi}} \left[\ln f_{\underline{y}|\underline{x}}(\underline{\rho} \mid \underline{\xi}) \right] \cdot f_{\underline{y}|\underline{x}}(\underline{\rho} \mid \underline{\xi} + \delta\underline{\xi}) \cdot d\underline{\rho} = -F \cdot \delta\underline{\xi} \tag{11.46}$$

Fisher's information multiplied by the deviation $\delta\underline{\xi}$ provides a measure for this deviation, which is mapped into another coordinate system (e.g., a deviation of a frequency mapped on a deviation in the voltage).

In figure 11.5 we see a gaussian pulse, which is shifted along the characteristic curve of the discriminator. At each time instant we compute the integral of the product of both functions. The resulting function at the end of this algorithm describes the cross-correlation between the gaussian pulse and the characteristic curve of the discriminator. The characteristic curve of the discriminator is generated from the gaussian pulse, which is symmetric to the ordinate. The resulting function achieves the value zero if the time shift of the gaussian pulse and the time shift of the characteristic curve of the discriminator are equal. In our case, this happens if the time shift vanishes.

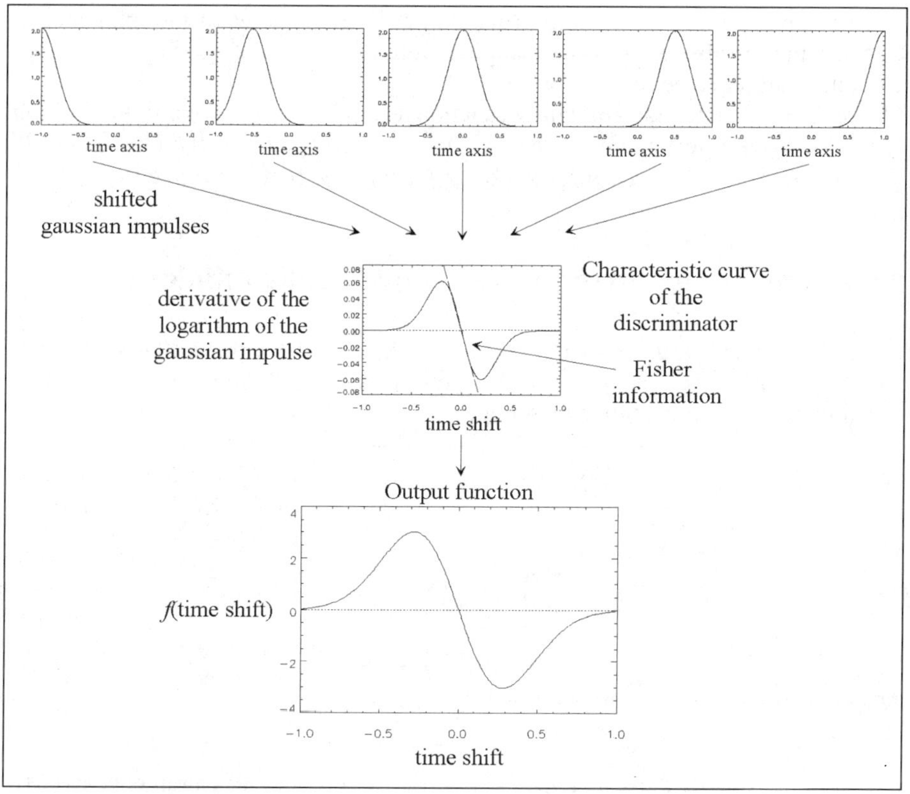

Fig. 11.5: Discriminator

This operation enables us to detect the shift of the gaussian pulse, entering the system, at the value we get from the calculation of the signal at the output of the system. If the output signal vanishes, the shift of the input signal is equal to the shift of the characteristic curve of the discriminator. Fisher's information determines the slope of the characteristic curve of the discriminator and thus the slope of the output signal. If we want to detect the non-shifted gaussian pulse as exactly as possible, we have to determine the zero point of the output function as exactly as possible. This becomes even more exactly, if the derivative of the output function in the zero point increases, i.e. if Fisher's information increases. An increasing Fisher information leads to an increasing information about the desired parameters of the input functions (observations). This is merely caused by an increase of the slope of the characteristic curve of the discriminator and thus we obtain a resulting improvement in the discrimination of different values in the coordinate system of the desired parameters.

This application of Fisher's information matrix reminds us of the fundamental tensor, which is also used in the mapping from one coordinate system to another coordinate system (see section 10.1.3).
Moreover we denoted the ambiguity function as the information about the position in the Kullback information, so that we here get a mapping of the uncertainty in the position $\delta\xi$ via Fisher's information, if we take the first derivative.

11.4 The ML estimation is asymptotically efficient

The starting point of this derivation is a state space model with a scalar state space variable x, which is mapped by the multi-dimensional nonlinear observation function $\underline{h}(x)$ to the contaminated output $\underline{y}$.

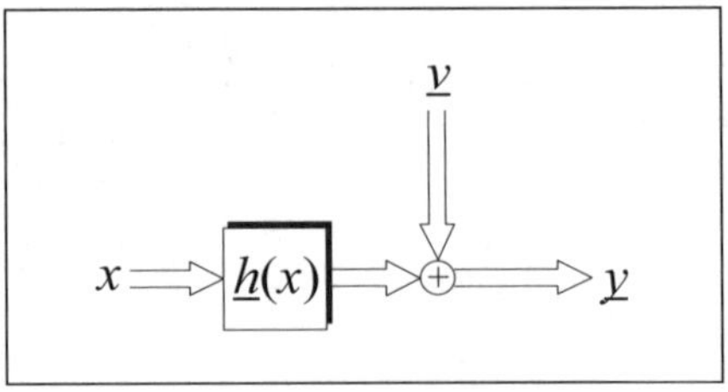

Fig. 11.6: Nonlinear observation equation

We want to estimate the scalar state variable x by repeated measurements and ML estimation. That is, we obtain measurements of the form:

$$y_i = h_i(x) + v_i \tag{11.47}$$

and the conditional distribution density $f_{\underline{y}|x}(\underline{\rho} \mid \xi)$ results from the sum of all observations. If we additionally assume independent measurements, we get:

$$f_{\underline{y}|x}(\underline{\rho} \mid \xi) = f_{y_1|x}(\rho_1 \mid \xi) \cdot f_{y_2|x}(\rho_2 \mid \xi) \cdot \ldots \cdot f_{y_m|x}(\rho_m \mid \xi)$$

$$f_{\underline{y}|x}(\underline{\rho} \mid \xi) = \prod_{i=1}^{m} f_{y_i|x}(\rho_i \mid \xi) \tag{11.48}$$

We want to find the ML estimate

$$L' = \frac{\partial}{\partial \xi} \ln f_{\underline{y}|x}(\underline{\rho} \mid \xi) \Big|_{\hat{\underline{x}}_{\mathrm{ML}}} \overset{!}{=} 0 \tag{11.49}$$

The independence directly leads to the equation

$$L' = \frac{\partial}{\partial \xi} \ln \prod_{i=1}^{m} f_{y_i|x}(\rho_i \mid \xi)\Big|_{\underline{\hat{x}}_{\mathrm{ML}}} \overset{!}{=} 0$$

$$L' = \frac{\partial}{\partial \xi} \sum_{i=1}^{m} \ln f_{y_i|x}(\rho_i \mid \xi)\Big|_{\underline{\hat{x}}_{\mathrm{ML}}} \overset{!}{=} 0$$

$$L' = \sum_{i=1}^{m} \frac{\partial}{\partial \xi} \ln f_{y_i|x}(\rho_i \mid \xi)\Big|_{\underline{\hat{x}}_{\mathrm{ML}}} \overset{!}{=} 0$$

$$L' = L'(x) = \sum_{i=1}^{m} \frac{1}{f_{y_i|x}(\rho_i \mid \xi)} \cdot \frac{\partial}{\partial \xi} f_{y_i|x}(\rho_i \mid \xi)\Big|_{\underline{\hat{x}}_{\mathrm{ML}}} \overset{!}{=} 0 \tag{11.50}$$

Now we generate a Taylor series expansion of this function $L'(x)$ at the ML estimate

$$L'(x) = L'(\hat{x}_{\mathrm{ML}}) + L''(\hat{x}_{\mathrm{ML}}) \cdot (x - \hat{x}_{\mathrm{ML}}) + \text{h.o.t.} \tag{11.51}$$

with

$$L''(x) = \frac{\partial}{\partial \xi} \sum_{i=1}^{m} \frac{1}{f_{y_i|x}(\rho_i \mid \xi)} \cdot \frac{\partial}{\partial \xi} f_{y_i|x}(\rho_i \mid \xi)\Big|_{\underline{\hat{x}}_{\mathrm{ML}}} \tag{11.52}$$

We cancel the first addend of this series expansion, as this addend describes the function at the ML estimate.

$$L'(x) = \underbrace{L'(\hat{x}_{\mathrm{ML}})}_{0} + L''(\hat{x}_{\mathrm{ML}}) \cdot (x - \hat{x}_{\mathrm{ML}}) + \text{h.o.t.} \tag{11.53}$$

At this point the function vanishes, so that we write

$$L'(x) = L''(\hat{x}_{\mathrm{ML}}) \cdot (x - \hat{x}_{\mathrm{ML}}) + \text{h.o.t.} \tag{11.54}$$

and we proceed with the calculation of the conditional density $f_{\underline{y}|x}(\underline{\rho} \mid \xi)$ from this relation. We neglect the higher order terms (h.o.t.)

$$L'(x) \cong L''(\hat{x}_{\mathrm{ML}}) \cdot (x - \hat{x}_{\mathrm{ML}}) \tag{11.55}$$

and integrate

$$L(x) \cong L''(\hat{x}_{\mathrm{ML}}) \cdot \int (x - \hat{x}_{\mathrm{ML}}) \cdot dx \tag{11.56}$$

$$L(x) \cong L''(\hat{x}_{\mathrm{ML}}) \cdot \frac{1}{2} \cdot (x - \hat{x}_{\mathrm{ML}})^2 + \text{const.} \tag{11.57}$$

To determine the constant, we examine this function at the ML estimate, where we find:

$$L(\hat{x}_{\mathrm{ML}}) \cong L''(\hat{x}_{\mathrm{ML}}) \cdot \frac{1}{2} \cdot (\hat{x}_{\mathrm{ML}} - \hat{x}_{\mathrm{ML}})^2 + \text{const.} = \text{const.} \tag{11.58}$$

We thus obtain:

$$L(x) = \ln f_{\underline{y}|x}(\underline{\rho} \mid \xi) \cong L''(\hat{x}_{\mathrm{ML}}) \cdot \frac{1}{2} \cdot (x - \hat{x}_{\mathrm{ML}})^2 + L(\hat{x}_{\mathrm{ML}})$$

$$f_{\underline{y}|x}(\underline{\rho} \mid \xi) \cong \exp\left\{L''(\hat{x}_{\mathrm{ML}}) \cdot \frac{1}{2} \cdot (x - \hat{x}_{\mathrm{ML}})^2 + l(\hat{x}_{\mathrm{ML}})\right\}$$

$$f_{\underline{y}|x}(\underline{\rho} \mid \xi) \cong \exp\{L(\hat{x}_{\mathrm{ML}})\} \cdot \exp\left\{L''(\hat{x}_{\mathrm{ML}}) \cdot \frac{1}{2} \cdot (x - \hat{x}_{\mathrm{ML}})^2\right\} \tag{11.59}$$

The likelihood function $f_{\underline{y}|x}(\underline{\rho} \mid \xi)$, or the conditional distribution density, is a gaussian distribution density, where the maximum is achieved for the expectation value (mean) $\hat{x}_{\mathrm{ML}}$. The variance of this distribution is equal to

$$\sigma^2 = -\frac{1}{L''(\hat{x}_{\mathrm{ML}})} \tag{11.60}$$

resulting from the comparison with

$$f_{\underline{y}|x}(\underline{\rho} \mid \xi) \cong \frac{1}{\sqrt{2\pi} \cdot \sigma} \cdot \exp\left\{-\frac{1}{2} \cdot \frac{(x - \hat{x}_{\mathrm{ML}})^2}{\sigma^2}\right\} \tag{11.61}$$

We now examine the variance in detail. We have

$$\sigma^2 = -\frac{1}{L''(\hat{x}_{\mathrm{ML}})}$$

with

$$L''(x) = \frac{\partial}{\partial \xi} \sum_{i=1}^{m} \frac{1}{f_{y_i|x}(\rho_i \mid \xi)} \cdot \frac{\partial}{\partial \xi} f_{y_i|x}(\rho_i \mid \xi) \Bigg|_{\hat{\underline{x}}_{\mathrm{ML}}}$$

$$L''(x) = \sum_{i=1}^{m} \frac{\partial}{\partial \xi} \left\{ \frac{1}{f_{y_i|x}(\rho_i \mid \xi)} \cdot \frac{\partial}{\partial \xi} f_{y_i|x}(\rho_i \mid \xi) \right\} \Bigg|_{\underline{\hat{x}}_{\mathrm{ML}}}$$

$$L''(x) = \sum_{i=1}^{m} \frac{\partial}{\partial \xi} \left\{ \frac{\partial}{\partial \xi} \ln f_{y_i|x}(\rho_i \mid \xi) \right\} \Bigg|_{\underline{\hat{x}}_{\mathrm{ML}}}$$

$$L''(x) = \sum_{i=1}^{m} \left\{ \frac{\partial^2}{\partial \xi^2} \ln f_{y_i|x}(\rho_i \mid \xi) \right\} \Bigg|_{\underline{\hat{x}}_{\mathrm{ML}}} \tag{11.62}$$

If we apply the weak law of the large numbers (Tschebycheff's inequality), which says that we may interchange the probabilities and the relative frequencies with a certain probability [vWE85]), we can substitute the summation by the integration, computing the expectation value, and thus we write:

$$l''(x) = \sum_{i=1}^{m} \left\{ \frac{\partial^2}{\partial \xi^2} \ln f_{y_i|x}(\rho_i \mid \xi) \right\} \Bigg|_{\underline{\hat{x}}_{\mathrm{ML}}} = m \cdot E\left\{ \frac{\partial^2}{\partial \xi^2} \ln f_{y_i|x}(\rho_i \mid \xi) \right\}$$

$$l''(x) = m \cdot (-F) \tag{11.63}$$

The variance is therefore given by the equation:

$$\sigma^2 = -\frac{1}{l''(\hat{x}_{\mathrm{ML}})} = -\frac{1}{m \cdot (-F)} = \frac{1}{m \cdot F}$$

$$\sigma^2 = \frac{1}{m \cdot F} \tag{11.64}$$

This term provides two statements about the variance of the desired parameter. The first statement is the familiar $\sqrt{n}$ law. The more measurements we get, the more m increases and the smaller the variance of the desired parameter becomes. The second statement is that the reciprocal Fisher information is a variance and that this variance describes the lowest limitation of the variance that can be achieved. As our result

$$f_{\underline{y}|x}(\underline{\rho} \mid \xi) \cong \exp\{l(\hat{x}_{\mathrm{ML}})\} \cdot \exp\left\{ l''(\hat{x}_{\mathrm{ML}}) \cdot \frac{1}{2} \cdot (x - \hat{x}_{\mathrm{ML}})^2 \right\}$$

$$f_{\underline{y}|x}(\underline{\rho} \mid \xi) \cong \exp\{l(\hat{x}_{\mathrm{ML}})\} \cdot \exp\left\{ -\frac{1}{2} \cdot m \cdot F \cdot (x - \hat{x}_{\mathrm{ML}})^2 \right\} \tag{11.65}$$

has been obtained by application of Tschebycheff's inequality (weak law of large numbers), this result is only achieved for large m, i.e. the distribution density is only achieved for large m, and the Cramér–Rao bound is also achieved for large m. Thus this estimation is asymptotically efficient, which means that the Cramér–Rao bound is asymptotically achieved (for large m).

Fisher's information is thus obviously connected to the maximum likelihood estimation, because it describes not only the width of the maximum but is also asymptotically achieved by an ML estimator.

11.5 Transition to the Akaike information criterion

The ambiguity function is – as we saw before – immediately connected to the maximum likelihood estimation. The equation

$$\gamma(\underline{x},\underline{x}_t)=E_{\underline{y}|\underline{x}_t}\left\{\ln\left[f_{\underline{y}|\underline{x}}(\underline{\rho}\mid\underline{x})\right]\right\}=\int_{\mathrm{P}}\ln\left[f_{\underline{y}|\underline{x}}(\underline{\rho}\mid\underline{x})\right]\cdot f_{\underline{y}|\underline{x}}(\underline{\rho}\mid\underline{x}_t)\cdot d\underline{\rho} \tag{11.66}$$

describes the expectation value of the log-likelihood function, where the expectation is taken with respect to another distribution density of the same family. The difference between these distribution densities is actually caused by the difference of the parameters $\underline{x}$ and $\underline{x}_t$.

If we calculate the maximum likelihood estimate from the log-likelihood function, we may expand the log-likelihood function into a series at the ML estimate

$$\begin{aligned}\ln\left[f_{\underline{y}|\underline{x}}(\underline{\rho}\mid\underline{\xi})\right]&=\ln\left[f_{\underline{y}|\underline{x}}(\underline{\rho}\mid\hat{\underline{x}}_{\mathrm{ML}})\right]+\frac{\partial}{\partial\underline{\xi}}\ln\left[f_{\underline{y}|\underline{x}}(\underline{\rho}\mid\underline{\xi})\right]\Bigg|_{\hat{\underline{x}}_{\mathrm{ML}}}\cdot(\underline{x}-\hat{\underline{x}}_{\mathrm{ML}})\\&+\frac{1}{2}\cdot\frac{\partial^2}{\partial\underline{\xi}^2}\ln\left[f_{\underline{y}|\underline{x}}(\underline{\rho}\mid\underline{\xi})\right]\Bigg|_{\hat{\underline{x}}_{\mathrm{ML}}}\cdot(\underline{x}-\hat{\underline{x}}_{\mathrm{ML}})^2+\text{h.o.t.}\end{aligned} \tag{11.67}$$

At the ML estimate the following condition holds

$$\frac{\partial}{\partial\underline{\xi}}\ln\left[f_{\underline{y}|\underline{x}}(\underline{\rho}\mid\underline{\xi})\right]\Bigg|_{\hat{\underline{x}}_{\mathrm{ML}}}=0 \tag{11.68}$$

and we get

$$\ln\left[f_{\underline{y}|\underline{x}}(\underline{\rho}\mid\underline{\xi})\right]=\ln\left[f_{\underline{y}|\underline{x}}(\underline{\rho}\mid\hat{\underline{x}}_{\mathrm{ML}})\right]+\frac{1}{2}\cdot\frac{\partial^2}{\partial\underline{\xi}^2}\ln\left[f_{\underline{y}|\underline{x}}(\underline{\rho}\mid\underline{\xi})\right]\Bigg|_{\hat{\underline{x}}_{\mathrm{ML}}}\cdot(\underline{x}-\hat{\underline{x}}_{\mathrm{ML}})^2+\text{h.o.t.} \tag{11.69}$$

We insert this function into the ambiguity function and obtain

$$\gamma(\underline{x},\underline{x}_t)=\int_{\mathrm{P}}\left[\ln\left[f_{\underline{y}|\underline{x}}(\underline{\rho}\mid\hat{\underline{x}}_{\mathrm{ML}})\right]+\frac{1}{2}\cdot\frac{\partial^2}{\partial\underline{\xi}^2}\ln\left[f_{\underline{y}|\underline{x}}(\underline{\rho}\mid\underline{\xi})\right]\Bigg|_{\hat{\underline{x}}_{ML}}\cdot(\underline{x}-\hat{\underline{x}}_{\mathrm{ML}})^2+\text{h.o.t.}\right]\cdot f_{\underline{y}|\underline{x}}(\underline{\rho}\mid\underline{x}_t)\cdot d\underline{\rho} \tag{11.70}$$

Neglecting the higher order terms provides

$$\gamma(\underline{x},\underline{x}_t)\approx\int_{\mathrm{P}}\left[\ln\left[f_{\underline{y}|\underline{x}}(\underline{\rho}\mid\hat{\underline{x}}_{\mathrm{ML}})\right]+\frac{1}{2}\cdot\frac{\partial^2}{\partial\underline{\xi}^2}\ln\left[f_{\underline{y}|\underline{x}}(\underline{\rho}\mid\underline{\xi})\right]\Bigg|_{\hat{\underline{x}}_{\mathrm{ML}}}\cdot(\underline{x}-\hat{\underline{x}}_{\mathrm{ML}})^2\right]\cdot f_{\underline{y}|\underline{x}}(\underline{\rho}\mid\underline{x}_t)\cdot d\underline{\rho} \tag{11.71}$$

$$\begin{aligned}\gamma(\underline{x},\underline{x}_t)\approx&\int_{\mathrm{P}}\ln\left[f_{\underline{y}|\underline{x}}(\underline{\rho}\mid\hat{\underline{x}}_{\mathrm{ML}})\right]\cdot f_{\underline{y}|\underline{x}}(\underline{\rho}\mid\underline{x}_t)\cdot d\underline{\rho}\\&+\int_{\mathrm{P}}\frac{1}{2}\cdot\frac{\partial^2}{\partial\underline{\xi}^2}\ln\left[f_{\underline{y}|\underline{x}}(\underline{\rho}\mid\underline{\xi})\right]\Bigg|_{\hat{\underline{x}}_{\mathrm{ML}}}\cdot(\underline{x}-\hat{\underline{x}}_{\mathrm{ML}})^2\cdot f_{\underline{y}|\underline{x}}(\underline{\rho}\mid\underline{x}_t)\cdot d\underline{\rho}\end{aligned} \tag{11.72}$$

The first term is constant, as both parameters, which may cause a variation of the densities, assume fixed values. We may thus also write

$$\begin{aligned}\gamma(\underline{x},\underline{x}_t)\approx&\ \text{const.}(\hat{\underline{x}}_{\mathrm{ML}},\underline{x}_t)\\&+\frac{1}{2}\cdot\int_{\mathrm{P}}\frac{\partial^2}{\partial\underline{\xi}^2}\ln\left[f_{\underline{y}|\underline{x}}(\underline{\rho}\mid\underline{\xi})\right]\Bigg|_{\hat{\underline{x}}_{\mathrm{ML}}}\cdot f_{\underline{y}|\underline{x}}(\underline{\rho}\mid\underline{x}_t)\cdot d\underline{\rho}\cdot(\underline{x}-\hat{\underline{x}}_{\mathrm{ML}})^2\end{aligned} \tag{11.73}$$

and the remaining term in the integral is merely the negative Fisher information with respect to the parameter $\underline{x}$ and we thus obtain

$$\gamma(\underline{x}, \underline{x}_t) \approx \text{const.}(\underline{\hat{x}}_{\text{ML}}, \underline{x}_t) - \frac{1}{2} \cdot F \cdot (\underline{x} - \underline{\hat{x}}_{\text{ML}})^2 \tag{11.74}$$

The ambiguity function, which is some kind of cross-information of a Shannon information and thus a part of Kullback's information, is sensitive with respect to alterations of the parameter $\underline{x}$, i.e. this functions serves to detect variations of the parameter $\underline{x}$.

Again we clearly notice the connection between Fisher's information and the metric tensor, because Fisher's information is the mapping from the variations of the parameters in the state space coordinate system to the variations of the observations $\underline{y}$ in the observation space coordinate system.

Akaike uses this connection between the ambiguity function and the likelihood estimation to derive an information criterion, which estimates the number of parameters of an AR model on the basis of an ambiguity function. This derivation will be the aim of the next chapter.

12 Akaike's information criterion

Akaike's information criterion is directly connected to Kullback's information measure and to the ambiguity function, which Akaike in [AKA73] [AKA74/2] calls (cross) entropy. On the basis of this ambiguity function we are able to derive a connection between Akaike's information criterion (AIC) and the maximum likelihood estimation. AIC is a criterion that allows the determination of the order of an AR model in system identification. This is done in the following way:

$$\mathrm{AIC}(n) \quad = \text{Akaike information of an AR model of order } n \tag{12.01}$$

The optimum order n_{optimal} of the model results from

$$\mathrm{AIC}(n_{\text{optimal}}) \quad = \min \{\mathrm{AIC}(n) \mid n = 1,2,\ldots,N\} \tag{12.02}$$

the minimum Akaike information occurring in the variation of the order of the model from 1 to N. And Akaike says that the minimization of the Kullback–Leibler distance is equal to the minimization of the AIC Akaike information criterion.

12.1 Akaike's information criterion and regression

To derive Akaike's information from the ambiguity function, we have to occupy ourselves with the problem of regression, which is an important part of this derivation. Regression means that we try to determine parameters that alter the given measurements, on the basis of these measurements. The most familiar regression is the following.

12.1.1 Least-squares regression

Here we have an observation

$$y_i = c_i^T \cdot \underline{x} + v_i \tag{12.03}$$

and we try to minimize the sum of the quadratic measurement errors. In our system we obtain a scalar observation, consisting of n parameters x_i. The equation that we have to minimize is

$$\sum_{i=1}^{m} v_i^2 = \sum_{i=1}^{m} \left(y_i - c_i^T \cdot \underline{x}\right)^2 \overset{!}{=} \min \tag{12.04}$$

This equation can also be denoted in another form. We therefore generate a measurement vector of the single observations and get

$$\underline{y} = C \cdot \underline{x} + \underline{v} \tag{12.05}$$

We minimize the sum of the quadratic errors

$$\underline{v}^T \cdot \underline{v} = \left(\underline{y} - C \cdot \underline{x}\right)^T \cdot \left(\underline{y} - C \cdot \underline{x}\right) \tag{12.06}$$

and this is equal to

$$\frac{\partial}{\partial \underline{x}} \left(\underline{y} - C \cdot \underline{x}\right)^T \cdot \left(\underline{y} - C \cdot \underline{x}\right) = 0$$

$$\frac{\partial}{\partial \underline{x}} \left(\underline{y}^T \cdot \underline{y} - \underline{y}^T \cdot C \cdot \underline{x} - (C \cdot \underline{x})^T \cdot \underline{y} + (C \cdot \underline{x})^T \cdot C \cdot \underline{x}\right) = 0$$

$$\frac{\partial}{\partial \underline{x}} \left(-2 \cdot \underline{y}^T \cdot C \cdot \underline{x} + \underline{x}^T \cdot C^T \cdot C \cdot \underline{x}\right) = 0$$

$$-2 \cdot C^T \cdot \underline{y} + 2 \cdot C^T \cdot C \cdot \underline{x} = 0$$

$$\underline{x}_{\mathrm{LS}} = \left(C^T \cdot C\right)^{-1} \cdot C^T \cdot \underline{y} \tag{12.07}$$

The least-squares estimate can thus be obtained from the measurement vector. To verify that this optimum is actually a minimum, we have to calculate the second order derivative, which provides

$$\frac{\partial^2}{\partial \underline{x} \cdot \partial \underline{x}^T} \left(\underline{y} - C \cdot \underline{x}\right)^T \cdot \left(\underline{y} - C \cdot \underline{x}\right) = 2 \cdot C^T \cdot C \tag{12.08}$$

If this matrix is positive definite, then the result obtained is actually a minimum of the variance. Positive definite means that the matrix meets the following condition for arbitrary vectors $\underline{\upsilon}$

$$\underline{\upsilon}^T \cdot C^T \cdot C \cdot \underline{\upsilon} = \underline{\gamma}^T \cdot \underline{\gamma} = \sum_{i=1}^{n} \gamma_i^2 \geq 0 \tag{12.09}$$

This term only vanishes, if all γ_i are equal to zero. The error between the measurements and the estimated measurements can then be denoted by the equation

$$\underline{y} = C \cdot \underline{x} + \underline{v}$$

$$\underline{x}_{\text{LS}} = \left(C^T \cdot C\right)^{-1} \cdot C^T \cdot \underline{y}$$

$$\underline{e} = \underline{y} - C \cdot \left(C^T \cdot C\right)^{-1} \cdot C^T \cdot \underline{y}$$

or

$$\underline{e} = \left[I_m - C \cdot \left(C^T \cdot C\right)^{-1} \cdot C^T\right] \cdot \underline{y} = M \cdot \underline{y} \tag{12.10}$$

The expectation value of the estimation error is

$$E\{\underline{e}\} = E\{M \cdot \underline{y}\}$$

$$E\{\underline{e}\} = E\{M \cdot (C \cdot \underline{x} + \underline{v})\}$$

$$E\{\underline{e}\} = E\left\{\left[I_m - C \cdot \left(C^T \cdot C\right)^{-1} \cdot C^T\right] \cdot (C \cdot \underline{x} + \underline{v})\right\} \tag{12.11}$$

With

$$\left[I_m - C \cdot \left(C^T \cdot C\right)^{-1} \cdot C^T\right] \cdot C = C - C \cdot \left(C^T \cdot C\right)^{-1} \cdot C^T \cdot C = C - C = 0 \tag{12.12}$$

we obtain

$$E\{\underline{e}\} = E\{M \cdot \underline{v}\} = M \cdot E\{\underline{v}\} = \underline{0} \tag{12.13}$$

and the variance of the estimation error is thus

$$E\left\{\underline{e}^T \cdot \underline{e}\right\} = E\left\{(M \cdot \underline{v})^T \cdot (M \cdot \underline{v})\right\}$$

$$E\left\{\underline{e}^T \cdot \underline{e}\right\} = E\left\{\underline{v}^T \cdot M^T \cdot M \cdot \underline{v}\right\}$$

$$E\left\{\underline{e}^T \cdot \underline{e}\right\} = E\left\{\text{tr}\left[\left(M^T \cdot M\right) \cdot \underline{v} \cdot \underline{v}^T\right]\right\}$$

$$E\left\{\underline{e}^T \cdot \underline{e}\right\} = \text{tr}\left[\left(M^T \cdot M\right) \cdot E\left\{\underline{v} \cdot \underline{v}^T\right\}\right] \tag{12.14}$$

As all measurements are carried out with the same accuracy we further transform the expectation value on the right-hand side

$$E\left\{\underline{e}^T \cdot \underline{e}\right\} = \text{tr}\left[\left(M^T \cdot M\right) \cdot \sigma^2 \cdot I_m\right]$$

$$E\left\{\underline{e}^T \cdot \underline{e}\right\} = \sigma^2 \cdot \text{tr}\left[\left(M^T \cdot M\right) \cdot I_m\right] = \sigma^2 \cdot \text{tr}\left[M^T \cdot M\right] \tag{12.15}$$

and now we examine the product of the two matrices

$$M^T \cdot M = \left[I_m - C \cdot \left(C^T \cdot C\right)^{-1} \cdot C^T\right]^T \cdot \left[I_m - C \cdot \left(C^T \cdot C\right)^{-1} \cdot C^T\right] \tag{12.16}$$

and this is equal to

$$\begin{aligned} &= I_m - \left(C \cdot \left(C^T \cdot C\right)^{-1} \cdot C^T\right)^T \\ &\quad - C \cdot \left(C^T \cdot C\right)^{-1} \cdot C^T + \left(C \cdot \left(C^T \cdot C\right)^{-1} \cdot C^T\right)^T \cdot C \cdot \left(C^T \cdot C\right)^{-1} \cdot C^T \end{aligned} \tag{12.17}$$

We get

$$\begin{aligned} \left(C \cdot \left(C^T \cdot C\right)^{-1} \cdot C^T\right)^T &= \left(C^T\right)^T \cdot \left[\left(C^T \cdot C\right)^{-1}\right]^T \cdot C^T \\ &= C \cdot \left[\left(C^T \cdot C\right)^{-1}\right]^T \cdot C^T \end{aligned} \tag{12.18}$$

and thus

$$\begin{aligned} M^T \cdot M = I_m &- C \cdot \left[\left(C^T \cdot C\right)^{-1}\right]^T \cdot C^T \\ &- C \cdot \left(C^T \cdot C\right)^{-1} \cdot C^T + C \cdot \left[\left(C^T \cdot C\right)^{-1}\right]^T \cdot \underbrace{C^T \cdot C \cdot \left(C^T \cdot C\right)^{-1}}_{I_n} \cdot C^T \end{aligned}$$

$$\begin{aligned} M^T \cdot M = I_m &- C \cdot \left[\left(C^T \cdot C\right)^{-1}\right]^T \cdot C^T \\ &- C \cdot \left(C^T \cdot C\right)^{-1} \cdot C^T + C \cdot \left[\left(C^T \cdot C\right)^{-1}\right]^T \cdot C^T \end{aligned}$$

$$M^T \cdot M = I_m - C \cdot \left(C^T \cdot C\right)^{-1} \cdot C^T = M \text{ (idempotent matrix)} \tag{12.19}$$

So we are able to simplify the previous equation

$$E\left\{\underline{e}^T \cdot \underline{e}\right\} = E\left\{\underline{v}^T \cdot M \cdot \underline{v}\right\}$$

$$E\left\{\underline{e}^T \cdot \underline{e}\right\} = E\left\{\mathrm{tr}\left[M \cdot \underline{v} \cdot \underline{v}^T\right]\right\}$$

$$= \mathrm{tr}\left[M \cdot E\left\{\underline{v} \cdot \underline{v}^T\right\}\right]$$

$$= \mathrm{tr}\left[M \cdot \sigma^2 \cdot I_m\right]$$

$$E\left\{\underline{e}^T \cdot \underline{e}\right\} = \mathrm{tr}\left[\sigma^2 \cdot M \cdot I_m\right] = \sigma^2 \cdot \mathrm{tr}[M] \tag{12.20}$$

M is a m×m-dimensional matrix. If we insert M, we obtain

$$E\left\{\underline{e}^T \cdot \underline{e}\right\} = \sigma^2 \cdot \mathrm{tr}[M] = \sigma^2 \cdot \mathrm{tr}\left[I_m - C \cdot \left(C^T \cdot C\right)^{-1} \cdot C^T\right]$$

$$E\left\{\underline{e}^T \cdot \underline{e}\right\} = \sigma^2 \cdot \left(\mathrm{tr}[I_m] - \mathrm{tr}\left[\left(C^T \cdot C\right)^{-1} \cdot C^T \cdot C\right]\right)$$

$$E\left\{\underline{e}^T \cdot \underline{e}\right\} = \sigma^2 \cdot \left(\mathrm{tr}[I_m] - \mathrm{tr}[I_n]\right)$$

$$E\left\{\underline{e}^T \cdot \underline{e}\right\} = \sigma^2 \cdot (m - n) \tag{12.21}$$

$\dim(\underline{y}) = m \qquad \dim(\underline{x}) = n$

The estimate of the variance of the measurement error on the basis of the least-squares estimates (which is sometimes called the model error, contrary to the parameter error) is obviously biased, but we easily eliminate this bias by rearranging the equation to

$$s^2 = \frac{E\left\{\underline{e}^T \cdot \underline{e}\right\}}{m - n} = \sigma^2 \tag{12.22}$$

and this is an unbiased estimator of the measurements error variance.

12.1.2 Application of the results to the ambiguity function

The ambiguity function

$$\text{Ambiguity Function} = AmbFct = \int_{\mathrm{P}} \ln f_{\underline{y}|\underline{x}}\left(\underline{\rho} \mid \underline{\xi}\right) \cdot g_{\underline{y}}\left(\underline{\rho}\right) \cdot d\underline{\rho} \tag{12.23}$$

is now to be maximized with respect to the multi-dimensional parameter $\underline{x}$ to obtain the optimum set of parameters. What is the advantage of this optimal set of parameters? From Kullback's information we know that in the case of a minimum discrimination information the conditional distribution density has a minimum difference to the density $g_{\underline{y}}$.

We generate a series expansion at the maximum likelihood estimate and get

$$f_{\underline{y}|\underline{x}}(\underline{\rho}\,|\,\underline{\xi}) = \ln f_{\underline{y}|\underline{x}}(\underline{\rho}\,|\,\underline{\hat{x}}_{\mathrm{ML}}) + \frac{\partial}{\partial\underline{\xi}} \ln f_{\underline{y}|\underline{x}}(\underline{\rho}\,|\,\underline{\xi})\Big|_{\underline{\xi}=\underline{\hat{x}}_{\mathrm{ML}}} \cdot (\underline{\xi} - \underline{\hat{x}}_{\mathrm{ML}})$$
$$+ \frac{1}{2}\cdot(\underline{\xi} - \underline{\hat{x}}_{\mathrm{ML}})^T \cdot \frac{\partial^2}{\partial\underline{\xi}\cdot\partial\underline{\xi}^T} \ln f_{\underline{y}|\underline{x}}(\underline{\rho}\,|\,\underline{\xi})\Big|_{\underline{\xi}=\underline{\hat{x}}_{\mathrm{ML}}} \cdot (\underline{\xi} - \underline{\hat{x}}_{\mathrm{ML}}) + \text{h.o.t.} \tag{12.24}$$

Neglecting the higher order terms (h.o.t.) we get the following formulation of the ambiguity function

$$AmbFct = \int_{\mathrm{P}} \Bigg[\ln f_{\underline{y}|\underline{x}}(\underline{\rho}\,|\,\underline{\hat{x}}_{\mathrm{ML}}) + \frac{\partial}{\partial\underline{\xi}} \ln f_{\underline{y}|\underline{x}}(\underline{\rho}\,|\,\underline{\xi})\Big|_{\underline{\xi}=\underline{\hat{x}}_{\mathrm{ML}}} \cdot (\underline{\xi} - \underline{\hat{x}}_{\mathrm{ML}}) +$$
$$+ \frac{1}{2}\cdot(\underline{\xi} - \underline{\hat{x}}_{\mathrm{ML}})^T \cdot \frac{\partial^2}{\partial\underline{\xi}\cdot\partial\underline{\xi}^T} \ln f_{\underline{y}|\underline{x}}(\underline{\rho}\,|\,\underline{\xi})\Big|_{\underline{\xi}=\underline{\hat{x}}_{\mathrm{ML}}} \cdot (\underline{\xi} - \underline{\hat{x}}_{\mathrm{ML}}) \Bigg] \cdot g_{\underline{y}}(\underline{\rho}) \cdot d\underline{\rho} \tag{12.25}$$

Because of the 'characteristic equation' of the ML estimation $\frac{\partial}{\partial\underline{\xi}} \ln f_{\underline{y}|\underline{x}}(\underline{\rho}\,|\,\underline{\xi})\Big|_{\underline{\xi}=\underline{\hat{x}}_{\mathrm{ML}}} = 0$ the remaining equation is

$$AmbFct = \int_{\mathrm{P}} \Bigg[\ln f_{\underline{y}|\underline{x}}(\underline{\rho}\,|\,\underline{\hat{x}}_{\mathrm{ML}})$$
$$+ \frac{1}{2}\cdot(\underline{\xi} - \underline{\hat{x}}_{\mathrm{ML}})^T \cdot \frac{\partial^2}{\partial\underline{\xi}\cdot\partial\underline{\xi}^T} \ln f_{\underline{y}|\underline{x}}(\underline{\rho}\,|\,\underline{\xi})\Big|_{\underline{\xi}=\underline{\hat{x}}_{\mathrm{ML}}} \cdot (\underline{\xi} - \underline{\hat{x}}_{\mathrm{ML}}) \Bigg] \cdot g_{\underline{y}}(\underline{\rho}) \cdot d\underline{\rho} \tag{12.26}$$

If we write

$$AmbFct = AmbFct_1 + AmbFct_2 \tag{12.27}$$

with

$$AmbFct_1 = \int_{\mathrm{P}} \ln f_{\underline{y}|\underline{x}}(\underline{\rho}\,|\,\underline{\hat{x}}_{\mathrm{ML}}) \cdot g_{\underline{y}}(\underline{\rho}) \cdot d\underline{\rho} \tag{12.28}$$

and

$$AmbFct_2 = \int_{\mathrm{P}} \frac{1}{2}\cdot(\underline{\xi} - \underline{\hat{x}}_{\mathrm{ML}})^T \cdot \frac{\partial^2}{\partial\underline{\xi}\cdot\partial\underline{\xi}^T} \ln f_{\underline{y}|\underline{x}}(\underline{\rho}\,|\,\underline{\xi})\Big|_{\underline{\xi}=\underline{\hat{x}}_{\mathrm{ML}}} \cdot (\underline{\xi} - \underline{\hat{x}}_{\mathrm{ML}}) \cdot g_{\underline{y}}(\underline{\rho}) \cdot d\underline{\rho} \tag{12.29}$$

then we have the result that we obtain from the application of the ambiguity function.

If we do not know the distribution density $g_{\underline{y}}(\underline{\rho})$ of the integrals, we cannot calculate the expectation values. We thus have to evade this calculation by determining the empirical expectation values, the averages, which can be computed via sample analysis.

From sample analysis we already know that the likelihood function is asymptotically efficient and that it provides a gaussian distribution for independent measurements

$$f_{\underline{y}|\underline{x}}\left(\underline{\rho} \mid \hat{\underline{x}}_{\mathrm{ML}}\right)$$

$$= \frac{1}{\sqrt{(2\pi)^m \cdot \left|\sigma^2 \cdot I_m\right|}} \cdot \exp\left\{-\frac{1}{2} \cdot \left(\underline{\rho} - C \cdot \hat{\underline{x}}_{\mathrm{ML}}\right)^T \cdot \left(\sigma^2 \cdot I_m\right)^{-1} \cdot \left(\underline{\rho} - C \cdot \hat{\underline{x}}_{\mathrm{ML}}\right)\right\} \tag{12.30}$$

The expectation value is known as

$$\underline{x}_{\mathrm{LS}} = \hat{\underline{x}}_{\mathrm{ML}} = \left(C^T \cdot C\right)^{-1} \cdot C^T \cdot \underline{y} \tag{12.31}$$

because for gaussian random variables these two estimates do not differ. The variance is thus also known as $\sigma^2 \cdot I_m$, as we carry out a sequence of independent measurements. The value of σ^2, however, has to be determined from sample analysis.

In the first part of the ambiguity function we calculate the expectation value of the logarithm of this distribution density

$$AmbFct_1 = \int_{\mathrm{P}} \ln f_{\underline{y}|\underline{x}}\left(\underline{\rho} \mid \hat{\underline{x}}_{\mathrm{ML}}\right) \cdot g_{\underline{y}}\left(\underline{\rho}\right) \cdot d\underline{\rho} \tag{12.32}$$

This calculation cannot be carried out if we do not know the distribution density $g_y(\underline{\rho})$. We have to fall back on the empirical expectation value, the average. With this empirical expectation we determine the expectation value and the variance of the gaussian density to obtain the expectation value of this conditional distribution density.

We find the expectation value

$$\hat{\underline{y}} = C \cdot \hat{\underline{x}}_{\mathrm{ML}} \tag{12.33}$$

and the covariance matrix for independent measurements

$$P = \hat{\sigma}^2 \cdot I_m \tag{12.34}$$

With these empirically determined values, the first part of the ambiguity function is equal to

$$AmbFct_1 = \ln\left[\frac{1}{\sqrt{(2\pi)^m \cdot \left|\hat{\sigma}^2 \cdot I_m\right|}} \cdot \exp\left\{-\frac{1}{2}\cdot\left(\underline{\hat{y}} - C\cdot\underline{\hat{x}}_{\text{ML}}\right)^T \cdot \left(\hat{\sigma}^2 \cdot I_m\right)^{-1} \cdot \left(\underline{\hat{y}} - C\cdot\underline{\hat{x}}_{\text{ML}}\right)\right\}\right]$$

and then

$$AmbFct_1 = -\frac{1}{2}\cdot\ln\left[(2\pi)^m \cdot \left|\hat{\sigma}^2 \cdot I_m\right|\right] - \frac{1}{2}\cdot\left(C\cdot\underline{\hat{x}}_{\text{ML}} - C\cdot\underline{\hat{x}}_{\text{ML}}\right)^T \cdot \left(\hat{\sigma}^2 \cdot I_m\right)^{-1} \cdot \left(C\cdot\underline{\hat{x}}_{\text{ML}} - C\cdot\underline{\hat{x}}_{\text{ML}}\right)$$

The second term vanishes, so that

$$AmbFct_1 = -\frac{1}{2}\cdot\ln\left[(2\pi)^m \cdot \left|\hat{\sigma}^2 \cdot I_m\right|\right]$$

$$AmbFct_1 = -\frac{m}{2}\cdot\ln(2\pi) - \frac{1}{2}\cdot\ln\hat{\sigma}^{2m}$$

$$AmbFct_1 = -\frac{m}{2}\cdot\ln(2\pi) - \frac{m}{2}\cdot\ln\hat{\sigma}^2 \tag{12.36}$$

In the second part of the ambiguity function we obtain

$$AmbFct_2 = \frac{1}{2}\cdot\int_{\text{P}}\left(\underline{x} - \underline{\hat{x}}_{\text{ML}}\right)^T \cdot \frac{\partial^2}{\partial\underline{\xi}\cdot\partial\underline{\xi}^T}\ln f_{\underline{y}|\underline{x}}\left(\underline{\rho}\mid\underline{\xi}\right)\Big|_{\underline{\xi}=\underline{\hat{x}}_{\text{ML}}} \cdot \left(\underline{x} - \underline{\hat{x}}_{\text{ML}}\right)\cdot g_{\underline{y}}\left(\underline{\rho}\right)\cdot d\underline{\rho} \tag{12.37}$$

with

$$\frac{\partial^2}{\partial\underline{\xi}\cdot\partial\underline{\xi}^T}\ln f_{\underline{y}|\underline{x}}\left(\underline{\rho}\mid\underline{\xi}\right)\Big|_{\underline{\xi}=\underline{\hat{x}}_{\text{ML}}} = -\frac{1}{\sigma^2}\cdot C^T \cdot I_m \cdot C = -\frac{1}{\sigma^2}\cdot C^T \cdot C \tag{12.38}$$

and this is the negative Fisher information for independent observations. Thus this part of the ambiguity function is equal to

$$AmbFct_2 = -\frac{1}{2}\cdot\int_{\text{P}}\left(\underline{x} - \underline{\hat{x}}_{\text{ML}}\right)^T \cdot \frac{1}{\sigma^2}\cdot C^T \cdot C\cdot\left(\underline{x} - \underline{\hat{x}}_{\text{ML}}\right)\cdot g_{\underline{y}}\left(\underline{\rho}\right)\cdot d\underline{\rho} \tag{12.39}$$

Now we do not calculate the expectation value, but we determine the averages, which means that we have to take into account the error in the parameters in the average of the sample analysis. We thus obtain the empirical expectation

$$AmbFct_2 = -\frac{1}{2} \cdot E\left\{ \left(\underline{x} - \hat{\underline{x}}_{ML}\right)^T \cdot \frac{1}{\sigma^2} \cdot C^T \cdot C \cdot \left(\underline{x} - \hat{\underline{x}}_{ML}\right) \right\}$$

$$AmbFct_2 = -\frac{1}{2} \cdot E\left\{ \frac{1}{\sigma^2} \cdot \mathrm{tr}\left[C^T \cdot C \cdot \left(\underline{x} - \hat{\underline{x}}_{ML}\right) \cdot \left(\underline{x} - \hat{\underline{x}}_{ML}\right)^T \right] \right\}$$

$$AmbFct_2 = -\frac{1}{2} \cdot \mathrm{tr}\left[C^T \cdot C \cdot E\left\{ \frac{1}{\sigma^2} \cdot \left(\underline{x} - \hat{\underline{x}}_{ML}\right) \cdot \left(\underline{x} - \hat{\underline{x}}_{ML}\right)^T \right\} \right]$$

$$AmbFct_2 = -\frac{1}{2} \cdot \mathrm{tr}\left[C^T \cdot C \cdot \frac{1}{\hat{\sigma}^2} \cdot \hat{\sigma}^2 \cdot \left(C^T \cdot C\right)^{-1} \right]$$

$$AmbFct_2 = -\frac{1}{2} \cdot \mathrm{tr}\left[C^T \cdot C \cdot \left(C^T \cdot C\right)^{-1} \right]$$

$$AmbFct_2 = -\frac{1}{2} \cdot \mathrm{tr}\left[I_n \right]$$

$$AmbFct_2 = -\frac{n}{2} \tag{12.40}$$

Summarizing, we get the complete ambiguity function

$$AmbFct = AmbFct_1 + AmbFct_2 = -\frac{m}{2} \cdot \ln(2\pi) - \frac{m}{2} \cdot \ln \hat{\sigma}^2 - \frac{n}{2}$$

$$AmbFct = -\frac{m}{2} \cdot \ln(2\pi) - \frac{m}{2} \cdot \ln \hat{\sigma}^2 - \frac{n}{2} = -\frac{\mathrm{AIC}}{2} \tag{12.41}$$

This term AIC is called ‘A information criterion’ from Akaike and it is generally known as Akaike’s information criterion.

$$\mathrm{AIC} = -2 \cdot AmbFct = -2 \cdot E\{\ln l\} = -2 \cdot E\{L\} \tag{12.42}$$

is directly related to the ambiguity function, because it denotes the expectation value (the empirical expectation value) of the likelihood function. We notice the use of this term when we think about this calculation again.

Consideration

We tried to determine an optimal set of n parameters $\underline{x}$ in such a way that the conditional distribution density $f_{\underline{y}|\underline{x}}(\underline{\rho} \mid \underline{\xi})$, consisting of these parameters, has a minimum discrimination information to the real but unknown distribution density function. The whole result from the computation of the Kullback information can be arranged into two parts. First it can be decomposed into the Shannon information of the real distribution density, which does not depend on the desired set of parameters. It is therefore constant in this examination. In the second part we obtained the negative ambiguity function, which we calculated. This part depends on the number n of the parameters $\underline{x}$ on the one hand, and on the other hand it depends on the ML estimate of the conditional distribution density, leading to the first two addends.

Thus we can use the result obtained to determine an optimum number n of parameters needed in the modeling. This is exactly the idea of Akaike. As the determination of the optimal number of parameters in an ARMA or an AR model is not a trivial task, Akaike searched for a method to determine this number of parameters on the basis of an information criterion. The criterion is the minimum Kullback–Leibler distance between the distribution density functions.

In general the likelihood ratio, as a quotient of two conditional distribution densities, contains all the information that is needed to determine the differences between the two distribution densities. And so it also contains all information that is needed to fit one distribution optimally to the other. This is exactly what we do when we optimize (minimize) the Kullback information by determining an optimal set of parameters.

12.2 BIC, SC or HQ

(Bayes's information criterion, Schwarz's criterion, Hannan–Quinn criterion)

With the examination of the Kullback information we obtain an extension of the maximum likelihood estimation. The maximum likelihood estimation contains the disadvantage that we do not consider any a priori information of the parameters. The consideration of the a priori information is achieved by the MAP (maximum a posteriori) estimation, where the ML estimation is expanded by the additional previous information, or with the Bayes estimation, which is a more general formulation of the MAP estimation [LOF90].

An extended approach thus has to consider the a priori information, i.e. the a priori distribution density $f_{\underline{x}}(\underline{\xi})$ of the parameters and then minimize the Kullback–Leibler distance between the real but unknown distribution density and the distribution, which we intend to fit optimally to this unknown density.

If we use this approach, we obtain the Bayes information criterion:

$$AmbFct = -\frac{m}{2}\cdot \ln(2\pi) - \frac{m}{2}\cdot \ln\hat{\sigma}^2 - \frac{n\cdot \ln m}{4} = -\frac{\text{BIC}}{2} = -\frac{\text{SC}}{2} \tag{12.43}$$

This formula has an additional factor ½ ln m and thus improves the Akaike information criterion. This criterion tends to estimate a smaller dimension of the models and for large values of m, i.e. a large number of samples, both criteria differ considerably from each other.

Another improvement is achieved by the Hannan–Quinn criterion [SAK90]

$$AmbFct = -\frac{m}{2}\cdot \ln(2\pi) - \frac{m}{2}\cdot \ln\hat{\sigma}^2 - \frac{n\cdot \ln\ln m}{2} = -\frac{\text{HQIC}}{2} \tag{12.44}$$

Summarizing, we have found the information criteria

$$\text{AIC} = m\cdot \ln\hat{\sigma}^2 + n \tag{12.45}$$

$$\text{BIC} = \text{SC} = m\cdot \ln\hat{\sigma}^2 + \frac{n\cdot \ln m}{2} \tag{12.46}$$

$$\text{HQIC} = \text{HQ} = m\cdot \ln\hat{s}^2 + n\cdot \ln\ln m \tag{12.47}$$

On one side these functions depend on the number of observation m, which have to be independent in these approaches. On the other side they depend on the number n of parameters. The minimum of each function indicates the desired value of n, i.e. the optimum number of parameters, needed to generate the observed data.

For a fixed number of samples the BIC tends to underestimate the dimension n, while the criterion of Akaike tends to overestimate the number n of the necessary parameters [TON90]. This consideration led to the development of the Hannan–Quinn criterion, providing values between the AIC and BIC and thus reducing the underestimation of the BIC. It is, of course, also possible to apply all criteria by choosing a cross-validation approach, where the available data are divided into several parts. Then each criterion is applied to another part of these data and thus provides its own estimate. The verification of the estimates can then be validated by additional data or by interchanging the sets of data.

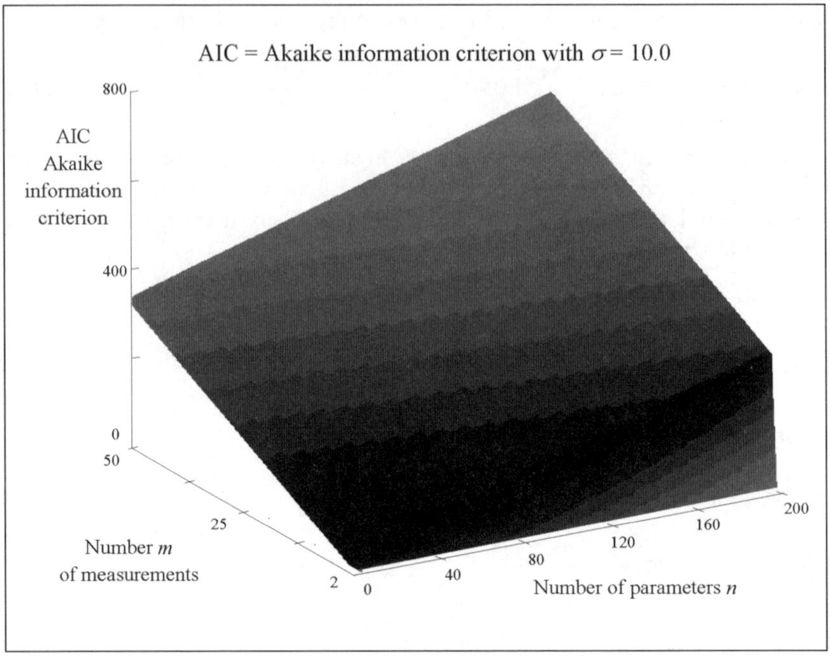

Fig. 12.1: Akaike information criterion with varying parameters m and n

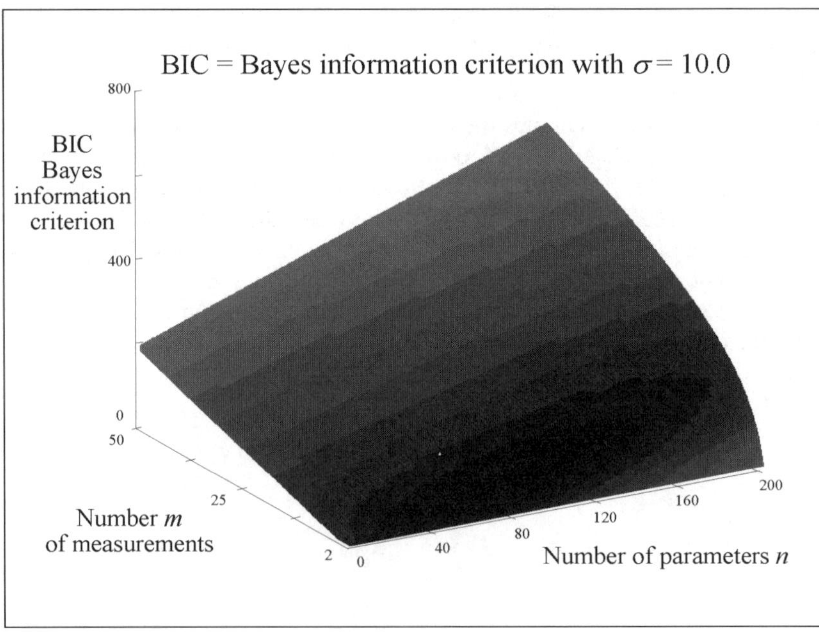

Fig. 12.2: Bayes information criterion with variation of m and n

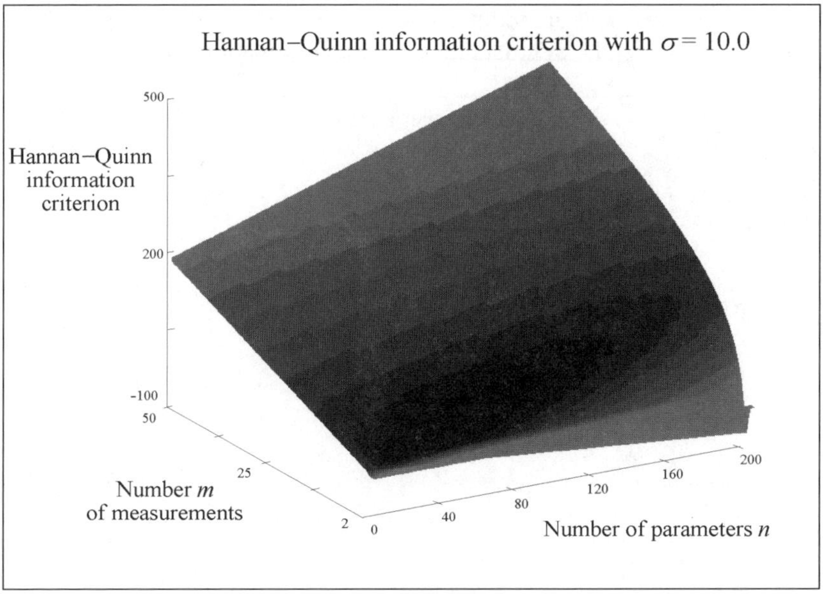

Fig. 12.3: Hannan–Quinn information criterion with variation of m and n

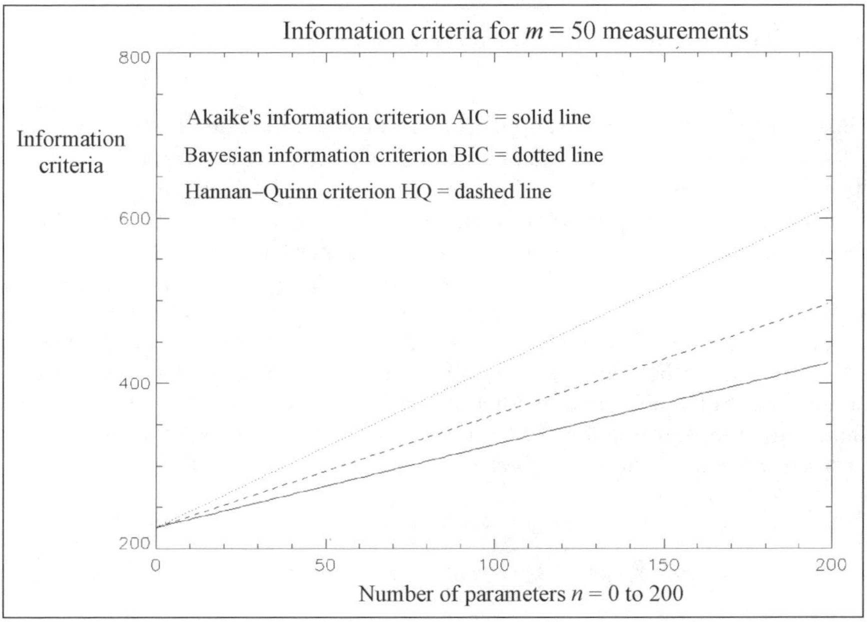

Fig. 12.4: Comparison of AIC, BIC and HQ for $\sigma = 10$

In the next figure we add an additional variation of the variance, which clearly shows the difference in the minima of these three information criteria.

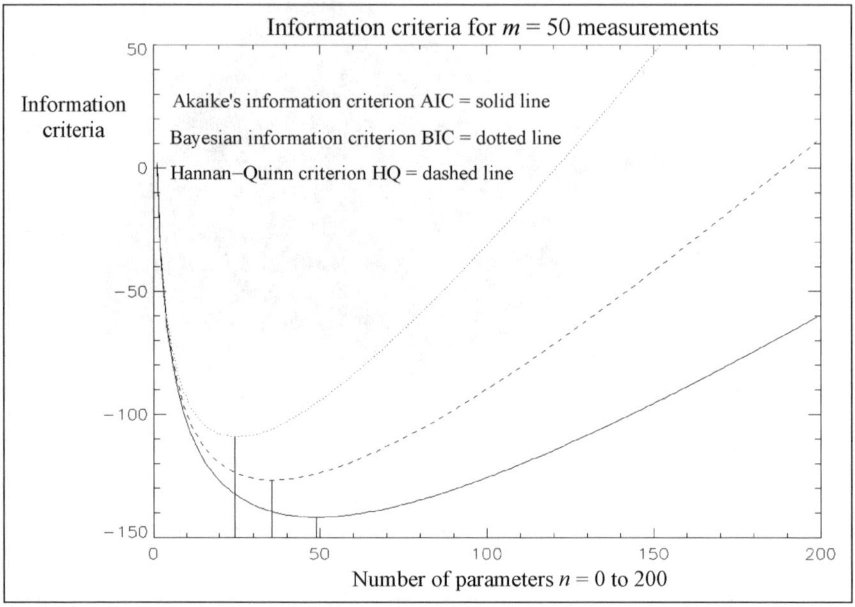

Fig. 12.5: Comparison of the complete AIC, BIC and HQ with $\sigma^2 = 1/\sqrt{n}$

For a fixed number of measurements and a constant likelihood estimate, figures 12.1 – 12.5 demonstrate that the minimum of the criterion of Hannan–Quinn lies between the minimum of BIC and the minimum of AIC, which we already stated. There are, of course, several other approaches, leading to criteria that allow us to choose a certain model or a certain set (number) of parameters [TON90].

A first variation of these information criteria occurs if we observe samples that are not independent of each other. In such a case we either change the criteria or we generate independent sets of data via a whitening filter (innovation model). Further criteria for the optimum determination of the number of parameters are for example the FPEs, the final prediction errors. These can be set into a direct relation to the information criteria, as they also contain the number of measurements and the number of necessary parameters. The FPE criterion is given by

$$\mathrm{FPE} = \frac{m+(n+1)}{m-(n+1)} \cdot \hat{\sigma}^2 \tag{12.48}$$

$\hat{\sigma}^2$ is again the sample variance
m = number of measurements (samples)
n = number of parameters

This criterion allows us to determine the optimum number n of parameters. Other criteria are described in [JUD85] [GRE90] so that we do not proceed with a more detailed description. We therefore finish the examination of the information criteria and focus our attention on the communication channel, the appropriate information measures and their connection.

13 Channel information

In order to classify the ambiguity function and several other information measures we examine the communication channel and the channel information, which is used in communication theory to describe the communication channel by means of information theory. Usually the following entropies are examined:

The entropy at the input of the communication channel, describing the entropy of the random variable X and their realizations x_i. This entropy is known as <u>source entropy</u>

$$H(X) = -\sum_{i} P(x_i) \cdot \mathrm{ld}[P(x_i)] \tag{13.01}$$

If the experiment is described by a continuous random variable $\underline{x}$, this entropy is denoted by the equation

$$S(\underline{x}) = -\int_{\underline{\Xi}}\int_{\underline{\mathrm{P}}} f_{\underline{x},\underline{y}}(\underline{\xi},\underline{\rho}) \cdot \ln\left[f_{\underline{x}}(\underline{\xi})\right] \cdot d\underline{\xi} \cdot d\underline{\rho}$$

$$S(\underline{x}) = -\int_{\underline{\Xi}} \ln\left[f_{\underline{x}}(\underline{\xi})\right] \cdot \int_{\underline{\mathrm{P}}} f_{\underline{x},\underline{y}}(\underline{\xi},\underline{\rho}) \cdot d\underline{\rho} \cdot d\underline{\xi}$$

$$S(\underline{x}) = -\int_{\underline{\Xi}} \ln\left[f_{\underline{x}}(\underline{\xi})\right] \cdot f_{\underline{x}}(\underline{\xi}) \cdot d\underline{\xi} \tag{13.02}$$

The entropy at the output of the channel, the entropy of the output, is given by

$$H(Y) = -\sum_{j} P(y_j) \cdot \mathrm{ld}[P(y_j)] \tag{13.03}$$

and

$$S(\underline{y}) = -\int_{\underline{\Xi}}\int_{\underline{\mathrm{P}}} f_{\underline{x},\underline{y}}(\underline{\xi},\underline{\rho}) \cdot \ln\left[f_{\underline{y}}(\underline{\rho})\right] \cdot d\underline{\xi} \cdot d\underline{\rho}$$

$$S(\underline{y}) = -\int_{\underline{\mathrm{P}}} \ln\left[f_{\underline{y}}(\underline{\rho})\right] \cdot \int_{\underline{\Xi}} f_{\underline{x},\underline{y}}(\underline{\xi},\underline{\rho}) \cdot d\underline{\xi} \cdot d\underline{\rho}$$

$$S(\underline{y}) = -\int_{\underline{\mathrm{P}}} \ln\left[f_{\underline{y}}(\underline{\rho})\right] \cdot f_{\underline{y}}(\underline{\rho}) \cdot d\underline{\rho} \tag{13.04}$$

The *equivocation* is described by

$$H(X \mid Y) = -\sum_{i,j} P(x_i, y_j) \cdot \mathrm{ld}\left[P(x_i \mid y_j)\right] \tag{13.05}$$

This entropy consists of the entropy

$$H(X \mid Y = y_j) = -\sum_{i} P(x_i \mid y_j) \cdot \mathrm{ld}\left[P(x_i \mid y_j)\right] \tag{13.06}$$

where an additional expectation for all possible realizations y_j is appended. In equation (13.06) it is merely the entropy that is contained in the random variable x, if the random variable y achieves a concrete realization y_j. This entropy actually varies with respect to the realizations y_j and is thus a random variable. To obtain a deterministic value, we compute the expectation value with respect to all possible realizations y_j. This leads to

$$H(X \mid Y) = \sum_{j} H(X \mid Y = y_j) \cdot P(y_j) \tag{13.07}$$

and this is the average conditional entropy $H(X|Y)$ or the expectation value of the conditional entropy. It is familiar to call it 'conditional entropy'. The continuous analog to this formulation is

$$S(\underline{x} \mid \underline{y}) = -\int_{\underline{\Xi}}\int_{\underline{\mathrm{P}}} f_{\underline{x},\underline{y}}(\underline{\xi}, \underline{\rho}) \cdot \ln\left[f_{\underline{x}|\underline{y}}(\underline{\xi} \mid \underline{\rho})\right] \cdot d\underline{\xi} \cdot d\underline{\rho} \tag{13.08}$$

The equivocation is thus the average of the uncertainty remaining after we received a signal. It thus describes the loss of the source information, as we can see in figure 13.1.

The *joint entropy*

$$H(X, Y) = -\sum_{i,j} P(x_i, y_j) \cdot \mathrm{ld}\left[P(x_i, y_j)\right] \tag{13.09}$$

$$S(\underline{x}, \underline{y}) = S(\underline{x} \cap \underline{y}) = -\int_{\underline{\Xi}}\int_{\underline{\mathrm{P}}} f_{\underline{x},\underline{y}}(\underline{\xi}, \underline{\rho}) \cdot \ln\left[f_{\underline{x},\underline{y}}(\underline{\xi}, \underline{\rho})\right] \cdot d\underline{\xi} \cdot d\underline{\rho} \tag{13.10}$$

describes the whole entropy of the emitted and received symbols. If the connection between input and output symbols is unique, the joint entropy is equal to the source entropy and it is equal to the entropy at the output of the channel. If the connection is not unique, the joint entropy is greater than the two other entropies.

$$H(X,Y) = H(X) = H(Y) \tag{13.11}$$

This equation describes a channel that is not contaminated with additional noise. Csiszár uses the notation $S(\underline{x} \cap \underline{y})$ to denote that this is the entropy of two random variables.

The *transinformation*, mutual information (Kolmogorov's information)

$$H(X;Y) = -\sum_{i,j} P(x_i, y_j) \cdot \mathrm{ld}\left[\frac{P(x_i) \cdot P(y_j)}{P(x_i, y_j)}\right] \tag{13.12}$$

$$S(\underline{x}; \underline{y}) = I(\underline{x}, \underline{y}) = -\iint_{\Xi\, \mathrm{P}} f_{\underline{x},\underline{y}}(\underline{\xi}, \underline{\rho}) \cdot \ln\left[\frac{f_{\underline{x},\underline{y}}(\underline{\xi}, \underline{\rho})}{f_{\underline{x}}(\underline{\xi}) \cdot f_{\underline{y}}(\underline{\rho})}\right] \cdot d\underline{\xi} \cdot d\underline{\rho} \tag{13.13}$$

The transinformation or the synentropy is equal to the average information content. It is – as we will notice – a measure for the correlation between the variables at the input and at the output of the channel, and it thus describes the similarity of these random variables. The greater the similarity between the two random variables is, the more the received symbols correspond to the emitted symbols (which is the aim of a correct transmission). The goal of an optimum transmission is thus to obtain a maximum of this mutual information.

The *dissipation* (see also the ambiguity function)

$$H(Y \mid X) = -\sum_{i,j} P(x_i, y_j) \cdot \mathrm{ld}\left[P(y_j \mid x_i)\right] \tag{13.14}$$

This entropy emerges from a conditional entropy, analogous to the equivocation.

$$H(Y \mid X = x_i) = -\sum_{j} P(y_j \mid x_i) \cdot \mathrm{ld}\left[P(y_j \mid x_i)\right] \tag{13.15}$$

Taking the expectation we find

$$H(Y \mid X) = \sum_{i} H(Y \mid X = x_i) \cdot P(x_i) \tag{13.16}$$

and this entropy describes an expected conditional entropy. In the continuous case we write

$$S(\underline{y} \mid \underline{x}) = -\iint_{\Xi\, \mathrm{P}} f_{\underline{x},\underline{y}}(\underline{\xi}, \underline{\rho}) \cdot \ln\left[f_{\underline{y}|\underline{x}}(\underline{\rho} \mid \underline{\xi})\right] \cdot d\underline{\xi} \cdot d\underline{\rho} \tag{13.17}$$

The *dissipation*, also known as context-entropy or the entropy depending on the perturbances, is the information that we receive at the output without being emitted. A special case of this dissipation is the already known ambiguity function, describing the ambiguities that may occur because of nonlinear transformations performed by the communication channel. Figure 13.1 shows the relations between these channel entropies. These connections are examined in detail in the following sections.

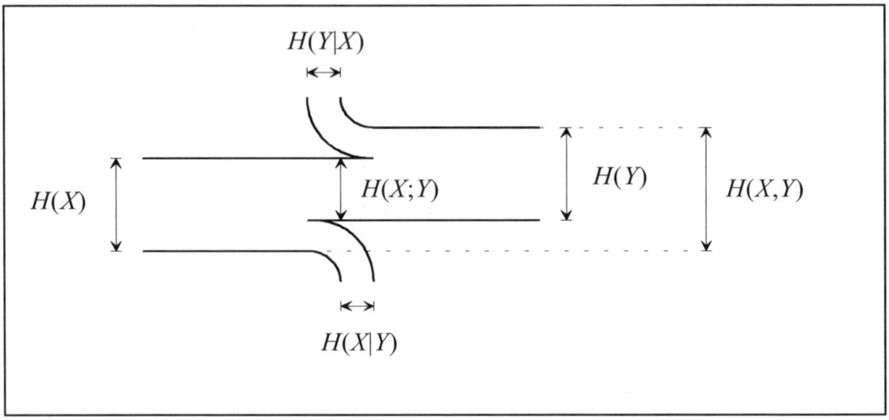

Fig. 13.1: Entropy measures in a communication channel

The connections between these entropies, which are also called extended entropies, are given by the equations

$$I(\underline{x},\underline{y}) = I(\underline{y},\underline{x}) = S(\underline{x}) - S(\underline{x} \mid \underline{y}) \tag{13.18}$$

$$I(\underline{x},\underline{y}) = S(\underline{y}) - S(\underline{y} \mid \underline{x}) \tag{13.19}$$

$$I(\underline{x},\underline{y}) = S(\underline{x}) + S(\underline{y}) - S(\underline{x},\underline{y}) \tag{13.20}$$

We again notice the symmetry of Kolmogorov's information. This symmetry can be extended to an arbitrary number of variables, if we start with the scalar connection

$$I(x_1,x_2) = S(x_1) + S(x_2) - S(x_1,x_2) \tag{13.21}$$

Here we write

$$I(x_1,x_2) = \sum_{i=1}^{2} S(x_i) - S(x_1,x_2) \tag{13.22}$$

Now we combine the two scalar random variables into one extended two-dimensional random variable and we introduce an additional random variable x_3. We get

$$I(x_1, x_2, x_3) = \sum_{i=1}^{3} S(x_i) - S(x_1, x_2, x_3) \tag{13.23}$$

In general we obtain

$$I(x_1, x_2, \ldots, x_n) = \sum_{i=1}^{n} S(x_i) - S(x_1, x_2, \ldots, x_n) \tag{13.24}$$

or

$$S(x_1, x_2, \ldots, x_n) = \sum_{i=1}^{n} S(x_i) - I(x_1, x_2, \ldots, x_n) \tag{13.25}$$

and we clearly see that the joint entropy is calculated from the sum of the single (marginal) entropies and the additional subtraction of the mutual information. If there is no mutual information, the random variables are independent and the joint entropy is equal to the sum of the single marginal entropies.

$$S(x_1, x_2, \ldots, x_n) = \sum_{i=1}^{n} S(x_i) \tag{13.26}$$

As $I(x_1, x_2, \ldots, x_n) \geq 0$ holds for arbitrary random variables, the joint entropy achieves its maximum when the random variables are mutually independent. The maximum of the joint entropy is thus determined by the sum of the marginal entropies. Kolmogorov's information is equal to the transinformation. It describes the amount of information about one random variable that is contained in another random variable. So it measures the mutual information of two random variables. The joint entropy

$$S(\underline{x}, \underline{y}) = -\int\limits_{\underline{\Xi}}\int\limits_{\underline{\mathrm{P}}} f_{\underline{x},\underline{y}}(\underline{\xi}, \underline{\rho}) \cdot \ln\left[f_{\underline{x},\underline{y}}(\underline{\xi}, \underline{\rho})\right] \cdot d\underline{\xi} \cdot d\underline{\rho} \tag{13.27}$$

$$S(\underline{x} \mid \underline{y}) = -\int\limits_{\underline{\Xi}}\int\limits_{\underline{\mathrm{P}}} f_{\underline{x},\underline{y}}(\underline{\xi}, \underline{\rho}) \cdot \ln\left[\frac{f_{\underline{x},\underline{y}}(\underline{\xi}, \underline{\rho})}{f_{\underline{y}}(\underline{\rho})}\right] \cdot d\underline{\xi} \cdot d\underline{\rho}$$

$$S(\underline{x}, \underline{y}) = S(\underline{x}) + S(\underline{y} \mid \underline{x})$$

$$S(\underline{x}, \underline{y}) = S(\underline{y}) + S(\underline{x} \mid \underline{y}) \tag{13.28}$$

is also symmetric.

If we restrict our examination to two random variables (we use the discrete formulations of these entropies) we obtain

$$H(X_1, X_2) = H(X_1) + H(X_2 \mid X_1) \tag{13.29}$$

$$H(X_1, X_2) = H(X_1) + H(X_2 \mid X_1)$$

$$H(X_2 \mid X_1) = H(X_1, X_2) - H(X_1)$$

The conditional entropy is equal to the joint entropy decreased by the marginal entropy of the variable, which determines the condition for the entropy. The extension to multi-dimensional random variables is

$$\begin{aligned} &H\left(X_{p+1}, X_{p+2}, \dots, X_{p+n} \mid X_1, X_2, \dots, X_p\right) \\ &= H\left(X_1, X_2, \dots, X_{n+p}\right) - H\left(X_1, X_2, \dots, X_p\right) \end{aligned} \tag{13.30}$$

or in vector notation

$$H(\underline{X}_2 \mid \underline{X}_1) = H(\underline{X}_1, \underline{X}_2) - H(\underline{X}_1) \tag{13.31}$$

The ambiguity function is the special case of the dissipation, where the true value $\underline{x}_t$ is known, or where the condition is generated on the basis of the true value. Under this condition (we know the true value) the entropy at the input vanishes

$$H(X) = -\sum_i P(x_i) \cdot \mathrm{ld}[P(x_i)] \tag{13.32}$$

as $P(\underline{x}_t) = 1$, which is equivalent to an infinite information with respect to the true value $\underline{x}_t$. Analogously we obtain $P(\underline{x}_t \mid \underline{y}_j) = 1$ and so the equivocation also vanishes, $H(X|Y) = 0$.
Because of

$$\frac{P(\underline{x}_t) \cdot P\left(\underline{y}_j\right)}{P\left(\underline{x}_t, \underline{y}_j\right)} = \frac{P\left(\underline{y}_j\right)}{P\left(\underline{x}_t, \underline{y}_j\right)} = \frac{1}{P\left(\underline{x}_t \mid \underline{y}_j\right)} = 1 \tag{13.33}$$

the mutual information is equal to zero, because the random variable $\underline{y}$ cannot add any information to the true value $\underline{x}_t$.

The remaining entropies are the entropy at the output $H(Y)$, the joint entropy $H(X,Y)$ and the dissipation $H(Y|X)$, which create some degenerated channel, presented in figure 13.2.

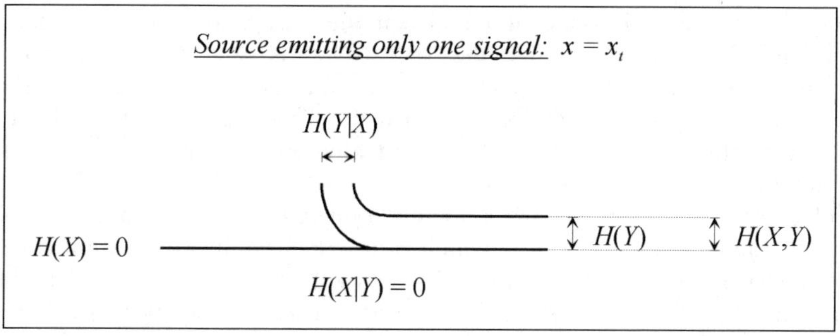

Figure no. 13.2: Channel information, without uncertainty at the input

An emitter that is merely able to send one signal, which it constantly does, achieves the same effect.

If we examine the differential form of the dissipation

$$S(\underline{y} \mid \underline{x}) = -\int_{\underline{\Xi}}\int_{\underline{\mathrm{P}}} f_{\underline{x},\underline{y}}(\underline{\xi},\underline{\rho}) \cdot \ln\left[f_{\underline{y}|\underline{x}}(\underline{\rho} \mid \underline{\xi})\right] \cdot d\underline{\xi} \cdot d\underline{\rho} \tag{13.34}$$

which is nothing other than the expectation of the logarithm of the conditional distribution density, we may write

$$S(\underline{y} \mid \underline{x}) = -E_{\underline{y},\underline{x}}\left\{\ln\left[f_{\underline{y}|\underline{x}}(\underline{\rho} \mid \underline{\xi})\right]\right\} \tag{13.35}$$

Application of the Bayesian rule provides

$$f_{\underline{x},\underline{y}}(\underline{\xi},\underline{\rho}) = f_{\underline{y}|\underline{x}}(\underline{\rho} \mid \underline{\xi}) \cdot f_{\underline{x}}(\underline{\xi}) \tag{13.36}$$

with

$$f_{\underline{x}}(\underline{\xi}) = \delta(\underline{\xi} - \underline{x}_t) \tag{13.37}$$

and thus

$$f_{\underline{x},\underline{y}}(\underline{\xi},\underline{\rho}) = f_{\underline{y}|\underline{x}}(\underline{\rho} \mid \underline{x}_t) \tag{13.38}$$

Then we the obtain the conditional expectation

$$-E_{\underline{y}|\underline{x}_t}\left\{\ln\left[f_{\underline{y}|\underline{x}}(\underline{\rho} \mid \underline{\xi})\right]\right\} = -\gamma(\underline{x},\underline{x}_t) \tag{13.39}$$

$$-\int_{\underline{\mathrm{P}}} \ln\left[f_{\underline{y}|\underline{x}}(\underline{\rho} \mid \underline{\xi})\right] \cdot f_{\underline{y}|\underline{x}}(\underline{\rho} \mid \underline{x}_t) \cdot d\underline{\rho} = -\gamma(\underline{x},\underline{x}_t) \tag{13.40}$$

and this is the desired connection between the dissipation and the ambiguity function for the special case that the 'true value' is sent. A maximum of the ambiguity function leads to a minimum of the dissipation, confirming the connection between the reciprocal information and the entropy, which we described in the introduction. A maximum ambiguity estimate is thus also a minimum dissipation estimate.

Such a communication channel is merely a mapping of one random variable X or $\underline{x}$ to the random variable Y or $\underline{y}$, which is described in the observation equation of a state space modeling, for instance. In this modeling the state space variable is mapped into the observation space. This means that the dissipation is caused not only by the mapping from input to output, but also by the additive noise or the already described ambiguities of nonlinear mappings, which lead to a loss of information during the transmission. Noise effects can be modeled, while the ambiguities are inevitable features of nonlinear mappings. Only in such cases where we have nonlinear mappings from the state space to the observation space, does the application of the ambiguity function make sense. These are the cases where ambiguities may occur.

The worse the signal-to-noise ratio becomes, the more the noise process influences the estimations. If we increase the additive noise in the observation, the uncertainty of the resulting estimate also increases (see section 11.4). The reason is that the maxima of the ambiguity function become more flat. As we already know, the spreading of these maxima, which is described by Fisher's information, is a measure for the uncertainty in the estimates and its increase is caused by stochastic part of the dissipation. With this consideration we finish the description of the channel entropies, as we do not need any further explanations.

Now we have discussed the familiar entropies in channel transmission, providing further descriptions of the ambiguity function, which has been identified as a special case of the dissipation. With this introduction of the information transmission over a communication channel we obtain an easier insight into the entropy measures, because we will switch to the state space description from the communication channel to describe the information transformation in state and observation space in the following sections. The observation equation

$$\underline{y}(k) = \underline{h}[\underline{x}(k), k] + \underline{v}(k) \tag{13.41}$$

is merely a transmission of the state space variable $\underline{x}(k)$ to the variable in the observation space $\underline{y}(k)$.

13.1 Redundancy

Redundancy is a concept that is mainly used in relation to channel coding, where we add a certain redundancy to a sequence of symbols to be able to detect and correct transmission errors. However, redundancy is in fact the part of the information that is completely superfluous to the receiver. Redundancy in the communication channel means that the knowledge of the output (a priori information) already provides information about certain parts of the input signal, which therefore become superfluous and thus redundant. The redundant part of the input signal is thus the part that can be predicted from the output. This equivocation of the communication channel describes the uncertainty that remains after the receipt of a message, i.e. the remaining uncertainty. If we diminish the uncertainty of the input by the remaining uncertainty that we have after we have received the signal, we get the information that we gain about the input via the transmission. This measure is called the rate of transmission, as it describes the information of the receiver with respect to the input.

$$R = H(X) - H(X \mid Y) = I(X \wedge Y) \tag{13.42}$$

Redundancy occurs if we are able to use this information to determine certain parts of the input signal. In channel coding the redundancy of a chosen code is determined by the equation

$$\text{Redundancy} = H_{\max} - H \tag{13.43}$$

which is the difference between the maximum entropy, achieved under optimum circumstances, and the real entropy of the coding.

Redundancy is for Brillouin merely another measure for the correlation. The entropy of a continuous random vector, occurring in this definition of the redundancy, can be computed by the equation

$$S = -\int_{-\infty}^{\infty} \ln f_{\underline{x}}(\underline{\xi}) \cdot f_{\underline{x}}(\underline{\xi}) \cdot d\underline{\xi} \tag{13.44}$$

Let $\underline{x} = (x_1, x_2)^T$ be a vector of the two zero-mean gaussian random variables x_1, x_2 with the variances σ_1^2, σ_2^2. Then we find

$$f_x(\xi) = \frac{1}{(2\pi)^{\frac{n}{2}} \cdot |P|^{\frac{1}{2}}} \cdot \exp\left\{-\frac{1}{2} \cdot \underline{\xi}^T \cdot P^{-1} \cdot \underline{\xi}\right\} \tag{13.45}$$

P is the covariance matrix $P = \begin{pmatrix} \sigma_1^2 & p_{12} \\ p_{12} & \sigma_2^2 \end{pmatrix}$

With the correlation coefficient

$$\rho = \frac{E\{x_1 \cdot x_2\}}{\sigma_1 \cdot \sigma_2} = \frac{p_{12}}{\sigma_1 \cdot \sigma_2} \quad ; \qquad p_{12} = \rho \cdot \sigma_1 \cdot \sigma_2$$

we get

$$P = \begin{pmatrix} \sigma_1^2 & \rho \cdot \sigma_1 \cdot \sigma_2 \\ \rho \cdot \sigma_1 \cdot \sigma_2 & \sigma_2^2 \end{pmatrix} \tag{13.46}$$

The determinant is thus

$$|P| = \sigma_1^2 \cdot \sigma_2^2 - (\rho \cdot \sigma_1 \cdot \sigma_2)^2 = \sigma_1^2 \cdot \sigma_2^2 \cdot (1-\rho^2)$$

and the inverse matrix

$$P^{-1} = \frac{1}{|P|} \cdot \begin{pmatrix} \sigma_2^2 & -\rho \cdot \sigma_1 \cdot \sigma_2 \\ -\rho \cdot \sigma_1 \cdot \sigma_2 & \sigma_1^2 \end{pmatrix}$$

$$P^{-1} = \begin{pmatrix} \frac{1}{\sigma_1^2 \cdot (1-\rho^2)} & \frac{-\rho}{\sigma_1 \cdot \sigma_2 \cdot (1-\rho^2)} \\ \frac{-\rho}{\sigma_1 \cdot \sigma_2 \cdot (1-\rho^2)} & \frac{1}{\sigma_2^2 \cdot (1-\rho^2)} \end{pmatrix} \tag{13.47}$$

We have thus described the inverse of the covariance matrix with the correlation coefficient and this enables us to verify the statement that redundancy and correlation coefficient are merely the same measures. We compute Shannon's entropy

$$S = -\int_{-\infty}^{\infty} \ln f_{\underline{x}}(\underline{\xi}) \cdot f_{\underline{x}}(\underline{\xi}) \cdot d\underline{\xi} \tag{13.48}$$

of the gaussian distribution

$$S = -\int_{-\infty}^{\infty} \left\{ -\ln\left((2\pi)^{\frac{n}{2}} \cdot |P|^{\frac{1}{2}} \right) - \frac{1}{2} \cdot \underline{\xi}^T \cdot P^{-1} \cdot \underline{\xi} \right\} \cdot f_x(\xi) \cdot d\xi \tag{13.49}$$

$$S = \ln\left((2\pi)^{\frac{n}{2}} \cdot |P|^{\frac{1}{2}} \right) + \frac{1}{2} \cdot \int_{-\infty}^{\infty} \underline{\xi}^T \cdot P^{-1} \cdot \underline{\xi} \cdot f_x(\xi) \cdot d\xi \tag{13.50}$$

with $n = 2$, as we assumed a two-dimensional random vector.

$$S = \ln(2\pi) + \frac{1}{2} \cdot \ln|P| + \frac{1}{2} \cdot \mathrm{tr}\left\{ P^{-1} \cdot \left(P - \underline{\mu}_x \cdot \underline{\mu}_x^T \right) \right\} \tag{13.51}$$

As the expectations are equal to zero, we obtain

$$S = \ln(2\pi) + \frac{1}{2} \cdot \ln|P| + 1$$

$$S = \ln(2\pi) + 1 + \frac{1}{2} \cdot \ln \sigma_1^2 \cdot \sigma_2^2 \cdot (1-\rho^2)$$

$$S = \ln(2\pi \cdot e) + \ln \sigma_1 \cdot \sigma_2 + \frac{1}{2} \cdot \ln\left(1-\rho^2\right) \tag{13.52}$$

The entropy is thus a function of the dimension of the variances or standard deviations of the single random variables and of the correlation coefficient. If the variances achieve fixed values, the only varying parameter is the correlation, which means that the entropy is a function of the correlation coefficient.

$$S = f(\rho) \tag{13.53}$$

and S_{max} can be computed as the zero of the derivative:

$$\frac{\partial}{\partial\rho} S = \frac{\partial}{\partial\rho}\left\{\ln(2\pi \cdot e) + \ln \sigma_1 \cdot \sigma_2 + \frac{1}{2} \cdot \ln\left(1-\rho^2\right)\right\}$$

$$\frac{\partial}{\partial\rho} S = \frac{1}{2} \cdot \frac{\partial}{\partial\rho} \ln\left(1-\rho^2\right) = \frac{-2 \cdot \rho}{2 \cdot \left(1-\rho^2\right)}$$

$$\frac{\partial}{\partial\rho} S = \frac{-\rho}{1-\rho^2} \overset{!}{=} 0 \tag{13.54}$$

$$\rho = 0 \qquad \text{for maximum entropy}$$

S_{min} can be obtained from the boundary condition $0 \leq \rho^2 \leq 1$, i.e. the bounded interval, which can be achieved by the correlation coefficient ($-1 \leq \rho \leq 1$). S_{min} is thus not obtained from a zero of the derivative, but from the boundaries, which we summarize by

$$\rho^2 = 1 \tag{13.56}$$

This leads to

$$S_{min} = \ln(2\pi \cdot e) + \ln \sigma_1 \cdot \sigma_2 + \frac{1}{2} \cdot \underbrace{\ln(1-1)}_{\to -\infty} \tag{13.57}$$

which is a negative infinite entropy, senseless for further computations. In this case both random variables x_1 and x_2 are exactly the same variable, and thus the knowledge of one variable provides complete knowledge about the other variable as well.

To obtain a complete description of the redundancy, we proceed with the maximum of the entropy at the point $\rho = 0$

$$S = \ln(2\pi \cdot e) + \ln \sigma_1 \cdot \sigma_2 + \frac{1}{2} \cdot \ln\left(1-\rho^2\right)$$

$$S_{\max} = \ln(2\pi \cdot e) + \ln \sigma_1 \cdot \sigma_2 + \frac{1}{2} \cdot \ln 1$$

$$S_{\max} = \ln(2\pi \cdot e) + \ln \sigma_1 \cdot \sigma_2 \tag{13.58}$$

The term $\frac{1}{2} \cdot \ln\left(1 - \rho^2\right)$ decreases the maximum entropy. Because of the boundary conditions $0 \le \rho^2 \le 1$, the argument in the logarithm is always less or equal 1, and thus the value of the logarithm is always negative if the correlation coefficient does not vanish.

Now Shannon defined redundancy as

$$\text{Redundancy} = S_{\max} - S \tag{13.59}$$

and with the appropriate values we get

$$\text{Redundancy} = \ln(2\pi \cdot e) + \ln \sigma_1 \cdot \sigma_2 - \left[\ln(2\pi \cdot e) + \ln \sigma_1 \cdot \sigma_2 + \frac{1}{2} \cdot \ln\left(1 - \rho^2\right)\right]$$

$$\text{Redundancy} = -\frac{1}{2} \cdot \ln\left(1 - \rho^2\right) \tag{13.60}$$

Thus redundancy is merely a logarithmic measure for the correlation. It satisfies the statements we made at the beginning of this chapter. The greater the redundancy is, the less new information is contained in the second random variable, as the redundancy increases with ρ^2. With an increasing similarity the correlation coefficient ρ^2 increases and with the correlation the redundancy and the mutual information increase. This, of course, diminishes the new information in the second random variable.

13.1.1 Knowledge, redundancy, utility

The utility of information and the resulting gain of knowledge always depend on the a priori knowledge of the receiver, according to our considerations. The possibility or ability to evaluate the information received implies a previous knowledge about the message that we obtain by transmission of the information. Because of this previous knowledge we expect information that diminishes the uncertainty regarding this knowledge.

The more previous knowledge we have, the less uncertainty can be diminished by the information received. Though a message is unchanged, it may have different information for two receivers, which differ from each other in their previous knowledge. When the message with the immanent information can only eliminate a small part of the uncertainty, the previous knowledge of the receiver already contained most parts of the information of the message, so that these parts are redundant to the receiver. As we saw in the previous computation, this can be explained by the similarity of the information contained in the message and the a priori knowledge of the receiver. If the a priori knowledge of the receiver increases, the redundancy of the message increases and the new information contained in the message decreases.

In this context we may also briefly discuss the relevance of a message. This relevance is a measure for the importance or for the usability of the information received. For our example, channel coding, this can be explained with the considerations described. If an emitter has some information about the channel, the receiver obtains the whole information in an optimum transmission. Now the following question arises: How far can the receiver reduce his uncertainty with the information received?

If the information stems from two completely different fields, the meaning of the information is not accessible to the receiver and thus the possibility of a utilization of the information vanishes. The fields of emitter and receiver obviously have to correlate to permit the receiver an appropriate understanding and reaction with respect to the information. The greater the correlation between the fields of sender and receiver, the greater is the efficiency, which allows a reduction of the uncertainty of the receiver. Completely uncorrelated fields lead to irrelevant information for the receiver, and he will not be able to use or even understand the information.

Thus transmission of information requires a certain correlation between sender and receiver. Uncorrelated fields do not provide any information for the receiver, who may only store the received message. Complete correlation means that the receiver already knows the signals that he is going to receive, and thus the message will not contain any information. Thus the transmission of information is bounded by the a priori knowledge of the receiver. In the same way the transmission of information is limited by the capacity of the channel. The maximum of the transmission limits the information that can be transferred.

13.2 Rate of transmission and equivocation

Shannon defined the rate of transmission of a communication channel

R = rate of transmission

$$R = H(X) - H(X \mid Y) \tag{13.61}$$

The rate of transmission is thus the knowledge that we gain about the input signal when the uncertainty at the input is diminished by the equivocation. In the same way the uncertainty at the output is diminished.

$$R = H(X) - [H(X,Y) - H(Y)]$$
$$R = H(X) + H(Y) - H(X,Y)$$
$$R = H(Y) - H(Y \mid X) \tag{13.62}$$

And this rate of transmission is equal to the uncertainty regarding x, reduced by the uncertainty regarding x when we have received y. The rate of transmission is thus

$$R = H(X) - H(X \mid Y)$$

$$R = -\sum_{j=1}^{m}\sum_{i=1}^{n} p(x_i, y_j) \cdot \ln p(x_i) + \sum_{j=1}^{m}\sum_{i=1}^{n} p(x_i, y_j) \cdot \ln p(x_i \mid y_j)$$

$$R = -\sum_{j=1}^{m}\sum_{i=1}^{n} p(x_i, y_j) \cdot [\ln p(x_i) - \ln p(x_i \mid y_j)]$$

$$R = \sum_{j=1}^{m}\sum_{i=1}^{n} p(x_i, y_j) \cdot \left[-\ln p(x_i) + \ln \frac{p(x_i, y_j)}{p(y_j)} \right]$$

$$R = \sum_{j=1}^{m}\sum_{i=1}^{n} p(x_i, y_j) \cdot \left[\ln \frac{p(x_i, y_j)}{p(x_i) \cdot p(y_j)} \right] = I(x, y) \tag{13.63}$$

= Kolmogorov's information

Kolmogorov's information (transinformation), known as mutual information of two random variables (which for gaussian distributions merely depends on the correlation coefficient), determines the rate of transmission of a communication channel or the flow of information through an algorithm.

$$\max(R) = \max(I(x,y)) = \text{channel capacity} = C \tag{13.64}$$

If we maximize Kolmogorov's information, for instance for a gaussian a posteriori estimate, to obtain the optimum weighting for the a priori information and the information that we get from some observation of gaussian distributed random variables, the resulting weighting matrix is the well-known Kalman gain [ARN97/1], used in optimal filtering.

The maximum of Kolmogorov's information determines the channel capacity or the capacity of an algorithm to transmit information. We write

$$I(x, y) \le C \tag{13.65}$$

If $I(x,y) > C$ should occur (C is a fixed channel capacity, which can be determined by transmission), then the message cannot be transferred without additional errors, even if the coding is perfect. The equivocation $H(X \mid Y)$ is always greater than zero and thus some part of the message, or of the information, 'disappears' (see figure 13.1).

The channel capacity will be part of further examinations in the following chapters, where we find that we cannot arbitrarily increase the channel capacity. There is a certain limit to this parameter, which cannot be exceeded without a loss of information.

Before we proceed with the examination of Kolmogorov's information, we first mention two interesting equations resulting from our previous considerations.

13.3 Hadamard's inequality and Gibbs's second theorem

We computed Shannon's information of a two-dimensional random variable as

$$S = \ln(2\pi) + \frac{1}{2} \cdot \ln|P| + 1 \tag{13.66}$$

$$S = \ln(2\pi \cdot e) + \ln \sigma_1 \cdot \sigma_2 + \frac{1}{2} \cdot \ln\left(1 - \rho^2\right) \tag{13.67}$$

We found that this entropy achieves its maximum if the correlation coefficient ρ vanishes. In the general result

$$S = \frac{n}{2} \cdot \ln(2\pi) + \frac{1}{2} \cdot \ln|P| + \frac{n}{2} = \frac{n}{2} \cdot \ln(2\pi \cdot e) + \frac{1}{2} \cdot \ln|P| \tag{13.68}$$

the entropy of a gaussian distribution merely depends on the covariance matrix, which is a symmetric, nonnegative definite matrix, and on the dimension. Hadamard's inequality states that for nonnegative definite matrices the determinant is always less than or equal to the product of the main diagonal elements

$$|P| = \det(P) \le \prod_{i=1}^{n} p_{ii} \tag{13.69}$$

Hadamard's inequality for nonnegative definite matrices

If $\underline{x}$ is a zero-mean random vector with covariance matrix P, we furthermore obtain the inequality

$$S(\underline{x}) \le \frac{n}{2} \cdot \ln(2\pi \cdot e) + \frac{1}{2} \cdot \ln|P| \quad \text{Gibbs's second theorem} \tag{13.70}$$

The entropy of a random vector with the given second order moments is always less than the entropy of a gaussian distribution density with equal second order moments. The gaussian distribution density maximizes this entropy if we know only the second order moments. The verification of this theorem can be carried out with Kullback's information, by comparing an arbitrary zero-mean distribution density function $f_{\underline{x}}(\underline{\xi})$ with the covariance matrix P to a zero-mean gaussian distribution function $g_{\underline{x}}(\underline{\xi})$ with the same covariance matrix P. Kullback's information is always nonnegative and thus

$$G_1 = \int_{-\infty}^{\infty} \ln \frac{f_{\underline{x}}(\underline{\xi})}{g_{\underline{x}}(\underline{\xi})} \cdot f_{\underline{x}}(\underline{\xi}) \cdot d\underline{\xi} \geq 0 \tag{13.71}$$

$$\int_{-\infty}^{\infty} \ln f_{\underline{x}}(\underline{\xi}) \cdot f_{\underline{x}}(\underline{\xi}) \cdot d\underline{\xi} - \int_{-\infty}^{\infty} \ln g_{\underline{x}}(\underline{\xi}) \cdot f_{\underline{x}}(\underline{\xi}) \cdot d\underline{\xi} \geq 0$$

$$-S(\underline{x}) \geq \int_{-\infty}^{\infty} \ln g_{\underline{x}}(\underline{\xi}) \cdot f_{\underline{x}}(\underline{\xi}) \cdot d\underline{\xi}$$

$$S(\underline{x}) \leq -\int_{-\infty}^{\infty} \ln g_{\underline{x}}(\underline{\xi}) \cdot f_{\underline{x}}(\underline{\xi}) \cdot d\underline{\xi}$$

$$S(\underline{x}) \leq \int_{-\infty}^{\infty} \ln g_{\underline{x}}(\underline{\xi}) \cdot f_{\underline{x}}(\underline{\xi}) \cdot d\underline{\xi}$$

$$S(\underline{x}) \leq -\int_{-\infty}^{\infty} \left\{ -\ln\left((2\pi)^{\frac{n}{2}} \cdot |P|^{\frac{1}{2}} \right) - \frac{1}{2} \cdot \underline{\xi}^T \cdot P^{-1} \cdot \underline{\xi} \right\} \cdot f_{\underline{x}}(\underline{\xi}) \cdot d\underline{\xi} \tag{13.72}$$

The solution of this integral leads – because of the same second order moments of $g_{\underline{x}}(\underline{\xi})$ and $f_{\underline{x}}(\underline{\xi})$ – to the inequality

$$S(\underline{x}) \leq \frac{n}{2} \cdot \ln(2\pi \cdot e) + \frac{1}{2} \cdot \ln|P| \tag{13.73}$$

which is Gibbs's second theorem.

Thus the gaussian distribution density is the density that maximizes Shannon's entropy if we know only the second order moments. If we are thus able to observe only the expectation value and the variance of some process, this inequality allows us to determine the maximum entropy of this process.

We now proceed with some further examinations regarding the channel entropies by using the correlation coefficient to describe the connections between information and the coupling of the random variables.

13.4 Kolmogorov's information

If we compute Kolmogorov's information

$$I(x_1,x_2)=I(x_2,x_1)=\int\limits_{\Xi_1}\int\limits_{\Xi_2} f_{x_1,x_2}(\xi_1,\xi_2)\cdot \ln\left[\frac{f_{x_1,x_2}(\xi_1,\xi_2)}{f_{x_1}(\xi_1)\cdot f_{x_2}(\xi_2)}\right]\cdot d\xi_1\cdot d\xi_2 \tag{13.74}$$

for two scalar gaussian random variables, whose joint distribution is also a gaussian distribution, the joint distribution can be written as

$$f_{\underline{x}}(\underline{\xi})=\frac{1}{(2\pi)^{\frac{n}{2}}\cdot|P_{xx}|^{\frac{1}{2}}}\cdot\exp\left\{-\frac{1}{2}\cdot\left(\underline{\xi}-\underline{\mu}_x\right)^T\cdot P_{xx}^{-1}\cdot\left(\underline{\xi}-\underline{\mu}_x\right)\right\} \tag{13.75}$$

where $\underline{x}^T=[x_1,x_2]$, $\quad \underline{\xi}^T=[\xi_1,\xi_2]$, $\quad \mu_{\underline{x}}^T=\lfloor\mu_1^T \mid \mu_2^T\rfloor$

$$P_{xx}=\begin{pmatrix}\sigma_1^2 & p_{12}\\ p_{21} & \sigma_2^2\end{pmatrix}=\begin{pmatrix}\sigma_1^2 & \rho\cdot\sigma_1\cdot\sigma_2\\ \rho\cdot\sigma_1\cdot\sigma_2 & \sigma_2^2\end{pmatrix}$$

with the correlation coefficient $\quad \rho=\dfrac{p_{12}}{\sigma_1\cdot\sigma_2}$

We get the determinant

$$|P_{xx}|=\det(P_{xx})=\left|\begin{pmatrix}\sigma_1^2 & \rho\cdot\sigma_1\cdot\sigma_2\\ \rho\cdot\sigma_1\cdot\sigma_2 & \sigma_2^2\end{pmatrix}\right|$$
$$=\sigma_1^2\cdot\sigma_2^2-\rho^2\cdot\sigma_1^2\cdot\sigma_2^2=\sigma_1^2\cdot\sigma_2^2\cdot(1-\rho^2)$$

and the inverse of the covariance matrix

$$\begin{aligned}P_{xx}^{-1}&=\frac{1}{|P_{xx}|}\cdot Adj\begin{pmatrix}\sigma_1^2 & \rho\cdot\sigma_1\cdot\sigma_2\\ \rho\cdot\sigma_1\cdot\sigma_2 & \sigma_2^2\end{pmatrix}\\ &=\frac{1}{\sigma_1^2\cdot\sigma_2^2\cdot(1-\rho^2)}\cdot\begin{pmatrix}\sigma_2^2 & -\rho\cdot\sigma_1\cdot\sigma_2\\ -\rho\cdot\sigma_1\cdot\sigma_2 & \sigma_1^2\end{pmatrix}\end{aligned} \tag{13.76}$$

The quadratic form in the exponential function becomes

$$\left(\underline{\xi}-\underline{\mu}_{\underline{x}}\right)^T \cdot P_{xx}^{-1} \cdot \left(\underline{\xi}-\underline{\mu}_{\underline{x}}\right)$$

$$=\frac{1}{\sigma_1^2\cdot\sigma_2^2\cdot\left(1-\rho^2\right)}\cdot\left(\underline{\xi}-\underline{\mu}_{\underline{x}}\right)^T\cdot\begin{pmatrix}\sigma_2^2 & -\rho\cdot\sigma_1\cdot\sigma_2\\ -\rho\cdot\sigma_1\cdot\sigma_2 & \sigma_1^2\end{pmatrix}\cdot\left(\underline{\xi}-\underline{\mu}_{\underline{x}}\right)$$

$$=\frac{1}{\sigma_1^2\cdot\sigma_2^2\cdot\left(1-\rho^2\right)}\cdot(\xi_1-\mu_1,\xi_2-\mu_2)\cdot\begin{pmatrix}\sigma_2^2 & -\rho\cdot\sigma_1\cdot\sigma_2\\ -\rho\cdot\sigma_1\cdot\sigma_2 & \sigma_1^2\end{pmatrix}\cdot\begin{pmatrix}\xi_1-\mu_1\\ \xi_2-\mu_2\end{pmatrix}$$

$$=\frac{1}{\sigma_1^2\cdot\sigma_2^2\cdot\left(1-\rho^2\right)}\cdot\Big[(\xi_1-\mu_1)\cdot\sigma_2^2-(\xi_2-\mu_2)\cdot\rho\cdot\sigma_1\cdot\sigma_2$$
$$-(\xi_1-\mu_1)\cdot\rho\cdot\sigma_1\cdot\sigma_2+(\xi_2-\mu_2)\cdot\sigma_1^2\Big]\cdot\begin{pmatrix}\xi_1-\mu_1\\ \xi_2-\mu_2\end{pmatrix}$$

$$=\frac{\left[(\xi_1-\mu_1)^2\cdot\sigma_2^2-2\cdot(\xi_2-\mu_2)\cdot(\xi_1-\mu_1)\cdot\rho\cdot\sigma_1\cdot\sigma_2,+(\xi_2-\mu_2)^2\cdot\sigma_1^2\right]}{\sigma_1^2\cdot\sigma_2^2\cdot\left(1-\rho^2\right)}$$

$$=\frac{\left[(\xi_1-\mu_1)^2\cdot\sigma_2^2-2\cdot(\xi_2-\mu_2)\cdot(\xi_1-\mu_1)\cdot\rho\cdot\sigma_1\cdot\sigma_2,+(\xi_2-\mu_2)^2\cdot\sigma_2^2\right]}{\sigma_1^2\cdot\sigma_2^2\cdot\left(1-\rho^2\right)} \tag{13.77}$$

and so the logarithm in Kolmogorov's information is equal to

$$\ln\left[\frac{f_{x_1,x_2}(\xi_1,\xi_2)}{f_{x_1}(\xi_1)\cdot f_{x_2}(\xi_2)}\right]=\ln\left[\frac{\sqrt{2\pi}\cdot\sigma_1\cdot\sqrt{2\pi}\cdot\sigma_2}{2\pi\cdot\sqrt{\sigma_1^2\cdot\sigma_2^2\cdot\left(1-\rho^2\right)}}\right]$$
$$-\frac{1}{2}\cdot\left\{\frac{1}{\left(1-\rho^2\right)}\cdot\left[\frac{(\xi_1-\mu_1)^2}{\sigma_1^2}-2\cdot(\xi_2-\mu_2)\cdot(\xi_1-\mu_1)\cdot\frac{\rho}{\sigma_1\cdot\sigma_2}+\frac{(\xi_2-\mu_2)^2}{\sigma_2^2}\right]\right.$$
$$\left.-\frac{(\xi_1-\mu_1)^2}{\sigma_1^2}-\frac{(\xi_2-\mu_2)^2}{\sigma_2^2}\right\}$$

$$=-\frac{1}{2}\cdot\ln\left(1-\rho^2\right)$$
$$-\frac{1}{2}\cdot\left\{\frac{1}{\left(1-\rho^2\right)}\cdot\left[\frac{(\xi_1-\mu_1)^2}{\sigma_1^2}-2\cdot(\xi_2-\mu_2)\cdot(\xi_1-\mu_1)\cdot\frac{\rho^2}{p_{12}}+\frac{(\xi_2-\mu_2)^2}{\sigma_2^2}\right]\right. \tag{13.78}$$
$$\left.-\frac{(\xi_1-\mu_1)^2}{\sigma_1^2}-\frac{(\xi_2-\mu_2)^2}{\sigma_2^2}\right\}$$

and we calculate the expectation value of this term

$$I(x_1,x_2)=\int_{\underline{\Xi}}\ln\left[\frac{f_{x_1,x_2}(\xi_1,\xi_2)}{f_{x_1}(\xi_1)\cdot f_{x_2}(\xi_2)}\right]\cdot f_{\underline{x}}(\underline{\xi})\cdot d\underline{\xi} \tag{13.79}$$

to obtain

$$I(x_1,x_2)=-\int_{\underline{\Xi}}\frac{1}{2}\cdot\ln\left(1-\rho^2\right)\cdot f_{\underline{x}}(\underline{\xi})\cdot d\underline{\xi}$$

$$-\frac{1}{2}\cdot\frac{1}{\left(1-\rho^2\right)}\cdot\int_{\underline{\Xi}}\left[\frac{(\xi_1-\mu_1)^2}{\sigma_1^2}-2\cdot(\xi_2-\mu_2)\cdot(\xi_1-\mu_1)\cdot\frac{\rho^2}{p_{12}}+\frac{(\xi_2-\mu_2)^2}{\sigma_2^2}\right]\cdot f_{\underline{x}}(\underline{\xi})\cdot d\underline{\xi}$$

$$+\frac{1}{2}\cdot\int_{\underline{\Xi}}\frac{(\xi_1-\mu_1)^2}{\sigma_1^2}+\frac{(\xi_2-\mu_2)^2}{\sigma_2^2}\cdot f_{\underline{x}}(\underline{\xi})\cdot d\underline{\xi}$$

$$I(x_1,x_2)=-\frac{1}{2}\cdot\ln\left(1-\rho^2\right)-\frac{1}{2}\cdot\frac{1}{\left(1-\rho^2\right)}\cdot\left[\frac{\sigma_1^2}{\sigma_1^2}-2\cdot p_{12}\cdot\frac{\rho^2}{p_{12}}+\frac{\sigma_2^2}{\sigma_2^2}\right]+\frac{1}{2}\cdot\left[\frac{\sigma_1^2}{\sigma_1^2}+\frac{\sigma_2^2}{\sigma_2^2}\right]$$

$$I(x_1,x_2)=-\frac{1}{2}\cdot\ln\left(1-\rho^2\right)-\frac{1}{2}\cdot\frac{1}{\left(1-\rho^2\right)}\cdot\left[2-2\cdot\rho^2\right]+1$$

$$I(x_1,x_2)=-\frac{1}{2}\cdot\ln\left(1-\rho^2\right)-\frac{2}{2}\cdot\frac{\left(1-\rho^2\right)}{\left(1-\rho^2\right)}+1$$

$$I(x_1,x_2)=-\frac{1}{2}\cdot\ln\left(1-\rho^2\right) \tag{13.80}$$

Kolmogorov's information is thus a function of the correlation coefficient ρ, assuming values in the interval $-1<\rho<1$. Thus the inequality $0\le\rho^2\le 1$ holds and the argument of the logarithm is always less than or equal to zero. So Kolmogorov's information merely achieves nonnegative values. We get

$$I(x_1,x_2)=S_1(x_1)-S_1(x_1\mid x_2)$$

$$I(x_1,x_2)=H(x_1)+H(x_2)-H(x_1,x_2) \tag{13.81}$$

and proceed with

$$H(x_1,x_2,\ldots,x_n)=\sum_{i=1}^{n}H(x_i)-I(x_1,x_2,\ldots,x_n) \tag{13.82}$$

If we now return to the conditional entropies

$$I(x_1, x_2) = S_1(x_1) - S_1(x_1 \mid x_2) = -\frac{1}{2} \cdot \ln\left(1 - \rho^2\right) \tag{13.83}$$

we may also extend this result to multi-dimensional random variables with dimension n. This leads to

$$I(x_{n+1}, \underline{x}) = S_1(x_{n+1}) - S_1(x_{n+1} \mid \underline{x}) = -\frac{1}{2} \cdot \ln\left(1 - R^2\right) \tag{13.84}$$

R is the multi-dimensional correlation coefficient describing the correlation between x_{n+1} and the components x_i $i = 1,\ldots,n$ of the random vector. R^2 is the scalar product of this vector.

13.5 Kullbacks divergence

$$D(\underline{x}, \underline{x}_s) = G(\underline{x}, \underline{x}_s) + G(\underline{x}_s, \underline{x}) = \int_{\underline{\Xi}} \ln\left(\frac{f_{\underline{x}}(\underline{\xi})}{f_{\underline{x}_s}(\underline{\xi})}\right) \cdot \left(f_{\underline{x}}(\underline{\xi}) - f_{\underline{x}_s}(\underline{\xi})\right) \cdot d\underline{\xi} \tag{13.85}$$

We start with the examination of the special case of two gaussian random variables x_1 and x_2, which we summarize in one vector $\underline{x}$.

$$f_{\underline{x}}(\underline{\xi}) = \frac{1}{(2\pi)^{\frac{n}{2}} \cdot |P_{xx}|^{\frac{1}{2}}} \cdot \exp\left\{-\frac{1}{2} \cdot \left(\underline{\xi} - \underline{\mu}_x\right)^T \cdot P_{xx}^{-1} \cdot \left(\underline{\xi} - \underline{\mu}_x\right)\right\} \tag{13.86}$$

$$\underline{x}^T = [x_1, x_2], \quad \underline{\xi}^T = [\xi_1, \xi_2], \quad \mu_{\underline{x}}^T = \left[\mu_1^T \mid \mu_2^T\right]$$

$$P_{xx} = \begin{pmatrix} \sigma_1^2 & p_{12} \\ p_{21} & \sigma_2^2 \end{pmatrix} = \begin{pmatrix} \sigma_1^2 & \rho \cdot \sigma_1 \cdot \sigma_2 \\ \rho \cdot \sigma_1 \cdot \sigma_2 & \sigma_2^2 \end{pmatrix}$$

with the correlation coefficient $\rho = \dfrac{p_{12}}{\sigma_1 \cdot \sigma_2}$

We obtain the determinant

$$|P_{xx}| = \det(P_{xx}) = \left|\begin{pmatrix} \sigma_1^2 & \rho \cdot \sigma_1 \cdot \sigma_2 \\ \rho \cdot \sigma_1 \cdot \sigma_2 & \sigma_2^2 \end{pmatrix}\right|$$

$$= \sigma_1^2 \cdot \sigma_2^2 - \rho^2 \cdot \sigma_1^2 \cdot \sigma_2^2 = \sigma_1^2 \cdot \sigma_2^2 \cdot \left(1 - \rho^2\right)$$

and the inverse matrix

$$P_{xx}^{-1} = \frac{1}{|P_{xx}|} \cdot Adj\begin{pmatrix} \sigma_1^2 & \rho \cdot \sigma_1 \cdot \sigma_2 \\ \rho \cdot \sigma_1 \cdot \sigma_2 & \sigma_2^2 \end{pmatrix}$$

$$= \frac{1}{\sigma_1^2 \cdot \sigma_2^2 \cdot (1-\rho^2)} \cdot \begin{pmatrix} \sigma_2^2 & -\rho \cdot \sigma_1 \cdot \sigma_2 \\ -\rho \cdot \sigma_1 \cdot \sigma_2 & \sigma_1^2 \end{pmatrix} \tag{13.87}$$

The quadratic form in the exponential function becomes

$$\left(\underline{\xi} - \underline{\mu}_x\right)^T \cdot P_{xx}^{-1} \cdot \left(\underline{\xi} - \underline{\mu}_x\right)$$

$$= \frac{1}{\sigma_1^2 \cdot \sigma_2^2 \cdot (1-\rho^2)} \cdot \left(\underline{\xi} - \underline{\mu}_x\right)^T \cdot \begin{pmatrix} \sigma_2^2 & -\rho \cdot \sigma_1 \cdot \sigma_2 \\ -\rho \cdot \sigma_1 \cdot \sigma_2 & \sigma_1^2 \end{pmatrix} \cdot \left(\underline{\xi} - \underline{\mu}_x\right)$$

$$= \frac{\left[(\xi_1 - \mu_1)^2 \cdot \sigma_2^2 - 2 \cdot (\xi_2 - \mu_2) \cdot (\xi_1 - \mu_1) \cdot \rho \cdot \sigma_1 \cdot \sigma_2 , + (\xi_2 - \mu_2)^2 \cdot \sigma_2^2\right]}{\sigma_1^2 \cdot \sigma_2^2 \cdot (1-\rho^2)} \tag{13.88}$$

Thus the divergence is

$$D\left(f_{\underline{x}}(\underline{\xi}), f_{x_1}(\xi_1) \cdot f_{x_2}(\xi_2)\right) = \int_{\Xi} \ln\left(\frac{f_{\underline{x}}(\underline{\xi})}{f_{x_1}(\xi_1) \cdot f_{x_2}(\xi_2)}\right) \cdot \left(f_{\underline{x}}(\underline{\xi}) - f_{x_1}(\xi_1) \cdot f_{x_2}(\xi_2)\right) \cdot d\underline{\xi} \tag{13.89}$$

and this is the divergence between the joint distribution density and the product of the marginal distribution densities. If the random variables are independent the divergence vanishes and we get $D = 0$, because in this case the joint distribution density function is equal to the product of the marginal distribution functions. If we decompose this divergence into two integrals, we obtain

$$D = \int_{\Xi} \ln\left(\frac{f_{\underline{x}}(\underline{\xi})}{f_{x_1}(\xi_1) \cdot f_{x_2}(\xi_2)}\right) \cdot f_{\underline{x}}(\underline{\xi}) \cdot d\underline{\xi} - \int_{\Xi} \ln\left(\frac{f_{\underline{x}}(\underline{\xi})}{f_{x_1}(\xi_1) \cdot f_{x_2}(\xi_2)}\right) \cdot f_{x_1}(\xi_1) \cdot f_{x_2}(\xi_2) \cdot d\underline{\xi}$$

$$D = I(x_1, x_2) - \int_{\Xi} \ln\left(\frac{f_{\underline{x}}(\underline{\xi})}{f_{x_1}(\xi_1) \cdot f_{x_2}(\xi_2)}\right) \cdot f_{x_1}(\xi_1) \cdot f_{x_2}(\xi_2) \cdot d\underline{\xi} \tag{13.90}$$

and the first term, the Kolmogorov information,

$$I(x_1,x_2)=I(x_2,x_1)=\int\limits_{\Xi_1}\int\limits_{\Xi_2} f_{x_1,x_2}(\xi_1,\xi_2)\cdot \ln\left[\frac{f_{x_1,x_2}(\xi_1,\xi_2)}{f_{x_1}(\xi_1)\cdot f_{x_2}(\xi_2)}\right]\cdot d\xi_1\cdot d\xi_2 \tag{13.91}$$

has already been solved

$$I(x_1,x_2)=-\frac{1}{2}\cdot\ln\left(1-\rho^2\right) \tag{13.92}$$

so that we proceed with the second term. We write

$$D_{\text{Term2}}=\int\limits_{\underline{\Xi}}\ln\left(\frac{f_{\underline{x}}(\underline{\xi})}{f_{x_1}(\xi_1)\cdot f_{x_2}(\xi_2)}\right)\cdot f_{x_1}(\xi_1)\cdot f_{x_2}(\xi_2)\cdot d\underline{\xi} \tag{13.93}$$

and obtain in the logarithm

$$\ln\left[\frac{f_{x_1,x_2}(\xi_1,\xi_2)}{f_{x_1}(\xi_1)\cdot f_{x_2}(\xi_2)}\right]=\ln\left[\frac{\sqrt{2\pi}\cdot\sigma_1\cdot\sqrt{2\pi}\cdot\sigma_2}{2\pi\cdot\sqrt{\sigma_1^2\cdot\sigma_2^2\cdot\left(1-\rho^2\right)}}\right]$$

$$-\frac{1}{2}\cdot\left\{\frac{1}{1-\rho^2}\cdot\left[\frac{(\xi_1-\mu_1)^2}{\sigma_1^2}-2\cdot(\xi_2-\mu_2)\cdot(\xi_1-\mu_1)\cdot\frac{\rho}{\sigma_1\cdot\sigma_2}+\frac{(\xi_2-\mu_2)^2}{\sigma_2^2}\right]\right.$$

$$\left.-\frac{(\xi_1-\mu_1)^2}{\sigma_1^2}-\frac{(\xi_2-\mu_2)^2}{\sigma_2^2}\right\}$$

$$=-\frac{1}{2}\cdot\ln\left(1-\rho^2\right)-\frac{1}{2}\cdot\left\{\frac{1}{1-\rho^2}\cdot\left[\frac{(\xi_1-\mu_1)^2}{\sigma_1^2}-2\cdot(\xi_2-\mu_2)\cdot(\xi_1-\mu_1)\cdot\frac{\rho^2}{p_{12}}+\frac{(\xi_2-\mu_2)^2}{\sigma_2^2}\right]-\right.$$

$$\left.-\frac{(\xi_1-\mu_1)^2}{\sigma_1^2}-\frac{(\xi_2-\mu_2)^2}{\sigma_2^2}\right\} \tag{13.94}$$

Inserting this into the integral, we find

$$D_{\text{Term2}} = -\int_{\underline{\Xi}} \frac{1}{2} \cdot \ln\left(1-\rho^2\right) \cdot f_{x_1}(\xi_1) \cdot f_{x_2}(\xi_2) \cdot d\underline{\xi} - \frac{1}{2} \cdot \frac{1}{1-\rho^2}$$

$$\cdot \int_{\underline{\Xi}} \left[\frac{(\xi_1-\mu_1)^2}{\sigma_1^2} - 2 \cdot (\xi_2-\mu_2) \cdot (\xi_1-\mu_1) \cdot \frac{\rho^2}{p_{12}} + \frac{(\xi_2-\mu_2)^2}{\sigma_2^2} \right] \cdot f_{x_1}(\xi_1) \cdot f_{x_2}(\xi_2) \cdot d\underline{\xi}$$

$$+ \frac{1}{2} \cdot \int_{\underline{\Xi}} \left[\frac{(\xi_1-\mu_1)^2}{\sigma_1^2} + \frac{(\xi_2-\mu_2)^2}{\sigma_2^2} \right] \cdot f_{x_1}(\xi_1) \cdot f_{x_2}(\xi_2) \cdot d\underline{\xi}$$

$$\begin{aligned} D_{\text{Term2}} = & -\frac{1}{2} \cdot \ln\left(1-\rho^2\right) \\ & + \frac{1}{1-\rho^2} \cdot \int_{\underline{\Xi}} \left[(\xi_2-\mu_2) \cdot (\xi_1-\mu_1) \cdot \frac{\rho^2}{p_{12}} \right] \cdot f_{x_1}(\xi_1) \cdot f_{x_2}(\xi_2) \cdot d\underline{\xi} \\ & - \frac{1}{2} \cdot \frac{1}{1-\rho^2} \cdot \int_{\underline{\Xi}} \left[\frac{(\xi_1-\mu_1)^2}{\sigma_1^2} + \frac{(\xi_2-\mu_2)^2}{\sigma_2^2} \right] \cdot f_{x_1}(\xi_1) \cdot f_{x_2}(\xi_2) \cdot d\underline{\xi} \\ & + \frac{1}{2} \cdot \int_{\underline{\Xi}} \left[\frac{(\xi_1-\mu_1)^2}{\sigma_1^2} + \frac{(\xi_2-\mu_2)^2}{\sigma_2^2} \right] \cdot f_{x_1}(\xi_1) \cdot f_{x_2}(\xi_2) \cdot d\underline{\xi} \end{aligned} \tag{13.95}$$

Now we combine the last two integrals to get

$$\begin{aligned} D_{\text{Term2}} = & -\frac{1}{2} \cdot \ln\left(1-\rho^2\right) \\ & + \frac{1}{1-\rho^2} \cdot \int_{\underline{\Xi}} \left[(\xi_2-\mu_2) \cdot (\xi_1-\mu_1) \cdot \frac{\rho^2}{p_{12}} \right] \cdot f_{x_1}(\xi_1) \cdot f_{x_2}(\xi_2) \cdot d\underline{\xi} \\ & + \left(-\frac{1}{2} \cdot \frac{1}{1-\rho^2} + \frac{1}{2} \right) \cdot \int_{\underline{\Xi}} \left[\frac{(\xi_1-\mu_1)^2}{\sigma_1^2} + \frac{(\xi_2-\mu_2)^2}{\sigma_2^2} \right] \cdot f_{x_1}(\xi_1) \cdot f_{x_2}(\xi_2) \cdot d\underline{\xi} \end{aligned} \tag{13.96}$$

and in the first integral we use the different random variables to obtain a decomposition into two integrals

$$D_{\text{Term2}} = -\frac{1}{2}\cdot \ln\left(1-\rho^2\right)$$

$$+\frac{1}{1-\rho^2}\cdot\frac{\rho^2}{p_{12}}\cdot\int_{\Xi_2}(\xi_2-\mu_2)\cdot\int_{\Xi_1}(\xi_1-\mu_1)\cdot f_{x_1}(\xi_1)\cdot d\xi_1\cdot f_{x_2}(\xi_2)\cdot d\xi_2$$

$$+\left(-\frac{1}{2}\cdot\frac{1}{1-\rho^2}+\frac{1}{2}\right)\cdot\left[\frac{\sigma_1^2}{\sigma_1^2}+\frac{\sigma_2^2}{\sigma_2^2}\right]$$

$$D_{\text{Term2}} = -\frac{1}{2}\cdot \ln\left(1-\rho^2\right)+\frac{1}{1-\rho^2}\cdot\frac{\rho^2}{p_{12}}\cdot\int_{\Xi_2}(\xi_2-\mu_2)\cdot 0\cdot f_{x_2}(\xi_2)\cdot d\xi_2$$

$$+\left(-\frac{1}{2}\cdot\frac{1}{1-\rho^2}+\frac{1}{2}\right)\cdot 2$$

$$D_{\text{Term2}} = -\frac{1}{2}\cdot \ln\left(1-\rho^2\right)-\frac{1}{1-\rho^2}+1$$

$$D_{\text{Term2}} = -\frac{1}{2}\cdot \ln\left(1-\rho^2\right)+\frac{1-\rho^2-1}{1-\rho^2}$$

$$D_{\text{Term2}} = -\frac{1}{2}\cdot \ln\left(1-\rho^2\right)-\frac{\rho^2}{1-\rho^2} \tag{13.97}$$

Now this term is to be inserted in the equation for the divergence, which leads to

$$D = I(x_1, x_2) - D_{\text{Term2}}$$

$$D = -\frac{1}{2}\cdot \ln\left(1-\rho^2\right)-\left[-\frac{1}{2}\cdot \ln\left(1-\rho^2\right)-\frac{\rho^2}{1-\rho^2}\right]$$

$$D = \frac{\rho^2}{1-\rho^2} \tag{13.98}$$

The divergence between the joint distribution density and the product of the marginal distribution densities is a function of the correlation coefficient. Contrary to Kullback's information the divergence is symmetric, which is caused by the summation of two Kullback information measures.

The divergence is, as we notice from our result, also a measure of the similarity or of the dissimilarity. The existing diversity of the two distribution densities examined arises because on one the hand we generate the joint distribution as a product of the marginal distribution densities of independent random variables and on the other hand we have the joint distribution density of correlated random variables. The difference between these distribution densities obviously lies in the correlation, which occurs as parameter in the result of our computation.

Let us examine the two extrema:

$\rho = 0$: The two random variables are not correlated and thus the divergence vanishes, because in this special case the joint distribution density is equal to the product of the marginal distribution densities.

$\rho = 1$: The divergence achieves an infinitely large value, because both random variables are identical. Thus there actually is no joint distribution density any more, because one of the two random variables is generated by a deterministic mapping from the other.

So the Kullback divergence is a symmetrical, nonnegative measure of the similarity of two distribution densities.

13.6 An example of a transmission

A random variable x, occuring at the input of a communication system, is received at the output with additional additive noise $y = x + n$. The additional noise is independent of the signal x. In this case we compute the conditional distribution density function of the output signal y as follows

$$f_{y|x}(\rho \mid \xi) \cdot d\rho = P\{\omega : y(\omega) \in (\rho, \rho + d\rho] \mid x(\omega) = \xi\} \tag{13.99}$$

This infinitesimally small interval describes the probability that there exists an event ω, with the realization $y(\omega)$ in the semi-closed interval $(\rho,\rho+d\rho]$ under the condition that the realization $x(\omega)$ is equal to ξ. From the transmission equation of the given channel we have

$$y(\omega) = x(\omega) + n(\omega) \tag{13.100}$$

and we insert this in the above equation to find

$$f_{y|x}(\rho \mid \xi) \cdot d\rho = P\{\omega : x(\omega) + n(\omega) \in (\rho, \rho + d\rho] \mid x(\omega) = \xi\} \tag{13.101}$$

Now we apply the condition and get

$$f_{y|x}(\rho \mid \xi) \cdot d\rho = P\{\omega : \xi + n(\omega) \in (\rho, \rho + d\rho] \mid x(\omega) = \xi\}$$

$$f_{y|x}(\rho \mid \xi) \cdot d\rho = P\{\omega : n(\omega) \in (\rho - \xi, \rho - \xi + d\rho] \mid x(\omega) = \xi\}$$

$$f_{y|x}(\rho \mid \xi)\cdot d\rho = P\{\omega : n(\omega) \in (\rho - \xi, \rho - \xi + d\rho] \mid x(\omega) = \xi\}$$
$$= f_{n|x}(\rho - \xi \mid \xi)\cdot d\rho \tag{13.102}$$

Because of the independence of the emitted signal x and the additive noise process n we finally obtain

$$f_{y|x}(\rho \mid \xi) = f_n(\rho - \xi) \tag{13.103}$$

and thus

$$f_{y,x}(\rho, \xi) = f_{y|x}(\rho \mid \xi)\cdot f_x(\xi) = f_n(\rho - \xi)\cdot f_x(\xi) \tag{13.104}$$

The joint distribution density of the input and output variables is thus generated by the product of the marginal distribution of the input variable and the shifted distribution density of the noise process. This result is achieved because of the assumption of independence of the input and the noise process.

This independence can be exploited in the determination of the correlation coefficient, which is thus obtained from the equation

$$\rho^2 = \frac{\sigma_x^2}{\sigma_y^2} = \frac{E\{(x - E\{x\})^2\}}{E\{(y - E\{y\})^2\}} = \frac{E\{(x - E\{x\})^2\}}{E\{((x+n) - E\{x+n\})^2\}}$$
$$= \frac{E\{(x - E\{x\})^2\}}{E\{(x - E\{x\})^2 + n^2\}} = \frac{\sigma_x^2}{\sigma_x^2 + \sigma_n^2} = \frac{S}{S+N} \tag{13.105}$$

Inserting this in the previous result, we get:

$$I(x,y) = -\frac{1}{2}\cdot \ln\left(1 - \rho^2\right) = -\frac{1}{2}\cdot \ln\left(1 - \frac{S}{S+N}\right) = -\frac{1}{2}\cdot \ln\left(\frac{N}{S+N}\right)$$

$$I(x,y) = \frac{1}{2}\cdot \ln\left(\frac{S+N}{N}\right)$$

$$I(x,y) = \frac{1}{2}\cdot \ln\left(1 + \frac{S}{N}\right) \tag{13.106}$$

This is the transinformation in a perturbed transmission, i.e. the amount of information that is transmitted from the sender to the adressee. Obviously this information depends on the logarithm of the signal-to-noise ratio, which is always nonnegative. The summation of 1 in the argument of the logarithm also ensures that the value of the logarithm is nonnegative. As soon as there is a signal with signal power S, the emitter transmits information to the receiver. If the signal power is too small, the detection of the signal at the output of the transmitter may not be unique, or it may not be possible at all.

The divergence is equal to

$$D = \frac{\rho^2}{1-\rho^2} = \frac{\dfrac{S}{S+N}}{1-\dfrac{S}{S+N}} = \frac{\dfrac{S}{S+N}}{\dfrac{S+N-S}{S+N}} = \frac{\dfrac{S}{S+N}}{\dfrac{N}{S+N}}$$

$$D = \frac{S}{N} \tag{13.107}$$

The divergence is equal to the signal-to-noise ratio. The greater this ratio becomes, the more similarity between the input and output variables we obtain. The more power we are able to provide for the emitter, the easier the detection of the signal is for the receiver. Thus an increasing noise process leads to a decreasing possibility to obtain the right assignment of the output signal to the input signal.

13.7 Communication channel and information processing

The communication channel allows us to explain the statements we made in the introduction. Information decreases the uncertainty that we have about a certain event. The evaluation of the information requires a correlation between the knowledge of emitter and receiver, which means that the uncertainties of the previous knowledge of sender and adressee are connected in some way. This can be shown with the transinformation, which we already evaluated as

$$I(x_1, x_2) = -\frac{1}{2} \cdot \ln\left(1-\rho^2\right) \tag{13.108}$$

The correlation coefficient ρ decides if there is a transmission of information, or if the random variables at the input and the output are not correlated (independent for gaussian random variables). Again we examine two special cases:

$\rho = 0$: Sender and receiver are completely uncorrelated and the flow of information or the mutual information completely vanishes.

$\rho = 1, -1$: Sender and receiver are completely correlated and thus the mutual information is infinite.

Such an infinite mutual information can be achieved via two different approaches. On the one hand we might achieve this infinite value by an exact transmission with an infinite transinformation (technically this is not possible, as we will see at the Shannon bound), leading to an exact correspondence of sender and receiver. On the other hand the receiver might already know the information, which also leads to an exact matching of sender and receiver. In this second case the transmitted information is completely redundant.

13.7.1 Semantic, syntactic and pragmatic information

The only part of the information that can be described with mathematical tools is the syntactic information, merely depending on the relative frequencies, or on the probabilities of the symbols. If, for instance, we send the message "I'll come later", the message "I'oeae llcmltr" contains exactly the same syntactic information while the semantic information has vanished. Even the pragmatic information is affected by the scrambling of the letters, because the desired effect of the message depends on the receiver. If the receiver tries to rearrange the letters, to find some meaningful expression, the message might have the desired effect. If he does not, the desired effect will not occur.

C. F. von Weizsäcker says: "Information is that (part of the message) which is understood", "Information is that which has an effect" and "Information is merely the part that generates information". All three statements describe the pragmatic information, because the adressee can only incorporate information if he is able to understand the message. This understanding requires a certain a priori information, a previous knowledge, and without this a priori information the message would be irrelevant and useless. A talk about quantum physics will not create any information for an auditor, who does not have a priori information, because the a priori information does not correlate with the field of quantum physics ($\rho << 1$) and thus the transinformation will be very small.

The second statement "Information is that which has an effect" requires the understanding on the side of the receiver. It also conditions a certain reaction of the receiver on the message. This reaction can be quantified by evaluating the actions of the agent who received the information, and can thus carry out conclusions regarding the information processing. So we may, for instance, configure the development of the agent in such a way that we choose the best agent from the different reactions of the agents on some received information.

In the third statement "Information is merely the part that generates information" the understanding of the information is the main goal. The understanding of the information is based on the a priori information, enabling us to classify the new information with our actual knowledge. The a priori information has to be such that the new information decreases an existing uncertainty. A priori information, which is not changed by the evaluation of a message, makes us think of the message as totally redundant. If someone measures a line of length 1.20 m and he gets the additional information that the length is somewhere between 1 and 2 m, the additional information will not influence his prior knowledge, because this information is totally redundant to him. Information has to change the a priori knowledge and it thus has to reduce the prior uncertainty to be useful. Furthermore there must be some similarity between the prior knowledge and the message, or some affinity of the receiver and the received information, because otherwise the information is classified as useless or irrelevant.

13.7.2 Information, first-time occurrence, confirmation

A reception of information certainly occurs if the message has an appropriate dosage between first-time occurrence and confirmation. In this case there exists a certain correlation between the prior information of the receiver and the received information and this enables the receiver to transform the information into knowledge (some authors denote knowledge as condensed information). The more we reduce the new part of the information, the better we are able to evaluate the information (the message), but the less information the message contains. The reduction of the gain of information certainly leads to a simplified evaluation, but it also limits the possible gain of information. Of course, there are also cases where the information cannot be received completely. Data compression is, for instance, used to compress the data of images in such a way that the information relevant to our eyes is maintained after the compression, even though the whole information of the image has been reduced by lossy compression algorithms such as JPG (joint photographic expert group) or GIF (graphics interchange format).

Obviously the adressee (the human eye) is used as a measure of quality, to determine the possible reduction of the transmission of information between the original image and the compressed image, without reducing the pragmatic information (the effect of the image on the receiver). The new part of the information is reduced in such a way that the receiver cannot detect any changes with his senses or his detectors.

Thus a message has to be a compromise between information as the new part of the message and redundancy as the already known part. Then the message contains relevant information, which can be evaluated on the basis of the redundant part.

13.8 Shannon's bound

How much information can we transmit over a channel? Are there any limitations? To answer these questions, we use a basic result of Nyquist's sampling theorem. This theorem provides the following statement:

A signal with the bandwidth B and duration T can be described by $2 \cdot B \cdot T$ samples. If each of these samples (on the average) has the power S, the total power of a sufficient number of samples is equal to

$$S_{\text{all}} = 2 \cdot B \cdot T \cdot S \tag{13.109}$$

We now define each signal as a point in an n-dimensional space. The configuration of these points shall be such that different signals do not affect each other (avoidance of intersymbol interference). The energy of each signal generates a hypersphere around each signal, which leads to hyperspheres of radius

$$r = \sqrt{2 \cdot B \cdot T \cdot S} \tag{13.110}$$

where the signal is situated. If there is an additional noise process, added to the deterministic signal, this radius becomes

$$r = \sqrt{2 \cdot B \cdot T(S+N)} \tag{13.111}$$

For pure noise we get

$$r = \sqrt{2 \cdot B \cdot T \cdot N}$$

As no signal affects another signal all, of these hyperspheres have to be arranged so that there are no intersections between two hyperspheres. If there are $n = 2 \cdot B \cdot T$ signals = samples, then the volume of the space where signals and noise occur is

$$V_{SN} = \left(\sqrt{2 \cdot B \cdot T \cdot (S+N)}\right)^{2 \cdot B \cdot T} \tag{13.112}$$

the volume of the noise signals is

$$V_N = \left(\sqrt{2 \cdot B \cdot T \cdot N}\right)^{2 \cdot B \cdot T} \tag{13.113}$$

and the ratio of these two volumes determines how many signals we may send in a finite time interval T across this channel.

$$N(B,T) = \frac{V_{SN}}{V_N} = \left(\frac{\sqrt{2 \cdot B \cdot T \cdot (S+N)}}{\sqrt{2 \cdot B \cdot T \cdot N}}\right)^{2 \cdot B \cdot T} = \left(\frac{S+N}{N}\right)^{B \cdot T}$$

$$N(B,T) = B \cdot T \cdot \ln\left(\frac{S+N}{N}\right) \tag{13.114}$$

As the signal duration is T seconds, we are able to send C bit per second across the channel

$$C = N(B,T) = B \cdot \ln\left(\frac{S+N}{N}\right) = B \cdot \ln\left(1 + \frac{S}{N}\right) \tag{13.115}$$

For a bandwidth B, a signal power S and a noise power N we may send a maximum of C [bit/s]. If we send more signals, the areas of the hyperspheres intersect and we obtain errors in the transmission. If the signal-to-noise ratio is equal to zero, i.e. the wanted signal is not available, we have $C = 0$ and we cannot transmit signals across the channel, as we do not have any signal power.

The examination of signal and noise power leads us to the number of signals that we are able to send per second. This is given by

$$C = B \cdot \ln\left(\frac{S+N}{N}\right) = B \cdot \ln\left(1 + \frac{S}{N}\right) \tag{13.116}$$

We recall the definitions of Nyquist and Hartley, who stated that

$$I_{\text{Hartley}} = \ln N = -\ln \frac{1}{p(x)} \tag{13.117}$$

N = number of possible sequences of symbols.

Combining single elements of a given alphabet, consisting of s characters, can generate these sequences. If

$$N = s^n \tag{13.118}$$

n = number of characters of the alphabet, used in the sequence.

We find

$$I_{\text{Hartley}} = \ln s^n = n \cdot \ln s \tag{13.119}$$

We have $n = 2 \cdot B$ symbols per second ($2 \cdot B \cdot T$ samples in the time interval T), leading to $B = n/2$. The rate of transmission is thus

$$C = \frac{n}{2} \cdot \ln\left(\frac{S+N}{N}\right) = n \cdot \ln\left(\sqrt{1 + \frac{S}{N}}\right) \tag{13.120}$$

Nyquist says that we can transmit

$$C = n \cdot \ln m \tag{13.121}$$

symbols per second if m is the number of the possible different symbols. Thus the quotient S/N allows us to determine directly the number of different signals that we may send across a channel. This is

$$C = n \cdot \ln m = n \cdot \ln\left(\sqrt{1 + \frac{S}{N}}\right)$$

$$m = \sqrt{1 + \frac{S}{N}} \tag{13.122}$$

For $S/N = 15$ we may send four different signals across a channel, which can be detected without any further signal processing.

$$C = B \cdot \ln\left(\frac{S+N}{N}\right) = B \cdot \ln\left(1+\frac{S}{N}\right) \tag{13.123}$$

To increase the number of possible signals, we may alter the parameters

B = increase the bandwidth
S = increase the power of the signal
N = decrease the power of the noise

In real applications N cannot be affected in a desired way, so that our interest is focussed on the remaining parameters, the power of the signal and the bandwidth.

Increasing the power of the signal by enhancing the amplitude only provides a logarithmic gain in the rate of transmission. Thus the increase of the bandwidth is the better choice, as it leads to a linear gain. So amplitude modulation, where we have to increase S, does not have the potential of phase or frequency modulation, which profits from the increasing bandwidth. But phase and frequency modulation are limited by other phenomena (nonlinear modulation and the related problems of nonlinear mappings especially with additive noise) and even the increase of the bandwidth cannot be carried out without limitation, as we will now explain.

Usually the power of the noise is the power of the thermic effects, known as Johnson noise or thermal noise. It is described by the equation

$$N = k \cdot T \cdot B \tag{13.124}$$

$k = 1.38 \cdot 10^{-23}$ J/K Boltzmann's constant

This is merely a simplified expression of

$$N = \frac{h \cdot f \cdot B}{\exp\left(\frac{h \cdot f}{k \cdot T}\right) - 1} \tag{13.125}$$

i.e. the noise, which also takes into account the effects of the quanta (see chapter 4).

$h = 6.625 \cdot 10^{-24}$ J/s Planck's constant

If we expand the exponential function into a series and neglect all higher order terms from the second term on, we obtain

$$N \cong \frac{h \cdot f \cdot B}{1 + \frac{h \cdot f}{k \cdot T} - 1} = \frac{h \cdot f \cdot B}{\frac{h \cdot f}{k \cdot T}} = k \cdot T \cdot B \tag{13.126}$$

for the thermal noise. The equation

$$N = \frac{h \cdot f \cdot B}{\exp\left(\frac{h \cdot f}{k \cdot T}\right) - 1} \tag{13.127}$$

is only necessary if the frequency is so great that $h \cdot f$ achieves values in the range of $k \cdot T$. This is true for

$$f = \frac{k \cdot T}{h} = 2.07 \cdot 10^{10} \cdot T \tag{13.128}$$

i.e. for frequencies in the range of 20 GHz/($k \cdot T$). For temperatures in the range of our room temperature, this leads to frequencies in the range of terahertz. With

$$N = k \cdot T \cdot B$$

in

$$C = B \cdot \ln\left(1 + \frac{S}{N}\right)$$

we find

$$C = B \cdot \ln\left(1 + \frac{S}{k \cdot T \cdot B}\right) \tag{13.129}$$

and if we build the limit for an infinite bandwidth, we obtain

$$\lim_{B \to \infty} C = \lim_{B \to \infty} B \cdot \ln\left(1 + \frac{S}{k \cdot T \cdot B}\right) = \infty \cdot 0 \tag{13.130}$$

which is an indefinite limit. We thus expand the logarithm into a series and get

$$\ln(1 + x) = x - \frac{x^2}{2} + \frac{x^3}{3} - \ldots = \sum_{n=1}^{\infty} (-1)^{n+1} \cdot \frac{x^n}{n} \tag{13.131}$$

and we obtain

$$\begin{aligned} \lim_{B \to \infty} C &= \lim_{B \to \infty} B \cdot \ln\left(1 + \frac{S}{k \cdot T \cdot B}\right) \\ &= \lim_{B \to \infty} B \cdot \left[\frac{S}{k \cdot T \cdot B} - \frac{\left(\frac{S}{k \cdot T \cdot B}\right)^2}{2} + \frac{\left(\frac{S}{k \cdot T \cdot B}\right)^3}{3} - \ldots \right] \end{aligned} \tag{13.132}$$

$$\lim_{B\to\infty} C = \lim_{B\to\infty}\left[\frac{S}{k\cdot T}-\frac{1}{2\cdot B}\left(\frac{S}{k\cdot T}\right)^2+\frac{1}{3\cdot B^2}\left(\frac{S}{k\cdot T}\right)^3-\ldots\right] \tag{13.133}$$

Only the first addend in the brackets does not have the bandwidth B in the denominator, i.e. only this term remains in the limit, and we get

$$\lim_{B\to\infty} C = \frac{S}{k\cdot T} \tag{13.134}$$

If we choose the dual logarithm instead of the natural logarithm, we have to divide by ln 2 and we thus obtain the result, which can be found in [LÜK85],

$$\lim_{B\to\infty} C = \frac{1}{\ln 2}\frac{S}{k\cdot T} = 1.44\cdot\frac{S}{k\cdot T} \tag{13.135}$$

Even if we increase the bandwidth to an arbitrary value, this bound of the channel capacity (rate of transmission) cannot be exceeded. This bound is known as the Shannon bound. Every signal processing in which the rate of transmission exceeds this Shannon bound loses information and thus leads to a lossy transmission.

13.9 Example of the channel capacity

Let us now examine a symmetric binary channel, shown in figure 13.3.

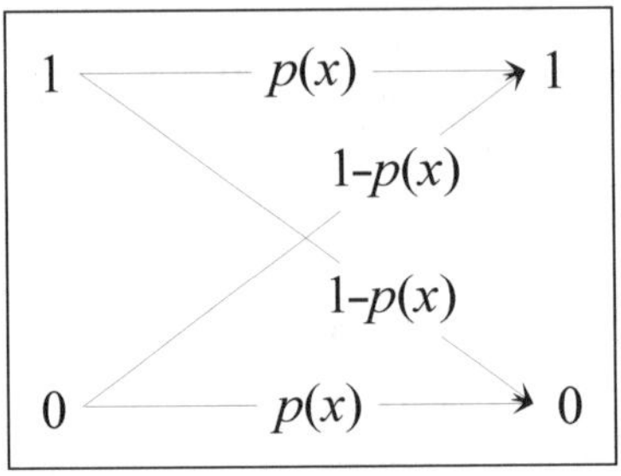

Fig. 13.3: Symmetric channel

We search for $R = H(X) - H(X\,|\,Y)$ or the maximum of this value, which is the channel capacity.

If we are limited to two symbols, our alphabet is also limited to two symbols (0,1). The entropy of the source thus achieves its maximum if both symbols occur with equal probability.

$$H(Y) = H(X) = 1 \text{ bit/symbol, as } p(0) = p(1) = \tfrac{1}{2} \tag{13.136}$$

Now we have to calculate the conditional entropy. We start with the following consideration:

The probability that we receive a 1 is equal to 1/2. If we receive a 1, then this symbol has been sent with a probability of p and with the probability $(1-p)$ the emitted symbol was a 0. Thus we have

$$H(X \mid y=1)=(-p \cdot \operatorname{ld} p-(1-p) \cdot \operatorname{ld}(1-p)) \tag{13.137}$$

and because of the symmetry of the channel

$$H(X \mid y=0)=(-p \cdot \operatorname{ld} p-(1-p) \cdot \operatorname{ld}(1-p)) \tag{13.138}$$

This leads to

$$\begin{aligned} H(X \mid Y) &= \frac{1}{2} \cdot H(X \mid y=1)+\frac{1}{2} \cdot H(X \mid y=0) \\ &= (-p \cdot \operatorname{ld} p-(1-p) \cdot \operatorname{ld}(1-p)) \end{aligned} \tag{13.139}$$

and thus we obtain the channel capacity

$$C=R=H(X)-H(X \mid Y) \tag{13.140}$$

$$C=1+p \cdot \operatorname{ld} p+(1-p) \cdot \operatorname{ld}(1-p) \tag{13.141}$$

This function is shown in figure 13.4

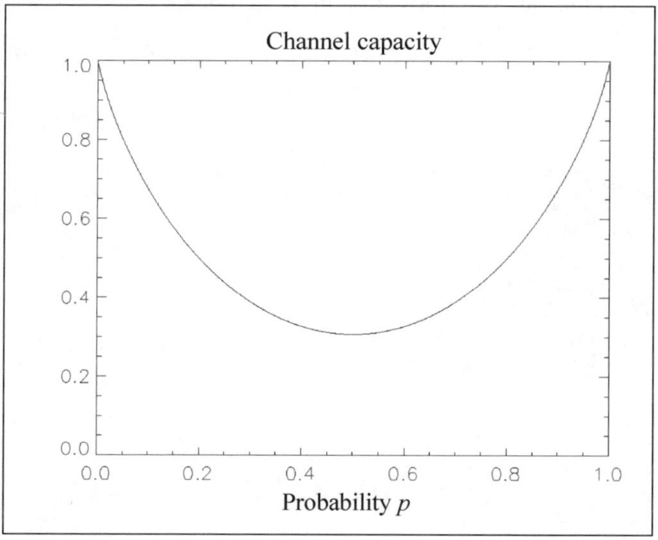

Fig. 13.4: Channel capacity

This function achieves a maximum at $p = 0$ and $p = 1$. In these cases we either receive the symbol that has been send ($p = 1$), or we receive the symbol that has not been sent ($p = 0$). In these cases the information that the receiver obtains about the signal at the input is optimal, because either he always receives the right symbol or he always receives the wrong symbol. There are no errors in transmission.

If we send only zeroes and we have an error in every 10th transmission, then we get $p = 0.9$ and this reduces the channel capacity to

$$C = 1 + 0.9 \cdot \mathrm{ld}\, 0.9 + 0.1 \cdot \mathrm{ld}\, 0.1 = 1 - 0.469 = 0.53 \tag{13.142}$$

If the error occurs in every 100th transmission, the channel capacity reduces merely to the value

$$C = 0.92 \tag{13.143}$$

It is important in this consideration to distinguish between the probabilities of the channel p and the probabilities $p(0)$, $p(1)$ of the symbols that are emitted by the source. The source emits the symbols with the source entropy, which is not affected by the channel. The transmission errors in the channel lead to the probability p, describing the number of symbols that are affected by the transmission and that are thus received as another symbol by the receiver. The probability of these errors in transmission is denoted by p!

With this example we finish the examination of the channel capacity and we now turn to entropies in state space models, to determine the transmission of information and the gain of information in these models.

14 'Deterministic' and stochastic information

14.1 Information in state space models

At the beginning of this chapter we want to carry out some basic considerations regarding the information in mappings with additional noise. Such mappings occur in state space models, when we model the relation between the state space variables and the observation space variables with the observation equation.

Signal or measurement data in general consist of an intended deterministic part and a second part, which is the noise that is generated by the superimposed disturbances. In state space models these two parts are described in the observation equation, where we add the noise process to the deterministic mapping. We thus start with the simplest observation equation:

$$\underline{y}(t) = \underline{x}(t) + \underline{v}(t) \tag{14.01}$$

If we assume all signals to be random variables, the corresponding equation for the distribution densities is

$$f_{\underline{y}(t)}(\underline{\rho}) = f_{\underline{x}(t)}(\underline{\rho}) * f_{\underline{v}(t)}(\underline{\rho}) \tag{14.02}$$

if the signal process $\underline{x}(t)$ and the noise process $\underline{v}(t)$ are statistically independent. If we additionally transform the input signal, we write

$$\underline{y}(t) = \underline{z}(t) + \underline{v}(t) = \underline{h}[\underline{x}(t)] + \underline{v}(t) \tag{14.03}$$

$$f_{\underline{y}(t)}(\underline{\rho}) = f_{\underline{z}(t)}(\underline{\rho}) * f_{\underline{v}(t)}(\underline{\rho}) \tag{14.04}$$

With the transformation of measure preserving mappings

$$f_{\underline{z}(t)}(\underline{\zeta}) \cdot d\underline{\zeta} = f_{\underline{x}(t)}(\underline{\xi}) \cdot d\underline{\xi} \tag{14.05}$$

and the connections

$$\underline{z}(t) = \underline{h}[\underline{x}(t)] \qquad \underline{x}(t) = \underline{h}^{-1}[\underline{z}(t)] \tag{14.06}$$

$$\underline{\zeta} = \underline{h}[\underline{\xi}] \qquad \underline{\xi} = \underline{h}^{-1}[\underline{\zeta}] \tag{14.07}$$

we find

$$f_{\underline{z}(t)}(\underline{\zeta})\cdot d\underline{\zeta} = f_{\underline{x}(t)}\left(\underline{h}^{-1}[\underline{\zeta}]\right)\cdot d\underline{h}^{-1}[\underline{\zeta}] \tag{14.08}$$

$$f_{\underline{z}(t)}(\underline{\zeta}) = f_{\underline{x}(t)}\left(\underline{h}^{-1}[\underline{\zeta}]\right)\cdot\left\|\frac{\partial \underline{h}^{-1}[\underline{\zeta}]}{\partial \underline{\zeta}}\right\| \tag{14.09}$$

and for the distribution density of the observed random variable we obtain the following equation

$$f_{\underline{y}(t)}(\underline{\rho}) = f_{\underline{x}(t)}\left(\underline{h}^{-1}[\underline{\rho}]\right)\cdot\left\|\frac{\partial \underline{h}^{-1}[\underline{\rho}]}{\partial \underline{\rho}}\right\| * f_{\underline{v}(t)}(\underline{\rho}) \tag{14.10}$$

This equation describes the general relation between the three random variables, the signal, the measurement noise and the measurement. It holds under the condition that the signal and the measurement noise are added and that they are independent random variables.

We are now interested in the information measures and values occurring in this connection, and we begin our examination with the information of the random variable $\underline{y}$ in observation space

$$S(\underline{y}) = -\int_{\underline{\mathrm{P}}} \ln\left[f_{\underline{y}(t)}(\underline{\rho})\right]\cdot f_{\underline{y}(t)}(\underline{\rho})\cdot d\underline{\rho} = E\left\{-\ln\left[f_{\underline{y}(t)}(\underline{\rho})\right]\right\} \tag{14.11}$$

$$S(\underline{x}) = -\int_{\underline{\Xi}} \ln\left[f_{\underline{x}(t)}(\underline{\xi})\right]\cdot f_{\underline{x}(t)}(\underline{\xi})\cdot d\underline{\xi} = E\left\{-\ln\left[f_{\underline{x}(t)}(\underline{\xi})\right]\right\} \tag{14.12}$$

$$S(\underline{v}) = -\int_{\underline{\Upsilon}} \ln\left[f_{\underline{v}(t)}(\underline{\upsilon})\right]\cdot f_{\underline{v}(t)}(\underline{\upsilon})\cdot d\underline{\upsilon} = E\left\{-\ln\left[f_{\underline{v}(t)}(\underline{\upsilon})\right]\right\} \tag{14.13}$$

$$S(\underline{y}) = -\int_{\underline{\mathrm{P}}} \ln\left[f_{\underline{x}(t)}\left(\underline{h}^{-1}[\underline{\rho}]\right)\cdot\left\|\frac{\partial \underline{h}^{-1}[\underline{\rho}]}{\partial \underline{\rho}}\right\| * f_{\underline{v}(t)}(\underline{\rho})\right]\cdot f_{\underline{x}(t)}\left(\underline{h}^{-1}[\underline{\rho}]\right)\cdot\left\|\frac{\partial \underline{h}^{-1}[\underline{\rho}]}{\partial \underline{\rho}}\right\| * f_{\underline{v}(t)}(\underline{\rho})\cdot d\underline{\rho} \tag{14.14}$$

Now the question arises how may we decompose the information $S(\underline{y})$ into the partial amounts of information $S(\underline{x})$ and $S(\underline{v})$.

Our first approach is to neglect the transformation $\underline{h}[\ ]$ and to start our examination with the two random variables $\underline{z}$ and $\underline{v}$, which create the random variable $\underline{y}$.

$$S(\underline{y}) = -\int_{\underline{P}} \ln\left[f_{\underline{z}(t)}(\underline{\rho}) * f_{\underline{v}(t)}(\underline{\rho})\right] \cdot f_{\underline{z}(t)}(\underline{\rho}) * f_{\underline{v}(t)}(\underline{\rho}) \cdot d\underline{\rho} \tag{14.15}$$

This computation is in fact easier to handle and easier to grasp on first sight. But the convolution in the logarithm provides some difficulties, so that we do not solve this equation, but we arrange another equation, based on conditional distribution densities.

14.2 The observation equation

We now ask where the information in an observation comes from and we again start our examination with the observation equation

$$\underline{y} = \underline{h}(\underline{x}) + \underline{v} \tag{14.16}$$

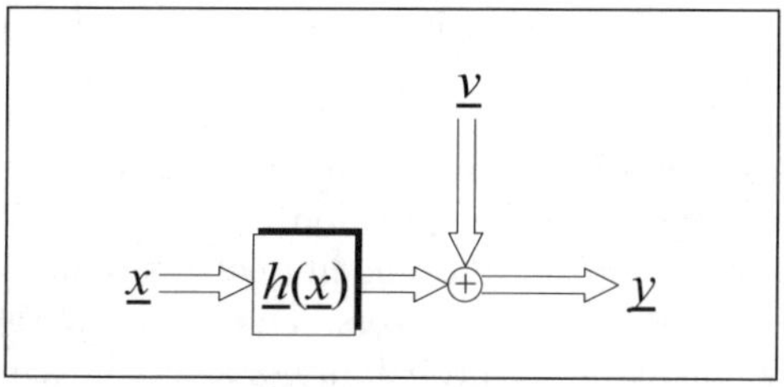

Fig. 14.1: Nonlinear observation equation

where we search for the information of the output signal

$$S(\underline{y}) = -\int_{P} \ln\left[f_{\underline{y}}(\underline{\rho})\right] \cdot f_{\underline{y}}(\underline{\rho}) \cdot d\underline{\rho} \tag{14.17}$$

which is an entropy of first order. It can also be written as

$$S(\underline{y}) = -\int_{\Xi}\int_{P} \ln\left[f_{\underline{y}}(\underline{\rho})\right] \cdot f_{\underline{y},\underline{x}}(\underline{\rho},\underline{\xi}) \cdot d\underline{\rho} \cdot d\underline{\xi} \tag{14.18}$$

The joint information function of the random variables $\underline{x}$ and $\underline{y}$, connected via the deterministic transformation and additive noise, is denoted as

$$S(\underline{x},\underline{y}) = -\int_{\Xi}\int_{P} \ln\left[f_{\underline{y},\underline{x}}(\underline{\rho},\underline{\xi})\right] \cdot f_{\underline{y},\underline{x}}(\underline{\rho},\underline{\xi}) \cdot d\underline{\rho} \cdot d\underline{\xi} \tag{14.19}$$

Application of the Bayesian rule leads to

$$S(\underline{x},\underline{y}) = -\int\limits_{\Xi}\int\limits_{\mathrm{P}} \ln\left[f_{\underline{y}|\underline{x}}(\underline{\rho}\mid\underline{\xi})\cdot f_{\underline{x}}(\underline{\xi})\right]\cdot f_{\underline{y}|\underline{x}}(\underline{\rho}\mid\underline{\xi})\cdot f_{\underline{x}}(\underline{\xi})\cdot d\underline{\rho}\cdot d\underline{\xi}$$

and

$$S(\underline{x},\underline{y}) = -\int\limits_{\Xi}\int\limits_{\mathrm{P}} \left\{\ln\left[f_{\underline{y}|\underline{x}}(\underline{\rho}\mid\underline{\xi})\right] + \ln\left[f_{\underline{x}}(\underline{\xi})\right]\right\}\cdot f_{\underline{y}|\underline{x}}(\underline{\rho}\mid\underline{\xi})\cdot f_{\underline{x}}(\underline{\xi})\cdot d\underline{\rho}\cdot d\underline{\xi}$$

$$S(\underline{x},\underline{y}) = -\int\limits_{\Xi}\int\limits_{\mathrm{P}} \ln\left[f_{\underline{y}|\underline{x}}(\underline{\rho}\mid\underline{\xi})\right]\cdot f_{\underline{y}|\underline{x}}(\underline{\rho}\mid\underline{\xi})\cdot f_{\underline{x}}(\underline{\xi})\cdot d\underline{\rho}\cdot d\underline{\xi}$$

$$-\int\limits_{\Xi}\int\limits_{\mathrm{P}} \ln\left[f_{\underline{x}}(\underline{\xi})\right]\cdot f_{\underline{y}|\underline{x}}(\underline{\rho}\mid\underline{\xi})\cdot f_{\underline{x}}(\underline{\xi})\cdot d\underline{\rho}\cdot d\underline{\xi}$$

We get:

$$S(\underline{x},\underline{y}) = -\int\limits_{\Xi}\int\limits_{\mathrm{P}} \ln\left[f_{\underline{y}|\underline{x}}(\underline{\rho}\mid\underline{\xi})\right]\cdot f_{\underline{y},\underline{x}}(\underline{\rho},\underline{\xi})\cdot d\underline{\rho}\cdot d\underline{\xi} + S(\underline{x}) = S(\underline{y}\mid\underline{x}) + S(\underline{x}) \tag{14.20}$$

This result is already familiar to us from the examination of the channel information. We want to rearrange this equation by using our knowledge of the model, which enables us to replace the conditional distribution density by the distribution density of the measurement errors. To do this, we at first denote the conditional distribution density by the probability of a certain semi-open interval

$$f_{\underline{y}|\underline{x}}(\underline{\rho}\mid\underline{\xi})\cdot d\underline{\rho} = P\left\{\omega \mid \underline{y}(\omega)\in\left(\underline{\rho},\underline{\rho}+d\underline{\rho}\right] \mid \underline{x}(\omega)=\underline{\xi}\right\} \tag{14.21}$$

and with the observation equation we get:

$$f_{\underline{y}|\underline{x}}(\underline{\rho}\mid\underline{\xi})\cdot d\underline{\rho} = P\left\{\omega \mid \underline{y}(\omega) = \underline{h}[\underline{x}(\omega)] + \underline{v}(\omega)\in\left(\underline{\rho},\underline{\rho}+d\underline{\rho}\right] \mid \underline{x}(\omega)=\underline{\xi}\right\}$$

$$f_{\underline{y}|\underline{x}}(\underline{\rho}\mid\underline{\xi})\cdot d\underline{\rho} = P\left\{\omega \mid \underline{y}(\omega) = \underline{h}[\underline{\xi}] + \underline{v}(\omega)\in\left(\underline{\rho},\underline{\rho}+d\underline{\rho}\right] \mid \underline{x}(\omega)=\underline{\xi}\right\}$$

$$f_{\underline{y}|\underline{x}}(\underline{\rho}\mid\underline{\xi})\cdot d\underline{\rho} = P\left\{\omega \mid \underline{v}(\omega)\in\left(\underline{\rho}-\underline{h}[\underline{\xi}],\underline{\rho}-\underline{h}[\underline{\xi}]+d\underline{\rho}\right] \mid \underline{x}(\omega)=\underline{\xi}\right\}$$

$$f_{\underline{y}|\underline{x}}(\underline{\rho}\mid\underline{\xi})\cdot d\underline{\rho} = f_{\underline{v}}\left(\underline{\rho}-\underline{h}[\underline{\xi}]\right)\cdot d\underline{\rho} \tag{14.22}$$

The conditioning on the random variable $\underline{x}$ vanishes, because we assume independence of $\underline{x}$ and $\underline{v}$. In this case we find

$$f_{\underline{v}|\underline{x}}(\underline{\upsilon}\mid\underline{\xi}) = \frac{f_{\underline{v},\underline{x}}(\underline{\upsilon},\underline{\xi})}{f_{\underline{x}}(\underline{\xi})} = \frac{f_{\underline{v}}(\underline{\upsilon})\cdot f_{\underline{x}}(\underline{\xi})}{f_{\underline{x}}(\underline{\xi})} = f_{\underline{v}}(\underline{\upsilon}) \tag{14.23}$$

and in the computation of the information this leads to

$$S(\underline{x},\underline{y}) = -\int_{\Xi}\left(\int_{\mathrm{P}} \ln\left[f_{\underline{y}|\underline{x}}(\underline{\rho} \mid \underline{\xi})\right] \cdot f_{\underline{y}|\underline{x}}(\underline{\rho} \mid \underline{\xi}) \cdot d\underline{\rho}\right) \cdot f_{\underline{x}}(\underline{\xi}) \cdot d\underline{\xi} + S(\underline{x}) \tag{14.24}$$

$$S(\underline{x},\underline{y}) = -\int_{\Xi}\left(\int_{\mathrm{P}} \ln\left[f_{\underline{v}}(\underline{\rho} - \underline{h}[\underline{\xi}])\right] \cdot f_{\underline{v}}(\underline{\rho} - \underline{h}[\underline{\xi}]) \cdot d\underline{\rho}\right) \cdot f_{\underline{x}}(\underline{\xi}) \cdot d\underline{\xi} + S(\underline{x}) \tag{14.25}$$

The term $-\left(\int_{\mathrm{P}} \ln\left[f_{\underline{v}}(\underline{\rho} - \underline{h}[\underline{\xi}])\right] \cdot f_{\underline{v}}(\underline{\rho} - \underline{h}[\underline{\xi}]) \cdot d\underline{\rho}\right)$ is a Shannon information of the random variable $\underline{v}$, which is shifted by $\underline{h}[\underline{x}]$. How does this shift influence Shannon's information?

From the gaussian distribution density we know that Shannon's information merely depends on the dimension of the random vector and on the covariance matrix. For

$$f_{\underline{v}}(\underline{\rho} - \underline{h}[\underline{\xi}]) = \frac{1}{(2\pi)^{\frac{m}{2}} \cdot |R|^{\frac{1}{2}}} \cdot \exp\left\{-\frac{1}{2} \cdot (\underline{\rho} - \underline{h}[\underline{\xi}])^T \cdot R^{-1} \cdot (\underline{\rho} - \underline{h}[\underline{\xi}])\right\} \tag{14.26}$$

we obtain

$$S(\underline{v}) = \frac{1}{2} \cdot \ln\left[(2\pi)^m \cdot |R|\right] + \frac{m}{2} \tag{14.27}$$

and for

$$f_{\underline{v}}(\underline{\rho}) = \frac{1}{(2\pi)^{\frac{m}{2}} \cdot |R|^{\frac{1}{2}}} \cdot \exp\left\{-\frac{1}{2} \cdot (\underline{\rho})^T \cdot R^{-1} \cdot (\underline{\rho})\right\} \tag{14.28}$$

$$S(\underline{v}) = \frac{1}{2} \cdot \ln\left[(2\pi)^m \cdot |R|\right] + \frac{m}{2} \tag{14.29}$$

Both information values are obviously equal, because the shift of the expectation value does not influence the information. In general we may proceed as follows: We know that

$$\underline{y} = \underline{h}(\underline{x}) + \underline{v} \tag{14.30}$$

We search for a connection of the two random variables $\underline{v}$ and $\underline{y}$. In order to denote the distribution density of one random variable by the distribution density of another random variable with an additional correction term, we apply a measure preserving mapping and obtain the formulation

$$f_{\underline{y}}(\underline{\rho}) \cdot d\underline{\rho} = f_{\underline{v}}(\underline{\upsilon}) \cdot d\underline{\upsilon}$$

$$f_{\underline{y}}(\underline{\rho}) = f_{\underline{v}}(\underline{\upsilon}) \cdot \left\| \frac{\partial \underline{\upsilon}}{\partial \underline{\rho}} \right\| \tag{14.31}$$

Now we use the functional connection

$$\underline{v} = \underline{y} - \underline{h}(\underline{x}) \quad \text{and} \qquad \underline{\upsilon} = \underline{\rho} - \underline{h}(\underline{\xi}) \tag{14.32}$$

leading to

$$f_{\underline{y}}(\underline{\rho}) = f_{\underline{v}}(\underline{\rho} - h(\underline{\xi})) \cdot \left\| \frac{\partial}{\partial \underline{\rho}} (\underline{\rho} - h(\underline{\xi})) \right\| \tag{14.33}$$

$$f_{\underline{y}}(\underline{\rho}) = f_{\underline{v}}(\underline{\rho} - h(\underline{\xi})) \cdot \left\| \frac{\partial \underline{\rho}}{\partial \underline{\rho}} \right\|$$

$$f_{\underline{y}}(\underline{\rho}) = f_{\underline{v}}(\underline{\rho} - h(\underline{\xi})) \tag{14.34}$$

The shift of the coordinate system (translation) does not affect the information, as the new distribution density function does not contain any additional correction term. We already demanded this as a basic condition from our first information functions.

As we are interested in the connection between measurement value, measurement noise and the state space variable, where we intend to describe the information, we again examine the integral

$$S(\underline{x}, \underline{y}) = -\int_{\Xi} \left(\int_{\mathrm{P}} \ln\left[f_{\underline{v}}(\underline{\rho} - \underline{h}[\underline{\xi}]) \right] \cdot f_{\underline{v}}(\underline{\rho} - \underline{h}[\underline{\xi}]) \cdot d\underline{\rho} \right) \cdot f_{\underline{x}}(\underline{\xi}) \cdot d\underline{\xi} + S(\underline{x}) \tag{14.35}$$

We now focus our attention on the term $-\left(\int_{\mathrm{P}} \ln\left[f_{\underline{v}}(\underline{\rho} - \underline{h}[\underline{\xi}]) \right] \cdot f_{\underline{v}}(\underline{\rho} - \underline{h}[\underline{\xi}]) \cdot d\underline{\rho} \right)$, which we rearrange appropriately. We thus substitute $\underline{\rho} - \underline{h}[\underline{\xi}] = \underline{\upsilon}$ and this leads to $d\underline{\rho} = d\underline{\upsilon}$ and for the limits we obtain the shift $\underline{\rho} = \mathrm{P} \;\Rightarrow\; \underline{\upsilon} = \mathrm{P} - \underline{h}[\underline{\xi}]$.

Thus the integral is given by

$$-\left(\int_{\mathrm{P} - \underline{h}[\underline{\xi}]} \ln\left[f_{\underline{v}}(\underline{\upsilon}) \right] \cdot f_{\underline{v}}(\underline{\upsilon}) \cdot d\underline{\upsilon} \right) \tag{14.36}$$

A remark on the range of the integration: The range of the integration, which would be necessary to make this integral a real Shannon information of the measurement noise would be Y = realization space of the measurement errors. But we have the range P–$\underline{h}[\xi]$ and this requires another examination of the observation equation, which we see in figure 14.2 .

$$\underline{y} = \underline{h}(\underline{x}) + \underline{v} \tag{14.37}$$

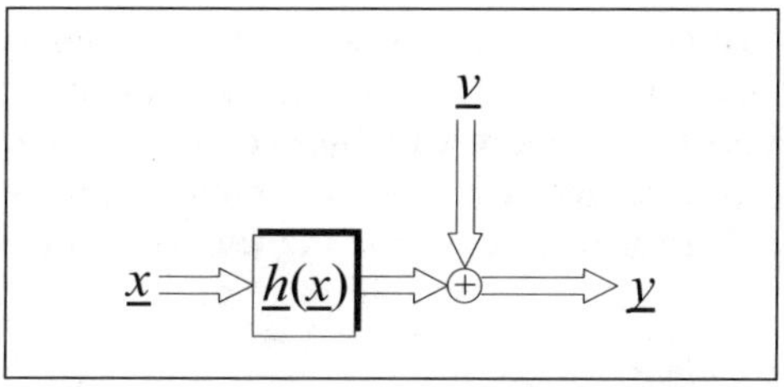

Fig. 14.2: Nonlinear observation equation

If we shift the realization space of the output variable $\underline{y}$ with the deterministic input $\underline{h}[\xi]$, we again arrive at the realization space of the measurement errors, described by the random variable $\underline{v}$. And this connection holds for arbitrary $\underline{h}[\xi]$. We therefore calculate the integral

$$-\left(\int_{\mathrm{P}-\underline{h}[\xi]} \ln\left[f_{\underline{v}}(\underline{\upsilon})\right] \cdot f_{\underline{v}}(\underline{\upsilon}) \cdot d\underline{\upsilon} \right) \tag{14.38a}$$

over the complete realization space Y of the measurement errors $\underline{v}$ and we thus write:

$$-\left(\int_{\mathrm{Y}} \ln\left[f_{\underline{v}}(\underline{\upsilon})\right] \cdot f_{\underline{v}}(\underline{\upsilon}) \cdot d\underline{\upsilon} \right) = S(\underline{v}) \tag{14.38}$$

This is the information of the measurement noise, or the uncertainty, which the measurement noise adds to the output. With this equation we return to the previous equation

$$S(\underline{x}, \underline{y}) = -\int_{\Xi} \left(\int_{\mathrm{P}} \ln\left[f_{\underline{v}}(\underline{\rho} - \underline{h}[\underline{\xi}])\right] \cdot f_{\underline{v}}(\underline{\rho} - \underline{h}[\underline{\xi}]) \cdot d\underline{\rho} \right) \cdot f_{\underline{x}}(\underline{\xi}) \cdot d\underline{\xi} + S(\underline{x}) \tag{14.39}$$

and we obtain

$$S(\underline{x},\underline{y}) = \int_{\Xi} S(\underline{v}) \cdot f_{\underline{x}}(\underline{\xi}) \cdot d\underline{\xi} + S(\underline{x})$$

$$S(\underline{x},\underline{y}) = S(\underline{v}) \cdot \int_{\Xi} f_{\underline{x}}(\underline{\xi}) \cdot d\underline{\xi} + S(\underline{x})$$

$$S(\underline{x},\underline{y}) = S(\underline{v}) + S(\underline{x}) \tag{14.40}$$

This solution can be verified intuitively, because the uncertainty that we find in the random variables $\underline{x}$ and $\underline{y}$ is obviously determined by the two random variables $\underline{x}$ and $\underline{v}$. As soon as these two random variables (their distribution densities) are determined, the statistical description of $\underline{y}$ is also found. Thus the information of these two parameters is sufficient to describe the joint information function of $\underline{x}$ and $\underline{y}$.

This leads us to two questions:

1. The first question is about the information of $\underline{v}$, i.e. $S(\underline{v})$.
2. The second question is about the nonlinear mapping $\underline{h}[\underline{x}]$, which does not occur in this joint information any more, though we know that Shannon's information depends on the transformations of the random variables. How can this be explained?

In the whole computation the mapping $\underline{h}[\underline{x}]$ occurs only once in the conditional distribution density $f_{\underline{y}|\underline{x}}(\underline{\rho} \mid \underline{\xi}) \cdot d\underline{\rho} = f_{\underline{v}}(\underline{\rho} - \underline{h}[\underline{\xi}]) \cdot d\underline{\rho}$. Otherwise this mapping is irrelevant, i.e. the deterministic transformation of the $\underline{x}$ information into the $\underline{y}$ coordinate system is not taken into account in this computation. Why? The reason is that this function merely provides a shift of the range of the integral.

$$-\left(\int_{\mathrm{P}-\underline{h}[\underline{\xi}]} \ln\left[f_{\underline{v}}(\underline{\upsilon})\right] \cdot f_{\underline{v}}(\underline{\upsilon}) \cdot d\underline{\upsilon} \right) \tag{14.41}$$

The deterministic function ensures that the range of the integration is equal to the range of the realization space of the measurement error random variable. It is, however, possible to change our approach slightly, which enables us to include the mapping in our considerations. Thus we introduce an additional parameter $\underline{z} = \underline{h}[\underline{x}]$ and we obtain the following observation equation, which is presented in figure 14.3.

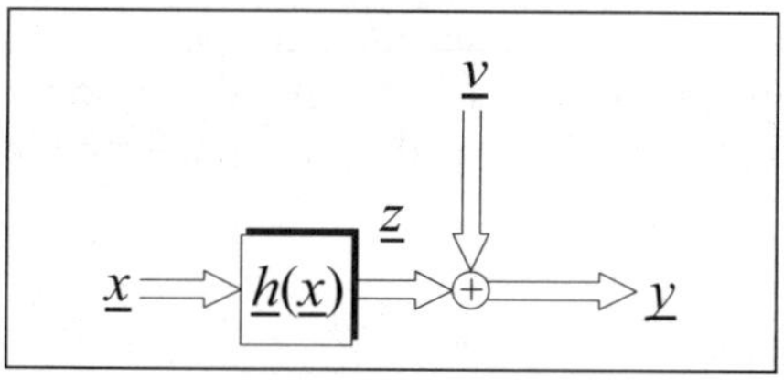

Fig. 14.3: Nonlinear observation equation

The related observation equation is

$$\underline{y} = \underline{h}(\underline{x}) + \underline{v} = \underline{z} + \underline{v} \tag{14.42}$$

where $\underline{z}$ and $\underline{v}$ are independent.

The joint distribution density of both variables is thus given by the equation

$$f_{\underline{v},\underline{z}}(\underline{\upsilon},\underline{\zeta}) = f_{\underline{v}}(\underline{\upsilon}) \cdot f_{\underline{z}}(\underline{\zeta}) \tag{14.43}$$

and it is thus equal to the product of the marginal distributions. For the information measures this leads to the following equation

$$S(\underline{v},\underline{z}) = S(\underline{v}) + S(\underline{z}) \tag{14.44}$$

The information of independent random variables simply adds. Now we proceed:

$$f_{\underline{y}}(\underline{\rho}) = f_{\underline{v}}(\underline{\rho}) * f_{\underline{z}}(\underline{\rho}) \tag{14.45}$$

$$f_{\underline{y}}(\underline{\rho}) = \int_{Z} f_{\underline{y},\underline{z}}(\underline{\rho},\underline{\zeta}) \cdot d\underline{\zeta} = \int_{Z} f_{\underline{y}|\underline{z}}(\underline{\rho} \mid \underline{\zeta}) \cdot f_{\underline{z}}(\underline{\zeta}) \cdot d\underline{\zeta} \tag{14.46}$$

and

$$S(\underline{y},\underline{z}) = -\int_{P}\int_{Z} \ln f_{\underline{y},\underline{z}}(\underline{\rho},\underline{\zeta}) \cdot f_{\underline{y},\underline{z}}(\underline{\rho},\underline{\zeta}) \cdot d\underline{\zeta} \cdot d\underline{\rho}$$

$$S(\underline{y},\underline{z}) = -\int_{P}\int_{Z} \ln f_{\underline{y}|\underline{z}}(\underline{\rho} \mid \underline{\zeta}) \cdot f_{\underline{y},\underline{z}}(\underline{\rho},\underline{\zeta}) \cdot d\underline{\zeta} \cdot d\underline{\rho} - \int_{Z} \ln f_{\underline{z}}(\underline{\zeta}) \cdot f_{\underline{z}}(\underline{\zeta}) \cdot d\underline{\zeta}$$

$$S(\underline{y},\underline{z}) = S(\underline{y} \mid \underline{z}) + S(\underline{z}) \tag{14.47}$$

and for this constellation we previously found

$$S(\underline{y},\underline{z}) = S(\underline{v}) + S(\underline{z}) \tag{14.48}$$

as the distribution density of $\underline{v}$ is only shifted. This shift leads to an equality of the range of the integral and the realization space of the measurement noise random variable, so that we obtain the result shown. As the random variable $\underline{z}$ is a nonlinear function of the random variable $\underline{x}$, they are connected by the equation

$$\underline{z} = \underline{h}(\underline{x}) \tag{14.49}$$

and for the distribution densities we obtain:

$$f_{\underline{z}}(\underline{\zeta}) \cdot d\underline{\zeta} = f_{\underline{x}}(\underline{\xi}) \cdot d\underline{\xi}$$

$$f_{\underline{z}}(\underline{\zeta}) = f_{\underline{x}}(\underline{\xi}) \cdot \left\| \frac{\partial \underline{\xi}}{\partial \underline{\zeta}} \right\| \tag{14.50}$$

Inserting

$$\underline{x} = \underline{h}^{-1}(\underline{z}) \quad \text{and} \quad \underline{\xi} = \underline{h}^{-1}(\underline{\zeta}) \tag{14.51}$$

leads to:

$$f_{\underline{z}}(\underline{\zeta}) = f_{\underline{x}}\left(\underline{h}^{-1}(\underline{\zeta})\right) \cdot \left\| \frac{\partial}{\partial \underline{\zeta}} \underline{h}^{-1}(\underline{\zeta}) \right\| \tag{14.52}$$

Thus

$$S(\underline{z}) = -\int_{Z} \ln f_{\underline{z}}(\underline{\zeta}) \cdot f_{\underline{z}}(\underline{\zeta}) \cdot d\underline{\zeta}$$

$$S(\underline{z}) = -\int_{Z} \ln\left[f_{\underline{x}}\left(\underline{h}^{-1}(\underline{\zeta})\right) \cdot \left\| \frac{\partial}{\partial \underline{\zeta}} \underline{h}^{-1}(\underline{\zeta}) \right\| \right] \cdot f_{\underline{x}}\left(\underline{h}^{-1}(\underline{\zeta})\right) \cdot \left\| \frac{\partial}{\partial \underline{\zeta}} \underline{h}^{-1}(\underline{\zeta}) \right\| \cdot d\underline{\zeta}$$

$$S(\underline{z}) = -\int_{Z} \left[\ln f_{\underline{x}}\left(\underline{h}^{-1}(\underline{\zeta})\right) + \ln \left\| \frac{\partial}{\partial \underline{\zeta}} \underline{h}^{-1}(\underline{\zeta}) \right\| \right] \cdot f_{\underline{x}}\left(\underline{h}^{-1}(\underline{\zeta})\right) \cdot \left\| \frac{\partial}{\partial \underline{\zeta}} \underline{h}^{-1}(\underline{\zeta}) \right\| \cdot d\underline{\zeta}$$

$$\begin{aligned} S(\underline{z}) = & -\int_{Z} \ln f_{\underline{x}}\left(\underline{h}^{-1}(\underline{\zeta})\right) \cdot f_{\underline{x}}\left(\underline{h}^{-1}(\underline{\zeta})\right) \cdot \left\| \frac{\partial}{\partial \underline{\zeta}} \underline{h}^{-1}(\underline{\zeta}) \right\| \cdot d\underline{\zeta} \\ & -\int_{Z} \ln \left\| \frac{\partial}{\partial \underline{\zeta}} \underline{h}^{-1}(\underline{\zeta}) \right\| \cdot f_{\underline{x}}\left(\underline{h}^{-1}(\underline{\zeta})\right) \cdot \left\| \frac{\partial}{\partial \underline{\zeta}} \underline{h}^{-1}(\underline{\zeta}) \right\| \cdot d\underline{\zeta} \end{aligned} \tag{14.53}$$

Substitution:

$$\underline{\xi} = \underline{h}^{-1}(\underline{\zeta}) \qquad d\underline{\xi} = d\underline{h}^{-1}(\underline{\zeta}) = \left\| \frac{\partial \underline{h}^{-1}(\underline{\zeta})}{\partial \underline{\zeta}} \right\| \cdot d\underline{\zeta} \tag{14.54}$$

$$S(\underline{z}) = -\int_{\Xi} \ln f_{\underline{x}}(\underline{\xi}) \cdot f_{\underline{x}}(\underline{\xi}) \cdot d\underline{\xi} - \int_{\Xi} \ln \left\| \frac{\partial \underline{\xi}}{\partial \underline{h}(\underline{\xi})} \right\| \cdot f_{\underline{x}}(\underline{\xi}) \cdot d\underline{\xi}$$

$$S(\underline{z}) = S(\underline{x}) - \int_{\Xi} \ln \left\| \frac{\partial \underline{h}(\underline{\xi})}{\partial \underline{\xi}} \right\|^{-1} \cdot f_{\underline{x}}(\underline{\xi}) \cdot d\underline{\xi} \tag{14.55}$$

and thus we have calculated the complete information in the observation equation of the state space representation.

We summarize:

The transformation, shown in figure 14.4, from the state space to the observation space

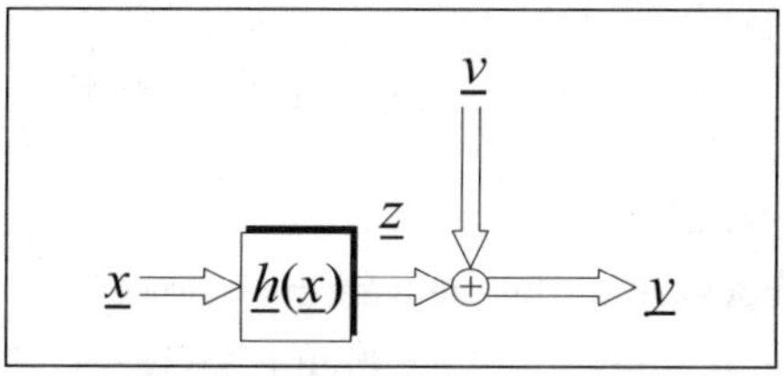

Fig. 14.4: Nonlinear observation equation

with the observation equations

$$\underline{y} = \underline{h}(\underline{x}) + \underline{v} = \underline{z} + \underline{v} \tag{14.56}$$

provides the information

$$S(\underline{y}, \underline{z}) = S(\underline{v}) + S(\underline{z}) \tag{14.57}$$

$$S(\underline{y}, \underline{z}) = S(\underline{v}) + S(\underline{x}) - \int_{\Xi} \ln \left\| \frac{\partial \underline{h}(\underline{\xi})}{\partial \underline{\xi}} \right\|^{-1} \cdot f_{\underline{x}}(\underline{\xi}) \cdot d\underline{\xi} \tag{14.58}$$

$$S(\underline{x}, \underline{y}) = S(\underline{v}) + S(\underline{x}) \tag{14.59}$$

$$S(\underline{x},\underline{y}) = S(\underline{y},\underline{z}) + \int_{\Xi} \ln \left\| \frac{\partial \underline{h}(\underline{\xi})}{\partial \underline{\xi}} \right\|^{-1} \cdot f_{\underline{x}}(\underline{\xi}) \cdot d\underline{\xi}$$

or

$$S(\underline{y},\underline{z}) = S(\underline{y} \mid \underline{z}) + S(\underline{z}) \tag{14.60}$$

In general we get:

$$S(\underline{y},\underline{x}) = S(\underline{y} \mid \underline{x}) + S(\underline{x}) \quad \text{or} \quad S(\underline{y},\underline{x}) = S(\underline{v}) + S(\underline{x}) \tag{14.61}$$

$$S(\underline{x},\underline{y}) = S(\underline{x} \mid \underline{y}) + S(\underline{y}) \tag{14.62}$$

$$S(\underline{y}) = S(\underline{x},\underline{y}) - S(\underline{x} \mid \underline{y}) \tag{14.63}$$

$$S(\underline{y}) = S(\underline{y} \mid \underline{x}) - S(\underline{x} \mid \underline{y}) + S(\underline{x}) \quad \text{or} \quad S(\underline{y}) = S(\underline{v}) - S(\underline{x} \mid \underline{y}) + S(\underline{x}) \tag{14.64}$$

If we examine the equation

$$S(\underline{y}) = S(\underline{v}) - S(\underline{x} \mid \underline{y}) + S(\underline{x}) \tag{14.65}$$

the uncertainty consists of three terms:

$S(\underline{x})$	Uncertainty at the input
$S(\underline{v})$	Uncertainty because of the measurement noise
$S(\underline{x} \mid \underline{y})$	Uncertainty at the input, if the output is known.

If the adressee receives a message at the output of the channel, then these three entropies influence the information which he is able to gain from this message. $S(\underline{x})$ and $S(\underline{v})$ increase the entropy at the output, while $S(\underline{x} \mid \underline{y})$ as equivocation decreases the entropy at the output. $S(\underline{x} \mid \underline{y})$ describes the uncertainty that remains when we receive the message. $I(\underline{x},\underline{y}) = S(\underline{x}) - S(\underline{x} \mid \underline{y})$, the transinformation or the mutual information, is the uncertainty that is eliminated by the receipt of the message. This 'deterministic' information, gained by the reception of a message, depends on the mutual information of both random variables $\underline{x}$ and $\underline{y}$ and this satisfies the already mentioned necessity of a correlation between emitter and receiver of the message.

$S(\underline{y}) = S(\underline{v}) - S(\underline{x} \mid \underline{y}) + S(\underline{x})$ is thus the uncertainty that can be eliminated by the reception of the message. Theoretically we are able to remove all uncertainty of the noise process, but only a part of the uncertainty of the real signal. This restriction is caused by $S(\underline{x} \mid \underline{y})$, because this term indicates that we obtain a remaining uncertainty, though we receive a certain message.

The two terms $S(\underline{x})$ and $S(\underline{v})$ ensure that there is an uncertainty at the output of the channel. But the effects of these entropies differ from each other. While $S(\underline{v})$ generates a pure stochastic uncertainty, caused by an additive noise process, the input uncertainty $S(\underline{x})$ is more an 'intended' uncertainty, which is transmitted to the receiver, who interprets this uncertain (new) part of the message as the information or the substance of the message. Only this part contains the new information, the surprising part of the message, which we do not know a priori. $S(\underline{x} \mid \underline{y})$ describes the uncertainty that remains after we have received the message. It thus is the part of the message that we are not able to classify.

$S(\underline{x})$, the message, and $S(\underline{v})$, the disturbance, of equation (14.64) ensure that there is an uncertainty at the output of the channel. Though both terms increase the uncertainty of the output, the effects of these two components are completely different. If $S(\underline{x})$ does not contain any uncertainty ($S(\underline{x}) = 0$), then the output is only generated by the noise process. Thus the uncertainty at the input has to be generated in some way, to increase the uncertainty at the output. Such an uncertainty at the input is, however, not to be generated by a deterministic function, but it is generated by sending certain events with certain probabilities. Only in this case is an uncertainty $S(\underline{x})$ achieved and thus we have information (a new part, a surprising part), which we transmit to the receiver. Moreover, $S(\underline{x} \mid \underline{y})$, the remaining uncertainty, can only come into existence if we have a nonzero input entropy $S(\underline{x})$.

If we do not have any uncertainty in the input signal $\underline{x}$, then we are able to determine this value with an arbitrary accuracy via repeated measurement of this deterministic value. This repeated measurement reduces the uncertainty added by the measurement noise process, and it allows us to approach the desired value $\underline{x}$ according to the $\sqrt{n}$ law (i.e. the standard deviation in the estimation of $\underline{x}$ decreases with a factor $1/\sqrt{n}$). So we are able to eliminate all the uncertainty of the additional measurement noise process $\underline{n}$.

This consideration shows the connection between the information and the surprising part of the message, which is important for the information that the message contains. If are able to reduce a message to this surprising part, then this message is built by an uncorrelated (unpredictable) sequence of signals. If we are able to gain all information out of the received sequence, the remaining uncertainty $S(\underline{x} \mid \underline{y}) = 0$ vanishes and we obtain a maximum amount of information, which is equal to

$$S(\underline{y}) = S(\underline{v}) + S(\underline{x}) \tag{14.66}$$

The basic idea of the innovation model, which uses an uncorrelated sequence (white noise) to generate measurement data, is to extract all the information about a signal from the completely uncertain sequence $S(\underline{x})$. Such an innovation sequence can, for instance, be found in the Kalman filter, where it is used to correct the predicted estimates. The information contained in the model of the Kalman filter and the information of the innovation sequence provide optimum state space estimates [LOF90] [MAY79] [ARN97/1]. As we are only able to describe the additive random process via stochastic parameters, the output signal consists of a deterministic part and an additional random process, contaminating the deterministic signal. If the estimated deterministic signal is optimally determined, the additive noise process is white. Such a white noise process does not have any preferred tendencies, which confirms that the estimation is optimal in the mean.

As we see in this consideration, it is important that there exists a correlation between emitter and receiver of a message, which enables the receiver to classify the information and to use the information (pragmatic information). Especially in the innovation approach we need some knowledge about the model of the process that we are observing, because we only get the innovations, the completely new and unknown part of the message, to determine the desired deterministic behavior of some model. Thus the innovations approach relies on the a priori knowledge of the model, which is merely 'adapted' by the incoming information.

We again interpret information according to Weizsäcker, i.e. information is the future reduction of the actual existing uncertainty. The more uncertainty the output of a system generates, the more degrees of freedom we get to make our decisions. The receipt of a concrete message reduces this uncertainty and generates the decision regarding a certain message. The measures of entropy, which we used in the description of the transmission, can so also be interpreted as measures of the degrees of freedom of the message space, where all possible messages are mapped.

14.3 Transmission faster than light

Regarding the discussion about the transmission of signals with velocities greater than the velocity of light, the transmission of information can clearly be limited to the velocity of light. If a system transmits information, this information – as we saw in the last chapter – has to be generated from the emitter by a 'production of uncertainty' or by a permanent creation of surprising messages. These permanent surprises, which we denote as the causal input signal $s(t)$ of such a transmission process, are then transmitted by the system, which is described by the impulse response $h(t)$. The output of the system can thus be denoted by the convolution integral and we obtain

$$g(t) = \int_{-\infty}^{\infty} s(\tau) \cdot h(t-\tau) \cdot d\tau \tag{14.67}$$

Because of the causality of the input signal we write

$$s(t) = 0 \qquad \text{for} \qquad t < 0 \tag{14.68}$$

and thus the lower limit of the integral is reduced, so that we now integrate over the range 0 to ∞. Moreover the impulse response $h(t)$ is causal, too

$$h(t) = 0 \qquad \text{for} \qquad t < 0 \tag{14.69}$$

because the impulse response is the reaction of a system to the Dirac impulse. If $h(t)$ were not to vanish for $t < 0$, the system would be able to anticipate the Dirac impulse, and thus it would be able to look into the future and to predict even surprising events. Such systems do not exist in reality and therefore the convolution integral becomes

$$g(t) = \int_{0}^{t} s(\tau) \cdot h(t-\tau) \cdot d\tau \tag{14.70}$$

Each new part of the input signal, each surprising part of $s(t)$, can thus at the earliest be detected at the output after some delay, caused by the system (speed of light as an upper limit). Only deterministic parts of the input signal $s(t)$ can be predicted at the output, if the receiver knows the input signal. But such deterministic parts of the input signal do not contain any information, because information lies in the changes in the signal and all deterministic parts of the signal are merely not interesting and irrelevant. The prediction of such deterministic signals, which artificially increase the speed of the transmission, is not suitable for the transmission of information, which is the actual aim of the transmission of messages.

14.4 Information about state space variables

We have two gaussian distributed random variables:

$$f_x(\xi) = \frac{1}{\sqrt{2\pi} \cdot \sigma_x} \cdot \exp\left\{-\frac{1}{2} \cdot \frac{\xi^2}{\sigma_x^2}\right\} \tag{14.71}$$

$$f_y(\rho) = \frac{1}{\sqrt{2\pi} \cdot \sigma_y} \cdot \exp\left\{-\frac{1}{2} \cdot \frac{\rho^2}{\sigma_y^2}\right\} \tag{14.72}$$

We do not consider mean values, as the information does not depend on the means but on the variances (uncertainties) of the random variables.

Our question is: How much information does the random variable $z = x + y$ contain about the random variable x? Or: How much information does the measurement $z = x + y$ contain about a desired state space random variable x, if y is an additive noise process?

The joint information $S(x,y) = S(x) + S(y)$ is equal to the sum of the marginal amounts of information, as we assume these random variables to be independent. The random variables z and x are not independent and thus there has to be some connection between the information of these variables. The distribution density of the summed random variable can be found by a short computation

$$f_z(\zeta) = \frac{1}{\sqrt{2\pi} \cdot \sigma_z} \cdot \exp\left\{ -\frac{1}{2} \cdot \frac{\zeta^2}{\sigma_z^2} \right\} \qquad \text{with} \qquad \sigma_z^2 = \sigma_x^2 + \sigma_y^2 \tag{14.73}$$

and:

$$f_{z|x}(\zeta \mid \xi) = f_y(\zeta - \xi) = \frac{1}{\sqrt{2\pi} \cdot \sigma_y} \cdot \exp\left\{ -\frac{1}{2} \cdot \frac{(\zeta - \xi)^2}{\sigma_y^2} \right\} \tag{14.74}$$

We thus directly compute the joint distribution density

$$f_{z,x}(\zeta, \xi) = f_{z|x}(\zeta \mid \xi) \cdot f_x(\xi) = f_y(\zeta - \xi) \cdot f_x(\xi) \tag{14.75}$$

and the conditional distribution density is given by the equation

$$f_{x|z}(\xi \mid \zeta) = \frac{f_{x,z}(\xi, \zeta)}{f_z(\zeta)} = \frac{f_y(\zeta - \xi) \cdot f_x(\xi)}{\int\limits_{-\infty}^{\infty} f_y(\zeta - \xi) \cdot f_x(\xi) \cdot d\xi} \tag{14.76}$$

This is exactly the formulation of the Bayesian estimation, which Jazwinski calls the filter equation. It is the result of the filter problem, where the state space variable has to be determined conditioned on the measurement variable, if state space variable and measurement noise are added. This exactly is our approach.

The conditional Shannon information, the equivocation of the conditional distribution density, is equal to

$$S(x \mid z) = \int\limits_{-\infty}^{\infty} \int\limits_{-\infty}^{\infty} \ln f_{x|z}(\xi \mid \zeta) \cdot f_{x,z}(\xi, \zeta) \cdot d\xi \cdot d\zeta$$

$$S(x \mid z)=\int_{-\infty}^{\infty}\int_{-\infty}^{\infty}\ln f_{x,z}(\xi,\zeta)\cdot f_{x,z}(\xi,\zeta)\cdot d\xi\cdot d\zeta-\int_{-\infty}^{\infty}\ln f_z(\zeta)\cdot f_z(\zeta)\cdot d\zeta$$

$$S(x \mid z)=S(x,z)-S(z) \tag{14.77}$$

Now we introduce the condition that the measurement is bounded to the interval $z\in[a,b]$, which reduces the information in the measurement variable because of this limitation to a limited interval (which may be caused by a limited area, where some vehicle is able to move, ...).

This is exactly the preliminary that Röhler, Sokolnikoff and Brillouin use in their definitions of the gain of information. The gain of information is the result of a reduction of the width of an interval. Thus the information of the measurement reduces the uncertainty of the state space random variable, which is the desired parameter in our system. The transmission of information can be described with the Fisher information (Discriminator, section 11.3.2.1). But how does the equivocation change?

The bounded interval reduces the integration range of z

$$S(x \mid z)=\int_{a}^{b}\int_{-\infty}^{\infty}\ln f_{x,z}(\xi,\zeta)\cdot f_{x,z}(\xi,\zeta)\cdot d\xi\cdot d\zeta-\int_{a}^{b}\ln f_z(\zeta)\cdot f_z(\zeta)\cdot d\zeta \tag{14.78}$$

and now we need the explicit formulation of the joint distribution density function

$$f_{z,x}(\zeta,\xi)=f_{z|x}(\zeta \mid \xi)\cdot f_x(\xi)=f_y(\zeta-\xi)\cdot f_x(\xi)$$

$$f_{z,x}(\zeta,\xi)=\frac{1}{\sqrt{2\pi}\cdot\sigma_y}\cdot\exp\left\{-\frac{1}{2}\cdot\frac{(\zeta-\xi)^2}{\sigma_y^2}\right\}\cdot\frac{1}{\sqrt{2\pi}\cdot\sigma_x}\cdot\exp\left\{-\frac{1}{2}\cdot\frac{\xi^2}{\sigma_x^2}\right\}$$

$$f_{z,x}(\zeta,\xi)=\frac{1}{2\pi\cdot\sigma_y\cdot\sigma_x}\cdot\exp\left\{-\frac{1}{2}\cdot\left[\frac{(\zeta-\xi)^2}{\sigma_y^2}+\frac{\xi^2}{\sigma_x^2}\right]\right\} \tag{14.79}$$

to use this in the calculation of the integral. Let us begin with Shannon's information of the summation random variable

$$S(z)=\int_{a}^{b}\ln f_z(\zeta)\cdot f_z(\zeta)\cdot d\zeta$$

$$S(z)=\int_{a}^{b}\ln\left[\frac{1}{\sqrt{2\pi}\cdot\sigma_z}\cdot\exp\left\{-\frac{1}{2}\cdot\left[\frac{\zeta^2}{\sigma_z^2}\right]\right\}\right]\cdot f_z(\zeta)\cdot d\zeta$$

$$S(z)=\int_a^b\left[-\frac{1}{2}\cdot\ln\left(2\pi\cdot\sigma_z^2\right)-\frac{1}{2}\cdot\left(\frac{\zeta^2}{\sigma_z^2}\right)\right]\cdot f_z(\zeta)\cdot d\zeta$$

$$S(z)=-\frac{1}{2}\cdot\ln\left(2\pi\cdot\sigma_z^2\right)\cdot\int_a^b f_z(\zeta)\cdot d\zeta-\frac{1}{2\cdot\sigma_z^2}\cdot\int_a^b\zeta^2\cdot f_z(\zeta)\cdot d\zeta \tag{14.80}$$

We now occupy ourselves with the first integral on the right-hand side of this equation, which is

$$\int_a^b f_z(\zeta)\cdot d\zeta=\int_{-\infty}^b f_z(\zeta)\cdot d\zeta-\int_{-\infty}^a f_z(\zeta)\cdot d\zeta \tag{14.81}$$

and to solve this integrals we use an arbitrary upper limit of the integration

$$\int_{-\infty}^x f_z(\zeta)\cdot d\zeta=\int_{-\infty}^x\frac{1}{\sqrt{2\pi}\cdot\sigma_z}\cdot\exp\left\{-\frac{1}{2}\cdot\frac{\zeta^2}{\sigma_z^2}\right\}\cdot d\zeta$$

$$=\frac{1}{\sqrt{2\pi}\cdot\sigma_z}\cdot\int_{-\infty}^x\exp\left\{-\frac{1}{2}\cdot\frac{\zeta^2}{\sigma_z^2}\right\}\cdot d\zeta \tag{14.82}$$

Now we substitute

$$\varphi=\frac{\zeta}{\sigma_z}\qquad\qquad \sigma_z\cdot d\varphi=d\zeta \tag{14.83}$$

The limits: $\zeta=-\infty\rightarrow\varphi=-\infty$

$\zeta=x\ \rightarrow\varphi=x\cdot\sigma_z$

and the integral is

$$\int_{-\infty}^x f_z(\zeta)\cdot d\zeta=\frac{1}{\sqrt{2\pi}\cdot\sigma_z}\cdot\int_{-\infty}^x\exp\left\{-\frac{1}{2}\cdot\frac{\zeta^2}{\sigma_z^2}\right\}\cdot d\zeta \tag{14.84}$$

$$=\frac{1}{\sqrt{2\pi}\cdot\sigma_z}\cdot\int_{-\infty}^{x\cdot\sigma_z}\exp\left\{-\frac{1}{2}\cdot\varphi^2\right\}\cdot\sigma_z\cdot d\varphi$$

$$=\frac{1}{\sqrt{2\pi}}\cdot\int_{-\infty}^{x\cdot\sigma_z}\exp\left\{-\frac{1}{2}\cdot\varphi^2\right\}\cdot d\varphi$$

$$\int_{-\infty}^{x} f_z(\zeta)\cdot d\zeta = \frac{1}{\sqrt{2\pi}}\cdot\Phi(x\cdot\sigma_z) \tag{14.85}$$

with

$$\Phi(x\cdot\sigma_z) = \int_{-\infty}^{x\cdot\sigma_z} \exp\left\{-\frac{1}{2}\cdot\varphi^2\right\}\cdot d\varphi \tag{14.86}$$

With this solution we obtain the desired solution of the integral

$$\int_{a}^{b} f_z(\zeta)\cdot d\zeta = \int_{-\infty}^{b} f_z(\zeta)\cdot d\zeta - \int_{-\infty}^{a} f_z(\zeta)\cdot d\zeta \tag{14.87}$$

and for the limited integration range we find

$$\int_{a}^{b} f_z(\zeta)\cdot d\zeta = \frac{1}{\sqrt{2\pi}}\cdot[\Phi(b\cdot\sigma_z)-\Phi(a\cdot\sigma_z)] \quad \text{with} \quad \sigma_z = \sqrt{\sigma_x^2+\sigma_y^2} \tag{14.88}$$

For Shannon's information of the limited integration range we find

$$S(z) = -\frac{1}{2}\cdot\ln\left(2\pi\cdot\sigma_z^2\right)\cdot\int_{a}^{b} f_z(\zeta)\cdot d\zeta - \frac{1}{2\cdot\sigma_z^2}\cdot\int_{a}^{b}\zeta^2\cdot f_z(\zeta)\cdot d\zeta$$

$$S(z) = -\frac{1}{2\cdot\sqrt{2\pi}}\cdot\ln\left(2\pi\cdot\sigma_z^2\right)\cdot[\Phi(b\cdot\sigma_z)-\Phi(a\cdot\sigma_z)] - \frac{1}{2\cdot\sigma_z^2}\cdot\int_{a}^{b}\zeta^2\cdot f_z(\zeta)\cdot d\zeta \tag{14.89}$$

The second integral remains

$$\int_{a}^{b}\zeta^2\cdot f_z(\zeta)\cdot d\zeta = \int_{a}^{b}\zeta^2\cdot\frac{1}{\sqrt{2\pi}\cdot\sigma_z}\cdot\exp\left\{-\frac{1}{2}\cdot\frac{\zeta^2}{\sigma_z^2}\right\}\cdot d\zeta$$

$$= \frac{1}{\sqrt{2\pi}\cdot\sigma_z}\cdot\int_{a}^{b}\zeta^2\cdot\exp\left\{-\frac{1}{2}\cdot\frac{\zeta^2}{\sigma_z^2}\right\}\cdot d\zeta \tag{14.90}$$

We substitute

$$\varphi = \frac{\zeta}{\sigma_z} \qquad \sigma_z\cdot\varphi = \zeta \qquad \sigma_z\cdot d\varphi = d\zeta \tag{14.91}$$

The limits:

$$\zeta = a \rightarrow \varphi = a \cdot \sigma_z$$

$$\zeta = b \rightarrow \varphi = b \cdot \sigma_z$$

And thus the integral is

$$\int_a^b \zeta^2 \cdot f_z(\zeta) \cdot d\zeta = \frac{1}{\sqrt{2\pi} \cdot \sigma_z} \cdot \int_{a \cdot \sigma_z}^{b \cdot \sigma_z} \sigma_z^2 \cdot \varphi^2 \cdot \exp\left\{-\frac{1}{2} \cdot \varphi^2\right\} \cdot \sigma_z \cdot d\varphi$$

$$= \frac{\sigma_z^2}{\sqrt{2\pi}} \cdot \int_{a \cdot \sigma_z}^{b \cdot \sigma_z} \varphi^2 \cdot \exp\left\{-\frac{1}{2} \cdot \varphi^2\right\} \cdot d\varphi \tag{14.92}$$

To solve this integral we examine the integral

$$\int_a^b \varphi^2 \cdot e^{-\frac{\varphi^2}{2}} \cdot d\varphi \tag{14.93}$$

which can be achieved via substitution.

$$e^{-\frac{\varphi^2}{2}} = \sum_{n=0}^{\infty} \frac{1}{n!} \cdot \left(-\frac{\varphi^2}{2}\right)^2$$

$$= \sum_{k=0}^{\infty} \frac{1}{(2k)!} \cdot \left(\frac{\varphi^2}{2}\right)^{2k} - \sum_{k=0}^{\infty} \frac{1}{(2k+1)!} \cdot \left(\frac{\varphi^2}{2}\right)^{2k+1} \tag{14.94}$$

With this substitution we get

$$\begin{aligned} \int_a^b \varphi^2 \cdot e^{-\frac{\varphi^2}{2}} \cdot d\varphi &= \int_a^b \varphi^2 \cdot \sum_{k=0}^{\infty} \frac{1}{(2k)!} \cdot \left(\frac{\varphi^2}{2}\right)^{2k} \cdot d\varphi \\ &\quad - \int_a^b \varphi^2 \cdot \sum_{k=0}^{\infty} \frac{1}{(2k+1)!} \cdot \left(\frac{\varphi^2}{2}\right)^{2k+1} \cdot d\varphi \end{aligned} \tag{14.95}$$

As the kernel exclusively consists of positive terms, we interchange summation and integration and obtain:

$$\int_a^b \varphi^2 \cdot e^{-\frac{\varphi^2}{2}} \cdot d\varphi = \sum_{k=0}^{\infty} \frac{1}{(2k)!} \cdot \int_a^b 2 \cdot \left(\frac{\varphi^2}{2}\right)^{2k+1} \cdot d\varphi - \sum_{k=0}^{\infty} \frac{1}{(2k+1)!} \cdot \int_a^b 2 \cdot \left(\frac{\varphi^2}{2}\right)^{2k+2} \cdot d\varphi \tag{14.96}$$

For the first integral on the right hand side we get:

$$\int_a^b 2 \cdot \left(\frac{\varphi^2}{2}\right)^{2k+1} \cdot d\varphi = 2 \cdot \left(\frac{1}{2}\right)^{2k+1} \cdot \int_a^b \varphi^{2\cdot(2k+1)} \cdot d\varphi$$

$$= \left(\frac{1}{2}\right)^{2k} \cdot \frac{1}{2\cdot(2k+1)+1} \cdot \left[\varphi^{2\cdot(2k+1)+1}\right]_a^b \tag{14.97}$$

Analogously we find for the remaining integral of equation (14.96)

$$\int_a^b 2 \cdot \left(\frac{\varphi^2}{2}\right)^{2k+2} \cdot d\varphi = 2 \cdot \left(\frac{1}{2}\right)^{2k+2} \cdot \int_a^b \varphi^{2\cdot(2k+2)} \cdot d\varphi$$

$$= \left(\frac{1}{2}\right)^{2k+1} \cdot \frac{1}{2\cdot(2k+2)+1} \cdot \left[\varphi^{2\cdot(2k+2)+1}\right]_a^b \tag{14.98}$$

Thus we obtain

$$\int_a^b \varphi^2 \cdot e^{-\frac{\varphi^2}{2}} \cdot d\varphi = \sum_{k=0}^{\infty} \frac{1}{(2k)!} \cdot \left(\frac{1}{2}\right)^{2k} \cdot \frac{1}{2\cdot(2k+1)+1} \cdot \left[\varphi^{2\cdot(2k+1)+1}\right]_a^b - \sum_{k=0}^{\infty} \frac{1}{(2k+1)!} \cdot \left(\frac{1}{2}\right)^{2k+1} \cdot \frac{1}{2\cdot(2k+2)+1} \cdot \left[\varphi^{2\cdot(2k+2)+1}\right]_a^b \tag{14.99}$$

and with summarized terms

$$\int_a^b \varphi^2 \cdot e^{-\frac{\varphi^2}{2}} \cdot d\varphi = \sum_{k=0}^{\infty} \frac{1}{(2k)!} \cdot \left(\frac{1}{2}\right)^{2k} \cdot \frac{1}{(4k+3)} \cdot \left[\varphi^{(4k+3)}\right]_a^b - \sum_{k=0}^{\infty} \frac{1}{(2k+1)!} \cdot \left(\frac{1}{2}\right)^{2k+1} \cdot \frac{1}{(4k+5)} \cdot \left[\varphi^{(4k+5)}\right]_a^b \tag{14.100}$$

To combine these two sums, we examine the exponents

$$4k+3 = 2\,(2k) + 3$$

$$4k+5 = 2\,(2k+1) + 3 \tag{14.101}$$

and write

$$\int_a^b \varphi^2 \cdot e^{-\frac{\varphi^2}{2}} \cdot d\varphi = \sum_{k=0}^{\infty} \frac{1}{(2k)!} \left(\frac{1}{2}\right)^{2k} \cdot \frac{1}{2\cdot(2k)+3} \cdot \left[\left(\varphi^2\right)^{2k} \cdot \varphi^3\right]_a^b$$
$$- \sum_{k=0}^{\infty} \frac{1}{(2k+1)!} \left(\frac{1}{2}\right)^{2k+1} \cdot \frac{1}{2\cdot(2k+1)+3} \cdot \left[\left(\varphi^2\right)^{2k+1} \cdot \varphi^3\right]_a^b \tag{14.102}$$

Now we are able to combine these two sums into one sum by introducing the summation variable n, which we replaced by $2k$ and $2k+1$ at the beginning of this calculation

$$\int_a^b \varphi^2 \cdot e^{-\frac{\varphi^2}{2}} \cdot d\varphi = \sum_{n=0}^{\infty} \frac{1}{n!} \left(\frac{1}{2}\right)^{n} \cdot \frac{1}{2n+3} \cdot \left[\left(-\varphi^2\right)^{n} \cdot \varphi^3\right]_a^b \tag{14.103}$$

and this is the solution of the integral, which we may also write as

$$\int_a^b \varphi^2 \cdot e^{-\frac{\varphi^2}{2}} \cdot d\varphi = \sum_{n=0}^{\infty} \frac{1}{n!} \left(\frac{1}{2}\right)^{n} \cdot \frac{1}{2n+3} \cdot \left[\left(-b^2\right)^{n} \cdot b^3 - \left(-a^2\right)^{n} \cdot a^3\right] \tag{14.104}$$

With our initial equation (14.89)

$$S(z) = -\frac{1}{2\cdot\sqrt{2\pi}} \cdot \ln\left(2\pi\cdot\sigma_z^2\right) \cdot \left[\Phi(b\cdot\sigma_z) - \Phi(a\cdot\sigma_z)\right] - \frac{1}{2\cdot\sigma_z^2} \cdot \int_a^b \zeta^2 \cdot f_z(\zeta) \cdot d\zeta \tag{14.105}$$

and with the substitution

$$S(z) = -\frac{1}{2\cdot\sqrt{2\pi}} \cdot \ln\left(2\pi\cdot\sigma_z^2\right) \cdot \left[\Phi(b\cdot\sigma_z) - \Phi(a\cdot\sigma_z)\right]$$
$$- \frac{1}{2\cdot\sigma_z^2} \cdot \frac{\sigma_z^2}{\sqrt{2\pi}} \cdot \int_{a\cdot\sigma_z}^{b\cdot\sigma_z} \varphi^2 \cdot \exp\left\{-\frac{1}{2}\cdot\varphi^2\right\} \cdot d\varphi$$

$$S(z) = -\frac{1}{2\cdot\sqrt{2\pi}} \cdot \left\{\ln\left(2\pi\cdot\sigma_z^2\right) \cdot \left[\Phi(b\cdot\sigma_z) - \Phi(a\cdot\sigma_z)\right] - \int_{a\cdot\sigma_z}^{b\cdot\sigma_z} \varphi^2 \cdot \exp\left\{-\frac{1}{2}\cdot\varphi^2\right\} \cdot d\varphi\right\} \tag{14.106}$$

we obtain

$$S(z)=-\frac{1}{2\cdot\sqrt{2\pi}}\cdot\left\{\ln\left(2\pi\cdot\sigma_z^2\right)\cdot\left[\Phi(b\cdot\sigma_z)-\Phi(a\cdot\sigma_z)\right]-\right.$$
$$\left.-\sum_{n=0}^{\infty}\frac{1}{n!}\cdot\left(\frac{1}{2}\right)^n\cdot\frac{1}{2n+3}\cdot\left[\left(-(b\cdot\sigma_z)^2\right)^n\cdot(b\cdot\sigma_z)^3-\left(-(a\cdot\sigma_z)^2\right)^n\cdot(a\cdot\sigma_z)^3\right]\right\} \tag{14.107}$$

To check this partial solution, we build the limit for $a \to -\infty$, $b \to \infty$ and we obtain

$$\lim_{\substack{a\to-\infty\\ b\to\infty}} S(z)=-\frac{1}{2\cdot\sqrt{2\pi}}\cdot\left\{\ln\left(2\pi\cdot\sigma_z^2\right)\cdot\left[0-\sqrt{2\pi}\right]-\int_{-\infty}^{\infty}\varphi^2\cdot\exp\left\{-\frac{1}{2}\cdot\varphi^2\right\}\cdot d\varphi\right\} \tag{14.108}$$

because of

$$\lim_{x\to\infty}\Phi(x)=\int_{-\infty}^{\infty}\exp\left\{-\frac{1}{2}\cdot\varphi^2\right\}\cdot d\varphi=\sqrt{2\pi} \tag{14.109}$$

The latter integral is an integral over a gaussian density with variance = 1, so that

$$\lim_{\substack{a\to-\infty\\ b\to\infty}} S(z)=-\frac{1}{2\cdot\sqrt{2\pi}}\cdot\left\{-\sqrt{2\pi}\cdot\ln\left(2\pi\cdot\sigma_z^2\right)-\sqrt{2\pi}\right\}$$

$$\lim_{\substack{a\to-\infty\\ b\to\infty}} S(z)=\frac{1}{2}\cdot\ln\left(2\pi\cdot\sigma_z^2\right)+\frac{1}{2} \tag{14.110}$$

and this is exactly the Shannon information of a scalar random variable z, which is described by a gaussian distribution with variance σ_z^2.

To compute completely the conditional entropy from equation (14.77), the calculation of the joint information remains. This is

$$S(x,z)=\int_{a}^{b}\int_{-\infty}^{\infty}\ln f_{x,z}(\xi,\zeta)\cdot f_{x,z}(\xi,\zeta)\cdot d\xi\cdot d\zeta \tag{14.111}$$

with

$$f_{z,x}(\zeta,\xi)=f_{z|x}(\zeta\mid\xi)\cdot f_x(\xi)=f_y(\zeta-\xi)\cdot f_x(\xi) \tag{14.112}$$

we get

$$S(x,z)=\int_{a}^{b}\int_{-\infty}^{\infty}\ln\left(f_y(\zeta-\xi)\cdot f_x(\xi)\right)\cdot f_y(\zeta-\xi)\cdot f_x(\xi)\cdot d\xi\cdot d\zeta$$

$$S(x,z)=\int_{a}^{b}\int_{-\infty}^{\infty}\ln f_y(\zeta-\xi)\cdot f_y(\zeta-\xi)\cdot f_x(\xi)\cdot d\xi\cdot d\zeta$$
$$+\int_{a}^{b}\int_{-\infty}^{\infty}\ln f_x(\xi)\cdot f_y(\zeta-\xi)\cdot f_x(\xi)\cdot d\xi\cdot d\zeta$$

$$S(x,z)=\int_{-\infty}^{\infty}\underbrace{\int_{a}^{b}\ln f_y(\zeta-\xi)\cdot f_y(\zeta-\xi)\cdot d\zeta}_{Int1}\cdot f_x(\xi)\cdot d\xi$$
$$+\int_{-\infty}^{\infty}\underbrace{\int_{a}^{b}f_y(\zeta-\xi)\cdot d\zeta}_{Int2}\cdot\ln f_x(\xi)\cdot f_x(\xi)\cdot d\xi \tag{14.113}$$

We start with the integral *Int*1 from equation (14.113)

$$\int_{a}^{b}\ln f_y(\zeta-\xi)\cdot f_y(\zeta-\xi)\cdot d\zeta$$

$$=\int_{a}^{b}\ln\left\{\frac{1}{\sqrt{2\pi}\cdot\sigma_y}\cdot\exp\left\{-\frac{1}{2}\cdot\frac{(\zeta-\xi)^2}{\sigma_y^2}\right\}\right\}\cdot\left\{\frac{1}{\sqrt{2\pi}\cdot\sigma_y}\cdot\exp\left\{-\frac{1}{2}\cdot\frac{(\zeta-\xi)^2}{\sigma_y^2}\right\}\right\}\cdot d\zeta$$

$$=-\frac{1}{2}\cdot\int_{a}^{b}\ln\left(2\pi\cdot\sigma_y^2\right)\cdot\left\{\frac{1}{\sqrt{2\pi}\cdot\sigma_y}\cdot\exp\left\{-\frac{1}{2}\cdot\frac{(\zeta-\xi)^2}{\sigma_y^2}\right\}\right\}\cdot d\zeta$$
$$-\frac{1}{2}\cdot\int_{a}^{b}\frac{(\zeta-\xi)^2}{\sigma_y^2}\cdot\left\{\frac{1}{\sqrt{2\pi}\cdot\sigma_y}\cdot\exp\left\{-\frac{1}{2}\cdot\frac{(\zeta-\xi)^2}{\sigma_y^2}\right\}\right\}\cdot d\zeta$$

$$= -\frac{1}{2} \cdot \frac{\ln\left(2\pi \cdot \sigma_y^2\right)}{\sqrt{2\pi} \cdot \sigma_y} \cdot \int_a^b \exp\left\{-\frac{1}{2} \cdot \frac{(\zeta - \xi)^2}{\sigma_y^2}\right\} \cdot d\zeta$$

$$-\frac{1}{2} \cdot \int_a^b \frac{(\zeta - \xi)^2}{\sigma_y^2} \cdot \frac{1}{\sqrt{2\pi} \cdot \sigma_y} \cdot \exp\left\{-\frac{1}{2} \cdot \frac{(\zeta - \xi)^2}{\sigma_y^2}\right\} \cdot d\zeta \tag{14.114}$$

Substitution:

$$\frac{\zeta - \xi}{\sigma_y} = x \qquad \zeta = x \cdot \sigma_y + \xi \qquad d\zeta = dx \cdot \sigma_y \tag{14.115}$$

The limits: $\zeta = a \rightarrow x = \dfrac{a - \xi}{\sigma_y}$

$\zeta = b \rightarrow x = \dfrac{b - \xi}{\sigma_y}$

and so

$$Int1 = -\frac{1}{2} \cdot \frac{\ln\left(2\pi \cdot \sigma_y^2\right)}{\sqrt{2\pi} \cdot \sigma_y} \cdot \int_{\frac{a-\xi}{\sigma_y}}^{\frac{b-\xi}{\sigma_y}} \exp\left\{-\frac{1}{2} \cdot x^2\right\} \cdot dx \cdot \sigma_y$$

$$-\frac{1}{2} \cdot \int_{\frac{a-\xi}{\sigma_y}}^{\frac{b-\xi}{\sigma_y}} x^2 \cdot \frac{1}{\sqrt{2\pi} \cdot \sigma_y} \cdot \exp\left\{-\frac{1}{2} \cdot x^2\right\} \cdot dx \cdot \sigma_y$$

$$= -\frac{\ln\left(2\pi \cdot \sigma_y^2\right)}{2 \cdot \sqrt{2\pi}} \cdot \int_{\frac{a-\xi}{\sigma_y}}^{\frac{b-\xi}{\sigma_y}} \exp\left\{-\frac{1}{2} \cdot x^2\right\} \cdot dx - \frac{1}{2 \cdot \sqrt{2\pi}} \cdot \int_{\frac{a-\xi}{\sigma_y}}^{\frac{b-\xi}{\sigma_y}} x^2 \cdot \exp\left\{-\frac{1}{2} \cdot x^2\right\} \cdot dx \tag{14.116}$$

This is not a final result, but we use this expression in the following examination to denote the term $Int1$.

This leads to

$$S(x,z)=-\frac{1}{2\cdot\sqrt{2\pi}}\cdot\int_{-\infty}^{\infty}\left[n\left(2\pi\cdot\sigma_y^2\right)\cdot\int_{\frac{a-\xi}{\sigma_y}}^{\frac{b-\xi}{\sigma_y}}\exp\left\{-\frac{1}{2}\cdot x^2\right\}\cdot dx\right.$$

$$\left.+\int_{\frac{a-\xi}{\sigma_y}}^{\frac{b-\xi}{\sigma_y}}x^2\cdot\exp\left\{-\frac{1}{2}\cdot x^2\right\}\cdot dx\right]\cdot f_x(\xi)\cdot d\xi$$

$$+\int_{-\infty}^{\infty}\underbrace{\int_a^b f_y(\zeta-\xi)\cdot d\zeta}_{Int2}\cdot\ln f_x(\xi)\cdot f_x(\xi)\cdot d\xi$$

$$S(x,z)=-\frac{1}{2\cdot\sqrt{2\pi}}\cdot\ln\left(2\pi\cdot\sigma_y^2\right)\cdot\int_{-\infty}^{\infty}\int_{\frac{a-\xi}{\sigma_y}}^{\frac{b-\xi}{\sigma_y}}\exp\left\{-\frac{1}{2}\cdot x^2\right\}\cdot dx\cdot f_x(\xi)\cdot d\xi$$

$$-\frac{1}{2\cdot\sqrt{2\pi}}\cdot\int_{-\infty}^{\infty}\int_{\frac{a-\xi}{\sigma_y}}^{\frac{b-\xi}{\sigma_y}}x^2\cdot\exp\left\{-\frac{1}{2}\cdot x^2\right\}\cdot dx\cdot f_x(\xi)\cdot d\xi \qquad (14.117)$$

$$+\int_{-\infty}^{\infty}\underbrace{\int_a^b f_y(\zeta-\xi)\cdot d\zeta}_{Int2}\cdot\ln f_x(\xi)\cdot f_x(\xi)\cdot d\xi$$

Now the calculation of *Int*2 from equation (14.113) remains

$$\int_a^b f_y(\zeta-\xi)\cdot d\zeta=\frac{1}{\sqrt{2\pi}\cdot\sigma_y}\cdot\int_a^b\exp\left\{-\frac{1}{2}\cdot\frac{(\zeta-\xi)^2}{\sigma_y^2}\right\}\cdot d\zeta \qquad (14.118)$$

Again we substitute:

$$\frac{\zeta-\xi}{\sigma_y}=x \qquad \zeta=x\cdot\sigma_y+\xi \qquad d\zeta=dx\cdot\sigma_y \qquad (14.119)$$

The limits:
$$\zeta = a \rightarrow x = \frac{a-\xi}{\sigma_y}$$
$$\zeta = b \rightarrow x = \frac{b-\xi}{\sigma_y}$$

This leads to:

$$Int2 = \int_{\frac{a-\xi}{\sigma_y}}^{\frac{b-\xi}{\sigma_y}} \frac{1}{\sqrt{2\pi} \cdot \sigma_y} \cdot \exp\left\{-\frac{1}{2} \cdot x^2\right\} \cdot dx \cdot \sigma_y$$

$$Int2 = \frac{1}{\sqrt{2\pi}} \cdot \int_{\frac{a-\xi}{\sigma_y}}^{\frac{b-\xi}{\sigma_y}} \exp\left\{-\frac{1}{2} \cdot x^2\right\} \cdot dx \tag{14.120}$$

and with this term we finish the calculation of the second integral. We thus obtain

$$\begin{aligned} S(x,z) = &-\frac{1}{2 \cdot \sqrt{2\pi}} \cdot \ln\left(2\pi \cdot \sigma_y^2\right) \cdot \int_{-\infty}^{\infty} \int_{\frac{a-\xi}{\sigma_y}}^{\frac{b-\xi}{\sigma_y}} \exp\left\{-\frac{1}{2} \cdot x^2\right\} \cdot dx \cdot f_x(\xi) \cdot d\xi \\ &-\frac{1}{2 \cdot \sqrt{2\pi}} \cdot \int_{-\infty}^{\infty} \int_{\frac{a-\xi}{\sigma_y}}^{\frac{b-\xi}{\sigma_y}} x^2 \cdot \exp\left\{-\frac{1}{2} \cdot x^2\right\} \cdot dx \cdot f_x(\xi) \cdot d\xi \\ &+\frac{1}{\sqrt{2\pi}} \cdot \int_{-\infty}^{\infty} \int_{\frac{a-\xi}{\sigma_y}}^{\frac{b-\xi}{\sigma_y}} \exp\left\{-\frac{1}{2} \cdot x^2\right\} \cdot dx \cdot \ln f_x(\xi) \cdot f_x(\xi) \cdot d\xi \end{aligned} \tag{14.121}$$

rearrange the terms and find

$$S(x,z) = -\frac{1}{2\cdot\sqrt{2\pi}} \cdot \int_{-\infty}^{\infty} \int_{\frac{a-\xi}{\sigma_y}}^{\frac{b-\xi}{\sigma_y}} \exp\left\{-\frac{1}{2}\cdot x^2\right\} \cdot \left[\ln\left(2\pi\cdot\sigma_y^2\right) + x^2 - 2\cdot\ln f_x(\xi)\right] \cdot dx \cdot f_x(\xi)\cdot d\xi \tag{14.122}$$

This is the joint information, if the random variable z is limited by the interval $[a,b]$.

Finally we summarize the calculated amounts of information

$$S(x \mid z) = S(x,z) - S(z) \tag{14.123}$$

by inserting the results from equations (14.122) and (14.107) into equation (14.123). This leads to the equation

$$\begin{aligned} S(x \mid z) = &-\frac{1}{2\cdot\sqrt{2\pi}} \cdot \int_{-\infty}^{\infty} \int_{\frac{a-\xi}{\sigma_y}}^{\frac{b-\xi}{\sigma_y}} \exp\left\{-\frac{1}{2}\cdot x^2\right\} \cdot \left[\ln\left(2\pi\cdot\sigma_y^2\right) + x^2 - 2\cdot\ln f_x(\xi)\right] \cdot dx \cdot f_x(\xi)\cdot d\xi \\ &+ \frac{1}{2\cdot\sqrt{2\pi}} \cdot \Big\{\ln\left(2\pi\cdot\sigma_z^2\right)\cdot\left[\Phi(b\cdot\sigma_z) - \Phi(a\cdot\sigma_z)\right] \\ &+ \sum_{n=0}^{\infty} \frac{1}{n!}\left(\frac{1}{2}\right)^n \cdot \frac{1}{2n+3} \cdot \left[\left(-(b\cdot\sigma_z)^2\right)^n \cdot (b\cdot\sigma_z)^3 - \left(-(a\cdot\sigma_z)^2\right)^n \cdot (a\cdot\sigma_z)^3\right]\Big\} \end{aligned} \tag{14.124}$$

with $\quad \Phi(x\cdot\sigma_z) = \int_{-\infty}^{x\cdot\sigma_z} \exp\left\{-\frac{1}{2}\cdot\varphi^2\right\}\cdot d\varphi$

This is the information we get about x if the measurement z is bounded by the interval $[a,b]$ and the connection of both random variables is $z = x + y$. This is the equivocation, describing the remaining uncertainty in the state space random variable x, if z is observed with the uncertainty σ_z^2. This can also be written as

$$S(x \mid z) = -\frac{1}{2\cdot\sqrt{2\pi}}$$

$$\cdot \int_{-\infty}^{\infty} \int_{\frac{a-\xi}{\sigma_y}}^{\frac{b-\xi}{\sigma_y}} \exp\left\{-\frac{1}{2}\cdot x^2\right\}\cdot\left[\ln\left(2\pi\cdot\sigma_y^2\right)+x^2-2\cdot\ln f_x(\xi)\right]\cdot dx\cdot f_x(\xi)\cdot d\xi$$

$$+\frac{1}{2\cdot\sqrt{2\pi}}\cdot\left\{\ln\left(2\pi\cdot\sigma_z^2\right)\cdot\left[\Phi(b\cdot\sigma_z)-\Phi(a\cdot\sigma_z)\right]\right.$$

$$\left. + \int_{a\cdot\sigma_z}^{b\cdot\sigma_z} \varphi^2\cdot\exp\left\{-\frac{1}{2}\cdot\varphi^2\right\}\cdot d\varphi\right\}$$

(14.125)

What happens if we increase the limits to $a \to -\infty$, $b \to \infty$, if there is no measurement and if we do not have any a priori information about the range of the expected measurements? We want to examine this limit to obtain some insight into the transmission of information. So we first calculate this limit.

$$\lim_{\substack{a\to-\infty \\ b\to\infty}} S(x \mid z)$$

$$= -\frac{1}{2\cdot\sqrt{2\pi}}\cdot \int_{-\infty}^{\infty} \int_{\frac{a-\xi}{\sigma_y}}^{\frac{b-\xi}{\sigma_y}} \exp\left\{-\frac{1}{2}\cdot x^2\right\}\cdot\left[\ln\left(2\pi\cdot\sigma_y^2\right)+x^2-2\cdot\ln f_x(\xi)\right]\cdot dx\cdot f_x(\xi)\cdot d\xi$$

$$\underbrace{+\frac{1}{2\cdot\sqrt{2\pi}}\cdot\left\{\ln\left(2\pi\cdot\sigma_z^2\right)\cdot\left[\Phi(b\cdot\sigma_z)-\Phi(a\cdot\sigma_z)\right]+\int_{a\cdot\sigma_z}^{b\cdot\sigma_z} \varphi^2\cdot\exp\left\{-\frac{1}{2}\cdot\varphi^2\right\}\cdot d\varphi\right\}}_{S_z}$$

(14.126)

We already verified $S(z)$ in equation (14.110), so we write

$$\lim_{\substack{a\to-\infty\\ b\to\infty}} S(x\,|\,z)=-\frac{1}{2\cdot\sqrt{2\pi}}$$

$$\cdot\int_{-\infty}^{\infty}\int_{-\infty}^{\infty}\exp\left\{-\frac{1}{2}\cdot x^2\right\}\cdot\left[\ln\left(2\pi\cdot\sigma_y^2\right)+x^2-2\cdot\ln f_x(\xi)\right]\cdot dx\cdot f_x(\xi)\cdot d\xi$$

$$+\frac{1}{2}\cdot\ln\left(2\pi\cdot\sigma_z^2\right)+\frac{1}{2}$$

rearrange the terms

$$\lim_{\substack{a\to-\infty\\ b\to\infty}} S(x\,|\,z)=-\frac{\ln\left(2\pi\cdot\sigma_y^2\right)}{2\cdot\sqrt{2\pi}}\cdot\int_{-\infty}^{\infty}\int_{-\infty}^{\infty}\exp\left\{-\frac{1}{2}\cdot x^2\right\}\cdot dx\cdot f_x(\xi)\cdot d\xi$$

$$-\frac{1}{2\cdot\sqrt{2\pi}}\cdot\int_{-\infty}^{\infty}\int_{-\infty}^{\infty}x^2\cdot\exp\left\{-\frac{1}{2}\cdot x^2\right\}\cdot dx\cdot f_x(\xi)\cdot d\xi$$

$$+\frac{1}{\sqrt{2\pi}}\cdot\int_{-\infty}^{\infty}\int_{-\infty}^{\infty}\exp\left\{-\frac{1}{2}\cdot x^2\right\}\cdot dx\cdot\ln f_x(\xi)\cdot f_x(\xi)\cdot d\xi$$

$$+\frac{1}{2}\cdot\ln\left(2\pi\cdot\sigma_z^2\right)+\frac{1}{2}$$

and proceed

$$\begin{aligned}\lim_{\substack{a\to-\infty\\ b\to\infty}} S(x\,|\,z)=&-\frac{\ln\left(2\pi\cdot\sigma_y^2\right)}{2}\cdot\int_{-\infty}^{\infty}\underbrace{\int_{-\infty}^{\infty}\frac{1}{\sqrt{2\pi}}\cdot\exp\left\{-\frac{1}{2}\cdot x^2\right\}\cdot dx\cdot f_x(\xi)\cdot d\xi}_{=1}\\ &-\frac{1}{2}\cdot\int_{-\infty}^{\infty}\underbrace{\int_{-\infty}^{\infty}\frac{1}{\sqrt{2\pi}}\cdot x^2\cdot\exp\left\{-\frac{1}{2}\cdot x^2\right\}\cdot dx}_{=1}\cdot f_x(\xi)\cdot d\xi\\ &+\int_{-\infty}^{\infty}\underbrace{\int_{-\infty}^{\infty}\frac{1}{\sqrt{2\pi}}\cdot\exp\left\{-\frac{1}{2}\cdot x^2\right\}\cdot dx}_{=1}\cdot\ln f_x(\xi)\cdot f_x(\xi)\cdot d\xi\\ &+\frac{1}{2}\cdot\ln\left(2\pi\cdot\sigma_z^2\right)+\frac{1}{2}\end{aligned}\tag{14.127}$$

We now write

$$\lim_{\substack{a\to-\infty\\ b\to\infty}} S(x\,|\,z) = -\frac{\ln\left(2\pi\cdot\sigma_y^2\right)}{2}\cdot\int_{-\infty}^{\infty} f_x(\xi)\cdot d\xi - \frac{1}{2}\cdot\int_{-\infty}^{\infty} f_x(\xi)\cdot d\xi$$

$$+\int_{-\infty}^{\infty} \ln f_x(\xi)\cdot f_x(\xi)\cdot d\xi + \frac{1}{2}\cdot\ln\left(2\pi\cdot\sigma_z^2\right)+\frac{1}{2}$$

and finally

$$\lim_{\substack{a\to-\infty\\ b\to\infty}} S(x\,|\,z) = -\frac{1}{2}\ln\left(2\pi\cdot\sigma_y^2\right)-\frac{1}{2}+\frac{1}{2}\cdot\ln\left(2\pi\cdot\sigma_x^2\right)+\frac{1}{2}+\frac{1}{2}\cdot\ln\left(2\pi\cdot\sigma_z^2\right)+\frac{1}{2} \tag{14.128}$$

with $\sigma_z^2 = \sigma_x^2 + \sigma_y^2$ we obtain

$$\lim_{\substack{a\to-\infty\\ b\to\infty}} S(x\,|\,z) = -\frac{1}{2}\ln\left(2\pi\cdot\sigma_y^2\right)-\frac{1}{2}+\frac{1}{2}\cdot\ln\left(2\pi\cdot\sigma_x^2\right)+\frac{1}{2}+\frac{1}{2}\cdot\ln\left(2\pi\cdot\left(\sigma_x^2+\sigma_y^2\right)\right)+\frac{1}{2}$$

$$\lim_{\substack{a\to-\infty\\ b\to\infty}} S(x\,|\,z) = \frac{1}{2}\cdot\ln\left(\frac{2\pi\cdot\sigma_x^2\cdot 2\pi\cdot\left(\sigma_x^2+\sigma_y^2\right)}{2\pi\cdot\sigma_y^2}\right)+\frac{1}{2}$$

$$\lim_{\substack{a\to-\infty\\ b\to\infty}} S(x\,|\,z) = \frac{1}{2}\cdot\ln\left(2\pi\cdot\frac{\sigma_x^2\cdot\left(\sigma_x^2+\sigma_y^2\right)}{\sigma_y^2}\right)+\frac{1}{2} \tag{14.129}$$

If we do not have any further information about the summation random variable z, apart from knowing that it is the sum of two gaussian random variables x and y, then this is the information we obtain about x on the basis of our knowledge about z. If we examine the unconditional information of the random variable x, we obtain

$$S(x) = \frac{1}{2}\cdot\ln\left(2\pi\cdot\sigma_x^2\right)+\frac{1}{2} \tag{14.130}$$

and compare this result with the information we gained in equation (14.129), we obtain the difference

$$\lim_{\substack{a\to-\infty\\ b\to\infty}} S(x\,|\,z) - S(x) = \frac{1}{2}\cdot\ln\left(\frac{\left(\sigma_x^2+\sigma_y^2\right)}{\sigma_y^2}\right)\geq 0 \tag{14.131}$$

This is the information about the random variable x, which we gain by the knowledge that the gaussian random variable z lies in the assumed range. It is exactly the gain of information that Norbert Wiener describes in his book 'Cybernetics' [WIE63].

The gain, which we obtain because of the knowledge of the output variable, depends on the quotient of the sum of the variance of the noise process and the variance of the signal process divided by the variance of the noise process. What does this mean? Well, firstly we notice that the gain of information is infinite if the variance of the noise process is equal to zero, $\sigma_y^2 = 0$. In this case we do not have any disturbance, but *y* is a deterministic variable [LIG66]. As soon as this deterministic variable becomes a stochastic (random) variable with $\sigma_y^2 \neq 0$, the gain of information is reduced to a finite value. With an increasing variance of the disturbance the gain of information decreases until it reaches the limit zero for $\sigma_y^2 \rightarrow \infty$, an infinite variance of the disturbance.

Obviously this examination again raises the problem of the transition from a deterministc variable to a random variable, because the difference in the gain of information, resulting from this transition, is infinite. A deterministic variable can only be generated by computation of the limit of a random variable, i.e. if the variance is equal to zero [LIG66]. As we are not able to provide exact measurements, this problem – which we already discussed in chapters 4 and 8 – does not occur in reality. But it affects the transition from the information of discrete random variables to the information of continuous variables in the same way. There we also have to compute limits, either for an infinite number of discrete values or for a lower resolution $d\underline{\xi}$ tending to zero. The limited accuracy of our measurements prevent such considerations in real systems, and it is also impossible to get an infinite number of initial states, because we always have some initial information (even if it is really inaccurate) regarding the system, which reduces the initial uncertainty to a finite number.

Another interesting point of view is the examination of the uncertainty of the signal σ_x^2, which is generated by an emitter. If there is no uncertainty in the signal, there is also no gain of information. This means that the emitter has to generate some uncertainty, containing information in the sense of a surprising part of the message or of the signal ('deterministic information'). If a signal does not contain any deterministic information (deliberately generated uncertainty), the knowledge of the output does not provide any gain of information.

With this consideration we decompose the information into two parts: Deterministic information and stochastic information. The second part, the stochastic information, describes influences of disturbances on the signal and the disturbances in the signal. The first part, the deterministic information, describes the actual information, which the emitter deliberately sends to the receiver (the new, surprising part of the message). Of course, this deterministic information cannot be described by a deterministic signal. Instead it is the intended part of the message, which contains a specific new knowledge for the receiver.

Such a part may be deterministic for the emitter, as he knows it. But it is an unexpected or unknown message for the receiver, who has no idea (prior information) that allows him to predict this information. A deterministic part of a signal is completely redundant and does not contain any information. This is the basic idea of the innovation approach, where only the surprising part (white noise) of the signal is used as an input signal to obtain a description of the system.

We now leave this chapter with the knowledge that only the non-predictable part of the signal contains information that the receiver classifies as interesting and relevant information. Deterministic parts of the signal are classified as redundant components, which may be used to reduce losses in the transformation. However, they do not contain any interesting components. They may allow the reconstruction of lost, interesting information, and such a loss of information is due to noise processes, which reduce the gain of information. In the limit these noise processes may even eliminate any gain of information.

15 Maximum entropy estimation

The maximum entropy estimation is a method that enables us to estimate the distribution density function of one or more random variables, if our previous knowledge is restricted by a limited number of samples or by other limitations. The basic idea of this method can be described with the Laplacian '*principle of insufficient reasoning*'. This principle states that we may only use a priori knowledge, which we get from measurements or other known restrictions. All other information, which is, for instance, implicated by an estimation algorithm, leads to a limitation of the results obtained.

Thus we should calculate the distribution density that implicates the least possible additional (unknown) information to the estimation algorithm, on the basis of the previous knowledge. This is the estimation that provides the maximum entropy estimate. The result of such an estimation allows us to add supplementary limitations to this outcome, which increase the information and decrease the entropy. But the derived result holds for the most general cases and can be improved by any additional information.

The informationentropy is the measure that is used to derive the desired distribution density function with maximum entropy.

$$S = -\sum_{i=1}^{N} p_i \cdot \ln p_i \tag{15.01}$$

or

$$S = -\int_{-\infty}^{\infty} \ln f_x(\xi) \cdot f_x(\xi) \cdot d\xi \tag{15.02}$$

The maximization of this measure certainly depends on the additional boundary conditions. If, for instance, we do not have any a priori information about the probabilities of the occurring events, we obtain the equation

$$-\sum_{i=1}^{N} p_i \cdot \ln p_i + \lambda \cdot \left(\sum_{i=1}^{N} p_i - 1 \right) = 0 \tag{15.03}$$

The only a priori information that we are able to include in this optimization is the knowledge that the distribution is a complete distribution

$$\sum_{i=1}^{N} p_i = 1 \tag{15.04}$$

Table 15.1: Distribution densities resulting from the maximum entropy estimation

RANGE	A PRIORI INFORMATION OF EXPECTATION VALUES	DISTRIBUTION DENSITY WITH MAXIMUM ENTROPY
$(0,1)$	no restrictions	$f_x(\xi)=1$ for $\xi \in (0,1)$ uniform distribution
(a,b) $0<a<b$	no restrictions	$f_x(\xi)=\frac{1}{b-a}\cdot rect\left(\frac{\xi-\frac{b+a}{2}}{b-a}\right)$ uniform distribution
$(0,1)$	$E\{\log(x)\}, E\{\log(1\text{-}x)\}$	$f_x(\xi)=\frac{\xi^{m-1}\cdot(1-\xi)^{n-1}}{B(m,n)}$ $B(m,n)$ = Beta function
$(0,\infty)$	$E\{x\}$	$f_x(\xi)=a\cdot e^{-a\cdot\xi}$ exponential density
$(0,\infty)$	$E\{x\}, E\{\log(x)\}$	$f_x(\xi)=\frac{a^p\cdot\xi^{p-1}\cdot e^{-a\cdot\xi}}{\Gamma(p)}$
$(-\infty,\infty)$	$\sigma^2 = \mathrm{E}\{x^2\}$	$f_x(\xi)=\frac{1}{\sqrt{2\pi}\cdot\sigma}\cdot\exp\left\{-\frac{1}{2}\cdot\frac{\xi^2}{\sigma^2}\right\}$ gaussian distribution density
$(-\infty,\infty)$	$\mu = \mathrm{E}\{x\}, \sigma^2 = E\{x^2\}$	$f_x(\xi)=\frac{1}{\sqrt{2\pi}\cdot\sigma}\cdot\exp\left\{-\frac{1}{2}\cdot\frac{(\xi-\mu)^2}{\sigma^2}\right\}$
$(-\infty,\infty)$	$E\{\lvert x\rvert\}$	$f_x(\xi)=\frac{a}{2}\cdot e^{-a\cdot\lvert\xi\rvert}$ Laplacian density

As a result (see section 3.2.3) we obtain the uniform distribution of the events, which also results from the principle of Laplace. This principle leads us to the conclusion: "If we do not have any a priori information, we must not prefer any event. Thus we have to assign the same probability to all events." This is exactly the assignment we get for a uniform distribution density function.

The distribution densities resulting from the maximization of the entropy depend on the boundary conditions. Different boundary conditions lead to different distribution densities. Some of them can be found in Table 15.1.

The optimum distribution density function resulting from the maximization of the entropy S obviously depends on the knowledge of the moments. All the conditions given have to be met by the distribution density, even if there are several conditions. The maximization is achieved via Shannon's entropy, which has been suggested by Edward Jaynes [JAY83] to be used for the maximum entropy estimation.

The maximization depends heavily on the chosen coordinate system. If, for instance, we determine the real and imaginary components in the complex plane on the basis of a given power $E\{x^2\}$, we obtain gaussian distribution densities for both components. If we switch to the cylindrical coordinate system, the amplitude and the phase of the very same signal do not at all have gaussian distribution densities. Consequently the question arises, which coordinate system should we choose to apply our a priori knowledge 'no prior information', if we do not have any previous knowledge. If we choose the 'wrong' coordinate system, we obtain a uniform distribution, but the transformation into the 'correct' coordinate system changes the uniform distribution into another distribution and thus it implicates additional information by changing the equal probabilities.

15.1 The difference between maximum entropy and minimum variance

Maximum entropy estimation is a method that not only enables us to determine certain parameters in order to meet some initial conditions, but also allows us to determine complete distribution densities on the basis of the previous knowledge, without implicating any additional restrictions during the optimization. Thus this method provides the most general solution on the basis of the available information.

If we have a priori information, describing certain parts of the realization space or of the sample space, we search for a complete description, which does not add any unnecessary restrictions. So we maximize the entropy under the condition of the previous information and we obtain a distribution density function (and thus the moments and other parameters), resulting merely from the a priori information. The further adaptation of this result can always be achieved via additional restrictions.

One problem in the Laplacian principle of insufficient reasoning remains. If we choose the uniform distribution density because of a complete lack of a priori information, we may do so in arbitrary coordinate systems. As soon as we change the coordinate system, certain events will be preferred, as the new distribution density will not be a uniform distribution any more. Thus we have to take care in the choice of the coordinate system.

The minimum variance estimation aims at the determination of certain parameters, which are immediately connected to the variance. The minimization of the variance provides optimum estimates for the desired parameters. This method cannot be used to determine complete distribution density functions (apart from gaussian distribution densities). The method merely allows the optimization of the central second order moment and in general it does not include additional information. And for a limited number of samples the variance can only be determined with a finite accuracy (sample analysis).

15.2 The difference from bootstrap or resampling methods

Bootstrap methods deal with a limited number of samples, which are used to determine the actual distribution density function. Because of this limited number of available samples this determination contains an error, which is explained by the law of large numbers. The empirical determination of a distribution density can only be achieved with a certain probability, increasing with the number of available samples. Because of the limited number of available samples the determination can only be achieved with a certain minimum error. The bootstrap method uses the available samples and generates additional sample sequences by random sampling from the original sampling sequence (resampling). This provides a smaller estimation error between the actual distribution density and the estimated distribution density.

This approach tries to include all available information via resampling from the given set of data. The resampled sets of data are all from the same type. If we thus measured an atypical sequence, all resampled sequences are atypical, too. That is, this method merely determines a distribution of the type of the original sampling sequence and not the distribution of the observed process. Both distribution densities are equal if the number of the originally observed samples tends to infinity, but this is the familiar law of the large number.

Contrary to the maximum entropy estimation the bootstrap method determines the distribution density function of the sampled sequence, which is used without any further restrictions. Concretely the bootstrap method determines the distribution of the sampling sequence, which should be (apart from atypical sequences) similar to the actual distribution density of the observed process. It is a statistical method, using a sample function to determine the distribution and arbitrary estimates.

The maximum entropy method determines a distribution density, which does not implicate any additional restrictions, apart from the a priori information. Moments and entropy functions are used as a priori information. It is a stochastic method, using the moments to determine the optimum distribution density. The moments can be determined via empirical methods.

15.3 A maximum entropy example

The maximum entropy method searches for the most probable state of the observed system. In other words, the maximum entropy method searches for the state that admits the greatest number of possible arrangements. An example is shown in Table 15.1: If we distribute four balls (numbered with 1,2,3,4) in two fields, we obtain 16 possible arrangements:

Table 15.2: Distribution of four balls in two fields

left field	right field	probabilities
1,2,3,4	empty	*P*(4 left) = 1/16
1,2,3	4	*P*(3 left) = 4/16
1,2,4	3	
1,3,4	2	
2,3,4	1	
1,2	3,4	*P*(2 left) = 6/16
1,3	2,4	
1,4	2,3	
2,3	1,4	
2,4	1,3	
3,4	1,2	
1	2,3,4	*P*(1 left) = 4/16
2	1,3,4	
3	1,2,4	
4	2,3,4	
empty	1,2,3,4	*P*(0 left) = 1/16

The probability that the left field contains two balls is the greatest probability in this table. This state has the maximum amount of entropy. For four balls we do not have a great difference in the probabilities. If we increase the number of the balls, the probabilities of the events '*i* balls in the left field' differ more clearly. This is shown in Table 15.3, where we see the distribution of N molecules in two fields. P is the number of all possible arrangements, while N_1 is the number of molecules in the left field and N_2 is the number of molecules in the right field.

Table 15.3: Distribution of N events in two fields

$N=$	2	$N=$	4	$N=$	6	$N=$	8	$N=$	10	$N=$	12
N_1/N_2	P	N_1/N_2	P	N_1/N_2	P	N_1/N_2	P	N_1/N_2	P	N_1/N_2	P
2/0	1	4/0	1	6/0	1	8/0	1	10/0	1	12/0	1
1/1	**2**	3/1	4	5/1	6	7/1	8	9/1	10	11/1	12
0/2	1	**2/2**	**6**	4/2	15	6/2	28	8/2	45	10/2	66
		1/3	4	**3/3**	**20**	5/3	56	7/3	120	9/3	220
		0/4	1	2/4	15	**4/4**	**70**	6/4	210	8/4	495
				1/5	6	3/5	56	**5/5**	**252**	7/5	792
				0/6	1	2/6	28	4/6	210	**6/6**	**924**
						1/7	8	3/7	120	5/7	792
						0/8	1	2/8	45	4/8	495
								1/9	10	3/9	220
								0/10	1	2/10	66
										1/11	12
										0/12	1

$$P = \frac{N!}{N_1! \cdot (N - N_1)!} \tag{15.05}$$

$$P = e^{\frac{S}{k}} \tag{15.06}$$

The greater the entropy becomes the more the probability increases that a state with higher probability is assumed in the system. And this state of higher probability is a state, where both fields have the same number of elements. The greater the entropy, the more regular is the distribution of the molecules in the fields.

15.4 Maximum entropy: The method

Entropy is first connected with Shannon's entropy or information. This is the basic approach in most explanations of the maximum entropy method. We therefore also start with this entropy, but we proceed with other entropies that can also be used for maximum entropy estimation.

15.4.1 Maximum Shannon entropy

For continuous random variables we have the following initial conditions:

$$S = -\int_{\Xi} \ln f_{\underline{x}}(\underline{\xi}) \cdot f_{\underline{x}}(\underline{\xi}) \cdot d\underline{\xi} \qquad \text{Shannon's informationentropy} \tag{15.07}$$

$$E\{\underline{g}_r(\underline{x})\} = a_r = \int_{\Xi} \underline{g}_r(\underline{\xi}) \cdot f_{\underline{x}}(\underline{\xi}) \cdot d\underline{\xi} \quad \text{for } r = 1,\ldots,m \quad \text{given moments} \tag{15.08}$$

$$\int_{\Xi} f_{\underline{x}}(\underline{\xi}) \cdot d\underline{\xi} = 1 \qquad \text{Condition for a distribution density function} \tag{15.09}$$

The optimization with additional boundary conditions is usually achieved by generating the Lagrangian function with the Lagrangian multipliers

$$\begin{aligned} LF = & -\int_{\Xi} \ln f_{\underline{x}}(\underline{\xi}) \cdot f_{\underline{x}}(\underline{\xi}) \cdot d\underline{\xi} + \sum_{r=1}^{m} \lambda_r \cdot \left[\int_{\Xi} \underline{g}_r(\underline{\xi}) \cdot f_{\underline{x}}(\underline{\xi}) \cdot d\underline{\xi} - a_r \right] \\ & + (\mu + 1) \cdot \left[\int_{\Xi} f_{\underline{x}}(\underline{\xi}) \cdot d\underline{\xi} - 1 \right] \end{aligned} \tag{15.10}$$

Now we first calculate the derivative with respect to the desired value, then we determine the zero of this derivative and this provides the determination equation for the Lagrangian multipliers μ, λ_r. In our case we search for the distribution density that maximizes the Shannon entropy on the basis of the given boundary conditions. We thus build

$$\frac{\partial}{\partial f_{\underline{x}}(\underline{\xi})} LF \stackrel{!}{=} 0 \tag{15.11}$$

$$\frac{\partial}{\partial f_{\underline{x}}(\underline{\xi})} LF = \int_{\Xi} \left\{ -\ln f_{\underline{x}}(\underline{\xi}) - 1 + \sum_{r=1}^{m} \left[\lambda_r \cdot \underline{g}_r(\underline{\xi}) \right] + \mu + 1 \right\} \cdot d\underline{\xi} \tag{15.12}$$

and demand

$$\int_{\Xi} \left\{ -\ln f_{\underline{x}}(\underline{\xi}) + \sum_{r=1}^{m} \left[\lambda_r \cdot \underline{g}_r(\underline{\xi}) \right] + \mu \right\} \cdot d\underline{\xi} \stackrel{!}{=} 0 \tag{15.13}$$

This term is only equal to zero when the kernel vanishes

$$-\ln f_{\underline{x}}(\underline{\xi}) + \sum_{r=1}^{m} \left[\lambda_r \cdot \underline{g}_r(\underline{\xi}) \right] + \mu \stackrel{!}{=} 0$$

$$\ln f_{\underline{x}}(\underline{\xi}) = \mu + \sum_{r=1}^{m} \left[\lambda_r \cdot \underline{g}_r(\underline{\xi}) \right]$$

$$f_{\underline{x}}(\underline{\xi}) = \exp\left\{\mu + \sum_{r=1}^{m}\left[\lambda_r \cdot \underline{g}_r(\underline{\xi})\right]\right\} \tag{15.14}$$

This is the desired distribution density. The remaining parameters (Lagrangian multipliers) can be determined with the boundary conditions

$$E\{\underline{g}_r(\underline{x})\} = a_r = \int_{\Xi} \underline{g}_r(\underline{\xi}) \cdot f_{\underline{x}}(\underline{\xi}) \cdot d\underline{\xi} \quad \text{for } r = 1,\ldots,m \quad \text{given moments} \tag{15.15}$$

$$\int_{\Xi} f_{\underline{x}}(\underline{\xi}) \cdot d\underline{\xi} = 1 \tag{15.16}$$

From the last boundary condition we obtain

$$\int_{\Xi} f_{\underline{x}}(\underline{\xi}) \cdot d\underline{\xi} = 1$$

and with

$$f_{\underline{x}}(\underline{\xi}) = \exp\left\{\mu + \sum_{r=1}^{m}\left[\lambda_r \cdot \underline{g}_r(\underline{\xi})\right]\right\}$$

this leads to

$$\int_{\Xi} \exp\left\{\mu + \sum_{r=1}^{m}\left[\lambda_r \cdot \underline{g}_r(\underline{\xi})\right]\right\} \cdot d\underline{\xi} = 1$$

$$\exp\{-\mu\} = \int_{\Xi} \exp\left\{\sum_{r=1}^{m}\left[\lambda_r \cdot \underline{g}_r(\underline{\xi})\right]\right\} \cdot d\underline{\xi} \tag{15.17}$$

and this term is the necessary standardization, leading to $\int_{\Xi} f_{\underline{x}}(\underline{\xi}) \cdot d\underline{\xi} = 1$. This ensures that the resulting function $f_{\underline{x}}(\underline{\xi})$ actually is a distribution density function.

For *discrete random variables* we obtain:

$$S = -\sum_{i=1}^{n} p_i \cdot \ln p_i \tag{15.18}$$

with the conditions

$$\sum_{i=1}^{n} p_i = 1 \tag{15.19}$$

and

$$\sum_{i=1}^{n} p_i \cdot g_r(x_i) = E\{g_r(x)\} = a_r \qquad \text{for } r = 1,\ldots,m \tag{15.20}$$

Again we have the moments and the general conditions for the distribution functions. The Lagrangian function is thus

$$LF = -\sum_{i=1}^{n} p_i \cdot \ln p_i + \sum_{r=1}^{m} \lambda_r \cdot \left[\sum_{i=1}^{n} p_i \cdot g_r(x_i) - a_r\right] + (\mu + 1) \cdot \left[\sum_{i=1}^{n} p_i - 1\right] \tag{15.21}$$

We now search for the probabilities

$$\frac{\partial LF}{\partial p_i} = -\ln p_i - 1 + \sum_{r=1}^{m} \lambda_r \cdot g_r(x_i) + (\mu + 1) \cdot [1]$$

$$\frac{\partial LF}{\partial p_i} = -\ln p_i + \mu + \sum_{r=1}^{m} \lambda_r \cdot g_r(x_i) = 0$$

$$\ln p_i = \mu + \sum_{r=1}^{m} \lambda_r \cdot g_r(x_i)$$

$$p_i = \exp\left\{\mu + \sum_{r=1}^{m} \lambda_r \cdot g_r(x_i)\right\} \qquad \text{for } i = 1,\ldots,n \tag{15.22}$$

This result is inserted into the boundary conditions, which again leads to the necessary standardization on the basis of the condition

$$\sum_{i=1}^{n} p_i = 1 \tag{15.23}$$

with

$$p_i = \exp\left\{\mu + \sum_{r=1}^{m} \lambda_r \cdot g_r(x_i)\right\}$$

we obtain

$$\sum_{i=1}^{n} \exp\left\{\mu + \sum_{r=1}^{m} \lambda_r \cdot g_r(x_i)\right\} = 1$$

$$\exp\{-\mu\} = \sum_{i=1}^{n} \exp\left\{\sum_{r=1}^{m} \lambda_r \cdot g_r(x_i)\right\} \tag{15.24}$$

This standardization ensures that the sum of all probabilities is equal to 1. All other Lagrangian multipliers can be determined from the condition

$$\sum_{i=1}^{n} p_i \cdot g_r(x_i) = E\{g_r(x)\} = a_r \qquad \text{for } r = 1,\ldots,m \tag{15.25}$$

by evaluating each equation.

This is the original maximum entropy approach, as Edward Jaynes describes it. With the measures we introduced in the previous chapters, we might ask if there are other useful information measures, which can be applied to some maximum entropy approach. How does an altered approach change the results? When do we use which entropy measure? All these questions cannot be answered in one general approach and we have to decide for each application which entropy measure we have to use. The different features of different entropy functions certainly change the desired results.

There is one generalization of the maximum entropy approach of Jaynes, which we are going to examine in the following section. This approach is based on the Kullback–Leibler distance. As we will see in this derivation, we may generate specific entropy measures for each distribution density function.

15.4.2 Minimum Kullback–Leibler distance

In order to generalize the maximum entropy estimation, we use the Kullback–Leibler distance to derive such an extended maximum entropy estimation approach. This estimation differs from the maximum entropy estimation, which is based on Shannons entropy, in one important point. While we maximize the Shannon entropy in the maximum entropy estimation only on the basis of the given moments, the minimization of the Kullback–Leibler distance additionally includes a prior knowledge about the distribution density. Both methods are equal, if the a priori distribution is a uniform distribution, which means that there actually is no a priori information in this density (apart from the limitation of the interval). Thus Kullback's information enables us to introduce an additional a priori information into the maximum entropy estimation. With this a priori density, we assume some basic form of the resulting density and the optimization only leads to an adaptation of this prior assumption. We now describe this method in detail:

At first the minimization of Kullback's information:

<u>*Continuous distribution density functions:*</u>

$$K = \int_{\Xi} \ln \frac{f_{\underline{x}}(\underline{\xi})}{g_{\underline{x}}(\underline{\xi})} \cdot f_{\underline{x}}(\underline{\xi}) \cdot d\underline{\xi} \qquad \text{Kullback's discrimination information} \quad (15.26)$$

$$E\{\underline{g}_r(\underline{x})\} = \int_{\Xi} \underline{g}_r(\underline{\xi}) \cdot f_{\underline{x}}(\underline{\xi}) \cdot d\underline{\xi} = a_r \quad \text{for } r = 1,...,m \quad \text{given moments} \quad (15.27)$$

$$\int_{\Xi} f_{\underline{x}}(\underline{\xi}) \cdot d\underline{\xi} = 1 \qquad \text{Condition for a distribution density} \quad (15.28)$$

In general we have another condition

$$\int_{\Xi} g_{\underline{x}}(\underline{\xi}) \cdot d\underline{\xi} = 1 \quad (15.29)$$

and we may ask for the moments of this second distribution density

$$E\{\underline{h}_n(\underline{x})\} = \int_{\Xi} \underline{h}_n(\underline{\xi}) \cdot g_{\underline{x}}(\underline{\xi}) \cdot d\underline{\xi} \quad \text{for } n = 1,...,l \quad \text{given moments} \quad (15.30)$$

The moments are not mentioned explicitly and we in fact do not need them explicitly, because the distribution density function $g_{\underline{x}}(\underline{\xi})$ contains all the moments. The appropriate Lagrangian function is

$$LF = \int_{\Xi} \ln \frac{f_{\underline{x}}(\underline{\xi})}{g_{\underline{x}}(\underline{\xi})} \cdot f_{\underline{x}}(\underline{\xi}) \cdot d\underline{\xi} + \sum_{r=1}^{m} \lambda_r \cdot \left[\int_{\Xi} \underline{g}_r(\underline{\xi}) \cdot f_{\underline{x}}(\underline{\xi}) \cdot d\underline{\xi} - a_r \right]$$
$$+ (\mu - 1) \cdot \left[\int_{\Xi} f_{\underline{x}}(\underline{\xi}) \cdot d\underline{\xi} - 1 \right] \quad (15.31)$$

The condition $\int_{\Xi} g_{\underline{x}}(\underline{\xi}) \cdot d\underline{\xi} = 1$ need not be considered in this optimization, because this is a basic condition for a distribution density. Additionally the derivative of the addend $\kappa_g \cdot \left[\int_\Xi g_{\underline{x}}(\underline{\xi}) \cdot d\underline{\xi} - 1\right]$ with respect to the desired distribution function $f_{\underline{x}}(\underline{\xi})$ would be equal to zero, because the prior distribution is a constant function.

We are now looking for a distribution density that minimizes the given Lagrangian function. We thus build the derivative with respect to this function and get

$$\frac{\partial}{\partial f_{\underline{x}}(\underline{\xi})} LF = \int_\Xi \left[\ln \frac{f_{\underline{x}}(\underline{\xi})}{g_{\underline{x}}(\underline{\xi})} + f_{\underline{x}}(\underline{\xi}) \cdot \frac{g_{\underline{x}}(\underline{\xi})}{f_{\underline{x}}(\underline{\xi})} \cdot \frac{1}{g_{\underline{x}}(\underline{\xi})} + \sum_{r=1}^{m} \lambda_r \cdot \underline{g}_r(\underline{\xi}) + (\mu - 1) \right] \cdot d\underline{\xi}$$

$$\frac{\partial}{\partial f_{\underline{x}}(\underline{\xi})} LF = \int_\Xi \left[\ln \frac{f_{\underline{x}}(\underline{\xi})}{g_{\underline{x}}(\underline{\xi})} + 1 + \sum_{r=1}^{m} \lambda_r \cdot \underline{g}_r(\underline{\xi}) + \mu - 1 \right] \cdot d\underline{\xi}$$

$$\frac{\partial}{\partial f_{\underline{x}}(\underline{\xi})} LF = \int_\Xi \left[\ln \frac{f_{\underline{x}}(\underline{\xi})}{g_{\underline{x}}(\underline{\xi})} + \mu + \sum_{r=1}^{m} \lambda_r \cdot \underline{g}_r(\underline{\xi}) \right] \cdot d\underline{\xi} \overset{!}{=} 0 \tag{15.32}$$

The derivative has to be equal to zero to indicate an extremum. This is achieved if the kernel of the integral vanishes

$$\ln \frac{f_{\underline{x}}(\underline{\xi})}{g_{\underline{x}}(\underline{\xi})} + \mu + \sum_{r=1}^{m} \lambda_r \cdot \underline{g}_r(\underline{\xi}) = 0$$

$$\ln f_{\underline{x}}(\underline{\xi}) = \ln g_{\underline{x}}(\underline{\xi}) - \mu - \sum_{r=1}^{m} \lambda_r \cdot \underline{g}_r(\underline{\xi})$$

$$f_{\underline{x}}(\underline{\xi}) = g_{\underline{x}}(\underline{\xi}) \cdot \exp\left\{ -\mu - \sum_{r=1}^{m} \lambda_r \cdot \underline{g}_r(\underline{\xi}) \right\} \tag{15.33}$$

Inserting this equation in the boundary condition $\int_\Xi f_{\underline{x}}(\underline{\xi}) \cdot d\underline{\xi} = 1$ leads to

$$\int_\Xi g_{\underline{x}}(\underline{\xi}) \cdot \exp\left\{ -\mu - \sum_{r=1}^{m} \lambda_r \cdot \underline{g}_r(\underline{\xi}) \right\} \cdot d\underline{\xi} = 1$$

$$\exp\{\mu\} = \int_{\Xi} g_{\underline{x}}(\underline{\xi}) \cdot \exp\left\{-\mu - \sum_{r=1}^{m} \lambda_r \cdot \underline{g}_r(\underline{\xi})\right\} \cdot d\underline{\xi} \tag{15.34}$$

This is the standardization required for a real distribution density function $f_{\underline{x}}(\underline{\xi})$.

Discrete distribution density functions:

Here we have the discrimination information

$$K = \sum_{i=1}^{n} p_i \cdot \ln \frac{p_i}{q_i} \tag{15.35}$$

with the boundary conditions $\sum_{i=1}^{n} p_i = 1$ and $\sum_{i=1}^{n} q_i = 1$ (15.36)

and

$$\sum_{i=1}^{n} p_i \cdot g_r(x_i) = E\{g_r(x)\} = a_r \qquad \text{for } r = 1,\ldots,m \tag{15.37}$$

The Lagrangian function is thus:

$$LF = \sum_{i=1}^{n} p_i \cdot \ln \frac{p_i}{q_i} + \sum_{r=1}^{m} \lambda_r \cdot \left[\sum_{i=1}^{n} p_i \cdot g_r(x_i) - a_r\right] + (\mu - 1) \cdot \left[\sum_{i=1}^{n} p_i - 1\right] \tag{15.38}$$

The derivative with respect to the desired probability p_i provides

$$\frac{\partial}{\partial p_i} LF = \ln \frac{p_i}{q_i} + p_i \cdot \frac{q_i}{p_i} \cdot \frac{1}{q_i} + \sum_{r=1}^{m} \lambda_r \cdot g_r(x_i) + (\mu - 1)$$

$$\frac{\partial}{\partial p_i} LF = \ln \frac{p_i}{q_i} + 1 + \sum_{r=1}^{m} \lambda_r \cdot g_r(x_i) + \mu - 1$$

$$\frac{\partial}{\partial p_i} LF = \ln \frac{p_i}{q_i} + \mu + \sum_{r=1}^{m} \lambda_r \cdot g_r(x_i) \overset{!}{=} 0 \tag{15.39}$$

This derivative has to be equal to zero, so that

$$\ln \frac{p_i}{q_i} = -\mu - \sum_{r=1}^{m} \lambda_r \cdot g_r(x_i)$$

$$\ln p_i = \ln q_i - \mu - \sum_{r=1}^{m} \lambda_r \cdot g_r(x_i)$$

$$p_i = q_i \cdot \exp\left\{-\mu - \sum_{r=1}^{m} \lambda_r \cdot g_r(x_i)\right\} \tag{15.40}$$

Again the standardization results from the first boundary condition

$$\sum_{i=1}^{n} p_i = 1 \tag{15.41}$$

where we insert the probability

$$\sum_{i=1}^{n} q_i \cdot \exp\left\{-\mu - \sum_{r=1}^{m} \lambda_r \cdot g_r(x_i)\right\} = 1$$

$$\exp\{\mu\} = \sum_{i=1}^{n} q_i \cdot \exp\left\{-\sum_{r=1}^{m} \lambda_r \cdot g_r(x_i)\right\} \tag{15.42}$$

Thus the minimization of the Kullback–Leibler distance also leads to an entropy estimate, which additionally contains a prior information in the form of a given distribution density function or a discrete probability distribution. We are now going to compare both methods, to explain the differences between the two entropy estimation approaches.

15.5 Maximum entropy and minimum discrimination information

In the discrete case we get the probabilities:

$$p_i = \exp\left\{\mu + \sum_{r=1}^{m} \lambda_r \cdot g_r(x_i)\right\} \qquad \text{Maximum entropy} \tag{15.43}$$

$$p_i = q_i \cdot \exp\left\{-\mu - \sum_{r=1}^{m} \lambda_r \cdot g_r(x_i)\right\} \tag{15.44}$$

Minimum discrimination information

In the continuous case we obtain:

$$f_{\underline{x}}(\underline{\xi}) = \exp\left\{\mu + \sum_{r=1}^{m} \lambda_r \cdot \underline{g}_r(\underline{\xi})\right\} \qquad \text{Maximum entropy} \tag{15.45}$$

$$f_{\underline{x}}(\underline{\xi}) = g_{\underline{x}}(\underline{\xi}) \cdot \exp\left\{-\mu - \sum_{r=1}^{m} \lambda_r \cdot \underline{g}_r(\underline{\xi})\right\} \tag{15.46}$$

Minimum discrimination information

Apart from the different sign of the Lagrangian multipliers, which has no effect on the results, we clearly notice that the extension made in Kullback's information provides a probability q_i or a distribution density $g_{\underline{x}}(\underline{\xi})$ which is merely weighted (Edgeworth expansion, Gram–Charlier expansion).

Before we proceed with further considerations, we carry out a more detailed examination of these two methods.

$$S = -\int_{\Xi} \ln f_{\underline{x}}(\underline{\xi}) \cdot f_{\underline{x}}(\underline{\xi}) \cdot d\underline{\xi} \qquad \text{Shannon's information} \tag{15.47}$$

$$S = -\sum_{i=1}^{n} p_i \cdot \ln p_i \tag{15.48}$$

$$K = \int_{\Xi} \ln \frac{f_{\underline{x}}(\underline{\xi})}{g_{\underline{x}}(\underline{\xi})} \cdot f_{\underline{x}}(\underline{\xi}) \cdot d\underline{\xi} \qquad \text{Kullback's discrimination information} \tag{15.49}$$

$$K = \sum_{i=1}^{n} p_i \cdot \ln \frac{p_i}{q_i} \tag{15.50}$$

$$K = \int_{\Xi} \ln \frac{f_{\underline{x}}(\underline{\xi})}{g_{\underline{x}}(\underline{\xi})} \cdot f_{\underline{x}}(\underline{\xi}) \cdot d\underline{\xi} = \int_{\Xi} \ln f_{\underline{x}}(\underline{\xi}) \cdot f_{\underline{x}}(\underline{\xi}) \cdot d\underline{\xi} - \int_{\Xi} \ln g_{\underline{x}}(\underline{\xi}) \cdot f_{\underline{x}}(\underline{\xi}) \cdot d\underline{\xi} \tag{15.51}$$

$$K = -S - \int_{\Xi} \ln g_{\underline{x}}(\underline{\xi}) \cdot f_{\underline{x}}(\underline{\xi}) \cdot d\underline{\xi} \tag{15.52}$$

$$K = \sum_{i=1}^{n} p_i \cdot \ln \frac{p_i}{q_i} = \sum_{i=1}^{n} p_i \cdot \ln p_i - \sum_{i=1}^{n} p_i \cdot \ln q_i$$

$$K = -S - \sum_{i=1}^{n} p_i \cdot \ln q_i \tag{15.53}$$

Kullback's information differs from Shannon's information first in the sign and then in an additional additive term. The reason for the different sign is that the maximization of Shannon's information can be achieved by the minimization of Kullback's information. The additive term introduces an additional precondition into the possibilities of the optimization of Kullback's information, which is not available in Shannon's information.

If we assume that there is no a priori knowledge about the precondition ('principle of insufficient reasoning'), we have to assign equal probabilities q_i or a uniform distribution $g_{\underline{x}}(\underline{\xi})$ to the events, as we must not prefer any event. This results in a uniform distribution density function, which leads to the Kullback information

$$K = -S - \int_{\Xi} \ln g_{\underline{x}}(\underline{\xi}) \cdot f_{\underline{x}}(\underline{\xi}) \cdot d\underline{\xi} \tag{15.54}$$

with $$\int_{\Xi} g_{\underline{x}}(\underline{\xi}) \cdot d\underline{\xi} = 1 \quad \text{and} \quad g_{\underline{x}}(\underline{\xi}) = \text{const. for all } \xi \in \Xi \tag{15.55}$$

$$K = -S - \int_{\Xi} \ln \text{const.} \cdot f_{\underline{x}}(\underline{\xi}) \cdot d\underline{\xi} = -S - \ln \text{const.} \cdot \int_{\Xi} f_{\underline{x}}(\underline{\xi}) \cdot d\underline{\xi}$$

$$K = -S - \ln \text{const.} \tag{15.56}$$

and discrete with $q_i = 1/n$

$$K = -S - \sum_{i=1}^{n} p_i \cdot \ln \frac{1}{n} = -S + \ln n \cdot \sum_{i=1}^{n} p_i$$

$$K = -S + \ln n \tag{15.57}$$

In this special case, where we use the uniform distribution density (no a priori information), Shannon's and Kullback's information differ only in the sign and in a constant shift. Thus a minimization of the Kullback–Leibler distance immediately leads to a maximization of Shannon's entropy.

So the maximum entropy method is merely a special case of the minimum discrimination estimation! Both methods lead to equal results if we do not have any a priori information.

This was the approach of the maximum entropy estimation, generating an optimization without any additional restrictions. We may certainly use other entropy measures to achieve maximum entropy estimates, but the basic approach of the computation is always the same. The problem in the applications is the use of the appropriate entropy function. The development of such appropriate entropy functions is the aim of the next section.

15.6 Generation of generalized entropy measures

Up to now the the optimization methods discussed have always contained an entropy function and certain preconditions in the form of moments. And we determined some optimal distribution density function, which had the maximum entropy with respect to the given boundary conditions. But it is also possible to do this optimization the other way around. We then get an inverse problem, where the distribution density function (and thus, of course, all moments, etc.) is known and we ask for the entropy function that is optimized with this preconditions. If we proceed with this question, we obtain other entropy functions, which may be generated for every problem that has to be solved.

In general we have preconditions:

- Entropy
- Distribution density function
- Moments
- A priori information (only for Kullback's information)

If we know two of these three quantities, the maximum entropy (ME) estimation allows us to calculate the third quantity without further restrictions. The MDI (minimum discrimination information) estimation allows us to calculate the fourth quantity if three quantities are known (the a priori information has to be known).

The standard problem (formulated by Jaynes) is the calculation of the distribution density from the moments and the entropy with the ME estimation.

Here we determine the entropy function from the distribution density and the moments (inverse problem). We start with the following considerations:

Preconditions:

<u>Continuous case:</u>

$$H = -\int_{\Xi} \Phi\left(f_{\underline{x}}\left(\underline{\xi}\right)\right) \cdot d\underline{\xi} \qquad \text{General entropy function} \tag{15.58}$$

$$E\left\{\underline{g}_r\left(\underline{x}\right)\right\} = \int_{\Xi} \underline{g}_r\left(\underline{\xi}\right) \cdot f_{\underline{x}}\left(\underline{\xi}\right) \cdot d\underline{\xi} = a_r \quad \text{for } r = 1,\dots,m \quad \text{given moments} \tag{15.59}$$

$$\int_{\Xi} f_{\underline{x}}\left(\underline{\xi}\right) \cdot d\underline{\xi} = 1 \qquad \text{Condition for a distribution density function} \tag{15.60}$$

We have to add two remarks to these conditions. Firstly the desired function $\Phi\left(f_{\underline{x}}\left(\underline{\xi}\right)\right)$ has to be a convex function, so that the entropy function $H = -\int_{\Xi} \Phi\left(f_{\underline{x}}\left(\underline{\xi}\right)\right) \cdot d\underline{\xi}$ is a concave function, which is a basic condition for entropy functions. Secondly the entropy function of this approach depends only on the distribution density function $f_{\underline{x}}\left(\underline{\xi}\right)$, so that we might ask whether Kullback's information can be determined with this configuration. As the a priori information has to be known, Kullback's information can be determined. Here we start again with the Lagrangian function

$$LF = -\int_{\Xi} \Phi\left(f_{\underline{x}}\left(\underline{\xi}\right)\right) \cdot d\underline{\xi} + \mu \cdot \left[\int_{\Xi} f_{\underline{x}}\left(\underline{\xi}\right) \cdot d\underline{\xi} - 1\right] + \sum_{r=1}^{m} \lambda_r \cdot \left[\int_{\Xi} \underline{g}_r\left(\underline{\xi}\right) \cdot f_{\underline{x}}\left(\underline{\xi}\right) \cdot d\underline{\xi} - a_r\right] \tag{15.61}$$

We optimize this equation with respect to the distribution density function

$$\frac{\partial}{\partial f_{\underline{x}}\left(\underline{\xi}\right)} LF = \int_{\Xi} \left[-\frac{\partial}{\partial f_{\underline{x}}\left(\underline{\xi}\right)} \Phi\left(f_{\underline{x}}\left(\underline{\xi}\right)\right) + \mu + \sum_{r=1}^{m} \lambda_r \cdot \underline{g}_r\left(\underline{\xi}\right)\right] \cdot d\underline{\xi} \overset{!}{=} 0 \tag{15.62}$$

Again the kernel has to be equal to zero

$$-\frac{\partial}{\partial f_{\underline{x}}\left(\underline{\xi}\right)} \Phi\left(f_{\underline{x}}\left(\underline{\xi}\right)\right) + \mu + \sum_{r=1}^{m} \lambda_r \cdot \underline{g}_r\left(\underline{\xi}\right) = 0$$

$$\frac{\partial}{\partial f_{\underline{x}}\left(\underline{\xi}\right)} \Phi\left(f_{\underline{x}}\left(\underline{\xi}\right)\right) = \mu + \sum_{r=1}^{m} \lambda_r \cdot \underline{g}_r\left(\underline{\xi}\right) \tag{15.63}$$

The derivative of the desired function $\Phi(\ )$ is thus determined by the standardization μ and by the functions of the dummy variables, weighted with the Lagrangian multipliers.

A useful rearrangement of the left-hand side is

$$\frac{\partial\Phi\left(f_{\underline{x}}\left(\underline{\xi}\right)\right)}{\partial f_{\underline{x}}\left(\underline{\xi}\right)}=\frac{\partial\Phi\left(f_{\underline{x}}\left(\underline{\xi}\right)\right)}{\partial\underline{\xi}}\frac{1}{\dfrac{\partial f_{\underline{x}}\left(\underline{\xi}\right)}{\partial\underline{\xi}}} \tag{15.64}$$

because the distribution density is known in this formulation of the problem and we search for the appropriate convex function Φ, fitting to the given distribution density.

Discrete:

$$H=-\sum_{i=1}^{n}\Phi(p_i) \tag{15.65}$$

with the boundary conditions $\sum_{i=1}^{n}p_i=1 \qquad \sum_{i=1}^{n}q_i=1$ (15.66)

and

$$\sum_{i=1}^{n}p_i\cdot g_r(x_i)=E\{g_r(x)\}=a_r \qquad \text{for } r=1,\ldots,m \tag{15.67}$$

With these preconditions (the desired function $\Phi(\ )$ has to be convex), we obtain the Lagrangian function

$$LF=-\sum_{i=1}^{n}\Phi(p_i)+\sum_{r=1}^{m}\lambda_r\cdot\left[\sum_{i=1}^{n}p_i\cdot g_r(x_i)-a_r\right]+\mu\cdot\left[\sum_{i=1}^{n}p_i-1\right] \tag{15.68}$$

The derivative is

$$\frac{\partial}{\partial p_i}LF=-\frac{\partial}{\partial p_i}\Phi(p_i)+\mu+\sum_{r=1}^{m}\lambda_r\cdot g_r(x_i)\overset{!}{=}0$$

$$\frac{\partial}{\partial p_i}\Phi(p_i)=\mu+\sum_{r=1}^{m}\lambda_r\cdot g_r(x_i) \tag{15.69}$$

So we are able to calculate entropy functions, which have to be optimized for each problem to derive the optimum estimate without any further boundary conditions.

15.6.1 Example: Gaussian distribution and Shannon's information

We now ask whether Shannon's information really is the optimum information measure when we deal with gaussian distribution densities. We start our examination with the equation

$$\frac{\partial}{\partial f_{\underline{x}}(\underline{\xi})}\Phi\left(f_{\underline{x}}(\underline{\xi})\right)=\mu+\sum_{r=1}^{m}\lambda_r\cdot\underline{g}_r(\underline{\xi}) \tag{15.70}$$

rearrange the left-hand side

$$\frac{\partial\Phi\left(f_{\underline{x}}(\underline{\xi})\right)}{\partial f_{\underline{x}}(\underline{\xi})}=\frac{\partial\Phi\left(f_{\underline{x}}(\underline{\xi})\right)}{\partial\underline{\xi}}\frac{1}{\dfrac{\partial f_{\underline{x}}(\underline{\xi})}{\partial\underline{\xi}}}=\mu+\sum_{r=1}^{m}\lambda_r\cdot\underline{g}_r(\underline{\xi})$$

$$\frac{\partial\Phi\left(f_{\underline{x}}(\underline{\xi})\right)}{\partial\underline{\xi}}=\left[\mu+\sum_{r=1}^{m}\lambda_r\cdot\underline{g}_r(\underline{\xi})\right]\cdot\frac{\partial f_{\underline{x}}(\underline{\xi})}{\partial\underline{\xi}} \tag{15.71}$$

and use our knowledge of the gaussian distribution density

$$f_{\underline{x}}(\underline{\xi})=\frac{1}{(2\pi)^{\frac{n}{2}}\cdot|P|^{\frac{1}{2}}}\cdot\exp\left\{-\frac{1}{2}\cdot(\underline{\xi}-\underline{m})^T\cdot P^{-1}\cdot(\underline{\xi}-\underline{m})\right\} \tag{15.72}$$

and of the derivative

$$\frac{\partial f_{\underline{x}}(\underline{\xi})}{\partial\underline{\xi}}=f_{\underline{x}}(\underline{\xi})\cdot\left\{-P^{-1}\cdot(\underline{\xi}-\underline{m})\right\} \tag{15.73}$$

This leads to

$$\frac{\partial\Phi\left(f_{\underline{x}}(\underline{\xi})\right)}{\partial\underline{\xi}}=\left[\mu+\sum_{r=1}^{m}\lambda_r\cdot\underline{g}_r(\underline{\xi})\right]\cdot f_{\underline{x}}(\underline{\xi})\cdot\left\{-P^{-1}\cdot(\underline{\xi}-\underline{m})\right\} \tag{15.74}$$

On the other hand Shannon's information has to be the optimum entropy function for gaussian distribution densities, so that we may write the function **Φ** and the derivative as

$$\frac{\partial\Phi\left(f_{\underline{x}}\left(\underline{\xi}\right)\right)}{\partial\underline{\xi}}=\frac{\partial}{\partial\underline{\xi}}\left[f_{\underline{x}}\left(\underline{\xi}\right)\cdot\ln f_{\underline{x}}\left(\underline{\xi}\right)\right]=\frac{\partial f_{\underline{x}}\left(\underline{\xi}\right)}{\partial\underline{\xi}}\cdot\ln f_{\underline{x}}\left(\underline{\xi}\right)+\frac{f_{\underline{x}}\left(\underline{\xi}\right)}{f_{\underline{x}}\left(\underline{\xi}\right)}\cdot\frac{\partial f_{\underline{x}}\left(\underline{\xi}\right)}{\partial\underline{\xi}}$$

$$\frac{\partial\Phi\left(f_{\underline{x}}\left(\underline{\xi}\right)\right)}{\partial\underline{\xi}}=\frac{\partial f_{\underline{x}}\left(\underline{\xi}\right)}{\partial\underline{\xi}}\cdot\left[1+\ln f_{\underline{x}}\left(\underline{\xi}\right)\right] \tag{15.75}$$

We now have to show that the following equations are equal

$$\frac{\partial\Phi\left(f_{\underline{x}}\left(\underline{\xi}\right)\right)}{\partial\underline{\xi}}=\left[\mu+\sum_{r=1}^{m}\lambda_r\cdot\underline{g}_r\left(\underline{\xi}\right)\right]\cdot\frac{\partial f_{\underline{x}}\left(\underline{\xi}\right)}{\partial\underline{\xi}} \tag{15.76}$$

$$\frac{\partial\Phi\left(f_{\underline{x}}\left(\underline{\xi}\right)\right)}{\partial\underline{\xi}}=\frac{\partial f_{\underline{x}}\left(\underline{\xi}\right)}{\partial\underline{\xi}}\cdot\left[1+\ln f_{\underline{x}}\left(\underline{\xi}\right)\right] \tag{15.77}$$

if we restrict our examination to second order moments. We thus write

$$\frac{\partial f_{\underline{x}}\left(\underline{\xi}\right)}{\partial\underline{\xi}}\cdot\left[1+\ln f_{\underline{x}}\left(\underline{\xi}\right)\right]=\left[\mu+\sum_{r=1}^{m}\lambda_r\cdot\underline{g}_r\left(\underline{\xi}\right)\right]\cdot\frac{\partial f_{\underline{x}}\left(\underline{\xi}\right)}{\partial\underline{\xi}}$$

$$1+\ln f_{\underline{x}}\left(\underline{\xi}\right)=\mu+\sum_{r=1}^{m}\lambda_r\cdot\underline{g}_r\left(\underline{\xi}\right) \tag{15.78}$$

To obtain a summation on the left-hand side of this equation, we generate a series expansion of the logarithm at the expectation value $E\{\underline{x}\}=\underline{m}$.

$$\begin{aligned}\ln f_{\underline{x}}\left(\underline{\xi}\right)=\ln f_{\underline{x}}\left(\underline{m}\right)&+\frac{\partial}{\partial\underline{\xi}}\ln f_{\underline{x}}\left(\underline{\xi}\right)\Big|_{\underline{\xi}=\underline{m}}\cdot\left(\underline{\xi}-\underline{m}\right)\\&+\frac{1}{2}\cdot\frac{\partial^2}{\partial\underline{\xi}\cdot\partial\underline{\xi}}\ln f_{\underline{x}}\left(\underline{\xi}\right)\Big|_{\underline{\xi}=\underline{m}}\cdot\left(\underline{\xi}-\underline{m}\right)\cdot\left(\underline{\xi}-\underline{m}\right)^T+\text{h.o.t.}\end{aligned} \tag{15.79}$$

This leads to

$$\begin{aligned}\ln f_{\underline{x}}\left(\underline{\xi}\right)=\ln f_{\underline{x}}\left(\underline{m}\right)&+\frac{1}{f_{\underline{x}}\left(\underline{\xi}\right)}\cdot\frac{\partial f_{\underline{x}}\left(\underline{\xi}\right)}{\partial\underline{\xi}}\Big|_{\underline{\xi}=\underline{m}}\cdot\left(\underline{\xi}-\underline{m}\right)\\&+\frac{1}{2}\cdot\frac{\partial}{\partial\underline{\xi}}\left[\frac{1}{f_{\underline{x}}\left(\underline{\xi}\right)}\cdot\frac{\partial f_{\underline{x}}\left(\underline{\xi}\right)}{\partial\underline{\xi}}\right]\Big|_{\underline{\xi}=\underline{m}}\cdot\left(\underline{\xi}-\underline{m}\right)\cdot\left(\underline{\xi}-\underline{m}\right)^T+\text{h.o.t.}\end{aligned}$$

$$\ln f_{\underline{x}}(\underline{\xi}) = \ln f_{\underline{x}}(\underline{m}) + \underbrace{\frac{1}{f_{\underline{x}}(\underline{\xi})} \cdot \left.\frac{\partial f_{\underline{x}}(\underline{\xi})}{\partial \underline{\xi}}\right|_{\underline{\xi}=\underline{m}}}_{=0} \cdot (\underline{\xi} - \underline{m})$$

$$+\frac{1}{2} \cdot \left.\left[\left(\underbrace{\frac{1}{f_{\underline{x}}(\underline{\xi})} \cdot \frac{\partial f_{\underline{x}}(\underline{\xi})}{\partial \underline{\xi}}}_{=0 \text{ für } \underline{\xi}=\underline{m}}\right)^2 + \frac{1}{f_{\underline{x}}(\underline{\xi})} \cdot \frac{\partial^2 f_{\underline{x}}(\underline{\xi})}{\partial \underline{\xi} \cdot \partial \underline{\xi}^T}\right]\right|_{\underline{\xi}=\underline{m}} \cdot (\underline{\xi} - \underline{m}) \cdot (\underline{\xi} - \underline{m})^T + \text{h.o.t.} \tag{15.80}$$

We get

$$\ln f_{\underline{x}}(\underline{\xi}) = \ln f_{\underline{x}}(m) + \frac{1}{2} \cdot \left.\left[\frac{1}{f_{\underline{x}}(\underline{\xi})} \cdot \frac{\partial^2 f_{\underline{x}}(\underline{\xi})}{\partial \underline{\xi} \cdot \partial \underline{\xi}^T}\right]\right|_{\underline{\xi}=\underline{m}} \cdot (\underline{\xi} - \underline{m}) \cdot (\underline{\xi} - \underline{m})^T + \text{h.o.t.} \tag{15.81}$$

and we insert this result into the equation

$$1 + \ln f_{\underline{x}}(\underline{\xi}) = \mu + \sum_{r=1}^{m} \lambda_r \cdot \underline{g}_r(\underline{\xi}) \tag{15.82}$$

Neglecting the higher order terms (h.o.t.) we obtain

$$1 + \ln f_{\underline{x}}(\underline{m}) + \frac{1}{2} \cdot \left.\left[\frac{1}{f_{\underline{x}}(\underline{\xi})} \cdot \frac{\partial^2 f_{\underline{x}}(\underline{\xi})}{\partial \underline{\xi} \cdot \partial \underline{\xi}^T}\right]\right|_{\underline{\xi}=\underline{m}} \cdot (\underline{\xi} - \underline{m}) \cdot (\underline{\xi} - \underline{m})^T = \mu + \sum_{r=1}^{m} \lambda_r \cdot \underline{g}_r(\underline{\xi}) \tag{15.83}$$

The additional restriction on second order moments in the precondition $g_r(\underline{\xi})$ provides

$$1 + \ln f_{\underline{x}}(\underline{m}) + \frac{1}{2} \cdot \left.\left[\frac{1}{f_{\underline{x}}(\underline{\xi})} \cdot \frac{\partial^2 f_{\underline{x}}(\underline{\xi})}{\partial \underline{\xi} \cdot \partial \underline{\xi}^T}\right]\right|_{\underline{\xi}=\underline{m}} \cdot (\underline{\xi} - \underline{m}) \cdot (\underline{\xi} - \underline{m})^T \tag{15.84}$$

$$= \mu + \lambda_1 \cdot \underline{\xi} + \lambda_2 \cdot (\underline{\xi} - \underline{\mu}) \cdot (\underline{\xi} - \underline{\mu})^T$$

Comparison of the coefficients leads to

$$\mu = 1 + \ln f_{\underline{x}}(\underline{m})$$ for the first Lagrangian multiplier (15.85)

(μ = multiplier, $\underline{m}$ = mean)

$$\mu = 1 - \frac{1}{2} \cdot \ln\left[(2\pi)^n \cdot |P|\right]$$ for the given gaussian distribution (15.86)

$$\lambda_1 = 0 \tag{15.87}$$

The a priori knowledge of the first order moment is not applied in the boundary conditions. This also does not occur in Shannon's information of gaussian distribution densities.

$$\lambda_2 = \frac{1}{2} \cdot \left[\frac{1}{f_{\underline{x}}(\underline{\xi})} \cdot \frac{\partial^2 f_{\underline{x}}(\underline{\xi})}{\partial \underline{\xi} \cdot \partial \underline{\xi}^T} \right]\Bigg|_{\underline{\xi}=\underline{m}} \tag{15.88}$$

The negative expectation value of this term is Fisher's information.

With $\dfrac{\partial f_{\underline{x}}(\underline{\xi})}{\partial \underline{\xi}} = f_{\underline{x}}(\underline{\xi}) \cdot \left\{ -P^{-1} \cdot (\underline{\xi} - \underline{m}) \right\}$ we obtain:

$$\lambda_2 = \frac{1}{2} \cdot \left[\frac{1}{f_{\underline{x}}(\underline{\xi})} \cdot \frac{\partial}{\partial \underline{\xi}} \left[f_{\underline{x}}(\underline{\xi}) \cdot \left\{ -P^{-1} \cdot (\underline{\xi} - \underline{m}) \right\} \right] \right]\Bigg|_{\underline{\xi}=\underline{m}}$$

$$\lambda_2 = \frac{1}{2} \cdot \left[\frac{1}{f_{\underline{x}}(\underline{\xi})} \cdot \frac{\partial}{\partial \underline{\xi}} \left[f_{\underline{x}}(\underline{\xi}) \cdot \left\{ -P^{-1} \cdot (\underline{\xi} - \underline{m}) \right\} + f_{\underline{x}}(\underline{\xi}) \cdot \left\{ -P^{-1} \right\} \right] \right]\Bigg|_{\underline{\xi}=\underline{m}}$$

$$\lambda_2 = -\frac{1}{2} \cdot P^{-1} \tag{15.89}$$

Thus Shannon's information is an appropriate information measure if we deal with gaussian random variables or if we restrict our attention to second order moments. But then we anyway restrict the available distribution densities to gaussian distribution density functions.

At the end of this chapter we want to extend the previous derivation to the maximum likelihood estimation. We therefore use the likelihood function

$$f_{\underline{y}|\underline{x}}(\underline{\rho} \mid \underline{\xi}) = \frac{1}{(2\pi)^{\frac{m}{2}} \cdot |R|^{\frac{1}{2}}} \cdot \exp\left\{ -\frac{1}{2} \cdot (\underline{\rho} - h(\underline{\xi}))^T \cdot R^{-1} \cdot (\underline{\rho} - h(\underline{\xi})) \right\} \tag{15.90}$$

and the derivative

$$\frac{\partial}{\partial \underline{\xi}} f_{\underline{y}|\underline{x}}(\underline{\rho} \mid \underline{\xi}) = f_{\underline{y}|\underline{x}}(\underline{\rho} \mid \underline{\xi}) \cdot \left(\frac{\partial}{\partial \underline{\xi}} h(\underline{\xi}) \right)^T \cdot R^{-1} \cdot (\underline{\rho} - h(\underline{\xi})) \tag{15.91}$$

Analogously to the previous derivation we get

$$1 + \ln f_{\underline{y}|\underline{x}}(\underline{\rho} \mid \underline{\xi}) = \mu + \sum_{r=1}^{m} \lambda_r \cdot \underline{g}_r(\underline{\xi}) \tag{15.92}$$

and expand the logarithm at an arbitrary point $\underline{x}_E$:

$$\begin{aligned} \ln f_{\underline{y}|\underline{x}}(\underline{\rho} \mid \underline{\xi}) &= \ln f_{\underline{y}|\underline{x}}(\underline{\rho} \mid \underline{x}_E) + \frac{\partial}{\partial \underline{\xi}} \ln f_{\underline{y}|\underline{x}}(\underline{\rho} \mid \underline{\xi}) \Big|_{\underline{\xi}=\underline{x}_E} \cdot (\underline{\xi} - \underline{x}_E) \\ &+ \frac{1}{2} \cdot \frac{\partial^2}{\partial \underline{\xi} \cdot \partial \underline{\xi}} \ln f_{\underline{y}|\underline{x}}(\underline{\rho} \mid \underline{\xi}) \Big|_{\underline{\xi}=\underline{x}_E} \cdot (\underline{\xi} - \underline{x}_E) \cdot (\underline{\xi} - \underline{x}_E)^T + \text{h.o.t.} \end{aligned} \tag{15.93}$$

This leads to

$$\begin{aligned} \ln f_{\underline{y}|\underline{x}}(\underline{\rho} \mid \underline{\xi}) &= \ln f_{\underline{y}|\underline{x}}(\underline{\rho} \mid \underline{x}_E) + \frac{1}{f_{\underline{y}|\underline{x}}(\underline{\rho} \mid \underline{\xi})} \cdot \frac{\partial f_{\underline{y}|\underline{x}}(\underline{\rho} \mid \underline{\xi})}{\partial \underline{\xi}} \Bigg|_{\underline{\xi}=\underline{x}_E} \cdot (\underline{\xi} - \underline{x}_E) \\ &+ \frac{1}{2} \cdot \left[\left(\frac{1}{f_{\underline{y}|\underline{x}}(\underline{\rho} \mid \underline{\xi})} \cdot \frac{\partial f_{\underline{y}|\underline{x}}(\underline{\rho} \mid \underline{\xi})}{\partial \underline{\xi}} \right)^2 + \frac{1}{f_{\underline{y}|\underline{x}}(\underline{\rho} \mid \underline{\xi})} \cdot \frac{\partial^2 f_{\underline{y}|\underline{x}}(\underline{\rho} \mid \underline{\xi})}{\partial \underline{\xi} \cdot \partial \underline{\xi}^T} \right] \Bigg|_{\underline{\xi}=\underline{x}_E} \\ &\cdot (\underline{\xi} - \underline{x}_E) \cdot (\underline{\xi} - \underline{x}_E)^T + \text{h.o.t.} \end{aligned} \tag{15.94}$$

Now we insert the maximum likelihood estimate

$$\ln f_{\underline{y}|\underline{x}}\left(\underline{\rho} \mid \underline{\xi}\right)=\ln f_{\underline{y}|\underline{x}}\left(\underline{\rho} \mid \underline{x}_{ML}\right)+\frac{1}{f_{\underline{y}|\underline{x}}\left(\underline{\rho} \mid \underline{\xi}\right)} \cdot \left.\underbrace{\frac{\partial f_{\underline{y}|\underline{x}}\left(\underline{\rho} \mid \underline{\xi}\right)}{\partial \underline{\xi}}}_{=0 \text{ for } \underline{\xi}=\underline{x}_{ML}}\right|_{\underline{\xi}=\underline{x}_{ML}} \cdot\left(\underline{\xi}-\underline{x}_{ML}\right)$$

$$+\frac{1}{2} \cdot\left.\left[\left(\frac{1}{f_{\underline{y}|\underline{x}}\left(\underline{\rho} \mid \underline{\xi}\right)} \cdot \underbrace{\frac{\partial f_{\underline{y}|\underline{x}}\left(\underline{\rho} \mid \underline{\xi}\right)}{\partial \underline{\xi}}}_{=0 \text{ for } \underline{\xi}=\underline{x}_{ML}}\right)^2+\frac{1}{f_{\underline{y}|\underline{x}}\left(\underline{\rho} \mid \underline{\xi}\right)} \cdot \frac{\partial^2 f_{\underline{y}|\underline{x}}\left(\underline{\rho} \mid \underline{\xi}\right)}{\partial \underline{\xi} \cdot \partial \underline{\xi}^T}\right]\right|_{\underline{\xi}=\underline{x}_{ML}}$$

$$\cdot\left(\underline{\xi}-\underline{x}_{ML}\right) \cdot\left(\underline{\xi}-\underline{x}_{ML}\right)^T+\text{h.o.t.} \tag{15.95}$$

This simplifies the equation

$$\ln f_{\underline{y}|\underline{x}}\left(\underline{\rho} \mid \underline{\xi}\right)=\ln f_{\underline{y}|\underline{x}}\left(\underline{\rho} \mid \underline{x}_{\mathrm{ML}}\right)$$
$$+\frac{1}{2} \cdot\left.\left[\frac{1}{f_{\underline{y}|\underline{x}}\left(\underline{\rho} \mid \underline{\xi}\right)} \cdot \frac{\partial^2 f_{\underline{y}|\underline{x}}\left(\underline{\rho} \mid \underline{\xi}\right)}{\partial \underline{\xi} \cdot \partial \underline{\xi}^T}\right]\right|_{\underline{\xi}=\underline{x}_{\mathrm{ML}}} \cdot\left(\underline{\xi}-\underline{x}_{\mathrm{ML}}\right) \cdot\left(\underline{\xi}-\underline{x}_{\mathrm{ML}}\right)^T+\text{h.o.t.} \tag{15.96}$$

and we again insert this equation in

$$1+\ln f_{\underline{y}|\underline{x}}\left(\underline{\rho} \mid \underline{\xi}\right)=\mu+\sum_{r=1}^{m} \lambda_r \cdot \underline{g}_r\left(\underline{\xi}\right) \tag{15.97}$$

and restrict our attention to second order moments. We find

$$1+\ln f_{\underline{y}|\underline{x}}\left(\underline{\rho} \mid \underline{x}_{\mathrm{ML}}\right)+\frac{1}{2} \cdot\left.\left[\frac{1}{f_{\underline{y}|\underline{x}}\left(\underline{\rho} \mid \underline{\xi}\right)} \cdot \frac{\partial^2 f_{\underline{y}|\underline{x}}\left(\underline{\rho} \mid \underline{\xi}\right)}{\partial \underline{\xi} \cdot \partial \underline{\xi}^T}\right]\right|_{\underline{\xi}=\underline{x}_{\mathrm{ML}}} \cdot\left(\underline{\xi}-\underline{x}_{\mathrm{ML}}\right) \cdot\left(\underline{\xi}-\underline{x}_{\mathrm{ML}}\right)^T$$

$$=\mu+\sum_{r=1}^{2} \lambda_r \cdot \underline{g}_r\left(\underline{\xi}\right)$$

$$1+\ln f_{\underline{y}|\underline{x}}\left(\underline{\rho}\mid\underline{x}_{\mathrm{ML}}\right)+\frac{1}{2}\cdot\left[\frac{1}{f_{\underline{y}|\underline{x}}\left(\underline{\rho}\mid\underline{\xi}\right)}\cdot\frac{\partial^2 f_{\underline{y}|\underline{x}}\left(\underline{\rho}\mid\underline{\xi}\right)}{\partial\underline{\xi}\cdot\partial\underline{\xi}^T}\right]\Bigg|_{\underline{\xi}=\underline{x}_{\mathrm{ML}}}\cdot\left(\underline{\xi}-\underline{x}_{\mathrm{ML}}\right)\cdot\left(\underline{\xi}-\underline{x}_{\mathrm{ML}}\right)^T$$

$$=\mu+\lambda_{r1}\cdot\underline{\xi}+\lambda_2\cdot\left(\underline{\xi}-\underline{x}_{\mathrm{ML}}\right)\cdot\left(\underline{\xi}-\underline{x}_{\mathrm{ML}}\right)^T \tag{15.98}$$

Comparison of the coefficients leads to

$$\mu=1+\ln f_{\underline{y}|\underline{x}}\left(\underline{\rho}\mid\underline{x}_{\mathrm{ML}}\right)$$

$$\mu=1-\frac{1}{2}\cdot\ln\left[(2\pi)^m\cdot|R|\right] \tag{15.99}$$

$$\lambda_1=0 \tag{15.100}$$

$$\lambda_2=\frac{1}{2}\cdot\left[\frac{1}{f_{\underline{y}|\underline{x}}\left(\underline{\rho}\mid\underline{\xi}\right)}\cdot\frac{\partial^2 f_{\underline{y}|\underline{x}}\left(\underline{\rho}\mid\underline{\xi}\right)}{\partial\underline{\xi}\cdot\partial\underline{\xi}^T}\right]\Bigg|_{\underline{\xi}=\underline{x}_{\mathrm{ML}}} \tag{15.101}$$

With $\frac{\partial}{\partial\underline{\xi}}f_{\underline{y}|\underline{x}}\left(\underline{\rho}\mid\underline{\xi}\right)=f_{\underline{y}|\underline{x}}\left(\underline{\rho}\mid\underline{\xi}\right)\cdot\left(\frac{\partial}{\partial\underline{\xi}}h\left(\underline{\xi}\right)\right)^T\cdot R^{-1}\cdot\left(\underline{\rho}-h\left(\underline{\xi}\right)\right)$ we obtain

$$\lambda_2=\frac{1}{2}\cdot\left[\frac{1}{f_{\underline{y}|\underline{x}}\left(\underline{\rho}\mid\underline{\xi}\right)}\cdot\frac{\partial}{\partial\underline{\xi}}\left[f_{\underline{y}|\underline{x}}\left(\underline{\rho}\mid\underline{\xi}\right)\cdot\left(\frac{\partial}{\partial\underline{\xi}}h\left(\underline{\xi}\right)\right)^T\cdot R^{-1}\cdot\left(\underline{\rho}-h\left(\underline{\xi}\right)\right)\right]\right]\Bigg|_{\underline{\xi}=\underline{x}_{\mathrm{ML}}}$$

$$\lambda_2=\frac{1}{2}\cdot\frac{1}{f_{\underline{y}|\underline{x}}\left(\underline{\rho}\mid\underline{\xi}\right)}\cdot\Bigg[\underbrace{f_{\underline{y}|\underline{x}}\left(\underline{\rho}\mid\underline{\xi}\right)\cdot\left(\frac{\partial}{\partial\underline{\xi}}h\left(\underline{\xi}\right)\right)^T\cdot R^{-1}\cdot\left(\underline{\rho}-h\left(\underline{\xi}\right)\right)}_{=\frac{\partial}{\partial\underline{\xi}}f_{\underline{y}|\underline{x}}\left(\underline{\rho}|\underline{\xi}\right)=0\ \text{for}\ \underline{\xi}=\underline{x}_{\mathrm{ML}}}$$

$$-f_{\underline{y}|\underline{x}}\left(\underline{\rho}\mid\underline{\xi}\right)\cdot\left(\frac{\partial}{\partial\underline{\xi}}h\left(\underline{\xi}\right)\right)^T\cdot R^{-1}\cdot\left(\frac{\partial}{\partial\underline{\xi}}h\left(\underline{\xi}\right)\right)\Bigg]\Bigg|_{\underline{\xi}=\underline{x}_{\mathrm{ML}}}$$

$$\lambda_2 = -\frac{1}{2} \cdot \left[\left(\frac{\partial}{\partial \underline{\xi}} h(\underline{\xi}) \right)^T \cdot R^{-1} \cdot \left(\frac{\partial}{\partial \underline{\xi}} h(\underline{\xi}) \right) \right] \Bigg|_{\underline{\xi} = \underline{x}_{\mathrm{ML}}} \tag{15.102}$$

This is Fisher's information with an additional factor −1/2. Fisher's information is the inverse covariance matrix of the distribution of the parameters in the maximum likelihood estimation (see sections 11.3 and 11.4).

We noticed in the previous calculations with the assumed gaussian distribution densities that Shannon's information is an appropriate entropy function for the application of the maximum entropy method to gaussian distribution densities and it is connected to the maximum likelihood estimation according to the latter derivation.

With this consideration we leave this chapter on the maximum entropy method and we add some further remarks to the concept of information, to information theory and to some fields of information theory.

Now we take a final look at Shannon's information of the likelihood function

$$S = -\int_{\mathrm{P}} \ln f_{\underline{y}|\underline{x}}(\underline{\rho} \mid \underline{\xi}) \cdot f_{\underline{y}|\underline{x}}(\underline{\rho} \mid \underline{\xi}) \cdot d\underline{\rho} \tag{15.103}$$

and we expand the logarithm into a Taylor series at the maximum likelihood estimate $\underline{x}_{\mathrm{ML}}$, which leads to

$$\begin{aligned} \ln f_{\underline{y}|\underline{x}}(\underline{\rho} \mid \underline{\xi}) &= \ln f_{\underline{y}|\underline{x}}(\underline{\rho} \mid \underline{x}_{\mathrm{ML}}) + \frac{\partial}{\partial \underline{\xi}} \ln f_{\underline{y}|\underline{x}}(\underline{\rho} \mid \underline{\xi}) \Bigg|_{\underline{x}_{\mathrm{ML}}} \cdot (\underline{\xi} - \underline{x}_{\mathrm{ML}}) \\ &+ \frac{1}{2} \cdot (\underline{\xi} - \underline{x}_{\mathrm{ML}})^T \cdot \frac{\partial}{\partial \underline{\xi} \cdot \partial \underline{\xi}^T} \ln f_{\underline{y}|\underline{x}}(\underline{\rho} \mid \underline{\xi}) \Bigg|_{\underline{x}_{\mathrm{ML}}} \cdot (\underline{\xi} - \underline{x}_{\mathrm{ML}}) + \text{h.o.t.} \end{aligned} \tag{15.104}$$

As we build this expansion at the ML estimate, the first order derivative vanishes and we obtain

$$\begin{aligned} \ln f_{\underline{y}|\underline{x}}(\underline{\rho} \mid \underline{\xi}) &= \ln f_{\underline{y}|\underline{x}}(\underline{\rho} \mid \underline{x}_{\mathrm{ML}}) \\ &+ \frac{1}{2} \cdot (\underline{\xi} - \underline{x}_{\mathrm{ML}})^T \cdot \frac{\partial}{\partial \underline{\xi} \cdot \partial \underline{\xi}^T} \ln f_{\underline{y}|\underline{x}}(\underline{\rho} \mid \underline{\xi}) \Bigg|_{\underline{x}_{\mathrm{ML}}} \cdot (\underline{\xi} - \underline{x}_{\mathrm{ML}}) + \text{h.o.t.} \end{aligned} \tag{15.105}$$

Neglecting the higher order terms, we get

$$S = -\int_{\mathrm{P}} \left[\ln f_{\underline{y}|\underline{x}}\left(\underline{\rho} \mid \underline{x}_{\mathrm{ML}}\right) + \frac{1}{2} \cdot \left(\underline{\xi} - \underline{x}_{\mathrm{ML}}\right)^T \cdot \frac{\partial}{\partial \underline{\xi} \cdot \partial \underline{\xi}^T} \ln f_{\underline{y}|\underline{x}}\left(\underline{\rho} \mid \underline{\xi}\right) \Bigg|_{\underline{x}_{\mathrm{ML}}} \cdot \left(\underline{\xi} - \underline{x}_{\mathrm{ML}}\right) \right] \cdot f_{\underline{y}|\underline{x}}\left(\underline{\rho} \mid \underline{\xi}\right) \cdot d\underline{\rho} \tag{15.106}$$

and thus

$$S = -\int_{\mathrm{P}} \ln f_{\underline{y}|\underline{x}}\left(\underline{\rho} \mid \underline{x}_{\mathrm{ML}}\right) \cdot f_{\underline{y}|\underline{x}}\left(\underline{\rho} \mid \underline{\xi}\right) \cdot d\underline{\rho} - \left(\underline{\xi} - \underline{x}_{\mathrm{ML}}\right)^T \cdot \int_{\mathrm{P}} \frac{1}{2} \cdot \frac{\partial}{\partial \underline{\xi} \cdot \partial \underline{\xi}^T} \ln f_{\underline{y}|\underline{x}}\left(\underline{\rho} \mid \underline{\xi}\right) \Bigg|_{\underline{x}_{\mathrm{ML}}} \cdot f_{\underline{y}|\underline{x}}\left(\underline{\rho} \mid \underline{\xi}\right) \cdot d\underline{\rho} \cdot \left(\underline{\xi} - \underline{x}_{\mathrm{ML}}\right) \tag{15.107}$$

with

$$f_{\underline{y}|\underline{x}}\left(\underline{\rho} \mid \underline{\xi}\right) = \frac{1}{(2\pi)^{\frac{m}{2}} \cdot |R|^{\frac{1}{2}}} \cdot \exp\left\{ -\frac{1}{2} \cdot \left(\underline{\rho} - \underline{h}(\underline{\xi})\right)^T \cdot R^{-1} \cdot \left(\underline{\rho} - \underline{h}(\underline{\xi})\right) \right\} \tag{15.108}$$

The second integral is the negative Fisher information and we therefore write

$$S = \int_{\mathrm{P}} \ln f_{\underline{y}|\underline{x}}\left(\underline{\rho} \mid \underline{x}_{\mathrm{ML}}\right) \cdot f_{\underline{y}|\underline{x}}\left(\underline{\rho} \mid \underline{\xi}\right) \cdot d\underline{\rho} + \frac{1}{2} \cdot \left(\underline{\xi} - \underline{x}_{\mathrm{ML}}\right)^T \cdot F \cdot \left(\underline{\xi} - \underline{x}_{\mathrm{ML}}\right) \tag{15.109}$$

with

$$F = \left[\left(\frac{\partial}{\partial \underline{\xi}} \underline{h}(\underline{\xi}) \right)^T \cdot R^{-1} \cdot \left(\frac{\partial}{\partial \underline{\xi}} \underline{h}(\underline{\xi}) \right) \right] \Bigg|_{\underline{\xi} = \underline{x}_{\mathrm{ML}}} \tag{15.110}$$

The first integral is the expectation value of the log-likelihood function, and in case of the gaussian distribution that we used, we obtain

$$S = -\int_{\mathrm{P}} \left[-\frac{1}{2} \cdot \ln\left[(2\pi)^m \cdot |R|\right] - \frac{1}{2} \cdot \left(\underline{\rho} - \underline{h}(\underline{x}_{\mathrm{ML}})\right)^T \cdot R^{-1} \cdot \left(\underline{\rho} - \underline{h}(\underline{x}_{\mathrm{ML}})\right) \right] \cdot f_{\underline{y}|\underline{x}}\left(\underline{\rho} \mid \underline{\xi}\right) \cdot d\underline{\rho} + \frac{1}{2} \cdot \left(\underline{\xi} - \underline{x}_{\mathrm{ML}}\right)^T \cdot F \cdot \left(\underline{\xi} - \underline{x}_{\mathrm{ML}}\right) \tag{15.111}$$

$$S = \frac{1}{2} \cdot \ln\left[(2\pi)^m \cdot |R|\right] + \frac{1}{2} \cdot \int\limits_{\mathrm{P}} \left(\underline{\rho} - \underline{h}(\underline{x}_{\mathrm{ML}})\right)^T \cdot R^{-1} \cdot \left(\underline{\rho} - \underline{h}(\underline{x}_{\mathrm{ML}})\right) \cdot f_{\underline{y}|\underline{x}}\left(\underline{\rho} \mid \underline{\xi}\right) \cdot d\underline{\rho}$$
$$+ \frac{1}{2} \cdot \left(\underline{\xi} - \underline{x}_{\mathrm{ML}}\right)^T \cdot F \cdot \left(\underline{\xi} - \underline{x}_{\mathrm{ML}}\right) \tag{15.112}$$

We now calculate the term

$$\frac{1}{2} \cdot \int\limits_{\mathrm{P}} \left(\underline{\rho} - \underline{h}(\underline{x}_{\mathrm{ML}})\right)^T \cdot R^{-1} \cdot \left(\underline{\rho} - \underline{h}(\underline{x}_{\mathrm{ML}})\right) \cdot f_{\underline{y}|\underline{x}}\left(\underline{\rho} \mid \underline{\xi}\right) \cdot d\underline{\rho} =$$

$$= \frac{1}{2} \cdot \mathrm{tr}\left\{ R^{-1} \cdot \int\limits_{\mathrm{P}} \left(\underline{\rho} - \underline{h}(\underline{x}_{\mathrm{ML}})\right) \cdot \left(\underline{\rho} - \underline{h}(\underline{x}_{\mathrm{ML}})\right)^T \cdot f_{\underline{y}|\underline{x}}\left(\underline{\rho} \mid \underline{\xi}\right) \cdot d\underline{\rho} \right\}$$

$$= \frac{1}{2} \cdot \mathrm{tr}\left\{ R^{-1} \cdot \int\limits_{\mathrm{P}} \left[\underline{\rho} \cdot \underline{\rho}^T - \underline{\rho} \cdot \underline{h}(\underline{x}_{\mathrm{ML}})^T - \underline{h}(\underline{x}_{\mathrm{ML}}) \cdot \underline{\rho}^T + \underline{h}(\underline{x}_{\mathrm{ML}}) \cdot \underline{h}(\underline{x}_{\mathrm{ML}})^T\right] \right.$$
$$\left. \cdot f_{\underline{y}|\underline{x}}\left(\underline{\rho} \mid \underline{\xi}\right) \cdot d\underline{\rho} \right\}$$

$$= \frac{1}{2} \cdot \mathrm{tr}\left\{ R^{-1} \cdot \left[R + \underline{h}(\underline{\xi}) \cdot \underline{h}(\underline{\xi})^T - \underline{h}(\underline{\xi}) \cdot \underline{h}(\underline{x}_{\mathrm{ML}})^T \right.\right.$$
$$\left.\left. - \underline{h}(\underline{x}_{\mathrm{ML}}) \cdot \underline{h}(\underline{\xi})^T + \underline{h}(\underline{x}_{\mathrm{ML}}) \cdot \underline{h}(\underline{x}_{\mathrm{ML}})^T \right]\right\}$$

$$= \frac{1}{2} \cdot \mathrm{tr}\left\{ I + R^{-1} \cdot \left(\underline{h}(\underline{\xi}) - \underline{h}(\underline{x}_{\mathrm{ML}})\right) \cdot \left(\underline{h}(\underline{\xi}) - \underline{h}(\underline{x}_{\mathrm{ML}})\right)^T \right\}$$

$$= \frac{m}{2} + \frac{1}{2} \cdot \left(\underline{h}(\underline{\xi}) - \underline{h}(\underline{x}_{\mathrm{ML}})\right)^T \cdot R^{-1} \cdot \left(\underline{h}(\underline{\xi}) - \underline{h}(\underline{x}_{\mathrm{ML}})\right) \tag{15.113}$$

and we obtain

$$S = \frac{1}{2} \cdot \ln\left[(2\pi)^m \cdot |R|\right] + \frac{m}{2} + \frac{1}{2} \cdot \left(\underline{h}(\underline{\xi}) - \underline{h}(\underline{x}_{\mathrm{ML}})\right)^T \cdot R^{-1} \cdot \left(\underline{h}(\underline{\xi}) - \underline{h}(\underline{x}_{\mathrm{ML}})\right)$$
$$+ \frac{1}{2} \cdot \left(\underline{\xi} - \underline{x}_{\mathrm{ML}}\right)^T \cdot F \cdot \left(\underline{\xi} - \underline{x}_{\mathrm{ML}}\right) \tag{15.114}$$

$$S = \frac{1}{2} \cdot \ln\left[(2\pi)^m \cdot |R|\right] + \frac{m}{2}$$
$$+ \frac{1}{2} \cdot \left(\underline{h}(\underline{\xi}) - \underline{h}(\underline{x}_{\mathrm{ML}})\right)^T \cdot R^{-1} \cdot \left(\underline{h}(\underline{\xi}) - \underline{h}(\underline{x}_{\mathrm{ML}})\right) + \frac{1}{2} \cdot \left(\underline{\xi} - \underline{x}_{\mathrm{ML}}\right)^T \cdot F \cdot \left(\underline{\xi} - \underline{x}_{\mathrm{ML}}\right) \tag{15.115}$$

The result obtained consists of two terms: The first term

$$S_{\text{term1}} = \frac{1}{2} \cdot \ln\left[(2\pi)^m \cdot |R|\right] + \frac{m}{2} \tag{15.116}$$

is the Shannon information of the gaussian distribution that we used in the calculation. The second term

$$S_{\text{term2}} = \frac{1}{2} \cdot \left(\underline{h}(\underline{\xi}) - \underline{h}(\underline{x}_{\text{ML}})\right)^T \cdot R^{-1} \cdot \left(\underline{h}(\underline{\xi}) - \underline{h}(\underline{x}_{\text{ML}})\right) + \frac{1}{2} \cdot \left(\underline{\xi} - \underline{x}_{\text{ML}}\right)^T \cdot F \cdot \left(\underline{\xi} - \underline{x}_{\text{ML}}\right) \tag{15.117}$$

describes the uncertainty in the second order moments in the two coordinate systems. The first addend in this second term describes the weighted quadratic uncertainty in the observation coordinate system where the weighting is achieved by the inverse of the covariance matrix describing the uncertainty (reciprocal information) of the observations. It includes possible ambiguities created by some nonlinear function $\underline{h}(\,)$.

The second addend does the same for the state space coordinate system, where the weighting matrix is Fisher's information describing the information (reciprocal uncertainty) of the state space variables.

16 Concluding remarks

16.1 Information, entropy and self-organization

These three concepts are closely related, which is the reason for us to carry out a more detailed examination in this chapter. Information and entropy have been subject to the considerations in the previous chapters and we noticed that we may roughly define information as organization and entropy as disorganization. Now this chapter adds the concept of self-organization, applied in biology and physics, which seems to present a contradiction to the second law of thermodynamics. This contradiction stems from the following consideration:

The non-living organisms on our earth disintegrate and thus they tend towards the Equilibrium State, the so-called 'heat death'.

Reason:

The increasing disorder in the disintegrating organisms leads to an increasing entropy until the maximum entropy is achieved. In this maximum entropy state the disorder achieves its maximum and cannot be increased any more. Therefore there are also no structures any more and as a consequence there is no information in the system. Moreover, and this is an important aspect regarding the comparison of entropy and energy (Carl F. von Weizsäcker uses this aspect), there is also no energy difference in the system.

Example:

A book (dead organism) disintegrates over the years to dust and thus loses all its information. The book includes a high degree of eliminated uncertainty (information), which vanishes during the disintegration.

The second law of thermodynamics states that the entropy has to increase in closed systems, which indicates that the disorder (disorganization) in closed systems has to increase.

In contrast to this second law of thermodynamics there exist living organisms, creating organized structures (self-organization) in the disintegrating environment and thus producing information, which enters these organizations. Thus they lose entropy (diminish their entropy) and so they seem to work against the second law of thermodynamics. They do not contradict the second law of thermodynamics, because the entropy increases only in closed systems. Self-organized systems are not closed systems, they are open systems, exchanging matter (air, food, ...), energy (heat, light, ...) and information with their environment. A constant emanation of heat to the environment increases the entropy of the environment and decreases the entropy of the system, providing the possibility of self-organization (generation of complex structures in an environment). This is the reason for the development of life on our earth [EBE90–94].

What are the preliminaries for life and for the generation of self-organization on our earth? At first the boundary conditions, such as temperature, light, climate, consistency of the atmosphere, have to be provided, to enable a survival of organisms. The still unexplained question about the genesis of life can likewise be answered with these preliminaries but not with a consistent scientific explanation. This is because the genesis of life out of dead matter is not comprehensible up to the present and it may not even be describable by random processes, as the probability of such an event would be very low.

At present it is suspected that there was some kind of 'soup' consisting of different materials (ammonia, ...). Then, a first, very primitive but living organism evolved, where the following rudimentary features may describe a living organism:

- Self-organization (generation of complex structures)
- Ability of reproduction

The concrete transition from this soup to such a primitive organism is still unexplained and may never be explained.

Man is able to clone, which has been demonstrated in several examples (e.g., the sheep Dolly), but man cannot create life. The 'raw material' of such a cloning process is always a living organism, containing the two rudimentary features and several other features, which we still do not understand completely. So there is an obvious difference between DNA in principle and the DNA of living cells. The DNA contains information about the structure, the construction and the 'mode of functioning' of a living being, but may not contain information about how the life of this living being came into existence.

The special preliminaries of self-organization on our earth are created by the sun, which permanently emits entropy in the form of radiation (heat, solar energy, ...) to the earth. This entropy is absorbed by the earth, which means that the entropy on the earth would have to increase permanently.

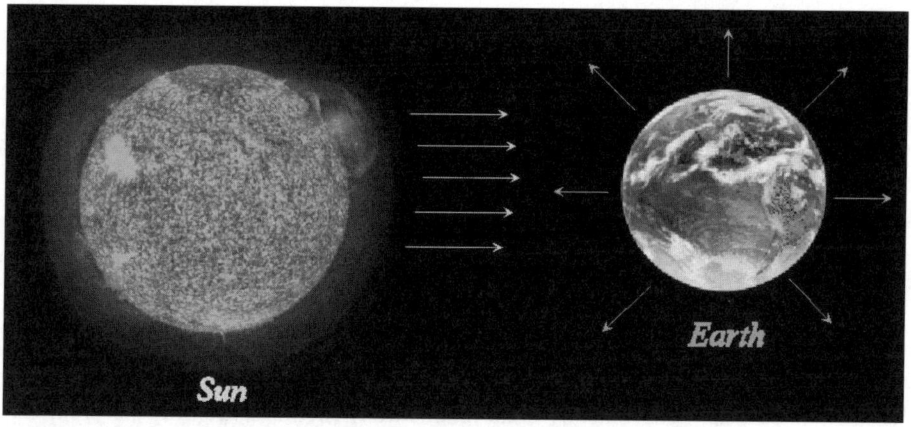

Fig. 16.1: Entropy emission from sun to earth

The sun emits a power (energy per time) of $P = 3.8 \cdot 10^{26}$ W into the universe. At the distance of our earth this leads to a radiant flux (intensity of the radiation = energy/time/area) of S = 1368 W/m^2. The area of cross-section of the earth is given by $A = \pi \cdot r^2$ and the incoming radiation power is obtained from the equation $P_{in} = S \cdot \pi \cdot r^2$. With the earths radius r = 6370 km this leads to $1.74 \cdot 10^{17}$ W or $1.74 \cdot 10^{17}$ J/s . It is estimated that the whole power demand of mankind is approximately $12 \cdot 10^{12}$ J/s = 12 000 GW (e.g., 10 000 large power plants of 1.2 GW each, or 2 kW per human being).

Now the earth not only absorbs energy from the sun, but it also emits energy into the universe. If we divide the heat energy by the temperature, obtaining the entropy, then the earth absorbs entropy from the sun and simultaneously emits entropy into the universe. Entropy is, according to Lord Kelvin, a measure for the quality of the energy. The greater the entropy of a system, the less structure and organization can be found. A system that is capable of generating self-organizing living structure has to emit entropy to be able to generate energy of higher quality.

The sun with its surface temperature of approximately 6000 K sends us an entropy $S_{in} = 1.74 \cdot 10^{17}$ W/5800 K and the earth simultaneously emits the entropy $S_{out} = 1.74 \cdot 10^{17}$ W/280 K, which leads to the difference of $-6 \cdot 10^{14}$ W/K . According to these thoughts each square meter of the earth exports approximately 1 W/K entropy. It is this export of entropy that provides the possibility for the generation of complex structures and for the development of life on earth. We thus live on negative entropy, i.e. we permanently generate energy of a higher quality, by emitting entropy [SCH89/1].

The generation of complex structures does not necessarily require a decrease of the entropy, as C. v. Weizsäcker explains. This generation diminishes the entropy because of the creation of organization of complex systems, but it increases the entropy as well, because it increases biodiversity and it thus again creates disorder.

Thus self-organization cannot be limited to the examination of the decrease of entropy. Instead we have to examine the complete system, which of course increases the complexity of the problem. The amalgamation of information is, however, closely related to the generation of complex structures. If we use the inverse conclusion, we may say that complex structures do contain information. This is exactly the approach of complexity theory.

16.2 Complexity theory

Andrej N. Kolmogorov and Alan Turing assign information to complex structures (increasing complexity → increasing information). Information eliminates – as we already discussed – the entropy (disorganization) in the same way as complex structures create organization in a disorganized environment.

So we may 'define' a random sequence by the feature that we are not able to find certain structures or patterns enabling us to reduce (compress) such a sequence. Non-random sequences consist of a sufficient number of patterns and can thus be compressed. This means that we may define the complexity of a data sequence by

Random sequence ↔ cannot be compressed

Non-random sequence ↔ can be compressed

The complexity thus describes the minimum number of data that we require to describe a given sequence. The number π can be denoted either by the Greek letter or by the specification of all numerals 3.1415.... In the same way we may write 1000 instead of 10^3 and there are several other possibilities to reduce the number of necessary symbols of a denotation. From this point of view complexity is the minimum number of symbols required for the description of a data sequence, a computer program, an algorithm, ...

The foundations of this concept of complexity are the considerations of Kolmogorov, v. Mises, v. Neumann, Church, Turing and others, regarding probability theory and the theory of random processes. The basic idea is:

If we assign weight to the identifiable patterns of a certain object, we are able to distinguish different objects by different patterns and we are additionally able to assign some kind of information to these object, by applying the weights (which may describe the peculiarity of the pattern). The more patterns some object has, the more information it provides, because the exact description of this object requires the complete enumeration of all patterns (characteristics). This assignment of information to such patterns can be extended to arbitrary objects and even to algorithms. We always assign weights to certain patterns and measure the information content (here: complexity) by means of these weights. This kind of 'information generation' is used in complexity theory to assign a complexity (information content) to data sequences.

Complexity is thus, in a simplified explanation, a measure for the detectable differences or for the number of distinguishable patterns. We do not want to deepen this consideration and – referring to further literature [LI93] – we turn to the runtime complexity, connected with the Halting Problem of Turing. The runtime complexity describes the duration required for the performance of an algorithm. If this duration tends to infinity, the program does not stop and we are confronted with the question, considered by Alan Turing as the Halting Problem.

Another point of view is presented in the approach of [YAS88]: In general the information functions do not contain any statements about the time required for the transmission of the information. This time can be considered in connection with the complexity, because the time required plays an important role in the transmission and processing of data. For example, encryption has to be optimized to obtain an optimum scrambling of the original message on the one hand and an optimized (minimized) transmission time on the other. Thus encryption requires optimum scrambling techniques and minimum coding length at the same time.

16.3 Data reduction

A statement made by Resnikoff [RES87]: "Theoretically we have two possibilities to process an increasing number of data. The first possibility – mainly used in technical applications up to the present – is the application of up to date technology to improve the processing and transmission of information. The aim is to proceed continually in this development to cope with the increasing number of data by an increased speed of the processing devices.

The second possibility – applied by nature because of its limited speed of development – is the reduction of the number of data. This data reduction is carried out in such a way that the important information is transmitted in a compact form, while irrelevant or redundant information is reduced to a minimum, or is even completely eliminated."

An example of this data reduction is also given in Resnikoff's book [RES87], where he examines the human eye. Our eye approximately contains 6 million light sensitive cells, suitable for the detection of different levels of luminance. If we now assume that these luminance values are encoded with 8 bit and that we need at least 50 images per second to perceive a continuous observation, then the product of these parameters $6 \cdot 10^{6} \cdot 8 \cdot 50$ leads to $2.4 \cdot 10^{9}$ bit/s, to be transported by the optic nerve. Such a rate of data (2.4 Gbit/s) exceeds even the capacity of fast computers.

In fact this throughput exceeds even the capacity of the cortical centers of our brain, which immediately implies that our brain has to perform some kind of data reduction to cope with this enormous amount of data. This data reduction is optimized by evolution in such a way that, for instance, rapid movement is sensed at the edges of our visual field much better than in the center.

Our present production of data is increasing so rapidly that data reduction is more and more used, in addition to the improvement of the devices, to cope with the amounts of information that we are currently producing. Data reduction can merely be applied without any losses (with respect to the information) if there are certain patterns. In this case the complexity is smaller than the number of data of the sequence and the number of data can be reduced without any loss of information. Lossy data reduction, applied in image reduction (e.g., Gif or Jpg), uses the information reception of the human eye to eliminate data with less information (higher frequencies) and thus to reduce the number of data. This is an intended elimination of irrelevant information.

16.4 Cryptology

The more data we transmit over the public internet, the more the security and connected encryption methods of these data become an important subject to users. The aim of the applied methods is the optimal encryption (κρυπτειν = to conceal) of the original message by algorithms, which have a very small probability to be decrypted by some unintended receiver. Absolute security will never be achieved, as it is (at least theoretically) possible to decrypt any code by brute force, i.e. by a computer with an appropriate power. The more time such a computer takes, the more security the coding provides in principle.

The generation of these codes is, for instance, achieved by application of mappings, which cannot be inverted without ambiguities. Such ambiguities complicate the decoding of the encrypted message to users who are not intended to read the original message. Because of the foreseen development of home banking, credit cards and other similar applications, this is currently a field of high interest to scientists.

16.5 Concluding considerations

Information is the pattern that is transported via physical, chemical, biological and other quantities and thus can be generally accessed. With this seizable definition we get at least a manageable idea of the concept of information. The resulting mathematical information measures describe the uncertainty that can be eliminated by a gain of information, i.e. some kind of potential concept of information. This is certainly caused by the character of these measures, describing the entropy and therefore the uncertainty.

The search for certain patterns, for information, has occupied man ever since he detected regularities in natural processes. It is in fact the aim of the natural sciences to describe these regularities with recurring patterns, to generate appropriate models and to obtain a theory of nature, which can be applied to take advantage of certain processes. This implies the detection and the knowledge of certain patterns, which we have to *re*cognize in nature.

The approach of pattern recognition in natural processes is applied in different scientific areas. Image processing takes advantage of concrete insight into the processes of our brain by creating detectors for edges, lines, different gradients and similar patterns. These patterns are detected, classified and their positions are located. Data analysis does not restrict the pattern recognition to the available data set. The given data set is transformed into other coordinate systems or in other forms of presentation, where the desired patterns are easier to identify. Such methods, e.g., subspace methods, search for the important pattern on the basis of the eigenvalues (PCA = principle component analysis) and they create a subspace of the eigenvectors. Thus the original signal space is transformed into the subspace of the interesting patterns.

A similar transformation is, for instance, the Fourier transformation, where the patterns are the frequency and the phase of the signals and the subspace is the frequency domain. In this context we may also mention some related transformations such as the Gabor transformation and wavelet transformations, which all search for special structures in the signals. The latter two transformations include the time of certain signal structures in the transformation, while the simple Fourier transformation merely describes the frequency and the phase of the signal.

Obviously the search for patterns is a basic element of scientific research. We are able to generate an organization of the observed system on the basis of these patterns and thus we are also able to describe the system. As we are not able to process the enormous amount of data that nature provides, or to obtain an insight into these process on the basis of these data, we decompose the systems into independent subsystems (canonical decomposition) and search for patterns. These patterns allow us to describe at least certain functions of these subsystems. If we combine these subsystems again and use the descriptions obtained for these subsystems, then we are able to describe the whole system by the connection of the subsystems. As, however, the whole system is more than the sum of his subsystems, the combination of the subsystems requires an additional examination of the mutual influences of the subsystems, because these are not included in the subsystems.

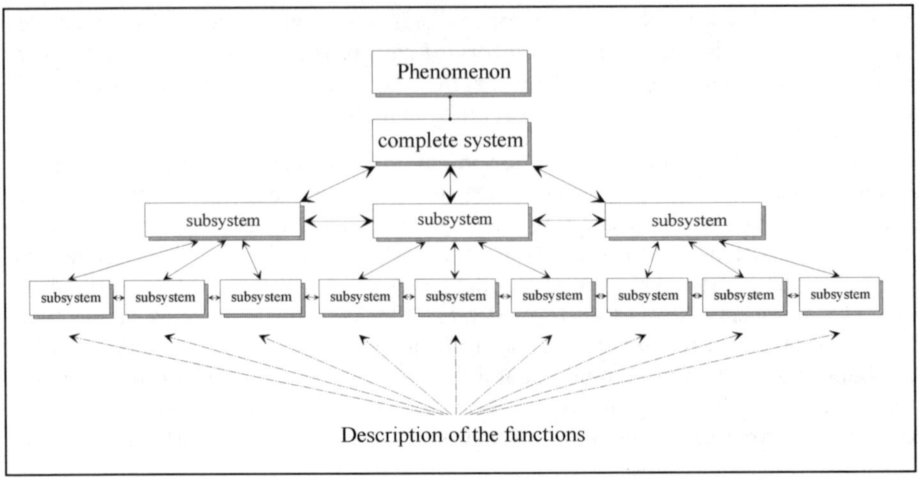

Fig. 16.2: System decomposition into subsystems

Figure 16.2 gives an impression of the exchange of information between the subsystems of one level and between the subsystems of different levels. An increasing complexity of the system leads to a more difficult description caused by the increasing dependence of different subsystems. In practical applications the description of such systems is usually restricted to the most important patterns, while the other patterns are combined into unimportant disturbances (effects that cannot be modeled). The determination of the important features can be achieved by such methods as PCA = principal component analysis, Karhunen–Loéve transformation or similar subspace methods (DOA, ESPRIT, ...). These methods determine the greatest eigenvalues, describing the most important features, and then they use these eigenvalues to determine the appropriate model or algorithm. In principle the combination of the subsystems into one system can be described as follows: The more subsystems we combine, the more we have to create simplifications to keep our knowledge of the system. These simplifications provide a loss of information, which should not change the important features of the system.

16.5.1 Information, entropy and probability

The concept of entropy as well as the concept of information are closely related to the probabilities of the events. The information functions are expectation values of distribution densities, which describe the probabilities of the possible events. These probabilities, which we assign to the events, have to be defined in some way before we use them. One possibility of a definition is the use of the relative frequency

$$h(A) = \frac{n(A)}{n} \tag{16.01}$$

describing the number $n(A)$ of the elements A occurring in n samples. The probability can be understood as the limit of this relative frequency for large n (number of samples) or as the expectation of the relative frequency. Both intuitive definitions of the probability have certain disadvantages, prohibiting a direct mathematical application of these heuristic denotations. The determination of the probability as the expectation of the relative frequencies already requires the definition of the probability in the computation of the expectation value

$$P(A) = E\{h(A)\} \tag{16.02}$$

and it thus leads to a circular argument, which is not allowed in a correct definition. The second approach computes the limit

$$P(A) = \lim_{n \to \infty} h(A) = \lim_{n \to \infty} \frac{n(A)}{n} \tag{16.03}$$

This limit cannot be computed with the common analytic methods and the observation of an infinite number of samples is not realistic. Moreover, the possibility that we are currently measuring an atypical sequence cannot be completely excluded. C. F. von Weizsäcker said: “The empirical verification of the theoretic probability can merely be achieved with a certain probability!” The empirical determination (measurement) of the probability can only be achieved by measuring the relative frequency and this relative frequency tends to the theoretical probability of the actual event only with a certain probability. The more samples we observe, the more the relative frequency should approach the theoretical probability (apart from atypical sequences).

Example: Throwing of a die

If we dice 20 times, we may not expect to obtain the probability 1/6 for each event from the sample space, because the quotient of the relative frequency already prohibits such a result.

$$h(A) = \frac{n(A)}{n} = \frac{n(A)}{20} \neq \frac{1}{6} \tag{16.04}$$

It is only possible to approach the true value 1/6 for a single event. Realistic results are

$$3/20 \approx 0.15$$

or

$$4/20 = 1/5 = 0.2$$

If we increase the number of samples to 100, the resulting relative frequency is closer to the true value and we get

16/100 = 0.16
17/100 = 0.17

The expected value is 0.16666.. .

The measurements verify our true value only with a limited accuracy and they contain errors with certain probabilities.

16.6 Information

How can we describe information? What is information?

We have, of course, not been able to answer these questions completely. Certain fundamental questions remain, which we may never be able to answer in a satisfying way. The nature of information can only be partially described by the objective descriptions that we applied, while other aspects of information are less amenable to scientific descriptions. Information is probably somewhat equivalent to the monads, used by the German philosopher Leibnitz, which do not have any extension and are mere substance. Weizsäcker uses exactly this term substance when he tries to find some unchanging part in the continuous flow of time. Information may thus be the steady component in nature, explaining the organization or the order. An appropriate example is DNA, containing the complete genetic information of a living being. The foundations of all complex processes, the arrangement and the functions of all organs as well as all typical features are included in this part of a cell. The DNA changes in the course of evolution, but this is an adaptation of nature, where the gain of information leads to a change of the information of the DNA. But what about our mind, elevating the human being above all other living beings? Is it also only a piece of matter, determined in the DNA? This question, which leads us back to the Laplacian demon and to other theories (theodize, etc.), cannot be answered in this context.

A principle of order or organization (deterministic pattern) was already known to the ancient Greek philosophers. Anaxagoras assumed that the universe began in a disorganized state, which he called chaos (χαος), and this chaos has been transformed from a spirit or esprit (νους) to a certain order, which he calls cosmos (κοσμος). (The esprit (νους) 'permeates' the cosmos).

Organization stems from the chaos, forms (patterns) occur, and something is 'informed'. The fundamental idea is thus very old and certainly 'rethought' in many different ways, but no closed definition has been found up to the present. Nevertheless, it is seems that information and its basic meaning is not only based on physically measurable quantities. Information is able to appear in different quantities and is able to change carrier very easily, which demonstrates a certain independence of the measurable quantities, which we are used to.

Nevertheless we are able to describe and apply a certain part of the information, the syntactic information, with the information functions described. The definition of the whole concept of information may be impossible for us, so that we have to rely on our familiar methods (see the foreword: Pascal). Though our applicable concept of information is based on a limited definition, the resulting applications are very extensive and the information age, which is now propagated, promises further applications of information processing and transmission. But this increasing information transmission refers to the syntactic level, which cannot guarantee pragmatic or useful information. The information transmitting devices are finally not responsible for the meaning or the sense of the transmitted messages. This responsibility is left to the users, to us. So it is up to us to extract the reasonable information out of the growing flood of information.

Appendix

A.1 Inequality for Kullback's information

We derive the inequality (Gibb's inequality) for discrete random variables.

We start with an estimate for the cross-entropy

$$\Psi = \sum_{i=1}^{n} p_i \cdot \ln q_i \tag{A.01}$$

$$\sum_{i=1}^{n} p_i = \sum_{i=1}^{n} q_i = 1 \tag{A.02}$$

by expanding this term and then by applying the Jensen inequality to obtain an upper limit for this expression.

Let $\quad q_i = p_i + u_i = p_i\left(1 + \frac{u_i}{p_i}\right) \quad$ with $\quad \sum_{i=1}^{n} u_i = 0 \quad$ (A.03)

If we substitute this into Ψ, we get:

$$\Psi = \sum_{i=1}^{n} p_i \cdot \ln q_i = \sum_{i=1}^{n} p_i \cdot \ln\left[p_i\left(1 + \frac{u_i}{p_i}\right)\right]$$

$$\Psi = \sum_{i=1}^{n} p_i \cdot \ln p_i + \sum_{i=1}^{n} p_i \cdot \ln\left(1 + \frac{u_i}{p_i}\right) \tag{A.04}$$

Application of the Jensen inequality $\;x \geq \ln(1+x)\;$ leads to

$$\Psi = \sum_{i=1}^{n} p_i \cdot \ln p_i + \sum_{i=1}^{n} p_i \cdot \ln\left(1 + \frac{u_i}{p_i}\right) \leq \sum_{i=1}^{n} p_i \cdot \ln p_i + \sum_{i=1}^{n} p_i \cdot \frac{u_i}{p_i}$$

$$\Psi = \sum_{i=1}^{n} p_i \cdot \ln p_i + \sum_{i=1}^{n} p_i \cdot \ln\left(1 + \frac{u_i}{p_i}\right) \le \sum_{i=1}^{n} p_i \cdot \ln p_i + \underbrace{\sum_{i=1}^{n} u_i}_{0}$$

$$\Psi = \sum_{i=1}^{n} p_i \cdot \ln q_i \le \sum_{i=1}^{n} p_i \cdot \ln p_i \tag{A.05}$$

Gibbs's inequality

We thus have an upper limit of the cross-entropy, which can also be achieved by using the nonnegativity of Kullback's information for discrete events.

A.2 The log-sum inequality

The log-sum inequality holds for convex or concave functions and results from the Taylor series expansion of these functions. The Taylor series expansion is:

$$f(x) = f(x_o) + f'(x_o) \cdot (x - x_0) + \frac{1}{2} \cdot f''(x_o) \cdot (x - x_0)^2 + \text{h.o.t.} \tag{A.06}$$

The approximation error of this Taylor series is expressed by the Lagrangian residuum, which is equal to the first term of the series, which is neglected in the approximation. If we truncate the series after the linear term, the error is determined by

$$F(x) = \frac{1}{2} \cdot f''(x_o) \cdot (x - x_0)^2 \tag{A.07}$$

As the term $(x - x_0)^2$ is always positive or equal to zero, the second derivative $f''(x_0)$ determines the sign of the error, and in general we distinguish

$$f''(x_o) = \left.\frac{d^2}{dx^2} f(x)\right|_{x=x_0} \ge 0 \qquad \text{convex function} \tag{A.08}$$

$$f''(x_o) = \left.\frac{d^2}{dx^2} f(x)\right|_{x=x_0} \le 0 \qquad \text{concave function} \tag{A.09}$$

Thus the Taylor series provides the inequalities

$$f(x) \ge f(x_o) + f'(x_o) \cdot (x - x_0) \qquad \text{convex}$$

or

$$f(x) \le f(x_o) + f'(x_o) \cdot (x - x_0) \qquad \text{concave} \qquad \text{(A.10)}$$

Now we examine the points on the line

$$x = x_2 + \lambda \cdot (x_1 - x_2) \qquad 0 \le \lambda \le 1$$

or

$$x = \lambda \cdot x_1 + (1 - \lambda) \cdot x_2 \qquad \text{(A.11)}$$

where the points x_1 and x_2 of this line are mapped on the values $f(x_1)$ and $f(x_2)$. If we now insert these two points x_1 and x_2 in the Taylor approximation at the point x_0 (in the inequality for convex functions), we find

$$x_0 = \lambda \cdot x_1 + (1 - \lambda) \cdot x_2$$

$$f(x_1) \ge f(x_o) + f'(x_o) \cdot (x_1 - x_0)$$

$$f(x_1) \ge f(x_o) + f'(x_o) \cdot (x_1 - \lambda \cdot x_1 - (1 - \lambda) \cdot x_2)$$

$$f(x_1) \ge f(x_o) + f'(x_o) \cdot (1 - \lambda) \cdot (x_1 - x_2) \qquad \text{(A.12)}$$

and

$$f(x_2) \ge f(x_o) + f'(x_o) \cdot (x_2 - x_0)$$

$$f(x_2) \ge f(x_o) + f'(x_o) \cdot (x_2 - \lambda \cdot x_1 - (1 - \lambda) \cdot x_2)$$

$$f(x_2) \ge f(x_o) + f'(x_o) \cdot (-\lambda) \cdot (x_1 - x_2) \qquad \text{(A.13)}$$

Now we multiply equation (A.12) by λ and equation (A.13) by $(1-\lambda)$, and add both results, to obtain

$$\lambda \cdot f(x_1) \ge \lambda \cdot f(x_o) + f'(x_o) \cdot \lambda \cdot (1 - \lambda) \cdot (x_1 - x_2)$$

$$(1 - \lambda) \cdot f(x_2) \ge (1 - \lambda) \cdot f(x_o) + f'(x_o) \cdot (-\lambda) \cdot (1 - \lambda) \cdot (x_1 - x_2)$$

$$\lambda \cdot f(x_1) + (1 - \lambda) \cdot f(x_2) \ge f(x_o)$$

$$\lambda \cdot f(x_1) + (1 - \lambda) \cdot f(x_2) \ge f(\lambda \cdot x_1 + (1 - \lambda) \cdot x_2) \qquad \text{(A.14)}$$

inequality for convex functions

For concave function we find the analogous inequality

$$\lambda \cdot f(x_1) + (1 - \lambda) \cdot f(x_2) \le f(\lambda \cdot x_1 + (1 - \lambda) \cdot x_2) \qquad \text{(A.15)}$$

inequality for concave functions

In the next step we want to extend this inequality for two weighted arguments to an inequality for n weighted arguments. We thus write

$$p_1 = \lambda \qquad \text{and} \qquad p_2 = 1-\lambda \qquad \sum_{i=1}^{n} p_i = 1 \qquad \text{with } n = 2$$

We already know that

$$p_1 \cdot f(x_1) + p_2 \cdot f(x_2) \geq f(p_1 \cdot x_1 + p_2 \cdot x_2) \tag{A.16}$$

Extension to three arguments: The left-hand side of the inequality is:

$$p_1 \cdot f(x_1) + p_2 \cdot f(x_2) + p_3 \cdot f(x_3) = \sum_{i=1}^{3} p_i \cdot f(x_i)$$

$$= p_3 \cdot f(x_3) + \sum_{i=1}^{2} p_i \cdot f(x_i)$$

$$= p_3 \cdot f(x_3) + (1-p_3) \cdot \sum_{i=1}^{2} p_i^* \cdot f(x_i) \tag{A.17}$$

with $p_i^* = \dfrac{p_i}{1-p_3}$

And we already know from the sum that

$$\sum_{i=1}^{2} p_i^* \cdot f(x_i) \geq f\left(\sum_{i=1}^{2} p_i^* \cdot x_i\right) \tag{A.18}$$

so that we find

$$\sum_{i=1}^{3} p_i \cdot f(x_i) \geq p_3 \cdot f(x_3) + (1-p_3) \cdot f\left(\sum_{i=1}^{2} p_i^* \cdot x_i\right) \tag{A.19}$$

and on the right-hand side of this inequality we have a term where we add two weighted values of a function. From the already derived inequality we know that for convex functions

$$\sum_{i=1}^{3} p_i \cdot f(x_i) \geq f\left(p_3 \cdot x_3 + (1-p_3) \cdot \sum_{i=1}^{2} p_i^* \cdot x_i\right) \tag{A.20}$$

If we now cancel the standardization of the weights

$$\sum_{i=1}^{3} p_i \cdot f(x_i) \geq f\left(p_3 \cdot x_3 + \sum_{i=1}^{2} p_i \cdot x_i \right)$$

we get

$$\sum_{i=1}^{3} p_i \cdot f(x_i) \geq f\left(\sum_{i=1}^{3} p_i \cdot x_i \right) \tag{A.21}$$

This step can now be repeated from n terms to $n+1$ terms, to show that this transition from two to three terms can be extended to an arbitrary number of addends. This is the general procedure in complete induction and it leads to the equation:

$$\sum_{i=1}^{n} p_i \cdot f(x_i) \geq f\left(\sum_{i=1}^{n} p_i \cdot x_i \right) \tag{A.22}$$

We may add the remark that at the beginning of the first derivation we had $p_1+p_2 = 1$, while in the last derivation we used the equation $p_1^* + p_2^* = 1/(1-p_3) < 1$. This can certainly be applied without further problems, because the weights $p_1+p_2 = 1$ merely ensure that the point x_0 lies on the determined line in the convex area, which was limited by the two points x_1 and x_2.

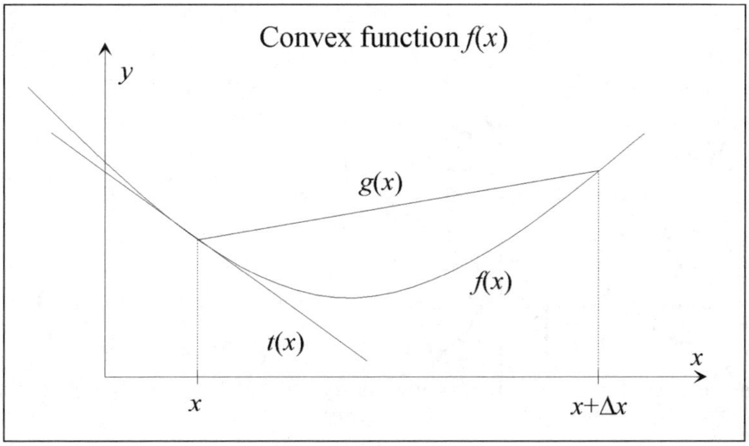

Fig. A.1: Convex function

All points in the interval $[x_1,x_2]$ meet this inequality, as we see in figure A.1. Thus it is also met by points with weights less than 1, which lead to the partial sum $p_1^* + p_2^* = 1/(1-p_3) < 1$, which is also less than 1.
As a consequence of this equation

$$\sum_{i=1}^{n} p_i \cdot f(x_i) \geq f\left(\sum_{i=1}^{n} p_i \cdot x_i\right) \tag{A.23}$$

and with application of the Tschebycheff inequality we proceed

$$E\{f(\underline{x})\} \geq f(E\{\underline{x}\}) \tag{A.24}$$

And now we come to the log-sum inequality, which is the aim of our derivation. We use

$$p_i \geq 0 \qquad \text{and} \qquad \sum_{i=1}^{n} p_i = 1$$

with

$$p_i = \frac{a_i}{\sum_{j=1}^{n} a_j} \qquad \text{and} \qquad x_i = \frac{b_i}{a_i} \tag{A.25}$$

and obtain

$$\sum_{i=1}^{n} \frac{a_i}{\sum_{j=1}^{n} a_j} \cdot f\left(\frac{b_i}{a_i}\right) \geq f\left(\sum_{i=1}^{n} \frac{a_i}{\sum_{j=1}^{n} a_j} \cdot \frac{b_i}{a_i}\right)$$

$$\frac{1}{\sum_{j=1}^{n} a_j} \sum_{i=1}^{n} a_i \cdot f\left(\frac{b_i}{a_i}\right) \geq f\left(\sum_{i=1}^{n} \frac{b_i}{\sum_{j=1}^{n} a_j}\right)$$

$$\sum_{i=1}^{n} a_i \cdot f\left(\frac{b_i}{a_i}\right) \geq \left(\sum_{j=1}^{n} a_j\right) \cdot f\left(\frac{\sum_{i=1}^{n} b_i}{\sum_{j=1}^{n} a_j}\right) \tag{A.26}$$

This is the log-sum inequality for convex functions.

For the special case of the negative logarithm, which is such a convex function, we obtain:

$$\sum_{i=1}^{n} a_i \cdot \ln\left(\frac{a_i}{b_i}\right) \geq \left(\sum_{i=1}^{n} a_i\right) \cdot \ln\left(\frac{\sum_{k=1}^{n} a_i}{\sum_{k=1}^{n} b_i}\right) \quad \text{if } f(\,) = -\ln(\,) \tag{A.27}$$

A.3 Generalized entropy, divergence and distance measures

Here we present only an enumeration of generalized measures, without any detailed examination of the presented measures, because this would be beyond the scope of this book.

A.3.1 Entropy measures

The most important properties of such entropy measures, which are met by Shannon's information, are:

Additivity:

$$H(P \cdot Q) = H(P) + H(Q) \tag{A.28}$$

For $P = \{p_1, p_2, \ldots, p_n\} \in \Delta_n^0$ and $Q = \{q_1, q_2, \ldots, q_m\} \in \Delta_m^0$ and for the product space $P \cdot Q = \{p_1 \cdot q_1, p_1 \cdot q_2, \ldots, p_1 \cdot q_m, p_2 \cdot q_1, p_2 \cdot q_2, \ldots, p_n \cdot q_m\} \in \Delta_{nm}^0$, where $P = \{p_1, p_2, \ldots, p_n\} \in \Delta_n^0$ = set of all probabilities of n events, where certain events may occur with a probability $p_i = 0$.

Recursivity:

$$H(p_1, p_2, \ldots, p_n) = H(p_1 + p_2, p_3, \ldots, p_n) + (p_1 + p_2) \cdot H\left(\frac{p_1}{p_1 + p_2}, \frac{p_2}{p_1 + p_2}\right) \tag{A.29}$$

$$p_1 + p_2 > 0 \ \text{ for all } \ P = \{p_1, p_2, \ldots, p_n\} \in \Delta_n^0$$

<u>*Summation property:*</u>

$$H(P) = \sum_{i=1}^{n} f(p_i) \qquad P = \{p_1, p_2, \ldots, p_n\} \in \Delta_n^0 \tag{A.30}$$

for Shannon's measure we get $f(p) = p \ln p$.

Rényi's entropy H_α is an extension with a generalized mean

$$\Phi_2(P) = H_\alpha(P) = \frac{1}{1-\alpha} \cdot \ln\left(\sum_{i=1}^{n} p_i^\alpha\right) \tag{A.31}$$

and it is an entropy of the order α. It meets the additivity, but it cannot be calculated with a recursive formulation and even the summation property does not hold.

Havrda and Charvát introduced the entropy H^s of grade s

$$\Phi_9(P) = H^s(P) = \frac{1}{2^{1-s} - 1} \cdot \left(\sum_{i=1}^{n} p_i^s - 1\right) \tag{A.32}$$

which does not meet the additivity, but is recursive in the grade s

$$\begin{aligned} H^s(p_1, p_2, \ldots, p_n) &= H^s(p_1 + p_2, p_3, \ldots, p_n) \\ &+ (p_1 + p_2)^s \cdot H^s\left(\frac{p_1}{p_1 + p_2}, \frac{p_2}{p_1 + p_2}\right) \end{aligned} \tag{A.33}$$

$$p_1 + p_2 > 0 \ \text{ for all } \ P = \{p_1, p_2, \ldots, p_n\} \in \Delta_n^0$$

and even the sum representation $H^s(P) = \sum_{i=1}^{n} f(p_i)$ is possible.

Moreover there is a third index, indicating an entropy ${}_tH$ of the kind t, and it has been presented by Arimoto with the entropy

$$\Phi_{12}(P) = {}_tH(P) = \frac{1}{2^{t-1}-1} \cdot \left(\left(\sum_{i=1}^{n} p_i^{\frac{1}{t}} \right)^t - 1 \right) \tag{A.34}$$

If we replace $1/t$ by α, we may easily derive Rényi's entropy.

Sharma and Mittai presented an entropy H^s_α of order α and grade s

$$\Phi_{13}(P) = H_1^s(P) = \frac{1}{2^{1-s}-1} \cdot \left[\exp_2 \left((s-1) \cdot \sum_{i=1}^{n} p_i \cdot \log p_i \right) - 1 \right] \tag{A.35}$$

$$\Phi_{14}(P) = H_\alpha^s(P) = \frac{1}{2^{1-s}-1} \cdot \left[\left(\sum_{i=1}^{n} p_i^\alpha \right)^{\frac{s-1}{\alpha-1}} - 1 \right] \tag{A.36}$$

Both entropies are neither additive nor recursive, and they do not even meet the sum-representation. They are generalizations of the entropies presented so far and can be reduced to the special forms.

Example: $s = 1$, this reduces to

$$\Phi_{14}(P) = H_\alpha^{s=0}(P) = \frac{1}{2^0-1} \cdot \left[\left(\sum_{i=1}^{n} p_i^\alpha \right)^0 - 1 \right] = \frac{0}{0} \tag{A.37}$$

calculating the limit according to de l'Hospital, we obtain Rényi's information.

For all $P = \{p_1, p_2, ..., p_n\} \in \Delta_n^0$ = set of all probabilities of n events, where certain events may occur with the probability $p_i = 0$, there exist several entropy (information) measures:

Shannon (1948):

$$\Phi_1(P) = -\sum_{i=1}^{n} p_i \cdot \ln p_i \tag{A.38}$$

Rényi (1961):

$$\Phi_2(P) = \frac{1}{1-\alpha} \cdot \ln \left(\sum_{i=1}^{n} p_i^\alpha \right) \qquad \alpha \neq 1,\ \alpha > 0 \tag{A.39}$$

Aczél and Daróczy (1963):

$$\Phi_3(P) = -\frac{\sum_{i=1}^{n} p_i^{\alpha} \cdot \ln p_i}{\sum_{i=1}^{n} p_i^{\alpha}} \qquad \alpha > 0 \tag{A.40}$$

$$\Phi_4(P) = \frac{1}{s-\alpha} \cdot \ln\left(\frac{\sum_{i=1}^{n} p_i^{\alpha}}{\sum_{i=1}^{n} p_i^{s}} \right) \qquad \alpha \neq s,\ \alpha \neq 1,\ \alpha > 0 \qquad \text{(see } \Phi_2\text{)} \tag{A.41}$$

$$\Phi_5(P) = \frac{1}{s} \cdot \arctan\left(\frac{\sum_{i=1}^{n} p_i^{\alpha} \cdot \sin(s \cdot \ln p_i)}{\sum_{i=1}^{n} p_i^{\alpha} \cdot \cos(s \cdot \ln p_i)} \right) \qquad s \neq 0,\ \alpha > 0 \tag{A.42}$$

Varma (1966):

$$\Phi_6(P) = \frac{1}{m-\alpha} \cdot \ln\left(\sum_{i=1}^{n} p_i^{\alpha-m+1} \right) \qquad m\text{-}1 < \alpha < m,\ m \geq 1 \tag{A.43}$$

$$\Phi_7(P) = \frac{1}{m \cdot (m-\alpha)} \cdot \ln\left(\sum_{i=1}^{n} p_i^{\frac{\alpha}{m}} \right) \qquad 0 < \alpha < m,\ m \geq 1 \tag{A.44}$$

Kapur (1967):

$$\Phi_8(P) = \frac{1}{1-t} \cdot \ln\left(\frac{\sum_{i=1}^{n} p_i^{t+s+1}}{\sum_{i=1}^{n} p_i^{s}} \right) \qquad t \neq 1,\ t > 0,\ s \geq 1 \quad \text{(see } \Phi_6\text{)} \tag{A.45}$$

Havrda and Charvát (1967):

$$\Phi_9(P) = \frac{1}{2^{1-s}-1} \cdot \left(\sum_{i=1}^{n} p_i^{s} - 1 \right) \qquad s \neq 1,\ s > 0 \tag{A.46}$$

Belis and Guiasu (1968):

$$\Phi_{10}(P) = -\frac{\sum_{i=1}^{n} p_i \cdot w_i \cdot \ln p_i}{\sum_{i=1}^{n} p_i \cdot w_i} \qquad w_i > 0,\ i = 1,2,3,\ldots,n \qquad (\text{see } \Phi_3) \tag{A.47}$$

Rathie (1970):

$$\Phi_{11}(P) = \frac{1}{1-\alpha} \cdot \ln\left(\frac{\sum_{i=1}^{n} p_i^{\alpha+s_i+1}}{\sum_{i=1}^{n} p_i^{s_i}}\right) \qquad \alpha \neq 1,\ \alpha > 0,\ s_i \geq 1,\ i = 1,2,3,\ldots,n \quad (\text{see } \Phi_8) \tag{A.48}$$

Arimoto (1971):

$$\Phi_{12}(P) = \frac{1}{2^{t-1}-1} \cdot \left(\left(\sum_{i=1}^{n} p_i^{\frac{1}{t}}\right)^{t} - 1\right) \qquad t \neq 1,\ t > 0 \tag{A.49}$$

Sharma and Mittal (1975):

$$\Phi_{13}(P) = \frac{1}{2^{1-s}-1} \cdot \left[\exp_2\left((s-1) \cdot \sum_{i=1}^{n} p_i \cdot \ln p_i\right) - 1\right] \qquad s \neq 1,\ s > 0 \tag{A.50}$$

$$\Phi_{14}(P) = \frac{1}{2^{1-s}-1} \cdot \left[\left(\sum_{i=1}^{n} p_i^{\alpha}\right)^{\frac{s-1}{\alpha-1}} - 1\right] \qquad s \neq 1,\ s > 0,\ \alpha \neq 1,\ \alpha > 0 \tag{A.51}$$

Taneja (1975):

$$\Phi_{15}(P) = -2^{\alpha-1} \cdot \sum_{i=1}^{n} p_i^{\alpha} \cdot \ln p_i \qquad \alpha > 0 \tag{A.52}$$

$$\Phi_{16}(P) = \frac{1}{2^{1-\alpha} - 2^{1-s}} \cdot \sum_{i=1}^{n} \left(p_i^{\alpha} - p_i^{s}\right) \qquad \alpha \neq s,\ \alpha > 0,\ s > 0 \tag{A.53}$$

$$\Phi_{17}(P) = -\frac{2^{\alpha-1}}{\sin(s)} \cdot \sum_{i=1}^{n} p_i^{\alpha} \cdot \sin(s \cdot \ln p_i) \quad s \neq k\pi, k = 0,1,2,\ldots \quad \alpha > 0 \qquad \text{(A.54)}$$

Picard (1979)

$$\Phi_{18}(P) = -\frac{\sum_{i=1}^{n} v_i \cdot \ln p_i}{\sum_{i=1}^{n} v_i} \qquad \text{(see } \Phi_3\text{)} \qquad \text{(A.55)}$$

$$\Phi_{19}(P) = \frac{1}{1-\alpha} \cdot \ln\left(\frac{\sum_{i=1}^{n} p_i^{\alpha-1} \cdot v_i}{\sum_{i=1}^{n} v_i}\right) \qquad \alpha \neq 1, \alpha > 0 \qquad \text{(see } \Phi_3\text{)} \qquad \text{(A.56)}$$

$$\Phi_{20}(P) = \frac{1}{2^{1-s}-1} \cdot \left[\frac{\exp_2\left((s-1) \cdot \sum_{i=1}^{n} v_i \cdot \ln p_i\right)}{\sum_{i=1}^{n} v_i} - 1\right] \qquad s \neq 1, s > 0 \text{ (see } \Phi_{13}\text{)}$$

(A.57)

$$\Phi_{21}(P) = \frac{1}{2^{1-s}-1} \cdot \left[\left(\frac{\sum_{i=1}^{n} v_i \cdot p_i^{\alpha-1}}{\sum_{i=1}^{n} v_i}\right)^{\frac{s-1}{\alpha-1}} - 1\right] \qquad s \neq 1, s > 0, \alpha \neq 1, \alpha > 0$$

(see Φ_{14}) (A.58)

exchanging the expectation over p_i by the expectation over v_i for all entropies. The weights are $v_i > 0$ for $i = 1,2,\ldots,n$ and $P = \{p_1, p_2, \ldots, p_n\} \in \Delta_n$, i.e. $p_i = 0$ is not allowed.

Ferreri (1980)

$$\Phi_{22}(P) = \left(1 + \frac{1}{\lambda}\right) \cdot \ln(1+\lambda) - \frac{1}{\lambda} \cdot \sum_{i=1}^{n} (1 + \lambda \cdot p_i) \cdot \ln(1 + \lambda \cdot p_i) \qquad \lambda > 0 \qquad \text{(A.59)}$$

Sant´anna and Taneja (1983):

$$\Phi_{23}(P) = -\sum_{i=1}^{n} p_i \cdot \ln\left(\frac{\sin(s \cdot p_i)}{2\sin\left(\frac{s}{2}\right)}\right) \qquad 0 < s < \pi \qquad \text{(A.60)}$$

$$\Phi_{24}(P) = -\sum_{i=1}^{n} \left(\frac{\sin(s \cdot p_i)}{2\sin\left(\frac{s}{2}\right)}\right) \cdot \ln\left(\frac{\sin(s \cdot p_i)}{2\sin\left(\frac{s}{2}\right)}\right) \qquad 0 < s < \pi \qquad \text{(A.61)}$$

$$\Phi_{25}(P) = -\sum_{i=1}^{n} \left(\frac{\sin(s \cdot p_i)}{2\sin\left(\frac{s}{2}\right)}\right) \qquad 0 < s < \pi \qquad \text{(A.62)}$$

The latter entropy cannot be reduced to Shannon's entropy, but it has certain properties similar to those of Shannon's entropy.

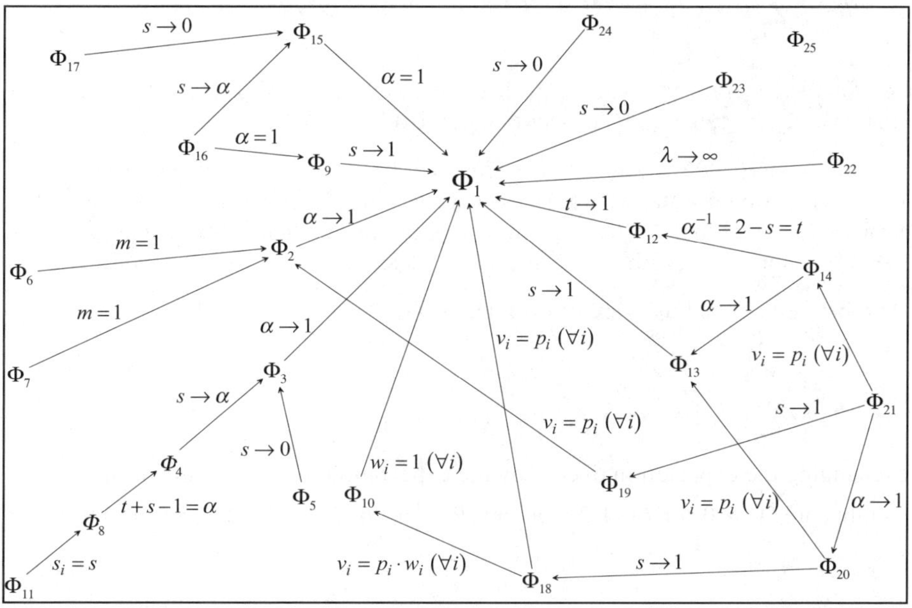

Fig. A.2: Connections between the information functions

Taneja uses these entropies to generate a *unified (α,s)-entropy* with

$$E_\alpha^s(P)=\begin{cases} H_\alpha^s(P) & \text{for } \alpha \neq 1, s \neq 1, \alpha > 0 \\ H_1^s(P) & \text{for } \alpha = 1, s \neq 1 \\ H_\alpha^1(P) & \text{for } \alpha \neq 1, s = 1, \alpha > 0 \\ H(P) & \text{for } \alpha = 1, s = 1 \end{cases} \tag{A.63}$$

and this most general form is then examined in detail.

During this examination Taneja defines certain properties (quasiconcave, pseudoconcave), which he uses for the description of the analytic and algebraic properties of the unified (α,s)-entropy. We do not want to describe Tanejas examinations. The complete explanation of this unified entropy can be found in [TAN89].

A.3.2 Generalized measures of distance

The quantity

$$G_\alpha^\rho=\left(\sum_{i=1}^{n} p_i^\alpha\right)^{\frac{1}{\rho}} \quad \alpha > 0 \text{ and } \rho \neq 0 \text{ for all } P=\{p_1,p_2,\ldots,p_n\}\in\Delta_n^0 \tag{A.64}$$

is called a generalized measure of distance [BOE79] [CAP88]. Another name of this quantity is 'generalized measure of security'.

This generalized measure of distance is not based on deterministic quantities or on the occurring events, where the distance might be of interest. It is based on the probabilities and thus on the uncertainty of the occurring events. Because of this use of the uncertainties this measure can be classified as a measure of information.

Another generalized measure of distance is presented in [CAP88]

$$T_\alpha^\rho=\left(\frac{\sum_{i=1}^{n} p_i^\alpha}{\sum_{i=1}^{n} p_i^\rho}\right)^{\frac{1}{\alpha-\rho}} \quad \rho\neq\alpha,\ \alpha > 0 \text{ and } \rho \neq 0 \ \text{ for all } P=\{p_1,p_2,\ldots,p_n\}\in\Delta_n^0 \tag{A.64}$$

These distance measures already contain other measures of distance, for example the measures of Trouborst et al. 1974, Györfi and Nemetz 1975, Devijver 1974 and Vajda 1968 (see [TAN89]).

Both measures have the following properties:

G_α^ρ : For all $P=\{p_1,p_2,...,p_n\}\in\Delta_n^0$ we have:

(i) G_α^ρ is a convex function of P for $\alpha>1$ and $\alpha\rho\geq 1$ or $0<\alpha<1$ $\rho<0$

(ii) G_α^ρ is a concave function of P for $0<\alpha<1$ $\rho>0$, $\alpha\rho\leq 1$

(i) G_α^ρ is a decreasing function of α ($\rho=$ const. and $\rho>0$)

(ii) G_α^ρ is an increasing function of α ($\rho=$ const. and $\rho<0$)

(iii) G_α^ρ is a decreasing function of ρ ($\alpha=$ const. and $\alpha>1$)

(iv) G_α^ρ is a decreasing function of ρ ($\alpha=$ const. and $0<\alpha<1$)

(i) $G_\alpha^\rho(1-p_{\max},p_{\max})\leq G_\alpha^\rho$ (A.65)

$$G_\alpha^\rho(1-p_{\max},p_{\max})\leq G_\alpha^\rho\left(\frac{1-p_{\max}}{n-1},...,\frac{1-p_{\max}}{n-1},p_{\max}\right)$$

$0<\alpha<1$ $\rho>0$ (or $\alpha>1$ $\rho<0$) (A.66)

(ii) $G_\alpha^\rho(1-p_{\max},p_{\max})\geq G_\alpha^\rho$ (A.67)

$$G_\alpha^\rho(1-p_{\max},p_{\max})\geq G_\alpha^\rho\left(\frac{1-p_{\max}}{n-1},...,\frac{1-p_{\max}}{n-1},p_{\max}\right)$$

$\alpha>1$ $\rho>0$ (or $0<\alpha<1$ $\rho<0$) (A.68)

T_α^ρ : For all $P=\left\{p_1,p_2,...,p_n\right\}\in\Delta_n^0$ we have:

(i) T_α^ρ is an increasing function of α ($\rho=$ const.)

(ii) T_α^ρ is an increasing function of ρ ($\alpha=$ const.)

(i) $T_\alpha^\rho\leq p_{\max}$ (A.69)

$$\text{(ii)}\quad T_\alpha^\rho(P) \geq \left(\frac{p_{\max}^\rho}{\sum_{i=1}^{n} p_i^\rho} \right)^{\frac{1}{\alpha-\rho}} \cdot p_{\max} \qquad \rho < \alpha \tag{A.70}$$

$$\text{(iii)}\quad T_\alpha^\rho(P) \geq \left(\frac{p_{\max}^\alpha}{\sum_{i=1}^{n} p_i^\alpha} \right)^{\frac{1}{\alpha-\rho}} \cdot p_{\max} \qquad \rho > \alpha \tag{A.71}$$

A.3.3 Generalized measures of the directed divergence

Kullback and Leibler (1951) were the first to introduce an information measure between two distribution density functions, the Kullback–Leibler divergence, discrimination information, relative information, directed divergence or Kullback's information.

$$D(P \| U) = \sum_{i=1}^{n} p_i \cdot \ln \frac{p_i}{u_i} \qquad \text{for all } P, U \in \Delta_n \tag{A.72}$$

This divergence can be extended by applying a generalized mean, which has been realized by Rényi (1961), who

$$D_\alpha^1(P \| U) = \frac{1}{1-\alpha} \cdot \ln\left(\sum_{i=1}^{n} p_i^\alpha \cdot u_i^{1-\alpha} \right) \qquad \alpha \neq 1,\ \alpha > 0 \ \text{ for all } P, U \in \Delta_n \tag{A.73}$$

Another generalization of the divergence is

$$D_s^s(P \| U) = \frac{1}{1-2^{1-s}} \cdot \left(\sum_{i=1}^{n} p_i^s \cdot u_i^{1-s} - 1 \right) \qquad s \neq 1,\ s > 0 \ \text{ for all } P, U \in \Delta_n \tag{A.74}$$

Both generalizations can be reduced to the Kullback–Leibler distance by calculation of the limit for

$$\lim_{\alpha \to 1} D_\alpha^1(P \| U) = \lim_{s \to 1} D_s^s(P \| U) = D(P \| U) \tag{A.75}$$

Furthermore there are two additional generalizations:

$$D_1^s(P \| U) = \frac{1}{1-2^{1-s}} \cdot \left(2^{(s-1)D(P\|U)} - 1\right) \quad s \neq 1 \quad \text{for all } P, U \in \Delta_n \tag{A.76}$$

and

$$D_\alpha^s(P \| U) = \frac{1}{1-2^{1-s}} \cdot \left[\left(\sum_{i=1}^{n} p_i^\alpha \cdot u_i^{1-\alpha}\right)^{\frac{s-1}{\alpha-1}} - 1\right] \quad s \neq 1,\ \alpha \neq 1,\ \alpha > 0 \tag{A.77}$$

for all P, $U \in \Delta_n$

And again we find the limit:

$$\lim_{s \to 1} D_\alpha^s(P \| U) = D_\alpha^1(P \| U) \tag{A.78}$$

$$\lim_{s \to 1} D_1^s(P \| U) = D(P \| U) \tag{A.79}$$

If $s = \alpha$, then both measures are equal

$$D_\alpha^s(P \| U) = D_s^s(P \| U) \tag{A.80}$$

All these generalized divergence measures can then be summarized by a unified directed divergence, defined by:

$$\boldsymbol{D}_\alpha^s(P \| U) = \begin{cases} D_\alpha^s(P \| U), & \alpha \neq 1, s \neq 1, \alpha > 0 \\ D_1^s(P \| U), & \alpha = 1, s \neq 1 \\ D_\alpha^1(P \| U), & \alpha \neq 1, s = 1, \alpha > 0 \\ D(P \| U), & \alpha = 1, s = 1 \end{cases} \tag{A.81}$$

for all P, $U \in \Delta_n$.

The special measure $D_s^s(P \| U)$ does not occur in this unified definition, as it is merely a special case of $D_\alpha^s(P \| U)$ for $s = \alpha$.

$\boldsymbol{D}_\alpha^s(P \| U)$ is thus called the unified (α, s)-directed divergence.

To circumvent difficulties, which may occur, when the probabilities are equal to zero, the probability space is assumed to be given by Δ_n, so that zero probabilities do not need further attention in the definition.

For all P, $U \in \Delta_n$ the unified (α, s)-directed divergence $\boldsymbol{D}_\alpha^s(P \| U)$ has the following properties:

(i) $\boldsymbol{D}_{\alpha}^{s}(P \parallel U) \geq 0$ for all $\alpha > 0$ and all s.

(ii) $\boldsymbol{D}_{\alpha}^{s}(P \parallel U)$ is a convex function of the pair $(P,U) \in \Delta_n \times \Delta_n$ for all $s \geq \alpha > 0$

(iii) $\boldsymbol{D}_{\alpha}^{s}(P \parallel U)$ is an increasing function of α (for constant s)

(iv)

$$\boldsymbol{D}_{\alpha}^{s}\left(\sum_{i=1}^{\sigma} p_i, 1-\sum_{i=1}^{\sigma} p_i \,\middle\|\, \sum_{i=1}^{\sigma} u_i, 1-\sum_{i=1}^{\sigma} u_i\right) \leq \boldsymbol{D}_{\alpha}^{s}(P \parallel U)$$

$$\leq \boldsymbol{D}_{\alpha}^{s}\left(p_1, p_2, ..., p_\sigma, \frac{1-\sum_{i=1}^{\sigma} p_i}{n-1}, ..., \frac{1-\sum_{i=1}^{\sigma} p_i}{n-1} \,\middle\|\, u_1, u_2, ..., u_\sigma, \frac{1-\sum_{i=1}^{\sigma} u_i}{n-1}, ..., \frac{1-\sum_{i=1}^{\sigma} u_i}{n-1}\right) \tag{A.82}$$

for $1 \leq \sigma \leq n$

(v) Let

$$P(c) = \left(\sum_{i=1}^{n} p_i c_{i1}, \sum_{i=1}^{n} p_i c_{i2}, ..., \sum_{i=1}^{n} p_i c_{in}\right) \in \Delta_n$$

$$U(c) = \left(\sum_{i=1}^{n} u_i c_{i1}, \sum_{i=1}^{n} u_i c_{i2}, ..., \sum_{i=1}^{n} u_i c_{in}\right) \in \Delta_n$$

with

$$\sum_{i=1}^{n} c_{ik} = \sum_{k=1}^{n} c_{ik} = 1 \qquad c_{ik} \geq 0 \text{ for } i, k = 1,2,...,n$$

Then we get

$$\boldsymbol{D}_{\alpha}^{s}(P(c) \parallel U(c)) \leq \boldsymbol{D}_{\alpha}^{s}(P \parallel U) \tag{A.83}$$

(vi)

$$\boldsymbol{D}_{\alpha}^{s}(P \parallel U)\begin{cases} \leq \boldsymbol{D}_{\alpha}^{1}(P \parallel U) & \text{for } s < 1 \\ \geq \boldsymbol{D}_{\alpha}^{1}(P \parallel U) & \text{for } s > 1 \end{cases} \tag{A.84}$$

$$\boldsymbol{D}_{1}^{s}(P \parallel U)\begin{cases} \leq \boldsymbol{D}(P \parallel U) & \text{for } s < 1 \\ \geq \boldsymbol{D}(P \parallel U) & \text{for } s > 1 \end{cases} \tag{A.85}$$

$$\boldsymbol{D}_{\alpha}^{s}(P \parallel U)\begin{cases} \leq \boldsymbol{D}_{1}^{s}(P \parallel U) & \text{for } 0 < \alpha < 1 \\ \geq \boldsymbol{D}_{1}^{s}(P \parallel U) & \text{for } \alpha > 1 \end{cases} \tag{A.86}$$

$$\boldsymbol{D}_{\alpha}^{1}(P \parallel U)\begin{cases} \leq \boldsymbol{D}(P \parallel U) & \text{for } 0 < \alpha < 1 \\ \geq \boldsymbol{D}(P \parallel U) & \text{for } \alpha > 1 \end{cases} \tag{A.87}$$

A.3.4 Generalized measures of divergence

In this section we present generalizations of two different measures of divergence. These two measures are

1) Information radius or Jensen's measure of divergence difference
2) *J*-divergence of Kullback and Leibler

A.3.4.1 Information radius and the J-divergence

Using the concavity of Shannon's information we may write:

$$\frac{H(P)+H(U)}{2} \leq H\left(\frac{P+U}{2}\right) \qquad \text{for all } P,\, U \in \Delta_n \tag{A.88}$$

The difference

$$R(P \parallel U) = H\left(\frac{P+U}{2}\right) - \frac{H(P)+H(U)}{2}$$

$$R(P \parallel U) = \sum_{i=1}^{n}\left[\frac{p_i \cdot \ln p_i + u_i \cdot \ln u_i}{2} - \left(\frac{p_i + u_i}{2}\right) \cdot \ln\left(\frac{p_i + u_i}{2}\right)\right] \tag{A.89}$$

is known as the information radius or as Jensen's measure of divergence difference for all P, $U \in \Delta_n$. Taneja calls this quantity the R-divergence. Another divergence is given by Kullback and Leibler, which we already know as the *J*-divergence. This is

$$J(P \| U) = D(P \| U) + D(U \| P) \qquad \text{for all } P,\, U \in \Delta_n. \tag{A.90}$$

with $\quad D(P \| U) = \sum_{i=1}^{n} p_i \cdot \ln \frac{p_i}{u_i}$

Inserting this, we obtain

$$J(P \| U) = \sum_{i=1}^{n} (p_i - u_i) \cdot \ln \frac{p_i}{u_i} \tag{A.91}$$

A relatively simple calculation leads to the connection between these two measures of divergence

$$R(P \| U) = \frac{1}{2} \cdot \left[D\left(P \| \frac{P+U}{2} \right) + D\left(U \| \frac{P+U}{2} \right) \right] \qquad \text{for all } P,\, U \in \Delta_n \tag{A.92}$$

A.3.4.2 Generalization of the R*-divergence*

The *R*-divergence can be generalized in three different ways.

<u>*1st way*:</u> Instead of Shannon's entropy *H*(*P*) we use the unified (α,s)-entropy $E_\alpha^s(P)$

$${}^{1}\boldsymbol{R}_\alpha^s(P \| U) = E_\alpha^s\left(\frac{P+U}{2} \right) - \frac{E_\alpha^s(P) + E_\alpha^s(U)}{2} \qquad \text{for all } P,\, U \in \Delta_n \tag{A.93}$$

In a more explicit way this means

$${}^{1}\boldsymbol{R}_\alpha^s(P \| U) = \begin{cases} {}^{1}R_\alpha^s(P \| U), & \alpha \neq 1, s \neq 1, \alpha > 0 \\ {}^{1}R_1^s(P \| U), & \alpha = 1, s \neq 1 \\ {}^{1}R_\alpha^1(P \| U), & \alpha \neq 1, s = 1, \alpha > 0 \\ {}^{1}R(P \| U), & \alpha = 1, s = 1 \end{cases} \tag{A.94}$$

where

$$^1R_\alpha^s(P\|U)=\frac{1}{1-2^{1-s}}\cdot\left\{\frac{1}{2}\cdot\left[\left(\sum_{i=1}^n p_i^\alpha\right)^{\frac{s-1}{\alpha-1}}+\left(\sum_{i=1}^n u_i^\alpha\right)^{\frac{s-1}{\alpha-1}}\right]-\left(\sum_{i=1}^n\left(\frac{p_i+u_i}{2}\right)^\alpha\right)^{\frac{s-1}{\alpha-1}}\right\}$$

for $\alpha \neq 1, s \neq 1, \alpha > 0$ (A.95)

$$^1R_1^s(P\|U)=\frac{1}{1-2^{1-s}}\cdot\left\{2^{(s-1)R(P\|U)}-1\right\} \quad \text{for } s\neq 1 \qquad \text{(A.96)}$$

$$^1R_\alpha^1(P\|U)=\frac{1}{\alpha-1}\cdot\ln\left\{\frac{\sum_{i=1}^n\left(\frac{p_i+u_i}{2}\right)^\alpha}{\sqrt{\left(\sum_{i=1}^n p_i^\alpha\right)\cdot\left(\sum_{i=1}^n u_i^\alpha\right)}}\right\} \quad \text{for } \alpha\neq 1, s=1, \alpha>0$$

(A.97)

If $s = \alpha$, we get:

$$^1R_\alpha^\alpha(P\|U)=\frac{1}{1-2^{1-\alpha}}\cdot\sum_{i=1}^n\left[\frac{p_i^\alpha+u_i^\alpha}{2}-\left(\frac{p_i+u_i}{2}\right)^\alpha\right] \quad \text{for } \alpha\neq 1, \alpha>0 \qquad \text{(A.98)}$$

This is the first possible generalization.

2nd way: Instead of $D(P\|U)$ = Kullback's divergence we use $\boldsymbol{D}_\alpha^s(P\|U)$, the unified (α, s)-directed divergence, which leads to:

$$^2\boldsymbol{R}_\alpha^s(P\|U)=\frac{1}{2}\cdot\left[\boldsymbol{D}_\alpha^s(P\|U)+\boldsymbol{D}_\alpha^s(U\|P)\right] \quad \text{for all } P, U\in\Delta_n \qquad \text{(A.99)}$$

Here we find

$$^2\boldsymbol{R}_\alpha^s(P\|U)=\begin{cases} ^2R_\alpha^s(P\|U), & \alpha\neq 1, s\neq 1, \alpha>0\\ ^2R_1^s(P\|U), & \alpha=1, s\neq 1\\ ^2R_\alpha^1(P\|U), & \alpha\neq 1, s=1, \alpha>0\\ ^2R(P\|U), & \alpha=1, s=1\end{cases} \quad \text{for all } P, U\in\Delta_n \qquad \text{(A.100)}$$

where

$$ {}^2R_\alpha^s(P\|U)=\frac{1}{2\cdot\left(1-2^{1-s}\right)} \cdot\left\{\left(\sum_{i=1}^{n} p_i^\alpha\cdot\left(\frac{p_i+u_i}{2}\right)^{1-\alpha}\right)^{\frac{s-1}{\alpha-1}}+\left(\sum_{i=1}^{n} u_i^\alpha\cdot\left(\frac{p_i+u_i}{2}\right)^{1-\alpha}\right)^{\frac{s-1}{\alpha-1}}-2\right\} \tag{A.101}$$

for $\alpha\neq 1, s\neq 1, \alpha>0$

$$ {}^2R_1^s(P\|U)=\frac{1}{2\cdot\left(1-2^{1-s}\right)}\cdot\left\{2^{(s-1)\cdot R(P\|P+U/2)}+2^{(s-1)\cdot R(U\|P+U/2)}-2\right\} \text{ for } s\neq 1 \tag{A.102}$$

$$ {}^2R_\alpha^1(P\|U)=\frac{1}{2\cdot(\alpha-1)} \cdot\left\{\ln\left(\sum_{i=1}^{n} p_i^\alpha\cdot\left(\frac{p_i+u_i}{2}\right)^{1-\alpha}\right)+\ln\left(\sum_{i=1}^{n} p_i^\alpha\cdot\left(\frac{p_i+u_i}{2}\right)^{1-\alpha}\right)\right\} \tag{A.103}$$

for $\alpha\neq 1, s=1, \alpha>0$

If $s=\alpha$, we get:

$$ {}^2R_\alpha^\alpha(P\|U)=\frac{1}{2\cdot\left(1-2^{1-\alpha}\right)}\cdot\left\{\sum_{i=1}^{n}\left(\frac{p_i^\alpha+u_i^\alpha}{2}\right)\cdot\left(\frac{p_i+u_i}{2}\right)^{1-\alpha}-1\right\}$$

for $\alpha\neq 1, \alpha>0$

(A.104)

This is the second possible generalization.

<u>*3rd way:*</u> We use the generalizations that we obtained in the second way and create another generalization:

$$ {}^{3}R_{\alpha}^{s}(P\|U)=\frac{1}{\left(1-2^{1-s}\right)}\cdot\left\{\left[\sum_{i=1}^{n}\left(\frac{p_i^{\alpha}+u_i^{\alpha}}{2}\right)\cdot\left(\frac{p_i+u_i}{2}\right)^{1-\alpha}\right]^{\frac{s-1}{\alpha-1}}-1\right\} \quad \text{(A.105)} $$

for $\alpha \neq 1, s \neq 1, \alpha > 0$

$$ {}^{3}R_{\alpha}^{1}(P\|U)=\frac{1}{\left(1-2^{1-s}\right)}\cdot\left\{\sum_{i=1}^{n}\left(\frac{p_i^{\alpha}+u_i^{\alpha}}{2}\right)\cdot\left(\frac{p_i+u_i}{2}\right)^{1-\alpha}\right\} \quad \text{(A.106)} $$

for $\alpha \neq 1, s = 1, \alpha > 0$

We obtain the limits:

$$ \lim_{s\to 1} {}^{3}R_{\alpha}^{s}(P\|U) = {}^{3}R_{\alpha}^{1}(P\|U) \quad \text{(A.107)} $$

$$ \lim_{\alpha\to 1} {}^{3}R_{\alpha}^{s}(P\|U) = {}^{3}R_{1}^{s}(P\|U) = {}^{1}R_{1}^{s}(P\|U) \quad \text{(A.108)} $$

and for $s = \alpha$ we have $\quad {}^{3}R_{s}^{s}(P\|U) = {}^{2}R_{s}^{s}(P\|U) \quad$ (A.109)

Thus we finally write

$$ {}^{3}\boldsymbol{R}_{\alpha}^{s}(P\|U)=\begin{cases} {}^{3}R_{\alpha}^{s}(P\|U), & \alpha \neq 1, s \neq 1, \alpha > 0 \\ {}^{3}R_{1}^{s}(P\|U), & \alpha = 1, s \neq 1 \\ {}^{3}R_{\alpha}^{1}(P\|U), & \alpha \neq 1, s = 1, \alpha > 0 \\ {}^{3}R(P\|U), & \alpha = 1, s = 1 \end{cases} \quad \text{for all } P, U \in \Delta_n \quad \text{(A.110)} $$

and this is the third possible generalization of the R-divergence.

For all P, $U \in \Delta_n$ we have:

(i) $\quad {}^{1}\boldsymbol{R}_{\alpha}^{s}(P\|U) \geq 0 \quad$ for $\alpha, s \in \Gamma_1$

$$ \Gamma_1 = \left\{(\alpha, s) : \alpha > 0 \text{ with } s \geq \alpha \text{ or } s \geq 2 - \frac{1}{\alpha}\right\} $$

(ii) $^{2}\boldsymbol{R}_{\alpha}^{s}(P \,\|\, U) \geq 0$ for all $\alpha > 0$ and all s

(iii) $^{3}\boldsymbol{R}_{\alpha}^{s}(P \,\|\, U) \geq 0$ for all $\alpha > 0$ and all s

(iv) $^{2}\boldsymbol{R}_{\alpha}^{s}(P \,\|\, U) \begin{cases} \leq {}^{3}\boldsymbol{R}_{\alpha}^{s}(P \,\|\, U), & s \leq \alpha \\ \geq {}^{3}\boldsymbol{R}_{\alpha}^{s}(P \,\|\, U), & s \geq \alpha \end{cases}$ for all $\alpha > 0$ and all s

A.3.4.3 Generalization of the J-divergence

Here we have two possibilities to generalize the J-divergence, which result in one or two scalar parameters. If we have only one parameter, either s or α, then the generalization of the divergence is presented by Rathie and Sheng, Burbea, Taneja (see [TAN89]):

$$J_{s}^{s}(P \,\|\, U) = \frac{1}{\left(1 - 2^{1-s}\right)} \cdot \left\{ \sum_{i=1}^{n} p_i^{s} \cdot u_i^{1-s} + \sum_{i=1}^{n} p_i^{1-s} \cdot u_i^{s} - 2 \right\} \quad \text{with } s \neq 1 \text{ and } s > 0 \tag{A.111}$$

$${}^{1}J_{\alpha}^{1}(P \,\|\, U) = \frac{1}{\alpha - 1} \cdot \left\{ \ln\left(\sum_{i=1}^{n} p_i^{\alpha} \cdot u_i^{1-\alpha} \right) + \ln\left(\sum_{i=1}^{n} p_i^{1-\alpha} \cdot u_i^{\alpha} \right) \right\}$$

with $\alpha \neq 1$ and $\alpha > 0$ (A.112)

and

$${}^{2}J_{\alpha}^{1}(P \,\|\, U) = \frac{1}{\alpha - 1} \cdot 2 \cdot \ln\left(\sum_{i=1}^{n} \frac{p_i^{\alpha} \cdot u_i^{1-\alpha} + p_i^{1-\alpha} \cdot u_i^{\alpha}}{2} \right)$$

with $\alpha \neq 1$ and $\alpha > 0$ as well as $P, U \in \Delta_n$ (A.113)

It can be shown that

$$\lim_{s \to 1} J_{s}^{s}(P \,\|\, U) = \lim_{\alpha \to 1} {}^{1}J_{\alpha}^{1}(P \,\|\, U) = 2 \cdot \lim_{\alpha \to 1} {}^{2}J_{\alpha}^{1}(P \,\|\, U) = J(P \,\|\, U) \tag{A.114}$$

with

$$J(P \,\|\, U) = \sum_{i=1}^{n} (p_i - u_i) \cdot \ln \frac{p_i}{u_i}$$

With two scalar parameters s and α we may find further generalizations:

$$
{}^{1}J_{\alpha}^{s}(P\,\|\,U)=\frac{1}{\left(1-2^{1-s}\right)}\cdot\left\{\left(\sum_{i=1}^{n}p_i^{\alpha}\cdot u_i^{1-\alpha}\right)^{\frac{s-1}{\alpha-1}}+\ln\left(\sum_{i=1}^{n}p_i^{1-\alpha}\cdot u_i^{\alpha}\right)^{\frac{s-1}{\alpha-1}}-2\right\}
$$

with $\alpha \neq 1$, $s \neq 1$ and $\alpha > 0$ (A.115)

and

$$
{}^{2}J_{\alpha}^{s}(P\,\|\,U)=\frac{1}{1-2^{1-s}}\cdot 2\cdot\left\{\left(\sum_{i=1}^{n}\frac{p_i^{\alpha}\cdot u_i^{1-\alpha}+p_i^{1-\alpha}\cdot u_i^{\alpha}}{2}\right)^{\frac{s-1}{\alpha-1}}-1\right\} \tag{A.116}
$$

with $\alpha \neq 1$, $s \neq 1$ and $\alpha > 0$

as logical extension of the parameters, which contain only one parameter.

The following limits are easy to verify:

$$
\lim_{s\to 1}{}^{1}J_{\alpha}^{s}(P\,\|\,U)={}^{1}J_{\alpha}^{1}(P\,\|\,U) \qquad \lim_{s\to 1}{}^{2}J_{\alpha}^{s}(P\,\|\,U)={}^{2}J_{\alpha}^{1}(P\,\|\,U) \tag{A.117a}
$$

$$
\lim_{\alpha\to 1}{}^{1}J_{\alpha}^{s}(P\,\|\,U)={}^{1}J_{1}^{s}(P\,\|\,U) \qquad \lim_{\alpha\to 1}{}^{2}J_{\alpha}^{s}(P\,\|\,U)={}^{2}J_{1}^{s}(P\,\|\,U) \tag{A.117b}
$$

Furthermore we have:

$$
{}^{1}J_{\alpha}^{1}(P\,\|\,U)=D_{\alpha}^{1}(P\,\|\,U)+D_{\alpha}^{1}(U\,\|\,P) \tag{A.118}
$$

and we summarize both J-divergences:

$$
{}^{3}\boldsymbol{R}_{\alpha}^{s}(P\,\|\,U)=\begin{cases}{}^{3}R_{\alpha}^{s}(P\,\|\,U), & \alpha\neq 1, s\neq 1, \alpha>0\\ {}^{3}R_{1}^{s}(P\,\|\,U), & \alpha=1, s\neq 1\\ {}^{3}R_{\alpha}^{1}(P\,\|\,U), & \alpha\neq 1, s=1, \alpha>0\\ {}^{3}R(P\,\|\,U), & \alpha=1, s=1\end{cases} \quad \text{for all } P, U\in\Delta_n \tag{A.119}
$$

The following properties hold:

(i) ${}^{r}\boldsymbol{J}_{\alpha}^{s}(P\,\|\,U)\geq 0$ ($r=1$ or 2) and for all $\alpha > 0$ and arbitrary s

(ii) ${}^{r}\boldsymbol{J}_{\alpha}^{s}(P\,\|\,U)$ is a convex function of the pair $(P, U) \in \Delta_n \times \Delta_n$ for all $s \geq \alpha > 0$

(iii) $$^{1}\boldsymbol{J}_{\alpha}^{s}(P \| U)\begin{cases}\leq {}^{2}\boldsymbol{J}_{\alpha}^{s}(P \| U), & \alpha \geq s \\ \geq {}^{2}\boldsymbol{J}_{\alpha}^{s}(P \| U), & s \geq \alpha > 0\end{cases}$$

Statistical applications of the generalized J-divergence can be found in [TAN87].

A.4 A short introduction to probability theory

In order to be able to define and handle probability in an objective mathematical manner, we first need a definition, a determination of the concept of probability, which can be used to derive an appropriate description. The (theoretical) probability is closely related to the heuristic relative frequency.

<u>Probability</u>	<u>Relative frequency</u>
Mathematical, axiomatic approach	Historic approach
Stochastics	Combinatorics

Relative frequency of the event A

N : Number of samples in an experiment

$N(A)$: Number of samples with the result A

The relative frequency of A: $$h(A) = \frac{N(A)}{N} \tag{A.120}$$

Taking the limit of the relative frequency for $N \to \infty$ carries out the transition to the probability

$$P(A) = \lim_{N \to \infty} \frac{N(A)}{N} = \lim_{N \to \infty} h(A) \tag{A.121}$$

or by interpreting the probability as an expectation value of the relative frequency.

The problem of the limit $N \to \infty$:

- When the number N of the samples occurring in an experiment is determined, this also determines the relative frequency with which atypical sequences or atypical events occur (We call an event 'atypical' if the relative frequency and the probability differ too much, which can be described via the Tschebycheff inequality).
- With an increasing number N the frequency of such atypical events decreases, but it never achieves the value zero (it does for $N \to \infty$, but this cannot be achieved).

- If we unfortunately observe such an atypical sequence of arbitrary length, the relative frequency will never converge to the probability.

The problem of the expectation $E\{h(A_i)\}$:

$$P(A_i) = E\{h(A_i)\} \tag{A.122}$$

- The calculation of the expectation value requires the knowledge of the probabilities, so that we have to use the probabilities in the definition of the probabilities. This leads to a circulus vituosus, a circular argument, not admissible in a definition.

Summarizing

The relative frequency converges with a certain probability to the probability. The theoretical probability can thus be identified by the relative frequency merely with a certain probability and the limit $N \to \infty$ cannot be achieved ('...in the long run we are all dead...'). Thus we define the probability in an axiomatic approach, first described by Andrej N. Kolmogorov.

A.4.1 Axiomatic definition of probability

Before we assign probabilities, we have to define a set that enables us to assign probabilities to the elements of this set.

A.4.1.1 Events, elementary events, sample space

Probability and events are inseparably connected to sets, because in every experiment we obtain events that are certain subsets of a given set and these events are described (weighted) by probabilities. (Example: In a lottery the event '3 hits' is independent of the order of the hits. All events '3 hits' are assigned one single probability, as the order of the hits is not important to the payout.) To be able to describe the sets with a mathematical formulation, we first define some terms:

Sample space, sample set	Ω	: basic set, event space
Elements of the sample set	ω	: elementary events

By elementary events we mean all possible results of an experiment. The sample space is described by the set definition

$$\Omega = \{\omega : \omega \in \Omega\} \tag{A.123}$$

Thus the sample space consists of all the elementary events that may occur in the observation. The next step in this definition is the introduction of events that consist of one or several elementary events and are subsets of the sample space.

$$A \subset \Omega : A\{\omega : \omega \in A : \omega \in \Omega\} \tag{A.124}$$

The events A are subsets of the sample space.

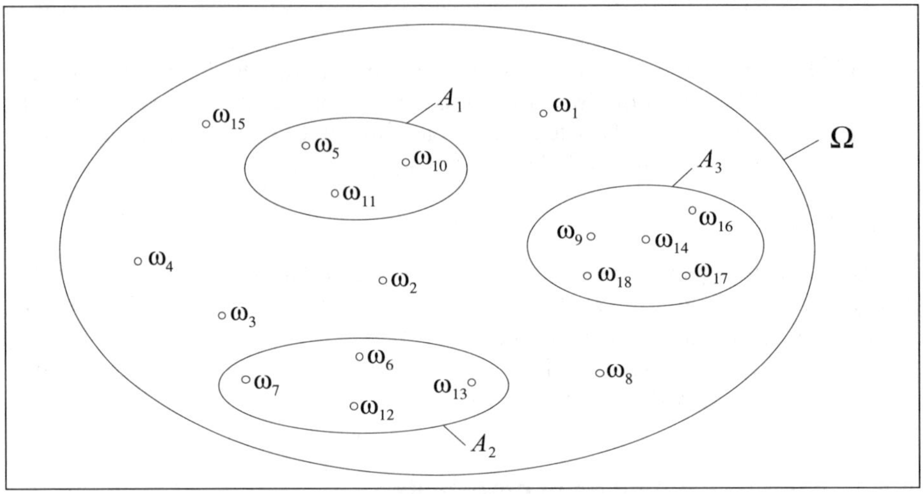

Fig. A.3: Sample space with events and elementary events

A are the observable subsets of the sample space and thus the subspaces that are of interest or useful to the observer.

A.4.1.2 Classes of subsets, fields

Which subspaces of the sample space Ω are events and which are not?

1. If $A \in \Omega$ is an event, we write

$$A \in \sigma_A = \text{is an element of the sigma-algebra} \tag{A.125}$$

and the complementary event $\overline{A} \in \sigma_A$ is an element of the sigma-algebra, too.

2. For two sets $A, B \in \sigma_A$, i.e. A, B are events, the combinations of this events are again events of the sigma-algebra.

$$A \cup B \in \sigma_A \quad \text{and} \quad A \cap B \in \sigma_A \tag{A.126}$$

analogously $\quad A \wedge B \in \sigma_A$ (A or B)

Generalized

$$A_i \in \sigma_A$$

$$\bigcup_i A_i \in \sigma_A \quad \text{and} \quad \bigcap_i A_i \in \sigma_A \tag{A.127}$$

and this has to hold for all countable intersections and union sets.

From the requirement that the complementary event has to be an event of the sigma-algebra, we conclude:

$$\bigcup_i A_i \in \sigma_A \Rightarrow \overline{\bigcup_i A_i} \in \sigma_A \Rightarrow \bigcap_i \overline{A_i} \in \sigma_A$$

according to de Morgan's rule.

3. $A \in \sigma_A \qquad \overline{A} \in \sigma_A$

$$\Rightarrow A \cap \overline{A} = \varnothing \in \sigma_A$$

$$\Rightarrow A \cup \overline{A} = \Omega \in \sigma_A \tag{A.128}$$

<u>*Summary*</u>:

The sigma-algebra is a class of sets (events) in the sample space, which are weighted with the probabilities (measures). The sigma-algebra contains the sample space Ω and the empty set $\varnothing$, as well as all complements, union sets and intersection of the interesting events.

All these sets can be observed, i.e. they can be distinguished in an experiment, which is why they are assigned with measures called probabilities. This observability of the events is an important criterion for the generation of an appropriate set organization (a generation of an appropriate algebra). Events we cannot observe or events that are not of any interest to us can be summarized in one event 'not interesting or not observable'.

Events A that do not have any further subsets (apart from the empty set), which create the smallest observable sets, are called atoms of the sigma-algebra.

<u>*Special algebras*</u>

$\Omega = \Re^n$ consists of all points of the n-dimensional euclidian space.

The set of the sample space is given by

$$A = \left\{ \underline{\omega} : \underline{\omega} \le \underline{a}, \underline{\omega} \in \Re^n \right\} \tag{A.129}$$

We have

$$\underline{\omega} \leq \underline{a} = \begin{pmatrix} \omega_1 \leq a_1 \\ \omega_2 \leq a_2 \\ \vdots \\ \omega_n \leq a_n \end{pmatrix} \tag{A.130}$$

and in the two-dimensional space we obtain an area of the form shown in figure A.4.

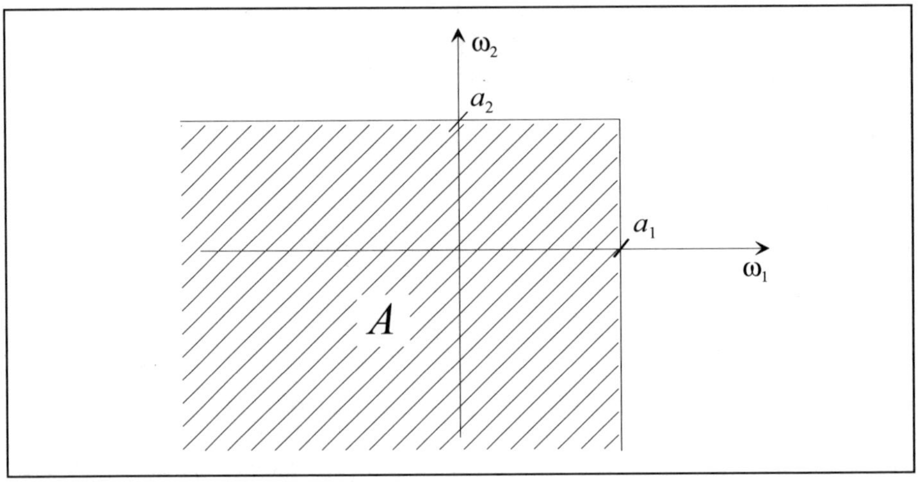

Fig. A.4: Two-dimensional Borel field

The sigma-algebra consisting of all intersections, union sets and complements of these sets, is called *Borel field* σ_B and contains all types of intervals (open, closed, semi-closed, points) and is of interest for many technical applications.

Example of a Borel field σ_B in the one-dimensional euclidian space:

$$A_1 = \{\omega : \omega \leq a_1, \omega \in \Re\} = (-\infty, a_1]$$
$$A_2 = \{\omega : \omega \leq a_2, \omega \in \Re\} = (-\infty, a_2] \tag{A.131}$$

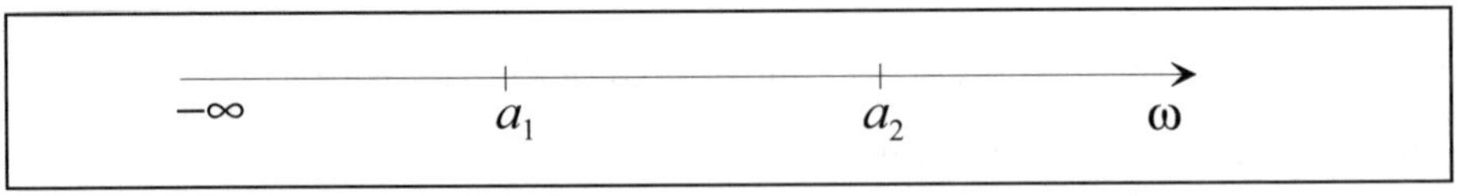

Fig. A.5: One-dimensional Borel field

$$A_1 \in \sigma_B \qquad \mathrm{A}_2 \in \sigma_B \tag{A.132}$$

A.4.1.3 Axiomatic definition of probability according to Kolmogorov

Intention:

Assignment of probabilities (measures) to the sets A of Ω, contained in the sigma-algebra, which are called events.

The probability function $P(\ )$ is a function, defined on the elements of the sigma-algebra, assigning probabilities $P(A_i)$ to all elements $A_i \in \sigma_A$. This function meets the following (Kolmogorov's) axioms:

1. $P(A) \geq 0$ (A.133)
2. $P(\Omega) = 1$ (A.134)
3. for disjoint events we get

$A_i \cap A_j \neq 0$ for $i \neq j$ mutually disjoint sets

$$P\left(\bigcup_i A_i\right) = \sum_i P(A_i) \text{ for all countable finite or infinite sets} \tag{A.135}$$

=> sigma-additivity of the probability

With the probability function $P(\ )$ the sigma-algebra is mapped on the interval [0,1].

Compatibility to the relative frequency

$P(\Omega) = 1$	$h(\Omega) = 1$
$P(\varnothing) = 0$	$h(\varnothing) = 0$

Besides $\overline{\varnothing} = \Omega$ and $\varnothing \cap \Omega = \varnothing$

$$\begin{aligned} P(\Omega) &= 1 = P(\Omega \cup \varnothing) \\ &= P(\Omega) + P(\varnothing) = 1 \end{aligned} \tag{A.136}$$

$$P(\varnothing) = 1 - P(\Omega) = 1 - 1 = 0 \tag{A.137}$$

Probability space

The triplet (Ω, σ_A, P) generates the probability space.

Ω	=	sample space	$\Omega = \{\omega : \omega \in \Omega\}$
σ_A	=	sigma-algebra	$\sigma_A = \{A : A \in \sigma_A\}$
$P()$	=	probability function	

A.4.1.4 Random variables

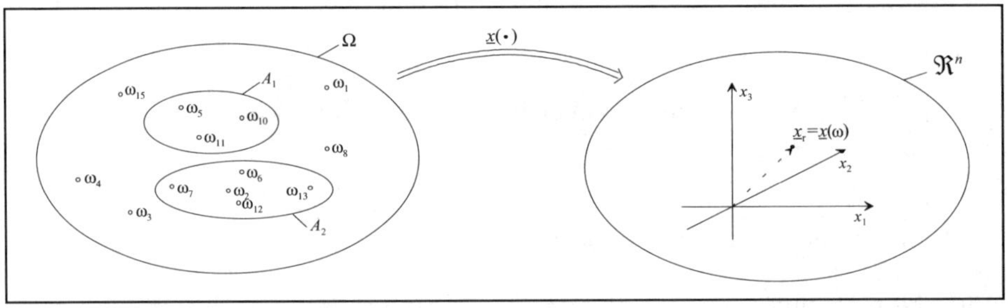

Fig. A.6: Random variables, mappings from sample space to realization space

The random variable $\underline{x}(\ )$ is a multi-dimensional, real mapping from the sample space Ω in the n-dimensional euclidian space $\Re^n$.

<u>*Scalar random variables*</u>

$x(\ )$ is a real function, assigning a realization $x_r = x(\omega) \in \Re$ to each event $\omega \in \Omega$ in the sample space in such a way that

$$A = \{\omega : x(\omega) \leq \xi\} \qquad \text{for arbitrary } \xi \in \Re \tag{A.138}$$

is an element of the sigma-algebra for $A \in \sigma_A$.

This transformation maps the sample space on the realization space of the real numbers $\Re$. The sigma-algebra of the sample space is thus mapped in a Borel field (special sigma-algebra), and the advantages of these Borel fields are evident in the description of distribution functions, as we will see later.

Concrete number $x_r = x(\omega)$ Realization of the random variable

Function $x(\ ) = x$ Random variable

Multi-dimensional random variables

$\underline{x}(\ )$ is a multi-dimensional real mapping from sample space Ω in the n-dimensional euclidian space $\Re^n$, so that

$$A = \{\omega : \underline{x}(\omega) \le \underline{\xi}\} \qquad \text{for arbitrary } \underline{\xi} \in \Re^n \text{ and for } A \in \sigma_A \tag{A.139}$$

A.4.1.5 Probability distribution

All sets of the form

$$A = \{\omega : \underline{x}(\omega) \le \underline{\xi}\} \qquad \text{are elements of the sigma-algebra (Borel field)} \tag{A.140}$$

$$A = \{\omega : x_1(\omega) \le \xi_1, x_2(\omega) \le \xi_2, \ldots, x_n(\omega) \le \xi_n\}$$

Now we define the probability distribution

$$F_{\underline{x}}(\underline{\xi}) = P(A) = P\{\omega : \underline{x}(\omega) \le \underline{\xi}\} \tag{A.141}$$

$$F_{\underline{x}}(\underline{\xi}) = P\{\omega : x_1(\omega) \le \xi_1, x_2(\omega) \le \xi_2, \ldots, x_n(\omega) \le \xi_n\}$$

$$F_{\underline{x}}(\underline{\xi}) = F_{x_1, x_2, \ldots, x_n}(\xi_1, \xi_2, \ldots, \xi_n)$$

This distribution function describes the probability of the set A, denoting the interval from $-\infty$ to the given limitation ξ (Borel field). A multi-dimensional distribution is then equal to a joint distribution function of scalar random variables.

Summary

The distribution function assigns probabilities to the sets of the Borel field. Thus this function is merely a mapping from realization space $\Re^n$, where the Borel fields and the appropriate algebra σ_B exist, in the probability space.

The triplet $(\Omega, \sigma_A, P(A))$ consisting of the sample space, the sigma-algebra and the probability describes the probability space as well as the triplet $\left(\Re^n, \sigma_B, F_{\underline{x}}(\underline{\xi})\right)$ consisting of the realization space, the Borel field (special sigma-algebra) and the distribution function.

A.4.1.6 Probability space, sample space, realization space

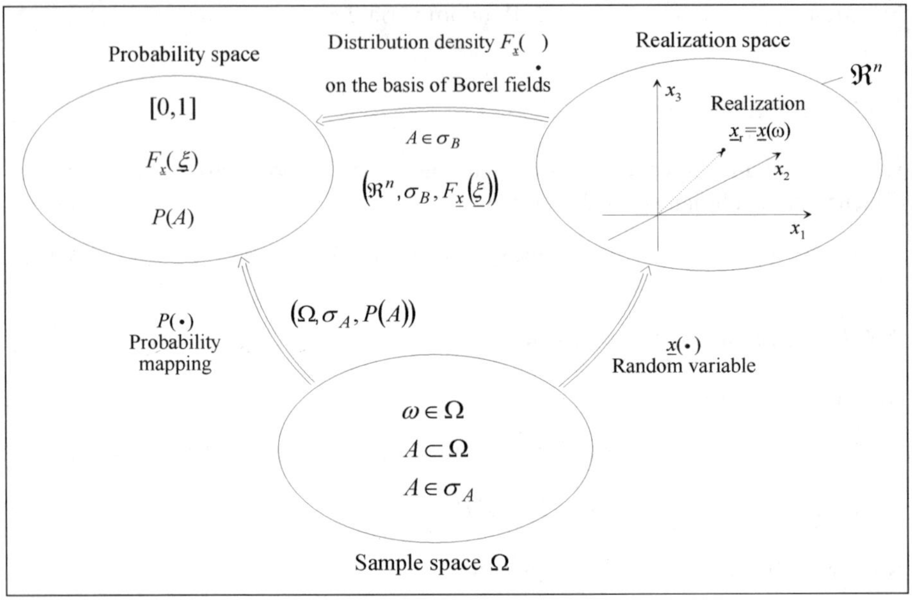

Fig. A.7: Probability space, sample space (event space), realization space

A.4.1.7 Probability distribution and distribution density function

Distribution function

$A = \{\omega : \underline{x}(\omega) \leq \underline{\xi}\}$ for arbitrary $\underline{\xi} \in \Re^n$ and for $A \in \sigma_A$

$$P(A) = F_{\underline{x}}(\underline{\xi}) \quad \text{distribution function} \tag{A.142}$$

Monotonic nature of the probability distribution

$$F_{\underline{x}}(\underline{\xi}_{i+1}) \geq F_{\underline{x}}(\underline{\xi}_i) \quad \text{for} \quad \underline{\xi}_{i+1} \geq \underline{\xi}_i \tag{A.143}$$

Verification:

$$A_i = \{\omega : \underline{x}(\omega) \leq \underline{\xi}_i\}$$

$$A_{i+1} = \{\omega : \underline{x}(\omega) \leq \underline{\xi}_{i+1}\}$$

$$= \{\omega : \underline{x}(\omega) \leq \underline{\xi}_i\} \cup \{\omega : \underline{\xi}_i < \underline{x}(\omega) \leq \underline{\xi}_{i+1}\} \quad \text{disjoint sets}$$

We thus obtain

$$F_{\underline{x}}(\underline{\xi}_{i+1}) = P(A_{i+1})$$

$$F_{\underline{x}}(\underline{\xi}_{i+1}) = P\{\omega : \underline{x}(\omega) \le \underline{\xi}_{i+1}\}$$
$$= P\{\omega : \underline{x}(\omega) \le \underline{\xi}_{i}\} + P\{\omega : \underline{\xi}_{i} < \underline{x}(\omega) \le \underline{\xi}_{i+1}\}$$

$$F_{\underline{x}}(\underline{\xi}_{i+1}) = F_{\underline{x}}(\underline{\xi}_{i}) + \underbrace{P\{\omega : \underline{\xi}_{i} < \underline{x}(\omega) \le \underline{\xi}_{i+1}\}}_{\ge 0 \text{ according to the definition}}$$

and therefore

$$F_{\underline{x}}(\underline{\xi}_{i+1}) \ge F_{\underline{x}}(\underline{\xi}_{i}) \qquad \text{q.e.d.}$$

Further properties of the distribution function: The limits

1) $F_{\underline{x}}(\infty, \infty, \dots, \infty) =$

$$= P\{\omega : x_1(\omega) \le \infty, x_2(\omega) \le \infty, \dots, x_2(\omega) \le \infty\} = P(\Omega) = 1 \qquad \text{(A.144)}$$

2) $F_{\underline{x}}(-\infty, -\infty, \dots, -\infty) =$

$$= P\{\omega : x_1(\omega) \le -\infty, x_2(\omega) \le -\infty, \dots, x_2(\omega) \le -\infty\} = P(\varnothing) = 0 \qquad \text{(A.145)}$$

Marginal distribution functions

$$F_{x_1, x_2, \dots, x_n}(\xi_1, \xi_2, \dots, \xi_k, \infty, \infty, \dots, \infty)$$

$$= P\left\{\omega : x_1(\omega) \le \xi_1, x_2(\omega) \le \xi_2, \dots, x_k(\omega) \le \xi_k, \underbrace{x_{k+1}(\omega) \le \infty, \dots, x_n(\omega) \le \infty}_{\text{is always met}}\right\}$$

(A.146)

As this condition always holds for the realizations of the random variables $k+1$ to n, the distribution function in this area is equal to 1. For the remaining random variables we find

$$= P\{\omega : x_1(\omega) \le \xi_1, x_2(\omega) \le \xi_2, \dots, x_k(\omega) \le \xi_k\}$$

$$F_{x_1, x_2, \dots, x_n}(\xi_1, \xi_2, \dots, \xi_k, \infty, \infty, \dots, \infty) = F_{x_1, x_2, \dots, x_k}(\xi_1, \xi_2, \dots, \xi_k) \qquad \text{(A.147)}$$

the marginal distribution function.

Special case $n=2$:

$$F_{x_1,x_2}(\xi_1,\infty)=F_{x_1}(\xi_1) \tag{A.148}$$

$$F_{x_1,x_2}(\infty,\xi_2)=F_{x_2}(\xi_2) \tag{A.149}$$

Furthermore we get

$$F_{x_1,x_2,\ldots,x_n}(\xi_1,\xi_2,\ldots,\xi_k,-\infty,-\infty,\ldots,-\infty)$$

$$=P\left\{\omega : x_1(\omega)\le\xi_1, x_2(\omega)\le\xi_2,\ldots,x_k(\omega)\le\xi_k,\underbrace{x_{k+1}(\omega)\le\infty,\ldots,x_n(\omega)\le\infty}_{\text{is never met}}\right\} \tag{A.150}$$

and therefore

$$F_{x_1,x_2,\ldots,x_n}(\xi_1,\xi_2,\ldots,\xi_k,-\infty,-\infty,\ldots,-\infty)=0 \tag{A.151}$$

As soon as only one limit is equal to $-\infty$, the distribution function of this event is equal to zero and the complete distribution function is also equal to zero.

A.4.1.8 Probability distribution density function (PDF)

If there exists a scalar real function $f_{\underline{x}}(\underline{\xi})$ for all $\underline{\xi}\in\Re^n$, which meets the equation

$$F_{x_1,x_2,\ldots,x_n}(\xi_1,\xi_2,\ldots,\xi_n)=\int_{-\infty}^{\xi_1}\int_{-\infty}^{\xi_2}\cdots\int_{-\infty}^{\xi_n} f_{x_1,x_2,\ldots,x_n}(\zeta_1,\zeta_2,\ldots,\zeta_n)\cdot d\zeta_1\cdot d\zeta_2\cdot\ldots\cdot d\zeta_n$$

$$F_{\underline{x}}(\underline{\xi})=\int_{-\underline{\infty}}^{\underline{\xi}} f_{\underline{x}}(\underline{\zeta})\cdot d\underline{\zeta} \tag{A.152}$$

then we call $f_{\underline{x}}(\,)$ probability distribution density function of the random variable $\underline{x}(\,)$. The inverse function

$$f_{\underline{x}}(\underline{\xi})=\frac{\partial}{\partial\underline{\xi}}F_{\underline{x}}(\underline{\xi})$$

$$f_{x_1,x_2,\ldots,x_n}(\xi_1,\xi_2,\ldots,\xi_n)=\frac{\partial^n}{\partial\xi_1\cdot\partial\xi_2\cdot\ldots\cdot\partial\xi_n}F_{x_1,x_2,\ldots,x_n}(\xi_1,\xi_2,\ldots,\xi_n) \tag{A.153}$$

generates the distribution density function as the derivative of the distribution function. The distribution density function does not always exist in the classical mathematical sense. Discontinuities in $F_{\underline{x}}(\underline{\xi})$ generate Dirac impulses in the distribution density function $f_{\underline{x}}(\underline{\xi})$.

In this case the distribution density function $f_{\underline{x}}(\underline{\xi})$ can be decomposed into a continuous and a discontinuous part. Discrete random variables in general lead to discontinuous distribution functions, which result in Dirac impulses in the distribution density functions.

Calculation of probabilities with the distribution density function

We found

$$P\{\omega : \underline{x}(\omega) \in (\underline{\xi}_1, \underline{\xi}_2]\} = F_{\underline{x}}(\underline{\xi}_2) - F_{\underline{x}}(\underline{\xi}_1) \tag{A.154}$$

and if we use the distribution density function

$$F_{\underline{x}}(\underline{\xi}_2) = \int_{-\infty}^{\underline{\xi}_2} f_{\underline{x}}(\underline{\zeta}) \cdot d\underline{\zeta} \tag{A.155}$$

$$F_{\underline{x}}(\underline{\xi}_1) = \int_{-\infty}^{\underline{\xi}_1} f_{\underline{x}}(\underline{\zeta}) \cdot d\underline{\zeta} \tag{A.156}$$

Inserting this, we obtain

$$\begin{aligned} P\{\omega : \underline{x}(\omega) \in (\underline{\xi}_1, \underline{\xi}_2]\} &= \int_{-\infty}^{\underline{\xi}_2} f_{\underline{x}}(\underline{\zeta}) \cdot d\underline{\zeta} - \int_{-\infty}^{\underline{\xi}_1} f_{\underline{x}}(\underline{\zeta}) \cdot d\underline{\zeta} \\ &= \int_{\underline{\xi}_1}^{\underline{\xi}_2} f_{\underline{x}}(\underline{\zeta}) \cdot d\underline{\zeta} \end{aligned} \tag{A.157}$$

The probability of the interval is equal to the area under the distribution density function in the given interval.

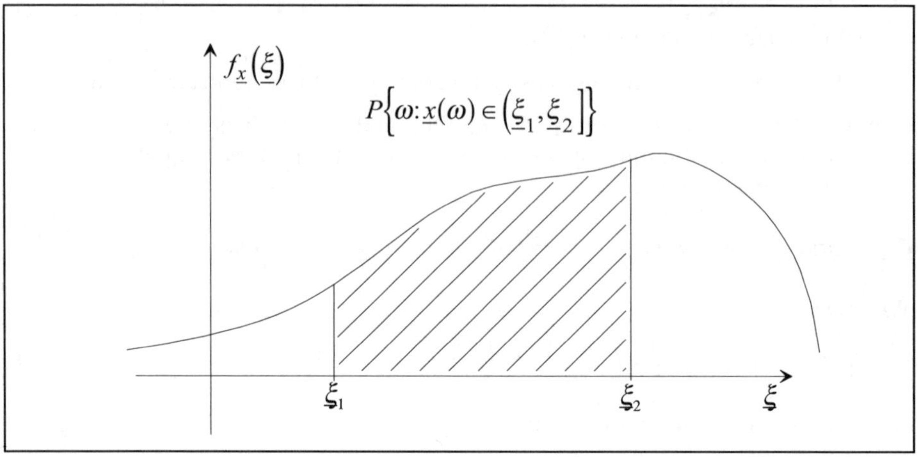

Fig. A.8: Probability on an interval

We may also write:

$$f_{\underline{x}}(\underline{\xi})\cdot d\underline{\xi} = dF_{\underline{x}}(\underline{\xi}) \tag{A.158}$$

And this is the infinitesimal probability of an interval of width $d\underline{\xi}$. The probability of a two-dimensional random variable requires a generalization:

$$\begin{aligned} P\{\omega : \underline{x}(\omega)\in A\} &= \int_A f_{\underline{x}}(\underline{\xi})\cdot d\underline{\xi} \\ &= \int\int_A ..\int f_{x_1,x_2,\ldots,x_n}(\xi_1,\xi_2,\ldots,\xi_n)\cdot d\xi_1\cdot d\xi_2\cdot\ldots\cdot d\xi_n \end{aligned} \tag{A.159}$$

To obtain the whole probability of the interval A, we integrate over all disjoint infinitesimal probability intervals

$$f_{\underline{x}}(\underline{\xi})\cdot d\underline{\xi} = dF_{\underline{x}}(\underline{\xi}) \tag{A.160}$$

and obtain the desired probability of the event A.

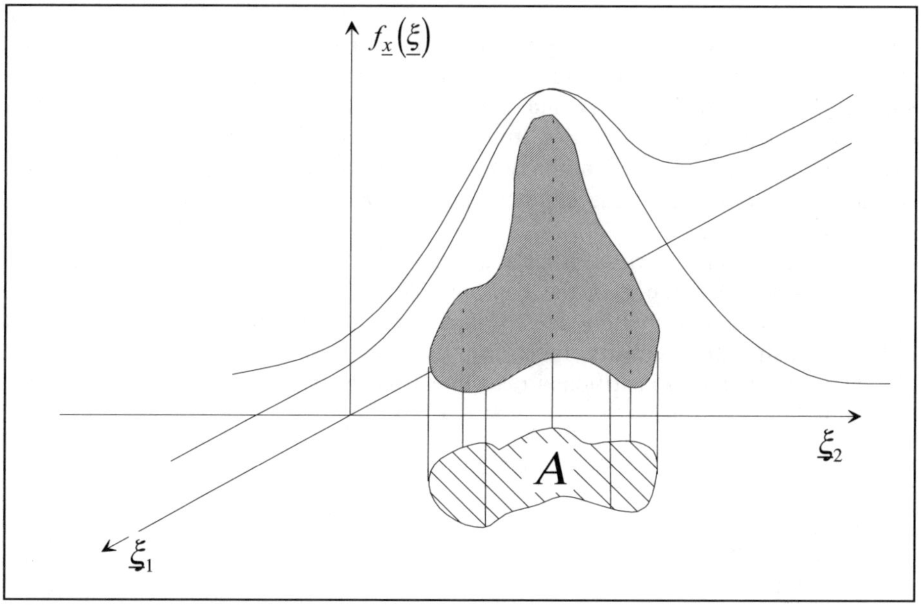

Fig. A.9: Probability of the area *A*

A.5 The regularity conditions

[KEN69/73]: The interchange of integration and differentiation is allowed if the limits of the integration are finite (i.e. the limits of the variation of the integration variable) and independent of the parameter ϑ that is used to generate the derivative. Even if the limitations are not finite, such an interchange is possible if the resulting integral is uniformly convergent for all ϑ, and the kernel is a continuous function of the integration variable ξ and the differentiation variable ϑ. These are the sufficient conditions for the interchange of integration and differentiation.

Regularity conditions:

The regularity conditions are the preliminaries for the existence of a maximum likelihood estimate. They are introduced from Blischke, Cramér (see for example section 33.3 in [KEN69/73]) and from Kullback (in the presentation of the connection between Kullback's information and Fisher's information) as preliminaries, under which a maximum likelihood estimate exists. This also includes the existence of the Cramér–Rao bound and the existence of Fisher's information matrix.

With these preliminaries Fisher's information is directly related to the ML estimate, whose existence is the basic condition for the existence of Fisher's information. If a distribution density function does not have an ML estimate (uniform distribution density function), then its information is equal to zero (Laplacian principle of insufficient reasoning), as all realizations do have the same probability of occurrence.

Cramér shows that the likelihood equation

$$\frac{\partial}{\partial \vartheta} \cdot \sum_{i=1}^{n} \ln f(\xi_i \mid \vartheta) = 0 \tag{A.161}$$

can be solved under the following conditions

(i) the partial derivatives

$$\frac{\partial}{\partial \vartheta} \ln f \quad ; \quad \frac{\partial^2}{\partial \vartheta^2} \ln f \quad ; \quad \frac{\partial^3}{\partial \vartheta^3} \ln f \tag{A.162}$$

exist for nearly all ξ and for each ϑ in a non-degenerated interval Θ.

(ii) For each $\vartheta \in \Theta$

$$\left| \frac{\partial}{\partial \vartheta} \ln f \right| < G_1(\xi) \tag{A.163}$$

$$\left| \frac{\partial^2}{\partial \vartheta^2} \ln f \right| < G_2(\xi) \tag{A.164}$$

$$\left| \frac{\partial^3}{\partial \vartheta^3} \ln f \right| < G_3(\xi) \tag{A.165}$$

where $G_1(\xi)$ and $G_2(\xi)$ can be integrated in the interval $(-\infty, +\infty)$ and $G_3(\xi)$ meets the equation

$$\int_{-\infty}^{\infty} G_3(\xi) \cdot f(\xi \mid \vartheta) \cdot d\xi < c \tag{A.166}$$

for an arbitrary constant c, which is independent of ϑ.

(iii) The inequality (see also the Cramér–Rao inequality)

$$0 < \int_{-\infty}^{\infty} \left[\frac{\partial}{\partial \vartheta} \ln f(\xi \mid \vartheta) \right]^2 \cdot f(\xi \mid \vartheta) \cdot d\xi = F < \infty \tag{A.167}$$

holds for all $\vartheta \in \Theta$.

Two statements of Blischke:

"... when the regularity conditions are violated, of course, other approaches to the investigation of asymptotic properties of maximum likelihood estimation must be used."
(Differential quotient instead of the derivative → Chapman–Robbins bound, ...)

"... typically in the non-regular case the maximum likelihood estimators are neither asymptotically normally distributed, nor asymptotically efficient."

These two facts are not unrelated.

With these regularity conditions the Cramér–Rao bound is asymptotically achieved by the so-called BAN (best asymptotically normal) estimators. In the non-regular case this bound does not exist or it cannot be achieved, not even asymptotically. That is, if Fishers information matrix diverges, the Cramér–Rao inequality degenerates to the trivial case

$$\operatorname{var}(\vartheta) \geq 0 \tag{A.167}$$

and a lower limit of the variance in the real sense (>0) does not exist any more.

Thus Fisher's information matrix is an indicator of the existence and the accessibleness of the Cramér–Rao bound.

A.6 State space description

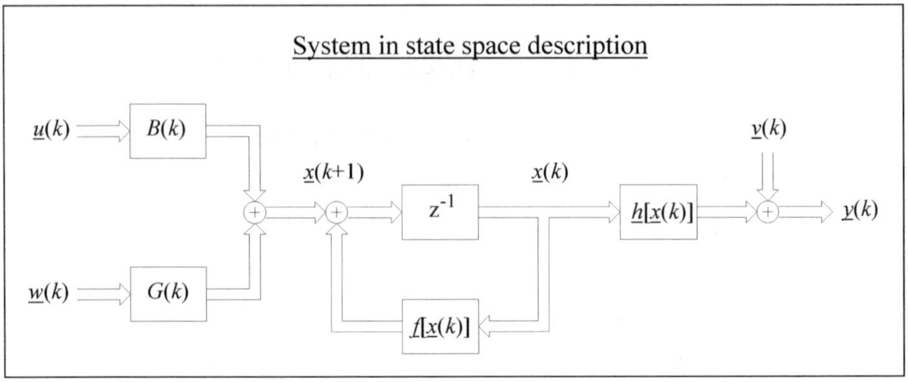

Fig. A.10: Discrete state space description

The vectors and mappings are:

$\underline{x}(k)$ = state space vector (at time k)

$\underline{y}(k)$ = measurement vector

$\underline{u}(k)$ = deterministic input vector

$\underline{w}(k)$ = stochastic input vector (white noise, driving noise)

$\underline{v}(k)$ = measurement noise

$B(k)$ = gain matrix of the deterministic input vector

$G(k)$ = gain matrix of the stochastic input vector

$\underline{f}[\underline{x}(k)]$ = nonlinear state transition function

$\underline{h}[\underline{x}(k)]$ = nonlinear observation function

In the linear state space model the nonlinear function $\underline{f}(\underline{x})$ and $\underline{h}(\underline{x})$ are replaced by the matrices

$A(k)$ = state transition matrix

$C(k)$ = observation matrix

The equalities of the discrete states space description are

Nonlinear denotation	Linear denotation
$\underline{x}(k+1)=\underline{f}[\underline{x}(k),k]+\underline{w}(k)$	$\underline{x}(k+1)=A(k)\cdot\underline{x}(k)+\underline{w}(k)$
$\underline{y}(k)=\underline{h}[\underline{x}(k),k]+\underline{v}(k)$	$\underline{y}(k)=C(k)\cdot\underline{x}(k)+\underline{v}(k)$

Bibliography

[ABD97] Abdel-Ghaffar, K. A. S.: A lower bound on the undetected error probability and strictly optimal codes, IEEE Transactions on Information Theory, Vol. IT-43, No. 5, 1997, pp. 1489-1502

[ABR93] Abrahams, Julia; Lipman, Marc J.: Relative uniformity of sources and the comparison of optimal code costs, IEEE Transactions on Information Theory, Vol. IT-39, No. 5, 1993, pp. 1695-1697

[ACZ75] Aczel, Daroscy: On Measures of Information and Their Characterization, Academic Press, New York, 1975

[AHM89] Ahmed, Nabil Ali; Gokhale, D. V.: Entropy expressions and their estimators for multivariate distributions, IEEE Transactions on Information Theory, Vol. IT-35, No. 3, 1989, pp. 688-695

[ALZ90] Alzer, H.: On weighted arithmetic, geometric and harmonic mean values, Glasnik mathematicki, Ser. 3, Vol. 25 (45), No. 2, 1990, pp. 279-285

[AKA69/1] Akaike, Hirotugu: Statistical predictor identification, Annals of the Institute of Statistical Mathematics, Vol. 21, 1969, pp. 203-217

[AKA69/2] Akaike, Hirotugu: A fundamental relation between predictor identification and power spectrum estimation, Annals of the Institute of Statistical Mathematics, Vol. 21, 1969, pp. 219-223

[AKA69/3] Akaike, Hirotugu: Fitting autoregressive models for prediction, Annals of the Institute of Statistical Mathematics, Vol. 21, 1969 pp. 243-247

[AKA69/4] Akaike, Hirotugu: Power spectrum estimation through autoregressive model fitting, Annals of the Institute of Statistical Mathematics, Vol. 21, 1969, pp. 407-419

[AKA73] Akaike, Hirotugu: Information theory and an extension of the maximum likelihood principles, 2nd International Symposium on Information Theory, Akademiai Kiado, Budapest, 1973, pp. 267-281

[AKA74/1] Akaike, Hirotugu: Stochastic theory of minimal realization, IEEE Transactions on Automatic Control, Vol. AC-19, No. 6, 1974, pp. 667-674

[AKA74/2] Akaike, Hirotugu: New look at the statistical model identification, IEEE Transactions on Automatic Control, Vol. AC-19, No. 6, 1974, pp. 716-723

[AKA75] Akaike, Hirotugu: Markovian representation of stochastic processes by canonical variables, SIAM Journal of Control, Vol. 13, No. 1, 1975, pp. 162-173

[AKA78] Akaike, Hirotugu: A bayesian analysis of the minimum AIC procedure, Annals of the institute of statistical mathematics , Vol. 30, Part A, 1978, pp. 9-14

[ANO96] Anoulova, Svetlana; Fisher, Paul; Pölt, Stefan; Simon, Hans Ulrich: Probably almost bayes decision, Information and Computation, Vol. 129, 1996, pp. 63-71

[AOK67] Aoki, Masanao: Optimization of Stochastic Systems, Topics in Discrete-Time Systems, Academic Press, New York, 1967

[ASH67] Ash, Robert: Information Theory, Interscience Publishing, New York, 1965

[ARI94] Aristoteles: Metaphysik, Rowohlt Verlag, Hamburg, 1994

[ARN96/1] Arndt, C.; Loffeld, O.: Information gained by pseudoredundant measurements, Europto, European Symposium on Laser, Optics, and Vision for Productivity in Manufacturing, Besançon, France, June 1996

[ARN96/2] Arndt, C.; Loffeld, O.: Information gained by Data-Fusion, Europto, European Symposium on Laser, Optics, and Vision for Productivity in Manufacturing, Besançon, France, June 1996

[ARN97/1] Arndt, Christoph : Informationsgewinnung und -verarbeitung in nichtlinearen dynamischen Systemen, Dissertation an der Universität Siegen, Shaker-Verlag, Aachen, 1997

[ARN97/2] Arndt, C.; Loffeld, O.: “Pseudoredundant Measurements in Nonlinear Mappings”, ISIT 97, IEEE International Symposium on Information Theory, Ulm, Germany, 29 June – 4 July 97

[ARN98/2] Arndt, C.; Hein, A.; Loffeld, O.: “Information Theory in Data Fusion”, Conference ‘Fusion of Earth data’, The European Association of Remote Sensing Laboratories (EARSeL), Société des Electriciens et Electroniciens de France (SEE), Ecole Nationale Supèrieure des Mines de Paris (ENSMP), 28–30 Januar 98, Sophia Antipolis, France

[ARN98/2] Arndt, C.; Loffeld, O.: "Information in SAR images with Special Emphasis on Phase Information", IASTED (The International Association of Science and Technologie for Development) International Conference Signal Processing and Communications, 11–14 February 1998, Canary Islands, Spain (http://www.gpds.ulpgc.es/icspc98)

[ARN92] Arnold, Ludwig: Stochastic Differential Equations: Theory and Applications, Krieger Publishing Company, Malabar, Fla., 1992

[ATH65] Athans, Michael; Schweppe, F. C.: Gradient Matrices and Matrix Calculations, Lincoln Laboratory, Lexington, Mass., Nov. 1965

[ATH66] Athans, Michael: Optimal Control, McGraw-Hill Book Company, New York, 1966

[AYR75] Ayres Frank Jr.: Differential- und Integralrechnung, McGraw-Hill Book Company, London, 1975

[AYR78] Ayres Frank Jr.: Matrizen, McGraw-Hill Book Company, London, 1978

[AYR91] Ayres Frank Jr.: Differentialgleichungen, McGraw-Hill Book Company, London, 1991

[BAH86] Baheti, R. H.: Efficient Approximation of Kalman filter for target tracking, IEEE Transanctions on Aerospace and Electronic Systems, New York, 1986

[BAJ96] Bajkova, Anisa T.: Reconstruction of ISAR images using generalized information methods, AEU International Journal of Electronic Communication, Vol. 50, No. 2, 1996, pp 165-168

[BAL97] Balcázar, J. L.; Gavaldá, R.: Computational power of neural networks: A characterization in terms of Kolmogorov complexity, IEEE Transactions on Information Theory, Vol. IT-43, No. 5, 1997, pp. 1175-1183

[BAL96] Balzer, D.; Arndt, C.; Loffeld, O.: Demodulation of phase modulated signals using an expanded observation model in Kalman Filters, Progress in Electromagnetic Research Symposium, PIERS, Innsbruck, Austria, July 1996

[BAR49] Barankin, E. W.: Locally best unbiased estimates, University of California, Berkeley, Annals of Mathematical Statistics, Vol. 20,1949, pp. 477 - 501

[BAR92] Barbour, Holst: Poisson Approximation, Clarendon Press, London, Barbour, 1992

[BAR78] Barndorff-Nielsen, O.: Information and Exponential Families in Statistical Theory, John Wiley&Sons, Chichester, 1978

[BAR91] Barron, A.; Cover, T.: Minimum complexity density estimation, , IEEE Transactions on Information Theory, Vol. IT-37, No. 4, 1991, pp. 1034-1054

[BAR66] Bartlett, Maurice: Stochastic Processes, Methods and Applications, Cambridge University Press, Cambridge, 1966

[BAS83] Basilevsky, Alexander: Applied Matrix Algebra in the Statistical Science, North-Holland, New York, 1983

[BAS74] Bass, J.: Éléments de Calcul de Probabilités, Masson et CIE, Paris, 1974

[BAS56] Basu, D.: On statistics of a complete sufficient statistic, Indian statistical Institute, Calcutta, Sankya : The Indian Journal of Statistics, Vol. 15, 1956, pp. 377-380

[BEL93] Bell, Mark R.: Information theory and radar waveform design, IEEE Transactions on Information Theory, Vol. IT-39, No. 5, 1993, pp. 1578-1597

[BER90] Berger, Toby; Ye, Zhongxing: ε-entropy and critical distortion of random fields, IEEE Transactions on Information Theory, Vol. IT-36, No. 4, 1990, pp. 717-725

[BHA45] Bhattacharyya, A.: On some sets of sufficient conditions leading to the normal bivariate distribution, Sankya : The Indian Journal of Statistics, Vol. 6, 1945, pp. 399-406

[BHA46] Bhattacharyya, A.: On some analogues of the amount of information and their use in statistical estimation Part 1, Sankya : The Indian Journal of Statistics, Vol. 8, 1946, pp. 1- 14

[BHA47] Bhattacharyya, A.: On some analogues of the amount of information and their use in statistical estimation Part 3, Sankya : The Indian Journal of Statistics, Vol. 8, 1947, pp. 201-217

[BHA48] Bhattacharyya, A.: On some analogues of the amount of information and their use in statistical estimation Part 4, Sankya : The Indian Journal of Statistics, Vol. 8, 1948, pp. 315 - 328

[BIE77] Bierman, Gerald J.: Factorization Methods for Discrete Sequential Estimation, Academic Press, New York, 1977

[BIR77] Birkenfeld, Wolfgang: Methoden zur Analyse von kurzen Zeitreihen, Birkhäuser Verlag, Basel and Stuttgart, 1977

[BLI69] Blischke, W. R.; Truelove, A. J.; Mundle, P. B.: On non-regular estimation. I. Variance bounds for estimators of location parameters, Control Data Corporation/C-E-I-R, American Statistical Association Journal, Sept. 1969, pp. 1056-1072

[BOB88] Bobrovski, Ben-Zion; Zakai, Moshe M.; Zeitouni, Ofer: Error bounds of the nonlinear filtering of signals with small diffusion coefficients, IEEE Transactions on Information Theory, IT-34, No. 4, 1988, pp. 710-721

[BOE79] Boekee, D. E.; van der Lubbe, J.: Some aspects of error bounds in feature selection, Pattern Recognition, No. 11, 1979, pp 353-360

[BOL81] Boltzmann, Ludwig: Vorlesungen über Gastheorie, Ambrosius Barth, Leipzig, 1981

[BON94] Bonavito, N. L.; et al.: Maximum entropy-driven bayesian reasoning, Telematics and Informatics, Vol. 11, No. 4, 1994, pp. 295-308

[BRA92] Brandt, Siegmund: Datenanalyse, Mit statistischen Methoden und Computerprogrammen, Bibliographisches Institut, Mannheim,Wien and Zürich, 1992

[BRI51/1] Brillouin, Leon: Maxwell´s Demon cannot operate: Information and entropy I, Journal of Applied Physics, 22, 1951, pp. 334-337

[BRI51/2] Brillouin, Leon: Information and entropy II, Journal of Applied Physics, Vol. 22, 1951, pp. 338-343

[BRI51/3] Brillouin, Leon: Information theory and most efficient codings for communication or memory devices, Journal of Applied Physics, Vol. 22, 1951, pp. 1108-1111

[BRI53] Brillouin, Leon: The negentropy principle of information, Journal of Applied Physics, Vol. 24, No. 9, 1953, pp. 1152-1163

[BRI54/1] Brillouin, Leon: Negentropy and information in telecommunications, writing, and reading, Journal of Applied Physics, Vol. 25, No. 5, 1954, pp. 595-599

[BRI54/2] Brillouin, Leon: Information theory and the uncertainty principle, Journal of Applied Physics, Vol. 25, No. 7, 1954, pp. 887-893

[BRI56] Brillouin, Leon: Science and Information Theory, Academic Press, New York, 1956

[BRI64] Brillouin, Leon: Scientific Uncertainty and Information, Academic Press, New York, 1964

[BRO87] Bronstein, I. N.; Semendjajev, K. A.: Taschenbuch der Mathematik, Verlag Harri Deutsch, Thun and Frankfurt/Main, 1987

[BRO93] Brown, Lawrence D.; Liu, Richard C.: Bounds on the Bayes and minimax risk for signal parameter estimation, IEEE Transactions on Information Theory, Vol. IT-39, No. 4, 1993, pp. 1386-1394

[BRO83] Brown, Robert Grover: Random Signal Analysis and Kalman Filtering, John Wiley & Sons, New York, 1983

[BUC91] Buck, Brian; Macaulay, Vincent A.: Maximum Entropy in Action, Clarendon Press, London, 1991

[CAM93] Cambel, A. B.: Applied Chaos Theory, Academic Press, London, 1993

[CAM85] Campbell, L. L.: The relation between information theory and differential geometry approach to statistics, Information Sciences, No. 25, 1985, pp. 199-210

[CAP88] Capocelli, Renato M; de Santis, Alfredo; Taneja, Inder Need: Bounds on the entropy series, IEEE Transactions on Information Theory, Vol. IT-34, No. 1, 1988, pp. 134-138

[CAR95] Carnot, Sadi: Über die bewegende Kraft des Feuers und die zur Entwicklung dieser Kraft geeigneten Maschinen, Verlag Harri Deutsch, Thun and Frankfurt am Main, 1995

[CHA95] Chaitin, G. J.: How to run algorithmic information theory on a computer, Lecture, 7.4,.Santa Fe institute, 1995 http://www.cs.auckland.ac.nz/CDMTCS/chaitin/howto.html

[CHA51] Chapman, Douglas G.; Robbins Herbert: Minimum variance estimation without regularity assumptions, University of Washington and University of North Carolina, Annals of Mathematical Statatistics, Vol. 22, 1951, pp. 581-586

[CHE67] Chen, C. H.: On information and distance measures, error bounds, and feature selection, Information Sciences, No. 10, 1967, pp. 159-171

[CHE90] Chen, G.: A simple treatment for suboptimal Kalman Filtering in case of data missing, IEEE AES Aerospace and Electronic Systems, Vol. 26, 1990, pp. 413-415

[CHI57] Chintschin, A. J.; Faddejew, D. K.;Kolmogoroff, A. N.; Rényi, A.; Balatoni, J.: Arbeiten zur Informationstheorie, VEB Verlag der Wissenschaften, Berlin, 1957

[CLA96] Clausius, Rudolf: Über die bewegende Kraft der Wärme...(1850), Harri Deutsch, Thun, 1996

[CLA90] Clarke, Bertrand S.; Barron, Andrew R.: Information-theoretic asymptotics of bayesian methods, IEEE Transactions on Information Theory, Vol. IT-36, No. 3, 1990, pp. 453-471

[CON87] Connes, A; Narnhofer, H; Thirringer, W.: Dynamical entropy of C^{**} algebras and von Neumann algebras, Communication in Mathematical Physics, Vol. 112, 1987, pp. 691-719

[COS84] Costa, M.; Cover, T.: On the similarity of the entropy power inequality, IEEE Transactions on Information Theory, Vol. IT-30, 1984, pp. 837-839

[COV92] Coveney, Peter; Highfield, Roger: Anti-Chaos, Der Pfeil der Zeit in der Selbstorganisation des Lebens, Rowohlt Verlag GmbH, Hamburg, 1994

[COV68] Cover, T.: Estimation by the nearest neighbour rule, IEEE Transactions on Information Theory, Vol. IT-14, No. 1, 1968, pp. 50-55

[COV81] Cover, T.; Van Campenhout, J.: Maximum entropy and conditional probability, IEEE Transactions on Information Theory, Vol. IT-27, No.4, 1981, pp. 483-489

[COV88] Cover, T.; Thomas, J.: Determinant inequalities via information theory, SIAM Journal of Matrix Analysis and Applications, Vol. 9, No.3, 1988, pp. 384-392

[COV91] Cover Thomas: Elements of Information Theory, John Wiley & Sons, New York, 1991

[COV97] Cover, Thomas M.; Ordentlich, Erik: The equivalence of optimal market gain and minimax regret universal portfolios, Proceedings of the International Symposium on Information theory, ISIT 1997, Ulm, p. 282

[CRA94] Cramer, Friedrich: Ordnung und Chaos, Die komplexe Struktur des Lebendigen, Insel Verlag, Frankfurt am Main, 1994

[CRA71] Cramér, Harald: Mathematical Methods of Statistics, University Press, Princeton, N. J., 1971

[CSI70] Csiszár, Imre: Channel Coding Theory, Springer-Verlag, Udine, 1970

[CSI74] Csiszár, Imre: On the general stochastic epidemic with immigration in large populations, Transactions on the 7th Prague Conference, Prague, Cechoslovakia, 1974, pp. 83-96

[CSI81] Csiszar, Imre; Körner, János: Information Theory, Coding theorem for Discrete Memoryless Systems, Academic Press, New York, 1981

[CSI91] Csiszár, Imre: Why least squares and maximum entropy? An axiomatic approach to inference for linear inverse problems, The Annals of Statistics, Vol. 19, No. 4, 1991, pp. 2032-2066

[CSI95/1] Csiszár, Imre: Generalized cutoff rates of Rényi information measures, IEEE Transactions on Information Theory, Vol. IT-41, No. 1, 1995, pp. 26-34

[CSI95/2] Csiszár, Imre, Narayan, Prakash: Channel Capacity for a given decoding metric, IEEE Transactions on Information Theory, Vol. IT-41, No. 1, 1995, pp. 35-43

[DAT91] Datta, Gauri Sankar; Ghosh, Malay: Bayesian prediction in linear models: Applications to small area estimation, The Annals of Statistics, Vol. 19, No. 4, 1991, 1748-1770

[DAV70] Davenport, Wilbur B. Jr.: Probability and Random Processes, McGraw-Hill Book Company, New York, 1970

[DAV87] Davenport, W. B.; Root W. L.: An Introduction to the Theory of Random Signals and Noise, IEEE Press, New York, 1987

[DEC96] Deco, Gustavo; Obradovic, Dragan: An Information-theoretic Approach to Neural Computing, Springer-Verlag, New York, 1996

[DEM91] Dembo, A.; Cover, T.: Information theoretic inequalities, IEEE Transactions on Information Theory, Vol. IT-37, No. 6, 1991, pp. 1501-1518

[DER56] Derman, Cyrus: Stochastic Approximation, Columbia University, Annals of Mathematical Statistics, Vol. 27, 1956, p.879-886

[DEV92] Devlin, Keith: Logic and Information, Cambridge University Press, Cambridge,1992

[DIE86] Dietrich, Paul: Lexikon der Nachrichtentechnik, Definitionen der Begriffe der Nachrichtentechnik in den NTG-Empfehlungen, VDE-Verlag, Berlin, 1986

[DIG97] Diggavi, Suhas N.; Cover, Thomas M.: Is maximum entopy noise the worst, Proceedings of the International Symposium on Information Theory, ISIT 1997, Ulm, p. 278

[DOO53] Doob, J. L.: Stochastic Processes, John Wiley & Sons, New York, 1953

[DRE75] Dreszer, Jerzy: Mathematik Handbuch, Verlag Harri Deutsch, Thun, 1975

[DUB78] Dubisson, Dernard: Information and systems, Proceedings of the IFAC Workshop, Pergamon Press, Oxford, 1978

[DWA70] Dwass, Meyer: Probability and Statistics, W. A. Benjamin, Menlo Park, California, 1970

[EBE90] Ebeling, Werner; Engel, Harald: Selbstorganisation in der Zeit, Akademischer Verlag, Berlin, 1990

[EBE91] Ebeling, Werner: Chaos - Ordnung - Information, Verlag Harri Deutsch, Thun and Frankfurt/Main, 1991

[EBE94] Ebeling, Werner: Chaos und Kosmos: Prinzipien der Evolution, Spektrum, Akademischer Verlag, Heidelberg, 1994

[EDW92] Edwards, Anthony W.: Likelihood, Johns Hopkins University Press, Cambridge, Mass. 1992

[EPH88] Ephraim, Yariv; Lec-Ari; Gray, Robert M.: Asymptotic minimum discrimination information measure for asymptotically weakly stationary processes, IEEE Transactions on Information Theory, Vol. IT-34, No. 5, 1988, pp. 1033-1040

[EPH89] Ephraim, Yariv; Dembo, Amit; Rabiner, Lawrence R.: A minimum discrimination information approach for hidden markov modeling, IEEE Transactions on Information Theory, Vol. IT-35, No. 5 1989, pp. 1001-1013

[ERK98] Erkip, E; Cover, T.: The efficiency of investment information, IEEE Transactions on Information Theory, Vol. IT-44, No. 3 1998, pp. 1026-1040

[FAL76] Falk, G.; Ruppel, W.: Energie und Entropie, Springer-Verlag, Berlin, 1976

[FAN66] Fano, Robert M.: Informationsübertragung ,Oldenbourg Verlag, München, 1966

[FED94] Feder, Meir; Merhav, Neri: Relations between entropy and error probability, IEEE Transactions on Information Theory, Vol. IT-40, No. 1, 1994, pp. 259-266

[FEI98] Feix, G.: Information and Entropie, 1998: http://www.phbib.ruhr-UniBochum.de/physik/htdocs/ inform/microw.html

[FEL91] Feldman, Israel: Constrained minimax estimation of the mean of the normal distribution with known variance, The Annals of Statistics, Vol. 19, No. 4, 1991, pp. 2259-2265

[FER80] Ferreri, C.: Hypoentropy and related heterogeneity, divergency and information measures, Statistica, Vol. 40, No. 2, 1980, pp. 155-168

[FEY72] Feynman, Richard Phillis: Lectures of Physics I/II/III, Addison-Wesley Publishing Company, Mass., 1972

[FIR77] Firchau, Volker: Wieviel sind Informationen maximal wert, Dissertation, Augsburg, 1977

[FLE72] Flechtner, Hans-Joachim: Grundbegriffe der Kybernetik, S. Hirzel Verlag, Stuttgart, 1972

[FRA89] Fraser, Andrew: Information and entropy in strange attractors, IEEE Transactions on Information Theory, Vol. IT-35, No. 2, 1989, pp. 245-262

[FRA57] Fraser, D.A.S and Irwin Guttman: Bhattacharrya bounds without regularity assumptions, Annals of Mathematical Statistics, Vol. 28, 1957, pp. 629-632

[FRE72] Freise, V.: Chemische Thermodynamik, Bibliographisches Institut, Mannheim, 1972

[FRI72] Frieden, Roy B.: Restoring with maximum likelihood and maximum entropy, Journal of the Optical Sciences of America, Vol. 62, No. 4, 1972, pp. 511-518

[FRI90] Frieden, B.: Fisher information, disorder, and the equilibrum distributions of physics, Physical Review A, Vol. 41, No. 8, 1990, pp. 4265-4276

[FRI95] Frieden, B.; Soffer, Berard, H.: Lagrangians of physics and the game of Fisher-information transfer, Physical Review E, Vol. 52, No. 3, 1995, pp.2274-2286

[FRI96] Frieden, B. Roy; Cocke, W. J.: Foundation for Fisher-information-based derivations of physical laws, Physical Review E, Vol. 54, No. 1, 1996, pp. 257-260

[FIR98] Frieden, Bernard Roy: Physics from Fisher Information, Cambridge University Press, Cambridge, 1998

[GAB46] Gabor, Dennis: Part 1. The analysis of information, Journal of the Institution of Electrical Engineers Part 3: Radio and Communication Engineering, 1946, pp. 429-457

[GAB47] Gabor, Dennis: New possibilities in speech transmission, Journal of the Institution of Electrical Engineers Part 3: Radio and Communication Engineering, No. 32, 1947, pp. 369-390

[GAB50] Gabor, Dennis: CIII. Communication theory in physics, The Philosophical Magazine, Vol. 41, 1950, pp. 1161-1187

[GAL68] Gallager, Robert: Information Theory and Reliable Communication, John Wiley, New York, 1968

[GEL68] Gelb, Arthur ; Wallace, E. V.: Multiple-Input Describing Functions and Nonlinear System Design, McGraw-Hill, New York, 1968

[GEL90] Gelb, Arthur (Editor): Applied Optimal Estimation, The M.I.T. Press, Mass., 1990

[GER91] Gerencsér, Lázló: AR(∞) estimation and nonparametric stochastic complexity, IEEE Transactions on Information Theory, Vol. IT-38, No. 6, 1991, pp. 1768-1778

[GIC71] Gichman, Skorochod: Stochastische Differentialgleichungen, Akademischer Verlag, Berlin, 1971

[GIC71] Gichman, Skorochod: The Theory of Stochastic Processes I/II/III, Springer-Verlag, 1971

[GOL68] Goldman, Stanford: Information Theory, Dover Publications, New York, 1968

[GOL87] Golic, Jovan DJ.: On the relationship between the information measures and the bayes probability of error, IEEE Transactions on Information Theory, IT-33, No. 5, 1987, pp. 681-693

[GRA77] Gray, Robert M. (Editor): Ergodic and Information theory, Dowden, Hutchinson & Ross, Stroudsburg, Pa., 1977

[GRA90] Gray, Robert M.: Entropy and Information Theory, Springer-Verlag, New York, 1990

[GRE90] Greene, William H.: Econometric Analysis, Macmillan Publishing, New York, 1990

[GRE87] Greiner, Walter; Neise, Ludwig; Stöcker, Horst: Thermodynamik und statistische Mechanik, Verlag Harri Deutsch, Thun, 1987

[GUI77] Guiasu, Silviu: Information Theory with Applications, McGraw-Hill Book Company, New York, 1977

[GUR71] Gura, I. A; Bierman, A. B.: On computational efficiency of linear filtering algorithms, Automatica, Vol. 7, 1971, pp. 299-314

[HAK88] Haken, Hermann: Information and Self-Organization, Springer Verlag-Berlin, 1988

[HAK95] Haken, Hermann: Erfolgsgeheimnisse der Natur, Synergetik: Die Lehre vom Zusammenwirken, Rowohlt Verlag, Hamburg, 1995

[HAM69] Hammad, Pierre: Shannon's information and Fisher's information for diffusion processes, American Statistical Association Journal, Sept. 1969

[HAR51] Hardy, G. H.; Littlewood, J. E., Pòlya G.: Inequalities, Cambridge University Press, Cambridge, 1951

[HAU81] Hauffe, Heinz: Informationsgehalt von Theorien, Springer-Verlag, Berlin, 1981

[HEI75] Heim, Roland: Zufall und Information, Dissertation an der Universität Tübingen, 1975

[HEL76] Helstrom, Carl W.: Quantum Detection and Estimation Theory, Academic Press, New York, 1976

[HEN74] Henze, Ernst: Einführung in die Informationstheorie, Vieweg, Braunschweig, 1974

[HEN80] Henning, Klaus: Die Entropie in der Systemtheorie, Dissertation an der Techn. Hochschule Aachen, 1980

[HEU86] Heuser, Harro: Lehrbuch der Analysis, Teil 1/2, B. G. Teubner, Stuttgart, 1986

[HO64] Ho, Y. C.: A Bayesian approach to problems in stochastic estimation and control, IEEE Transactions on Automatic Control, Vol. 9, 1964, pp. 333-339

[HOF91] Hofstadter, Douglas R.: Gödel, Escher, Bach, Deutscher Taschenbuch Verlag, Stuttgart, 1991

[HON94] Honlo, Y; Wen-Tsung, L.: Application of Akaike information criterion statistics to geotechnical inverse analysis, Structural Safety, Vol. 14, No. 1/2, 1994, pp. 5-29

[HOT97/1] Hotz, Günter: Algorithmische Informationstheorie, Vorlesungsskript der Universität des Saarlandes, Saarbrücken, 1997

[HOT97/2] Hotz, Günter: Algorithmische Informationstheorie, Teubner-Verlag, Stuttgart, 1997

[HUR90] Hurvich, Clifford M.; Beltrâo, Kaizô I.: Cross-validation choice of a spectrum estimate and ist connections with AIC, Journal of Time Series Analysis, Vol. 11, No. 2, 1990, pp. 121-137

[HUR93] Hurvich, Clifford M.; Tsai, Chin-Ling: A corrected Akaike information criterion for vector autoregressive model selection, Journal of Time Series Analysis, Vol. 14, No. 3, 1993, pp. 271-279

[ING97] Ingarden, R. S.; Kossakowski, A.; Ohya, M.: Information Dynamics and Open Systems, Kluwer Academic Publishers, Dordrecht, 1997

[ISA73] Isaacson, E.; Keller, H. B.: Analyse numerischer Verfahren, Verlag Harri Deutsch, Zürich and Frankfurt am Main, 1973

[JAC94] Jacovitti, G.; Neri, A.: Estimation of the autocorrelation function of complex gaussian stationary processes by amplitude clipped signals, IEEE Transactions on Information Theory, Vol. IT-40, No. 1, 1994, pp. 239-245

[JAY83] Jaynes, E. T. (Editor: R. D. Rosenkrantz): Papers on Probability and Statistics, D. Reidel Publishing Company, Dordrecht, 1983

[JAZ70] Jazwinski, A. H.: Stochastic processes and filtering, Academic Press, New York, 1970

[JOH84] Johnson, Rodney W.; Shore, John E.: Which is the better entropy for speech processing: -SlogS or logS ?, IEEE Transactions on Acoustics, Speech and Signal Processing, Vol. ASSP-32, No. 1, 1984, pp.

[JON79] Jones, D. S.: Elementary Information Theory, Clarendon Press, Oxford, 1979

[JUD85] Judge, GeorgeG.; Griffiths, Hill: The Theory and Practice of Econometrics, John Wiley and Sons., New York, 1985

[JUM90] Jumarie, Guy: Relative Information, Theory and Application, Springer-Verlag Berlin, 1990

[KAI67] Kailath, Thomas: The divergence and Bhattacharrya distance measures in signal selection, IEEE Transactions on Communication Technology, Vol. COM-15, No. 1, 1967, pp.52-60

[KAI76] Kailath, Thomas: Lectures on Linear Least Squares Estimation, Prentice-Hall, Englewood Cliffs, N.J., 1976

[KAI80] Kailath, Thomas: Linear Systems, Prentice-Hall, Englewood Cliffs, N.J., 1980

[KAM85] Kammer Hans-Werner, Schwabe, Kurt: Thermodynamik irreversibler Prozesse, Akademie Verlag, Berlin, 1985

[KAN95] Kanaya, Fumio; Han, Te Sun: The asymptotic of posterior entropy and error probability for bayesian estimation, IEEE Transactions on Information Theory, Vol. IT-41, No. 6, 1995, pp. 1988-1992

[KAN93] Kanazawa, Yuichiro: Hellinger distance and Akaike's information criterion for the histogram, Statistics and Probability Letters, Vol. 17, 1993, pp. 293-298

[KAP67] Kapur, J. N.: Generalized entropy of order α and type β, The Mathematical Seminar, Vol. 4, No. 4, 1967, pp. 78-94

[KAP88] Kapur, Jagat Narain: Mathematical Modelling, John Wiley & Sons. New York, 1988

[KAP89] Kapur, J. N.: Maximum Entropy Models in Science and Engineering, John Wiley & Sons, New York, 1989

[KAP92] Kapur, Jagat Narain, Kesavan, H. K.: Entropy optimization principles with applications, Academic Press, Boston, Mass., 1992

[KAP94] Kapur, J. N.: Measures of Information and their Applications, John Wiley & Sons, New York, 1994

[KAU93] Kauffman, Stuart A.: The Origins of Order, Self-Organization and Selection in Evolution, Oxford University Press, New York, 1993

[KAS68] Kashyap, R. L.; Claydon, Colin, C.: Estimation of probability density and distribution functions, IEEE Transactions on Information Theory, Vol. IT-14, No. 4, 1968, pp. 549-556

[KEN69/73] Kendall & Stuart: The Advanced Theory of Statistics, Vol. I (1969) / Vol. II (1973), Charles Griffin & Company Limited, London and High Wycombe

[KES89] Kesavan, H. K.; Kapur, J. N.: The generalized maximum entropie principle, IEEE Transactions on Systems, Man and Cybernetics, Vol. 19. No. 5, 1989, pp.1042-1052

[KIE52] Kiefer, J.: On minimum variance estimators, Cornell University, Annals of Mathematical Statistics, Vol. 23, 1952, pp. 627-629

[KIE56] Kiefer, J.; Wolfowitz, J.: Consistency of the maximum likelihood estimator in the presence of infinitely many incidental parameters, Cornell University, Annals of Mathematical Statistics, Vol. 27, 1956, pp. 887-906

[KIE57] Kiefer, J; Weiss, Lionel: Some properties of generalized sequential probability ratio tests, Cornell University and University of Oregon, Annals of Mathematical Statistics, Vol. 28, 1957, pp. 57-74

[KNI89] Knight, Keith: Consistency of Akaike´s information criterion for infinite variance autoregresive processes, The Annals of Statistics, Vol. 17, No. 2, 1989, pp. 824-840

[KOD94] Koda, H.; Tanaka, H.: Formulation of the modified Akaike information criterion (AIC), Systems and Computers in Japan, Vol. 25, 1994, pp. 61-72

[KOH89] Kohonen, Teuvo: Self-organization and Associative Memory, Springer-Verlag, Berlin, 1989

[KOL30] Kolmogorov, Andrej N.: Sur la notion de la moyenne, Atti della R. Academia Nazionale dei Lincei, 12, 1930, pp. 388-391

[KOL56] Kolmogorov, Andrej N.: Foundations of the Theory of Probability, Chelsea Publishing, New York, 1956

[KOL65] Kolmogorov, Andrej N.: Three approaches to the quantitative definition of information, Problemy Peredachi Informatsii, Vol. 1, No. 1, 1965, pp. 3-11

[KOL68] Kolmogorov, Andrej N.: Logical basis for information theory and probability theory, IEEE transactions on Information Theory, Vol. IT-14, No. 5, 1968, pp. 662-665

[KOL75] Kolmogorov, Andrej N.: Reelle Funktionen und Funktionalanalysis, Verlag der Wissenschaft, Berlin, 1975

[KOL77] Kolmogorov, Andrej N.: Grundbegriffe der Wahrscheinlichkeitsrechnung, Springer, Berlin, 1977

[KOL93] Kolmogorov, Andrej N.: Selected works 1/2/ 3, Kluwer Academic Publishing, Dordrecht, 1993

[KOU94] Kouritzin, Michael A.: On almost-shure bounds for the LMS algorithm, IEEE Transactions on Information Theory, Vol. IT-40, No. 2, 1994, pp. 372-383

[KRA93] Kravtsov, Yurii A.: Limits of Predictability, Springer-Verlag Berlin, 1993

[KRE77] Kress, Dieter: Theoretische Grundlagen der Signal- und Informationsübertragung, Vieweg Verlag, Braunschweig, 1977

[KUL87] Kullback, Solomon: Topics in Statistical Information Theory, Springer-Verlag, Berlin, 1987

[KUM86] Kumar, P. R.; Varaiya, Pravin: Stochastic Systems Estimation, Identification, and Adaptive Control, Prentice Hall, Englewood Cliffs, N.J., 1986

[LAR79] Larson, Harold: Random Noise, Signals and Dynamic Systems, Vol. I/II, John Wiley&Sons, New York, 1979

[LAC87] Laccoume, Jean-Louis; Durrani Tariq Salim; Stora, Raymond: Traitement du Signal (Signal Processing), North-Holland, Amsterdam, 1987

[LAI98] Lai, T. L.: Information bounds and quick detection of parameter changes in stochastic systems, IEEE Transactions on Information Theory, IT-44, No. 7, November 1998, pp. 2917-2929

[LEE88] Lee, Y. C. (Editor): Evolution, Learning and Cognition, World Scientific Publishing, Singapore, 1988

[LEH51] Lehmann, E. L.: A general concept of unbiasedness, University of California, Berkeley, and Princeton University, Annals of Mathematical Statistics, Vol. 22, 1951, pp. 587-591

[LEV89] Lev-Ari, Hanoch; Parkerm Sydney R.; Kailath, Thomas: Multidimensional maximum-entropy covariance extension, IEEE Transactions on Information Theory, Vol. IT-35, No. 3, 1989, pp. 497-508

[LI93] Li, Ming; Vitányi, Paul: An introduction to Kolmogorov complexity and its applications, Springer-Verlag, New York, 1993

[LIG66] Lighthill, Michael J.: Einführung in die Theorie der Fourieranalysis und der verallgemeinerten Funktionen, Bibliographisches Institut, Mannheim, 1966

[LIN91] Lin, Jianhua: Divergence measures based on the Shannon entropy, IEEE Transactions on Information Theory, Vol. IT-37, No. 1, 1991, pp. 145-151

[LIN94] Linder, Tamás; Zamir, Ram: On the asymptotic tightness of the Shannon lower bound, IEEE Transactions on Information Theory, Vol. IT-40, No. 6, 1994, pp. 2026-2031

[LIN68] Lindgren, B. W.: Statistical Theory, Collier-Macmillan Limited, London, 1968

[LIP74] Lipcer, R. S.; Shiryayev, A. N.: Statistics of Random Processes, Springer Verlag, New York, 1974

[LOE63] Loève, Michel: Probability Theory, Van Nostrand Reinhold Company, New York, 1963

[LOF90] Loffeld, Otmar: Estimationstheorie I/II, Oldenbourg Verlag, München, 1990

[LÜK85] Lüke, H. D.: Signalübertragung, Springer-Verlag, Berlin, 1985

[LOU89] Luo, Ren C.; Kay, Michael G.: Multisensor integration and fusion of intelligent systems, IEEE Transactions on Systems, Man and Cybernetics, Vol. 19, No. 5, 1989, pp. 901-931

[MAH71] Mahalanabis, A. K.; Farooq, M.: A second-order method for state estimation of nonlinear dynamical systems, International Journal of Control, Vol. 14, No. 4, 1971, pp. 631-639

[MAK75] Makhoul, John: Linear prediction: A tutorial review, Proceedings of the IEEE, Vol. 63, No. 4, April 1975, pp. 561-580

[MAS83] Masry, Elias: Probability density estimation from sampled data, IEEE Transactions on Information Theory, Vol. IT-29, No. 5, 1983, pp. 696-709

[MAY79] Maybeck, Peter S.: Stochastic Models, Estimation and Control I/II/III, Academic Press, New York, 1979

[MCE77] McEliece, Robert J.: The Theory of Information and Coding, Encyclopedia of mathematics and its applications, Addison-Wesley Publishing Company, London, 1977

[MCC87] McCullagh, Peter: Tensor Methods in Statistics, Chapman and Hall, London, 1987

[MEA84] Mead, Lawrence R.; Papanicolaou, N.: Maximum entropy in the problem of moments, Journal of Mathematical Physics, Vol 25, No. 8, 1984, pp. 2404-2417

[MEN92] Menendez, M. L.: Some statistical applications of generalized Jensen's divergence measures, Fuzzy Sets and ystems, Vol. 52, No. 2, 1992, pp. 169-180

[MEY93] Meyer, M. Eugene; Gokhale, D. V.: Kullback–Leibler information measure for studying convergence rates of densities and distributions, IEEE Transactions on Information Theory, Vol. IT-39, Vol. 4, 1993, pp. 1401-1404

[MID87/1] Middleton, David: Topics in Communication Theory, Peninsula Publishing, Los Altos, Calif., 1987

[MID87/2] Middleton, David: Introduction to Statistical Communication Theory, Peninsula Publishing, Los Altos, Calif., 1987

[MUT94] Mutua, Francis M.: The use of the Akaike information criterion in the identification of an optimum flood frequency model, Hydroligocal Sciences Journal, Vol. 39, No. 3, 1994, pp. 235-244

[NAK93] Nakamori, Seiichi: Optimal filtering algorithm using covariance information in linear discrete-time distributed parameter systems, Signal Processing, Vol. 33, 1993, pp. 44-56

[NEI64] Neidhardt, Peter: Informationstheorie und automatische Informationsverarbeitung, Berliner Union, Stuttgart, 1964

[NIC91] Nicolis, John S.: Chaos and Information Processing, A Heuristic Outline, World Scientific Publishing, Singapore, 1991

[ÖST93] Österreicher, F.; Vajda, I: Statistical information and discrimination, IEEE Transactions on Information Theory, Vol. IT-39, Vol. 3, 1993, pp. 1036-1039

[OHY83] Ohya, M.: On compound state and mutual information in quantum information theory, IEEE Transactions on Information Theory, Vol. IT-29, No. 5, 1983, pp. 770-774

[ORN93/1] Ornstein, Donald Samuel; Weiss, Benjamin: Entropy and data compression schemes, IEEE Transactions on Information Theory, Vol. IT-39, Vol. 1, 1993, pp. 78-83

[ORN93/2] Ornstein, Robert; Thompson, Richard F.: Unser Gehirn, Das lebendige Labyrinth, Rowohlt Verlag, 1993

[PAL97/1] Palus, Milan: On entropy rates of dynamical systems and gaussian processes, Institute of Computer Sciences, Academy of Sciences of the Czech Republic, Jan. 1997

[PAL97/2] Palus, Milan: Kolmogorov entropy from time series using information-theoretic functionals, Neural Network World, Vol. 3, 1997, pp. 269-292

[PAP81] Papoulis, Athanasios: Systems and Transforms with Application in Optics, Robert E. Krieger Publishing Company, Malabar, Florida, 1981

[PAP84] Papoulis, Athanasios: Probability, Random Variables and Stochastic Processes, McGraw-Hill, Singapore, 1984

[PAR55] Pardo, L.; Morales, D.; Taneja, I. J.: Generalized Jensen difference divergence measures and Fisher measure of information, Kybernetes, Vol. 24, No. 2, 1955, pp. 15-28

[PAS77] Pascal, Blaise: Pensées, Èditions Gallimard, 1977

[PAZ93] Pazman, Andrej: Nonlinear Statistical Models, Kluwer Academic Publishers, Bratislava, 1993

[PEA74] Pearson, Carl E.: Handbook of Applied Mathematics, Van Nostrand Reinhold Company, London, 1974

[PEC92] Pecaric, Josip E.; Proschan, Frank; Tong, Y. L.: Convex Functions, Partial Orderings and Statistical Applications, Academic Press, Boston, Mass., 1992

[PEN89] Penrose, Roger: The emperors new mind, Oxford University Press, New York, 1989

[PET67] Peters, Johannes: Einführung in die allgemeine Informationstheorie, Springer-Verlag, Berlin, 1967

[PIC79] Picard, Claude-Francois: Weighted probablistic information measures, Journal of Combinatorics, Information and System Sciences, Vol. 4, No. 4, 1979, pp. 343-356

[PIE80] Pierce, John Robinson: An Introduction to Information Theory, Dover Publications, New York, 1980

[PIN92] Pina, R. K.; Puetter, R. C.: Incorporation of spatial information in bayesian image reconstruction: The maximum residual likelihood criterion, Publications of the Astronomical Society of the Pacific, Vol. 104, No. 681, 1992, pp. 1069-1103

[PIN92] Pinna, G. G. et al.: On the use of the Akaike information criterion in AR spectral analysis of cardiovascular variability signals: A case report, Computers and Cardiology, 1993, pp. 471- 474

[PÖT80] Pötschke, Dieter; Sobik, Fred: Mathematische Informationstheorie, Akademie Verlag, Berlin, 1980

[POL93] Politis, Dimitris Nicolas: Nonparametric maximum entropy, IEEE Transactions on Information Theory, Vol. IT-39, No. 4, 1993, pp. 1409-1413

[PON86] Pontryagin, Lev Semenovitch: The Mathematical Theory of Optimal Processes, Gordon and Breach Science Publishers, Montreux, 1986

[POP94] Popper, Karl: Logik der Forschung, J. C. B. Mohr, Tübingen, 1994

[POP97] Popper, Karl; Eccles, John: Das Ich und sein Gehirn, Piper Verlag, München, 1997

[POS84] Posthoff, C.;Woschni, E. G.: Funktionaltransformationen in der Informationstechnik, Akademie Verlag Berlin, 1984

[PRI68] Price, Charles F.: An analysis of the divergence problem in the Kalman filter, IEEE Transactions on Automatic Control, December 1968, pp. 699-702

[PRI90] Prigogine, Ilya; Stengers, Isabelle: Dialog mit der Natur, Neue Wege wissenschaftlichen Denkens, Piper Verlag, München, 1990

[PRI92] Prigogine, Ilya: Vom Sein zum Werden, Piper Verlag, München, 1992

[PUE95] Puetter, R. C.: Pixon-based multiresolution image reconstruction and the quantification of picture information content, The International Journal of Image Systems and Technology, April 1995

[RAI80] Raisbeck, Gordon: Informationstheorie, Oldenbourg Verlag, München, 1970

[RAT70] Rathie, Pushpa N.: On a generylized entropy and a coding theorem, Journal of Applied Probability, Vol. 7, 1970, pp. 124-133

[REI87] Reif, Frederick: Statistische Physik und Theorie der Wärme, de Gruyter, Berlin, 1987

[REI90] Reif, Frederick: Statistische Physik, Vieweg Verlag, Braunschweig, 1990

[REN55] Rényi, Alfréd: On a new Axiomatic Approach of Probability, Acta Hungarica, Vol. 6, 1955, pp. 285-333,

[REN73] Rényi, Alfréd: Wahrscheinlichkeitstheorie mit einem Anhang über Informationstheorie, VEB Verlag der Wissenschaften, Berlin, 1973

[REN76/1] Rényi, Alfréd: Selected Papers (I: -1955), Akad. Kiado, Budapest, 1976

[REN76/2] Rényi, Alfréd: Selected Papers (II: 1956-1961)), Akad. Kiado, Budapest, 1976

[REN76/3] Rényi, Alfréd: Selected Papers (III: 1962-1970)), Akad. Kiado, Budapest, 1976

[REN82] Rényi, Alfréd: Tagebuch der Informationstheorie, Birkhäuser, Basel, 1982

[RES87] Resnikoff, Howard: The Illusion of Reality, Springer-Verlag, Berlin, 1987

[RIB94] Riba, Jaume; Vézquez, Gregori: Bayesian recursive estimation and timing exploiting the cyclostationarity property, Signal Processing, Vol. 40, 1994, pp. 21-37

[RIC88] Rice, John A.: Mathematical Statistics and Data Analysis, Pacific Grove, Calif., 1988

[ROB56] Robbins, Herbert: Mixture of distributions, Department of Mathematical Statistics, University of North Carolina, Annals of Mathematical Statistics, Vol. 27, 1956, pp. 360 - 369

[RÖH67] Röhler, Rainer: Informationstheorie in der Optik, Wissenschaftliche Verlagsgesellschaft, Stuttgart, 1967

[RUN66] Rund, Hanno: The Hamilton–Jacobi Theory in the Calculus of Variations, van Nostrand Company (Canada), Toronto, 1966

[SAK90] Sakai, Hideaki: An application of a BIC-Type method to harmonic analysis and a new criterion for order determination of an AR process, IEEE Transactions on Acoustics, Speech and Signal Processing, Vol. 38, No. 6, 1990, pp. 999-1004

[SAL97] Salm, Wolfgang: Entropie und Information – Naturwissenschaftliche Schlüsselbegriffe, Aulis Verlag Deubner & Co. KG, Köln, 1997

[SAN85] Sant´anna, Annibal, P.; Taneja, Inder Jeet: Trigonometric entropies, Jensen difference divergence measures, and error bounds, Information Sciences, Vol. 35, 1985, pp. 145-155

[SAU92] Sauvageot, Jean-Luc; Thouvenot, Jean-Paul: Une nouvelle définition de l´entrope dynamique des systèmes non commutatifs, Communication in Mathematical Physics, 145, 1992, pp. 411-423

[SCH75] Schadach, Dieter J.: Grundlagen und Anwendungen einer nicht-boolschen Informationstheorie auf Teilraumverbänden von Tensorprodukten separabler Hilbert-Räume, Dissertation an der Christian-Albrecht-Universität Kiel, 1975

[SCH68] Schalkwijk, J. Pieter M.: Center-of-gravity information feedback, IEEE Transactions on Information Theory, Vol. IT-14, No. 2, 1968, pp. 324-330

[SCH93] Scharf, Louis L.; McWhorter, L. T.: Geometry of the Cramér–Rao bound, Signal Processing, Vol. 31, 1993, pp. 301-311

[SCH89/1] Schrödinger, Erwin: Was ist Leben, Piper Verlag, München, 1989

[SCH89/2] Schrödinger, Erwin: Die Natur und die Griechen, Diogenes Verlag, Zürich, 1989

[SCH78] Schwarz, Gideon: Estimation the dimension of a model, The Annals of Statistics, Vol. 6, No. 2, 1978, pp. 461-464

[SCH73] Schweppe, Fred C.: Uncertain Dynamic Systems, Prentice Hall,Englewood Cliffs, N.J., 1973

[SCH90] Schwartz, Mischa: Information transmission, modulation and noise, McGraw-Hill Book Company, New York, 1990

[SHA71] Shannon, C. E; Waver, W.: The Mathematical Theory of Communication, University of Illinois Press, Urbana, 1971

[SHA93] Shannon, Claude E.: Collected Papers Ed. by N. J. A. Sloane, IEEE Press, Piscataway, N.J., 1993

[SHA98/1] Shannon, Claude E.: Commemorative Issue, IEEE Transactions on Information Theory, Vol. IT-44, No. 6, 1998

[SHA98/2] Shamai, S. (Shitz); Wyner, A. D.: Information-theoretic considerations for symmetric, cellular, multiple-access fading channels - Part I: IEEE Transactions on Information Theory, Vol. IT-43, No. 6, 1998, pp. 1877-1894

[SHA98/3] Shamai, S. (Shitz); Wyner, A. D.: Information-theoretic considerations for symmetric, cellular, multiple-access fading channels - Part II: IEEE Transactions on Information Theory, Vol. IT-43, No. 6, 1998, pp. 1895-1911

[SHA73] Sharma, B. D.; Mittal, D. P.: New non-additive measures of entropy for discrete probability distributions, Journal of Mathematical Sciences, Vol. 10, 1975, pp. 28-40

[SHA92] Sharov, Alexei, A.: Biosemiotics: Functional-evolutionary approach to the analysis orf the sense of information, Biosemiotics, T-A. Sebeok; J. Umiker-Sebeok (eds.), The Semiotic Web 1991, Mouton de Gruyter, New York, 1992, pp. 345-373

[SHI76] Shibata, Ritei: Selection of the order of an autoregressive model by Akaike´s information criterion, Biometrika, Vol. 63, No. 1, 1976, pp. 117-126

[SHI87] Shields, Paul C.: The ergodic and entropy theorems revisited, IEEE Transactions on Information Theory, Vol. IT-33, No. 2, 1987, pp. 263-266

[SIM94] Simmonds, J. G.: A Brief on Tensor Analysis, Springer-Verlag, New York, 1994

[SLE74] Slepian, David [Hrsg.]: Key Papers in the Development of Information Theory, IEEE Press, New York, 1974

[SOK64] Sokolnikow, I. S.: Tensor Analysis, Theory and Applications to Geometry and Mechanics of Continua, John Wiley & Sons, New York, 1964

[SOR85] Sorenson, H. W.: Kalman Filtering, Theory and Applications, IEEE Press, New York, 1985

[STE93] Steward, Ian: Spielt Gott Roulette, Insel Verlag, Frankfurt am Main, 1993

[STO91] Stonier, Tom: Information und die innere Struktur des Universums, Springer- Verlag, Berlin, 1991

[SZA96] Szabo, Nick: Introduction to algorithmic information theory, http://www.best.comTILDEszabo/kolmogorov.html, 1996

[SZI28] Szilard, L.: Über die Entropieverminderung ind einem thermodynamischen System bei Eingriffen intelligenter Wesen, Zeitschrift für die Physik, Vol. 53, 1928, pp. 840-856

[TAN85] Taneja, Inder Neet: Generalized error bounds in pattern recognition, Pattern Recognition Letters, Amsterdam, Vol. 6, No. 3, 1985, pp. 361-368

[TAN89] Taneja, Inder Neet: On generalized information measures and their applications, Advances in Electronics and Electron Physics, Vol. 76, 1989, pp. 327-413

[TAY65] Taylor, Agnus, E.: General Theory of Functions and Integration, Blaisdell Publishing Company, Waltham, Mass., 1965

[TEB93] Teboulle, Marc; Vajda, Igor: Convergence of best *Φ*-entropy estimates, IEEE Transactions on Information Theory, Vol. IT-39, No. 1, 1993, pp. 297-301

[THO85] Thomopoulos, Stelios, C. A.: Minimum error entropy estimation and entropic prediction filtering: An optimal predictive coding scheme, IEEE Transactions on Information Theory, Vol. IT-31, No. 5, 1985, pp. 697-703

[TIE93] Tietze, Hans: Das philosophische Gesamtwerk, Band 4: Theorie der Information, Schäuble Verlag, Berlin, 1993

[TON90] Tong, Howell: Non-linear Time series, a dynamical system approach, Clarendon Press, London, 1990

[TOP74] Topsoe, Flemming: Informationstheorie, Teubner-Verlag, Stuttgart, 1974

[VAA89] Vaart, Aad, van der: On the asymptotic information bound, The Annals of Statistics, Vol. 17, No. 4, 1989, pp 1487-1500

[VAN81] Van Campenhout, Jan, M.; Cover, Thomas M.: Maximum entropy and conditional probability, IEEE Transactions on Information Theory, Vol. IT-27, No. 4, 1981, pp. 483-489

[VAR66] Varma, R. S.: Generalizations of Rényi´s entropy of order α, Journal of Mathematical Sciences, Vol. 67, No. 1, 1966, pp. 34-48

[VÖL90] Völz, Horst: Computer und Kunst, Urania Verlag, Leipzig - Jena - Berlin, 2. Auflage 1990

[VÖL94] Völz, Horst: Information verstehen, Vieweg Verlag, Braunschweig, 1994

[VON95] Von der Linden, W.: Maximum entropy data analysis, Applied Physics A, Materials Sciences and Processing, Vol. 60, No. 2, 1995, pp. 155-165

[vMI64/1] von Mises, Richard: Selected Papers I: Geometry, Mechanics, Analysis, Verl. Univ. Microfilms Internat., Ann Arbor, Mich., 1964

[vMI64/2] von Mises, Richard: Selected Papers II: Probability and Statistics, General, Verl. Univ. Microfilms Internat., Ann Arbor, Mich., 1964

[vMI64/3] von Mises, Richard: Mathematical Theory of Probability and Statistics, Academic Press, New York, 1964 [vMI72] von Mises, Richard: Wahrscheinlichkeit Statistik und Wahrheit (Probability, Statistics and Truth), Springer-Verlag, Berlin, 1972

[vWE85] von Weizsäcker, Carl Friedrich: Aufbau der Physik, Carl Hanser Verlag, München, 1985

[vWE92] von Weizsäcker, Carl Friedrich: Zeit und Wissen, Carl Hanser Verlag, München, 1992

[WAN93] Wang, Xiaobao: An AIC type estimator for the number of cosinusoids, Journal of Time Series Analysis, Vol. 14, No. 4, 1993, pp. 433-440

[WAT60] Watanabe, Satosi: Information theoretical analysis of multivariate correlation, IBM Journal of Research and Development, Vol. 4, 1960, pp. 66-82

[WHA71] Whalen, Anthony D.: Detection of Signals in Noise, Academic Press, San Diego, Calif., 1971

[WIE63] Wiener, Norbert: Kybernetik, Econ Verlag GmbH, Düsseldorf, 1963

[WIE50] Wiener, Norbert: Extrapolation, Interpolation and Smoothing of Stationary Time Series, M.I.T. Press, Cambridge, Mass., 1950

[WOL56] Wolfowitz , J.: The minimum distance method, Cornell University, Annals of Mathematical Statistics, 1956, pp. 75 - 88

[WOL90] Wolkenstein, Michail W.: Entropie und Information, Harri Deutsch Verlag, Frankfurt am Main, 1990

[WOO80] Woodward, Philpi M.: Probability and Information Theory with Application, Artech House Books, Dedham, Mass., 1980

[WON71] Wong, Eugene: Stochastic Processes in Information and Dynamical Systems, McGraw Hill Book Company, New York, 1971

[WOS88] Woschni, E. G.: Informationstechnik, Alfred Hüthig Verlag, Heidelberg, 1988

[WYN94] Wyner, Aaron D.: 1994 Shannon Lecture, Typical Sequences and all that: Entropy, pattern matching, and data compression, AT&T Bell Laboratories, Murray Hill, N.J., 1994

[YAS88] Yaser S. Abu-Mustafa (ed.): Complexity in Information theory, Springer-Verlag, New York, 1988

[YEU91] Yeung, Raymond W.: A new outlook on Shannon's information measure, IEEE Transactions on Information Theory, Vol. IT-37, No. 3, 1991, pp. 466-474

[YEU97] Yeung, R. W.: A framework for information inequalities, Proceedings on the International Symposium on Information Theory, ISIT 1997, Ulm, 1997, pp. 268

[YEU34] Yeung, R. W.: A framework for linear information inequalities, IEEE Transactions on Information Theory, Vol. IT-43, No. 6, October 1998, pp. 1924-1934

[YUA99] Yuan, A; Clarke, B.: An information criterion for likelihood selection, IEEE Transactions on Information Theory, Vol. IT-45, No. 2, March 1999

[ZAM93] Zamir, Ram; Feder, Meir: A generalization of the entropy power inequality with applications, IEEE Transactions on Information Theory, Vol. IT-39, No. 5, 1993, pp. 1723-1728

Index